Introduction to Health and Safety at Work

Introduction to Health and Safety at Work covers the fundamentals of occupational safety and health for the thousands of students who complete the NEBOSH National General Certificate in Occupational Health and Safety each year. This seventh edition closely follows the NEBOSH National General Certificate syllabus which was updated in 2019 and comes into use in 2020.

The highly illustrated content covers all of the essential elements of health and safety management, the legal framework, risk assessment and control standards and also includes checklists, report forms and record sheets to supplement learning. It also has an extensive summary of current health and safety legislation.

▶ Aligned to the NEBOSH National General Certificate in Occupational Health and Safety
▶ Practice questions and answers to test knowledge and increase understanding

The book is suitable for all students following a level 3 Health and Safety course and a source of reference and guidance for managers at work in the UK. Written by renowned authors, this book is often provided as part of the Certificate course and is essential reading for a student.

Seventh Edition

Introduction to Health and Safety at Work

for the NEBOSH
National General Certificate in Occupational
Health and Safety

Phil Hughes MBE, MSc, CFIOSH

Chairman NEBOSH 1995–2001, President of IOSH 1990–1991

Ed Ferrett PhD, BSc (Hons Eng), CEng, MIMechE, MIET, CMIOSH

Vice Chairman NEBOSH 1999–2008

Routledge
Taylor & Francis Group

LONDON AND NEW YORK

Seventh edition published 2021
by Routledge
2 Park Square, Milton Park, Abingdon, Oxon, OX14 4RN

and by Routledge
52 Vanderbilt Avenue, New York, NY 10017

Routledge is an imprint of the Taylor & Francis Group, an informa business

First edition published 2003
Sixth edition published by Routledge 2016

British Library Cataloguing-in-Publication Data
A catalogue record for this book is available from the British Library

Library of Congress Cataloging-in-Publication Data
Names: Hughes, Phil (Phillip W.), 1946- author. | Ferrett, Ed, author.
Title: Introduction to health and safety at work : for the NEBOSH national general certificate in occupational health and safety / Phil Hughes, MBE, MSc, CFIOSH, Chairman NEBOSH 1995-2001, President of IOSH 1990-1991, Ed Ferrett PhD, BSc (Hons Eng), CEng, MIMechE, MIET, CMIOSH, Vice Chairman NEBOSH 1999-2008.
Description: Seventh edition. | Abingdon, Oxon ; New York, NY : Routledge, 2021. | Includes bibliographical references and index.
Identifiers: LCCN 2020028678 (print) | LCCN 2020028679 (ebook) | ISBN 9780367638351 (hbk) | ISBN 9780367482886 (pbk) | ISBN 9781003039075 (ebk)
Subjects: LCSH: Industrial hygiene--Great Britain--Handbooks, manuals, etc. | Industrial safety--Great Britain--Handbooks, manuals, etc.
Classification: LCC T55 .H84 2021 (print) | LCC T55 (ebook) | DDC 658.3/82–dc23
LC record available at https://lccn.loc.gov/2020028678
LC ebook record available at https://lccn.loc.gov/2020028679

ISBN: 978-0-367-63835-1 (hbk)
ISBN: 978-0-367-48288-6 (pbk)
ISBN: 978-1-003-03907-5 (ebk)

Typeset in Univers LT
by KnowledgeWorks Global Ltd.

Contents

List of illustrations

Figures

Tables

Boxes

Preface to the seventh edition

Since 2002, *Introduction to Health and Safety at Work* has established itself as the foremost text for students taking the NEBOSH National General Certificate (NGC). It has become a significant work of reference for managers with health and safety responsibilities, and it is therefore a matter of primary importance that it should be kept up to date, as far as is possible, with new legislation and recent developments. The National General Certificate in Occupational Health and Safety is the most widely held health and safety qualification of its kind in the UK.

It is suitable for managers, supervisors and staff from all types of organisations, making day-to-day decisions at work, who need a broad understanding of health and safety issues and to be able to manage risks effectively. Many people take the NEBOSH National General Certificate as a first step in a career in health and safety. It provides a valuable overview, and is a sound basis for further professional study. This book has been designed to help candidates who are studying on the National General Certificate course and to provide them with a useful reference text in their subsequent careers.

Many larger organisations choose the NEBOSH National General Certificate as a key part of their supervisor or management development programme. By ensuring that line managers have a sound understanding of the principles of risk management, they build an effective safety culture in the company. Smaller organisations, operating in lower risk environments, often choose the NEBOSH National General Certificate as the appropriate qualification for the manager taking the lead on health and safety issues.

This seventh edition has been produced to include all the recent syllabus changes. There have been very few legislative changes since the last edition but any small changes have been included. Some details of the new Occupational Management System ISO 45001 have also been included.

The qualification is divided into two distinct units each of which is assessed separately. These are a written management examination (NGC1) and a risk assessment of the candidate's workplace (NGC2).

The NGC course consists of 11 elements but it was felt the element 3 'Managing risk – understanding people and processes' is too large to include as a single chapter. It has, therefore, been split into two chapters – Chapter 3 Health and safety management of people and processes and Chapter 4 Health and safety management of risk and controls.

We have included our own questions at the end of each chapter using the NEBOSH format. Candidates that can successfully answer these questions unaided should have no problems in the examination or the risk assessment.

As mentioned in previous editions, it is the policy of NEBOSH to examine new relevant legislation 6 months after its introduction whether it is specifically mentioned in the syllabus document or not.

Finally, one of the objectives of the book is to provide a handbook for the use of any person who has health and safety as part of his/her responsibilities. We thought that it would be helpful, therefore, to add a few useful topics which are outside the syllabus; these are included at the end of some chapters and an additional chapter covers other aspects of health and safety and related topics that many readers will find helpful on completion of the course – environmental considerations, international issues and health and safety in the home.

We hope that you find this new edition to be useful.

Phil Hughes
Ed Ferrett

Acknowledgements

Throughout the book, definitions used by the relevant legislation and the Health and Safety Executive and advice published in Approved Codes of Practice or various Health and Safety Commission/Executive publications have been utilised. Most of the references produced at the end of each Act or Regulation summary in Chapter 13 are drawn from the HSE Books range of publications.

The authors' grateful thanks go to Liz Hughes and Jill Ferrett for proof reading and patience and their administrative help during the preparation of this edition. The authors are particularly grateful to Liz for the excellent study guide that she has written for all NEBOSH students, which is included at the end of this book, for the section on report writing in Chapter 5 and the sections on home safety and cycle safety in Chapter 4. Liz gained an honours degree in psychology at the University of Warwick, later going on to complete a Master's degree at the same university. She taught psychology in further and higher education, where most of her students were either returning to education after a gap of many years, or were taking a course to augment their existing professional skills. She went on to qualify as a social worker specialising in mental health, and later moved into the voluntary sector where she managed development for a number of years. Liz then helped to set up and manage training for the National Schizophrenia Fellowship (now called Rethink) in the Midlands.

We would also like to acknowledge the additional contribution made by Jill Ferrett for the help that she gave during the research for the book and with some of the word processing. Given her background in economics and higher education, her advice on certain legal and economic issues has been particularly valuable.

We would like to thank various NEBOSH and HSE staff for their generous help and advice. Finally we would like to thank Stephen Vickers, a past Chief Executive of NEBOSH, for his encouragement at the beginning of the project and Guy Loft and all the production team at Taylor and Francis who have worked hard to translate our dream into reality.

About the authors

Phil Hughes MBE is a well-known UK safety professional with over 50 years worldwide experience as Head of Environment, Health and Safety at two large multinationals: Courtaulds and Fisons. Phil started work in health and safety in the Factory Inspectorate at the Derby District UK in 1969 and moved to Courtaulds in 1974. He joined IOSH in that year and became Chairman of the Midland Branch, then National Treasurer and was President in 1990–1991. Phil has been very active on the NEBOSH Board for over 10 years and served as Chairman from 1995 to 2001. He was also a Professional Member of the American Society of Safety Engineers for many years and has lectured widely throughout the world. Phil received the RoSPA Distinguished Service Award in May 2001 and was a Director and Trustee of RoSPA from 2003 to 2010. He received an MBE in the New Year Honours List 2005 for services to health and safety. Phil is a Chartered Fellow of IOSH.

Ed Ferrett is an experienced health and safety consultant who has practised for over 35 years. With a PhD and an honours degree in mechanical engineering from Nottingham University, Ed spent 30 years in higher and further education, retiring as the Head of the Faculty of Technology of Cornwall College in 1993. Since then he has been an independent consultant to several public and private sector organisations including construction businesses, the Regional Health and Safety Adviser for the Government Office (West Midlands), and was Chair of West of Cornwall Primary Care NHS Trust for 6 years until 2006.

Ed was a member of the NEBOSH Board from 1995 until 2010 and Vice Chair from 1999 to 2008. He has delivered many health and safety courses and has been a lecturer in NEBOSH courses for various course providers. He has been an External Examiner for an MSc course and BSc course in health and safety at two UK universities, a Reporting Inspector for Independent Further and Higher Education with the British Accreditation Council and a NEBOSH Ambassador. Ed is a Chartered Engineer and a Chartered Member of IOSH.

How to use this book and what it covers

Introduction to Health and Safety at Work Seventh Edition is basically designed to:

1. cover the syllabus of the NEBOSH National General Certificate (NG1 and NG2) September 2019 editions in Occupational Health and Safety and other level 3 OSH awards;

2. go beyond the NEBOSH syllabus in covering some construction, climate change, environmental, home safety and international aspects;

3. provide a good basis in OSH for students who wish to progress to the NEBOSH Diploma or a University first or second degree;

4. provide a text which more than covers the IOSH Managing Safely syllabus or other similar awards;

5. give summaries of UK OSH legislation relevant to the NG1 and NG2;

6. help students study, revise and sit the examinations;

7. provide brief guidance to students who carry out the practical assessment;

8. provide brief guidance for searching the internet;

9. supply a range of significant OSH websites;

10. provide a good updated reference text for managers with OSH responsibilities and OSH practitioners in industry and commerce; and

11. provide numerous templates for typical fire and safety forms.

We expect the book to be used as a basis for training, and as further reference when students are back in their own workplaces. We believe that all questions can be answered from the material in the book but we would also urge students to study some of the documents given as reference sources at the end of each chapter. It would be helpful to visit some of the websites where further detailed guidance is available. The websites featured in the text were found to be correct at the time of writing in March 2020.

Poor

Corrected

Poor

Corrected

Animated inspection exercises

For more detail see the NEBOSH National General Certificate in Occupational Health and Safety Qualification guide for Learning Partners

https://www.nebosh.org.uk/qualifications/national-general-certificate/#resources

Usable images

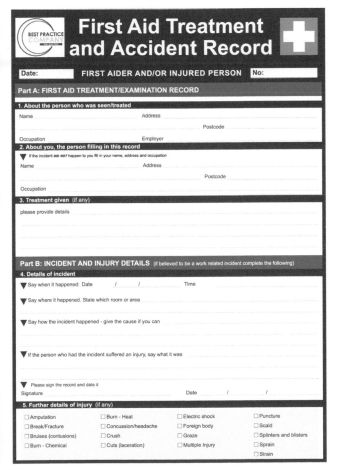

Forms in Word that can be downloaded for use at work (see Chapter 17 for full set)

Principal abbreviations

Most abbreviations are defined within the text. Abbreviations are not always used if it is not appropriate within the particular context of the sentence. The most commonly used ones are as follows:

ACGIH American Conference of Governmental Industrial Hygienists
ACL Approved Carriage List
ACM asbestos-containing material
ACoP Approved Code of Practice
ADR Accord dangereux routier (European agreement concerning the international carriage of dangerous goods by road)
AFNOR French Standards Association
AFSSET French Agency for Environmental and Occupational Health Safety
AIB Asbestos Insulation Board
ALARP as low as reasonably practicable
AND European provisions concerning the international carriage of dangerous goods by inland waterways
ANSI American National Standards Institute
ASCC Australian Safety and Compensation Council
ASEAN Association of Southeast Asian Nations
ASSE American Society of Safety Engineers
ASTM American Society for Testing and Materials (now ASTM International)
ATDSDR Agency for Toxic Substances and Diseases Registry (USA)
ATEX Atmosphere Explosive (used in the context of two European Directives, 94/9/EC and 1999/92/EC)
BA breathing apparatus
BAT best available techniques
BEBOH British Examining Board in Occupational Hygiene
BIOH British Institute of Occupational Hygiene
BLR blue light radiation
BPM best practicable means
BRE Building Research Establishment
BSI British Standards Institution
CAR Control of Asbestos Regulations (UK)
CAS Chemical Abstracts Service (USA)
CBI Confederation of British Industry
CD consultative document

CDM Construction (Design and Management) Regulations (UK)
CEN Comité Européen de Normalisation
CENELEC Comité Européen de Normalisation Électrotechnique
CIB Chartered Institute of Building
CIRC Centre International de Recherche sur le Cancer (France)
CIS International Occupational Safety and Health Information Centre
CISDOC International Labour Organization database available on OSHROM
CISPR Comité International Spécial des Perturbations Radioélectriques
CLAW Control of Lead at Work Regulations
CLP Classification Packaging and Labelling
CNS Comité de Normalisation de la Soudure (France)
CO carbon monoxide
CONIAC Construction Industry Advisory Committee
COSHH Control of Substances Hazardous to Health Regulations (UK)
CPR Cardiopulmonary resuscitation
CSA Canadian Standards Association
CTS carpal tunnel syndrome
CVD cardiovascular disease
dB decibels
DB dry bulb
dB(A) decibel (A-weighted)
dB(C) decibel (C-weighted)
DSE display screen equipment
DSEAR Dangerous Substances and Explosive Atmospheres Regulations (UK)
E&W England and Wales
EA Environment Agency
EAV exposure action value
EC European Commission
EEF Engineering Employers Federation
ELV exposure limit value
EMAS Employment Medical Advisory Service
EPA Environmental Protection Act 1990 (UK)
EU European Union
EU-OSH European Agency for Safety and Health at Work

FAO	Food and Agriculture Organization of the United Nations
FOPS	falling-object protective structure(s)
FPO	Fire Prevention Officer
GATT	General Agreement on Tariffs and Trade
GHGB	Good Health is Good Business
GHS	Globally Harmonised System of Classification and Labelling of Chemicals
GTAW	gas tungsten arc welding
HACCP	hazard analysis critical control point
HAI	hospital acquired infections
HAM	hand-held monitor
HASAC	Health and Safety Advice Centre
HAV	hand–arm vibration
HGV	Heavy Goods Vehicle
HOPE	Healthcare, Occupational and Primary for Employees
HSCER	Health and Safety (Consultation with Employees) Regulations (UK)
HSE	Health and Safety Executive (UK)
HSG	Health and Safety Guidance Booklet
HSW Act	Health and Safety at Work etc. Act 1974 (UK)
HWL	Healthy Working Lives
IAC	Industry Advisory Committee
IChemE	Institution of Chemical Engineers
IEA	International Ergonomics Association
IEC	International Electrotechnical Commission
IEE	Institution of Electrical Engineers
ILO	International Labour Organization
INDG	Industry Guidance
IOH	Institution of Occupational Hygienists
IOSH	Institution of Occupational Safety and Health
IPCS	International Programme on Chemical Safety
IPMS	Institution of Professionals, Managers and Specialists
IPPC	integrated pollution prevention and control
IPPR	Institute for Public Policy Research
IPR	integrated pollution regulation
IRM	Institute of Risk Management
IRSM	International Institute of Risk and Safety Management
ISBN	International Standard Book Number(ing)
ISO	International Organization for Standardization
LD50	lethal dose fifty
LDLo	lethal dose low
LEA	Local Enterprise Agency
LEAL	Lower Exposure Action Level
LEL	lower exposure limit
Leq	equivalent continuous sound level
Leq(8)hr	equivalent continuous sound level (normalised to 8 hours)
LEV	local exhaust ventilation
LNG	liquefied natural gas
LOLER	Lifting Operations and Lifting Equipment Regulations (UK)
LPG	liquefied petroleum gas
MEL	maximum exposure limit
mg/m^3	milligrams per cubic metre
MHOR	Manual Handling Operations Regulations (UK)
MHSWR	Management of Health and Safety at Work Regulations (UK)
MORR	Management of Occupational Road Risk
MOT	Ministry of Transport (still used for vehicle tests in UK)
MSD	musculoskeletal disorder
MSDS	Material Safety Data Sheet(s)
NEBOSH	National Examination Board in Occupational Safety and Health
NIOSH	National Institute for Occupational Safety and Health (USA)
NVQ	National Vocational Qualification
OBE	open book examination
OECD	Organisation for Economic Co-operation and Development
OEL	occupational exposure limit
OES	occupational exposure standard
OHS	Occupational Health Service
OHSAS	Occupational Health and Safety Assessment Series
OHSLB	Occupational Health and Safety Lead Body
OIAC	Oil Industry Advisory Committee
OSH	Occupational Safety and Health or Occupational Health and Safety
OSHA	Occupational Safety and Health Administration (USA)
PPE	personal protective equipment
ppm	parts per million
PTFE	polytetrafluoroethylene
PUWER	Provision and Use of Work Equipment Regulations (UK)
PVC	polyvinyl chloride
RCD	residual current device
RCE	Report of the Committee of Experts
REACH	Registration Evaluation and Authorisation and restriction of Chemicals
RIDDOR	Reporting of Injuries, Diseases and Dangerous Occurrences Regulations (UK)
ROES	representative(s) of employee safety
RoSPA	Royal Society for the Prevention of Accidents
RPE	respiratory protective equipment
RRFSO	Regulatory Reform (Fire Safety) Order (UK)
RTA	road traffic accident
SaHW	Safe and Healthy Working
sfairp	so far as is reasonably practicable
SMEs	small and medium-sized enterprises
SPL	sound pressure level
SQA	Scottish Qualification Authority (Scotland)
STEL	Short-term Exposure Limit
SWL	safe working load
SWP	safe working pressure

TLV	threshold limit value		VRO	Vocationally Related Qualifications (England, Wales, Northern Ireland)
TUC	Trades Union Congress			
TWA	time-weighted average		WAHR	Work at Height Regulations (UK)
UEAL	Upper Exposure Action Level		WBV	whole-body vibration
UK	United Kingdom		WEL	workplace exposure limit
ULD	upper limb disorder		WHO	World Health Organization
UNEP	United Nations Environment Programme		WRULD	work-related upper limb disorder
UNESCO	United Nations Educational, Scientific and Cultural Organization			
VAWR	Vibration at Work Regulations (UK)			

See ILO for more information on abbreviations and acronyms at www.ilo.org/legacy/english/protection/safework/cis/products/safetytm/acronym.htm

Coronavirus (COVID-19): working safely

Since we completed this seventh edition, the world has been shaken by the Covid-19 pandemic. This has changed the way in which we use places of work and our ability to mix with other people both outside in the open air and more particularly inside buildings. The UK, Scottish, Welsh and Northern Ireland Governments have all played their part in developing regulations and guidance. In addition, Public Health bodies and the HSE have been involved in developing, communicating and enforcing the requirements.

All places where people from more than one household unit may come in contact have been subject to restrictions which vary across the UK. In the second wave of the virus in the Autumn of 2020, restrictions vary from one area/county to another. This includes: private homes; gardens; public outside spaces; a wide variety of public buildings; private offices; factories; other workplaces; and most forms of transport.

For workplaces, the HSE's guidance is very accessible and regularly kept up to date. For details on making workplaces Covid-secure see: https://www.hse.gov.uk/coronavirus/index.htm

Coronavirus information for people and businesses can be found on the following websites:

- England: https://www.gov.uk/coronavirus
- Scotland: https://www.gov.scot/coronavirus-covid-19/
- Wales: https://gov.wales/coronavirus
- Northern Ireland: https://www.nidirect.gov.uk/campaigns/coronavirus-covid-19

The summary contained here comes from the HSE guidance. Employers have a legal duty to ensure the health and safety of their employees and those who may be affected by their activities. This includes the protection from health risks such as Covid-19.

The Actions require:

- A risk assessment to be prepared and reviewed when conditions or requirements change. The HSE have a template for a suitable and sufficient risk assessment: https://www.hse.gov.uk/coronavirus/assets/docs/risk-assessment.pdf;
- Homeworking so as to avoid contact between people to be considered, wherever possible;
- Travel to be limited and carefully controlled to maintain social distancing on public transport and/or private vehicles;
- Maintain social distancing at all times, particularly inside buildings, or where this is not possible reduce the number of people in the workplace;
- Personal hygiene, especially hand washing and the use of sanitiser, to be regular and thorough;
- Cleaning of work areas (especially surfaces which are frequently touched), sanitation facilities, canteens, common areas and vehicles to be effective and frequent. Sanitiser to be provided at entrance and exit points;
- Assigning and keeping people in shift teams;
- Using screens and barriers to separate people from each other;
- Providing adequate ventilation;
- Clearly mark out distancing and travel directions (one-way systems if possible);
- Provide adequate information and guidance to employees on the standards and procedures to be followed;
- Ensure that the standards in workplaces and facilities are properly maintained by regular inspections and audits;
- Ensure face coverings are worn indoors and in accordance with regulations and guidance. However, face coverings are mainly intended to protect others and not the wearer. When used correctly, they cover the nose and mouth, which are the main sources of transmitting coronavirus (COVID-19). There is more advice on face coverings at https://www.gov.uk/government/publications/face-coverings-when-to-wear-one-and-how-to-make-your-own/face-coverings-when-to-wear-one-and-how-to-make-your-own. This includes guidance on:
  - ▷ When and where you must wear them and enforcement measures;
  - ▷ How to wear them correctly;
  - ▷ Different arrangements in England, Scotland and Wales;

▷ The use of face coverings as a public health protection measure falls under specific regulations. Face coverings are not classed as personal protective equipment (PPE) because:
  ▷ there is currently no need to conform to a manufacturing standard;
  ▷ they do not provide protection for work risks such as dust and spray.

▶ Surgical face masks are designed to be worn in medical settings to limit the spread of infection. They are not considered to be PPE when worn outside of healthcare activities. Unlike face coverings, they:
  ▷ are manufactured to a recognised standard;
  ▷ are resistant to droplets of fluids and splashes.

▶ First Aid during the pandemic needs to be carefully reassessed to ensure that first aiders are properly protected and there is sufficient first aid cover with a reduced workforce. For detailed guidance see https://www.hse.gov.uk/coronavirus/first-aid-and-medicals/first-aid-certificate-coronavirus.htm.

Accommodation

Employers who provide accommodation for their workers should consider the following:

▶ Identify measures to keep workers safe while they are staying in accommodation and working, in the Covid-19 risk assessment;
▶ Minimise numbers of people living in shared accommodation;
▶ Treat each accommodation living unit (e.g. caravan) as a household;
▶ How you will ensure that, if one person in the household shows symptoms, all will self-isolate in line with the government guidance, see: https://www.gov.uk/government/publications/covid-19-stay-at-home-guidance;
▶ Workers should not live or stay in more than one household;
▶ Keep people who live in the same household together in the same work group;
▶ Try not to mix households while they are working;
▶ Try to keep each household socially distanced from other households;
▶ Check workers' health before they start work each day;
▶ If a worker develops symptoms of coronavirus while working they should return to their household. The individual, and anyone else highlighted by government guidance should then follow the government guidelines on isolation;
▶ Provide fire safety precautions as normal.

Arriving and leaving work

The precautions necessary include:

▶ Reduce the chance of anyone with coronavirus entering the premises, for example, display signs asking people not to enter if they have COVID symptoms;

▶ Limit the number of people entering a workplace, or parts of a workplace, to avoid overcrowding. If possible arrange partial working from home for some employees;
▶ Where possible, introduce shift working to limit the number of people in the workplace at any one time. This may also reduce the burden on public transport if employees use it to travel to work;
▶ Consider introducing suitable barriers or screens where people regularly interact, such as service desks or reception areas. If used, ensure they are cleaned and disinfected regularly;
▶ Consider these other control measures for when people are arriving or leaving work:
  ▷ Stagger arrival and departure times to reduce crowding into and out of the workplace;
  ▷ Provide additional parking or facilities such as bike racks to help people walk, run, or cycle to work where possible;
  ▷ Provide more entry points to reduce congestion;
  ▷ Provide more storage for workers for clothes and bags. Encourage storage of personal items and clothing in personal storage spaces, for example lockers during shifts;
  ▷ Use markings and introduce one-way flow at entry and exit points;
  ▷ Provide handwashing facilities at entry and exit points, or hand sanitiser where handwashing is not possible;
  ▷ Where possible, avoid using touch-based security devices such as keypads. If you need to use them, make sure they're cleaned regularly.

Deliveries

Put in place procedures for dealing with deliveries and visitors. Allocate time slots for customers. See https://www.hse.gov.uk/coronavirus/drivers-transport-delivery.htm.

Vehicle cleaning

Where company vehicles have different users ensure they are cleaned regularly. Clean commonly touched areas in vehicles including:

▶ handles (inside and out);
▶ steering wheel and starter button;
▶ centre touchscreen and stereo;
▶ all control switches, and levers;
▶ keys and key fob.

Ensure that people practice good hygiene before and after using the vehicle. Provide hand sanitiser where necessary.

Talk to your workers and provide information

Talk to all workers and encourage them to co-operate with procedures and cleaners. They can make sure that surfaces are left clear at the end of the day so that deep cleaning can be done more effectively. Papers or items left on surfaces might make cleaning less effective.

Keeping people informed about any changes to cleaning, and the reasons for it, reduces the chance of uncooperative staff. It may also reassure workers that the employer is doing what they can to keep them safe.

Use posters to inform people that regular cleaning will take place and that they should co-operate with cleaners.

Where users are expected to clean equipment after use as part of the cleaning regime, posters can help to remind people of their responsibilities.

Proect home workers

An employer has the same health and safety responsibilities for home workers as for any other workers.

When someone is working from home, permanently or temporarily, an employer should consider:

▶ How will they keep in touch with them?
▶ What work activity will they be doing (and for how long)?
▶ Can it be done safely?
▶ Do control measures need to be put in place to protect them?

Home working can cause work-related stress and affect people's mental health. Being away from managers and colleagues could make it difficult to get proper support. Put procedures in place so you can keep in direct contact with home workers so you can recognise signs of stress as early as possible.

It is also important to have an emergency point of contact and to share this so people know how to get help, if they need it.

The moral, financial and legal reasons for health and safety management

Chapter summary

This chapter covers the following topics in Element 1 of the NEBOSH NGC syllabus:

Morals and money

The force of law – punishment and compensation

The most important legal duties for employers and workers

Managing contractors effectively

Learning outcomes

These are the NEBOSH learning outcomes covered by this chapter:

The learner will be able to:

1. Justify health and safety improvements using moral, financial and legal arguments
2. Advise on the main duties for health and safety in the workplace
3. Help their organisation manage contractors

Assessment criteria

1.1 Discuss the moral, financial and legal reasons for managing health and safety in the workplace

1.2 Explain how the law works and the consequences of non-compliance

1.3 Summarise the main health and safety duties of employers and workers in HSW Act 1974 and MHSWR 1999

1.4 Explain how contractors should be selected, monitored and managed

Introduction

Occupational health and safety is relevant to all branches of industry, business and commerce including traditional industries, information technology companies, the National Health Service, care homes, schools, universities, leisure facilities and offices.

The purpose of this chapter is to introduce the foundations on which appropriate health and safety management systems may be built. Occupational health and safety affects all aspects of work. In a low hazard organisation, health and safety may be supervised by a single competent manager. In a high hazard manufacturing plant, many different specialists, such as engineers (electrical, mechanical and civil), lawyers, medical doctors and nurses, trainers, work planners and supervisors may be required to assist the professional health and safety practitioner in ensuring that there are satisfactory health and safety standards within the organisation.

There are many barriers to the achievement of good standards. The pressure of production or performance targets, financial constraints and the complexity of the organisation are typical examples of such obstacles. In difficult economic times, organisations need to reduce their costs, and this often impacts on the management of health and safety such as the maintenance of complex machinery and the training of workers in safe systems of work. However, there are some powerful incentives for organisations to strive for high health and safety standards. These incentives are moral, legal and financial.

Corporate responsibility, a term used extensively in the 21st century world of work, covers a wide range of issues. It includes the effects that an organisation's business has on the environment, human rights and poverty in the least developed nations. Health and safety in the workplace is an important corporate responsibility issue.

Corporate responsibility has various definitions. However, broadly speaking, it covers the ways in which organisations manage their core business to add social, environmental and economic value in order to produce a positive, sustainable impact on both society and the business itself. Terms such as 'corporate social responsibility', 'socially responsible business' and 'corporate citizenship' all refer to this concept.

The UK Health and Safety Executive's (HSE) mission is to ensure that the risks to health and safety of workers are properly controlled. In terms of corporate responsibility, it is working to encourage organisations to:

▶ improve health and safety management systems to reduce injuries and ill health;

▶ demonstrate the importance of health and safety issues at board level;

▶ report publicly on health and safety issues within their organisation, including their performance against targets.

The HSE believes that effective management of health and safety:

▶ is vital to employee well-being;

▶ has a role to play in enhancing the reputation of businesses and helping them achieve high-performance teams;

▶ is financially beneficial to business.

This chapter covers the legal responsibilities that exist between people who control premises and those who use them, and between contractors and those who hire them; and the duties of suppliers, manufacturers and designers of articles and substances for use at work.

The chapter also describes the legal responsibilities that exist between duty-holders under the Construction (Design and Management) Regulations (CDM) to ensure that health and safety is fully integrated into the management of any construction project and to encourage everyone involved with the project (the client, designer and principal contractor) to work together effectively.

Some basic definitions

Before a detailed discussion of health and safety issues can take place, some basic occupational health and safety definitions are required.

▶ **Health** – The protection of the bodies and minds of people from illness resulting from the materials, processes or procedures used in the workplace.

▶ **Safety** – The protection of people from physical injury. The borderline between health and safety is ill-defined and the two words are normally used together to indicate concern for the physical and mental well-being of the individual at the place of work.

▶ **Welfare** – The provision of facilities to maintain the health and well-being of individuals at the workplace. Welfare facilities include washing and

sanitation arrangements, the provision of drinking water, heating, lighting, accommodation for clothing, seating (when required by the work activity or for rest), eating and rest rooms. First-aid arrangements are also considered as welfare facilities.

▶ **Occupational or work-related ill health** – This is concerned with those illnesses or physical and mental disorders that are either caused or triggered by workplace activities. Such conditions may be induced by the particular work activity of the individual, or by activities of others in the workplace. The time interval between exposure and the onset of the illness may be short (e.g. asthma attacks) or long (e.g. deafness or cancer).

▶ **Environmental protection** – These are the arrangements to cover those activities in the workplace which affect the environment (in the form of flora, fauna, water, air and soil) and, possibly, the health and safety of employees and others. Such activities include waste and effluent disposal and atmospheric pollution.

▶ **Accident** – This is defined by the Health and Safety Executive (HSE) as 'any unplanned event that results in injury or ill health of people, or damage or loss to property, plant, materials or the environment or a loss of a business opportunity'. Other authorities

Figure 1.1 At work

define an accident more narrowly by excluding events that do not involve injury or ill health. This book will always use the HSE definition.

▶ **Near miss** – This is any incident that could have resulted in an accident. Knowledge of near misses is very important as research has shown that, approximately, for every 10 'near miss' events at a particular location in the workplace, a minor accident will occur.

▶ **Dangerous occurrence** – This is a 'near miss' which could have led to serious injury or loss of life. Dangerous occurrences are defined in the Reporting of Injuries, Diseases and Dangerous Occurrences Regulations (often known as RIDDOR) and are always reportable to the enforcement authorities. Examples include the collapse of a scaffold or a crane or the failure of any passenger-carrying equipment.

▶ **Hazard and risk**
  ▷ A **hazard** is something with the *potential* to cause harm (this can include articles, substances, plant or machines, methods of working, the working environment and other aspects of work organisation). Hazards take many forms including, for example, chemicals, electricity and working from a ladder. A hazard can be ranked relative to other hazards or to a possible level of danger.
  ▷ A **risk** is the *likelihood* of potential harm from that hazard being realised. Risk (or strictly the level of risk) is also linked to the severity of its consequences. A risk can be reduced and the hazard controlled by good management.

It is very important to distinguish between a *hazard* and a *risk* – the two terms are often confused and activities such as construction work are frequently called high risk when they are high hazard. Although the hazard will continue to be high, the risks will be reduced as controls are implemented. The level of risk remaining when controls have been adopted is known as the **residual risk**. There should only be high residual risk where there is poor health and safety management and inadequate control measures.

1.1 The moral and financial reasons for promoting good standards of health and safety

The first concern of most managers when they start work at a new organisation is to understand the implications of their new role and to form good relationships with other members of the team. Concerns about health and safety are often not a first or even second consideration. So why bother about health and safety?

There are three basic drivers for good health and safety management; these are moral, legal and financial reasons. The moral reasons are centred on the need to protect people from injury and disease while they are at work. The legal reasons are embodied in the criminal and civil law, and the financial reasons come as a consequence of infringements of health and safety law with the consequent fines, compensation payments, associated financial costs and even, in extreme cases, imprisonment. Each of these reasons will now be discussed in turn.

1.1.1 Moral reasons

The moral reasons are supported by the occupational accident and disease rates.

Accidents

Accidents at work can lead to serious injury and even death. Over recent years, between 130 and 190 workers have been fatally injured each year at their place of work and over 20,000 others suffered major injuries such as amputations, fractures and burns. Statistics are collected on all people who are injured at workplaces – not just employees – and over 350 members of the public have been fatally injured at places of work. Since 1995 suicides and trespassers on the railways have been included in the HSE figures – this has led to a significant increase in the overall fatality figures.

Table 1.1 shows a typical annual accident breakdown between employees, self-employed and members of the public. These figures give an indication of the scale of the problem even though the actual figures for any given year may be slightly higher or lower. The industries with the highest fatality and major accident rates (per 100,000 employees) are agriculture, construction, transport, waste and recycling and manufacturing, and the most common causes are slipping or tripping (41%), and falls from a height (16%). A further large number of injuries to employees caused an absence from work of over seven days. Of these less serious injuries, the most common causes were handling, lifting or carrying (36%), and slipping or tripping (24%).

Accident statistics are published each year by HSE and indicate that there is a need for health and safety awareness even in occupations which many would consider very low hazard, such as the health services and hotels. In fact over 70% of all deaths occur in the service sector, and manufacturing is considerably safer than construction and agriculture. These latter two industries accounted for almost half of all fatal injuries to workers. Some of the most common causes of deaths and serious injuries in the agriculture sector include handling livestock and using tractors, quad bikes and chainsaws. Finally, a further significant proportion of work-related accidents occur while travelling on roads and not at the workplace.

These injury figures show that there is clearly a very strong moral case for improvement in health and safety performance.

Table 1.1 Annual accidents for different groups of people

	Fatalities	Major
Total	520	46,000
Employees	130	20,000
Self-employed	40	1,000
Members of the public	350	25,000

Disease

Work-related ill health and occupational disease can lead to absence from work and, in some cases, to death. Such occurrences may also lead to costs to the State (the Industrial Injuries Scheme) and to individual employers (sick pay and, possibly, compensation payments). Each year thousands of people die from work-related diseases mainly due to past working conditions. The industry sectors having ill-health rates that consistently have been higher than the rate for all industries are health and social work, public administration and education.

Table 1.2 Approximate proportions (%) of cases of work-related ill health reported by General Practitioners in any year

Type of illness	Percentage
Musculoskeletal disorders	53
Mental ill health (stress, anxiety)	31
Dermatitis and other skin disorders	10
Other diagnosis including infections	4
Respiratory disease	2
Hearing loss	0.1

Stress and musculoskeletal disorders are the largest causes of work-related ill health. There are on average over 400,000 workers suffering from stress-related ill health each year of which 50% have suffered for over one year or longer causing over 10 million working days lost each year. Over 400,000 workers suffer from musculoskeletal disorders (mainly back pain and upper limb disorders) and 60% of these people suffer for one year or longer causing over 7 million working days lost each year. Data from 300 General Practitioners (GPs) (Table 1.2) confirms that musculoskeletal disorders are the most common type of work-related illness, but that mental ill health (usually caused by stress) accounts for more working days lost (Table 1.3).

Occupational asthma is the UK's fastest growing workplace disease and affects between 1,500 and 3,000 people each year. Every year in the UK, 7,000 people are thought to contract asthma that is either caused by their work or is made worse by it. In some cases people are left disabled and unable to work. Other work-related respiratory diseases include chronic

obstructive pulmonary disease (COPD), pneumoconiosis and silicosis. According to GPs, there are 8,500 new cases of work-related respiratory diseases each year.

Work-related cancer is another serious problem and results in 8,000 deaths and 13,500 new cases each year. The leading cause of these deaths is occupational exposure to asbestos which accounts for least 4,500 deaths each year although this figure is expected to increase in the future. The most common forms of such cancers are lung cancer and mesothelioma.

Recent research has shown that one in five people who are on sickness leave from work for 6 weeks will stay off work and leave paid employment.

1.1.2 Societal expectations of good standards of health and safety

Societal expectations are not static and tend to rise over time, particularly in a wealthy nation like the UK. For example, the standards of safety accepted in a motor car 50 years ago would be considered to be totally inadequate at the beginning of the 21st century. People expect safe, quiet, comfortable cars that do not break down and which retain their appearance for many thousands of miles. Industry should strive to deliver these same high standards for the health and safety of employees or service providers. The question is whether societal expectations are as great an influence on workplace safety standards as they are on product safety standards. Society can influence standards through:

▶ people only working for good employers. This is effective in times of low unemployment;
▶ national and local news media highlighting good and bad employment practices;
▶ schools teaching good standards of health and safety;
▶ the purchase of fashionable and desirable safety equipment, such as trendy crash helmets for mountain bikes;
▶ buying products only from responsible companies (the difficulty of defining what is responsible has been partly overcome through ethical investment criteria, but this is possibly not widely enough understood to be a major influence);
▶ watching TV and other programmes which improve safety knowledge and encourage safe behaviour from an early age.

1.1.3 Financial reasons

Costs of accidents

Any accident or incidence of ill health will cause both direct and indirect costs and incur an insured and an uninsured cost. It is important that all of these costs are taken into account when the full cost of an accident is calculated. In a study undertaken by the HSE, it was shown that indirect costs or hidden costs could be

36 times greater than direct costs of an accident. In other words, the direct costs of an accident or disease represent the tip of the iceberg when compared to the overall costs (Figure 1.2). Annually over 27 million days are lost in the UK due to injuries and work-related ill health and Table 1.3 shows the causes of these lost days. The total cost of illness and injury to UK industry has been estimated as £15 billion – 60% on ill health and about 40% injuries and fatalities.

Clearly ill health has a serious effect on the economy of the UK due to lost productivity, reduction in income tax receipts, increases in long-term sickness and increased healthcare costs. Injuries and ill health cost the UK an estimated £15 billion, with £10 billion from illness and £5 billion in injury each year.

Table 1.3 Causes of working days lost in the UK

Cause	Percentage
Stress and anxiety	40
Musculoskeletal disorders	28
Injury	17
Other illnesses	15

Source: HSE.

1.1.4 The business case for health and safety

The business case for health and safety is centred on the potential costs of poor standards of health and safety. Fines in excess of £250,000 and even higher levels of compensation payments are not uncommon. As mentioned earlier, the costs may be direct or indirect and insured or uninsured. Some examples of these follow.

Direct costs

These are costs which are directly related to the accident and may be insured or uninsured.

Insured direct costs normally include:

- claims on employers and public liability insurance;
- damage to buildings, equipment or vehicles;
- any attributable production and/or general business loss;
- the absence of employees.

Uninsured direct costs include:

- fines resulting from prosecution by the enforcement authority;
- sick pay;
- some damage to product, equipment, vehicles or process not directly attributable to the accident (e.g. caused by replacement staff);
- increases in insurance premiums resulting from the accident;
- any compensation not covered by the insurance policy due to an excess agreed between the employer and the insurance company;
- legal representation following any compensation claim.

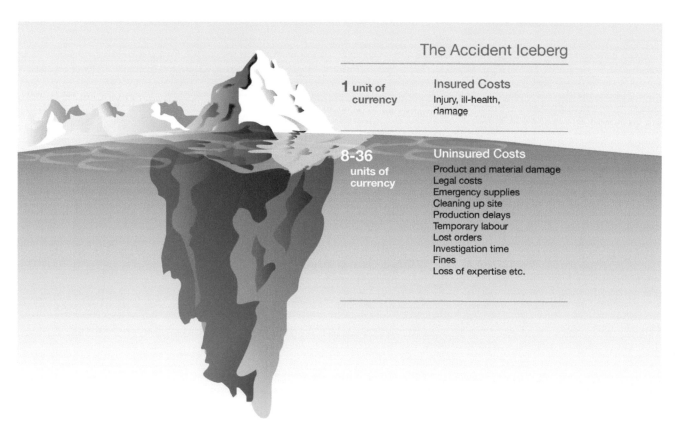

Figure 1.2 Insured and uninsured costs

Indirect costs

These are costs which may not be directly attributable to the accident but may result from a series of accidents. Again these may be insured or uninsured.

Insured indirect costs include:

▶ a cumulative business loss;
▶ product or process liability claims;
▶ recruitment of replacement staff.

Uninsured indirect costs include:

▶ loss of goodwill and a poor corporate image;
▶ accident investigation time and any subsequent remedial action required;
▶ production delays;
▶ extra overtime payments;
▶ lost time for other employees, such as a first-aider, who tend to the needs of the injured person;
▶ the recruitment and training of replacement staff;
▶ additional administration time incurred;
▶ first-aid provision and training;
▶ lower employee morale possibly leading to reduced productivity.

Some of these items, such as business loss, may be uninsurable or too prohibitively expensive to insure. Therefore, insurance policies can never cover all of the costs of an accident or disease because either some items are not covered by the policy or the insurance excess is greater than the particular item cost.

1.1.5 Employers' Liability Compulsory Insurance

The Employers' Liability (Compulsory Insurance) Act makes it a legal requirement for all employers to have employers' liability insurance. This ensures that any employee, who successfully sues his/her employer following an accident, is assured of receiving compensation irrespective of the financial position of the employer. For employers, the insurance covers the cost of legal fees and compensation in the event of a claim by a worker. Only very few businesses are not required to have employers' liability insurance.

There is a maximum penalty of up to £2,500 for every day without appropriate cover for employers who do not have such insurance. In addition, one or more copies of the current certificate must be displayed at each place of business and be 'reasonably protected' from being defaced or damaged. Recently, the rules requiring an employer to display the certificate have changed, so that the requirement will be satisfied if the certificate is made available in electronic format and is reasonably accessible to relevant employees.

1.1.6 Legal reasons

The legal reasons concerning the employer's duty of care in criminal and civil law will be covered later in this chapter.

Some statistics on legal enforcement indicate the legal consequences resulting from breaches in health and safety law. There have been some very high compensation awards for health and safety cases in the Civil Courts and fines in excess of £100,000 in the Criminal Courts. Table 1.4 shows the typical number of enforcement notices served each year in Great Britain. Most notices are served in the manufacturing, construction and agricultural sectors. Local authorities serve 40% of all improvement notices and 20% of all prohibition notices. A small number of enforcement notices are also issued by the Office of Rail Regulation (ORR).

Table 1.4 also indicates the typical number of prosecutions by the HSE and local authorities each year. HSE (together with the Procurator Fiscal in Scotland) present 80% of the prosecutions and the remainder is presented by Local Authority Environmental Health Officers. These prosecutions result in approximately £15m in fines each year. Most of these prosecutions were for infringements of various Construction Regulations (including the Work at Height Regulations) and the Provision and Use of Work Equipment Regulations.

There are clear legal reasons for effective health and safety management systems.

There is clearly a strong moral, economic and legal case to do more to reduce this unacceptable level of injury and ill health in the workplace.

1.2 The legal framework for the regulation of health and safety including sources and types of law

1.2.1 Sources of law (England and Wales)

There are two sources of law – statute law and common law.

Table 1.4 Typical recent annual health and safety enforcement activity in Great Britain

	Improvement notices	Deferred prohibition	Immediate prohibition	Offences prosecuted
HSE	5,700	17	3,000	950*
Local Authorities	3,300	13	1,300	200
Total	9,000	30	4,300	1,150

Note

* Includes 30 prosecutions by the Procurator Fiscal in Scotland.

Statute law

Statute law is a law which has been laid down by Parliament as Acts of Parliament and other legislation. In health and safety, an Act of Parliament, the HSW Act 1974, lays down a general legal framework. Specific health and safety duties are, however, covered by Regulations or Statutory Instruments – these are also examples of statute law. If there is a conflict between statute and common law, statute law takes precedence. However, as with common law, judges interpret statute law usually when it is new or ambiguous. Although for health and safety, statute law is primarily the basis of criminal law, there is a tort of breach of statutory duty. However the application of this tort has been considerably reduced by the Enterprise and Regulatory Reform Act 2013 (see later under 1.2.4).

Common law

Common law dates from the 11th century when William I set up Royal Courts to apply a uniform (common) system of law across the whole of England. Prior to that time, there was a variation in law, or the interpretation of the same law, from one town or community to another. Common law is based on judgements made by courts (or strictly judges in courts). In general, courts are bound by earlier judgements on any particular point of law – this is known as 'judicial precedent'. Lower courts must follow the judgements of higher courts. Hence judgements made by the Justices in the Supreme Court form the basis of most of the common law currently in use.

In health and safety, the legal definition of negligence, duty of care and terms such as 'practicable' and 'as far as is reasonably practicable' are all based on legal judgements and form part of the common law. Common law also provides the foundation for most civil claims made on health and safety issues.

1.2.2 Sub-divisions or types of law

There are two sub-divisions of the law that apply to health and safety issues: criminal law and civil law.

Criminal law

Criminal law consists of rules of behaviour laid down by the Government or the State and, normally, enacted by Parliament through Acts of Parliament. These rules or Acts are imposed on the people for the protection of the people. Criminal law is enforced by several different Government Agencies who may prosecute individuals and organisations for contravening criminal laws. It is important to note that, except for very rare cases, only these agencies are able to decide whether to prosecute an individual or not.

An individual or organisation who breaks criminal law is deemed to have committed an offence or crime and, if he/she is prosecuted, the court will determine whether he/she is guilty or not. If the individual is found guilty, the court could sentence him/her to a fine or imprisonment. Owing to this possible loss of liberty, the level of proof required by a Criminal Court is very high and is known as proof 'beyond reasonable doubt', which is as near certainty as possible. Although the prime object of a Criminal Court is the allocation of punishment, the court can award compensation to the victim or injured party. One example of criminal law is the Road Traffic Act, which is enforced by the police. However, the police are not the only criminal law enforcement agency. The Health and Safety at Work (HSW) etc. Act is another example of criminal law and this is enforced either by the HSE or Local Authority Environmental Health Officers (EHOs). Other agencies which enforce criminal law include the Fire Authority, the Environment Agency, Trading Standards and Customs and Excise.

There is one important difference between procedures for criminal cases in general and criminal cases involving health and safety. The prosecution in a criminal case has to prove the guilt of the accused beyond reasonable doubt. Although this obligation is not totally removed in health and safety cases, Section 40 of the HSW Act 1974 transferred, where there is a duty to do something 'so far as is reasonably practicable' or 'so far as is practicable' or 'use the best practicable means', the onus of proof is on the accused to show that there was no better way to discharge his/her duty under the Act. However, when this burden of proof is placed on the accused, they only need to satisfy the court on the balance of probabilities that what they are trying to prove has been done.

Civil law

Civil law concerns disputes between individuals or individuals and companies. An individual sues another individual or company to address a civil wrong or tort (or delict in Scotland). The individual who brings the complaint to court is known as the claimant or plaintiff (pursuer in Scotland), and the individual or company who is being sued is known as the defendant (defender in Scotland).

The Civil Court is concerned with liability and the extent of that liability, rather than guilt or non-guilt. Therefore, the level of proof required is based on the 'balance of probability', which is a lower level of certainty than that of 'beyond reasonable doubt' as required by the Criminal Court. If a defendant is found to be liable, the court would normally order him/her to pay compensation and possibly costs to the plaintiff. However, the lower the balance of probability, the lower the level of compensation awarded. In extreme cases, where the balance of probability is just over 50%, the plaintiff may 'win' the case but lose financially because costs may not be awarded and the level of compensation is low. The level of compensation may also be reduced through the defence of **contributory negligence**, which is discussed later under 1.2.4. For cases involving health and safety, civil disputes usually follow accidents or illnesses and concern negligence or a breach of statutory duty. The vast majority of cases are settled 'out of court'. Although actions are often between individuals, where the defendant is an employee who was acting in the course of his/her employment during the alleged incident, the defence of the action is transferred to his/her employer – this is known as **vicarious liability**. The civil action then becomes one between the individual and the employer.

The Supreme Court has stated that vicarious liability can be established in 'employment like relationships' (e.g. voluntary workers) and not only in employee/employer relationships.

This liability could be reduced by demonstrating that all reasonable steps have been taken to prevent the alleged perpetrator from undertaking the alleged wrongful act. This might include any of the following actions:

- ▶ the training of staff in equality and diversity principles;
- ▶ the establishment and communication of appropriate policies to all staff;
- ▶ the highlighting clearly to all staff what is deemed acceptable and unacceptable conduct and the consequences of the latter.

The relationship between the sub-divisions and sources of law

The two sub-divisions of law may use either of the two sources of law (as shown in Figure 1.3). For example, murder is a common law crime. In terms of health and safety, however, criminal law is only based on statute law, whereas civil law may be based on either common law or statute law.

In summary, criminal law seeks to protect everyone in a society whereas civil law seeks to protect and recompense any individual citizen or organisation.

1.2.3 Criminal law liabilities

The role, functions and powers of Health and Safety Executive (HSE)/HSE Northern Ireland (HSENI); Procurator Fiscal (Scotland) and local authorities Health and Safety Executive (HSE)

The Health and Safety at Work etc. Act 1974 (HSW Act) established the Health and Safety Commission and gave it the responsibility to draft new Regulations and to enforce them either through its executive arm,

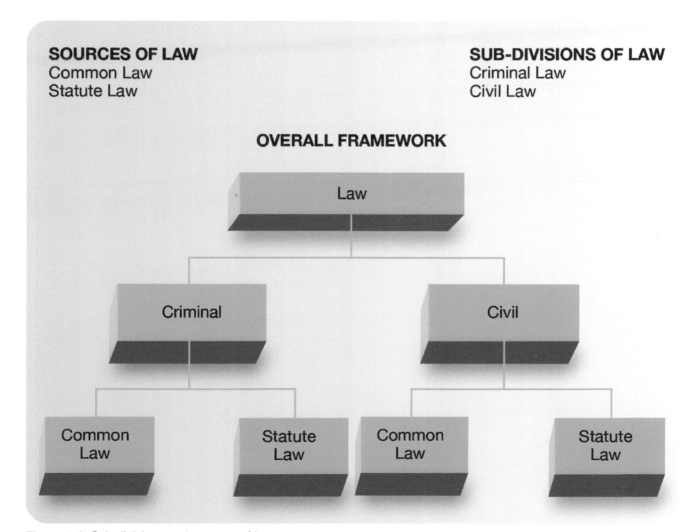

SOURCES OF LAW
Common Law
Statute Law

SUB-DIVISIONS OF LAW
Criminal Law
Civil Law

OVERALL FRAMEWORK

Law

Criminal

Civil

Common Law

Statute Law

Common Law

Statute Law

Figure 1.3 Sub-divisions and sources of law

known as the HSE, or through the Local Authority Environmental Health Officer (EHO).

In April 2008, the Health and Safety Commission and HSE merged into a single unified body called the HSE. The merger increased the outside, or non-executive, input to its work. This improved communication, accountability and the oversight of longer term strategy. The merger meant that the size of the Board of the Executive is no more than 12 members. The Board of the Executive assumed responsibility for running all aspects of the organisation, including setting the overall strategic direction, financial and performance management and prioritisation of resources.

The HSE retains its independence, reflecting the interests of employers, employees and local authorities and is committed to maintaining its service delivery.

In Northern Ireland, the **Health and Safety Executive of Northern Ireland (HSENI)** has a very similar role to that of the HSE.

In England and Wales, the **Crown Prosecution Service (CPS)** is the main prosecuting authority for criminal cases prepared by the police and other investigators. CPS prosecutors present cases in both the Magistrates' Courts and the higher courts. The head of the CPS is the Director of Public Prosecutions (DPP). In Northern Ireland, the main prosecuting authority is the **Public Prosecution Service for Northern Ireland (PPSNI)** and has a similar role to the CPS. Both organisations are independent of the police and decide whether a prosecution is to proceed to court. For health and safety offences, the CPS decides on manslaughter and corporate manslaughter cases.

Although occupational health and safety in Scotland is a reserved issue (it was not devolved to the Scottish Executive), there are some differences in some areas of activity. A summary of the concordat between the HSE and the Scottish Executive is given in Chapter 13 and covers several of these differences.

The **Procurator Fiscal** is the public prosecutor in Scotland with a similar role to that of the Crown Prosecution Service in England and Wales. They investigate all sudden and suspicious deaths (similar to a coroner), conduct Fatal Accident Inquiries (similar to an inquest in England and Wales) and handle criminal complaints against the police. They also receive reports from specialist reporting agencies such as Her Majesty's Revenue and Customs.

For the majority of crimes in Scotland the Procurators Fiscal present cases for the prosecution in the Sheriff and District Courts and have the discretion not to prosecute and pursue alternatives free from political interference. However, they are always subject to the directions of Crown Office and the Lord Advocate.

Procurators Fiscal make preliminary investigations into criminal cases, take written statements from witnesses (known as precognition) and are responsible for the investigation and prosecution of crime. This includes the power to direct the police in their investigation, but except for serious crimes such as murder the police normally complete their enquiries before involving the Procurator Fiscal. They have the power to offer fixed penalties instead of prosecution (a fiscal fine) and to give warnings, compensation orders, work orders and road traffic fixed penalties. They can also divert offenders from prosecution to social workers, psychological counselling or psychiatric treatment.

All suspicious, sudden and accidental deaths must be reported to the Procurator Fiscal, who prosecutes on behalf of the HSE. They have a responsibility to identify if any criminal action has occurred and, where appropriate, prosecute. Where a criminal offence is suspected to have occurred, the Procurator Fiscal will instruct the local police to investigate. The Procurator Fiscal also investigates and, where appropriate, prosecutes cases of homicide and corporate homicide involving health and safety issues. However, if the police decide that there is not a case of homicide or a corporate homicide, the investigation will be handed over to the HSE to investigate. As in the rest of the UK, the HSE (or Local Authorities) in Scotland will investigate non-fatal work-related incidents. Finally it is important to note that no offence in Scottish law can be proved unless there is evidence from more than one source – this is known as corroboration.

The prime function of the HSE remains the same – to monitor, review and enforce health and safety law and to produce codes of practice and guidance. However, the HSE also undertakes many other activities, such as the compilations of health and safety statistics, leading national health and safety campaigns, investigations into accidents or complaints, visiting and advising employers and the production of a very useful website.

The HSE proposes, as part of its strategic plan, to:

▶ lead and engage with others to improve workplace health and safety;
▶ provide an effective regulatory framework;
▶ secure effective management and control of risk; and
▶ reduce the likelihood of low-frequency, high-impact catastrophic incidents.

Fee for Intervention (FFI)

The Health and Safety (Fees) Regulations introduced the Fee for Intervention (FFI) scheme which places a duty on HSE to recover its costs for carrying out its regulatory functions from those found to be in **material breach** of health and safety law. A **material breach** is when, in the opinion of the HSE inspector, there is or has been a contravention of health and safety law that requires them to issue a notice in writing of that opinion to the duty-holder.

Written notification is given by the inspector in either a simple notification of the particular contravention or an improvement or prohibition notice, or a prosecution. The notification will include the following information:

▷ the law that the inspector's opinion relates to;
▷ the reasons for their opinion; and
▷ notification that a fee is payable to HSE.

Any organisation that is served such a notification is liable for fines to cover HSE-related costs, including call-outs, inspections and investigations and the taking of enforcement action. Breaches of health and safety law ranging from slips, trips and falls to not providing enough toilets or washing facilities have resulted in HSE fines. The two sectors that received the most fines in the first year of FFI were manufacturing (38%) and construction (36%).

Figure 1.4 Employees at work

Powers of inspectors

Inspectors under the Act work either for the HSE or the Local Authority. Local authorities are responsible for retail and service outlets and the HSE is responsible for all other work premises including the local authorities themselves. Both groups of inspectors have the same powers.

Table 1.5 shows the breakdown of responsibilities between the HSE and the Local Authorities.

In 2020, following the Grenfell Tower fire, the UK Government announced the establishment of a Building Safety Regulator within the HSE. This is intended to raise building safety and performance standards particularly for higher risk buildings.

Table 1.5 Premises inspected by HSE and Local Authorities

HSE inspect:	Local Authorities inspect:
factories	offices (except government offices)
farms	shops
building sites	hotels
nuclear installations	restaurants
mines	leisure premises
schools and colleges	nurseries and playgroups
fairgrounds	pubs and clubs
gas, electricity and water systems	museums (privately owned)
hospitals and nursing homes	places of worship
central and local government premises	sheltered accommodation
offshore installations	care homes

The detailed powers of inspectors are given in Chapter 13. In summary, an inspector has the right to:

▷ enter premises at any reasonable time, accompanied by a police officer, if necessary;
▷ examine, investigate and require the premises to be left undisturbed;
▷ take samples, photographs and, if necessary, dismantle and remove equipment or substances;
▷ require the production of books or other relevant documents and information;
▷ seize, destroy or render harmless any substance or article;
▷ issue enforcement notices and initiate prosecutions.

An inspector may issue a formal caution when an offence has been committed, but it is deemed that the public interest does not require a prosecution. Formal cautions are not normally considered if the offender has already had a formal caution.

Amongst many other duties the **Office of Rail Regulation (ORR)** has responsibility for the independent health and safety regulation of the railway industry in the UK. It is a statutory body and is the combined economic and safety regulatory authority for the UK's railway network.

Its role in health and safety on the railways was originally undertaken by the HM Railway Inspectorate (HMRI). HMRI, which began in 1840, was responsible for overseeing safety on Britain's railways and tramways. From 1990 to 2006, it became part of the Health and Safety Executive and then was transferred to the Office of Rail Regulation.

In 2009, a single rail regulatory body covering both safety and economic issues, the Safety Directorate, was created within the ORR but the individual inspectors continue to be known as Her Majesty's Railway Inspectors.

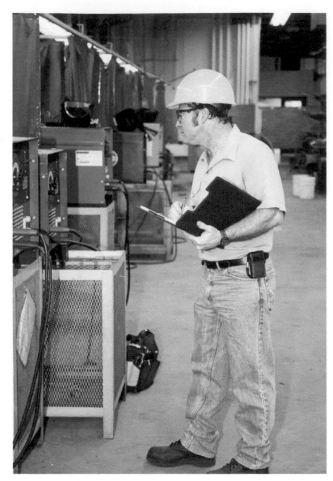

Figure 1.5 The inspector inspects

Enforcement notices

There are two types of enforcement notice.

Improvement Notice – This identifies a specific breach of the law and specifies a date by which the situation is to be remedied. An appeal may be made to the Employment Tribunal and must be made within 21 days. The notice is then suspended until the appeal is either heard or withdrawn. There are five main grounds for an appeal:

1. The inspector interpreted the law incorrectly.
2. The inspector exceeded his/her powers.
3. The breach of the law is admitted but the proposed solution is not practicable or reasonably practicable.
4. The time allowed to comply is too short.
5. The breach of the law is admitted but the breach is so insignificant that the notice should be cancelled.

Prohibition Notice – This is used to halt an activity which the inspector feels could lead to a serious personal injury. The notice will identify which legal requirement is being or is likely to be contravened. The notice takes effect as soon as it is issued. As with the improvement notice, an appeal may be made to the Employment Tribunal but, in this case, the notice remains in place during the appeal process.

The Court of Appeal in the *Chilcott v. Thermal Transfer Ltd* case has ruled that the Employment Tribunal is 'required to consider and decide whether on the facts which were or should have been known to the inspector at the time, a prohibition notice was justified'. Hindsight should not be taken into account when considering such appeals. The ruling is applicable to challenges to improvement notices as well as prohibition notices.

There are two forms of prohibition notice:

▶ an immediate prohibition notice – this stops the work activity immediately until the specified risk is reduced;
▶ a deferred prohibition notice – this stops the work activity within a specified time limit.

Summary of the actions available to an inspector

Following a visit by an inspector to premises, the following actions are available to an inspector:

▶ take no action;
▶ give verbal advice;
▶ give written advice;
▶ formal or simple caution;
▶ serve an improvement notice;
▶ serve a prohibition notice;
▶ prosecute.

In any particular situation, more than one of these actions may be taken.

The decision to prosecute will be made either when there is sufficient evidence to provide a realistic possibility of conviction or when prosecution is in the public interest.

Prosecution and penalties

Indictable and summary offences

Most minor offences, including the majority of health and safety offences, are called summary offences and are tried without a jury only in a Magistrates' Court (or Sheriff Court in Scotland). The most serious offences are called indictable offences (solemn in Scotland) and are tried only by the Crown Court (or the High Court of Justiciary in Scotland). If the defendant is to be tried on indictment, the Magistrates' Court (or lower court) will conduct committal proceedings.

Disqualification of directors

Under Section 2(1) of the Company Directors Disqualification Act 1986, a court is empowered to make a disqualification order against a person convicted of an indictable offence in connection with the promotion, formation or management of a company. Disqualified persons must not, without the leave of the court, be a director, liquidator or administrator of a company, or receiver and manager of a company's property, or in any way, directly or indirectly, be concerned or take part in the promotion, formation or management of a company for a specified period.

The maximum period of disqualification is 5 years when an order is made by a Magistrates' Court, or 15 years by the Crown Court or any other court.

Penalties

In 2008, the Health and Safety (Offences) Act raised the maximum penalties available to the courts in respect of certain health and safety offences by amending Section 33 of the HSW Act. Further, under the Legal Aid, Sentencing and Punishment of Offenders Act 2012 Section 85(1) from 12 March 2012, Magistrates' Courts can issue unlimited fines and imprison for up to 12 months.

The sentencing guidelines that direct judges to set penalties for occupational health and safety regulatory breaches and corporate manslaughter offences were increased significantly in 2016. This has resulted in an increase in penalties of £1m or more. The major factors behind the increase are the guidelines' instruction to courts to weigh the potential for harm in a regulatory breach rather than just the actual harm caused, and the explicit link between fine level and annual turnover.

Magistrates' Court (or lower court)

The Magistrates' Court deals with summary offences.

It can impose an unlimited fine and/or up to 12 months' imprisonment. The sentencing option of 12 months applies in Scotland but only applies in England and Wales when Section 154(1) of the Criminal Justice Act 2003 is enacted.

Crown Court (or higher court)

The Crown Court deals with indictable (solemn in Scotland) offences and some summary cases referred to it by the lower court which may feel that its powers are too limited.

Fines are unlimited in the Crown Court for all health and safety offences. Possible imprisonment for up to 2 years may also be imposed by the court.

The courts also have the power to disqualify directors who are convicted of health and safety offences for up to 5 years at the lower court and 15 years at the higher court.

Imprisonment for health and safety offences has been rare. Over a 10-year period from 1996, only five people were sent to prison for health and safety offences. With the changes introduced in the Health and Safety Offences Act, this number may increase. Imprisonment is a possible penalty for most health and safety offences in both the Magistrates' Court and the Crown Court and a possible penalty for an offence committed by a corporate body. Additionally, certain offences, which were triable only in the lower courts, such as the failure

to comply with an improvement notice, are now triable in either court. This could lead to the offender receiving a higher sentence if their case was referred to the higher court. Courts (or the Crown Prosecution Service in England and Wales or the Procurator Fiscal's Office in Scotland) may refer cases to the next highest court if the case is complex or the powers of the lower court are thought to be too limited.

Work-related deaths

Work-related deaths are investigated by the police jointly with the HSE or Local Authority to ascertain whether a charge of manslaughter (culpable homicide in Scotland) or corporate manslaughter is appropriate. An inquest by a coroner may be made into any workplace death whether due to an accident or industrial disease. Before the inquest, the coroner will decide whether to order a post-mortem. At the inquest, the coroner sits with a jury and decides which witnesses will be called to give evidence. Decisions are usually made on the balance of probability rather than 'beyond reasonable doubt'. The appropriate enforcement agency, such as the HSE, may attend the inquest either to give evidence or to decide whether the evidence presented warrants a prosecution.

If the police decide not to pursue a manslaughter charge, the HSE or Local Authority will continue the investigation under health and safety law. Recent procedural changes now allow prosecution to take place before the inquest where manslaughter or homicide charges are not relevant – if it is considered appropriate and in the interest of justice.

The Corporate Manslaughter and Corporate Homicide Act 2007

The Act created a new statutory offence of corporate manslaughter which replaced the common law offence of manslaughter by gross negligence where corporations and similar entities were concerned. In Scotland, the new offence is called 'corporate homicide'. The Act introduced a new offence but did not introduce any new health and safety duties. An organisation will have committed the new offence if:

▶ it owes a duty of care to another person in defined circumstances;
▶ there is a management failure by its senior managers; and
▶ it is judged that its actions or inaction amount to a gross breach of that duty resulting in a person's death.

The health and safety duties relevant to the Act are:

▶ employer and occupier duties including the provision of safe systems of work and training on any equipment used;

duties connected with:
- ▷ the supply of goods and services to customers;
- ▷ the operation of any activity on a commercial basis;
- ▷ any construction and maintenance work;
- ▷ the use or storage of plant, vehicles or any other item.

A breach of duty will be gross if the management of health and safety at the organisation falls far below that which would reasonably be expected. It is, therefore, a judgement on the health and safety culture of the organisation (as described in Chapter 3) at the time of the incident. The emphasis of the Act is on the responsibility of the senior management of the organisation rather than a single individual. A conviction is only likely if the management and organisation of health and safety by senior management is a substantial element in the breach of the duty of care.

On conviction, the offence is punishable by an unlimited fine and the courts are able to make:

- ▶ a remedial order requiring the organisation to take steps to remedy the management failure concerned;
- ▶ a publicity order requiring the organisation to publicise details of its conviction and fine.

If a court makes a publicity order, it will stipulate in which medium a public announcement should be made and will consider requiring a statement to be displayed on the company's website. The prosecution will provide the court (and serve on the defendant) a draft copy of the suggested order, which the judge will then personally endorse.

The publicity order will probably require an organisation convicted of corporate manslaughter to publicise:

- ▶ the fact that it has been convicted;
- ▶ the particulars of the offence;
- ▶ the amount of any fine;
- ▶ the terms of any remedial order.

The convicted organisation will normally be ordered to pay the prosecution's costs. Normally orders for compensation are considered by the Civil Courts, but it is possible that occasional cases may arise where a Criminal Court considers a compensation order in respect of bereavement and funeral expenses.

It is important to note that the Act did not create a new individual liability. Individuals may still be charged with the existing offence of manslaughter by gross negligence and/or a breach of Section 37 of the HSW Act. Crown immunity does not apply to the offence, although a number of public bodies and functions are exempt from it (such as military operations and the response of emergency services).

The police will investigate the death with the technical support of the relevant enforcement agency (often but not always the HSE). However, the Crown Prosecution Service (in Scotland the Procurator Fiscal) decides whether to initiate a prosecution after the police investigation.

Sentencing guidelines have been published to help courts deal with bodies found guilty of corporate manslaughter. Such bodies can be given up to a seven-figure fine and seldom less than £500,000. However, although the level of the fine should not be linked to the turnover or profit of the body, the financial circumstances of the body and the effect on those innocent parties affected by the undertaking must be considered.

A **defence** is that the failings were exclusively at a junior level. The failings of senior management must have formed a substantial element in the breach for the offence of corporate manslaughter. However, the failings at senior management level do not of themselves have to amount to a gross breach of duty. Liability for the offence is assessed by looking at the failings of the organisation as a whole.

The prosecution must also prove that the organisation owed a relevant duty of care to the deceased and that the breach of duty was the cause of the death. The breach must have made a more than minimal contribution to the death.

1.2.4 Civil law liabilities

Common law torts and duties

Negligence

The only tort (civil wrong) of real significance in health and safety is negligence. Negligence is the lack of reasonable care or conduct which results in injury, damage (or financial loss) of or to another. Whether the act or omission was reasonable is usually decided as a result of a court action or out-of-court settlement.

There have been two important judgements that have defined the legal meaning of negligence. In 1856, negligence was judged to involve actions or omissions and the need for reasonable and prudent behaviour. In 1932, Lord Atkin said the following:

> 'You must take reasonable care to avoid acts or omissions which you reasonably foresee would be likely to injure your neighbour. Who then, in law is my neighbour? The answer seems to be persons who are so closely and directly affected by my act that I ought reasonably to have them in contemplation as being so affected when I am directing my mind to the acts or omissions which are called in question.'

This judgement followed the case of *Donoghue v. Stevenson*, in which Mrs Donoghue, the claimant, consumed part of a bottle of ginger beer that contained a decomposed snail. The snail was not visible, as the bottle was opaque. Neither her friend, who bought it for her, nor the shopkeeper who sold it was aware of its presence. Mrs Donoghue became ill and sued the manufacturer, Mr Stevenson, for negligence. The case worked its way through the court system up to the House of Lords where the judges agreed that Mrs Donoghue had a valid claim, but disagreed on the remedy. One judge thought that it was a product liability case but Lord Atkin argued that the law should recognise a unifying principle that everyone owes a duty of reasonable care to their neighbour. He quoted the Bible in support of his argument, specifically the general biblical principle that 'Thou shalt love thy neighbour.' In framing this landmark judgement, he created the doctrine that nobody should harm their neighbour.

It can be seen, therefore, that for negligence to be established, it must be reasonable and foreseeable that the injury could result from the act or omission. In practice, the court may need to decide whether the injured party is the neighbour of the perpetrator. A collapsing scaffold could easily injure a member of the public, who could be considered a neighbour to the scaffold erector. A recent court case has ruled that hindsight after an incident cannot be used to establish negligence. The Supreme Court has decided in recent cases that foreseeability must play a part in determining whether a place is or was safe. The law does not aim to create an environment that is entirely risk-free. It is directed at situations where there is a material risk to health and safety, which any reasonable person would appreciate and take steps to guard against. If a risk cannot be reasonably foreseen, then it is not material.

An employee who is suing his/her employer for negligence needs to establish the following three criteria:

1. A duty was owed to the employee by the employer and the incident took place during the course of his/her employment.
2. There was a breach of that duty because the event was foreseeable and all reasonable precautions had not been taken.
3. The breach resulted in the specific injury, disease, damage and/or loss suffered.

These tests should also be used by anyone affected by the employer's undertaking (such as contractors and members of the public), who is suing the employer for negligence.

If the employer is unable to defend against the three criteria, two further partial defences are available. It could be argued that the employee was fully aware of the risks that were taken by not complying with safety instructions (known as *volenti non fit injuria* or 'the risk was willingly accepted'). This defence is unlikely to be totally successful because courts have ruled that employees have not accepted the risk voluntarily as economic necessity forces them to work.

The second possible defence is that of '**contributory negligence**' where the employee is deemed to have contributed to the negligent act. This partial defence, if successful, can significantly reduce the level of compensation (by up to 80% in some cases). There are two other defences that are not often used – Act of God

and a disagreement on the facts of the case. An Act of God is a legal term that describes an accident that could not have been prevented by human foresight but occurred due to some form of unpredictable natural disaster.

Any negligence claim must be made within a set time period (currently 3 years).

Duty of care

Several judgements have established that employers owe a duty of care to each of their employees. This duty cannot be assigned to others, even if a consultant is employed to advise on health and safety matters, or if the employees are sub-contracted to work with another employer. This duty may be sub-divided into five groups. Employers must:

1. provide a safe place of work, including access and egress;
2. provide safe plant and equipment;
3. provide a safe system of work;
4. provide safe and competent fellow employees;
5. provide adequate levels of supervision, information, instruction and training.

Employer duties under common law are often mirrored in statute law as in the HSW Act 1974. This, in effect, makes them both common law and statutory duties.

These duties apply even if the employee is working at a third-party premises or if he/she has been hired by the employer to work for another employer.

Duty of care is relative, not absolute. The greater the foreseeability and the greater the risk of harm, the greater the duty of care becomes. Organisations cannot abdicate their duty of care to consultants or contractors. For example, recent cases involving work-related stress have established that once an employer knows that an employee has a stress condition, the duty of care obligation is greater towards that employee than to an employee who appears to be coping well with their workload.

Breach of statutory duty

Although for health and safety, statute law is primarily the basis of criminal law, there is a tort of breach of statutory duty which can be used when a person is seeking compensation following an accident or illness.

In the past when there was a breach of statutory duty, such as the duties laid down by the HSW Act or its Regulations, an individual who was affected by the breach could, subject to certain conditions, pursue a civil action to obtain compensation.

However, the Enterprise and Regulatory Reform Act 2013 in Section 69 amended Section 47(2) of the HSW Act 1974 as follows:

'(2) Breach of a duty imposed by a statutory instrument containing (whether alone or with other provision) health and safety regulations shall not be actionable except to the extent that regulations under this section so provide.'

This means that breaches of statutory duty, as described by Regulations made under the HSW Act, no longer give rise to civil liability unless there is a specific provision to the contrary. One specific provision is given in the Health and Safety at Work etc. Act 1974 (Civil Liability) (Exceptions) Regulations 2013 which allows workers who are pregnant, or new or expectant mothers (including agency workers), to continue to bring claims for breach of statutory duty in relation to breaches of Sections 16–17A of the Management of Health and Safety at Work Regulations to the extent that the breach causes damage. These sections relate to requirements to carry out risk assessments and make particular arrangements for new and expectant mothers to protect their health and safety. These employees may sue their employer both for negligence and the breach of a statutory duty. Such an action is termed a **double-barrelled action**. However, it is important to stress that double damages would not result because the compensation is awarded for the actual harm suffered.

The requirement to provide competent fellow employees includes the provision of adequate supervision, instruction and training. As mentioned earlier, employers are responsible for the actions of their employees (**vicarious liability**) provided that the action in question took place during the normal course of their employment.

In response to the growing 'compensation culture', the Compensation Act became law in 2006. It contains provisions relating to the law of negligence and breach of statutory duty. It expects courts not to prevent or impede a desirable activity from taking place when they consider the particular steps to be taken to meet a reasonable standard of care following a claim for negligence or breach of statutory duty. The Act attempts to ensure that normal activities are not prevented because of a fear of litigation and excessive risk aversion. The emphasis needs to be on risk management rather than risk aversion. It states that in claims for negligence or breach of statutory duty, an apology, offer of treatment or other redress shall not of itself amount to an admission of liability. The Compensation Act also contains provisions on the damages for mesothelioma and the regulation of claims management services.

Levels of statutory duty

There are three principal levels of statutory duty which form a hierarchy of duties. These levels are used extensively in health and safety statutory (criminal) law, but have been defined by judges under common law. The three levels of duty are absolute, practicable and reasonably practicable.

Absolute

This is the highest level of duty and often occurs when the risk of injury is so high that injury is inevitable unless safety precautions are taken. It is a rare requirement regarding physical safeguards, although it was more common before 1992 when certain sections of the Factories Act 1961 were still in force. The duty is absolute and the employer has no choice but to undertake the duty. The verbs used in the Regulations are 'must' and 'shall'.

An example of this is Regulation 11(1a) of the Provision and Use of Work Equipment Regulations which requires that every employer shall ensure that effective measures are taken to prevent access to any dangerous part of machinery or rotating stock bar.

For certain health and safety Regulations, such as the Electricity at Work Regulations and the Control of Substances Hazardous to Health Regulations, an absolute duty may be defended using the argument that 'all reasonable precautions and all due diligence' were taken to comply with the law. To rely on this defence, proof is needed that a control system for the risks outlined in the particular Regulation is in place. Proof of due diligence requires that this control system has suitable safeguards and control measures and that they operate effectively.

Many of the health and safety management requirements contained in health and safety law place an absolute duty on the employer. Examples of this include the need for written safety policies and risk assessments when employee numbers rise above a basic threshold.

Practicable

This level of duty is more often used than the absolute duty as far as the provision of safeguards is concerned and, in many ways, has the same effect.

A duty that 'the employer ensures, so far as is practicable, that any control measure is maintained in an efficient state' means that if the duty is technically possible or feasible then it must be done irrespective of any difficulty, inconvenience or cost. Examples of this duty may be found in the Provision and Use of Work Equipment Regulations [Regulation 11(2) (a and b)].

Reasonably practicable

This is the most common level of duty in health and safety law and was defined by Judge Asquith in *Edwards v. the National Coal Board* (1949) as follows:

> '"Reasonably practicable" is a narrower term than "physically possible," and seems to me to imply that computation must be made by the owner in which the quantum of risk is placed on one scale and the sacrifice involved in the measures necessary for averting the risk (whether in time, money or trouble) is placed in the other, and that, if it be shown that there is a gross disproportion between them – the risk being insignificant in relation to the sacrifice – the defendants discharge the onus on them.'

In other words, if the risk of injury is very small compared with the cost, time and effort required

Figure 1.6 Diagrammatic view of 'reasonably practicable'

to reduce it, no action is necessary. It is important to note that money, time and trouble must 'grossly outweigh', not balance (Figure 1.6), the risk. This duty requires judgement on the part of the employer (or his/her adviser) and clearly needs a risk assessment to be undertaken with conclusions noted. Continual monitoring is also required to ensure that risks do not increase. There are numerous examples of this level of duty, including the HSW Act [e.g. Section 2(3)], the Manual Handling Operations Regulations (see Chapter 13) and the Control of Substances Hazardous to Health Regulations (see Chapter 13). If organisations follow 'good practice', defined by the HSE as 'those standards which satisfy the law', then they are likely to satisfy the 'reasonably practicable' test. Approved Codes of Practice and Guidance published by the HSE describe and define many of these standards.

The term 'suitable and sufficient' is used to define the scope and extent required for health and safety risk assessment and may be interpreted in a similar way to 'reasonably practicable'. More information is given on this definition in Chapter 4.

1.3 The main requirements of the Health and Safety at Work etc. Act 1974 and the Management of Health and Safety at Work Regulations 1999

1.3.1 The Health and Safety at Work Act 1974 (HSW Act)

Background to the Act

The HSW Act resulted from the findings of the Robens Report, published in 1972. Earlier legislation had tended to relate to specific industries or workplaces. This resulted in over 5 million workers being unprotected by any health and safety legislation. Contractors and members of the public were generally ignored. The law was more concerned with the requirement for plant and equipment to be safe rather than the development of

parallel arrangements for raising the health and safety awareness of employees.

A further serious problem was the difficulty that legislation had in keeping pace with developments in technology. For example, following a court ruling in 1955 which, in effect, banned the use of grinding wheels throughout industry, it then took 15 years to produce the Abrasive Wheels Regulations 1970 to address the problem raised by the 1955 court judgement (*John Summers and Sons Ltd. v. Frost*). In summary, health and safety legislation before 1974 tended to be reactive rather than proactive.

Lord Robens was asked, in 1970, to review the provisions made for the health and safety of people at work. His report produced conclusions and recommendations upon which the HSW Act 1974 was based. The principal recommendations were as follows:

▶ There should be a single Act that covers all workers and that Act should contain general duties which should 'influence attitudes'.
▶ The Act should cover all those affected by the employer's undertaking, such as contractors, visitors, students and members of the public.
▶ There should be an emphasis on health and safety management and the development of safe systems of work. This would involve the encouragement of employee participation in accident prevention. (This was developed many years later into the concept of the health and safety culture.)
▶ Enforcement should be targeted at 'self-regulation' by the employer rather than reliance on prosecution in the courts.

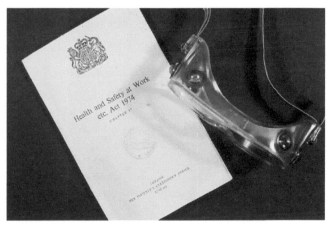

Figure 1.7 HSW Act

These recommendations led directly to the introduction of the HSW Act in 1974.

Regulations

The HSW Act is an **Enabling Act** which allows the Secretary of State to make further laws (known as

Regulations) without the need to pass another Act of Parliament. Regulations are laws, approved by Parliament. These are usually made under the HSW Act, following proposals from the HSE. This applies to Regulations based on EU Directives as well as 'home-grown' ones. It is a criminal offence to breach a Regulation and any breaches may result in enforcement action as explained later in this chapter.

The HSW Act, and general duties in the Management Regulations, aim to help employers to set goals, but leave them free to decide how to control hazards and risks which they identify. Guidance and Approved Codes of Practice give advice, but employers are free to take other routes to achieving their health and safety goals, so long as they do what is reasonably practicable. But some hazards are so great, or the proper control measures so expensive, that employers cannot be given discretion in deciding what to do about them. Regulations identify these hazards and risks and set out specific action that must be taken. Often these requirements are absolute – employers have no choice but to follow them and there is no qualifying phrase of 'reasonably practicable' included.

Some Regulations apply across all organisations – the Manual Handling Regulations would be an example. These apply wherever things are moved by hand or bodily force. Equally, the Display Screen Equipment Regulations apply wherever visual display units are used at work. Other Regulations apply to hazards unique to specific industries, such as mining or construction.

Wherever possible, the HSE will set out the Regulations as goals and describe what must be achieved, but not how it must be done.

Sometimes it is necessary to be *prescriptive*, and to spell out what needs to be done in detail, as some standards are absolute. For example, all mines should have two exits; contact with live electrical conductors should be avoided. Sometimes European law requires prescription.

Some activities or substances are so dangerous that they have to be licensed, for example explosives and asbestos. Large, complex installations or operations require 'safety cases', which are large-scale risk assessments subject to scrutiny by the Regulator. An example would be the privatised railway companies who are required to produce safety cases for their operations.

Approved Code of Practice (ACoP)

An ACoP is produced for many sets of Regulations by the HSE and attempts to give more details on the requirements of the Regulations. It also attempts to give the level of compliance needed to satisfy the Regulations. ACoPs are aimed at what the HSE calls 'experienced health and safety audiences' to explain the law and enable the control of more complex risks. It's important to note that although ACoPs are not law,

they do have a special legal status (sometimes referred to as quasi-legal), which means if the advice in an ACoP is followed properly a duty-holder can be reasonably assured they are complying with the law.

The relationship of an ACoP to a Regulation is similar to the relationship of the Highway Code to the Road Traffic Act. A person is never prosecuted for contravening the Highway Code but can be prosecuted for contravening the Road Traffic Act. If a company is prosecuted for a breach of health and safety law and it is proved that it has not followed the relevant provisions of the ACoP, a court can find it at fault, unless the company can show that it has complied with the law in some other way.

As most health and safety prosecutions take place in a Magistrates' Court, it is likely that the lay magistrates will consult the relevant ACoP as well as the Regulations when dealing with a particular case. Therefore, in practice, an employer must have a good reason for not adhering to an ACoP.

Approved Codes of Practice generally are only directly legally binding if:

▶ the Regulations or Act indicates that they are so, for example the Safety Signs and Signals Regulations Schedule 2 specify British Standard Codes of Practice for alternative hand signals;
▶ they are referred to in an Enforcement Notice.

Guidance

Guidance, which has no formal legal standing, comes in two forms – legal and best practice. The Legal Guidance series of booklets is issued by the HSE to cover the technical aspects of Health and Safety Regulations. These booklets generally include the Regulations and the ACoP, where one has been produced.

Best practice guidance is normally published in the HSG series of publications by the HSE. Examples of best practice guidance books include *Health and Safety in Construction* HSG150 and *Lighting at Work* HSG38.

An example of the relationship between these forms of requirement/advice can be shown using a common problem found throughout industry and commerce – minimum temperatures in the workplace.

Regulation 7 of the Workplace (Health, Safety and Welfare) Regulations states '(1) During working hours, the temperature in all workplaces inside buildings shall be reasonable'.

The ACoP states 'The temperature in workrooms should normally be at least 16 degrees Celsius unless much of the work involves severe physical effort in which case the temperature should be at least 13 degrees Celsius. ...'

It would, therefore, be expected that employers would not allow their workforce to work at temperatures below those given in the ACoP unless it was

reasonable for them to do so (if, for example, the workplace was a refrigerated storage unit).

Best practice guidance to cover this example is given in HSG194 *Thermal Comfort in the Workplace* in which possible solutions to the maintenance of employee welfare in low-temperature environments are given.

General duties and key sections of the Act

A summary of the HSW Act is given in Chapter 13 – only an outline is given here.

Section 2 Duties of employers to employees

To ensure, so far as is reasonably practicable, the health, safety and welfare of all employees. In particular:

- safe plant and systems of work;
- safe use, handling, transport and storage of substances and articles;
- provision of information, instruction, training and supervision;
- safe place of work, access and egress;
- safe working environment with adequate welfare facilities;
- a written safety policy together with organisational and other arrangements (if there are five or more employees);
- consultation with safety representatives and formation of safety committees where there are recognised trade unions.

Section 3 Duties of employers to others affected by their undertaking

A duty to safeguard those not in their employment but affected by the undertaking. This includes members of the public, contractors, patients, customers, visitors and students.

Section 4 Duties of landlords, owners and those in control of premises

This includes a duty to ensure that means of access and egress are safe for those using the premises. Those in control of non-domestic premises also have a duty to ensure, so far as is reasonably practicable, that the premises, the means of access and exit, and any plant (such as boilers and air conditioning units) or substances are safe and without risks to health.

Section 6 Duties of suppliers

Persons who design, manufacture, import or supply any article or substance for use at work must ensure, so far as is reasonably practicable, that it is safe and without risk to health. Articles must be safe when they are set, cleaned, used and maintained. Substances must be without risk to health when they are used, handled, stored or transported. This requires that information must be supplied on the safe use of the articles and substances. There may be a need to guarantee the required level of safety by undertaking tests and examinations.

Section 7 Duties of employees

Two main duties are:

- to take reasonable care for the health and safety of themselves and others affected by their acts or omissions;
- to cooperate with the employer and others to enable them to fulfil their legal obligations.

Section 8

No person is to misuse or interfere with safety provisions (sometimes known as the 'horseplay section').

Section 9

Employees cannot be charged for health and safety requirements such as personal protective equipment.

Section 36 Offences due to the fault of another person

Where an offence is committed by a person that was caused by the act of a second person, the second person may also be charged with the first person. This situation may arise during the trial of the first person when it becomes apparent that the action was the fault of a second person. The HSE have the power under this section to prosecute either both persons or just one. The first person is usually the employer and the second person could be, for example, a manufacturer of a faulty product or a self-employed contractor.

Section 37 Personal liability of directors

Where an offence is committed by a corporate body with the consent or connivance of, or is attributable to any neglect of, a director or other senior officer of the body, both the corporate body and the person are liable to prosecution.

HSE figures show that there has been an increase over the past few years in prosecutions of senior managers and directors under Section 37. Recent case law has confirmed that directors cannot avoid a charge of neglect under Section 37 by arranging their organisation's business so as to leave them ignorant of circumstances which would trigger their obligation to address health and safety breaches. In 2017, 40 directors were successfully prosecuted under Section 37 of the Health and Safety at Work etc. Act 1974 and six were convicted of corporate manslaughter on a personal basis.

1.3.2 Employers' duties

Employers have duties under both criminal and civil law. The general duties of employers under the HSW Act relate to:

- the health, safety and welfare at work of employees and others whether part-time, casual, temporary or homeworkers, on work experience, government training schemes or on site as contractors – that is anyone working under their control or direction;

▶ the health and safety of anyone who visits or uses the workplace and anyone who is allowed to use the organisation's equipment;

▶ the health and safety of those affected by the work activity, for example neighbours, and the general public.

Employers must protect the health and safety of employees and others who might be affected by their work activities. Health and safety is about sensible, proportionate actions that protect people – not unnecessary bureaucracy and paperwork.

This 10-point list shows some of the key duties required by the HSW Act and other legislation that apply to nearly every organisation:

1. Obtain Employers' Liability Compulsory Insurance and display the certificate.
2. Ensure that the organisation has competent health and safety advice available. This does not have to be an external consultant.
3. Develop a health and safety policy that outlines the health and safety management system.
4. Undertake risk assessments on the business activities that includes details of the required control measures.
5. Ensure that relevant actions are taken following the risk assessment to prevent accidents and ill health.
6. Provide basic welfare facilities, such as toilets, washing facilities and drinking water.
7. Provide free health and safety training for workers.
8. Consult workers on health and safety.
9. Display the health and safety law poster or give workers a leaflet with the information.
10. Report certain work-related accidents, diseases and dangerous occurrences.

Failure to comply with these duties can have serious consequences – for both organisations and individuals. As has already been stated, sanctions include fines, imprisonment and disqualification.

Employers are required to do a risk assessment of the work activities carried out by homeworkers. It may be necessary for employers to visit the home of the worker to carry out a risk assessment, although homeworkers can also help in identifying the hazards for their employers themselves.

The Working Time Regulations set a limit on the time that a worker can be required to work, which averages at 48 hours a week (over a 17-week rolling reference period), although workers can voluntarily opt out to work longer. Young workers must not work more than eight hours per day and 40 hours per week. Night workers are limited to an average of eight hours' work in a 24-hour period and entitled to a free health assessment. Records and monitoring of working times, rest breaks and holiday entitlement are also covered by the Regulations. Working time will not usually include time spent travelling to and from the workplace and time during rest breaks. The Regulations apply to 'workers', including employees, temporary workers and freelancers, but not the genuinely self-employed. Workers are entitled to regular breaks in the working day and rest periods between working days. Employers must provide that rest periods can be taken, but do not need to ensure they are actually taken. The rest period is in addition to annual leave and can be paid or unpaid.

Visitors and general public

Employers have a duty to ensure the health and safety of the public while on their premises, even if the individuals concerned, like children, are not supposed to be there. Two cases reported in 2008 have highlighted how far this liability can extend.

CASE STUDIES

Two companies pleaded guilty to charges of breaching the HSW Act 1974 following the drowning of a 9-year-old girl, who was playing with other children in the company car park when they strayed onto nearby reservoirs. At the time of the accident, in 2004, the main gates to the factory were off their hinges because work was being carried out on the site and a second gate, which led to the reservoir, was only secured with a nylon rope.

In a separate case, a construction company was found guilty of failing to prevent unauthorised persons, including children, from gaining access to an area where construction material and equipment were stored. A child was seriously injured by falling paving stones while playing on a partly built housing estate where materials were being stored during construction work.

Misunderstandings over who is responsible for monitoring and protecting contractors and sub-contractors – leading to the failure to carry out responsibilities – is often at the heart of high-profile health and safety cases. But situations such as these could be avoided with a clearer understanding of employers' and contractors' responsibilities.

Visitors to a site whether authorised or not are often more at risk than employees because:

▶ they are unfamiliar with the workplace processes, the hazards and associated risks they present;
▶ they may not have the appropriate personal protective equipment (PPE);
▶ they will have a lack of knowledge of the site or premises layout;

- walkways are often inadequate, unsigned or poorly lit;
- they are not familiar with the emergency procedures or means of escape;
- they may be particularly vulnerable if they suffer from a disability or are very young.

Many of these problems with visitors can be overcome by, for example:

- visitors signing in and being provided with a site escort;
- providing appropriate PPE and identity badges;
- providing simple induction procedures with a short video and information on site rules, hazards and emergency procedures;
- clear marking of walkways and areas where unauthorised people are not permitted.

Night working

Employers should ascertain whether they employ people who would be classified as night workers. If so, they should check:

- the number of hours that the night workers normally work;
- if night workers work more than eight hours per day on average, whether the number of hours can be reduced and if any exceptions apply;
- that a health assessment is made on each night worker;
- that a regular programme of health checks is made on each night worker;
- that proper records of night workers are maintained, including details of health assessments; and
- that night workers are not involved in work which is particularly hazardous.

It is important to note that the Working Time Regulations state that night-shifts must not exceed eight hours in every 24. There are several risks to which night workers are exposed due to fatigue, lack of sleep, insufficient rest breaks and an increase in possible accidents due to tiredness. Research is currently being undertaken whether there is a link between night working and the development of breast cancer in women. More than 3.5 million people are employed as shift workers in the UK. They work in a wide variety of industries including the emergency services, health care, the utilities, transport, manufacturing, entertainment and retail. Poorly designed shift-working arrangements and long working hours that do not balance the demands of work with time for rest and recovery can result in fatigue, accidents, injuries and ill health. Fatigue results in slower reactions, reduced ability to process information, memory lapses, absent-mindedness, decreased awareness, lack of attention, underestimation of risk, reduced coordination, etc. Fatigue can lead to errors and accidents, ill health

and injury, and reduced productivity and is often a root cause of major accidents. Fatigue has also been implicated in 20% of accidents on major roads.

Temporary workers Directive adopted by EU Parliament

The European Parliament has adopted a Directive on temporary agency work which enables temporary workers to be treated equally with those of the employer company. However, following an agreement reached in 2008, agency workers should receive the same pay and conditions as permanent staff after being employed for 12 weeks.

The agreement was made between the UK Government, the CBI and the TUC, and the following points were agreed:

- After 12 weeks in a given job, there will be entitlement to equal treatment.
- Equal treatment will be defined to mean 'at least the basic working and employment conditions that would apply to the workers concerned if they had been recruited directly by that undertaking to occupy the same job'. This does not cover occupational social security schemes.
- The Government will consult regarding the implementation of the Directive, in particular how disputes will be resolved regarding the definition of equal treatment and how both sides of the industry can reach appropriate agreements on the treatment of agency workers.
- The new arrangements will be reviewed at an appropriate point 'in the light of experience'.

1.3.3 Duties of managers and supervisors

Managers and supervisors are essential communication links between the Board and the workforce. They should ensure that the agreed health and safety standards are communicated to the workers and adherence to those standards monitored effectively. They have a responsibility to deliver the duty of care of the employer to the workforce.

Therefore managers should:

- familiarise themselves with the health and safety management system of the organisation;
- ensure that there is an adequate and appropriate level of supervision for all workers;
- ensure that supervisors are aware of:
  - the health and safety standards of the organisation;
  - the specific hazards within their area of supervision;
  - the need to set a good example on health and safety issues;

▷ the need to monitor the health and safety performance of their workforce; and
▷ the training needs, including induction training, of their workforce;

► ensure that sufficient resources are available to allow tasks to be completed safely and without risks to health; and
► communicate to the Chief Executive and the Board the adherence or otherwise of the health and safety standards agreed by the Board.

Effective supervision of the workforce is one of the important keys to successful health and safety management. It will enable managers to monitor the effectiveness of the training that workers have received, and whether they have the necessary capacity and competence to do the job assigned to them. Supervisors should lead by example in health and safety matters (for example in the wearing of PPE).

Hence supervisors should ensure that:

► workers in their department understand the risks associated with their workplace and the measures available to control them;
► the risk control measures are up to date and are being properly used, maintained and monitored;
► particular attention is paid to new, inexperienced or young people and those whose first language is not English;
► workers are encouraged to raise concerns over any shortcomings in health and safety provision;
► arrangements are in place to supervise the work of contractors.

1.3.4 Duties of employees and others

Employees and agency workers

They have specific duties under the HSW Act, which are:

► to take reasonable care for the health and safety of themselves and of other persons who may be affected by their acts or omissions at work. This involves the same wide group that the employer has to cover, not just the people on the next desk or bench;
► to cooperate with employers in assisting them to fulfil their statutory duties;
► not to interfere with deliberately or misuse anything provided, in accordance with health and safety legislation, to further health and safety at work. See also Chapter 13.

Self-employed

There are 4.2 million self-employed workers in the UK of whom 3 million do not employ anyone. Approximately half of the self-employed use their home as a workplace.

The duties of the self-employed under the HSW Act are fairly limited. Under the HSW Act the self-employed are:

► responsible for their own health and safety;
► responsible to ensure that others who may be affected are not exposed to risks to their health and safety.

These responsibilities are extended by the Management of Health and Safety at Work Regulations, which requires self-employed people to:

► carry out risk assessment;
► cooperate with other people who work in the premises and, where necessary, in the appointment of a health and safety coordinator;
► provide comprehensible information to other employees working in their premises.

Section 3(2) of HSW Act is to be amended in order to exempt self-employed persons from the general duty in respect of themselves and other persons (not being their employees), except those undertaking activities on a prescribed list.

An undertaking will be prescribed if one of the following four criteria is met:

► there are high numbers of self-employed in a particular industry and high rates of injuries and/or fatalities (e.g. agriculture);
► there is a significant risk to members of the public (e.g. fairgrounds); or
► there is the potential for mass fatalities (e.g. explosives).

The occupational sectors for which no exemption is proposed currently (December 2019) include: agricultural activities; construction; quarries; mining; diving; sites that are subject to the Control of Major Accident Hazards (COMAH) Regulations and sub-COMAH sites; offshore activities; nuclear installations; explosives; and gas-fitting and installation.

1.3.5 Persons in control of premises

Section 4 of the HSW Act requires that 'Persons in control of **non-domestic** premises' take such steps as are reasonable in their position to ensure that there are no risks to the health and safety of people who are not employees but use the premises. This duty extends to:

► people entering the premises to work;
► people entering the premises to use machinery or equipment, for example a launderette;
► access to and exit from the premises;
► corridors, stairs, lifts and storage areas.

Those in control of premises are required to take a range of steps depending on the likely use of the premises and the extent of their control and knowledge of the actual use of the premises.

Figure 1.8 NEBOSH is in control here

1.3.6 Duties of manufacturers and others in the supply chain

Introduction

Market leaders in every industry are increasing their grip on the chain of supply. They do so by monitoring rather than managing, and also by working more closely with suppliers. The result of this may be that suppliers or contractors are absorbed into the culture of the dominant firm, while avoiding the costs and liabilities of actual management. Powerful procurement departments emerge to define and impose the necessary quality standards and guard the lists of preferred suppliers.

The trend in many manufacturing businesses is to involve suppliers in a greater part of the manufacturing process so that much of the final production is the assembly of prefabricated subassemblies. This is particularly true of the automotive and aircraft industries. This is good practice as it:

- involves the supplier in the design process;
- reduces the number of items being managed within the business;
- reduces the number of suppliers;
- improves quality management by placing the onus on suppliers to deliver fully checked and tested components and systems.

In retail, suppliers are even given access to daily sales figures and forecasts of demand which would normally be considered as highly confidential information. In the process, the freedom of local operating managers to pick and choose suppliers is reduced. Even though the responsibility to do so is often retained, it is strongly qualified by centrally imposed rules and lists, and assistance or oversight.

Suppliers have to be:

- trusted;
- treated with fairness in a partnership;
- given full information to meet the demands being placed on them.

Under these conditions, suppliers and contractors looking for business with major firms need greater flexibility and wider competence than earlier. This often implies increased size and perhaps mergers, though in principle bids could be, and perhaps are, made by loose partnerships of smaller firms organised to secure such business.

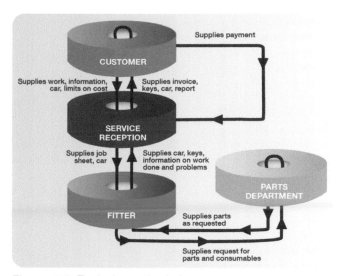

Figure 1.9 Typical supply chain

Advantages of good supply chain management

Reduction of waste

This is an important objective of any business and involves not only waste of materials but also that of time. Examples of waste are:

- unwanted materials due to over-ordering, damage or incorrect specifications;
- extraneous activities like double handling, for example between manufacturer, builders merchant and the site;
- re-working and re-fitting due to poor quality, design, storage or manufacture;
- waste of time such as waiting for supplies due to excessive time from ordering to delivery or early delivery long before they are needed.

Faster reaction

A well-managed supply chain should be able to respond rapidly to changing requirements. Winter conditions require very different materials than those used during warmer or drier seasons. A contractor may have to modify plans rapidly and suppliers may need to ramp up or change production at short notice.

Reduction in accidents

A closer relationship between client, designers, principal contractors and suppliers of services and products can result not only in a safer finished product but, in construction, a safer method of erection. If more products are pre-assembled in ideal factory conditions and then fixed in place on site, it is often safer than utilising a full assembly approach in poor weather conditions on temporary work platforms. Examples are made-up roof trusses and prefabricated doors and windows already fitted to their frames.

Legislation

The HSW Act Section 6 places a duty on everyone in the supply chain, from the designer to the final installer, of articles of plant or equipment for use at work or any article of fairground equipment to:

▶ ensure that the article will be safe and without risk to health at all times when it is being set, used, cleaned or maintained;

▶ carry out any necessary testing and examination to ensure that it will be safe;

▶ provide adequate information about its safe setting, use, cleaning, maintenance, dismantling and disposal.

There is an obligation on designers or manufacturers to do any research necessary to prove safety in use. Erectors or installers have special responsibilities to make sure, when handed over, that the plant or equipment is safe to use.

Similar duties are placed on manufacturers and suppliers of substances for use at work to ensure that the substance is safe when properly used, handled, processed, stored or transported and to provide adequate information and do any necessary research, testing or examining.

Where articles or substances are imported, the suppliers' obligations outlined above attach to the importer, whether they are a separate importing business or the user personally who does the importing.

Often items are obtained through hire purchase, leasing or other financing arrangements with the ownership of the items being vested with the financing organisation. Where the financing organisation's only function is to provide the money to pay for the goods, the supplier's obligations do not attach to them.

Information for customers

The quality movement has drawn attention to the need to ensure that there are processes in place which ensure quality, rather than just inspecting and removing defects when it is too late. In much the same way, organisations need to manage health and safety rather than acting when it is too late.

Customers need information and specifications from the manufacturer or supplier – especially where there is a potential risk involved for them. When deciding what the supplier needs to pass on, careful thought is required about the health and safety factors associated with any product or service.

This means focusing on four key questions and then framing the information supplied so that it deals with each one. The questions are as follows:

▶ Are there any inherent dangers in the product or service being passed on – what could go wrong?

▶ What can the manufacturer or supplier do while working on the product or service to reduce the chance of anything going wrong later?

▶ What can be done at the point of handover to limit the chances of anything going wrong?

▶ What steps should customers take to reduce the chances of something going wrong? What precisely would they need to know?

Depending on what is being provided to customers, the customer information may need to comply with the following legislation:

▶ Supply of Machinery (Safety) Regulations;

▶ Provision and Use of Work Equipment Regulations;

▶ Control of Substances Hazardous to Health Regulations; and

▶ CLP Regulations (Classification, labelling and packaging of substances).

This list is not exhaustive. See Chapter 13 for summaries.

Buying problems

Examples of problems that may arise when purchasing include:

▶ second-hand equipment which does not conform to current safety standards such as an office chair which does not provide adequate back support or have five feet/castors;

▶ starting to use new substances which do not have safety data sheets;

▶ machinery which, while well-guarded for operators, may pose risks for a maintenance engineer.

Figure 1.10 Inadequate chair – take care when buying especially if second-hand. It should have five feet and an adjustable backrest

A risk assessment should be done on any new product, taking into account the likely life expectancy (delivery, installation, use, cleaning, maintenance, disposal, etc.). The supplier should be able to provide the information needed to do this. This will help the purchaser make an informed decision on the total costs because the risks will have been identified as will the precautions needed to control those risks. A risk assessment will still be needed for a CE-marked product. The CE marking signifies the manufacturer's declaration that the product conforms to relevant European Directives. Declarations from reputable manufacturers will normally be reliable. However, purchasers should be alert to fake or inadequate declarations and technical standards which may affect the health and safety of the product despite the CE marking. The risk assessment is still necessary to consider how and where the product will be used, what effect it might have on existing operations and what training will be required.

Employers have some key duties when buying plant and equipment:

▶ They must ensure that work equipment is safe and suitable for its purpose and complies with the relevant legislation. This applies equally to equipment which is adapted to be used in ways for which it was not originally designed.

▶ When selecting work equipment, they must consider existing working conditions and health and safety issues.

▶ They must provide adequate health and safety information, instructions, training and supervision for operators. Manufacturers and suppliers are required by law to provide information that will enable safe use of the equipment, substances, etc. and without risk to health.

Some of the issues that will need to be considered when buying in product or plant include:

▶ ergonomics – risk of work-related upper limb disorders (WRULDs);
▶ manual handling needs;
▶ access/egress;
▶ storage, for example of chemicals;
▶ risk to contractors when decommissioning old plant or installing new plant;
▶ hazardous materials – provision of extraction equipment or personal protective equipment;
▶ waste disposal;
▶ safe systems of work;
▶ training;
▶ machinery guarding;
▶ emissions from equipment/plant, such as noise, heat or vibration.

Appendix 1.1 gives a useful checklist for supply chain health and safety management.

1.3.7 Role and functions of external agencies

The HSE and the Local Authorities (a term used to cover county, district and unitary councils) are all external agencies that have a direct role in the monitoring and enforcement of health and safety standards. There are,

Figure 1.11 Diagram showing the main external agencies that impact on the workplace

however, three other external agencies which have a regulatory influence on health and safety standards in the workplace.

Fire and Rescue Authority

The Fire and Rescue Authority is situated within a single or group of local authorities and is normally associated with rescue, firefighting and the offering of general advice. It has also been given powers to enforce fire precautions within places of work under fire safety law. The powers of the Authority are very similar to those of the HSE on health and safety matters. The Authority issues Alteration Notices to workplaces and conducts routine and random fire inspections (often to examine fire risk assessments). It should be consulted during the planning stage of proposed building alterations when such alterations may affect the fire safety of the building (e.g. means of escape).

The Fire and Rescue Authority can issue both Enforcement and Prohibition Notices. The Authority can prosecute for offences against fire safety law.

The Environment Agency (Scottish Environment Protection Agency)

The Environment Agency was established in 1995 and was given the duty to protect and improve the environment. It is the regulatory body for environmental matters and has an influence on health and safety issues.

It is responsible for authorising and regulating emissions from industry.

Other duties and functions of the Agency include:

▶ ensuring effective controls of the most polluting industries;
▶ monitoring radioactive releases from nuclear sites;
▶ ensuring that discharges to controlled waters are at acceptable levels;
▶ setting standards and issuing permits for collecting, transporting, processing and disposal of waste (including radioactive waste);
▶ enforcement of the Producer Responsibility Obligations (Packaging Waste) Regulations. These resulted from an EU Directive which seeks to minimise packaging and recycle at least 50% of it;
▶ enforcement of the Waste Electrical and Electronic Equipment (WEEE) Directive and its associated Directives.

The Agency may prosecute in the criminal courts for the infringement of environmental law – in one case a fine of £4 million was imposed.

Insurance companies

Insurance companies play an important role in the improvement of health and safety standards. Since 1969, it has been a legal requirement for employers

to insure their employees against liability for injury or disease arising out of their employment. This is called employers' liability insurance and was covered earlier in this chapter. Certain public sector organisations are exempted from this requirement because any compensation is paid from public funds. Other forms of insurance include fire insurance and public liability insurance (to protect members of the public).

Figure 1.12 Good standards prevent harm and save money

Premiums for all these types of insurance are related to level of risks which is related to standards of health and safety. In recent years, there has been a considerable increase in the number and size of compensation claims and this has placed further pressure on insurance companies.

Insurance companies are becoming effective health and safety regulators by weighing the premium offered to an organisation according to its safety and/or fire precaution record.

1.3.8 The Management of Health and Safety at Work Regulations

On 1 January 1993, following an EC Directive, the Management of Health and Safety at Work Regulations became law in the UK. These Regulations were updated in 1999 and are described in detail in Chapter 13. In many ways, the Regulations were not introducing new concepts or replacing the 1974 Act – they simply reinforced or amended the requirements of the HSW Act. Some of the duties of employers and employees were redefined. The duties of employers, employees and others are summarised below but the influence of these duties on health and safety management is discussed in detail in Chapters 3 and 4.

Employers' duties

Employers must:

▶ undertake suitable and sufficient written risk assessments when there are five or more employees;

- preventative and protective measures must be on the basis of the principles of prevention specified in Schedule 1 to the regulations;
- put in place effective arrangements for the planning, organisation, control, monitoring and review of health and safety measures in the workplace (including health surveillance). Such arrangements should be recorded if there are five or more employees;
- employ (to be preferred) or contract competent persons to help them comply with health and safety duties;
- develop suitable emergency procedure and ensure that employees and others are aware of these procedures and can apply them;
- provide health and safety information to employees and others, such as other employers, the self-employed and their employees who are sharing the same workplace and parents of child employees or those on work experience;
- cooperate in health and safety matters with other employers who share the same workplace;
- provide employees with adequate and relevant health and safety training;
- provide temporary workers and their contract agency with appropriate health and safety information;
- protect new and expectant mothers and young persons from particular risks;
- under certain circumstances, as outlined in Regulation 6, provide health surveillance for employees.

The information that should be supplied by employers under the Regulations is:

- risks identified by any risk assessments including those notified to the employer by other employers sharing the same workplace;
- the preventative and protective measures that are in place;
- the emergency arrangements and procedures and the names of those responsible for the implementation of the procedures.

Regulation 19 requires employers to protect young persons from risks resulting from their immaturity.

The protection of young persons (under the age of 18) from risks associated with their immaturity is covered in detail in Chapter 4. This immaturity is often associated with the inability to perceive hazards and may be intellectual and/or physical. The influence of peer pressure is also an important factor amongst young people.

Finally, it is important to stress, as noted above, that Schedule 1 of the Regulations outlines the principles of prevention which employers and the self-employed need to apply so that health and safety risks are addressed and controlled. These principles are discussed in detail in Chapter 4.

Employees' duties

Employees must:

- use any equipment or substance in accordance with any training or instruction given by the employer;
- report to the employer any serious or imminent danger;
- report any shortcomings in the employer's protective health and safety arrangements.

Joint occupation of premises

The Management of Health and Safety at Work Regulations specifically state that where two or more employers share a workplace – whether on a temporary or a permanent basis – each employer shall:

- cooperate with other employers;
- take reasonable steps to coordinate between other employers to comply with legal requirements;
- take reasonable steps to inform other employers where there are risks to health and safety.

All employers and self-employed people involved should satisfy themselves that the arrangements adopted are adequate. Where a particular employer controls the premises, the other employers should help to assess the shared risks and coordinate any necessary control procedures. Where there is no controlling employer, the organisations present should agree joint arrangements to meet regulatory obligations, such as appointing a health and safety coordinator.

1.4 Legal and organisational roles and duties for managing contractors and construction projects

1.4.1 Introduction

A contractor is anyone who is brought in to work and is not an employee. People on work experience or on labour-only contracts or temporary staff are considered to be employees. Contractors are used for maintenance, repairs, installation, construction, demolition, computer work, cleaning, security, health and safety and many other tasks. Sometimes there are several contractors on site at any one time. Clients need to think about how their work may affect each other's and how they interact with the normal site occupier.

1.4.2 Legal considerations

The HSW Act applies to all work activities. It requires employers to ensure, so far as is reasonably practicable, the health and safety of:

- their employees;
- other people at work on their site, including contractors;
- members of the public who may be affected by their work.

All parties to a contract have specific responsibilities under health and safety law, and these cannot be passed on to someone else:

▶ Employers are responsible for protecting people from harm caused by work activities. This includes the responsibility not to harm contractors and sub-contractors on site.

▶ Employees and contractors have to take care not to endanger themselves, their colleagues or others affected by their work.

▶ Contractors also have to comply with the HSW Act and other health and safety legislation. Clearly, when contractors are engaged, the activities of different employers do interact. So cooperation and communication are needed to make sure all parties can meet their obligations.

▶ Employees have to cooperate with their employer on health and safety matters, and not do anything that puts them or others at risk.

▶ Employees must be trained and clearly instructed in their duties.

▶ Self-employed people must not put themselves in danger, or others who may be affected by what they do.

▶ Suppliers of chemicals, machinery and equipment have to make sure their products or imports are safe, and provide information on this.

The Management of Health and Safety at Work Regulations apply to everyone at work and encourage employers to take a more systematic approach to dealing with health and safety by:

▶ assessing the risks which affect employees and anyone who might be affected by the site occupier's work, including contractors;

▶ setting up emergency procedures;

▶ providing training;

▶ cooperating with others on health and safety matters, for example contractors who share the site with an occupier;

▶ providing temporary workers, such as contractors, with health and safety information.

Most contracted work is governed by a legally binding contract and it is, therefore, very important that the contract covers all parts of the work – for example, competent workers, safe access when working at height, fire precautions and safe waste disposal. The principles of cooperation, coordination and communication between organisations underpin the Management of Health and Safety at Work Regulations and the CDM Regulations.

1.4.3 Construction (Design and Management) (CDM 2015) Regulations

Businesses often engage contractors for construction projects at one time or another to build plant, convert or extend premises and demolish buildings. The CDM 2015 Regulations apply to all construction projects. Larger projects which are notifiable (see details in Chapter 13) have more extensive requirements.

All projects require the following:

▶ clients to check the competence of all their appointees; ensure there are suitable management arrangements for the project; allow sufficient time and resources for all stages and provide pre-construction information to designers and contractors. There are special arrangements under the regulations for domestic clients (see later in this section);

▶ designers to eliminate hazards and reduce risks during the design stage and provide information about remaining risks;

▶ contractors to plan, manage and monitor their own work and that of employees; check the competence of all their appointees and employees; train their own employees; provide information to their employees; comply with the requirements for health and safety on site detailed in Part 4 of the Regulations and other Regulations such as the Work at Height Regulations; and ensure there are adequate welfare facilities for their employees;

▶ everyone to assure their own competence; cooperate with others and coordinate work so as to ensure the health and safety of construction workers and others who may be affected by the work; report obvious risks; take account of the general principles of prevention in planning or carrying out construction work; and comply with the requirements in Part 4 of CDM 2015 and other relevant regulations for any work under their control.

For even small projects, clients should ensure that contractors provide:

▶ information regarding the contractor's health and safety policy;

▶ information on the contractor's health and safety organisation detailing the responsibilities of individuals;

▶ information on the contractor's procedures and standards of safe working;

▶ the method statements for the project in hand;

▶ details on how the contractor will audit and implement its health and safety procedures;

▶ procedures for investigating incidents and learning the lessons from them.

Smaller contractors may need some guidance to help them produce suitable method statements. While they do not need to be lengthy, they should set out those features essential to safe working, for example access arrangements, PPE, control of chemical risks and fire precautions.

Copies of relevant risk assessments for the work to be undertaken should be requested. These need not be very detailed but should indicate the risk and the control methods to be used.

The client, principal designer, designer, principal contractor and other contractors all have specific roles under CDM 2015 Regulations.

1.4.4 Explanation of terms used and duties in CDM 2015

(a) **The Client** is an organisation or individual for whom a construction project is undertaken. This can include, for example, local authorities, school governors, insurance companies and project originators for private finance initiative (PFI) projects. If it is likely that more than one contractor will be working on a project at any one time, the client must appoint a principal designer and a principal contractor. The client has the following duties:

- make arrangements for managing the project to ensure that construction work is carried out safely so far as is reasonably practicable;
- ensure that the welfare requirements in Schedule 2 of the Regulations are followed;
- provide pre-construction information to each designer and each contractor engaged on the project;
- take reasonable steps to ensure that the health and safety arrangements are maintained and reviewed throughout the project;
- ensure that the principal designer complies with their duties under the Regulations;
- ensure that the principal contractor complies with their duties under the Regulations;
- ensure before the construction phase begins that the principal contractor (or contractor if only one is used) draws up a construction phase plan;
- ensure that the principal designer prepares an appropriate health and safety file for the project which:
  - ▷ includes information on asbestos as required under the Control of Asbestos Regulations;
  - ▷ is revised from time to time;
  - ▷ is kept available for inspection; and
  - ▷ is passed on if interest in the project passes to another client.

Domestic clients are a special case and do not have specific duties under CDM 2015 (see the note at the end of this section).

(b) **The Principal Designer** must plan, manage, monitor and coordinate the pre-construction phase of a project, taking into account the general principles of prevention to ensure:

- that the project is carried out safely so far as is reasonably practicable;
- that assistance is given to the client for the provision of pre-construction information;
- health and safety risks are identified, eliminated or controlled;

- the cooperation of everyone working on the project;
- the preparation and revision as necessary of the health and safety file;
- the provision of pre-construction information to every designer and contractor; and
- liaison with the principal contractor for the duration of the project, particularly over the preparation of the construction phase plan.

The primary function of the principal designer is to identify and eliminate through design foreseeable risk at the pre-construction and construction phases. The principal designer has to ensure that the architects consider buildability, maintenance and use of the building after it has been handed over. He/she does not have to be a designer in the traditional sense but must be a professional who can control the design process, with the skills, knowledge and experience to ensure safety and health compliance.

(c) **Designers** are those who have a trade or a business which involves them in:

- preparing designs for construction work, including variations. This includes preparing drawings, design details, specifications, bills of quantities and the specification (or prohibition) of articles and substances, as well as all the related analysis, calculations and preparatory work; and
- arranging for their employees or other people under their control to prepare designs relating to a structure or part of a structure.

It does not matter whether the design is recorded (e.g. on paper or a computer) or not (e.g. it is communicated only orally).

Designers may include:

- architects, civil and structural engineers, building surveyors, landscape architects, other consultants, manufacturers and design practices (of whatever discipline) contributing to, or having overall responsibility for, any part of the design, for example drainage engineers;
- anyone who specifies or alters a design, or who specifies the use of a particular method of work or material, such as a design manager, quantity surveyor who insists on specific material, or a client who stipulates a particular layout for a new building;
- building service designers, engineering practices or others designing plant which forms part of the permanent structure (including lifts, heating, ventilation and electrical systems), for example a specialist provider of permanent fire extinguishing installations;
- those purchasing materials where the choice has been left open, for example those purchasing building blocks and so deciding the weights that a bricklayer must handle;
- contractors carrying out design work as part of their contribution to a project, such as an engineering

contractor providing design, procurement and construction management services;

- temporary works engineers, including those designing auxiliary structures, such as formwork, falsework, facade retention schemes, scaffolding and sheet piling;
- interior designers, including shop-fitters who also develop the design;
- heritage organisations who specify how work is to be done in detail, for example providing detailed requirements to stabilise existing structures; and
- those determining how buildings and structures are altered, for example during refurbishment, where this has the potential for partial or complete collapse.

In summary, therefore, under CDM, a designer includes architects, consulting engineers, quantity surveyors, building service engineers and temporary works engineers.

The duties of the designer are;

- before starting work be satisfied that the client is aware of their duties under CDM 2015;
- to take into account the general principles of prevention to eliminate risks for anyone:
  - ▷ working on the project;
  - ▷ maintaining or cleaning a structure; or
  - ▷ using a structure designed as a workplace;
- if elimination is not possible the designer must so far as is reasonably practicable:
  - ▷ take steps to reduce and control health and safety risks through the design process;
  - ▷ provide information about health and safety risks to the principal designer;
  - ▷ ensure appropriate information is contained in the health and safety file; and
- provide with the design sufficient information to help clients, other designers and contractors to fulfil their duties under CDM.

(d) **The Principal Contractor** is the contractor appointed by the client. The principal contractor can be an organisation or an individual, and is the main or managing contractor. The principal contractor must consult and engage with workers on the project. A principal contractor's key duties are to coordinate and manage the construction phase of the project and to ensure the health and safety of everybody involved with the construction work, or who are affected by it. Other duties are to ensure that:

- construction work is carried out safely so far as is reasonably practicable;
- a construction phase plan is prepared prior to setting up a site and reviewed and revised as necessary;
- there is coordination of the implementation of legal requirements and the construction phase plan for all people on site;
- necessary site health and safety rules are drawn up;

- suitable site induction is provided;
- access to unauthorised persons is controlled;
- the welfare requirements in Schedule 2 of the Regulations are provided throughout the construction phase;
- there is liaison with the principal designer throughout the project particularly regarding information for the health and safety file and management of the construction phase plan; and
- the health and safety file is appropriately updated, reviewed and revised to take account of changes.

(e) **The Contractor** is any person who, in the course or furtherance of a business, undertakes or manages construction work. A contractor has the following duties:

- not to undertake construction work on a project unless satisfied that the client is aware of their duties under CDM 2015;
- to plan, manage and monitor the way construction work is done to ensure it is safe so far as is reasonably practicable;
- where there is no principal contractor, to ensure that a construction phase plan is drawn up prior to setting up the site;
- to provide any employees or persons under their control, any information and instruction to ensure the safety of the project including:
  - ▷ suitable site induction;
  - ▷ emergency procedures;
  - ▷ information on risks to their health and safety either identified by their risk assessments or from another contractor's operations;
- to provide any other health and safety training required by the Management of Health and Safety at Work Regulations;
- to not start work until access to unauthorised persons is controlled;
- to ensure that the welfare requirements in Schedule 2 of the Regulations are provided for employees and others under their control;
- to comply with any directions given by the principal designer or principal contractor and any site rules; and
- to consult with their workers.

(f) **Pre-construction health and safety information**
The client must provide designers and contractors with specific health and safety information needed to identify the hazards and risks associated with the design and construction work. The information must be identified, assembled and supplied in good time, so that those who need it during the preparation of a bid for construction work or the planning of such work may estimate the resources required to enable the design, planning and construction work to be properly organised and carried out. The topics which should be addressed in the pre-construction health and safety information are given in Appendix 1.2.

1

(g) **Construction phase health and safety plan**

The principal contractor must define the way in which the construction phase will be managed and the key health and safety issues for a particular project must be recorded. The health and safety plan should outline the organisation and arrangements required to manage risks and coordinate the work on site. It should not be a repository for detailed generic risk assessments, records of how decisions were reached or detailed method statements; but it may, for example, set out when such documents will need to be prepared. It should be well focused, clear and easy for contractors and others to understand – emphasising key points and avoiding irrelevant material. It is crucial that all relevant parties are involved and cooperate in the development and implementation of the plan as work progresses.

The plan must be tailored to the particular project and relevant photographs and sketches will simplify and shorten explanations. The plan should be organised so that relevant sections can easily be made available to designers and contractors.

The construction phase health and safety plan for the initial phase of the construction work must be prepared before any work begins. Further details may be added as completed designs become available and construction work proceeds. Appendix 1.3 gives details of what should be included in the health and safety plan.

(h) **Health and safety file**

This is a record of health and safety information for the client to retain. It alerts those who are responsible for the structure and equipment within it to the significant health and safety risks that will need to be addressed during subsequent use, construction, maintenance, cleaning, repair, alterations, refurbishment and demolition work. Appendix 1.4 gives details of what should be included in the health and safety file.

(i) **Method statement**

This is a formal document that describes the sequences of operations for safe working that will ensure health and safety during the performance of a task. It results from the risk assessment carried out for the task or operation and the identified control measures. If the risk is low, a verbal statement may normally suffice.

(j) **Notifiable work**

Construction work is notifiable to the HSE if it lasts longer than 30 working days and has more than 20 workers simultaneously or will involve more than 500 person days of work (e.g. 50 people working for 10 days). Holidays and weekends are not counted if there is no construction work on those days. See Figures 1.13 and 1.14. The client is responsible for making the notification as soon as possible after

Figure 1.13 Domestic client: CDM applies but it is not notifiable due to short duration project

their appointment to the particular project. The notice must be displayed where it can be read by people working on the site and must be updated as necessary. (See Chapter 13 for the details required.)

Notification does not give rise to any additional duties.

(k) **Domestic Clients**

Unless there is a written agreement between the domestic client and the principal designer, the duties of the client must be carried out by the

Figure 1.14 Large-scale contract: CDM applies and it is notifiable

contractor or principal contractor where there is more than one.

Where there is more than one contractor, the domestic client must appoint a principal designer and principal contractor in writing. Where no appointment is made, the first designer and contractor appointed are deemed to be the principal one in each case.

1.4.5 The principles of assessing and managing contractors

Scale of contractor use

The use of contractors is increasing as many companies turn to outside resources to supplement their own staff and expertise. Contractors are often used for information technology projects, maintenance, repairs, installation, construction, demolition and many other jobs. Organisations have increasingly used contractors over recent years rather than employ more full-time employees. This may be to supplement their staff particularly for specialist tasks, or to undertake non-routine activities. There may also be other reasons for the increased use of contractors:

▶ Demand for products or services is uncertain.
▶ Contractors can be used when more flexibility is required.
▶ Contractors usually supply their own tools and equipment associated with the contract.
▶ There are no permanent staff available to perform the work.
▶ The financial overheads and legal employment obligations are lower.
▶ Most of the costs associated with increasing and reducing employee numbers as product demand varies do not relate to contractors.
▶ Permanent staff can concentrate on the core business of the organisation.

The advantage of the flexibility of contractors in uncertain economic times is the main reason for the increased use of them over recent years. There are, however, some disadvantages in the use of independent contractors. These include:

▶ Contractors/sub-contractors may cost more than the equivalent daily rate for employing a worker.
▶ By relying on contractors, the skills of permanent staff are not developed.
▶ There is less control over contractors than permanently employed staff. This can be a significant problem if the contractor sub-contracts some of the work.
▶ The control of the contractor and the quality of the work is crucially dependent on the terms of the contract

Contractor selection

The selection of the right contractor for a particular job is probably the most important element in ensuring that the risks to the health and safety of everybody involved in the activity and people in the vicinity are reduced as far as possible. Ideally, selection should be made from a list of approved contractors who have demonstrated that they are able to meet the client's requirements.

The selection of a contractor has to be a balanced judgement with a number of factors being taken into account. Fortunately, a contractor who works well and meets the client's requirements in terms of the quality and timeliness of the work is also likely to have a better than average health and safety performance. Cost, of course, will have to be part of the judgement but may not provide any indication of which contractor is likely to give the best performance in health and safety terms. In deciding which contractor should be chosen for a task, the following should be considered:

▶ Do they have an adequate health and safety policy?
▶ Can they demonstrate that the person responsible for the work is competent?
▶ Can they demonstrate that competent safety advice will be available?
▶ Do they monitor the level of accidents at their work site?
▶ Do they have a system to assess the hazards of a job and implement appropriate control measures?
▶ Will they produce a method statement which sets out how they will deal with all significant risks?
▶ Do they have guidance on health and safety arrangements and procedures to be followed?
▶ Do they have effective monitoring arrangements?
▶ Do they use trained and skilled staff who are qualified where appropriate? (Judgement will be required, as many construction workers have had little or no training except training on the job.) Can the company demonstrate that the employees or other workers used for the job have had the appropriate training and are properly experienced and, where appropriate, qualified?
▶ Can they produce good references indicating satisfactory performance?

Figure 1.15 Contractors at work from a lorry-mounted loading grab

Management and authorisation of contractors

The management of contractors is essential since accidents tend to happen more easily when the work of contractors is not properly supervised or the hazards of their job have not been identified and suitable controls introduced. It is important that contractors are made aware of:

- the health and safety procedures and rules of the organisation;
- the hazards on your site, particularly those associated with the project;
- any special equipment or personal protective equipment that they need to use;
- the emergency procedures and the sound of the alarm, and
- the safe disposal of waste.

Good communication and supervision of contractors is essential with a named point of contact from the organisation known to the contractors. Regular liaison is important so that the progress of the contract to a satisfactory completion may be monitored.

It is important that the activities of the contractors within the organisation of the client are effectively planned and coordinated so there is minimal interference with the normal activities of the organisation and the health and safety of its staff is not put at risk. This means that staff should be made aware of the contractor activity and any particular hazards associated with it, such as dust and noise.

Contractors, their employees, sub-contractors and their employees, should not be allowed to commence work on any client's site without authorisation signed by the company contact. The authorisation should clearly define the range of work that the contractor can carry out and set down any special requirements, for example protective clothing, fire exits to be left clear, and isolation arrangements.

Permits will be required for operations such as hot work. All contractors should keep a copy of their authorisation at the place of work. A second copy of the authorisation should be kept at the site and be available for inspection.

The company contact signing the authorisation will be responsible for all aspects of the work of the contractor. The contact will need to check as a minimum the following:

- that the correct contractor for the work has been selected;
- that the contractor has made appropriate arrangements for supervision of staff;
- that the contractor has received and signed for a copy of the contractor's safety rules;
- that the contractor is clear what is required, the limits of the work and any special precautions that need to be taken;
- that the contractor's personnel are properly qualified for the work to be undertaken.

The company contact should check whether sub-contractors will be used. They will also require authorisation, if deemed acceptable. It will be the responsibility of the company contact to ensure that sub-contractors are properly supervised.

Appropriate supervision will depend on a number of factors, including the risk associated with the job, experience of the contractor and the amount of supervision the contractor will provide. The responsibility for ensuring there is proper supervision lies with the person signing the contractor's authorisation.

The company contact will be responsible for ensuring that there is adequate and clear communication between different contractors and company personnel where this is appropriate.

Safety rules for contractors

In the conditions of contract, there should be a stipulation that the contractor and all of their employees adhere to the contractor's safety rules. Contractors' safety rules should contain as a minimum the following points:

- **Health and safety** – that the contractor operates to at least the minimum legal standard and conforms to accepted industry good practice;
- **Supervision** – that the contractor provides a good standard of supervision of their own employees;
- **Sub-contractors** – that they may not use sub-contractors without prior written agreement from the organisation;
- **Authorisation** – that each employee must carry an authorisation card issued by the organisation at all times while on site.

Example of rules for contractors

Contractors engaged by the organisation to carry out work on its premises will:

- familiarise themselves with so much of the organisation's health and safety policy as affects them and will ensure that appropriate parts of the policy are communicated to their employees, and any sub-contractors and employees of sub-contractors who will do work on the premises;
- cooperate with the organisation in its fulfilment of its health and safety duties to contractors and take the necessary steps to ensure the like cooperation of their employees;
- comply with their legal and moral health, safety and food hygiene duties;
- ensure the carrying out of their work on the organisation's premises in such a manner as not to put either themselves or any other persons on or about the premises at risk;

ensure that where they wish to avail themselves of the organisation's first-aid arrangements/facilities while on the premises, written agreement to this effect is obtained prior to first commencement of work on the premises;

supply a copy of their statement of policy, organisation and arrangements for health and safety written for the purposes of compliance with the Management of Health and Safety at Work Regulations and Section 2(3) of the HSW Act where applicable and requested by the organisation;

abide by all relevant provisions of the organisation's safety policy, including compliance with health and safety rules and CDM 2015;

ensure that on arrival at the premises, they and any other persons who are to do work under the contract report to reception or their designated organisation contact.

Without prejudice to the requirements stated above, contractors, sub-contractors and employees of contractors and sub-contractors will, to the extent that such matters are within their control, ensure:

the safe handling, storage and disposal of materials brought onto the premises;

that the organisation is informed of any hazardous substances brought onto the premises and that the relevant parts of the Control of Substances Hazardous to Health Regulations in relation thereto are complied with;

that fire prevention and fire precaution measures are taken in the use of equipment which could cause fires;

that steps are taken to minimise noise and vibration produced by their equipment and activities;

that scaffolds, ladders and other such means of access, where required, are erected and used in accordance with Work at Height Regulations and good working practice;

that any welding or burning equipment brought onto the premises is in safe operating condition and used in accordance with all safety requirements;

that any lifting equipment brought onto the premises is adequate for the task and has been properly tested/certified;

that any plant and equipment brought onto the premises is in safe condition and used/operated by competent persons;

that for vehicles brought onto the premises, any speed, condition or parking restrictions are observed;

that compliance is made with the relevant requirements of the Electricity at Work Regulations 1989;

that connection(s) to the organisation's electricity supply is from a point specified by its management and is by proper connectors and cables;

that they are familiar with emergency procedures existing on the premises;

that welfare facilities provided by the organisation are treated with care and respect;

that access to restricted parts of the premises is observed and the requirements of food safety legislation are complied with;

that any major or lost-time accident or dangerous occurrence on the organisation's premises is reported as soon as possible to their site contact;

that where any doubt exists regarding health and safety requirements, advice is sought from the site contact.

The foregoing requirements do not exempt contractors from their statutory duties in relation to health and safety, but are intended to assist them in attaining a high standard of compliance with those duties.

1.5 Additional useful background information that is not part of the NEBOSH NGC syllabus

1.5.1 The legal system in England and Wales

The description that follows applies to England and Wales (and with a few minor differences to Northern Ireland). Only the court functions concerning health and safety are mentioned. Figure 1.16 shows the court hierarchy in schematic form.

Criminal law

Magistrates' Courts

Most criminal cases begin and end in the Magistrates' Courts. Health and safety cases are brought before the court by enforcement officers (Health and Safety Executive or Local Authority Environmental Health Officers) and they are tried by a bench of three lay magistrates (known as Justices of the Peace) or a single district judge. The lay magistrates are members of the public, usually with little previous experience of the law, whereas the district judge is legally qualified.

The Magistrates' Court has limited powers with a maximum fine of £5,000 for employees to £20,000 for employers or for those who ignore prohibition notices. Magistrates are also able to imprison for up to 12 months. The vast majority of health and safety criminal cases are dealt with in the Magistrates' Court. See Table 13.1 in Chapter 13 for details of penalties under the Health and Safety Offences Act 2008.

Crown Court

The Crown Court hears the more serious cases (indictable), which are passed to them from the Magistrates' Court – normally because the sentences available to the magistrates are felt to be too lenient. Cases are heard by a judge and jury, although some cases are heard by a judge alone. The penalties available to the Crown Court are an unlimited fine and up to 2 years' imprisonment for breaches of enforcement notices. The Crown Court also hears appeals from the Magistrates' Court.

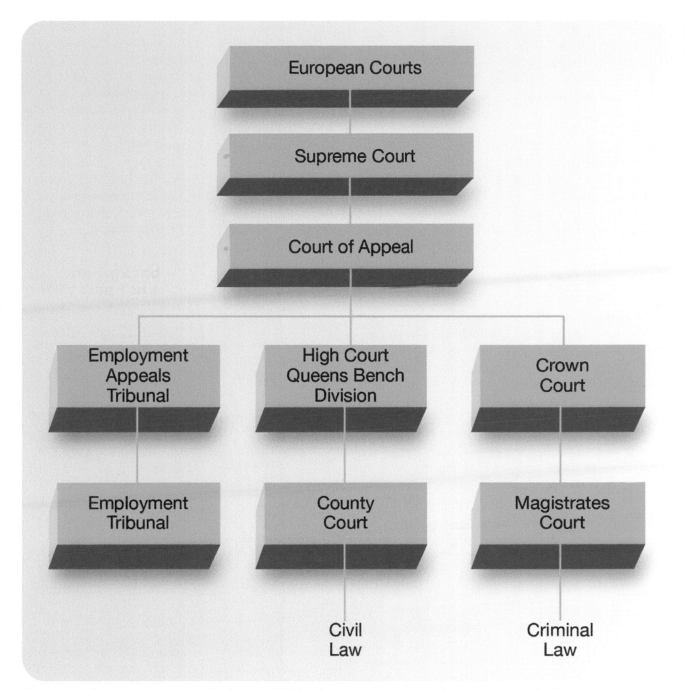

Figure 1.16 The court system in England and Wales for health and safety showing the principal courts

Appeals from the Crown Court are made to the Court of Appeal (Criminal Division), who may then give leave to appeal to the most senior court in the country – the Supreme Court. The most senior judge at the Court of Appeal is the Lord Chief Justice.

Civil law

County Court

The lowest court in civil law is the County Court, which only deals with minor cases (for compensation claims of up to £50,000 if the High Court agrees). Cases are normally heard by a judge sitting alone. For personal injury claims of less than £5,000, a small claims court is also available.

High Court

Many health and safety civil cases are heard in the High Court (Queen's Bench Division) before a judge only. It deals with compensation claims in excess of £50,000 and acts as an appeal court for the County Court.

Appeals from the High Court are made to the Court of Appeal (Civil Division). The Supreme Court receives appeals from the Court of Appeal or, on matters of law interpretation, directly from the High Court. The most senior judge at the Court of Appeal is the Master of the Rolls.

The Justices in the Supreme Court are sometimes called upon to make judgements on points of law, which are then binding on lower courts. Such judgements form the basis of common law, which was covered in Section 1.2 above.

The Supreme Court

From October 2009, the Supreme Court of the United Kingdom was established and assumed the judicial functions of the House of Lords. It is an independent institution, presided over by 12 independently appointed judges, known as Justices of the Supreme Court, and is housed in the Middlesex Guildhall on Parliament Square in London. The Supreme Court is the final court of appeal in the UK for civil cases and hears appeals in criminal cases from England, Wales and Northern Ireland. It plays an important role in the development of United Kingdom common law, as was the case previously with the House of Lords. As an appeal court, the Supreme Court cannot consider a case unless a relevant order has been made in a lower court.

Other courts – Employment Tribunals

These were established in 1964 and primarily deal with employment and conditions of service issues, such as unfair dismissal. However, they also deal with appeals over health and safety enforcement notices, disputes between recognised safety representatives and their employers and cases of unfair dismissal involving health and safety issues. There are usually three members who sit on a tribunal. These members are appointed and are often not legally qualified. Appeals from the tribunal may be made to the Employment Appeal Tribunal or, in the case of enforcement notices, to the High Court. Appeals from tribunals can only deal with the clarification of points of law.

1.5.2 The legal system in Scotland

Scotland has both criminal and civil courts but prosecutions are initiated by the Procurator Fiscal rather than the Health and Safety Executive. The lowest Criminal Court is called the District Court and deals with minor offences. The Sheriff Court has a similar role to that of the Magistrates' Court (for criminal cases) and the County Court (for civil cases), although it can deal with more serious cases involving a sheriff and jury.

The High Court of Justiciary, in which a judge and jury sit, has a similar role to the Crown Court and appeals are made to the Court of Criminal Appeal. The High Court of Justiciary is the supreme criminal court of Scotland. The High Court is both a court of first instance and a court of appeal. The Outer House of the Court of Session deals with civil cases in a similar way to the English High Court. The Inner House of the Court of Session is the appeal court for civil cases.

For both appeal courts, the Supreme Court is the final court of appeal. There are Employment Tribunals in Scotland with the same role as those in England and Wales.

1.5.3 European Courts

There are two European Courts – the European Court of Justice and the European Court of Human Rights.

The European Court of Justice, based in Luxembourg, is the highest court in the European Union (EU). It deals primarily with community law and its interpretation. It is normally concerned with breaches of community law by member states and cases may be brought by other member states or institutions. Its decisions are binding on all member states. There is currently no right of appeal.

The European Court of Human Rights, based in Strasbourg, is not directly related to the EU – it covers most of the countries in Europe including the EU member states. As its title suggests, it deals with human rights and fundamental freedoms. With the introduction of the Human Rights Act 1998 in October 2000, many of the human rights cases are now heard in the UK.

1.5.4 The influence of the European Union on health and safety

Much of the UK health and safety law originates in Europe. Proposals from the European Commission may be agreed by member states. The member states are then responsible for making them part of their national law.

In Britain itself and in much of the EU, health and safety law is based on the principle of risk assessment described in Chapter 4. The main role of the EU in health and safety is to harmonise workplace and legal standards and remove barriers to trade across member states. A Directive from the EU is legally binding on each member state and must be incorporated into the national law of each member state. Directives set out specific minimum aims which must be covered within the national law. Some member states incorporate Directives more speedily than others.

Directives are proposed by the European Commission, comprising commissioners, who are citizens of each of the member states. The proposed Directives are sent to the European Parliament, which is directly elected from the member states. The European Parliament may accept, amend or reject the proposed Directives. The proposed Directives are then passed to the Council of Ministers, who may accept the proposals on a qualified majority vote, unless the European Parliament has rejected them, in which case they can only be accepted on a unanimous vote of the Council. The Council of Ministers consists of one senior government minister from each of the member states.

The powers of the EU in health and safety law are derived from the Treaty of Rome 1957 and the Single European Act 1986. For health and safety, the Single European Act added two additional Articles – 100A and 118A. The 1997 Treaty of Amsterdam renumbered them as 95A and 138A, respectively. Article 95A is concerned with health and safety standards of equipment and plant and its Directives are implemented in the UK by the Department for Business, Energy and Industrial Strategy.

Article 138A is concerned with minimum standards of health and safety in employment and its Directives are implemented by the Health and Safety Executive.

The objective of the Single European Act 1986 is to produce a 'level playing field' for all member states so that goods and services can move freely around the EU without any one state having an unfair advantage over another. The harmonisation of health and safety requirements across the EU is one example of the 'level playing field'.

The first introduction of an EU Directive into UK health and safety law occurred on 1 January 1993 when a Framework Directive on Health and Safety Management and five daughter Directives were introduced using powers contained in the HSW Act 1974. These Directives, known as the European Six Pack, covered the following areas:

- management of health and safety at work;
- workplace;
- provision and use of work equipment;
- manual handling;
- personal protective equipment;
- display screen equipment.

Since 1993, several other EU Directives have been introduced into UK law. Summaries of the more common UK Regulations are given in Chapter 13.

The UK Government has issued its draft proposals for minor changes to health and safety regulations to ensure they still function as intended after the country exits the European Union. Minor amendments will be made to regulations to remove EU references but legal requirements will remain the same. Health and safety standards will be maintained. It has also reiterated that the level of protection afforded to workers under UK employment law which originated from the European Union will continue post-Brexit. See the Health and Safety (Amendment) (EU Exit) Regulations 2018.

1.6 Further information

The Health and Safety at Work etc. Act 1974

The Enterprise and Regulatory Reform Act 2013 (Section 69)

The Health and Safety at Work etc. Act 1974 (Civil Liabilities) (Exceptions) Regulations 2013

The Interpretation Act 1978

The Management of Health and Safety at Work Regulations 1999 (as amended)

The Construction (Design and Management) Regulations 2015

Corporate Manslaughter and Corporate Homicide Act 2007

Health and Safety (Fees) Regulations 2012 (Regulations 23–25)

Guidance on the application of Fees for Intervention (FFI), HSE47, HSE Books www.hse.gov.uk/pubns/hse47.pdf

Managing for health and safety, HSG65 third edition 2013, HSE Books, ISBN: 978-0-7176-6456-6 www.hse.gov.uk/pubns/priced/hsg65.pdf

HSE, Managing for health and safety www.hse.gov.uk/managing/

HSE, The Health and Safety Toolbox: how to control risks at work www.hse.gov.uk/toolbox/

Leading health and safety at work (INDG417) HSE Books, ISBN 978 0 7176 6267 8 www.hse.gov.uk/pubs/indg417.htm

Managing Health and Safety in Construction (ACoP) (L144), HSE Books, ISBN 978 0 7176 6223 4 www.hse.gov.uk/pubns/books/l144.htm

Health and Safety in Construction (Guidance) (HSG150rev), HSE Books, ISBN 978 0 7176 6182 4 www.hse.gov.uk/pubns/books/hsg150.htm

Health and safety management standard ISO 45001: 2018 https://www.iso.org/standard/63787.html

HSE Statistics www.hse.gov.uk/statistics/index.htm

1.7 Practice questions

1. In relation to occupational health and safety, **explain**, using an example in **EACH** case, the meaning of the following terms:
   (a) hazard
   (b) risk
   (c) welfare
   (d) work-related ill health
   (e) near miss.

2. (a) **Explain** the reasons that organisations are required to have employer liability insurance.
   (b) **Outline SIX** possible direct **AND SIX** possible indirect costs to an organisation following a serious accident in the workplace. Identify those costs that are insurable and those that are not.

3. (a) **Outline** the key differences between civil law and criminal law.
   (b) **Identify** the main differences between common law and statute law and **explain** by giving **ONE** example how these two sources of law are applied in occupational health and safety cases.

4. In relation to health and safety, **outline** the role of the following:
   (a) employment tribunals
   (b) civil courts
   (c) criminal courts.

5. (a) **Identify** the courts in which summary and indictable health and safety offences are heard.
   (b) Giving an example in each case, **outline** the common law duties of care that employers

have towards their employees for their health, safety and welfare.

6. (a) **Explain** the meaning of the term 'negligence' in health and safety matters.

   (b) **Outline** the three absolute defences and two partial defences available to an employer in a case of alleged negligence brought by an employee.

   (c) **Explain** the significance of vicarious liability for an employer following an injury to an employee resulting from the negligence of another employee.

7. (a) **Explain**, giving an example in each case, the differences between the levels of statutory duty – absolute, practicable and reasonably practicable.

   (b) **Outline** the meaning of the terms statute-barred breaches of statutory duty and double-barrelled actions.

   (c) **Identify** the general duties placed on employees by:
      (i) the Health and Safety at Work etc. Act 1974
      (ii) the Management of Health and Safety at Work Regulations 1999.

8. (a) **Outline** the general duties placed on employers by the Health and Safety at Work etc. Act 1974 in Sections 2, 3 and 6.

   (b) Identify the possible defences that an employer may give following a prosecution under the Health and Safety at Work etc. Act 1974.

   (c) **Identify FOUR** enforcement agencies who have the power to prosecute companies for non-compliance with health and safety legislation.

   (d) **Explain**, using relevant examples, the circumstances under which a health and safety inspector may serve an improvement notice or a prohibition notice on an employer. Briefly outline the appeal procedure in each case.

9. While visiting a company, a health and safety enforcement officer observes a contractor operating a dangerous part of a machine that is unguarded.

   (a) **Describe** the powers available to the officer under the Health and Safety at Work etc. Act 1974

   (b) **Identify** the sections of the Health and Safety at Work etc. Act 1974 that may have been breached.

   (c) **Identify** the actions available and the likely one that was taken by the officer.

10. Giving an example in each case, **explain** the meaning, legal status and roles of:
    (a) health and safety regulations
    (b) HSE Approved Codes of Practice
    (c) HSE guidance.

11. **Describe** the possible penalties that a court could impose on a company that is convicted of an offence under the Corporate Manslaughter and Corporate Homicide Act 2007.

12. (a) **Outline** the measures that directors of corporations can take to promote good health and safety standards within their organisations.

    (b) **Outline** ways by which the health and safety of visitors to the workplace may be protected.

    (c) **Describe** the issues that should be addressed when assessing risks to night workers.

    (d) **Describe** the important factors that will help to achieve cooperation and coordination in jointly occupied workplace.

13. **Describe** the possible health and safety issues that may arise in the management of a supply chain.

14. **Outline FIVE** duties of employers to their employees under the Management of Health and Safety at Work Regulations 1999.

15. Following an investigation of an accident to a contractor while on an employer's premises, it was found that the employer had failed to provide relevant health and safety information to the contractor. The employer was also unable to demonstrate any source of competent health and safety advice.

    (a) **Outline** the duties that the employer owes to the contractor under the Health and Safety at Work etc. Act 1974.

    (b) **Outline** the duties that the employer owes under the Management of Health and Safety at Work Regulations 1999 that may not have been discharged.

    (c) **Outline** the health and safety information that should have been provided to and from the contractor before work commenced.

16. **Outline FOUR** duties of each of the following persons under the Construction (Design and Management) Regulations:
    (a) the client
    (b) the principal designer
    (c) the principal contractor.

17. Several construction contractors are to be employed by a printing company to refurbish its building.
    (a) **Outline** the health and safety duties of the self-employed.

    (b) **Outline** the main issues to be considered during the selection of the contractors by the printing company.

    (c) **Outline** the matters to be addressed by the company while supervising the contractors and the contractors in supervising its employees.

APPENDIX 1.1 Checklist for supply chain health and safety management

This checklist is a reminder of the topics that might need to be discussed with people with whom individual contractors may be working.

It is not intended to be exhaustive and not all questions will apply at any one time, but it should help people to get started.

1. Responsibilities

▷ What are the hazards of the job?
▷ Who is to assess particular risks?
▷ Who will coordinate action?
▷ Who will monitor progress?

2. The job

▷ Where is it to be done?
▷ Who with?
▷ Who is in charge?
▷ How is the job to be done?
▷ What other work will be going on at the same time?
▷ How long will it take?
▷ What time of day or night?
▷ Do you need any permit to do the work?

3. The hazards and risk assessments

Site and location

Consider the means of getting into and out of the site and the particular place of work – are they safe? – and:

▷ Will any risks arise from environmental conditions?
▷ Will you be remote from facilities and assistance?
▷ What about physical/structural conditions?
▷ What arrangements are there for security?

Substances

▷ What suppliers' information is available?
▷ Is there likely to be any microbiological risk?
▷ What are the storage arrangements?
▷ What are the physical conditions at the point of use? Check ventilation, temperature, electrical installations, etc.
▷ Will you encounter substances that are not supplied, but produced in the work, for example fumes from hot work during dismantling plant? Check how much, how often, for how long, method of work, etc.
▷ What are the control measures? For example, consider preventing exposure, providing engineering controls, using personal protection (in that order of choice).

▷ Is any monitoring required?
▷ Is health surveillance necessary, for example for work with sensitisers? (Refer to health and safety data sheet.)

Plant and equipment

▷ What are the supplier/hirer/manufacturer's instructions?
▷ Are any certificates of examination and test needed?
▷ What arrangements have been made for inspection and maintenance?
▷ What arrangements are there for shared use?
▷ Are the electrics safe to use? Check the condition of power sockets, plugs, leads and equipment. (Don't use damaged items until they have been repaired.)
▷ What assessments have been made of noise levels?

4. People

▷ Is information, instruction and training given, as appropriate?
▷ What are the supervision arrangements?
▷ Are members of the public/inexperienced people involved?
▷ Have any disabilities/medical conditions been considered?

5. Emergencies

▷ What arrangements are there for warning systems in case of fire and other emergencies?
▷ What arrangements have been made for fire/emergency drills?
▷ What provision has been made for first-aid and fire-fighting equipment?
▷ Do you know where your nearest fire exits are?
▷ What are the accident reporting arrangements?
▷ Are the necessary arrangements in place for availability of rescue equipment and rescuers?

6. Welfare

Who will provide:

▷ shelter
▷ food and drinks
▷ washing facilities
▷ toilets (male and female)
▷ clothes changing/drying facilities?

There may be other pressing requirements which make it essential to rethink health and safety as the work progresses.

APPENDIX 1.2 Pre-construction information

This is taken from Appendix 2 of the ACoP *Managing Health and Safety in Construction*, L144.

When drawing up the pre-construction information, each of the following topics should be considered.

Information should be included where the topics are relevant to the work proposed. The pre-construction information provides information for those bidding for or planning work, and for the development of the construction phase plan. The level of detail in the information should be proportionate to the risks involved in the project.

Pre-construction information

1. Description of project

(a) Project description and programme details including:
- key dates (including planned start and finish of the construction phase), and
- the minimum time to be allowed between appointment of the principal contractor and instruction to commence work on site.

(b) Details of client, principal designer, designers and other consultants.

(c) Whether or not the structure will be used as a workplace – if so, the finished design will need to take account of the relevant requirements of the Workplace (Health, Safety and Welfare) Regulations.

(d) Extent and location of existing records and plans.

2. Client's considerations and management requirements

(a) Arrangements for:
- planning for and managing the construction work, including any health and safety goals for the project
- communications and liaison between client and others
- security of the site
- welfare provisions.

(b) Requirements relating to the health and safety of the client's employees or customers or those involved in the project such as:
- site hoarding requirements
- site transport arrangements or vehicle movement restrictions
- client permit-to-work systems
- fire precautions
- emergency procedures and means of escape
- 'no-go' areas or other authorisation requirements for those involved in the project
- any areas the client has designated as confined spaces
- smoking and parking restrictions.

3. Environmental restrictions and existing on-site risks

(a) Safety hazards, including:
- boundaries and access, including temporary access, e.g. narrow streets, lack of parking, turning or storage

- any restrictions on deliveries or waste collection or storage
- adjacent land uses, e.g. schools, railway lines or busy roads
- existing storage of hazardous materials
- location of existing services particularly those that are concealed – water, electricity, gas, etc.
- ground conditions, underground structures or water courses where these might affect the safe use of plant, e.g. cranes, or the safety of ground work
- information about existing structures – stability, structural form, fragile or hazardous materials, anchorage points for fall arrest systems (particularly where demolition is involved)
- previous structural modifications, including weakening or strengthening of the structure (particularly where demolition is involved)
- fire damage, ground shrinkage, movement or poor maintenance which may have adversely affected the structure
- any difficulties relating to plant and equipment in the premises, such as overhead gantries whose height restricts access
- health and safety information contained in earlier design, construction or 'as-built' drawings, such as details of pre-stressed or post-tensioned structures.

(b) Health hazards, including:
- asbestos, including results of surveys (particularly where demolition is involved)
- existing storage of hazardous material
- contaminated land, including results of surveys
- existing structures containing hazardous materials
- health risks arising from client's activities.

4. Significant design and construction hazards

(a) Significant design assumptions and suggested work methods, sequences or other control measures.

(b) Arrangements for coordination of ongoing design work and handling design changes.

(c) Information on significant risks identified during design.

(d) Materials requiring particular precautions.

5. The health and safety file

Description of its format and any conditions relating to its content.

APPENDIX 1.3 Construction phase plan

This is taken from Appendix 3 of the ACoP *Managing Health and Safety in Construction*, L144.

When drawing up the construction phase plan, employers should consider each of the following topics: information should be included in the plan where the topic is relevant to the work proposed; the plan sets out how health and safety is to be managed during the construction phase. The level of detail should be proportionate to the risks involved in the project.

Construction phase plan

1. Description of project

(a) Project description and programme details including key dates.
(b) Details of client, principal designer, designers, principal contractor and other consultants.
(c) Extent and location of existing records and plans that are relevant to health and safety on site, including information about existing structures when appropriate.

2. Management of the work

(a) Management structure and responsibilities.
(b) Health and safety goals for the project and arrangements for monitoring and review of health and safety performance.
(c) Arrangements for:
   - regular liaison between parties on site
   - consultation with the workforce
   - exchange of design information between the client, principal designer, designers and contractors on site
   - handling design changes during the project
   - the selection and control of contractors
   - the exchange of health and safety information between contractors
   - site security
   - site induction
   - on-site training
   - welfare facilities and first aid
   - the reporting and investigation of accidents and incidents including near misses
   - the production and approval of risk assessments and written systems of work
   - site rules
   - fire and emergency arrangements.

3. Arrangements for controlling significant site risks

(a) Safety risks, including:
   - delivery and removal of materials (including waste) and work equipment taking account of any risks to the public, e.g. during access to and egress from the site
   - dealing with services – water, electricity and gas, including overhead power lines and temporary electrical installations
   - accommodating adjacent land use
   - stability of structures whilst carrying out construction work, including temporary structures and existing unstable structures
   - preventing falls
   - work with or near fragile materials
   - control of lifting operations
   - the maintenance of plant and equipment
   - work on excavations and work where there are poor ground conditions
   - work on wells, underground earthworks and tunnels
   - work on or near water where there is a risk of drowning
   - work involving diving
   - work in a caisson or compressed air working
   - work involving explosives
   - traffic routes and segregation of vehicles and pedestrians
   - storage of materials (particularly hazardous materials) and work equipment
   - any other significant safety risks.
(b) health risks, including:
   - the removal of asbestos
   - dealing with contaminated land
   - manual handling
   - use of hazardous substances, particularly where there is a need for health monitoring
   - reducing noise and vibration
   - working with ionising radiation; any other significant health risks.

4. The health and safety file

(a) layout and format
(b) arrangements for the collation and gathering of information
(c) storage of information.

APPENDIX 1.4 The health and safety file

The contents of the health and safety file are likely to remain the same as follows:

- a brief description of the work carried out;
- any residual hazards which remain and how they have been dealt with (e.g. surveys or other information concerning asbestos, contaminated land, water-bearing strata and buried services);
- key structural principles (e.g. bracing, sources of substantial stored energy – including pre- or post-tensioned members) and safe working loads (SWL)

for floors and roofs, particularly where these may preclude placing scaffolding or heavy machinery there;

▶ hazardous material used (e.g. lead paint, pesticides and special coatings which should not be burnt off);

▶ information regarding the removal or dismantling of installed plant and equipment (e.g. any special arrangements for lifting, order of or other special instructions for dismantling);

▶ health and safety information about equipment provided for cleaning or maintaining the structure;

▶ the nature, location and markings of significant services, including underground cables, gas supply equipment and firefighting services;

▶ information and as-built drawings of the structure, its plant and equipment (e.g. the means of safe access to and from service voids, fire doors and compartmentalisation).

CHAPTER 2

Health and safety management systems and policy

Chapter summary

This chapter covers the following topics in Element 2 of the NEBOSH NGC syllabus:

What they are and the benefits they bring

What good health and safety management systems look like

Learning outcomes

These are the NEBOSH learning outcomes covered by this chapter:

The learner will be able to:

Work within a health and safety management system, recognising what effective general policy, organisation and arrangements should look like

Assessment criteria

2.1 Give an overview of the elements of a health and safety management system and the benefits of having a formal/certified system

2.2 Discuss the main ingredients of health and safety management systems that make it effective – general policy, organisation, arrangements

---

2.1 Key elements and benefits of a health and safety management system

2.1.1 Introduction

Every organisation should have a clear policy for the systematic management of health and safety so that health and safety risks may be effectively addressed and controlled and for those with five or more employees, it is a legal requirement that this policy must be in writing. A good health and safety policy will indicate the goals that the adopted health and safety management system hopes to achieve. The health and safety policy and management system will also complement those policies in areas such as quality, the environment and human resources. As for those areas, for the health and safety policy and its associated management system to be successful, it must have realistic goals and the active support and involvement of all levels of management within the organisation.

Planning to implement the health and safety policy is the key element of any health and safety management system. An effective planning system for health and safety requires organisations to set up, operate and maintain a management system that can detect, eliminate and control hazards and risks. This process is particularly important when there are long latency periods before a health problem becomes apparent, such as asbestosis or lung cancer caused by asbestos fibre inhalation.

Since the main objective of any health and safety plan is to prevent people being harmed at work, workplace precautions must be provided and maintained at the point of risk. Risks are created in the workplace as resources and information are used to create products and services in the manufacturing process. This business activity model also applies to other industries including construction, education, leisure, hospitals and local authorities. Workplace precautions to match the hazards and risks are needed at each stage of the business activity. They can include machine guards,

electrical safeguards, flammable liquid storage, dust controls, safety instructions and systems of work.

This chapter is concerned with the planning for managing occupational health and safety within an organisation. The preparation of an effective health and safety policy is an essential first step in the formulation of the health and safety management system. The policy will only remain as words on paper, however good the intentions, until there is an effective organisation set up to implement and monitor its requirements. Some policies are written so that most of the wording concerns strict requirements laid on employees and only a few vague words cover managers' responsibilities. Generally, such policies do not meet the requirements of the HSW Act or the Management of Health and Safety at Work Regulations, which require an effective policy with a robust organisation and arrangements to be set up. For a policy to be effective, it must be honoured in the spirit as well as the letter. A good health and safety policy together with an effective occupational health and safety management system will also enhance the performance of the organisation in areas other than health and safety, help with the personal development of the workforce and reduce financial losses.

2.1.2 Key elements of a health and safety management system

Most of the key elements required for effective health and safety management are very similar to those required for good quality, finance and general business management. Commercially successful organisations usually have good health and safety management systems in place. The principles of good and effective management provide a sound basis for the improvement of health and safety performance.

HSG65

The HSE document HSG65, Managing for health and safety, describes the occupational health and safety management system used extensively in the UK. In 2013, the HSE's guidance in HSG65 moved from using the POPMAR model (Policy, Organising, Planning, Measuring performance, Auditing and Review) to a 'Plan, Do, Check, Act' approach. Figure 2.1 illustrates this model.

The HSE have argued that the move towards Plan, Do, Check, Act achieves a better balance between the systems and behavioural aspects of management. It also treats health and safety management as an integral part of good management generally, rather than as a stand-alone system. A description of this four-step approach to occupational health and safety management is:

PLAN – establish standards for health and safety management based on risk assessment and legal requirements.

DO – implement plans to achieve objectives and standards.

CHECK – measure progress with plans and compliance with standards.

ACT – review against objectives and standards and take appropriate action.

The **Plan–Do–Check–Act** model for occupational health and safety management also forms the basis of the two occupational health and safety management systems: ISO 45001 2018 and ILO-OSH 2001.

All recognised occupational health and safety management systems, including HSG65, have some basic and common elements. These are:

▶ a planning phase;
▶ a performance phase;
▶ a performance assessment phase; and
▶ a performance improvement phase.

The planning phase – PLAN

The planning phase always includes a **policy statement** which outlines the health and safety aims, objectives and commitment of the organisation and lines of responsibility. **Hazard identification and risk assessment** takes place during this phase and the significant hazards may well be included in the policy statement. It is important to note that in some reference texts, in particular those in languages other than English, the whole process of hazard identification, risk determination and the selection of risk reduction or control measures is termed 'risk assessment'. However, all three occupational health and safety management systems described in this chapter refer to the individual elements of the process separately and use the term 'risk assessment' for the determination of risk only.

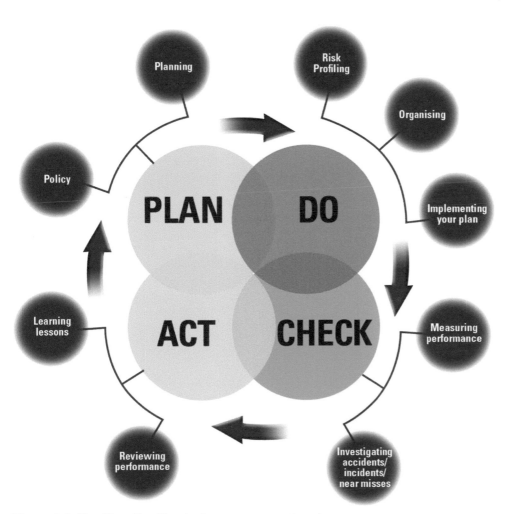

Figure 2.1 The Plan, Do, Check, Act management cycle

At the **planning stage**, emergency procedures should be developed and relevant health and safety legal requirements and other standards identified together with appropriate benchmarks from similar industries. An **organisational structure** must be defined so that health and safety responsibilities are allocated at all levels of the organisation and issues such as competent persons and health and safety training are addressed. Realistic targets should be agreed within the organisation and be published as part of the policy.

The performance phase – DO

The performance phase (Do) will only be successful if there is good **communication** at and between all levels of the organisation. This implies employee participation as both **worker representatives** and on **safety committees**. Effective communication with the workforce, for example with clear **safe systems of work** and other health and safety procedures, will not only aid the implementation and operation of the plan but also produce continual improvement of performance – a key requirement of all occupational health and safety, quality and environmental management systems. There should also be effective communication with other stakeholders, such as regulators, contractors, customers and trade unions. The performance phase must be monitored on a regular basis since this will indicate whether there is an effective occupational health and safety management system and a good **health and safety culture** within the organisation.

The performance assessment phase – CHECK

The performance assessment phase (Check) may be either **active or reactive** or, ideally, a mixture of both (see also Leading and Lagging Indicators in Section 5.1). Active assessment includes work-based **inspections and audits**, regular health and safety committee meetings, feedback from training sessions and a constant review of risk assessments. Reactive assessment relies on records of accident, work-related injuries and ill health as well as near miss and any enforcement notices. Any recommended remedial or preventative actions, following an investigation, must be implemented immediately and monitored regularly.

The performance improvement phase – ACT

The performance improvement phase (Act) involves a **review** of the effectiveness of the health and safety management system and the identification of any weaknesses. The review, which should be undertaken by the management of the organisation, will assess whether targets have been met and the reasons for any under-performance. Issues such as the level of resources made available, the vigilance

of supervisors and the level of cooperation of the workforce should be considered at the review stage. When recommendations are made, the review process must define a timescale by which any improvements are implemented and this part of the process must also be monitored. **Continual improvement** implies a commitment to improve performance on a proactive, continuous basis without waiting for a formal review to take place. Most management systems include an **audit** requirement, which may be either internal or external, or both. The audit process examines the effectiveness of the whole management process and may act as a control on the review process. Many enquiry reports into health and safety management issues have asserted that health and safety performance should be subject to audit in the same way that financial performance must be audited.

The NEBOSH National General Certificate NGC1 syllabus follows the Plan, Do, Check and Act model and this is reflected in Chapters 2–5 of this book as shown in Table 2.1.

Table 2.1 Location and contents of the key elements of a health and safety management system in Chapters 2–5

Plan, Do, Check, Act	Topics covered	Chapter
PLAN	Key elements of a health and safety management system	2
	Purpose and importance of setting a policy for health and safety	
	Key features and appropriate content of an effective health and safety policy	
DO	Organisational health and safety roles and responsibilities of employers, directors, managers and supervisors	3 & 4
	Concept of health and safety culture and its significance in the management of health and safety in an organisation	
	Human factors which influence behaviour at work	
	How health and safety behaviour at work can be improved	
	Importance of planning	
	Principles and practice of risk assessment	
	General principles of control and hierarchy of risk reduction measures	
	Sources of health and safety information	
	Safe systems of work	
	Permits to work	
	Emergency procedures and arrangement for contacting the emergency services	
	Requirements for, and effective provision of, first aid in the workplace	
CHECK	Active and reactive monitoring	5
	Investigating incidents	
	Recording and reporting incidents	
ACT	Health and safety auditing	5
	Review of health and safety performance	

2.1.3 Major occupational health and safety management systems

There are two other major occupational health and safety management systems that are in use globally in addition to HSG65. These are as follows:

▶ ISO 45001 2018 – this replaces OHSAS 18001 – 2007 which was developed in conjunction with the ISO 9000 series for quality management and the ISO 14000 series for environmental management.

▶ ILO-OSH 2001 was developed by the International Labour Organization (ILO) after an extensive study of many occupational health and safety management systems used across the world. It was established as an international system following the publication of *Guidelines on occupational safety and health management systems* in 2001. It is very similar to ISO 45001 2018.

ISO 45001 2018

The ISO 45001 occupational health and safety management system elements are based on the Plan, Do, Check, Act model as follows:

▶ policy (PLAN)
▶ planning (PLAN)
▶ implementation and operation (DO)
▶ checking and corrective action (CHECK)
▶ management review (ACT)
▶ continual improvement (ACT).

It uses risk-based thinking as a method of identifying risks and opportunities in all parts of the cycle to improve performance and minimise negative outcomes.

ISO 45001 replaced OHSAS 18001 as the definitive occupational health and safety standard in 2016. Its purpose is to enable an organisation to proactively improve its occupational health and safety performance in preventing injury and ill health. As with HSG65 it applies the Plan, Do, Check, Act management model. It places more emphasis on seeking continual improvement by addressing occupational health and safety risks and other initiatives such as health, education and training. It also stresses the importance of risk management and continuing risk assessment.

One of the key changes from earlier standards (such as ISO 18001) is that ISO 45001 requires demonstrable leadership from the very top of organisations. Those who control or direct an organisation must take an active role in promoting a positive culture and communicating this throughout the organisation. They need to demonstrate their active involvement in the integration of the occupational health and safety management system into the overall business processes.

ISO 45001 places more emphasis in the following areas:

▶ worker participation
▶ continual improvement
▶ hierarchy of control – particularly applied during the planning and operational stages
▶ risk management
▶ compliance status – requires relevant legal and other requirements to be kept up to date and compliance status checked regularly
▶ contractors, procurement and outsourcing – to be effectively managed by effectively extending the management system as far into the supply chain as the organisation has control and influence
▶ performance evaluation – of occupational health and safety performance and also the efficacy of the system.

The key benefits of ISO 45001 to an organisation include:

▶ Easier integration with other ISO standards because of the Annex SL structure.
▶ Increased awareness of its occupational health and safety risks.
▶ Continual improvement of its occupational health and safety performance.
▶ Improves managerial involvement and supervision.
▶ Reputation for being one of the safest places to work by improving individual safety.
▶ Improved ability to respond to regulatory compliance issues.
▶ Reduction in the overall costs of incidents.
▶ Reduction in downtime and the costs of disruption to operations.
▶ Reduction in the cost of insurance premiums.
▶ Reduced absenteeism and employee turnover rates.
▶ Increases return on financial investment.
▶ Recognition of achievement of an international benchmark.

Reasons for introducing health and safety management systems

The reasons include the need to:

▶ **maintain compliance with legal requirements**;
▶ ensure that the organisation's health and safety policy is undertaken in an effective and efficient manner;
▶ define the health and safety duties and responsibilities of everyone in the organisation;
▶ plan and set health and safety standards;
▶ measure performance against those standards;
▶ ensure that arrangements are in place to address health and safety legislative requirements;
▶ ensure that arrangements are in place to control and minimise any workplace risks to the health, safety and welfare of employees of the organisation;

▶ investigate promptly any accidents or incidents of ill health;
▶ establish effective communication on health and safety issues between all levels of the organisation;
▶ reduce employee turnover;
▶ improve productivity in the organisation through continual improvement; and
▶ improve the reputation and image of the organisation and gain more business.

2.1.4 Benefits of having a formal/ certified health and safety management system

Formal or certified occupational health and safety management systems have many benefits, of which the principal ones are:

▶ It is much easier to achieve and demonstrate legal compliance. Enforcement authorities have more confidence in organisations which have a health and safety management system in place.
▶ They ensure that health and safety is given the same emphasis as other business objectives, such as quality and finance. They will also aid integration, where appropriate, with other management systems.
▶ They enable significant health and safety risks to be addressed in a systematic manner.
▶ They can be used to show legal compliance with terms such as 'practicable' and 'so far as is reasonably practicable'.
▶ They indicate that the organisation is prepared for an emergency.
▶ They illustrate that there is a genuine commitment to health and safety throughout the organisation.
▶ They increase the confidence of stakeholders and help the organisation win contracts and gain a competitive advantage.
▶ An effective ISO 45001 management system can help the organisation save money in the longer term after the initial costs of certification have been recouped.

There are, however, several problems associated with occupational health and safety management systems, although most of them are soluble because they are caused by poor implementation of the system. The main problems are:

▶ The arrangements and procedures are not apparent at the workplace level and the audit process is only concerned with a desktop review of procedures.
▶ The documentation is excessive and not totally related to the organisation due to the use of generic procedures.
▶ Other business objectives, such as production targets, lead to ad hoc changes in procedures.
▶ Integration, which should really be a benefit, can lead to a reduction in the resources and effort applied to health and safety.

▶ A lack of understanding by supervisors and the workforce leads to poor system implementation.
▶ The performance review is not implemented seriously thus causing cynicism throughout the organisation.

For there to be a valid certification process, the following must be in place:

▶ A documented standard for the health and safety management system with which the organisation requiring certification must comply.
▶ An external audit by an independent and competent body of the health and safety management system to check that the organisation is complying with the standard.
▶ An authoritative external body that ensures that the auditors remain independent and competent.

Benefits and limitations of integrated quality, environmental, and health and safety management systems

Benefits

The objectives and processes of the three management systems are essentially the same. A well-planned integrated management system (IMS) is likely to:

▶ operate more cost-effectively than separate systems;
▶ facilitate decision-making that best reflects the overall needs of the organisation;
▶ offer more rewarding career opportunities for specialists in each discipline;
▶ lead to the avoidance of duplication (e.g. in personnel, meetings, software and paperwork) and shared expertise between each system;
▶ identify a positive function in one system that may be integrated into the other systems;
▶ link with management arrangements for other activities, such as product safety and security;
▶ reduce the possibility of addressing problems in one system that create difficulties in other systems;
▶ involve timely overall IMS reviews thus avoiding any incompatibilities developing in individual systems;
▶ reduce distortions in resource allocations in the separate systems.

Limitations

▶ the existing systems currently work well as separate systems and the process of integration may undermine this success;
▶ relevant specialists in one of the systems may continue to concentrate in that area of their expertise;
▶ there may be conceptual difficulties between the systems that are difficult to reconcile such as different approaches to audits;
▶ an IMS could become over-centralised and unnecessarily complex;

▶ the period of time during which an organisation is planning and implementing an IMS could become a period of organisational vulnerability;

▶ the degree of complexity for each individual system within the IMS may vary considerably;

▶ some of the systems are certifiable whereas others are not and some are based on law whereas others are not. This could lead to distortions in the coherence of the IMS;

▶ rivalries within the organisation may damage the operation of the IMS;

▶ regulators and single topic auditors may have difficulty evaluating their part of the IMS; and

▶ a negative or poor culture in one system may influence the other systems in the IMS.

2.2 Purpose and key features of an effective health and safety policy

Every organisation should have a clear policy for the management of health and safety so that everybody associated with the organisation is aware of its health and safety aims and objectives and how they are to be achieved. For a policy to be effective, it must be honoured in the spirit as well as the letter. A good health and safety policy will also enhance the performance of the organisation in areas other than health and safety, help with the personal development of the workforce and reduce financial losses. A clear health and safety policy contributes to business efficiency and continuous improvement throughout the operation. The demonstration of senior management involvement provides evidence to all stakeholders that responsibilities to people and the environment are taken seriously. The policy should state the intentions of the business in terms of clear aims, objectives, organisation, arrangements and targets for all health and safety issues.

2.2.1 Legal duties concerning health and safety policy

Section 2(3) of the Health and Safety at Work (HSW) Act and the Employers' Health and Safety Policy Statements (Exception) Regulations 1975 require employers, with **five or more** employees, to prepare and review on a regular basis a written health and safety policy together with the necessary organisation and arrangements to carry it out and to bring the policy and any revision of it to the notice of their employees.

This does not mean that organisations with four or fewer employees do not need to have a health and safety policy – it simply means that it does not have to be written down. The number of employees is the maximum number at any one time whether they are full time, part time or seasonal.

This obligation on employers was introduced for the first time by the HSW Act and is related to the employers' reliance on the Act on self-regulation to improve health

and safety standards rather than on enforcement alone. A good health and safety policy involves the development, monitoring and review of the standards needed to address and reduce the risks to health and safety produced by the organisation.

The law requires that the written health and safety policy should include the following three sections:

▶ a health and safety policy statement of intent which includes the health and safety aims and objectives of the organisation;

▶ the health and safety organisation detailing the people with specific health and safety responsibilities and their duties;

▶ the health and safety arrangements in place in terms of systems and procedures.

The Management of Health and Safety at Work Regulations also require the employer to 'make and give effect to such arrangements as are appropriate, having regard to the nature of their activities and the size of their undertaking, for the effective planning, organisation, control, monitoring and review of the preventative and protective measures'. They further require that these arrangements must be recorded when there are five or more employees.

When an inspector from the Health and Safety Executive (HSE) or Local Authority visits an establishment, it is very likely that they will wish to see the health and safety policy as an initial indication of the management attitude and commitment to health and safety. There have been instances of prosecutions being made due to the absence of a written health and safety policy. (Such cases are, however, usually brought before the courts because of additional concerns.)

Figure 2.2 Well-presented policy documents

2.2.2 Key features and appropriate content of an effective health and safety policy

Policy statement of intent

The health and safety policy statement of intent is often referred to as the health and safety policy statement or sometimes, incorrectly, as the health and safety policy. It should contain the aims (which are not measurable) and objectives (which are measurable) of the organisation or company. Aims will probably remain unchanged during policy revisions whereas objectives will be reviewed and modified or changed every year. The statement should be written in clear and simple language so that it is easily understandable. It should also be fairly brief and broken down into a series of smaller statements or bullet points.

The statement should be signed and dated by the most senior person in the organisation. This will demonstrate management commitment to health and safety and give authority to the policy. It will indicate where ultimate responsibility lies and the frequency with which the policy statement is reviewed.

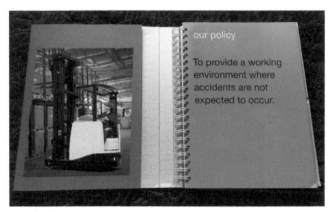

Figure 2.3 Part of a policy commitment

The most senior manager is normally the Chief Executive Officer (CEO) or the Managing Director. It is the responsibility of the CEO, under the HSW Act, to ensure that the health and safety policy is developed and communicated to all employees in the organisation. They will need to ensure the following:

▶ key functions of health and safety management, such as monitoring and audit, accident investigation and training, are included in the organisational structure;
▶ adequate resources are available to manage health and safety effectively;
▶ the production of various health and safety arrangements in terms of rules and procedures;
▶ arrangements for the welfare of employees;
▶ the regular review and, if necessary, updating of the health and safety policy.

The policy statement should be written by the organisation and not by external consultants, as it needs to address the specific health and safety issues and hazards within the organisation. In large organisations, it may be necessary to have health and safety policies for each department and/or site with an overarching general policy incorporating the individual policies.

Such an approach is often used by local authorities and multinational companies.

The following points should be included or considered when a health and safety policy statement is being drafted:

▶ the aims, which should cover health and safety, welfare and relevant environmental issues;
▶ the position of the senior person in the organisation or company who is responsible for health and safety (normally the Chief Executive);
▶ the names of the Health and Safety Adviser and any safety representatives or other competent health and safety persons;
▶ a commitment to the basic requirements of the HSW Act (access, egress, risk assessments, safe plant and systems of work, use, handling, transport and handling of articles and substances, information, training and supervision);
▶ a commitment to the additional requirements of the Management of Health and Safety at Work Regulations (risk assessment, emergency procedures, health surveillance and employment of competent persons);
▶ duties towards the wider general public and others (contractors, customers, students, etc.);
▶ the principal hazards in the organisation;
▶ specific policies of the organisation (e.g. smoking policy, violence to staff, etc.);
▶ a commitment to employee consultation possibly using a safety committee or staff council;
▶ duties of employees (particularly those defined in the Management of Health and Safety at Work Regulations);
▶ specific health and safety performance targets for the immediate and long-term future;
▶ a commitment to provide the necessary resources to achieve the objectives outlined in the policy statement.

Health and safety **performance targets** are an important part of the statement of intent because:

▶ they indicate that there is management commitment to improve health and safety performance;
▶ they motivate the workforce with tangible goals resulting, perhaps, in individual or collective rewards;
▶ they offer evidence during the monitoring, review and audit phases of the management system.

The type of target chosen depends very much on the areas which need the greatest improvement in the organisation. The following list, which is not exhaustive, shows common health and safety performance targets:

▶ a specific reduction in the number of accidents, incidents (not involving injury) and cases of work-related ill health (perhaps to zero);
▶ a reduction in the level of sickness absence;
▶ a specific increase in the number of employees trained in health and safety;

- an increase in the reporting of minor accidents and 'near miss' incidents;
- a reduction in the number of civil claims;
- no enforcement notices from the HSE or Local Authority;
- a specific improvement in health and safety audit scores;
- the achievement of a nationally recognised health and safety management standard such as HSG65.

A typical health and safety policy statement would state the following: 'The health and safety policy of the organisation is to:

- prevent accidents and cases of work-related ill health;
- manage health and safety risks in the workplace;
- provide clear instructions and information, and adequate training, to ensure that employees are competent to do their work;
- provide personal protective equipment;
- consult with all employees on matters affecting their health and safety;
- provide and maintain safe plant and equipment;
- ensure safe handling and use of substances;
- maintain safe and healthy working conditions;
- implement emergency procedures, including evacuation in case of fire or other significant incident; and
- review and revise this policy regularly.'

The policy statement of intent should be posted on prominent noticeboards throughout the workplace and brought to the attention of all employees at induction and refresher training sessions. It can also be communicated to the workforce during team briefing sessions, at 'toolbox' talks which are conducted at the workplace or directly by email, intranet, newsletters or booklets. It should be a permanent item on the agenda for health and safety committee meetings where it and its related targets should be reviewed at each meeting.

Setting health and safety objectives

A health and safety plan is necessary to guide the organisation – setting out the objectives for a specified time period. The speed with which the objectives will be achieved will depend on the resources available, the current state of health and safety compliance and the policy aims of the organisation. If the policy is to achieve excellence, the objectives will need to be tougher than those wishing simply to comply with minimum legal standards. It is essential to decide on priorities, with the emphasis on effective and adequate workplace precautions that meet legal requirements. High hazard/risk activities should receive priority. In some cases short-term measures may be needed quickly to minimise risks while longer-term solutions are designed and

implemented.

There are two basic methods of setting objectives within an organisation – from the top or from the bottom – and either has its merits. Many policies are set from the board room and it is here that the resources for health and safety and the associated standards can be agreed. This is the 'top-down' approach.

However, at times directors may be somewhat remote from the workplaces where most accidents occur. In some world leading businesses, objectives are set at the workplace by those who are exposed to the significant risks. These are then approved at higher levels and coordinated with other workplaces within the organisation. The overall aims may be set centrally and then the details organised at site or workplace levels. This is the 'bottom-up' approach and helps to achieve commitment from people at the workplace.

Whatever approach is used, it is vitally important to ensure that safety representatives and/or representatives of employee safety are fully involved through consultation when setting the objectives and during their implementation. The objectives should be properly documented and may differ in detail at each functional level.

Health and safety objectives need to be specific, measurable, agreed with those who deliver them, realistic and set against a suitable timescale (SMART). Both short- and long-term objectives should be set and prioritised against business needs. Objectives at different levels or within different parts of an organisation should be aligned so they support the overall policy objectives. Personal targets can also be agreed with individuals to secure the attainment of objectives of the organisation, and an example of such objectives is given in Box 2.1.

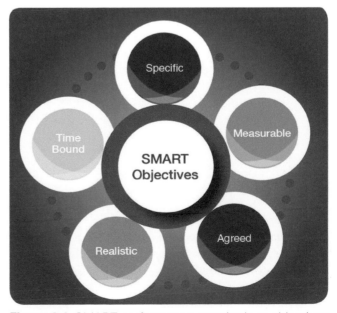

Figure 2.4 SMART performance standards or objectives

Box 2.1 Example of objectives

An organisation might wish to minimise the risk to workers of working at height and aim to prevent all serious injury from falls when working at height. The objectives at each functional level may be as follows:

1. Board room – objective is to eliminate all serious injuries from working at height within one year. The Board may set one of its members as the champion to make this happen and provide advice from in-house expertise or external consultants. The budget to be approved within the first month.

2. The board champion may set the following objectives/plan to achieve the Board's aim in consultation with the organisation's central health and safety steering committee:
   a) All sites to reduce serious accidents by 50% within 6 months and 100% within 12 months.
   b) All site managers to audit existing work at height operations; determine what can be done to avoid working at height; and report in 1 month with a draft budget;
   c) Safety adviser to establish existing accident performance; produce best practice guidance with safety representatives for those work-at-height operations which are considered unavoidable; identify suitable levels of competency and training requirements for those working at height.

3. Site managers may set the following in consultation with safety representatives and the site health and safety committee:
   a) All departmental supervisors to ensure immediately that work-at-height operations are inspected and risk assessed before work starts, all equipment to be assessed for suitability and checked for condition.
   b) All people working at height to be assessed for competence and specific training given as needed, for example working in mobile elevating work platforms.
   c) Toolbox talks on general work-at-height awareness to be provided.

There are three complementary outputs from the planning process:

- health and safety plans with objectives for developing, maintaining and improving the health and safety management system, such as:
  - ▷ requiring each site of a multi-site organisation to have an annual health and safety plan and an accident and incident investigation system;
  - ▷ establishing a reliable risk assessment process for COSHH;
  - ▷ involving employees in preparing workplace precautions;
  - ▷ completing all manual handling assessments by the end of the current year;
  - ▷ providing a new guard for a particular machine;
- specifications for management arrangements, risk control and workplace precautions; and
- performance standards for implementing the health and safety management system, identifying the contribution of individuals to implementing the system (this is essential to building a positive health and safety culture).

Health and safety targets will need to be set so that the agreed performance standards are approached and met. Since there may well be many standards, it will be necessary to prioritise important standards and associated targets. These prioritised targets are known as **Key Performance Indicators (KPIs)**. The targets must be realistic and a procedure known as benchmarking will help to ascertain whether this is the case.

Benchmarking

Benchmarking compares the performance of the organisation with that of similar organisations. A small construction company will benchmark itself against the performance of the UK construction industry perhaps using statistics supplied by the HSE. Such benchmarks, or examples of good practice, are defined by comparison with the health and safety performance of other parts of the organisation or the national performance of the occupational group of the organisation. The HSE publish an annual report, statistics and a bulletin, all of which may be used for this purpose. Typical benchmarks include accident rates per employee and accident or disease causation. It is important that the correct benchmarks are chosen and the organisation is assured of the accuracy of the data. The advantages of benchmarking are that:

- the key performance indicators for a particular organisation may be easily identified;
- it helps with continuous improvement;
- it focuses attention on weaker performance areas;
- it gives confidence to various stakeholders; and
- it is useful feedback for boards, chief executives and managers.

The relationship of health and safety performance to other business activities

The best health and safety policies do not separate health and safety and human resources management since people are a key resource. Organisations want a fit, competent and committed workforce rather than simply preventing accidents and ill health. This integrated approach will extend to people outside the organisation with policies for the control of environmental pollution and product safety. Organisations often fail to manage health and safety effectively because they see it as something distinct from other managerial tasks. The principles and approach to managing health and safety are exactly the same as those required for managing quality or the environment.

Well-managed health and safety is important in business terms because stakeholders (such as shareholders, customers and the general public) are influenced by any bad publicity resulting from poor health and safety standards. The identification, assessment and control of health and safety and other risks is a managerial responsibility and of equal importance to production and quality. There is good evidence that good health and safety management also produces good quality products at a reasonable cost. However, the practical implications of health and safety must be carefully thought through to avoid conflict between the demands of policy and other operational requirements. Insufficient attention to health and safety can lead to problems with production – for example, work schedules that fail to take into account the problems of fatigue or inadequate resources allocated to training.

The considerable advances in information technology have been very helpful in the development of health and safety management systems. It has enabled the rapid identification of data that is critical to the management of health and safety (e.g. accident and ill-health trends, health and safety performance variations between departments within the organisation). It has also simplified the collection and analysis of essential data.

Research by HSE found that 'in many organisations the visible leadership and emphasis on continual improvement with respect to health and safety lagged behind that for quality of a product or service'.

Organisation of health and safety

This section of the policy defines the names, positions and responsibilities of those within the organisation or company who have a specific responsibility for health and safety. Therefore, it identifies those health and safety responsibilities and the reporting lines through the management structure. This section will include the following groups together with their associated responsibilities:

▶ directors and senior managers (responsible for setting policy, objectives and targets);
▶ supervisors (responsible for checking day-to-day compliance with the policy);

▶ health and safety advisers (responsible for giving advice during accident investigations and on compliance issues);
▶ other specialists, such as an occupational nurse, chemical analyst and an electrician, responsible for giving specialist advice on particular health and safety issues;
▶ health and safety representatives (responsible for representing employees during consultation meetings on health and safety issues with the employer);
▶ employees (responsible for taking reasonable care of the health and safety of themselves and others who may be affected by their acts or omissions);
▶ fire marshals (responsible for the safe evacuation of the building in an emergency);
▶ First-aiders (responsible for administering first aid to injured persons).

For smaller organisations, some of the specialists mentioned above may well be employed on a consultancy basis.

For the health and safety organisation to work successfully it must be supported from the top (preferably at board level) and some financial resource made available.

It is also important that certain key functions are included in the organisation structure. These include:

▶ accident investigation and reporting;
▶ health and safety training and information;
▶ health and safety monitoring and audit;
▶ health surveillance;
▶ monitoring of plant and equipment, their maintenance and risk assessment;
▶ liaison with external agencies;
▶ management and/or employee safety committees – the management committee will monitor day-to-day problems and any concerns of the employee health and safety committee.

The role of the health and safety adviser is to provide specialist information to managers in the organisation and to monitor the effectiveness of health and safety procedures. The adviser is not 'responsible' for health and safety or its implementation; that is the role of the line managers.

Finally the job descriptions, which define the duties of each person in the health and safety organisational structure, must not contain responsibility overlaps or blur chains of command. Each individual must be clear about his/her responsibilities and the limits of those responsibilities.

More detailed information is given on organisational health and safety roles and responsibilities of employers, directors, managers, supervisors and health and safety practitioners under Section 2.3 in this chapter.

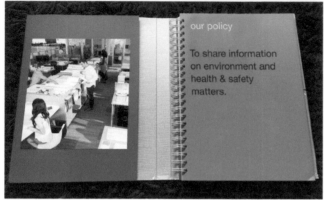

Figure 2.5 Good information, training and working with employees is essential

Arrangements for health and safety

The arrangements section of the health and safety policy gives details of the specific systems and procedures used to assist in the implementation of the policy statement. This will include health and safety rules and procedures and the provision of facilities such as a first-aid room and washrooms. It is common for risk assessments (including COSHH, manual handling and personal protective equipment (PPE) assessments) to be included in the arrangements section, particularly for those hazards referred to in the policy statement. It is important that arrangements for fire and other emergencies and for information, instruction, training and supervision are also covered. Local codes of practice (e.g. for fork-lift truck drivers) should be included.

The following list covers the more common items normally included in the arrangements section of the health and safety policy:

▶ employee health and safety code of practice;
▶ employee consultation procedures;
▶ accident and illness reporting and investigation procedures;
▶ emergency procedures, first aid;
▶ fire drill procedure;
▶ procedures for undertaking risk assessments;
▶ control of exposure to specific hazards (noise, vibration, radiation, manual handling, hazardous substances, etc.);

▶ machinery safety (including safe systems of work, lifting and pressure equipment);
▶ electrical equipment (maintenance and testing);
▶ maintenance procedures;
▶ permits to work procedures;
▶ use of PPE;
▶ monitoring procedures including health and safety inspections and audits;
▶ procedures for the control and safety of contractors and visitors;
▶ provision of welfare facilities;
▶ training procedures and arrangements;
▶ catering and food hygiene procedures;
▶ arrangements for consultation with employees;
▶ terms of reference and constitution of the safety committee;
▶ procedures and arrangements for waste disposal.

The three sections of the health and safety policy are usually kept together in a health and safety manual and copies distributed around the organisation.

Review of health and safety policy

It is important that the health and safety policy is monitored and reviewed on a regular basis. For this to be successful, a series of benchmarks needs to be established.

Figure 2.6 The policy might be good but is it put into practice – unsafe use of a ladder

There are several reasons for a review of the health and safety policy. The more important reasons are that:

▶ significant organisational changes may have taken place;
▶ there have been changes in key personnel;
▶ there have been changes in legislation and/or guidance;
▶ new work methods have been introduced;
▶ there have been alterations to working arrangements and/or processes;
▶ there have been changes following consultation with employees;
▶ the monitoring of risk assessments or accident/incident investigations indicates that the health and safety policy is no longer totally effective;
▶ information from manufacturers has been received;
▶ advice from an insurance company has been received;
▶ the findings of an external health and safety audit have been received;
▶ enforcement action has been taken by the HSE or Local Authority;
▶ a sufficient period of time has elapsed since the previous review.

A positive promotion of health and safety performance will achieve far more than simply preventing accidents and ill health. It will:

▶ support the overall development of personnel;
▶ improve communication and consultation throughout the organisation;
▶ minimise financial losses due to accidents and ill health and other incidents;
▶ directly involve senior managers in all levels of the organisation;
▶ improve supervision, particularly for young persons and those on occupational training courses;
▶ improve production processes;
▶ improve the public image of the organisation or company.

It is apparent, however, that some health and safety policies appear to be less than successful. There are many reasons for this. The most common are:

▶ the statements in the policy and the health and safety priorities not understood by or properly communicated to the workforce;
▶ minimal resources made available for the implementation of the policy;
▶ too much emphasis on rules for employees and too little on management policy;
▶ a lack of parity with other activities of the organisation (such as finance and quality control) due to mistaken concerns about the costs of health and safety and the effect of those costs on overall performance. This attitude produces a poor health and safety culture;
▶ lack of senior management involvement in health and safety, particularly at board level;

▶ employee concerns that their health and safety issues are not being addressed or that they are not receiving adequate health and safety information. This can lead to low morale among the workforce and, possibly, high absenteeism;
▶ high labour turnover;
▶ inadequate or no PPE;
▶ unsafe and poorly maintained machinery and equipment;
▶ a lack of health and safety monitoring procedures.

In summary, a successful health and safety policy is likely to lead to a successful organisation or company. A checklist for assessing any health and safety policy has been produced by the HSE and has been reproduced in Appendix 2.1.

2.3 Additional useful background information that is not part of the NEBOSH NGC syllabus

2.3.1 Planning a health and safety management system

The planning (**PLAN**) of a health and safety management system involves the development and implementation of suitable management arrangements to control risks by the introduction of workplace preventative measures. Such measures must be proportionate and appropriate to the hazards and risks in the organisation. Any management system must be able to accommodate changes in the activities of the organisation, its personnel or legislative requirements.

There are three key questions that need to be answered during the planning process:

▶ Where is the organisation at the moment in terms of the management of health and safety?
▶ Where does it want to be?
▶ How does it get there?

The answers to these questions at site level will be different to those at the headquarters in a large multi-site organisation. Therefore they should be asked at all levels in an organisation to obtain a complete understanding of the effectiveness of health and safety management across the whole organisation. For planning to be effective it must be properly coordinated throughout the organisation to avoid overlaps and omissions.

The health and safety planning process comprises three sections:

▶ correct information about the existing situation;
▶ suitable benchmarks against which to make comparisons (see earlier in this chapter);
▶ competent people to carry out the analysis and make judgements.

The organisation has to compare the current situation against both the health and safety management framework described earlier (Figure 2.1) and specific legal requirements. The key requirements of this framework are as follows:

1. **A clear health and safety policy** – Evidence shows that a well-considered policy contributes to business efficiency and continual improvement throughout the organisation. It helps to minimise financial losses arising from avoidable accidents and demonstrates to the workforce that accidents are not necessarily the fault of any individual member of the workforce. Such a management attitude could lead to an increase in workforce cooperation, job satisfaction and productivity. This demonstration of senior management involvement offers evidence to all stakeholders that responsibilities to people and the environment are taken seriously by the organisation. A good health and safety policy helps to ensure that there is a systematic approach to risk assessment and sufficient resources, in terms of people and money, have been allocated to protect the health and safety and welfare of the workforce. It can also support quality improvement programmes which are aimed at continual improvement.

2. **A well-defined health and safety organisation** – The shared understanding of the organisation's values and beliefs, at all levels of the organisation, is an essential component of a positive health and safety culture. For a positive health and safety culture to be achieved, an organisation must have clearly defined health and safety responsibilities so that there is always management control of health and safety throughout the organisation. The formal organisational structure should be such that the promotion of health and safety becomes a collaborative activity between the workforce, safety representatives and the managers. An effective organisation is essential for the **DO** stage of the management system and will be noted for good staff involvement and participation, high-quality communications, the promotion of competency and the empowerment of all employees to make informed contributions to the work of the organisation.

3. **A clear health and safety plan** – This involves the setting and implementation of performance standards and procedures using an effective occupational health and safety management system. The plan is based on risk assessment methods to decide on priorities and set objectives for controlling or eliminating hazards and reducing risks. Measuring success requires the establishing of performance standards against which achievements can be identified.

4. **The measurement of health and safety performance** – This includes both active (sometimes called proactive) and reactive monitoring to see how effectively the occupational health and safety management system is working. It forms part of the **CHECK** stage of the management system. Active monitoring involves looking at the premises, plant and substances plus the people, procedures and systems. Reactive monitoring discovers through investigation of accidents and incidents why controls have failed. It is also important to measure the organisation against its own long-term goals and objectives.

5. **Reviewing performance** – The results of monitoring and independent audits should be systematically reviewed to evaluate the performance of the management system against the objectives and targets established by the health and safety policy. It is at the review or **ACT** stage of the management system that the objectives and targets set in the health and safety policy may be changed. Changes in the health and safety environment in the organisation, such as an accident, should also trigger a performance review – this is discussed in more detail later in this chapter. Performance reviews are not only required by the HSW Act but are part of any organisation's commitment to continuous improvement. Comparisons should be made with internal performance indicators and the external performance indicators of similar organisations with exemplary practices and high standards.

6. **Auditing** – An independent and structured audit of all parts of the health and safety management system reinforces the review process. Such audits may be internal or external (the differences are discussed in Chapter 5). The audit assesses compliance with the health and safety management arrangements and procedures. If the audit is to be really effective, it must assess both the compliance with stated procedures and the performance in the workplace. It will identify weaknesses in the health and safety policy and procedures and identify unrealistic or inadequate standards and targets.

Other aspects of health and safety management systems

Any occupational health and safety management system will fail unless there is a positive health and safety culture within the organisation and the active involvement of internal and external stakeholders.

A structured and well-organised occupational health and safety management system is essential for the maintenance of high health and safety standards within all organisations and countries. Some systems, such as ISO 45001 2018, offer the opportunity for integration with quality and environmental management systems. This enables a sharing of resources although it is important that technical activities, such as health and safety risk assessment, are only undertaken by persons trained and competent in that area.

For an occupational health and safety management system to be successful, it must address workplace risks and be 'owned' by the workforce. It is, therefore, essential that the audit process examines shop floor health and safety behaviour to check that it mirrors that required by the health and safety management system.

Finally, whichever system is adopted, there must be continual improvement in health and safety performance if the application of the occupational health and safety management system is to succeed in the long term.

Key characteristics of a health and safety management system

The four basic elements common to all occupational health and safety management systems, as described earlier in this chapter, contain the different activities of the system together with the detailed arrangements and activities required to deliver those activities. However, there are four key characteristics of a **successful** occupational health and safety management system:

▶ a positive health and safety culture;
▶ the involvement of all stakeholders;
▶ an effective audit; and
▶ continual improvement.

A positive health and safety culture

In Chapter 3, the essential elements for a successful health and safety culture are detailed and discussed. In summary, they are:

▶ leadership and commitment to health and safety throughout the organisation;
▶ an acceptance that high standards of health and safety are achievable;
▶ the identification of all significant hazards facing the workforce and others;
▶ a detailed assessment of health and safety risks in the organisation and the development of appropriate control and monitoring systems;
▶ a health and safety policy statement outlining short- and long-term health and safety objectives. Such a policy should also include national codes of practice and health and safety standards;
▶ relevant communication and consultation procedures and training programmes for employees at all levels of the organisation;
▶ systems for monitoring equipment, processes and procedures and the prompt rectification of any defects found;
▶ the prompt investigation of all incidents and accidents and reports made detailing any necessary remedial actions.

Some of these essential elements form part of the health and safety management system, but unless all are present within the organisation, it is unlikely that occupational health and safety will be managed

successfully no matter which system is introduced. The chosen management system must be effective in reducing risks in the workplace else it will be nothing more than a paper exercise.

The involvement of stakeholders

There are a number of internal and external stakeholders of the organisation who will have an interest and influence on the introduction and development of the occupational health and safety management system.

The **internal stakeholders** include:

▶ **Directors and trustees of the organisation** – Following several national and international reports on corporate governance in recent years (such as the Combined Code of Corporate Governance 2003), the measurement of occupational health and safety performance and the attainment of health and safety targets have been recognised as being as important as other measures of business performance and targets. A report on health and safety performance should be presented at each Board meeting and be a periodic agenda item for sub-committees of the Board such as audit and risk management.

▶ **The workforce** – Without the full cooperation of the workforce, including contractors and temporary employees, the management of health and safety will not be successful. The workforce is best qualified to ensure and provide evidence that health and safety procedures and arrangements are actually being implemented at the workplace. So often this is not the case even though the occupational health and safety management system is well designed and documented. Worker representatives can provide useful evidence on the effectiveness of the management system at shop floor level and they can also provide a useful channel of communication between the senior management and the workforce. One useful measure of the health and safety culture is the enthusiasm with which workers volunteer to become and continue to be occupational health and safety representatives. For a representative to be really effective, some training is essential. Organisations which have a recognised trade union structure will have fewer problems with finding and training worker representatives. More information on the duties of worker representatives is given in Chapter 3.

▶ **Health and safety professionals** – Such professionals will often be appointed by the organisation to manage the occupational health and safety management system and monitor its implementation. They will, therefore, have a particular interest in the development of the system, the design of objectives and the definition of targets or goals. Unless they have a direct-line management responsibility, which is not very common, they can only act as advisers, but still have a positive influence on the health and safety culture of the organisation. The appointed health and safety professional(s) also liaises with other associated competent persons, such as for electrical appliance and local exhaust ventilation testing.

The **external stakeholders** include:

▶ **Insurance companies** – As compensation claims increase, insurance companies are requiring more and more evidence that health and safety is being effectively managed. There is increasing evidence that insurance companies are becoming less prepared to offer cover to organisations with a poor health and safety record and/or occupational health and safety management system.

▶ **Investors** – Health and safety risks will, with other risks, have an effect on investment decisions. Increasingly, investment organisations require evidence that these risks are being addressed before investment decisions are made.

▶ **Regulators** – In many parts of the world, particularly in the Far East, national regulators and legislation require certification to a recognised international occupational health and safety management system standard. In many countries, regulators use such a standard to measure the awareness of a particular organisation of health and safety issues.

▶ **Customers** – Customers and others within the supply chain are increasingly insisting on some form of formal occupational health and safety management system to exist within the organisation. The construction industry is a good example of this trend. Much of this demand is linked to the need for corporate social responsibility and its associated guidelines on global best practice.

▶ **Neighbours** – The extent of the interest of neighbours will depend on the nature of the activities of the organisation and the effect that these activities have on them. The control of noise, and dust and other atmospheric contaminants are examples of common problem areas which can only be addressed on a continuing basis using a health, safety and environmental management system.

▶ **International organisations** – The United Nations, the ILO, the International Monetary Fund, the World Bank and the World Health Organization are all examples of international bodies which have shown a direct or indirect interest in the management of occupational health and safety. In particular, the ILO is keen to see minimum standards of health and safety established around the world. The ILO works to ensure for everyone the right to work in freedom, dignity and security – which includes the right to a safe and healthy working environment. More than 70 ILO Conventions and Recommendations relate to questions of health and safety. In addition the ILO has issued more than 30 Codes of Practice on Occupational Health and Safety. For more information see their website www.ilo.org/safework.

There is a concern of many of these international organisations that as production costs are reduced by relocating operations from one country to another, there is also a lowering in occupational health and safety standards. The introduction of internationally recognised occupational health and safety management systems will help to alleviate such fears.

An effective audit

An effective audit is the final step in the occupational health and safety management system control cycle. The use of audits enables the reduction of risk levels and the effectiveness of the occupational health and safety management system to be improved. The auditing process is discussed in detail in Chapter 5, as is the important choice that must be made between the use of internal or external auditors.

Continual improvement

Continual improvement is recognised as a vital element of all occupational health and safety management systems if they are to remain effective and efficient as internal and external changes affect the organisation. Internal changes may be caused by business reorganisation, such as a merger, new branches and changes in products, or by new technologies, employees, suppliers or contractors. External changes could include new or revised legislation, guidance or industrial standards, new information regarding hazards or campaigns by regulators.

Continual improvement is part of the **ACT** stage of the management system and need not necessarily be done at high cost or add to the complexity of the management system. Continual improvement is discussed in detail in Chapter 5.

2.3.2 Organisational health and safety roles and responsibilities of employers, directors, managers and supervisors

Control of the health and safety organisation

Like all management functions, establishing control and maintaining it day in day out is crucial to effective health and safety management. Managers, particularly at senior levels, must take proactive responsibility for controlling issues that could lead to ill health, injury or loss. A nominated senior manager at the top of the organisation needs to oversee policy implementation and monitoring. The nominated person will need to report regularly to the most senior management team and will be a director or principal of the organisation.

Health and safety responsibilities will need to be assigned to line managers and expertise must be made available, either inside or outside the enterprise, to help them achieve the requirements of the HSW Act and the Regulations made under the Act. The purpose of the health and safety organisation is to harness the collective enthusiasm, skills and effort of the entire workforce with managers taking key responsibility and providing clear direction. The prevention of accidents and ill health through management systems of control becomes the focus rather than looking for individuals to blame after the incident occurs.

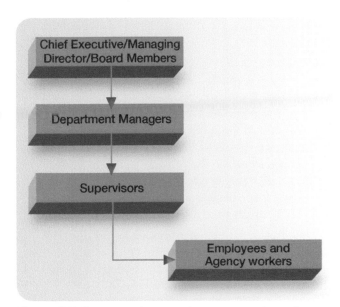

Figure 2.7 Everyone from senior manager down has health and safety responsibilities

The control arrangements should be part of the written health and safety policy. Performance standards will need to be agreed and objectives set which link the outputs required to specific tasks and activities for which individuals are responsible. For example, the objective could be to carry out a workplace inspection once a week to an agreed checklist and rectify faults within three working days. The periodic, say annual, audit would check to see if this was being achieved and if not, what the reasons for non-compliance with the objective were.

People should be held accountable for achieving the agreed objectives through existing or normal procedures such as:

▶ job descriptions, which include health and safety responsibilities;
▶ performance appraisal systems, which look at individual contributions;
▶ arrangements for dealing with poor performance;
▶ where justified, the use of disciplinary procedures.

Such arrangements are only effective if health and safety issues achieve the same degree of importance as other key management concerns, and a good performance is considered to be an essential part of career and personal development.

Organisational health and safety responsibilities – employers

The organisational health and safety responsibilities of employers are closely linked to their statutory duties which are covered in detail in Chapters 1 and 13.

Organisational health and safety responsibilities – directors

There is an absence of explicit legal duties that deal with directors' responsibilities for health and safety, but there is voluntary guidance to inform directors of the requirements for good management of health and safety within their organisations. The HSE and the Institute of Directors have published a guide INDG417 Leading health and safety at work. This guide is based on a plan that will deliver, monitor and review the management concept and the following information is closely based on this guide.

The following are quotations from health and safety leaders in the public and private sectors taken from the guide:

> 'Health and safety is integral to success. Board members who do not show leadership in this area are failing in their duty as directors and their moral duty, and are damaging their organisation.'

> 'An organisation will never be able to achieve the highest standards of health and safety management without the active involvement of directors. External stakeholders viewing the organisation will observe the lack of direction.'

For many organisations, health and safety is a corporate governance issue. The Board should integrate health and safety into its main governance structures, including board sub-committees, such as for risk, quality, remuneration and audit. The Turnbull guidance on the Combined Code on Corporate Governance requires listed companies to have robust systems of internal control, covering not just 'narrow' financial risks but also risks relating to the environment, business reputation and health and safety.

Effective health and safety performance comes from the top; members of the Board have both collective and individual responsibility for health and safety. Directors and boards need to examine their own behaviours, both individually and collectively, against the guidance given by HSE. If they fall short of the standards it sets them, then they must change so that they become effective leaders in health and safety.

Directors and board members need to take action so that:

▶ the health and safety of employees and others, such as members of the public, may be protected;
▶ risk management includes health and safety risks and becomes a key business risk in board decisions; and
▶ health and safety duties imposed by legislation are followed.

In INDG417, the Institute of Directors and the HSE state that:

'Protecting the health and safety of employees or members of the public who may be affected by the activities of an organisation is an essential part of risk management and must be led by the Board.'

The starting points for these two organisations are the following essential principles. These principles are intended to underpin the actions in this guidance and so lead to good health and safety performance.

- ▶ Strong and active leadership from the top:
  - ▷ visible, active commitment from the Board;
  - ▷ establishing effective 'downward' communication systems and management structures;
  - ▷ integration of good health and safety management with business decisions.
- ▶ Worker involvement:
  - ▷ engaging the workforce in the promotion and achievement of safe and healthy conditions;
  - ▷ effective 'upward' communication;
  - ▷ providing high-quality training.
- ▶ Assessment and review:
  - ▷ identifying and managing health and safety risks;
  - ▷ accessing (and following) competent advice;
  - ▷ monitoring, reporting and reviewing performance.

The HSE in its guidance on directors' and board responsibility for health and safety recommends the following five action points:

- ▶ The Board needs to accept formally and publicly its collective role in providing health and safety leadership in its organisation.
- ▶ Each member of the Board needs to accept their individual role in providing health and safety leadership for their organisation.
- ▶ The Board needs to ensure that all board decisions reflect its health and safety intentions, as articulated in the health and safety policy statement.
- ▶ The Board needs to recognise its role in engaging the active participation of workers in improving health and safety.
- ▶ The Board needs to ensure that it is kept informed of, and alert to, relevant health and safety risk management issues. The Health and Safety Executive recommends that boards appoint one of their number to be the 'Health and Safety Director'.

Directors need to ensure that the Board's health and safety responsibilities are properly discharged. To achieve this, the Board will need to:

- ▶ carry out an annual review of health and safety performance;
- ▶ keep the health and safety policy statement up to date with current board priorities and review the policy at least every year;
- ▶ ensure that there are effective management systems for monitoring and reporting on the organisation's health and safety performance;

- ▶ ensure that any significant health and safety failures and their investigation are communicated to board members;
- ▶ ensure that when decisions are made the health and safety implications are fully considered;
- ▶ ensure that regular audits are carried out to check that effective health and safety risk management systems are in place.

By appointing a Health and Safety Director there will be a board member who can ensure that these health and safety risk management issues are properly addressed, both by the Board and more widely throughout the organisation.

The Chairman and/or Chief Executive have a critical role to play in ensuring risks are properly managed and that the Health and Safety Director has the necessary competence, resources and support of other board members to carry out their functions. Indeed, some boards may prefer to see all the health and safety functions assigned to their Chairman and/or Chief Executive. As long as there is clarity about the health and safety responsibilities and functions, and the Board properly addresses the issues, this is acceptable.

The health and safety responsibilities of all board members should be clearly articulated in the organisation's statement of health and safety policy and arrangements. It is important that the role of the Health and Safety Director should not detract either from the responsibilities of other directors for specific areas of health and safety risk management or from the health and safety responsibilities of the Board as a whole. Some form of health and safety training will probably be required for directors.

INDG417 states that there are four elements that boards need to incorporate into their management of health and safety. These are:

- ▶ planning the direction of health and safety;
- ▶ delivering the plan for health and safety;
- ▶ monitoring health and safety performance; and
- ▶ reviewing health and safety performance.

Details of these four elements are given in Appendix 2.2.

INDG417 was revised in 2019. It continued to emphasise the same essential principles but simplified the core actions as follows: to take responsibility and 'ownership' of health and safety, members of the Board must ensure that:

- ▶ health and safety arrangements are adequately resourced;
- ▶ they obtain competent health and safety advice;
- ▶ risk assessments are carried out; and
- ▶ employees or their representatives are involved in decisions that affect their health and safety.

The guidance sets out the following:

- ▶ a four-point agenda for embedding the essential health and safety principles;

- ▶ a summary of legal liabilities;
- ▶ a checklist of key questions for leaders;
- ▶ a list of resources and references for implementing this guidance in detail.

Typical managers' organisational responsibilities

In addition to the legal responsibilities on management, there are many specific responsibilities imposed by each organisation's health and safety policy. A summary of the organisational responsibilities for health and safety of typical line managers and their accountability of each level of the line organisation is given below. Many organisations will not fit this exact structure but most will have those who direct, those who manage or supervise and those who have no line responsibility, but have responsibilities to themselves and fellow workers. A more detailed list of the responsibilities is given in Appendix 2.3.

Managing Directors/Chief Executives

Managing Directors/Chief Executives are responsible for the health, safety and welfare of all those who work or visit the organisation. In particular, they:

1. are responsible and accountable for health and safety performance within the organisation;
2. must ensure that adequate resources are available for the health and safety requirements within the organisation including the appointment of a senior member of the senior management with specific responsibility for health and safety;
3. appoint one or more competent persons and adequate resources to provide assistance in meeting the organisation's health and safety obligations including specialist help where appropriate;
4. establish, implement and maintain a formal, written health and safety programme for the organisation that encompasses all areas of significant health and safety risk;
5. approve, introduce and monitor all site health and safety policies, rules and procedures; and
6. review annually the effectiveness and, if necessary, require revision of the health and safety programme.

Departmental Managers

The principal Departmental Managers may report to the Site Manager, Managing Director or Chief Executive. In particular, they:

1. are responsible and accountable for the health and safety performance of their department;
2. are responsible for the engagement and management of contractors and that they are properly supervised;
3. must ensure that any machinery, equipment or vehicles used within the department are maintained, correctly guarded and meet agreed health and safety standards. Copies of records of all maintenance, statutory and insurance inspections must be kept by the Departmental Manager;
4. develop a training plan that includes specific job instructions for new or transferred employees and follow up on the training by supervisors. Copies of records of all training must be kept by the Departmental Manager; and
5. personally investigate all lost workday cases and dangerous occurrences and report to their line manager. Progress any required corrective action.

Supervisors

The supervisors are responsible to and report to their Departmental Manager. In particular, they:

1. are responsible and accountable for their team's health and safety performance;
2. enforce all safe systems of work procedures that have been issued by the Departmental Manager;
3. instruct employees in relevant health and safety rules, make records of this instruction and enforce all health and safety rules and procedures;
4. supervise any contractors that are working within their area of supervision; and
5. enforce personal protective equipment requirements, make spot checks to determine that protective equipment is being used and periodically appraise condition of equipment. Record any infringements of the personal protective equipment policy.

Role and functions of health and safety practitioners and other advisers

Competent person

One or more competent persons must be appointed to help managers comply with their duties under health and safety law. The essential point is that managers should have access to expertise to help them fulfil their legal requirements. However, they will always remain as advisers and will not assume responsibility in law for health and safety matters. This responsibility always remains with line managers and cannot be delegated to an adviser whether inside or outside the organisation. The appointee could be:

- ▶ the employer themselves if they are competent. This may be appropriate in a small, low hazard business;
- ▶ one or more employees, provided that they have sufficient time and other resources to undertake the task properly;
- ▶ a person(s) from outside the organisation who has sufficient expertise to help.

The HSE have produced a leaflet entitled 'Getting specialist help with health and safety' (INDG420). This gives simple guidance for those looking for specialist health and safety help.

When there is an employee with the ability to do the job, it is better for them to be appointed than to use outside specialists. Many health and safety issues can be tackled by people with an understanding of current best practice and the ability to judge and solve problems. Some specialist help is needed on a permanent basis whilst other help may only be required for a short period. There is a wide range of specialists available for different types of health and safety problem, these include:

▶ engineers for specialist ventilation or chemical processes;
▶ occupational hygienists for assessment and practical advice on exposure to chemical (dust, gases, fumes, etc.), biological (viruses, fungi, etc.) and physical (noise, vibration, etc.) agents;
▶ occupational health professionals for medical examinations and diagnosis of work-related disease, pre-employment and sickness advice, health education;
▶ ergonomists for advice on suitability of equipment, comfort, physical work environment, work organisation;
▶ physiotherapists for prevention and treatment of musculoskeletal disorders;
▶ radiation protection advisers for advice on compliance with the Ionising Radiation Regulations;
▶ health and safety practitioners for general advice on implementation of legislation, health and safety management, risk assessment, control measures and monitoring performance.

Health and safety practitioner

Status and **competence** are essential for the role of health and safety practitioners and other advisers. They must be able to advise management and employees or their representatives with authority and independence. They need to be able to advise on:

▶ creating and developing health and safety policies. These will be for existing activities in addition to new acquisitions or processes;
▶ the promotion of a positive health and safety culture. This includes helping managers to ensure that an effective health and safety policy is implemented;
▶ health and safety planning. This will include goal-setting, deciding priorities and establishing adequate systems and performance standards. Short- and long-term objectives need to be realistic;
▶ day-to-day implementation and monitoring of policy and plans. This will include accident and incident investigation, reporting and analysis;
▶ performance reviews and audit of the whole health and safety management system.

To do this properly, health and safety practitioners need to:

▶ have proper training and be suitably qualified – for example NEBOSH Diploma or competence-based IOSH membership, relevant degree and, where appropriate, a Chartered Safety and Health Practitioner or a NEBOSH Certificate in small to medium-sized low hazard premises, like offices, call centres, warehouses and retail stores;
▶ keep up-to-date information systems on such topics as civil and criminal law, health and safety management and technical advances;
▶ know how to interpret the law as it applies to their own organisation;
▶ actively participate in the establishment of organisational arrangements, systems and risk control standards relating to hardware and human performance. Health and safety practitioners will need to work with management on matters such as legal and technical standards;
▶ undertake the development and maintenance of procedures for reporting, investigating, recording and analysing accidents and incidents;
  ▷ develop and maintain procedures to ensure that senior managers get a true picture of how well health and safety is being managed (where a benchmarking role may be especially valuable). This will include monitoring, review and auditing; and
▶ be able to present their advice independently and effectively.

Figure 2.8 Safety practitioner at the front line

A new national register of occupational safety consultants has been established by the HSE to help employers access good quality, proportionate advice. The Occupational Safety Consultants Register (OS-CR) provides firms with details of consultants who have met the highest qualification standard of recognised professional bodies and who are bound by a code of conduct that requires them to only give advice which is sensible and proportionate.

Relationships within the organisation

Health and safety practitioners:

▶ support the provision of authoritative and independent advice;
▶ report directly to directors or senior managers on matters of policy and have the authority to stop work if it contravenes agreed standards and puts people at risk of injury;
▶ are responsible for professional standards and systems.

They may also have line management responsibility for other health and safety practitioners, in a large group of companies or on a large and/or high hazard site.

Relationships outside the organisation

Health and safety practitioners also have a function outside their own organisation. They provide the point of liaison with a number of other agencies including the following:

▶ environmental health officers and licensing officials;
▶ architects and consultants;
▶ HSE and the Fire and Rescue Authorities;
▶ the Police;
▶ HM Coroner or the Procurator Fiscal;
▶ local authorities;
▶ insurance companies;
▶ contractors;
▶ clients and customers;
▶ the public;
▶ equipment suppliers;
▶ the media;
▶ General Practitioners;
▶ IOSH and occupational health specialists and services.

Many safety advisers think it unlikely they will ever be investigated and if they are they will be covered by insurance. This may not necessarily be the case. If a safety adviser is being investigated, the normal procedure is to claim under the insurance cover that is in place, either the professional indemnity (PI) if self-employed, or under the employer's policy if employed in-house. It is not unusual to find that cover for advice and representation in criminal investigations and proceedings is excluded from these policies. Therefore, health and safety advisers, particularly those that are self-employed, should always check their insurance cover.

2.4 Further information

The Health and Safety at Work etc. Act 1974

Employers' Health and Safety Policy Statements (Exception) Regulations 1975

The Management of Health and Safety at Work Regulations 1999 (as amended)

Managing for health and safety, HSG65 third edition 2013, HSE Books, ISBN: 978-0-7176-6456-6 www.hse.gov.uk/pubns/priced/hsg65.pdf

HSE, 'Managing for health and safety' www.hse.gov.uk/managing/

HSE, The Health and Safety Toolbox: how to control risks at work www.hse.gov.uk/toolbox/

Health and safety made simple, INDG449, HSE Books, www.hse.gov.uk/pubns/indg449.htm

Plan, Do, Check, Act – An introduction to managing for health and safety, INDG275 (rev 1), HSE books www.hse.gov.uk/pubns/indg275.htm

BS 45002-0:2018 Occupational health and safety management systems Part 0: General guidelines for the application of ISO 45001, BSI Standards Publications

BS ISO 45001:2018, Occupational health and safety management systems – Requirements with guidance for use, BSI Standards Publications

IOSH briefing on ISO 45001 www.iosh.co.uk/45001

2.5 Practice questions

1. (a) **Outline** the key elements of a health and safety management system.
    (b) **Outline** the actions involved in delivering effective arrangements as described in the HSE publication 'Managing for health and safety' (HSG65 third edition).

2. Review (evaluation) is one of the key elements of a health and safety management system.
    Identify three components of this element **AND give** an example in each case.

3. (a) **Identify** the economic benefits that an organisation may obtain by implementing a successful health and safety management system.
    (b) **Outline** factors that should be considered during the planning stage of a safety management system.
    (c) **Discuss** the possible results of a poor implementation of a health and safety management system.

4. (a) **State** the legal requirements for an employer to prepare a written health and safety policy.
    (b) **Identify** the three main sections of a health and safety policy document and explain the purpose and general content of each section.
    (c) **Outline** the responsibilities of a Chief Executive in relation to the health and safety policy.

5. (a) **Explain** the purpose of the 'statement of intent' section of a health and safety policy.
   (b) **Outline** the **SIX** topics that may be addressed in this section of the health and safety policy.
   (c) **Explain** why a health and safety policy should be signed by the most senior person in an organisation.
   (d) **Outline** the indications that the objectives within the 'statement of intent' of the health and safety policy are not reflected in the health and safety performance of an organisation.

6. (a) **Outline** the key functions that should be included in the health and safety organisational structure of a company.
   (b) **Identify** the categories of persons that would be found in the health and safety organisational structure.
   (c) **Identify SIX** employee responsibilities which could be included in the 'organisation' section of a health and safety policy.

7. The 'arrangements' section is an important part of a health and safety policy.
   (a) **Identify** the information that should be contained in it.
   (b) **Outline** the key areas that should be addressed within it.

8. (a) **Outline** the occasions when a health and safety policy should be reviewed.
   (b) **Describe FOUR** external **AND FOUR** internal influences that might initiate a health and safety policy review.

9. (a) **Explain** why it is important for an organisation to set performance targets in terms of its health and safety performance.
   (b) **Outline SEVEN** types of performance target that an organisation might typically set in relation to health and safety.

10. Health and safety performance targets can be measured using benchmarking.
    (a) **Explain** the term 'benchmarking'.
    (b) **Give TWO** examples of health and safety performance information that can be used for benchmarking.

11. (a) **Describe** the four elements that boards of directors should take into account when overseeing the management of health and safety within their organisation.
    (b) **Outline** the responsibilities that Departmental Managers have to ensure the health, safety and welfare of those working in the department.
    (c) **Outline** the factors that will determine the level of supervision that a new employee should receive during their initial period of employment within an organisation.

APPENDIX 2.1 Health and Safety Policy checklist

The following checklist is intended as an aid to the writing and review of a health and safety policy. It is derived from HSE Information.

General policy and organisation

▶ Does the statement express a commitment to health and safety and are your obligations towards your employees made clear?

▶ Does it say which senior manager is responsible for seeing that it is implemented and for keeping it under review, and how this will be done?

▶ Is it signed and dated by you or a partner or senior director?

▶ Have the views of managers and supervisors, safety representatives and the safety committee been taken into account?

▶ Were the duties set out in the statement discussed with the people concerned in advance, and accepted by them, and do they understand how their performance is to be assessed and what resources they have at their disposal?

▶ Does the statement make clear that cooperation on the part of all employees is vital to the success of your health and safety policy?

▶ Does it say how employees are to be involved in health and safety matters, for example by being consulted, by taking part in inspections and by sitting on a safety committee?

▶ Does it show clearly how the duties for health and safety are allocated and are the responsibilities at different levels described?

▶ Does it say who is responsible for the following matters (including deputies where appropriate)?
  ▷ reporting investigations and recording accidents
  ▷ fire precautions, fire drill and evacuation procedures
  ▷ first aid
  ▷ safety inspections
  ▷ the training programme
  ▷ ensuring that legal requirements are met, for example regular testing of lifts and notifying accidents to the health and safety inspector.

Arrangements that need to be considered

▶ Keeping the workplace, including staircases, floors, ways in and out, washrooms, etc. in a safe and clean condition by cleaning, maintenance and repair;

▶ the requirements of the Work at Height Regulations;

▶ any suitable and sufficient risk assessments.

Plant and substances

▶ Maintenance of equipment such as tools, ladders, etc. (are they in a safe condition?);

▶ maintenance and proper use of safety equipment such as helmets, boots, goggles, respirators, etc.;

▶ maintenance and proper use of plant, machinery and guards;

▶ regular testing and maintenance of lifts, hoists, cranes, pressure systems, boilers and other dangerous machinery, emergency repair work, and safe methods of carrying out these functions;

▶ maintenance of electrical installations and equipment;

▶ safe storage, handling and, where applicable, packaging, labelling and transport of flammable and/or hazardous substances;

▶ controls of work involving harmful substances such as lead and asbestos;

▶ the introduction of new plant, equipment or substances into the workplace by examination, testing and consultation with the workforce;

▶ exposure to ultraviolet and optical radiation; and

▶ exposure to non-ionising and ionising radiation.

Other hazards and issues

▶ Mental ill health at work including the home–work interface;

▶ noise problems – wearing of hearing protection, and control of noise at source;

▶ vibration problems – hand–arm and whole-body control techniques and personal protection;

▶ preventing unnecessary or unauthorised entry into hazardous areas;

▶ lifting of heavy or awkward loads;

▶ protecting the safety of employees against assault when handling or transporting the employer's money or valuables;

▶ special hazards to employees when working on unfamiliar sites, including discussion with site manager where necessary;

▶ control of works transport, for example fork-lift trucks, by restricting use to experienced and authorised operators or operators under instruction (which should deal fully with safety aspects);

▶ driving on public roads while at work;

▶ substance abuse at work.

Figure 2.9 Emergency procedures

Figure 2.10 Fork-lift reach truck in warehouse

Emergencies

▶ Ensuring that fire exits are marked, unlocked and free from obstruction;
▶ maintenance and testing of firefighting equipment, fire drills and evacuation procedures;
▶ first aid, including name and location of person responsible for first aid and deputy, and location of first-aid box.

Communication

▶ Giving employees information about the general duties under the HSW Act and specific legal requirements relating to their work;
▶ giving employees necessary information about substances, plant, machinery and equipment with which they come into contact;
▶ discussing with contractors, before they come on site, how they plan to do their job, whether they need any equipment from your organisation to help them, whether they can operate in a segregated area or only when part of the plant is shut down and, if not, what hazards they may create for your employees and vice versa.

Training

▶ Training employees, supervisors and managers to enable them to work safely and to carry out their health and safety responsibilities efficiently.

Supervising

▶ Supervising employees so far as necessary for their safety – especially young workers, new employees and employees carrying out unfamiliar tasks.

Keeping check

▶ Regular inspections and checks of the workplace, machinery appliances and working methods.

APPENDIX 2.2 Leadership actions for directors and board members

This guidance, taken from INDG417, sets out an agenda for the effective leadership of health and safety by all directors, governors, trustees, officers and their equivalents in the private, public and third sectors. It applies to organisations of all sizes.

There are four elements that boards need to incorporate into their management of health and safety. These are:

▶ planning the direction of health and safety;
▶ delivering the plan for health and safety;
▶ monitoring health and safety performance; and
▶ reviewing health and safety performance.

1. Planning the direction of health and safety

The Board should set the direction for effective health and safety management. Board members need to establish a health and safety policy that is much more than a document – it should be an integral part of the organisation's culture, its values and performance standards.

All board members should take the lead in ensuring the communication of health and safety duties and benefits throughout the organisation. Executive directors must develop the policy since they have responsibility for the day-to-day operation of the organisation and must respond quickly where difficulties arise or new risks are introduced; non-executive directors must make sure that health and safety is properly addressed.

Essential actions

For a policy to be effective, all board members must be aware of the significant risks faced by their organisation. The policy should include the Board's role and that of individual board members in leading the management of health and safety in the organisation. The Board needs to 'own' and understand the key issues involved and decide on the best methods to communicate, promote and champion health and safety.

The health and safety policy is a 'living' document and it should evolve over time, for example, during major organisational changes such as restructuring or a significant acquisition.

Good practice guide

- ▶ Health and safety should appear regularly on the agenda for board meetings.
- ▶ The Chief Executive should give the clear leadership, but it may also be useful to name one of their members as the health and safety 'champion'.
- ▶ The presence on the Board of a Health and Safety Director gives a strong signal that the issue is being taken seriously and that its strategic importance is understood.
- ▶ The setting of targets helps the Board to measure performance.
- ▶ A non-executive director can act as a scrutiniser to ensure that robust processes are in place to support boards that face significant health and safety risks.

2. Delivering the plan for health and safety

Delivery depends on an effective management system to ensure, so far as is reasonably practicable, the health and safety of employees, customers and members of the public. Organisations should aim to protect people by introducing management systems and practices that ensure risks are dealt with sensibly, responsibly and proportionately.

Essential actions

To take responsibility and 'ownership' of health and safety, members of the Board must ensure that:

- ▶ health and safety arrangements are adequately resourced;
- ▶ they obtain competent health and safety advice;
- ▶ risk assessments are carried out;
- ▶ employees or their representatives are involved in decisions that affect their health and safety.

The Board should consider the health and safety implications of introducing new processes, working

practices or personnel and dedicate adequate resources to the task and seek competent advice when necessary. All board decisions, particularly when implementing change, must be made in the context of the organisation's health and safety policy.

Good practice guide

- ▶ Leadership is more effective if visible – board members can reinforce health and safety policy by being seen on the 'shop floor', following all safety measures themselves and addressing any breaches immediately.
- ▶ Health and safety should be considered when senior management appointments are made.
- ▶ Having good procurement standards for goods, equipment and services should prevent the introduction of expensive health and safety hazards.
- ▶ The health and safety arrangements of partners, key suppliers and contractors should be assessed regularly since their performance could adversely affect director's own performance.
- ▶ A separate risk management or health and safety sub-committee of the Board, chaired by a senior executive, should ensure that key issues are addressed and guard against time and effort being wasted on trivial risks and unnecessary bureaucracy.
- ▶ Health and safety training to some or all board members will promote understanding and knowledge of the key health and safety issues in the organisation.
- ▶ Worker involvement in health and safety, above the legal duty to consult worker representatives, will improve participation and indicate the commitment of senior management.

3. Monitoring health and safety performance

Monitoring and reporting are vital parts of a health and safety system. Management systems must allow the Board to receive both specific (e.g. incident-led) and routine reports on the performance of health and safety policy. Much day-to-day health and safety information need be reported only at the time of a formal review. But only a strong system of monitoring can ensure that the formal review will proceed as planned – and that relevant events in the interim are brought to the Board's attention.

Core actions

The Board should ensure that:

- ▶ appropriate weight is given to reporting both preventative information (such as progress of training and maintenance programmes) and incident data (such as accident and sickness absence rates);

▶ periodic audits of the effectiveness of management structures and risk controls for health and safety are carried out;

▶ the impact of changes such as the introduction of new procedures, work processes or products, or any major health and safety failure, is reported as soon as possible to the Board;

▶ there are procedures to implement new and changed legal requirements and to consider other external developments and events.

Good practice guide

▶ Effective monitoring of sickness absence and workplace health will alert the Board to underlying problems that could seriously damage performance or result in accidents and long-term illness.

▶ The collection of workplace health and safety data will allow the Board to benchmark the organisation's performance against others in its sector.

▶ Appraisals of senior managers should include an assessment of their contribution to health and safety performance.

▶ Boards should receive regular reports on the health and safety performance and actions of contractors.

▶ Organisations will find that they win greater support for health and safety from workers by involving them in the monitoring process.

4. Reviewing health and safety performance

A formal boardroom review of health and safety performance is essential. It allows the Board to establish whether the essential health and safety principles – strong and active leadership, worker involvement, and assessment and review – have been embedded in the organisation. It tells senior managers whether their system is effective in managing risk and protecting people.

Core actions

The Board should review health and safety performance at least once a year. The review process should:

▶ examine whether the health and safety policy reflects the organisation's current priorities, plans and targets;

▶ examine whether risk management and other health and safety systems have been effectively reported to the Board;

▶ report health and safety shortcomings, and the effect of all relevant board and management decisions;

▶ decide actions to address any weaknesses and a system to monitor their implementation;

▶ instigate immediate reviews in the light of major shortcomings or events.

Good practice guide

▶ Performance on health and safety and well-being should be reported in the organisation's annual report to investors and stakeholders.

▶ Board members should make extra 'shop floor' visits to gather information for the formal review.

▶ Good health and safety performance should be celebrated at central and local level.

INDG417 was revised in 2019. It continued to emphasise the same essential principles but simplified the core actions as follows: to take responsibility and 'ownership' of health and safety, members of the Board must ensure that:

▶ health and safety arrangements are adequately resourced;

▶ they obtain competent health and safety advice;

▶ risk assessments are carried out; and

▶ employees or their representatives are involved in decisions that affect their health and safety.

The 2019 guidance sets out the following:

▶ a four-point agenda for embedding the essential health and safety principles;

▶ a summary of legal liabilities;

▶ a checklist of key questions for leaders;

▶ a list of resources and references for implementing this guidance in detail.

APPENDIX 2.3 Detailed health and safety responsibilities

Managing Directors/Chief Executives

1. Are responsible and accountable for health and safety performance at the organisation;

2. Develop a strong, positive health and safety culture throughout the company and communicate it to all managers. This should ensure that all managers have a clear understanding of their health and safety responsibilities;

3. Provide guidance and leadership on health and safety matters to their management team;

4. Establish minimum acceptable health and safety standards within the organisation;

5. Ensure that adequate resources are available for the health and safety requirements within the organisation and authorise any necessary major health and safety expenditures;

6. Evaluate, approve and authorise health and safety related projects developed by the organisation's health and safety advisers;

7. Review and approve health and safety policies, procedures and programmes developed by the organisation's managers;

8. Ensure that a working knowledge of the areas of health and safety that are regulated by various governmental agencies, particularly the HSE, Local

Authorities and the Environment Agency, are maintained;

9. Ensure that health and safety is included as an agenda topic at all formal senior management meetings;

10. Review and act upon major recommendations submitted by outside loss prevention consultants and insurance companies;

11. Ensure that health and safety is included in any tours such as fire inspections of the organisation's sites and note any observed acts or conditions that fall short of or exceed agreed health and safety standards;

12. Ensure that all fatalities, major property losses, serious lost workday injuries and dangerous occurrences are investigated;

13. Establish, implement and maintain a formal, written health and safety programme for the organisation that encompasses all areas of significant health and safety risk;

14. Establish controls to ensure uniform adherence to the health and safety programme across the organisation;

15. Attend the health and safety committee meetings at the organisation;

16. Review, on a regular basis, all health and safety activity reports and performance statistics;

17. Review health and safety reports submitted by outside agencies and determine that any agreed actions have been taken;

18. Review annually the effectiveness of, and, if necessary, require revision of the site health and safety programme;

19. Appraise the performance of the health and safety advisers and provide guidance or training where necessary;

20. Monitor the progress of managers and others towards achieving their individual health and safety objectives.

Departmental Managers

1. Are responsible and accountable for the health and safety performance of their department;

2. Contact each supervisor frequently (daily) to monitor the health and safety standards in the department;

3. Hold departmental health and safety meetings for supervisors and employee representatives at least once a month;

4. Ensure that any machinery, equipment or vehicles used within the department are maintained and correctly guarded and meet agreed health and safety standards. Copies of records of all maintenance, statutory and insurance inspections must be kept by the Departmental Manager;

5. Ensure that all fire and other emergency equipment is properly maintained on a regular basis with all faults rectified promptly and that all departmental staff are aware of fire and emergency procedures;

6. Ensure that there is adequate first-aid cover on all shifts and all first-aid boxes are adequately stocked;

7. Ensure that safe systems of work procedures are in place for all jobs and that copies of all procedures are submitted to the site Managing Director for approval;

8. Review all job procedures on a regular basis and require each supervisor to check that the procedures are being used correctly;

9. Approve and review, annually, all departmental health and safety risk assessments, rules and procedures, maintain strict enforcement and develop plans to ensure employee instruction and re-instruction;

10. Ensure that all health and safety documents (such as the organisation's health and safety manual, risk assessments, rules and procedures) are easily accessible to all departmental staff;

11. Establish acceptable housekeeping standards, defining specific areas of responsibility, and assign areas to supervisors; make a weekly spot check across the department, hold a formal inspection with supervisors at least monthly and submit written reports of the inspections to the health and safety adviser with deadlines for any required actions;

12. Authorise purchases of tools and equipment necessary to attain compliance with the organisation's specifications and relevant statutory Regulations;

13. Develop a training plan that includes specific job instructions for new or transferred employees and follow up on the training by supervisors. Copies of records of all training must be kept by the Departmental Manager;

14. Review the health and safety performance of their department each quarter and submit a report to the Managing Director/Chief Executive;

15. Personally investigate all lost workday cases and dangerous occurrences and report to the Managing Director/Chief Executive. Progress any required corrective action;

16. Adopt standards for assigning PPE to employees, insist on strict enforcement and make spot field checks to determine compliance;

17. Evaluate the health and safety performances of supervisors;

18. Develop in each supervisor strong health and safety attitudes and a clear understanding of their specific duties and responsibilities;

19. Instil, by action, example and training, a positive health and safety culture among all departmental staff;

20. Instruct supervisors in site procedures for the care and treatment of sick or injured employees;
21. Ensure that the names of any absentees, written warnings and all accident reports are submitted to the Human Resources Manager.

Supervisors

1. Are responsible and accountable for their team's health and safety performance;
2. Conduct informal health and safety meetings with their employees at least monthly;
3. Enforce all safe systems of work procedures that have been issued by the Departmental Manager;
4. Report to the Departmental Manager any weaknesses in the safe system of work procedures or any actions taken to revise such procedures. These weaknesses may be revealed by either health and safety risk assessments or observations;
5. Report any jobs that are not covered by safe systems of work procedures to the Departmental Manager;
6. Review any unsafe acts and conditions and either eliminate them or report them to the Departmental Manager;
7. Instruct employees in relevant health and safety rules, make records of this instruction and enforce all health and safety rules and procedures;
8. Make daily inspections of assigned work areas and take immediate steps to correct any unsafe or unsatisfactory conditions, report to the Departmental Manager those conditions that cannot be immediately corrected and instruct employees on housekeeping standards;
9. Instruct employees that tools/equipment are to be inspected before each use and make spot checks of tools'/equipment's condition;
10. Instruct each new employee personally on job health and safety requirements in assigned work areas;
11. Provide on-the-job instruction on safe and efficient performance of assigned jobs for all employees in the work area;
12. Report any apparent employee health problems to the Departmental Manager;
13. Enforce PPE requirements; make spot checks to determine that protective equipment is being used and periodically appraise condition of equipment. Record any infringements of the PPE policy;
14. In the case of a serious injury, ensure that the injured employee receives prompt medical attention, isolate the area and/or the equipment as necessary and immediately report the incident to the Departmental Manager. In case of a dangerous occurrence, the Supervisor should take immediate steps to correct any unsafe condition and, if necessary, isolate the area and/or

the equipment. As soon as possible, details of the incident and any action taken should be reported to the Departmental Manager;
15. Investigate all accidents, serious incidents and cases of ill health involving employees in assigned work areas. Immediately after an accident, complete an accident report form and submit it to the Departmental Manager for onward submission to the health and safety adviser. A preliminary investigation report and any recommendations for preventing a recurrence should be included on the accident report form;
16. Check for any changes in operating practices, procedures and other conditions at the start of each shift/day and before relieving the 'on duty' Supervisor (if applicable). A note should be made of any health and safety related incidents that have occurred since their last working period;
17. At the start of each shift/day, make an immediate check to determine any absentees. Report any absentees to the Departmental Manager;
18. Make daily spot checks and take necessary corrective action regarding housekeeping, unsafe acts or practices, unsafe conditions, job procedures and adherence to health and safety rules;
19. Attend all scheduled and assigned health and safety training meetings;
20. Act on all employee health and safety complaints and suggestions;
21. Maintain, in their assigned area, health and safety signs and noticeboards in a clean and legible condition.

Employees and agency workers

1. Are responsible for their own health and safety;
2. Ensure that their actions will not jeopardise the safety or health of other employees;
3. Obey any safety rules, particularly regarding the use of PPE or other safety equipment;
4. Learn and follow the operating procedures and health and safety rules and procedures for the safe performance of the assigned job;
5. Must correct, or report to their Supervisor, any observed unsafe practices and conditions;
6. Maintain a healthy and safe place to work and cooperate with managers in the implementation of health and safety matters;
7. Make suggestions to improve any aspect of health and safety;
8. Maintain an active interest in health and safety;
9. Follow the established procedures if accidents occur by reporting any accident to the Supervisor;
10. Report any absence from the company caused by illness or an accident.

CHAPTER 3

Health and safety management of people and processes

Introduction

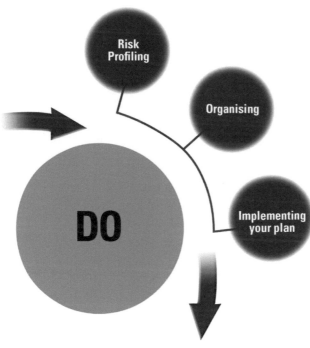

Figure 3.1 Do part of the management cycle

This chapter is about managers in businesses, or other organisations, setting out clear responsibilities and lines of communications for everyone in the enterprise. The policy sets the direction for health and safety within the enterprise and forms the written intentions of the principals or directors of the business. The organisation needs to be clearly communicated and people need to know what they are responsible for in the day-to-day operations. A vague statement that 'everyone is responsible for health and safety' is misleading and fudges the real issues. Everyone is responsible, but management has particular responsibilities. There is no equality of responsibility under law between those who provide direction and create policy and those who are employed to follow. Principals, or employers in terms of the Health and Safety at Work (HSW) Act, have substantially more responsibility than employees.

In 1972, the Robens Report recognised that the introduction of health and safety management systems was essential if the ideal of self-regulation of health and safety by industry was to be realised. It further recognised that a more active involvement of the workforce in such systems was essential if self-regulation was to work. Self-regulation and the implicit need for health and safety management systems and employee involvement were incorporated into the Health and Safety at Work (HSW) Act.

Since the introduction of the HSW Act, health and safety standards have improved considerably but there have been some catastrophic failures. One of the worst was the fire on the offshore oil platform, Piper Alpha, in 1988 when 167 people died. At the subsequent enquiry, the concept of a safety culture was defined by the Director General of the Health and Safety Executive (HSE) at that time, J. R. Rimington. This definition has remained as one of the key points for a successful health and safety management system.

3.1 Health and safety culture

3.1.1 Definition of a health and safety culture

The health and safety culture of an organisation may be described as the development stage of the organisation in health and safety management at a particular time. HSG65 gives the following definition of a health and safety culture:

'The safety culture of an organisation is the product of individual and group values, attitudes, perceptions, competencies and patterns of behaviour that determine the commitment to, and the style and proficiency of, an organisation's health and safety management.

Organisations with a positive safety culture are characterised by communications founded on mutual trust, by shared perceptions of the importance of safety and by confidence in the efficacy of preventive measures.'

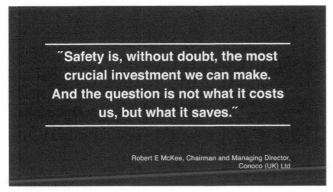

"Safety is, without doubt, the most crucial investment we can make. And the question is not what it costs us, but what it saves."

Robert E McKee, Chairman and Managing Director, Conoco (UK) Ltd

Figure 3.2 Safety investment

There is concern among some health and safety professionals that many health and safety cultures are developed and driven by senior managers with very little input from the workforce. Others argue that this arrangement is sensible because the legal duties are placed on the employer. A positive health and safety culture needs the involvement of the whole workforce just as a successful quality system does. There must be a joint commitment in terms of attitudes and values. The workforce must believe that the safety measures put in place will be effective and followed even when financial and performance targets may be affected.

3.1.2 Safety culture and safety performance

The relationship between health and safety culture and health and safety performance

The following elements are the important components of a **positive** health and safety culture:

▶ leadership and commitment to health and safety throughout the organisation, which is demonstrated in a genuine and visible way;

▶ health and safety treated as seriously as other corporate aims and adequately resourced;

▶ health and safety must be a line-management responsibility;

▶ acceptance that high standards of health and safety are achievable as part of a long-term strategy formulated by the organisation requiring sustained effort and interest;

▶ a detailed assessment of health and safety risks in the organisation and the development of appropriate control and monitoring systems;

▶ a health and safety policy statement that conveys a sense of optimism and outlines short- and long-term health and safety objectives. Such a policy should

also include codes of practice and required health and safety standards;

▶ realistic and achievable targets and performance measured against them;

▶ relevant employee training programmes, communication and consultation procedures to ensure ownership and participation in health and safety throughout the organisation;

▶ systems for monitoring equipment, processes, behaviour and procedures and the prompt rectification of any defects; and

▶ the prompt investigation of all incidents and accidents and reports made detailing any necessary remedial actions.

If the organisation adheres to these elements, then a basis for a good performance in health and safety will have been established. However, to achieve this level of performance, sufficient financial and human resources must be made available for the health and safety function at all levels of the organisation.

Managers at all levels of the organisation must show their commitment to health and safety by leading through example and ensuring that adequate resources in terms of time, money and people, are allocated to health and safety management. Good managers appear regularly on the shop floor, talk about health and safety and visibly demonstrate their commitment by their actions – such as stopping production to resolve issues. If employees perceive that managers are not seriously committed to high health and safety standards, they will assume that they are expected to put commercial interests first, and safety initiatives or programmes will be undermined as a result.

All managers, supervisors and members of the governing body (e.g. directors) should receive training in health and safety and be made familiar during training sessions with the health and safety targets of the organisation. The depth of training undertaken will depend on the level of competence required of the particular manager. Managers should be accountable for health and safety within their departments and be rewarded for significant improvements in health and safety performance. They should also be expected to discipline employees within their departments who infringe health and safety policies or procedures.

Important indicators of a health and safety culture

As the safety culture of an organisation develops, increasing use should be made of specific leading indicators and the setting and monitoring of specific goals. The main indicators for the development of health and safety must:

▶ be objective and easy to measure and collect;

▶ be relevant to the organisation;

▶ provide immediate and reliable indication of the performance level;

▶ be cost-effective in terms of equipment, personnel and additional technology required to gather the information; and

▶ be understood and owned by the organisation.

The goals must be relevant and improve safety performance over a reasonable time period and be specific to the key risks and activities of the organisation.

There are several outputs or indicators of the state of the health and safety culture of an organisation. The most important are the numbers of accidents, near misses and occupational ill-health cases occurring within the organisation.

Although the number of accidents may give a general indication of the health and safety culture, a more detailed examination of accidents and accident statistics is normally required. A calculation of the rate of accidents enables health and safety performance to be compared between years and organisations.

The simplest measure of accident rate is called the incidence rate and is defined as:

$$\frac{\text{Total number of accidents}}{\text{Number of persons employed}} \times 1{,}000$$

or the total number of accidents per 1,000 employees.

A similar measure (per 100,000) is used by the HSE in its annual report on national accident statistics and enables comparisons to be made within an organisation between time periods when employee numbers may change. It also allows comparisons to be made with the national occupational or industrial group relevant to the organisation.

There are four main problems with this measure which must be borne in mind when it is used. These are:

▶ there may be a considerable variation over a time period in the ratio of part-time to full-time employees;

▶ the measure does not differentiate between major and minor accidents and takes no account of other incidents, such as those involving damage but no injury (although it is possible to calculate an incidence rate for a particular type or cause of accident);

▶ there may be significant variations in work activity during the periods being compared;

▶ under-reporting of accidents will affect the accuracy of the data.

Some industries prefer the accident frequency rate per million hours worked. This method, by counting hours worked rather than the number of employees, avoids distortions by including part-time as well as full-time employees and by overtime worked.

The calculation is:

$$\frac{\text{Number of accidents in the period}}{\text{Total hours worked in the period}} \times 1{,}000{,}000$$

There is more information on these two measures in Chapter 5.

Indications of a poor health and safety culture

Subject to the above limitations, an organisation with a high accident incidence or frequency rate is likely to have a negative or poor health and safety culture.

There are other indications of a poor health and safety culture or climate. These include:

▶ a high sickness, ill-health and absentee rate among the workforce;

▶ the perception of a blame culture;

▶ high staff turnover leading to a loss of momentum in making health and safety improvements;

▶ no resources (in terms of budget, people or facilities) made available for the effective management of health and safety;

▶ management decisions that consistently put production or cost before health and safety considerations;

▶ a lack of compliance with relevant health and safety law and the safety rules and procedures of the organisation;

▶ regular procedural violations;

▶ poor selection procedures and management of contractors;

▶ poor levels of communication, cooperation and control;

▶ a weak health and safety management structure;

▶ either a lack, or poor levels, of health and safety competence; and

▶ high insurance premiums.

A positive health and safety culture is not enforceable through legislation. However there can be enforcement that addresses the outcomes of a poor culture. For example if a company has unsuccessfully relied on procedural controls to avoid major accidents, there could be enforcement action to ensure compliance with relevant health and safety legislation or for the provision of appropriate safeguards.

In summary, a poor health and safety performance within an organisation is an indication of a negative health and safety culture.

Factors affecting a health and safety culture

The most important factor affecting the culture is the commitment to health and safety from the top of an organisation. This commitment may be shown in many different ways. It needs to have a formal aspect in terms of an organisational structure, job descriptions and a health and safety policy, but it also needs to be apparent during crises or other stressful times. The health and safety procedures may be circumvented or simply forgotten when production or other performance targets are threatened.

Structural reorganisation or changes in market conditions will produce feelings of uncertainty among the workforce which, in turn, will affect the health and safety culture.

Poor levels of supervision, health and safety information and training are very significant factors in reducing health and safety awareness and, therefore, the culture.

The influence of peers can also have a very significant effect on the health and safety culture in an organisation. For example, if several workers in a workplace are not wearing appropriate personal protective equipment, then this is likely to influence other workers in the workplace not to wear the personal protective equipment even if it is a mandatory requirement.

Finally, the degree of consultation and involvement with the workforce in health and safety matters is crucial for a positive health and safety culture. Most of these factors may be summed up as human factors.

A safety culture questionnaire is given in Appendix 3.1.

3.2 Improving health and safety culture

Safe behaviour should be encouraged and strengthened through positive reinforcement by good example and supervision. Workers should be encouraged to set their own safety targets. Positive feedback on behaviour can sometimes reverse normal peer pressure so that safe working becomes the accepted social norm. However, safe methods of working need to be identified and communicated to workers. This requires a significant amount of time and resources, but will result in an improvement in health and safety performance. The success of behavioural change is very dependent on the commitment of management.

No single section or department of an organisation can develop a positive health and safety culture on its own. In addition to the commitment by the management, there needs to be the promotion of health and safety standards, effective communication within the organisation, cooperation from and with the workforce and an effective and developing training programme. Each of these topics will be examined in turn to show their effect on improving the health and safety culture in the organisation.

3.2.1 Commitment of management

As mentioned earlier, there needs to be a commitment from the very top of the organisation and this commitment will, in turn, produce higher levels of motivation and commitment throughout the organisation. Probably, the best indication of this concern for health and safety is shown by the status given to health and safety and the amount of resources (money, time and people) allocated to health and safety.

The management of health and safety should form an essential part of a manager's responsibility and they should be held to account for their performance on health and safety issues. Specialist expertise should be made available when required (e.g. for noise assessment), either from within the workforce or by the employment of external contractors or consultants. Health and safety should be discussed on a regular basis at management meetings at all levels of the organisation. If the organisation employs sufficient people to make direct consultation with all employees difficult, there should be a health and safety committee at which there is employee representation. In addition, there should be recognised routes for anybody within the organisation to receive health and safety information or have their health and safety concerns addressed.

The health and safety culture is enhanced considerably when senior managers appear regularly at all levels of an organisation whether it be the shop floor, the hospital ward or the general office and are willing to discuss health and safety issues with staff. A visible management is very important for a positive health and safety culture.

Finally, the positive results of a management commitment to health and safety will be the active involvement of all employees in health and safety, the continuing improvement in health and safety standards and the subsequent reduction in accident and occupational ill-health rates. This will lead, ultimately, to a reduction in the number and size of compensation claims.

HSG48 makes some interesting suggestions to managers on the improvements that may be made to health and safety which will be seen by the workforce as a clear indication of their commitment. The suggestions are:

▶ review the status of the health and safety committees and health and safety practitioners. Ensure that any recommendations are acted upon or implemented;
▶ ensure that senior managers receive regular reports on health and safety performance and act on them;
▶ ensure that any appropriate health and safety actions are taken quickly and are seen to have been taken;
▶ any action plans should be developed in consultation with employees, based on a shared perception of hazards and risks, be workable and continually reviewed.

3.2.2 Promotion of health and safety standards

For a positive health and safety culture to be developed, everyone within the organisation needs to understand the standards of health and safety expected by the organisation and the role of the individual in achieving and maintaining those standards.

Such standards are required to control and minimise health and safety risks.

Standards should clearly identify the actions required of people to promote health and safety. They should also specify the competencies needed by employees and should form the basis for measuring health and safety performance.

Health and safety standards cover all aspects of the organisation. Typical examples include:

▶ the design and selection of premises;
▶ the design and selection of plant and substances (including those used on site by contractors);
▶ the recruitment of employees and contractors;
▶ the control of work activities, including issues such as risk assessment;
▶ competence, maintenance and supervision;
▶ emergency planning and training;
▶ the transportation of the product and its subsequent maintenance and servicing.

Having established relevant health and safety standards, it is important that they are actively promoted within the organisation by all levels of management. The most effective method of promotion is by leadership and example. There are many ways to do this such as:

▶ the involvement of managers in workplace inspections and accident investigations;
▶ the use of PPE (e.g. goggles and hard hats) by all managers and their visitors in designated areas;
▶ ensuring that employees attend specialist refresher training courses when required (e.g. first aid and fork-lift truck driving);
▶ full cooperation with fire drills and other emergency training exercises;
▶ comprehensive accident reporting and prompt follow-up on recommended remedial actions.

The benefit of good standards of health and safety will be shown directly in less lost production, accidents and compensation claims, and lower insurance premiums. It may also be shown in higher product quality and better resource allocation.

An important and central necessity for the promotion of high health and safety standards is health and safety competence. What is meant by 'competence'?

Competence

The word 'competence' is often used in health and safety literature. One definition, made during a civil case in 1962, stated that a competent person is:

'a person with practical and theoretical knowledge as well as sufficient experience of the particular machinery, plant or procedure involved to enable them to identify defects or weaknesses during plant and machinery examinations, and to assess their importance in relation to the strength and function of that plant and machinery.'

This definition concentrates on a manufacturing rather than service industry requirement of a competent person.

Regulation 7 of the Management of Health and Safety at Work Regulations requires that:

'every employer shall employ one or more competent persons to assist him in undertaking the measures he needs to take to comply with the requirements and prohibitions imposed upon him by or under the relevant statutory provisions.'

In other words, competent persons are required to assist the employer in meeting his/her obligations under health and safety law. This may mean a health and safety adviser in addition to, say, an electrical engineer, an occupational nurse and a noise assessment specialist. The number and range of competent persons will depend on the nature of the business of the organisation.

It is recommended that competent employees are used for advice on health and safety matters rather than external specialists (consultants). It is recognised, however, that if employees, competent in health and safety, are not available in the organisation, then an external service may be enlisted to help. The key is that management and employees need access to health and safety expertise.

The regulations do not define 'competence' but do offer advice to employers when appointing health and safety advisers who should have:

▶ a knowledge and understanding of the work involved, the principles of risk assessment and prevention and current health and safety applications;
▶ the capacity to apply this to the task required by the employer in the form of problem and solution identification, monitoring and evaluating the effectiveness of solutions and the promotion and communication of health and safety and welfare advances and practices.

Such competence does not necessarily depend on the possession of particular skills or qualifications. It may only be required to understand relevant current best practice, be aware of one's own limitations in terms of experience and knowledge and be willing to supplement existing experience and knowledge. However, in more complex or technical situations, membership of a relevant professional body and/or the possession of an appropriate qualification in health and safety may be necessary. It is important that any competent person employed to help with health and safety has evidence of relevant knowledge, skills and experience for the tasks involved. The appointment of a competent person as an adviser to an employer does not absolve the employer from their responsibilities under the HSW Act and other relevant statutory provision.

Finally, it is worth noting that the requirement to employ competent workers is not restricted to those having a health and safety function but covers the whole workforce.

Competent workers must have sufficient training, experience, knowledge and other qualities to enable them to properly undertake the duties assigned to them.

3.2.3 Identifying and keeping up to date with legal requirements

The comparison with legal standards and requirements is an essential part of the setting of health and safety objectives. The organisation should have systems to identify legal requirements and keep up to date with changes to the law. This might be achieved by:

▶ regular checking of the HSE website looking at the 'news' section. This website gives detailed guidance on compliance – many published guides are now free to be downloaded as PDF files;
▶ free subscription to HSE bulletins on the internet;
▶ reading various health and safety periodicals;
▶ members of trade associations being kept up to date through their journals or websites;
▶ the text of legislation which can be obtained free of charge online.

There is more information on OSH websites and guidance on searching the internet in Chapter 17. Today there are so many ways of keeping up to date through the internet that the comment 'ignorance of the law is no excuse' is more poignant than ever.

3.2.4 Types of communication

Many problems in health and safety arise due to poor communication. It is not just a problem between management and workforce – it is often a problem the other way or indeed at the same level within an organisation. It arises from ambiguities or, even, accidental distortion of a message.

There are three basic methods of communication in health and safety – verbal, written and graphic.

Verbal communication is the most common. It is communication by speech or word of mouth. Verbal communication should only be used for relatively simple pieces of information or instruction. It is most commonly used in the workplace, during training sessions or at meetings.

There are several potential problems associated with verbal communication. The speaker needs to prepare the communication carefully so that there is no confusion about the message. It is very important that the recipient is encouraged to indicate their understanding of the communication. There have been many cases of accidents occurring because a verbal instruction has not been understood. There are several barriers to this understanding from the point of view of the recipient, including language and dialect, the use of

technical language and abbreviations, background noise and distractions, hearing problems, ambiguities in the message, mental weaknesses and learning disabilities, and lack of interest and attention.

Having described some of the limitations of verbal communication, it does have some merits. It is less formal, enables an exchange of information to take place quickly and the message to be conveyed as near to the workplace as possible. Training or instructions that are delivered in this way are called toolbox talks and can be very effective.

Written communication takes many forms from the simple memo to the detailed report. A memo should contain one simple message and be written in straightforward and clear language. The title should accurately describe the contents of the memo. In recent years, emails have largely replaced memos, as it has become a much quicker method to ensure that the message gets to all concerned (although a recent report has suggested that many people are becoming overwhelmed by the number of emails which they receive!). The advantage of memos and emails is that there is a record of the message after it has been delivered. The disadvantage is that they can be ambiguous or difficult to understand or, indeed, lost within the system.

Reports are more substantial documents and cover a topic in greater detail. The report should contain a detailed account of the topic and any conclusions or recommendations. The main problem with reports is that they are often not read properly due to the time constraints on managers. It is important that all reports have a summary attached so that the reader can decide whether it needs to be read in detail (see Chapter 5).

The most common way in which written communication is used in the workplace is the **noticeboard**. For a noticeboard to be effective, it needs to be well positioned within the workplace and there needs to be a regular review of the notices to ensure that they are up to date and relevant. The use of noticeboards as a means of communicating health and safety information to employees has some limitations that include:

▶ the information may not be read;
▶ the noticeboards may not be accessible;
▶ the information may become outdated or defaced;
▶ some employees may not be able to read while others may not understand what they have read;
▶ there may be language barriers;
▶ the information is mixed in with other non-health and safety information; and
▶ there is no opportunity offered for feedback.

The following alternative methods could be used for the communication of essential health and safety information:

▶ memos, emails and company intranet;
▶ toolbox talks and team briefings;

- induction training and any further back-up training sessions;
- newsletters, bulletins and payslips;
- digital video media including DVDs;
- a staff handbook; and
- through safety committees, safety representatives, and representatives of employee safety.

Graphic communication is communication by the use of drawings, photographs or DVDs. It is used to impart either health and safety information (e.g. fire exits) or health and safety propaganda. The most common forms of health and safety propaganda are the poster and the DVD. Both can be used very effectively as training aids, as they can retain interest and impart a simple message. Their main limitation is that they can become out of date fairly quickly or, in the case of posters, become largely ignored.

There are many sources of health and safety information which may need to be consulted before an accurate communication can be made. These include regulations, judgements, Approved Codes of Practice, guidance, British and European standards, periodicals, case studies and HSE publications.

3.2.5 The health and safety poster

The Health and Safety (Information for Employees) Regulations require that the approved poster entitled 'Health and Safety Law – what you should know' is displayed or the approved leaflet is distributed. This information tells employees in general terms about the requirements of health and safety law. Employers must also inform employees of the local address of the enforcing authority (either the HSE or the Local Authority) and the Employment Medical Advisory Service (EMAS). This should be marked on the poster or supplied with the leaflet.

Research by the HSE showed that the earlier versions of the law poster and law leaflet were visually unappealing and rarely read. They were redesigned to be more readable and engaging. The poster and leaflet are available in a range of formats as part of our commitment to make health and safety information more accessible.

The Health and Safety Information (Amendment) Regulations allows the HSE to approve and publish new posters and leaflets which do not need organisations to update or add enforcing authority and Employment Medical Advisory Service (EMAS) contact information. However, the poster will still need to be displayed and provide employees with basic health and safety information. The HSE is concerned that the poster needs to be understood by employees:

- who have visual and/or learning difficulties;
- who have poor English reading skills;
- who work in an environment where the risk of being denied employment rights is high.

Figure 3.3 Health and safety poster – must be displayed or brochure given to employees

The poster (Figure 3.3) is a simplified version of the previous one and contains a single Incident Contact Centre number for the HSE helpline. The leaflets that employers can give to workers are replaced with pocket cards.

Health and Safety Law pocket cards are published by the HSE and are a legal alternative to the HSE-approved Health and Safety Law poster. The pocket cards are suited for issuing to workers. This provides them with a personal copy of relevant health and safety law.

In addition to the health and safety poster, the following types of health and safety information could be displayed on a workplace noticeboard:

- a copy of the Employer's Liability Insurance Certificate;
- details of first-aid arrangements;
- emergency evacuation and fire procedures;
- minutes of the last health and safety committee meeting;
- details of health and safety targets and performance against them;
- health and safety posters and campaign details.

There are many other examples of written communications in health and safety, such as employee handbooks, company codes of practice, minutes of safety committee meetings and health and safety procedures.

3.2.6 Consultation with the workforce

General

It is important to gain the cooperation of all employees if a successful health and safety culture is to become established. This cooperation is best achieved by consultation. Joint consultation can help businesses be more efficient and effective by reducing the number of accidents and work-related ill health and also to motivate staff by making them aware of health and safety problems.

There are two pieces of legislation that cover health and safety consultation with employees and both are summarised in Chapter 13. They are the Safety Representatives and Safety Committees Regulations 1977 and the Health and Safety (Consultation with Employees) Regulations 1996. These Regulations require employers to consult with their employees on health and safety matters and make provision for both trade-union-appointed safety representatives and **representatives of employee safety (ROES)** elected by the workforce.

Employers must consult their workforce on a range of health and safety issues, including:

▶ any measures at the workplace which may substantially affect their health and safety, for example changes in systems of work, types of equipment or chemicals being used;
▶ arrangements for getting competent people to help them comply with health and safety requirements;
▶ the information that must be given to employees about risks to health and safety and any preventative measures;
▶ planning and organising health and safety training;

▶ the health and safety consequences of using new technology or substances.

Also, in workplaces in which the employer recognises a trade union, the employer must consult with trade-union-appointed safety representatives on health and safety matters affecting the employees they represent.

Trade-union-appointed safety representatives may:

▶ investigate possible dangers at work, the causes of accidents there and general complaints by employees on health and safety issues and take these up with the employer;
▶ carry out inspections of the workplace;
▶ represent employees in discussions with the HSE inspectors and receive information from them;
▶ attend safety committee meetings.

In workplaces in which trade unions are not recognised:

▶ employees must be consulted on health and safety, either directly or through their elected representatives.

Elected ROES may:

▶ take up with employers concerns about possible risks and dangerous events in the workplace that may affect the employees they represent;
▶ take up with employers general matters affecting the health and safety of the employees they represent;
▶ represent employees who elected them in consultations with health and safety inspectors.

Finally, an employer should ensure that all representatives receive reasonable training in health and safety at the employer's expense, are allowed time during working hours to perform their duties, and provide other facilities and assistance that the representatives should reasonably require.

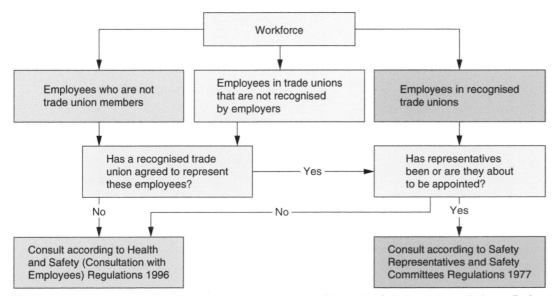

Figure 3.4 The law on consulting employees about health and safety in your workplace. References to the Regulations are colour-coded to help find the parts that are most relevant to a particular organisation: for workplaces where the Safety Representatives and Safety Committees Regulations 1977 apply; for workplaces where the Health and Safety (Consultation with Employees) Regulations 1996 apply

The Safety Representatives and Safety Committees Regulations

These Regulations apply only to those organisations that have recognised trade unions for collective bargaining purposes. The recognised trade union may appoint safety representatives from among the employees and notify the employer in writing. In some workplaces, safety representatives have agreed to represent the entire workforce.

The Regulations give safety representatives several functions not duties (a failure to carry out a function is not a breach of law as it would be if they were legal duties). Functions include:

(a) representing employees in consultation with the employer;
(b) investigating potential hazards and dangerous occurrences;
(c) being involved with risk assessment procedures;
(d) investigating the causes of accidents, cases of work-related diseases or ill health and dangerous occurrences;
(e) investigating employee complaints relating to health, safety and welfare;
(f) making representations to the employer on health, safety and welfare matters;
(g) carrying out inspections of the workplace. They must also be allowed to inspect the workplace at least once a quarter or sooner if there has been a substantial change in the conditions of work;
(h) representing employees at the workplace in consultation with enforcing inspectors;
(i) receiving information from health and safety inspectors;
(j) attending safety committee meetings.

Representatives must be allowed time off with pay to fulfil these functions and to undergo health and safety training. They must have access to suitable facilities and assistance to carry out their functions.

Safety committees

If two or more safety representatives have requested in writing that a safety committee be set up, the employer has 3 months to comply.

In medium to large organisations the easiest and often the most effective method of consultation is the health and safety committee. It will realise its full potential if its recommendations are seen to be implemented and both management and employee concerns are freely discussed. It will not be so successful if it is seen as a talking shop.

The committee should have stated objectives which mirror the objectives in the organisation's health and safety policy statement, and its own terms of reference. Terms of reference should include the following:

► the study of accident and notifiable disease statistics to enable reports to be made of recommended remedial actions;

► the examination of health and safety audits and statutory inspection reports;
► the consideration of reports from the external enforcement agency;
► the review of new legislation, Approved Codes of Practice and guidance and their effect on the organisation;
► the monitoring and review of all health and safety training and instruction activities in the organisation;
► the monitoring and review of health and safety publicity and communication throughout the organisation;
► development of safe systems of work and safety procedures;
► reviewing risk assessments;
► considering reports from safety representatives;
► continuous monitoring of arrangements for health and safety and revising them whenever necessary.

There are no fixed rules on the composition of the health and safety committee except that it should be representative of the whole organisation. It should have representation from the workforce and the management including at least one senior manager (other than the Health and Safety Adviser). Managers and representatives should agree who should chair the meetings, how often meetings should be held and what they hope to achieve.

Accident and ill-health investigations

Properly investigated accidents and ill health can reveal weaknesses which need to be remedied. A joint investigation with safety representatives is more likely to inspire confidence with workers so that they cooperate fully with the investigation, as in many cases those involved may be concerned about being blamed for the accident.

Safety representatives are entitled to contact enforcing authority inspectors. If this is just for information, they can be contacted directly. If it is a formal complaint against the employer, the inspector will need to know if the employer has been informed. The inspectors can be contacted anonymously. They will keep the person's identity secret in such circumstances.

Training, facilities and assistance

Safety representatives are legally entitled to paid time off for training, which is usually freely available from their trade union or the Trades Union Congress (TUC).

Training courses topics often include:

► the role and functions of the safety representative;
► health and safety legislation;
► how to identify and minimise hazards;
► how to carry out a workplace inspection and accident investigation;
► employer's health and safety arrangements, including emergency procedures, risk assessments and health and safety policies;
► further information on training courses.

The employer is also required to provide facilities and assistance for safety representatives. Depending on the circumstances, these could include:

▶ noticeboard;
▶ telephone;
▶ lockable filing cabinet;
▶ access to an office to meet workers in private;
▶ camera;
▶ key health and safety information;
▶ access to specialist assistance and support in understanding technical issues.

The Health and Safety (Consultation with Employees) Regulations

In non-unionised workplaces where there are no safety representatives, or in a workplace that has trade union recognition but either the trade union has not appointed a safety representative or they do not represent the whole workforce, the Health and Safety (Consultation with Employees) Regulations will apply.

Employers must consult either on an individual basis (e.g. in very small companies) or through ROES elected by the workforce. See Table 3.1 for a comparison of functions with union-appointed safety representatives.

The guidance to these Regulations emphasises the difference between informing and consulting. Consultation involves listening to the opinion of employees on a particular issue and taking it into account before a decision is made. Informing employees means providing information on health and safety issues such as risks, control systems and safe systems of work.

The functions of the ROES are to:

▶ represent the interest of workers on health and safety matters to the employer;
▶ approach the employer regarding potential hazards and dangerous occurrences at the workplace;
▶ approach the employer regarding general matters affecting the health and safety of the people they represent;
▶ to speak for the people they represent in consultation with inspectors.

The employer must consult the ROES on the following:

▶ the introduction of any measure or change which may substantially affect employees' health and safety;
▶ the arrangements for the appointment of competent persons to assist in following health and safety law;
▶ any information resulting from risk assessments or their resultant control measures which could affect the health, safety and welfare of employees;
▶ the planning and organisation of any health and safety training required by legislation;
▶ the health and safety consequences to employees of the introduction of new technologies into the workplace.

However, the employer is not expected to disclose information if:

▶ it violates a legal prohibition;
▶ it could endanger national security;

Table 3.1 A comparison of the functions of health and safety representatives

Safety Representatives and Safety Committees Regulations 1977	Health and Safety (Consultation with Employees) Regulations 1996
Representatives	
Appointed in writing by a trade union recognised for collective bargaining purposes.	Elected by the workforce, where the employer has decided not to consult directly.
Known as	
Safety representatives.	Representatives of employee safety (ROES).
Functions	
Investigate potential hazards and dangerous occurrences at the workplace, and complaints by an employee relating to health, safety and welfare at work, and examine causes of workplace accidents.	
Representation to the employer on the above investigations, and on general matters affecting the health and safety of employees they represent.	Representation to the employer on: ▶ potential hazards and dangerous occurrences ▶ general matters affecting the health and safety of employees they represent ▶ specific matters on which the employer must consult.
Inspect the workplace	
Represent employees in dealings with health and safety inspectors.	Represent employees in dealings with health and safety inspectors.
Receive certain information from inspectors.	
Attend health and safety committee meetings.	

Source: HSE INDG232(rev) page 4 HSE Web.

▶ it relates specifically to an individual without his/her consent;

▶ it could harm substantially the business of the employer or infringe commercial security;

▶ it was obtained in connection with legal proceedings.

There are several benefits that accrue to organisations that consult with their employees. These include:

▶ healthier and safer workplaces because employees can help to identify hazards and develop relevant procedures to eliminate, reduce or control risks;

▶ a stronger commitment to implementing procedures and decisions because employees have been involved proactively in the decision-making process;

▶ the development of better work practices;

▶ a reduction in workplace accidents; and

▶ greater cooperation and trust between employers and employees so that the overall performance of the organisation improves.

Consultation does not remove the employer's right and duty to manage since the employer will always make the final decision following the consultation. But talking to employees and involving them in the decision-making process is a vital part of successfully managing health and safety.

3.2.7 Health and safety training

The provision of information and training for employees will develop their awareness and understanding of the specific hazards and risks associated with their jobs and working environment. It will inform them of the control measures that are in place and any related safe procedures that must be followed. Apart from satisfying legal obligations, several benefits will accrue to the employer by the provision of sound information and training to employees. These benefits include:

▶ a reduction in accident severity and frequency;

▶ a reduction in injury and ill-health related absence;

▶ a reduction in compensation claims and, possibly, insurance premiums;

▶ an improvement in the health and safety culture of the organisation;

▶ improved staff morale and retention.

Health and safety training is a very important part of the health and safety culture and it is also a legal requirement, under the Management of Health and Safety at Work Regulations and other regulations (see Appendix 3.2), for an employer to provide such training. Training is required on recruitment, at induction or on being exposed to new or increased risks due to:

▶ being transferred to another job or given a change in responsibilities;

▶ the introduction of new work equipment or a change of use in existing work equipment;

▶ the introduction of new technology;

▶ the introduction of a new system of work or the revision of an existing system of work;

▶ an increase in the employment of more vulnerable employees (young or disabled persons);

▶ particular training required by the organisation's insurance company (e.g. specific fire and emergency training).

Additional training may well be needed following a single or series of accidents or near misses, the introduction of new legislation, the issuing of an enforcement notice or as a result of a risk assessment or safety audit.

It is important during the development of a training course that the target audience is taken into account. If the audience are young persons, the chosen approach must be capable of retaining their interest and any illustrative examples used must be within their experience. The trainer must also be aware of external influences, such as peer pressures, and use them to advantage. For example, if everybody wears PPE then it will be seen as the thing to do. Levels of literacy and numeracy are other important factors.

Figure 3.5 Health and safety training needs and opportunities

The way in which the training session is presented by the use of DVDs, PowerPoint slides, case studies, lectures or small discussion groups needs to be related to the material to be covered and the backgrounds of the trainees. Supplementary information in the form of copies of slides and additional background reading is often useful. The environment used for the training sessions is also important in terms of room layout and size, lighting and heating.

Attempts should be made to measure the effectiveness of the training by course evaluation forms issued at the time of the session, by a subsequent refresher session and by checking for improvements in health and safety performance (such as a reduction in specific accidents).

There are several different types of training, which include: induction, job specific, supervisory and management, and specialist. Informal sessions held at the place of work are known as 'toolbox talks'. Such sessions should only be used to cover a limited number of issues. They can become a useful route for employee consultation.

Induction training

Workers are as likely to have an accident in the first six months at a workplace as they are during the whole of the rest of their working life. Therefore induction training is very important.

Induction training should always be provided to new employees, trainees and contractors. While such training covers items such as pay, conditions and quality, it must also include health and safety. It is useful if the employee signs a record to the effect that training has been received. This record may be required as evidence should there be a subsequent legal claim against the organisation.

New workers are most vulnerable to accidents during their first six months at a new workplace. In the construction industry, eight out of 16 fatal accidents happened during the first 10 days on site, half of them on the very first day. The causes of these accidents were:

▶ lack of experience of working in a new industry or workplace;
▶ lack of familiarity with the job and the work environment;
▶ reluctance to raise concerns (or not knowing how to); and
▶ eagerness to impress workmates and managers.

They may not recognise hazards as a potential source of danger, possibly due to a lack of familiarity with the workplace, and may ignore warning signs and rules, or cut corners.

A key requirement for all new workers is adequate supervision particularly during the first few weeks. They must know how and with whom they can raise any concerns about any health and safety issues. It is important that they have understood the information, instruction and training needed to work safely. They must also have understood any emergency arrangements or procedures. Young persons and migrant workers are two groups of potentially vulnerable workers who may require particular attention at induction.

Most induction training programmes would include the following topics:

▶ the health and safety policy of the organisation including a summary of the organisation and arrangements including employee consultation;
▶ a brief summary of the health and safety management system including the name of the employee's direct supervisor, safety representative and source of health and safety information;
▶ the employee responsibility for health and safety including any general health and safety rules (e.g. smoking prohibitions);
▶ the accident reporting procedure of the organisation, the location of the accident book and the location of the nearest first-aider;
▶ the fire and other emergency procedures including the location of the assembly point;
▶ the hazards that are specific to the workplace;
▶ a summary of any relevant risk assessments and safe systems of work;
▶ the location of welfare, canteen facilities and rest rooms;
▶ procedures for reporting defects or possible hazards and the name of the responsible person to whom the report should be made;
▶ details of the possible disciplinary measures that may be enacted for non-compliance with health and safety rules and procedures.

Additional items which are specific to the organisation may need to be included, such as:

▶ internal transport routes and pedestrian walkways (e.g. fork-lift truck operations);
▶ the correct use of PPE and maintenance procedures;
▶ manual handling techniques and procedures;
▶ details of any hazardous substances in use and any procedures relating to them (e.g. health surveillance).

There should be some form of follow-up with each new employee after 3 months to check that the important messages have been retained. This is sometimes called a refresher course, although it is often better done on a one-to-one basis.

It is very important to stress that the content of the induction course should be subject to constant review and updated following an accident investigation, new legislation, changes in the findings of a risk assessment or the introduction of new plant or processes. A well-designed induction programme may help to encourage a positive health and safety culture and reduce the effect of negative peer pressure.

Job-specific training

Job-specific training ensures that employees undertake their job in a safe manner. Such training, therefore, is a form of skill training and is often best done 'on the job' – sometimes known as 'toolbox training'. Details of the safe system of work or, in more hazardous jobs, a permit-to-work system, should be covered. In addition to normal safety procedures, emergency procedures and the correct use of PPE also need to be included. The results of risk assessments are very useful in the development of this type of training. It is important that any common causes of human errors (e.g. discovered as a result of an accident investigation), any standard safety checks or maintenance requirements, are addressed.

It is common for this type of training to follow an operational procedure in the form of a checklist which the employee can sign on completion of the training. The new employee will still need close supervision for some time after the training has been completed.

Supervisory and management training

Supervisory and management health and safety training follows similar topics to those contained in an induction training course but will be covered in more depth. There will also be a more detailed treatment of health and safety law. There has been considerable research over the years into the failures of managers that have resulted in accidents and other dangerous incidents. These failures have included:

▶ lack of health and safety awareness, enforcement and promotion (in some cases, there has been encouragement to circumvent health and safety rules);
▶ lack of consistent supervision of and communication with employees;
▶ lack of understanding of the extent of the responsibility of the supervisor.

It is important that all levels of management, including the Board, receive health and safety training. This will not only keep everybody informed of health and safety legal requirements, accident prevention techniques and changes in the law, but also encourage everybody to monitor health and safety standards during visits or tours of the organisation.

Specialist training

Specialist health and safety training is normally needed for activities that are not related to a specific job but more to an activity. Examples include first aid, fire prevention, fork-lift truck driving, overhead crane operation, scaffold inspection and statutory health and safety inspections. These training courses are often provided by specialist organisations and successful participants are awarded certificates. Details of two of these courses will be given here by way of illustration.

Fire prevention training courses include the causes of fire and fire spread, fire and smoke alarm systems, emergency lighting, the selection and use of fire extinguishers and sprinkler systems, evacuation procedures, high-risk operations and good housekeeping principles.

A fork-lift truck drivers' course would include the general use of the controls, loading and unloading procedures, driving up or down an incline, speed limits, pedestrian awareness (particularly in areas where pedestrians and vehicles are not segregated), security of the vehicle when not in use, daily safety checks and defect reporting, refuelling and/or battery charging and emergency procedures.

Details of other types of specialist training appear elsewhere in the text.

Training is a vital part of any health and safety programme and needs to be constantly reviewed and updated. Many health and safety Regulations require specific training (e.g. manual handling, PPE and display screens). Additional training courses may be needed when there is a major reorganisation, a series of similar accidents or incidents, or a change in equipment or a process. Finally, the methods used to deliver training must be continually monitored to ensure that they are effective.

A list of typical legislation that requires some form of health and safety training is given in Appendix 3.2. The list has been compiled to cover the needs of most small and medium-sized organisations.

3.3 How human factors influence behaviour at work both positively and negatively

3.3.1 Human factors

Over the years, there have been several studies undertaken to examine the link between various accident types, graded in terms of their severity, and near misses. One of the most interesting was conducted in the USA by H. W. Heinrich in 1950. He looked at over 300 accidents/incidents and produced the ratios illustrated in Figure 3.6.

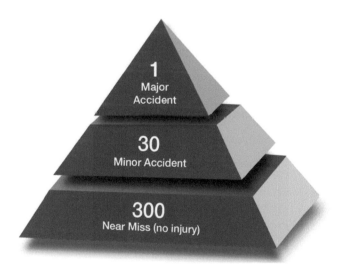

Figure 3.6 Heinrich's accidents/incidents ratios

This study indicated that for every 10 near misses, there will be an accident. Although the accuracy of this study may be debated and other studies have produced different ratios, it is clear that if near misses are continually ignored, an accident will result. Further, the HSE Accident Prevention Unit has suggested that 90% of all accidents are due to human error and 70% of all accidents could have been avoided by earlier (proactive) action by management. It is clear from many research projects that the major factors in most accidents are human factors.

While Heinrich's triangle is useful in many ways, it does not suggest that minor and major incidents have the same causes as has been suggested by some technical papers over the years. Neither does it imply that the causes of incident frequency are the same as the causes of severe (major) injuries. The causes of severe injuries are often different to those of the causes of minor injuries. It is not reasonable, therefore, to assume that the majority of major injuries or incidents will occur where the majority of incidents take place. The hazards in a construction company's main office will not present the same major hazards as those at the company's various construction sites.

The HSE has defined human factors as, 'environmental, organisational and job factors, and human and individual characteristics which influence behaviour at work in a way which can affect health and safety'.

In simple terms, in addition to the environment, the health and safety of people at work are influenced by:

- the organisation;
- the job; and
- individual (personal) factors.

These are known as human factors as they each have a human involvement. The individual factors which differentiate one person from another are only one part of those factors – and not always the most important.

Each of these elements will be considered in turn.

The organisation

The organisation is the company or corporate body and has the major influence on health and safety. It must have its own positive health and safety culture and produce an environment in which it:

- manages health and safety throughout the organisation, including the setting and publication of a health and safety policy and the establishment of a health and safety organisational structure;
- measures the health and safety performance of the organisation at all levels and in all departments. The performance of individuals should also be measured. There should be clear health and safety targets and standards and an effective reporting procedure for accidents and other incidents so that remedial actions may be taken;
- motivates managers within the organisation to improve health and safety performance in the workplace in a proactive rather than reactive manner.

The HSE has recommended that an organisation needs to provide the following elements within its management system:

- a clear and evident commitment from the most senior manager downwards, which provides a climate for safety in which management's objectives and the need for appropriate standards are communicated and in which constructive exchange of information at all levels is positively encouraged;
- an analytical and imaginative approach identifying possible routes to human factor failure. This may well require access to specialist advice;
- procedures and standards for all aspects of critical work and mechanisms for reviewing them;
- effective monitoring systems to check the implementation of the procedures and standards;
- incident investigation and the effective use of information drawn from such investigations;
- adequate and effective supervision with the power to remedy deficiencies when found.

It is important to recognise that there are often reasons for these elements not being present, resulting in weak management of health and safety. The most common reason is that individuals within the management organisation do not understand their roles – or their roles have never been properly explained to them. The higher a person is within the structure the less likely it is that he/she has received any health and safety training. Such training at board level is rare.

Objectives and priorities may vary across and between different levels in the structure, leading to disputes which affect attitudes to health and safety. For example, a warehouse manager may be pressured to block walkways so that a large order can be stored prior to dispatch.

Motivations can also vary across the organisation, which may cause health and safety to be compromised. The production controller will require that components of a product are produced as near simultaneously as possible so that their final assembly is performed as quickly as possible. However, the health and safety adviser will not want to see safe systems of work compromised.

In an attempt to address some of these problems, the HSE has produced guidance on the safety duties of company directors. Each director and the Board, acting collectively, will be expected to provide health and safety leadership in the organisation. The Board will need to ensure that all its decisions reflect its health and safety intentions and that it engages the workforce actively in the improvement of health and safety. The Board will also be expected to keep itself informed of changes in health and safety risks. (See Chapter 2 Section 2.3.2 for more details on directors' responsibilities.)

The following simple checklist may be used to check any organisational health and safety management structure.

Does the structure have:

- an effective health and safety management system?
- a positive health and safety culture?
- arrangements for the setting and monitoring of standards?
- adequate supervision?
- effective incident reporting and analysis?
- learning from experience?
- clearly visible health and safety leadership?

▶ suitable team structures?
▶ efficient communication systems and practices?
▶ adequate staffing levels?
▶ suitable work patterns?

HSG48 'Reducing Error and Influencing Behaviour' gives the following causes for failures in organisational and management structures:

▶ poor work planning leading to high work pressure;
▶ lack of safety systems and barriers;
▶ inadequate responses to previous incidents;
▶ management based on one-way communications;
▶ deficient coordination and responsibilities;
▶ poor management of health and safety;
▶ poor health and safety culture.

Organisational factors play a significant role in the health and safety of the workplace. However, this role is often forgotten when health and safety is being reviewed after an accident or when a new process or piece of equipment is introduced.

The job

Jobs may be highly dangerous or present only negligible risk of injury. Health and safety is an important element during the design stage of the job and any equipment, machinery or procedures associated with the job. Method study helps to design the job in the most cost-effective way and ergonomics helps to design the job with health and safety in mind. Ergonomics is the science of matching equipment, machines and processes to people rather than the other way round. An ergonomically designed machine will ensure that control levers, dials, meters and switches are sited in a convenient and comfortable position for the machine operator. Similarly, an ergonomically designed workstation will be designed for the comfort and health of the operator. Chairs, for example, will be designed to support the back properly throughout the working day (Figure 3.7).

Figure 3.7 Well-designed workstation for sitting or standing

Physically matching the job and any associated equipment to the person will ensure that the possibility of human error is minimised. It is also important to ensure that there is cognitive matching of the person's information and decision-making requirements. A person must be capable, either through past experience or through specific training, of performing the job with the minimum potential for human error.

The major considerations in the design of the job, which would be undertaken by a specialist, have been listed by the HSE as follows:

▶ the identification and detailed analysis of the critical tasks expected of individuals and the appraisal of any likely errors associated with those tasks;
▶ evaluation of the required operator decision making and the optimum (best) balance between the human and automatic contributions to safety actions (with the emphasis on automatic whenever possible);
▶ application of ergonomic principles to the design of man–machine interfaces, including displays of plant and process information, control devices and panel layout;
▶ design and presentation of procedures and operating instructions in the simplest terms possible;
▶ organisation and control of the working environment, including the workspace, access for maintenance, lighting, noise and heating conditions;
▶ provision of the correct tools and equipment;
▶ scheduling of work patterns, including shift organisation, control of fatigue and stress and arrangements for emergency operations;
▶ efficient communications, both immediate and over a period of time.

For some jobs, particularly those with a high risk of injury, a job safety analysis should be undertaken to check that all necessary safeguards are in place. All jobs should carry a job description and a safe system of work for the particular job. The operator should have sight of the job description and be trained in the safe system of work before commencing the job. More information on both these latter items is given in Chapter 4.

The following simple checklist may be used to check that the principal health and safety considerations of the job have been taken into account:

▶ Have the critical parts of the job been identified and analysed?
▶ Have the employee's decision-making needs been evaluated?
▶ Has the best balance between human and automatic systems been evaluated?
▶ Have ergonomic principles been applied to the design of equipment displays, including displays of plant and process information, control information and panel layouts?
▶ Has the design and presentation of procedures and instructions been considered?

► Has the guidance available for the design and control of the working environment, including the workspace, access for maintenance, lighting, noise and heating conditions, been considered?

► Have the correct tools and equipment been provided?

► Have the work patterns and shift organisation been scheduled to minimise their impact on health and safety?

► Has consideration been given to the achievement of efficient communications and shift handover?

HSG48 gives the following causes for failures in job health and safety:

► illogical design of equipment and instruments;
► constant disturbances and interruptions;
► missing or unclear instructions;
► poorly maintained equipment;
► high workload;
► noisy and unpleasant working conditions.

It is important that health and safety monitoring of the job is a continuous process. Some problems do not become apparent until the job is started. Other problems do not surface until there is a change of operator or a change in some aspect of the job.

It is very important to gain feedback from the operator on any difficulties experienced because there could be a health and safety issue requiring further investigation.

Individual (personal) factors

Individual or personal factors, which affect health and safety, may be defined as any condition or characteristic of an individual which could cause or influence him/her to act in an unsafe manner. They may be physical, cognitive or psychological in nature. Individual factors, therefore, include issues such as attitude, motivation, training and human error and their interaction with the physical, mental and perceptual capability of the individual. Typical individual factors could include:

► Self-interest in the case of incentive or bonus schemes to maximise earnings;
► loss of/or poor memory;
► loss of/or poor hearing;
► language issues and communication problems;
► physical health;
► experience and competence;
► personality and attitude;
► age;
► the position in team; and
► training and competence.

Language and communication issues may be particular problems for some migrant workers. The law requires that employers provide workers with comprehensible and relevant information about risks and about the procedures they need to follow to ensure they can work safely and without risk to health. This does not

have to be in English. The employer may make special arrangements, which could include translation, using interpreters or replacing written notices with clearly understood symbols or diagrams. Any health and safety training provided must take into account the worker's capabilities, including language skills.

Figure 3.8 Poor working conditions

The following **negative** factors can also affect the individual:

► low skill and competence levels;
► tired colleagues;
► bored or disinterested colleagues;
► individual medical or mental problems;
► complacency from repetitive tasks and lack of awareness training;
► inexperience, especially if the employee is new or a young person; and
► peer pressure to conform to the 'group' or an individual's perception of how a task should be completed.

These factors have a significant effect on health and safety. Some of them, normally involving the personality of the individual, are unchangeable but others, involving skills, attitude, perception and motivation, can be changed, modified or improved by suitable training or other measures. In summary, the person needs to be matched to the job.

Studies have shown that the most common individual factors which contribute to accidents are low skill and competence levels, tiredness, boredom, low morale and individual medical problems. The setting of realistic timescales and work tasks schedules will lead to considerable improvements in these factors. However, the key issue is effective and relevant training. This can help to improve attitudes and the understanding of risks and safe working. It will also enhance or improve skills

It is difficult to separate all the physical, mental or psychological factors because they are interlinked. However, the three most common factors are psychological factors – attitude, motivation and perception.

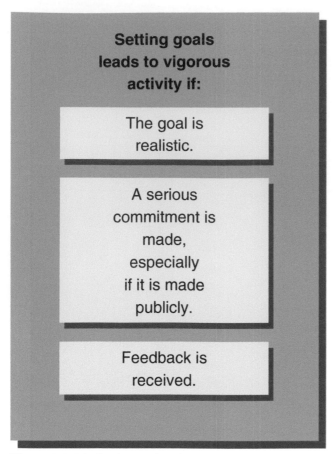

Setting goals leads to vigorous activity if:

The goal is realistic.

A serious commitment is made, especially if it is made publicly.

Feedback is received.

Figure 3.9 Motivation and activity

Attitude is the tendency to behave in a particular way in a certain situation. One of the principal aims of a good safety culture is to change behaviour. But, if this change is to become permanent, then attitude will have to change. However, good intentions are not enough and attitudes and behaviour do not always coincide. Attitudes are influenced by the prevailing health and safety culture within the organisation, the commitment of the management, the experience of the individual and the influence of the peer group. Peer group pressure is a particularly

important factor among young people, and health and safety training must be designed with this in mind by using examples or case studies that are relevant to them. Behaviour may be changed by training, the formulation and enforcement of safety rules and meaningful consultation – attitude change often follows.

Motivation is the driving force behind the way a person acts or the way in which people are stimulated to act. Involvement in the decision-making process in a meaningful way will improve motivation as will the use of incentive schemes. However, there are other important influences on motivation such as recognition and promotion opportunities, job security and job satisfaction. Self-interest, in all its forms, is a significant motivator and individual factor.

Perception is the way in which people interpret the environment or the way in which a person believes or understands a situation (Figure 3.10). In health and safety, the perception of hazards is an important concern. Many accidents occur because people do not perceive that there is a risk. There are many common examples of this, including the use of personal protective equipment (such as hard hats) and guards on drilling machines and the washing of hands before meals. It is important to understand that when perception leads to an increased health and safety risk, it is not always caused by a conscious decision of the individual concerned. The stroboscopic effect caused by the rotation of a drill at certain speeds under fluorescent lighting will make the drill appear stationary. It is a well-known phenomenon, especially among illusionists, that people will often see what they expect to see rather than reality. Routine or repetitive tasks will reduce attention levels, leading to the possibility of accidents.

Other individual factors which can affect health and safety include physical stature, age, experience, health, hearing, intelligence, language, skills, level of competence and qualifications.

(a)

(b)

(c)

Figure 3.10 Visual perceptions: (a) Are the lines of the same length? (b) Faces or vase? (c) Face or saxophone player?

Memory is an important individual factor, as it is influenced by training and experience. The efficiency of memory varies considerably between people and during the lifetime of an individual. The overall health of a person can affect memory as can personal crises. Owing to these possible problems with memory, important safety instructions should be available in written as well as verbal form.

Finally, it must be recognised that some employees do not follow safety procedures either due to peer pressure or a wilful disregard of those procedures.

The following checklist is modelled on HSG48 and may be used to check that the relevant individual factors have been covered:

▶ Has the job specification been drawn up and does it include age, physique, skill, qualifications, experience, aptitude, knowledge, intelligence and personality?
▶ Have the skills and aptitudes been matched to the job requirements?
▶ Have the personnel selection policies and procedures been set up to select appropriate individuals?
▶ Has an effective training system been implemented?
▶ Have the needs of special groups of employees been considered?
▶ Have the monitoring procedures been developed for the personal safety performance of safety critical staff?
▶ Have fitness for work and health surveillance been provided where it is needed?
▶ Have counselling and support for ill health and stress been provided?

Individual factors are the attributes that employees bring to their jobs and may be strengths or weaknesses. Negative individual factors cannot always be neutralised by improved job design. It is, therefore, important to ensure that personnel selection procedures should match people to the job. This will reduce the possibility of accidents or other incidents.

3.3.2 Human errors and violations

In 1985, a report by the HSE found that senior management failures resulted in 61% of all accidents, and 47% of those accidents had human causes. Human failures in health and safety are classified either as errors or violations. An error is an unintentional deviation from an accepted standard, whereas a violation is a deliberate deviation from the standard (Figure 3.11).

Human errors

Human errors fall into three groups – slips, lapses and mistakes, which can be further sub-divided into rule-based and knowledge-based mistakes.

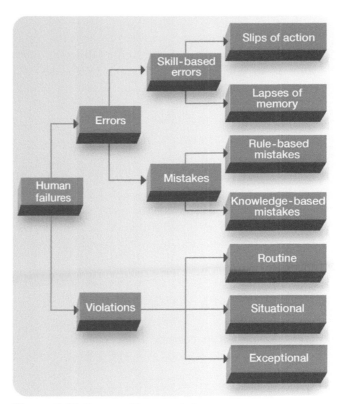

Figure 3.11 Types of human failure

Slips and lapses

These are very similar in that they are caused by a momentary memory loss often due to lack of attention or loss of concentration. They are not related to levels of training, experience or motivation and they can usually be reduced by redesigning the job or equipment or minimising distractions.

Slips are failures to carry out the correct actions of a task. Examples include the use of the incorrect switch, reading the wrong dial or selecting the incorrect component for an assembly. A slip also describes an action taken too early or too late within a given working procedure.

Lapses are failures to carry out particular actions which may form part of a working procedure. A fork-lift truck driver leaving the keys in the ignition lock of his truck is an example of a lapse as is the failure to replace the petrol cap on a car after filling it with petrol. Lapses may be reduced by redesigning equipment so that, for example, an audible horn indicates the omission of a task. They may also be reduced significantly by the use of detailed checklists.

Mistakes

Mistakes occur when an incorrect action takes place but the person involved believes the action to be correct. A mistake involves an incorrect judgement. There are two types of mistake – rule-based and knowledge-based.

Rule-based mistakes occur when a rule or procedure is remembered or applied incorrectly. These mistakes usually happen when, due to an error, the rule that is normally used no longer applies. For example, a particular job requires the counting of items into groups of ten followed by the adding together of the groups so that the total number of items may be calculated. If one of the groups is miscounted, the final total will be incorrect even though the rule has been followed.

Knowledge-based mistakes occur when well-tried methods or calculation rules are used inappropriately. For example, the depth of the foundations required for a particular building was calculated using a formula. The formula, which assumed a clay soil, was used to calculate the foundation depth in a sandy soil. The resultant building was unsafe.

The HSE has suggested the following points to consider when the potential source of human errors is to be identified:

▶ What human errors can occur with each task? There are formal methods available to help with this exercise.
▶ What influences are there on performance? Typical influences include time pressure, design of controls, displays and procedures, training and experience, fatigue and levels of supervision.
▶ What are the consequences of the identified errors?
▶ What are the significant errors?
▶ Are there opportunities for detecting each error and recovering it?
▶ Are there any relationships between the identified errors? Could the same error be made on more than one item of equipment due, for example, to the incorrect calibration of an instrument?

Errors and mistakes can be reduced by the use of instruction, training and relevant information. However, communication can also be a problem, particularly at shift handover times. Environmental and organisational factors involving workplace stress will also affect error levels.

The following steps are suggested to reduce the likelihood of human error:

▶ Examine and reduce the workplace stressors (e.g. noise, poor lighting) which increase the frequency of errors.
▶ Examine and reduce any social or organisational stressors (e.g. insufficient staffing levels, peer pressure).
▶ Design plant and equipment to reduce error possibilities (poorly designed displays, ambiguous instructions).
▶ Ensure that there are effective training arrangements.
▶ Simplify any complicated or complex procedures.
▶ Ensure that there is adequate supervision, particularly for inexperienced or young trainees.

▶ Check that job procedures, instructions and manuals are kept up to date and are clear.
▶ Include the possibility of human error when undertaking the risk assessment.
▶ Isolate the human error element of any accident or incident and introduce measures to reduce the risk of a repeat.
▶ Monitor the effectiveness of any measures taken to reduce errors.

Violations

There are three categories of violation – routine, situational and exceptional.

Routine violation occurs when the breaking of a safety rule or procedure is the normal way of working. It becomes routine not to use the recommended procedures for tasks. An example of this is the regular speeding of fork-lift trucks in a warehouse so that orders can be fulfilled on time.

There are many reasons given for routine violations, for example:

▶ taking short cuts to save time and energy;
▶ a belief that the rules are unworkable or too restrictive;
▶ lack of knowledge of the procedures;
▶ perception that the rules are no longer applied;
▶ poor supervision and a lack of enforcement of the rules;
▶ new workers thinking that routine violations are the norm and not realising that this was not the safe way of working.

Finally, it must be recognised that there are some situations where peer pressure or simply a wilful disregard for procedures or other people's safety may result in routine violations. Routine violations can be reduced by regular monitoring, ensuring that the rules are actually necessary, or redesigning the job.

The following features are very common in many workplaces and often lead to routine violations:

▶ poor working posture due to poor ergonomic design of the workstation or equipment;
▶ equipment difficult to use and/or slow in response;
▶ equipment difficult to maintain or pressure on time available for maintenance;
▶ procedures unduly complicated and difficult to understand;
▶ unreliable instrumentation and/or warning systems;
▶ high levels of noise and other poor aspects to the environment (fumes, dusts, humidity);
▶ associated PPE either inappropriate, difficult and uncomfortable to wear or ineffective due to lack of maintenance.

Situational violations occur when particular job pressures at particular times make rule compliance difficult. They may happen when the correct equipment is not available or weather conditions are adverse. A common example is the use of a ladder rather than a scaffold for working at height to replace window frames in a building. Situational violations may be reduced by improving job design, the working environment and supervision.

Exceptional violations rarely happen and usually occur when a safety rule is broken to perform a new task. Good examples of this are the violations which can occur during the operations of emergency procedures such as for fires or explosions. These violations should be addressed in risk assessments and during training sessions for emergencies (e.g. fire training).

Everybody is capable of making errors. It is one of the objectives of a positive health and safety culture to reduce them and their consequences as much as possible.

3.4 Additional useful background information that is not part of the NEBOSH NGC syllabus

3.4.1 Internal influences on health and safety performance

There are many influences on health and safety standards, some are positive and others negative. No business, particularly small businesses, is totally divorced from their suppliers, customers and neighbours. This section considers the internal influences on a business, including management commitment, production demands, communication, competence and employee relations.

Figure 3.12 Internal influences on safety culture

Management commitment

Managers, particularly senior managers, can give powerful messages to the workforce by what they do for health and safety. Managers can achieve the level of health and safety performance that they demonstrate they want to achieve. Employees soon get the negative message if directors disregard safety rules and ignore written policies to get urgent production to the customer or to avoid personal inconvenience. It is what they do that counts not what they say. In Chapter 2, managers' organisational roles were listed showing the ideal level of involvement for senior managers. Depending on the size and geography of the organisation, senior managers should be personally involved in:

▶ health and safety inspections or audits;
▶ meetings of the central health and safety or joint consultation committees;
▶ involvement in the investigation of accidents, ill health and incidents. The more serious the incident the more senior the manager who takes an active part in the investigation.

Production/service demands

Managers need to balance the demands placed by customers with the action required to protect the health and safety of their employees. How this is achieved has a strong influence on the standards adopted by the organisation. The delivery driver operating to near impossible delivery schedules and the manager agreeing to the strident demands of a large customer regardless of the risks to employees involved are typical dilemmas facing workers and managers alike. The only way to deal with these issues is to be well organised and agree with the client/ employees ahead of the crisis as to how they should be prioritised.

Rules and procedures should be intelligible, sensible and reasonable. They should be designed to be followed under normal production or service delivery conditions. If following the rules involves very long delays or impossible production schedules, they should be revised rather than ignored by workers and managers alike, until it is too late and an accident occurs. Sometimes the safety rules are simply used in a court of law, in an attempt to defend the company concerned, following an accident.

Managers who plan the impossible schedule or ignore a safety rule to achieve production or service demands are held responsible for the outcomes. It is acceptable to balance the cost of the action against the level of risk being addressed but it is never acceptable to ignore safety rules or standards simply to get the work done. Courts are not impressed by managers who put profit considerations ahead of safety requirements.

Communication

Communications were covered in depth earlier in the chapter and clearly will have significant influence on health and safety issues. This will include:

▶ poorly communicated procedures that will not be understood or followed;
▶ poor verbal communications which will be misunderstood and will demonstrate a lack of interest by senior managers;
▶ missing or incorrect signs which may cause accidents rather than prevent them;
▶ managers who are nervous about face-to-face discussions with the workforce on health and safety issues, which will have a negative effect.

Managers and supervisors should plan to have regular discussions to learn about the problems faced by employees and discuss possible solutions. Some meetings, like that of the safety committee, are specifically planned for safety matters, but this should be reinforced by discussing health and safety issues at all routine management meetings. Regular one-to-one talks should also take place in the workplace, preferably to a planned theme or safety topic, to get specific messages across and get feedback from employees.

Competence

Competent people, who know what they are doing and have the necessary skills to do the task correctly and safely, will make the organisation more effective. Competence can be bought in through recruitment or consultancy but it is often much more effective to develop it among employees. It demonstrates commitment to health and safety and a sense of security for the workforce. The loyalty that it creates in the workforce can be a significant benefit to safety standards. Earlier in this chapter, there is more detail on how to achieve competence.

Employee representation

Given the resources and freedom to fulfil their function effectively, enthusiastic, competent employee safety representatives can make a major contribution to good health and safety standards. They can provide the essential bridge between managers and employees. People are more willing to accept the restrictions that some precautions bring if they are consulted and feel involved, either directly in small workforces or through their safety representatives.

3.4.2 External influences on health and safety performance

The role and function of some external organisations is set out in Chapter 1. Here the influence they can have

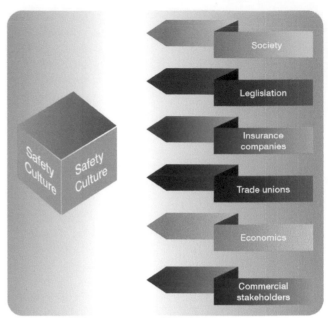

Figure 3.13 External influences on safety culture

is briefly discussed, including societal expectations, legislation and enforcement, insurance companies, trade unions, economics, commercial stakeholders (Figure 3.13).

Societal expectations

Societal expectations have been covered in Chapter 1. In general society assumes that health and safety considerations have been considered in all aspects of their lives and becomes concerned when this does not seem to be the case.

Legislation and enforcement

Good comprehensible legislation should have a positive effect on health and safety standards. Taken together, legislation and enforcement can affect standards by:

▶ providing a level to which every employer has to conform;
▶ insisting on minimum standards which also enhance people's ability to operate and perform well;
▶ providing a tough, visible threat of getting shut down or a heavy fine;
▶ stifling development by being too prescriptive, for example woodworking machines did not develop quickly in the 20th century, partly because the Regulations were so detailed in their requirements that new designs were not feasible;
▶ providing well-presented and easily read guidance for specific industries at reasonable cost or free.

On the other hand, a weak enforcement regime can have a powerful negative effect on standards.

Insurance companies

Insurance companies can influence health and safety standards mainly through financial incentives. Employers' liability insurance is a legal requirement in the UK and therefore all employers have to obtain this type of insurance cover. Insurance companies can influence standards through:

- discounting premiums to those in the safest sectors or best individual companies;
- insisting on risk reduction improvements to remain insured. This is not very effective where competition for business is fierce;
- encouraging risk reduction improvements by bundling services into the insurance premium;
- providing guidance on standards at reasonable cost or free.

Trade unions

Trade unions can influence standards by:

- providing training and education for members;
- providing guidance and advice cheaply or free to members;
- influencing governments to regulate, enhance enforcement activities and provide guidance;
- influencing employers to provide high standards for their members. This is sometimes confused with financial improvements with health and safety getting a lower priority;
- encouraging members to work for safer employers;
- helping members to get proper compensation for injury and ill health if it is caused through their work.

Economics

Economics can play a major role in influencing health and safety standards. The following ways are the most common:

- lack of orders and/or money can cause employers to try to ignore health and safety requirements;
- if employers were really aware of the actual and potential cost of accidents and fires, they would be more concerned about prevention. The HSE believes that the ratio between insured and uninsured costs of accidents is between 1:8 and 1:36;
- perversely, when the economy is booming activity increases and, particularly in the building industry, accidents can sharply increase. The pressures to perform and deliver for customers can be safety averse;
- businesses that are only managed on short-term performance indicators seldom see the advantage of the long-term gains that are possible with a happy, safe and fit workforce.

The cost of accidents and ill health, in both human and financial terms, needs to be visible throughout the organisation so that all levels of employee are encouraged to take preventative measures.

Commercial stakeholders

A lot can be done by commercial stakeholders to influence standards. This includes:

- insisting on proper arrangements for health and safety management at supplier companies before they tender for work or contracts;
- checking on suppliers to see if the workplace standards are satisfactory;
- encouraging ethical investments;
- considering ethical as well as financial standards when banks provide funding;
- providing high-quality information for customers;
- insisting on high standards to obtain detailed planning permission (where this is possible);
- providing low-cost guidance and advice.

3.5 Further information

The Health and Safety at Work etc. Act 1974

The Management of Health and Safety at Work Regulations 1999 (as amended)

The Safety Representatives and Safety Committees Regulations 1977

The Health and Safety Information for Employees Regulations 1989

The Health and Safety (Consultation with Employees) Regulations 1996

Corporate Manslaughter and Corporate Homicide Act 2007

Occupational Health and Safety Assessment Series (OHSAS 18000): Occupational Health and Safety Management Systems OHSAS 18001:2007 ISBN: 978-0-5805-9404-5; OHSAS18002:2008 ISBN: 978-0-5806-2686-9

Reducing Error and Influencing Behaviour (HSG48), HSE Books, ISBN: 978-0-7176-2452-2 www.hse.gov.uk/pubns/priced/hsg48.pdf

First-aid at Work, The Health and Safety (First-Aid) Regulations 1981 (L74, third edition 2013), HSE Books, ISBN: 978-0-7176-6560-0 www.hse.gov.uk/pubns/priced/l74.pdf

Leading health and safety at work (INDG417 rev 1 2013) HSE Books www.hse.gov.uk/pubns/indg417.htm

Human factors and ergonomics www.hse.gov.uk/humanfactors/

Consulting workers on health and safety, Safety Representatives and Safety Committee Regulations 1977 (as amended) and Health and Safety (Consultation with Employees) Regulations 1996 (as amended), ACoP, L146, HSE Books, ISBN: 978-0-7176-6461-0 www.hse.gov.uk/pubns/priced/l146.pdf

3.6 Practice questions

1. (a) **Explain** the meaning of the term 'health and safety culture'.
   (b) **Describe FIVE** requirements of a positive health and safety culture.
   (c) **Identify** ways in which the health and safety culture of an organisation might be improved.
   (d) **Outline EIGHT** possible indications of a poor health and safety culture in an organisation.

2. (a) The number of accidents is a measure of the health and safety culture in a company. **State** the definition of 'accident incidence rate'.
   (b) **Describe** other measures of health and safety culture. **Outline** how information on accidents could be used to promote health and safety in the workplace.

3. (a) **Explain** the meaning of the term 'competent person' when related to health and safety.
   (b) **Outline** the factors that the employer should consider when selecting an individual for the role of health and safety competent person within the organisation.
   (c) **Outline** the organisational factors that may cause a person to work unsafely even though they are competent.
   (d) **Identify FOUR** types of specialist help that may be required to support the health and safety function.

4. (a) **Identify FIVE** topics for which there would be health and safety standards in a company.
   (b) **Outline** the possible actions that a chief executive and senior managers could take to improve the health and safety standards in the workplace.

5. Following a significant increase in accidents, a health and safety campaign is to be launched within a manufacturing company to encourage safer working by employees.
   (a) **Outline** how the company might ensure that the nature of the campaign is effectively communicated to, and understood by, employees.
   (b) **Identify** a range of methods that an employer can use to provide health and safety information directly to individual employees.
   (c) **Outline** reasons why the safety procedures may not have been followed by all employees.

6. **Outline** the advantages and disadvantages of communicating health and safety messages to employees:
   (a) Verbally
   (b) In writing
   (c) Graphically.

7. Noticeboards are often used to display health and safety information in the workplace.
   (a) **Outline** the limitations of relying on this method to communicate to employees and how these limitations may be addressed.
   (b) **Identify FOUR** types of health and safety information that might usefully be displayed on a noticeboard.
   (c) **Identify FOUR** other methods which the employer could use to communicate essential health and safety information to their employees.

8. (a) **Explain** the circumstances in which an employer must form a health and safety committee under the Safety Representatives and Safety Committees Regulations.
   (b) **Outline SIX** reasons why a health and safety committee may prove to be ineffective in practice;

9. With reference to the Health and Safety (Consultation with Employees) Regulations 1996:
   (a) **explain** the difference between consulting and informing and the ways in which an employer can consult with the workforce;
   (b) **outline FOUR** health and safety matters on which employers must consult their employees;
   (c) **outline** the rights and functions of representatives of employee safety elected under the Regulations;
   (d) **identify FOUR** types of information that an employer is not obliged to disclose to an employee representative.

10. (a) **Outline** the importance of induction training for new employees in reducing the number of accidents in a workplace.
    (b) **Outline** reasons for additional refresher health and safety training at a later stage of employment within a workplace.
    (c) **Identify** the reasons for a review of the health and safety training programme within an organisation.

11. A building contractor is to undertake building maintenance work in the roof space of a busy warehouse. **Outline** the issues that should be covered in an induction programme for the contractor's employees.

12. Following an accident at a manufacturing company, an investigation concluded that there had been a poor record of specialist and supervisory training.
    (a) **Identify THREE** types of specialist training that may have been relevant to the company.

(b) **Outline** the possible supervisory failures that could have caused an accident.

13. (a) **Explain**, using an example, the meaning of the term 'attitude'.

(b) **Outline THREE** influences on the attitude towards health and safety of employees within an organisation.

14. (a) **Explain** the meaning of the term 'motivation'.

(b) Other than lack of motivation, **outline FOUR** reasons why employees may fail to comply with safety procedures at work.

(c) **Outline** ways in which employers may motivate their employees to comply with health and safety procedures.

15. (a) **Explain** the meaning of the term 'perception'.

(b) **Outline** the factors relating to the individual that may influence a person's perception of an occupational risk.

(c) **Outline** ways in which employees' perceptions of hazards in the workplace might be improved.

16. The human factors which influence behaviour in the workplace are the organisation, the job and personal factors.

(a) **Describe** the essential elements in the organisation and the job that will ensure good health and safety behaviour.

(b) **Outline** personal or individual factors that may contribute to human errors occurring at work.

17. (a) **Explain** the meaning of the term 'human error'.

(b) **Describe** the difference between human errors and violations.

(c) **Describe**, using practical examples, **THREE** types of human error and **THREE** types of violation that can lead to accidents in the workplace.

18. **Describe TWO** internal and **TWO** external influences on the health and safety culture of an organisation.

APPENDIX 3.1 Safety culture questionnaire

The HSE have suggested that the following questionnaire can be used to inspect the safety culture of an organisation.

1. Management commitment

Where is safety perceived to be in management's priorities (senior/middle/first line)?

How do they show this?

How often are they seen in the workplace?

Do they talk about safety when in the workplace and is this visible to the workforce?

Do they 'walk the talk'?

Do they deal quickly and effectively with safety issues raised?

What balance do their actions show between safety and production?

Are management trusted over safety?

2. Communication

Is there effective two-way communication about safety?

How often are safety issues discussed:
- ▶ with line manager/subordinate?
- ▶ with colleagues?

What is communicated about the safety programme of the company?

How open are people about safety?

3. Employee involvement

How are people (all levels, especially operators) involved in safety?

How often are individual employees asked for their input on safety issues?

How often do operators report unsafe conditions or near misses, etc.?

Is there active, structured operator involvement, e.g. workshops, projects, safety circles?

Is there a continuous improvement/total quality approach?

Whose responsibility is safety regarded to be?

Is there genuine cooperation over safety – a joint effort between all in the company?

4. Training/information

Do employees feel confident that they have all the training that they need

How accurate are employees' perceptions of hazards and risks?

How effective is safety training in meeting needs (including managers'!)?

How are needs identified?

How easily available is safety information?

5. Motivation

Do managers give feedback on safety performance (and how)?

Are they likely to notice unsafe acts?

Do managers (all levels – senior/middle/first line) always confront unsafe acts?

How do they deal with them?

Do employees feel they can report unsafe acts?

How is discipline applied to safety?

What do people believe are the expectations of managers?

Do people feel that this is a good place to work (why/why not)?

Are they proud of their company?

6. Compliance with procedures

What are written procedures used for?

What decides whether a particular task will be captured in a written procedure?

Are they read? Are they helpful?

What other rules are there?

Are there too many procedures and rules?

How well are people trained in them?

Are they audited effectively?

Are they written by users?

Are they linked to risks?

7. Learning organisation

Does the company really learn from accident history, incident reporting, etc.?

Do employees feel confident in reporting incidents or unsafe conditions?

Do they report them?

Do reports get acted upon?

Do they get feedback?

APPENDIX 3.2 List of typical legislation requiring health and safety training

	Legislation
1.	Management of Health and Safety at Work Regulations
2.	Regulatory Reform (Fire Safety) Order 2005
3.	Provision and Use of Work Equipment Regulations
4.	Control of Substances Hazardous to Health Regulations
5.	Health and Safety (First-aid) Regulations
6.	Health and Safety (Display Screen Equipment) Regulations
7.	Manual Handling Operations Regulations
8.	Control of Noise at Work Regulations
9.	Personal Protective Equipment at Work Regulations
10.	Health and Safety (Safety Signs and Signals) Regulations
11.	Control of Asbestos Regulations
12.	Confined Spaces Regulations
13.	Health and Safety (Consultation with Employees) Regulations
14.	Control of Vibration at Work Regulations

3

Health and safety management of risk and controls

Chapter summary

This chapter covers the following topics in Element 3 of the NEBOSH NGC syllabus:

Assessing risk

Management of change

Safe systems of work for general work activities

Permit-to-work systems

Emergency procedures

Learning outcomes

These are the NEBOSH learning outcomes covered by this chapter:

The learner will be able to:

1. Do a general risk assessment in their own workplace – profiling and prioritising risks, inspecting the workplace, recognising a range of common hazards, evaluating risks (taking account of current controls), recommending further control measures, planning actions

2. Recognise workplace changes that have significant health and safety impacts and effective ways to minimise those impacts

3. Develop basic safe systems of work (including taking account of typical emergencies) and knowing when to use permit-to-work systems for special risks

Assessment criteria

3.4 Explain the principles of the risk assessment process

and

5–11 Produce a risk assessment of a workplace which considers a wide range of identified hazards (drawn from elements 5–11) and meets best practice standards ('suitable and sufficient')

3.5 Discuss typical workplace changes that have significant health and safety impacts and ways to minimise those impacts

3.6 Describe what to consider when developing and implementing a safe system of work for general activities

3.7 Explain the role, function and operation of a permit-to-work system

3.8 Discuss typical emergency procedures (including training and testing) and how to decide what level of first aid is needed in the workplace

Introduction

Risk assessment and minimum levels of risk prevention or control are regulated by health and safety legal requirements. Risk assessment is an essential part of the planning stage of any health and safety management system. Risk assessment methods are used to decide on priorities and to set objectives for eliminating hazards and reducing risks. Wherever possible, risks should be eliminated through the selection and design of facilities, equipment and processes. If risks cannot be eliminated, they should be minimised by the use of physical controls or, as a last resort, through systems of work and personal protective equipment. The control of risks is essential to secure and maintain a healthy and safe workplace, which complies with the relevant legal requirements. In this chapter, hazard identification and risk assessment are covered together with appropriate risk control measures. A hierarchy of control methods is discussed that gives a preferred order of approach to risk control.

This chapter also concerns the principles of prevention that should be adopted when deciding on suitable measures to eliminate or control both acute and chronic risks to the health and safety of people at work. The principles of control can be applied to both health

risks and safety risks although health risks have some distinctive features that require a special approach. The chapter concludes with emergency procedures and the treatment of first aid.

4.1 Assessing risk

4.1.1 Introduction

There are several descriptions of the meaning of 'risk assessment'. NEBOSH defines risk assessment as:

'the identification of preventative and protective measures by the evaluation of the risk(s) arising from a hazard(s), taking into account the adequacy of any existing controls, and deciding whether or not the risk(s) is acceptable.'

4.1.2 Legal aspects of risk assessment

The general duties of employers to their employees in Section 2 of the HSW Act 1974 imply the need for risk assessment. This duty was also extended by Section 3 of the Act to anybody else affected by activities of the employer – contractors, visitors, customers or members of the public. However, the Management of Health and Safety at Work Regulations are much more specific concerning the need for risk assessment. The following requirements are laid down in those Regulations:

▶ the risk assessment shall be 'suitable and sufficient' and cover both employees and non-employees affected by the employer's undertaking (e.g. contractors, members of the public, students, patients, customers); every self-employed person shall make a 'suitable and sufficient' assessment of the risks to which they or those affected by the undertaking may be exposed;

▶ any risk assessment shall be reviewed if there is reason to suspect that it is no longer valid or if a significant change has taken place;

▶ where there are five or more employees, the significant findings of the assessment shall be recorded and any specially at-risk group of employees identified. (This does not mean that employers with four or less employees need not undertake risk assessments.)

The term 'suitable and sufficient' is important as it defines the limits to the risk assessment process. A suitable and sufficient risk assessment should:

▶ identify the significant risks and ignore the trivial ones;

▶ identify and prioritise the measures required to comply with any relevant statutory provisions;

▶ remain appropriate to the nature of the work and valid over a reasonable period of time;

▶ identify the risk arising from or in connection with the work. The level of detail should be proportionate to the risk.

The significant findings that should be recorded include a detailed statement of the hazards and risks, the preventative, protective or control measures in place and any further measures required to reduce the risks present.

When assessing risks under the Management of Health and Safety at Work Regulations, reference to other Regulations may be necessary even if there is no specific requirement for a risk assessment in those Regulations. For example, reference to the legal requirements of the Provision and Use of Work Equipment Regulations will be necessary when risks from the operation of machinery are being considered. However, there is no need to repeat a risk assessment if it is already covered by other Regulations (e.g. a risk assessment considering personal protective equipment is required under the COSHH Regulations so there is no need to undertake a separate risk assessment under the Personal Protective Equipment Regulations).

Apart from the duty under the Management of Health and Safety at Work Regulations to undertake a health and safety assessment of the risks to any person (employees, contractors or members of the public) who may be affected by the activities of the organisation, the following Regulations require a specific risk assessment to be made:

▶ Ionising Radiation Regulations;

▶ Control of Asbestos Regulations;

Figure 4.1 Reducing the risk – finding an alternative to fitting a wall-mounted boiler

▶ Control of Noise at Work Regulations;

▶ Manual Handling Operations Regulations;

▶ Health and Safety (Display Screen Equipment) Regulations;

▶ Personal Protective Equipment at Work Regulations;

▶ Confined Spaces Regulations;

▶ Work at Height Regulations;

▶ Regulatory Reform (Fire Safety) Order (not under HSW Act);

▶ Control of Vibration at Work Regulations;

▶ Control of Lead at Work Regulations;

▶ Control of Substances Hazardous to Health Regulations.

See www.hse.gov.uk/risk/, the HSE's risk management site.

4.1.3 Forms of risk assessment

There are two basic forms of risk assessment.

A **quantitative** risk assessment attempts to measure the risk by relating the probability of the risk occurring to the possible severity of the outcome and then giving the risk a numerical value. This method of risk assessment is used in situations where a malfunction could be very serious (e.g. aircraft design and maintenance or the petrochemical industry).

The more common form of risk assessment is the **qualitative** assessment, which is based purely on personal judgement and is normally defined as high, medium or low. Qualitative risk assessments are usually satisfactory as the definition (high, medium or low) is normally used to determine the time frame over which further action is to be taken.

The term '**generic**' risk assessment is sometimes used, and describes a risk assessment which covers similar activities or work equipment in different departments, sites or companies. Such assessments are often produced by specialist bodies, such as trade associations. If used, they must be appropriate to the particular job and they will need to be extended to cover additional hazards or risks.

4.1.4 **Some definitions**

Some basic definitions were introduced in Chapter 1 and those relevant to risk assessment are reproduced here.

Hazard and risk

A hazard is something with the **potential** to cause harm (this can include articles, substances, plant or machines, methods of working, the working environment and other aspects of work organisation). Hazards take many forms including, for example, chemicals, electricity or the use of a ladder. A hazard can be ranked relative to other hazards or to a possible level of danger.

A risk is the **likelihood** of potential harm from that hazard being realised. Risk (or strictly the level of risk) is also linked to the severity of its consequences. A risk can be reduced and the hazard controlled by good management.

It is very important to distinguish between a hazard and a risk – the two terms are often confused and activities often called high risk are in fact high hazard. There should only be high residual risk where there is poor health and safety management and inadequate control measures.

Electricity is an example of a high hazard as it has the potential to kill a person. The risk associated with electricity – the likelihood of being killed on coming into contact with an electrical device – is, hopefully, low.

A **risk profile** is an analysis of the types of threats an organisation faces by assigning numerical values to the different types of threats and the danger that they pose. It informs all aspects of the approach to leading and managing its health and safety risks.

A risk profile examines:

▶ the nature and level of the threats faced by an organisation;
▶ the likelihood of adverse effects occurring;
▶ the level of disruption and costs associated with each type of *risk;* and
▶ the effectiveness of the control measures in place.

The risks may be immediate *safety* or health-related hazards (or both). The outcome of risk profiling will be that the right risks have been identified and prioritised for action, and minor risks will not have been given too much priority. It also informs decisions about the risk control measures that are needed.

Occupational or work-related ill health

This is concerned with those acute and chronic illnesses or physical and mental disorders that are either caused or triggered by workplace activities. Such conditions may be induced by the particular work activity of the individual or by activities of others in the workplace. The time interval between exposure and the onset of the illness may be short (e.g. acute asthma attacks) or long (e.g. chronic deafness or cancer).

Accident

This is defined by the Health and Safety Executive (HSE) as *'any unplanned event that results in injury or ill health of people, or damage or loss to property, plant, materials or the environment or a loss of a business opportunity'*. Other authorities define an accident more narrowly by excluding events that do not involve injury or ill health.

Figure 4.2 Accident at work

Figure 4.3 Bird's well-known accident triangle

Incident and near miss

HSE states that an *'Incident includes all undesired circumstances and "near misses" which could cause accidents'*. Knowledge of near misses is very important as research has shown that, approximately, for every 10 'near miss' events at a particular location in the workplace, a minor accident will occur.

Dangerous occurrence

This is a 'near miss' or 'damage incident' which could have led to serious injury or loss of life. Dangerous occurrences are defined in the Reporting of Injuries, Diseases and Dangerous Occurrences Regulations (often known as RIDDOR) and are always reportable to the enforcement authorities. Examples include the collapse of a scaffold or a crane or the failure of any passenger-carrying equipment.

In 1969, F. E. Bird collected a large quantity of accident data and produced a well-known triangle (Figure 4.3). It can be seen that damage and near miss accidents occur much more frequently than injury accidents and are, therefore, a good indicator of risks. The study also shows that most accidents are predictable and avoidable.

4.1.5 The objectives of risk assessment

The *main* objective of risk assessment is to determine the measures required by the organisation to comply with relevant health and safety legislation and, thereby, reduce the level of occupational injuries and ill health. The purpose is to help the employer or self-employed person to determine the measures required to comply with their legal statutory duty under the HSW Act 1974 or its associated Regulations. The risk assessment will need to cover all those who may be at risk, such as customers, contractors and members of the public. In the case of shared workplaces, an overall risk assessment may be needed in partnership with other employers.

In Chapter 1, the moral, legal and financial arguments for health and safety management were discussed in detail. The important distinction between the direct and indirect costs of accidents is reiterated here.

Any accident or incidence of ill health will cause both direct and indirect costs and incur an insured and an uninsured cost. It is important that all of these costs are taken into account when the full cost of an accident is calculated. In a study undertaken by the HSE, it was shown that indirect or hidden costs could be 36 times greater than direct costs of an accident. In other words, the direct costs of an accident or disease represent the tip of the iceberg when compared with the overall costs.

Direct costs are costs that are directly related to the accident. They may be insured (claims on employers'

and public liability insurance, damage to buildings, equipment or vehicles) or uninsured (fines, sick pay, damage to product, equipment or process).

Indirect costs may be insured (business loss, product or process liability) or uninsured (loss of goodwill, extra overtime payments, accident investigation time, production delays).

There are many reasons for the seriousness of a hazard not to be obvious to the person exposed to it. It may be that the hazard is not visible (radiation, certain gases and biological agents) or have no short-term effect (work-related upper limb disorders). Some common causes of accidents include lack of attention, lack of experience, not wearing appropriate PPE, sensory impairment and inadequate information, instruction and training.

4.1.6 Accident categories

There are several categories of accident, all of which will be dealt with in more detail in later chapters. The principal categories are as follows:

▶ contact with moving machinery or material being machined;
▶ struck by moving, flying or falling object;
▶ hit by a moving vehicle;
▶ struck against something fixed or stationary;
▶ injured while handling, lifting or carrying;
▶ slips, trips and falls on the same level;
▶ falls from a height;
▶ trapped by something collapsing;
▶ drowned or asphyxiated;
▶ exposed to, or in contact with, a harmful substance;
▶ exposed to fire;
▶ exposed to an explosion;
▶ contact with electricity or an electrical discharge;
▶ injured by an animal;
▶ physically assaulted by a person;
▶ other kind of accident.

4.1.7 Health risks

Risk assessment is not only concerned with injuries in the workplace but also needs to consider the possibility of occupational ill health. Health risks fall into the following four categories:

1. chemical (e.g. paint solvents, exhaust fumes);
2. biological (e.g. bacteria, pathogens);
3. physical (e.g. noise, vibrations);
4. psychological (e.g. occupational stress).

There are two possible health effects of occupational ill health.

They may be **acute**, which means that they occur soon after the exposure and are often of short duration, although in some cases emergency admission to hospital may be required.

They may be **chronic**, which means that the health effects develop with time. It may take several years for the associated disease to develop and the effects may be slight (mild asthma) or severe (cancer).

Health risks are discussed in more detail in Chapter 6.

4.1.8 The management of risk assessment

Risk assessment is part of the planning and implementation stage of the health and safety management system recommended by the HSE in its publication HSG65. All aspects of the organisation, including health and safety management, need to be covered by the risk assessment process. This will involve the assessment of risk in areas such as maintenance procedures, training programmes and supervisory arrangements. A general risk assessment of the organisation should reveal the significant hazards present and the general control measures that are in place. Such a risk assessment should be completed first and then followed by more specific risk assessments that examine individual work activities.

Risk assessors

It is important that the risk assessment team is selected on the basis of its competence to assess risks in the particular areas under examination in the organisation. The Team Leader or Manager should have health and safety experience and relevant training in risk assessment. It is sensible to involve the appropriate line manager, who has responsibility for the area or activity being assessed, as a team member. Other members of the team will be selected on the basis of their experience, their technical and/or design knowledge and any relevant standards or Regulations relating to the activity or process. At least one team member must have communication and report writing skills. A positive attitude and commitment to the risk assessment task are also important factors. It is likely that team members will require some basic training in risk assessment.

4.1.9 The practice of risk assessment

HSE has produced a free leaflet entitled 'Risk assessment – A brief guide to controlling risks in the workplace', INDG163. It gives practical advice on assessing risks and recording the findings, and is aimed at small and medium-sized companies in the service and manufacturing sectors. There are five steps in risk assessment (Figure 4.4). These are:

1. look for the hazards;
2. decide who might be harmed, and how;
3. evaluate the risks and decide whether existing precautions are adequate or more should be done;
4. record the significant findings;
5. review the assessment and revise it if necessary.

Each of these steps will be examined in turn in the following sections

Figure 4.4 Five steps to risk assessment

Step 1 – Look for the hazards

The essential first step in risk assessment is to seek out and identify hazards. Relevant sources of information include:

- legislation and supporting Approved Codes of Practice which give practical guidance and include basic minimum requirements;
- process information;
- product information provided under Section 6 of the HSW Act;
- relevant British, European and international standards;
- industry or trade association guidance;
- the personal knowledge and experience of managers and employees;
- accident, ill health and incident data from within the organisation, from other organisations or from central sources; and
- expert advice and opinion and relevant research.

Other possible sources health and safety information include:

- absentee records;
- inspection and audit reports undertaken by the organisation and by external organisations such as the HSE;
- maintenance, risk assessment (including COSHH) and training records;
- any equipment examination or test reports;
- information and data from manufacturers and suppliers;
- the internet and encyclopaedias; and
- health and safety magazines and journals and specialist technical and legal publications.

See Chapter 17 for more information and website addresses. Many of these sources of information will be referred to throughout this book.

Figure 4.5 Checking the label for health risks

There should be a critical appraisal of all routine and non-routine business activities. People exposed may include not just employees, but also others such as members of the public, contractors and users of the products and services. Employees and safety representatives can make a useful contribution in identifying hazards.

In the simplest cases, hazards can be identified by observation and by comparing the circumstances with the relevant information (e.g. single-storey premises will not present any hazards associated with stairs). In more complex cases, measurements such as air sampling or examining the methods of machine operation may be necessary to identify the presence of hazards from chemicals or machinery. In the most complex or high-risk cases (for example, in the chemical or nuclear industry) special techniques and systems may be needed such as hazard and operability studies (HAZOPS) and hazard analysis techniques such as event or fault tree analysis. Specialist advice may be needed to choose and apply the most appropriate method.

Only significant hazards, which could result in serious harm to people, should be identified. Trivial hazards are a lower priority.

A tour of the area under consideration by the risk assessment team is an essential part of hazard identification as is consultation with the relevant section of the workforce.

A review of accident, incident and ill-health records will also help with the identification. Other sources of information include safety inspection, survey and audit reports, job or task analysis reports, manufacturers' handbooks or data sheets and Approved Codes of Practice and other forms of guidance.

Hazards will vary from workplace to workplace but the checklist in Appendix 4.1 shows the common hazards that are significant in many workplaces. Many questions in the NEBOSH examinations involve several common hazards found in most workplaces.

It is important during hazard identification that unsafe conditions are not confused with hazards. Unsafe conditions should be rectified as soon as possible after observation. Examples of unsafe conditions include missing machine guards, faulty warning systems and oil spillage on the workplace floor.

Step 2 – Decide who might be harmed and how

Employees and contractors who work full time at the workplace are the most obvious groups at risk and it will be a necessary check that they are competent to perform their particular tasks. However, there may be other groups who spend time in or around the workplace. These include young workers, trainees, new and expectant mothers, cleaners, contractor and maintenance workers and members of the public. Members of the public will include visitors, patients, students or customers as well as passers-by.

The risk assessment must include any additional controls required due to the vulnerability of any of these groups, perhaps caused by inexperience or disability. It must also give an indication of the numbers of people from the different groups who come into contact with the hazard and the frequency of these contacts.

Step 3 – Evaluating the risks and the adequacy of current controls

This step is really two steps – evaluating the risks and evaluating the adequacy of current controls.

Evaluating the risks

During most risk assessments it will be noted that some of the risks posed by the hazard have already been addressed or controlled. The purpose of the risk assessment, therefore, is to reduce the remaining risk. This is called the **residual risk**.

The goal of risk assessment is to reduce all residual risks to as low a level as reasonably practicable. In a relatively complex workplace, this will take time so that **a system of ranking risk** is required – the higher the risk level the sooner it must be addressed and controlled.

For most situations, a **qualitative** risk assessment will be perfectly adequate. (This is certainly the case for NEBOSH Certificate candidates and is suitable for use during the risk assessment.) During the risk assessment, a judgement is made as to whether the risk level is high, medium or low in terms of the risk of somebody being injured. This designation defines a timetable for remedial actions to be taken thereby reducing the risk. High-risk activities should normally be addressed in days, medium risks in weeks and low risks in months or in some cases no action will be required. It will usually be necessary for risk assessors to receive some training in risk level designation.

A **quantitative** risk assessment attempts to quantify the risk level in terms of the likelihood of an incident and its subsequent severity. Clearly the higher the likelihood and severity, the higher the risk will be. The likelihood depends on such factors as the control measures in place, the frequency of exposure to the hazard and the category of person exposed to the hazard. The severity will depend on the magnitude of the hazard (voltage, toxicity, etc.).

Likelihood of occurrence	Likelihood level
Harm is certain or near certain to occur	High 3
Harm will often occur	Medium 2
Harm will seldom occur	Low 1
Severity of harm	**Severity level**
Death or major injury (as defined by RIDDOR)	Major 3
7-day injury or illness (as defined by RIDDOR)	Serious 2
All other injuries or illnesses	Slight 1
Risk = Severity × Likelihood	

The HSE has suggested a simple 3 x 3 matrix to determine risk levels.

Likelihood	Severity		
	Slight 1	Serious 2	Major 3
Low 1	Low 1	Low 2	Medium 3
Medium 2	Low 2	Medium 4	High 6
High 3	Medium 3	High 6	High 9

Thus: 6–9 High Risk, 3–4 Medium Risk, 1–2 Low Risk

It is possible to apply such methods to organisational risk or to the risk that the management system for health and safety will not deliver in the way in which it was expected or required. Such risks will add to the activity or occupational risk level. In simple terms, poor supervision of an activity will increase its overall level of risk. A risk management matrix has been developed which combines these two risk levels, as shown here.

RISK MANAGEMENT MATRIX		Occupational risk level		
		Low	Medium	High
Organisational risk level	Low	L	L	M
	Medium	L	M	H
	High	M	H	Unsatisfactory

L Low Risk, **M** Medium Risk, **H** High Risk

Whichever type of risk evaluation method is used, the level of risk simply enables a timetable of risk reduction to an acceptable and tolerable level to be formulated. The legal duty requires that all risks should be reduced to as low as is reasonably practicable.

In established workplaces, some control of risk will be in place already. The effectiveness of these controls needs to be assessed so that an estimate of the residual risk may be made. Many hazards have had specific Acts, Regulations or other recognised standards developed to reduce associated risks. Examples of such hazards are fire, electricity, lead and asbestos. The relevant legislation and any accompanying Approved Codes of Practice or guidance should be consulted first and any recommendations implemented. Advice on control measures may also be available from trade associations, trade unions or employers' organisations.

Where there are existing preventative measures in place, it is important to check that they are working properly and that everybody affected has a clear

understanding of the measures. It may be necessary to strengthen existing procedures, for example by the introduction of a permit-to-work system. More details on the principles of control below.

Principles to consider when controlling risk

The control of risks is essential to secure and maintain a healthy and safe workplace which complies with the relevant legal requirements. Hazard identification and risk assessment are covered earlier in this chapter and these together with appropriate risk control measures form the core of the HSG65 'PLAN' section of the management model.

In industry today safety is controlled through a combination of **engineered measures** such as the provision of safety protection (e.g. guarding and warning systems), and **operational measures** in training, safe work practices, operating procedures and method statements, along with **management supervision**.

These measures (collectively) are commonly known in health and safety terms as **control measures**. Some of these more common measures will be explained in more detail later.

This section concerns the principles that should be adopted when deciding on suitable measures to eliminate or control both acute and chronic risks to the health and safety of people at work (Figure 4.6). The principles of control can be applied to both health risks and safety risks, although health risks have some distinctive features that require a special approach.

Figure 4.6 When controls breakdown

Chapters 6–12 deal with specific workplace hazards and controls, subject by subject. The principles of prevention now enshrined in the Management of Health and Safety at Work (MHSW) Regulations need to be used jointly with the hierarchy of control methods which give the preferred order of approach to risk control.

When risks have been analysed and assessed, decisions can be made about workplace precautions.

All final decisions about risk control methods must take into account the relevant legal requirements, which establish minimum levels of risk prevention or control.

Some of the duties imposed by the HSW Act and the relevant statutory provisions are **absolute** and must be complied with. Many requirements are, however, qualified by the words **so far as is reasonably practicable**, or **so far as is practicable**. These require an assessment of cost, along with information about relative costs, effectiveness and reliability of different control measures. Further guidance on the meaning of these three expressions is provided in Chapter 1.

General principles of prevention

The MHSW Regulations Schedule 1 specifies the general principles of prevention which are set out in Article 6(2) of the European Council Directive 89/391/EEC. For the first time, the principles have been enshrined directly in Regulations which state, at Regulation 4, that *'Where an employer implements any preventative measures he shall do so on the basis of the principles specified in Schedule 1'*. These principles are:

1. Avoiding risks

This means, for example, trying to stop doing the task or using different processes or doing the work in a different, safer way.

2. Evaluating the risks which cannot be avoided

This requires a risk assessment to be carried out.

3. Combating the risks at source

This means that risks, such as a dusty work atmosphere, are controlled by removing the cause of the dust rather than providing special protection; or that slippery floors are treated or replaced rather than putting up a sign.

4. Adapting the work to the individual

This involves the design of the workplace, the choice of work equipment and the choice of working and production methods, with a view, in particular, to alleviating monotonous work and work at a predetermined work rate and to reducing their effect on health.

This will involve consulting those who will be affected when workplaces, methods of work and safety procedures are designed. The control individuals have over their work should be increased, and time spent working at predetermined speeds and in monotonous work should be reduced where it is reasonable to do so.

5. Adapting to technical progress

It is important to take advantage of technological and technical progress, which often gives designers and employers the chance to improve both safety and working methods. With the internet and other international information sources available, very wide knowledge, going beyond what is happening in the UK or Europe, will be expected by the enforcing authorities and the courts.

6. Replacing the dangerous by the non-dangerous or the less dangerous

This involves substituting, for example, equipment or substances with non-hazardous or less hazardous substances.

7. Developing a coherent overall prevention policy

This covers technology, organisation of work, working conditions, social relationships and the influence of factors relating to the working environment. Health and safety policies should be prepared and applied by reference to these principles.

8. Giving collective protective measures priority over individual protective measures

This means giving priority to control measures which make the workplace safe for everyone working there so giving the greatest benefit, for example removing hazardous dust by exhaust ventilation rather than providing a filtering respirator to an individual worker. This is sometimes known as a 'Safe Place' approach to controlling risks.

9. Giving appropriate instruction to employees

This involves making sure that employees are fully aware of company policy, safety procedures, good practice, official guidance, any test results and legal requirements. This is sometimes known as a 'Safe Person' approach to controlling risks where the focus is on individuals. A properly set-up health and safety management system should cover and balance both a Safe Place and Safe Person approach.

Evaluating the controls

1. Hierarchy of risk control

When assessing the adequacy of existing controls or introducing new controls, a hierarchy of risk controls should be considered. The health and safety management system ISO 45001 (para. 8.1.2) states that the organisation shall establish a process for achieving risk reduction based upon the following hierarchy:

(a) eliminate the hazard;
(b) substitute with less hazardous materials, processes, operations or equipment;
(c) use engineering controls;
(d) use safety signs, markings and warning devices and administrative controls;
(e) use personal protective equipment.

The organisation shall ensure that the Occupational Health and Safety risks and determined controls are taken into account when establishing, implementing and maintaining its Occupational Health and Safety management system.

The hierarchy reflects that risk elimination and risk control by the use of physical engineering controls and safeguards can be more reliably maintained than those which rely solely on people. These concepts are now written into the Control of Substances Hazardous to Health (COSHH) Regulations and the Management of Health and Safety at Work (MHSW) Regulations.

Where a range of control measures are available, it will be necessary to weigh up the relative costs of each against the degree of control each provides, both in the short and long term. Some control measures, such as eliminating a risk by choosing a safer alternative substance or machine, provide a high degree of control and are reliable. Physical safeguards such as guarding a machine or enclosing a hazardous process need to be maintained. In making decisions about risk control, it will therefore be necessary to consider the degree of control and the reliability of the control measures along with the costs of both providing and maintaining the measure. The ISO 45001 hierarchy will now be discussed in detail.

(a) Elimination of the hazard

The best and most effective way of reducing risks is by avoiding a hazard and its associated risks. For example avoid working at height by installing a permanent working platform with stair access; avoid entry into a confined space by, for example, using a sump pump in a pit which is removed by a lanyard for maintenance; eliminate the fire risks from tar boilers by using bitumen which can be applied cold.

(b) Substitution

Substitution describes the use of a less hazardous form of a substance or process. There are many examples of substitution such as the use of water-based rather than oil-based paints; the use of asbestos substitutes and the use of compressed air as a power source rather than electricity to reduce both electrical and fire risks; and the use of mechanical excavators instead of hand digging.

In some cases it is possible to change the method of working so that risks are reduced. For example use rods to clear drains instead of strong chemicals; use a long-handled water hose brush to clean windows instead of ladders. Sometimes the pattern of work can be changed so that people can do things in a more natural way, for example when placing components for packing consider whether people are right- or left-handed; encourage people in offices to take breaks from computer screens by getting up to photocopy, fetch files or print documents.

Care must be taken to consider any additional hazards which may be involved and thereby introduce additional risks, as a result of a substitution.

(c) Engineering controls

This describes the control of risks by means of engineering design rather than a reliance on preventative actions by the employee. There are several ways of achieving such controls:

1. Control the risks at the source (e.g. the use of more efficient dust filters or the purchase of less noisy equipment).
2. Control the risk of exposure by:
   - ▶ **isolating** the equipment by the use of an enclosure, a barrier or guard;
   - ▶ **insulating** any electrical or temperature hazard;
   - ▶ **ventilating** away any hazardous fumes or gases either naturally or by the use of extractor fans and hoods (Figure 4.7).

Figure 4.7 Proper control of gases and vapours in a laboratory

(d) Safety signs, markings and warning devices and administrative controls

Safety signs, markings and warning devices

A summary of the Safety Signs and Signals Regulations is given in Chapter 13. All general health and safety signs used in the workplace must include a pictorial symbol categorised by shape, colour and graphic image (Figure 4.8).

Prohibition
A red circular band with diagonal crossbar on a white background, the symbol within the circle to be black denoting a safety sign that indicates that a certain behaviour is prohibited.

Warning
A yellow triangle with black border and symbol within the yellow area denoting a safety sign that gives warning of a hazard.

Mandatory
A blue circle with white symbol denoting a sign that indicates that a specific course of action must be taken.

Safe condition
A green oblong or square with symbol or text in white denoting a safety sign providing information about safe conditions.

Fire equipment
A red oblong or square with symbol in white denoting a safety sign that indicates the location of fire-fighting equipment.

Figure 4.8 Colour categories and shapes of signs

Overhead load

Safety helmet must be worn

Safety harness must be worn

Not drinkable

No access for pedestrians

Figure 4.9 Examples of warning, mandatory and prohibition signs

All workplaces need to display safety signs of some kind, but deciding what is required can be confusing. Presented here are the basic requirements for the majority of small premises or sites, such as small construction sites, canteens, shops, small workshop units and offices. This section does not cover any signs which food hygiene law may require.

Most requirements are covered by the Health and Safety (Safety Signs and Signals) Regulations. These require signs wherever a risk has not been controlled by other means. For example if a wet area of floor is cordoned off, a warning sign will not be needed, because the barrier will keep people out of the danger area. Signs are not needed where the sign would not reduce the risk or the risk is insignificant.

The following signs are typical of some of the ones most likely to be needed in these premises. Others may be necessary, depending on the hazards and risks present.

(i) Overhead obstacles, construction site and prohibition notices (Figures 4.9 and 4.10)

Figure 4.10 Falling object and construction site entrance signs

(ii) Wet floors

These need to be used wherever a slippery area is not cordoned off. Lightweight stands holding double-sided signs are readily available (Figure 4.11).

(iii) Chemical storage

Where hazardous cleaning chemicals are stored, apart from keeping the store locked, a suitable warning notice should be posted if it is considered this would help to reduce the risk of injury (Figure 4.12).

(iv) Fire safety signs

The Regulations apply in relation to general fire precautions. The guidance under the Fire Safety Order

Figure 4.11 Wet floor signs

Toxic

Corrosive

Figure 4.12 Examples of chemical warning signs

requires signs to comply with BS5499-4 and 5 and the Safety Signs and Signals Regulations.

Since 1998 the older, text-only 'fire exit' signs should have been supplemented or replaced with pictogram signs. Fire safety signs complying with BS5499-4 and 5 already contain a pictogram and do not require changing (Figure 4.13).

Fire exit

Figure 4.13 Examples of fire safety signs

(v) Fire action signs

These and other fire safety signs, such as fire extinguisher location signs, will be needed (Figure 4.14).

Hose reel   Fire extinguisher   Fire alarm call point

Figure 4.14 Examples of fire action signs

(vi) First aid

Signs showing the location of first-aid facilities will be needed. Advice on the action to take in the case of electric shock is no longer a legal requirement but is recommended (see Figure 12.2).

First-aid post   First-aid stretcher   Eyewash

Figure 4.15 Examples of first-aid signs

(vii) Gas pipes and LPG cylinder stores

Liquefied petroleum gas (LPG) cylinder stores should have the sign shown in Figure 4.16.

Figure 4.16 LPG sign

(viii) No smoking

Areas substantially enclosed should have the sign shown in Figure 4.17 under smoke-free legislation. See Chapters 11 and 13 for more details.

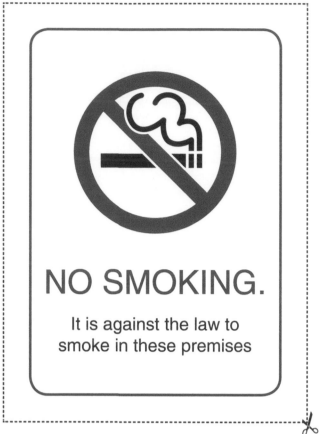

Figure 4.17 Smoke-free – no smoking sign

(ix) Fragile roofs

Signs should be erected at roof access points and at the top of outside walls where ladders may be placed (Figure 4.18).

Figure 4.18 Fragile roof signs

(x) Obstacles or dangerous locations

Signs that warn of obstacles or dangerous locations, such as low head height or slippery floors, should have alternating yellow and black stripes.

(xi) Other signs and posters

▶ Health and safety law – What you should know (there is a legal requirement to display this poster or distribute equivalent leaflet).

▶ Certificate of Employer's Liability Insurance (there is a legal requirement to display this).

▶ Scalds and burns are common in kitchens. A poster showing recommended action is advisable, for example 'First-aid for Burns'.

(xii) Sign checklist

Existing signs should be checked to ensure that:

▶ they are correct and up to date;

▶ they carry the correct warning symbol where appropriate;

▶ they are relevant to the hazard;

▶ they are easily understood;

▶ they are suitably located and not obscured;

▶ they are clean, durable and weatherproof where necessary;

▶ illuminated signs have regular lamp checks;

▶ they are used when required (e.g. 'Caution wet floor' signs);

▶ they are obeyed and effective.

Administrative controls

Reduced time exposure

This involves reducing the time during the working day that the employee is exposed to the hazard, by giving the employee either other work or rest periods. It is normally only suitable for the control of health hazards associated with, for example, noise, vibration, excessive heat or cold, display screens and hazardous substances. However, it is important to note that for many hazards there are short-term exposure limits as well as normal occupational exposure limits (OELs) over an 8-hour period (see Chapter 8). Short-term limits must not be exceeded during the reduced time exposure intervals.

It cannot be argued that a short time of exposure to a dangerous part of a machine is acceptable. However, it is possible to consider short bouts of intensive work with rest periods when employees are engaged in heavy labour, such as manual digging when machines are not permitted due to the confines of the space or buried services.

Isolation/segregation

Controlling risks by isolating them or segregating people and the hazard is an effective control measure and is used in many instances; for example separating vehicles and pedestrians on factory sites, providing separate walkways for the public on road repairs, providing warm rooms on sites or noise refuges in noisy processes.

The principle of isolation is usually followed with the storage of highly flammable liquids or gases which are put into open, air-ventilated compounds away from other hazards such as sources of ignition, or away from people who may be at risk from fire or explosion.

Safe systems of work

Operating procedures or safe systems of work are probably the most common form of control measure used in industry today and may be the most economical and, in some cases, the only practical way of managing a particular risk. They should allow for methodical execution of tasks. The development of safe operating procedures should address the hazards that have been identified in the risk assessment. The system of work describes the safe method of performing the job or activity. A safe system of work is a requirement of the Health and Safety at Work Act and is dealt with in detail later.

If the risks involved in the task are high or medium, the details of the system should be in writing and should be communicated to the employee formally in a training session. Details of systems for low-risk activities may be conveyed verbally. There should be records that the employee (or contractor) has been trained or instructed in the safe system of work and that they understand it and will abide by it.

Training

Training helps people acquire the skills, knowledge and attitudes to make them competent in the health and safety aspects of their work. There are generally two types of safety training:

▶ Specific safety training (or on-the-job training) which aims at tasks where training is needed due to the specific nature of such tasks. This is usually a job for supervisors, who by virtue of their authority and close daily contact, are in a position to convert safety generalities to the everyday safe practice procedures that apply to individual tasks, machines, tools and processes;

▶ Planned training, such as general safety training, induction training, management training, skill training or refresher courses that are planned by the organisation, and relate to managing risk through policy, legislative or organisational requirements that are common to all employees.

Before any employee can work safely, they must be shown safe procedures for completing their tasks. The purpose of safety training should be to improve the safety awareness of employees and show them how to perform their jobs while employing acceptable, safe behaviour.

See Chapter 3 for more detail on health and safety training.

Information

Organisations need to ensure that they have effective arrangements for identifying and receiving relevant health and safety information from outside the organisation including:

▶ ensuring that pertinent health and safety information is communicated to all people in the organisation who need it;

▶ ensuring that relevant information is communicated to people outside the organisation who require it;

▶ encouraging feedback and suggestions from employees on health and safety matters.

Anyone who is affected by what is happening in the workplace will need to be given safety information. This does not only apply to staff. It can also apply to visitors, members of the public and contractors.

Information to be provided for people in a workplace includes:

► who is at risk and why;
► how to carry out specific tasks safely;
► correct operation of equipment;
► emergency action;
► accident and hazard reporting procedures;
► the safety responsibilities of individual people.

Information can be provided in a variety of ways. These include safety signs, posters, newsletters, memos, emails, personal briefings, meetings, toolbox talks, formal training, written safe systems of work and written health and safety arrangements.

Welfare

Welfare facilities include general workplace ventilation, lighting and heating and the provision of drinking water, sanitation and washing facilities. There is also a requirement to provide eating and rest rooms. Risk control may be enhanced by the provision of eye washing and shower facilities for use after certain accidents (Figure 4.19).

Figure 4.19 Emergency shower and eye wash station where there is a serious risk of chemical contamination

Good housekeeping is a very cheap and effective means of controlling risks. It involves keeping the workplace clean and tidy at all times and maintaining good storage systems for hazardous substances and other potentially dangerous items. The risks most likely to be influenced by good housekeeping are fire and slips, trips and falls.

See Chapter 9 for more information on the work environment.

Monitoring and supervision

All risk control measures, whether they rely on engineered or human behavioural controls, must be monitored for their effectiveness, with supervision to ensure that they have been applied correctly. Competent people who have a sound knowledge of the equipment or process should undertake this monitoring. Checklists are useful to ensure that no significant factor is forgotten. Any statutory inspection or insurance company reports should be checked to see whether any areas of concern were highlighted and if any recommendations were implemented. Details of any accidents, illnesses or other incidents will give an indication on the effectiveness of the risk control measures. Any emergency arrangements should be tested during the monitoring phase including first-aid provision.

It is crucial that the operator should be monitored to ascertain that all relevant procedures have been understood and followed. The operator may also be able to suggest improvements to the equipment or system of work. The supervisor is an important source of information during the monitoring process.

Where the organisation is involved with shift work, it is essential that the risk controls are monitored on all shifts to ensure the uniformity of application.

The effectiveness and relevance of any training or instruction given should be monitored.

Periodically the risk control measures should be reviewed. Monitoring and other reports are crucial for the review to be useful. Reviews often take place at safety committee and/or at management meetings. A serious accident or incident should lead to an immediate review of the risk control measures in place.

(e) Personal protective equipment

Personal protective equipment (PPE) should only be used as a last resort. There are many reasons for this. The most important limitations are that PPE:

► only protects the person wearing the equipment, not others nearby;
► relies on people wearing the equipment at all times;
► must be used properly;
► must be replaced when it no longer offers the correct level of protection. This is particularly relevant when respiratory protection is used.

The benefits of PPE are:

▶ it gives immediate protection to allow a job to continue while engineering controls are put in place;

▶ in an emergency it can be the only practicable way of effecting rescue or shutting down plant in hazardous atmospheres;

▶ it can be used to carry out work in confined spaces where alternatives are impracticable. But it should never be used to allow people to work in dangerous atmospheres, which are, for example, enriched with oxygen or potentially explosive.

See Chapter 13 for more details on PPE.

Figure 4.21 Respiratory protection and disposable overalls are needed when working in high levels of asbestos dust

Figure 4.20 Good dust control for a chasing operation. A dust mask is still required for complete protection

2. Other hierarchies of risk control

There are several other similar hierarchies of risk control, which have been used over many years. A typical example is as follows:

▶ elimination:
▶ substitution;
▶ changing work methods/patterns;
▶ reduced or limited time exposure;
▶ engineering controls (e.g. isolation, insulation and ventilation);
▶ good housekeeping;
▶ safe systems of work;
▶ training and information;
▶ personal protective equipment;

▶ welfare;
▶ monitoring and supervision; and
▶ review.

3. Prioritisation of risk control

The prioritisation of the implementation of risk control measures will depend on the risk rating (high, medium and low) but the timescale in which the measures are introduced will not always follow the ratings. It may be convenient to deal with a low-level risk at the same time as a high-level risk or before a medium-level risk. It may also be that work on a high-risk control system is delayed due to a late delivery of an essential component – this should not halt the overall risk reduction work. It is important to maintain a continuous programme of risk improvement rather than slavishly following a predetermined priority list.

Step 4 – Recording significant findings

It is very useful to keep a written record of the risk assessment even if there are less than five employees in the organisation. For an assessment to be 'suitable and sufficient', only the significant hazards and conclusions need be recorded. The record should also include details of the groups of people affected by the hazards and the existing control measures and their effectiveness. The conclusions should identify any new controls required and a review date. The HSE booklet *'Five Steps to Risk Assessment'* provides a very useful guide and examples of the detail required for most risk assessments.

There are many possible layouts which can be used for the risk assessment record. Examples are given in Appendices 4.2 and 4.3 and Chapter 17. It should be noted that in Appendices 4.2 and 4.3, the initial qualitative risk level at the time of the risk assessment is given – the residual risk level when all the additional controls have been implemented will be 'low'. This should mean that an annual review will be sufficient.

The written record provides excellent evidence to a health and safety inspector of compliance with the law. It is also useful evidence if the organisation should become involved in a civil action.

The record should be accessible to employees and a copy kept with the safety manual containing the safety policy and arrangements.

Step 5 – Monitoring and review

As mentioned earlier, the risk controls should be reviewed periodically. This is equally true for the risk assessment as a whole. Review and revision may be necessary when conditions change as a result of the introduction of new machinery, processes or hazards. There may be new information on hazardous substances or new legislation. There could also be changes in the workforce, for example the introduction of trainees. The risk assessment needs to be revised only if significant changes have taken place since the last assessment was done. An accident or incident or a series of minor ones provides a good reason for a review of the risk assessment. This is known as the post-accident risk assessment.

Examples of completed risk assessments are given in Appendices 4.2 and 4.3. They show how small and medium-sized businesses have approached risk assessment.

4.1.10 Cost-benefit analysis

In recent years, risk assessment has been accompanied by a cost-benefit analysis that attempts to evaluate the costs and benefits of risk control and reduction. The costs could include capital investment, maintenance and training and produce benefits such as reduced insurance premiums, higher productivity and better product quality. The pay-back period for most risk reduction projects (other than the most simple) has been shown to be between two and five years. Although the benefits are often difficult to quantify, cost-benefit analysis does help to justify the level of expenditure on a risk reduction project.

4.1.11 Special cases

There are several groups of persons who require an additional risk assessment due to their being more 'at risk' than other groups. Three such groups will be considered: young persons, expectant and nursing mothers and disabled workers.

Young persons

There are about 20 fatalities of young people at work each year. Therefore, any risk assessment involving young people needs to consider the particular vulnerability of young persons in the workplace. Young workers clearly have a lack of experience and awareness of risks in the workplace, a tendency to be subject to peer pressure and a willingness to work hard. Many young workers will be trainees or on unpaid work experience. Young people are not fully developed, have poor perception of risk and are more vulnerable to physical, biological and chemical hazards than adults.

An amendment to the HSW Act enables trainees on government-sponsored training schemes to be treated as employees as far as health and safety is concerned. The Management of Health and Safety at Work Regulations define a young person as anybody under the age of 18 years and stipulate that a special risk assessment must be completed which takes into account their immaturity and inexperience. The assessment must be completed before the young person starts work. If the young person is of school age (16 years or less), the parents or guardian of the child should be notified of the outcome of the risk assessment and details of any safeguards which will be used to protect the health and safety of the child.

The following key elements should be covered by the risk assessment:

▶ details of the work activity, including any equipment or hazardous substances;
▶ details of any prohibited equipment or processes;
▶ details of health and safety training to be provided;
▶ details of supervision arrangements.

The extent of the risks identified in the risk assessment will determine whether employers should restrict the work of the people they employ. Except in special circumstances, young people should not be employed to do work which:

▶ is beyond their physical or psychological capacity;
▶ exposes them to substances chronically harmful to human health, for example toxic or carcinogenic substances;
▶ exposes them to radiation;
▶ involves a risk of accidents which they are unlikely to recognise because of, for example, their lack of experience, training or insufficient attention to safety;
▶ involves a risk to their health from extreme heat, noise or vibration.

These restrictions will not apply in **special circumstances** where young people over the minimum school leaving age are doing work necessary for their training, under proper supervision by a competent person, and providing the risks are reduced to the lowest level, so far as is reasonably practicable. Under no circumstances can children of compulsory school age do work involving these risks, whether they are employed or under training such as work experience.

Induction training is important for young workers and such training should include site rules, restricted areas, prohibited machines and processes, fire precautions, emergency procedures, welfare arrangements and details of any further training related to their particular job. At induction, they should be introduced to their mentor and given close supervision, particularly during the first few weeks of their employment.

If work-experience trainees are not at school on any days during the placement, then they should not work more than 8 hours a day or 40 hours a week (Working Time Directive) and no night work or overtime. They should also have a 20-minute rest within a 4.5-hour working period.

A guide is available to employers who organise site visits for young people. The guide 'Guidance for employers: Getting involved in work-experience and work-related learning' is available from LANTRA (www.lantra.co.uk).

Expectant and nursing mothers

The Management of Health and Safety at Work Regulations 1999 incorporate the Pregnant Workers Directive from the EU. Pregnant women and nursing mothers are more at risk from certain types of hazard. In most of these instances the hazard presents a risk not only to the woman but also to the baby. Many of these hazards can cause miscarriage, birth defects or ill health in the baby. If any type of work could present a particular risk to expectant or nursing mothers, the risk assessment must include an assessment of such risks. Should these risks be unavoidable, then the woman's working conditions or hours must be altered to avoid the risks. The alternatives for her are to be offered other work or be suspended from work on full pay. The woman must notify the employer in writing that she is pregnant, or has given birth within the previous six months and/or is breastfeeding.

Pregnant workers should not be exposed to chemicals, such as pesticides and lead, or to biological hazards, such as hepatitis, or effects likely to be passed on genetically or likely to harm the unborn child. Female agricultural workers, veterinaries or farmers who are pregnant should not assist with lambing, so that any possible contact with ovine chlamydia is avoided.

Other work activities that may present a particular risk to pregnant women at work are radiography, involving possible exposure to ionising radiation, and shop work when long periods of standing are required during shelf filling or stock-taking operations.

Other hazards which might affect such women are:

▶ manual handling especially later in pregnancy;
▶ chemical or biological agents (e.g. lead and the rubella virus);
▶ passive smoking;
▶ lack of rest room facilities;
▶ temperature variations;
▶ ergonomic issues related to prolonged standing, sitting or the need for awkward body movement;
▶ whole-body vibration;
▶ issues associated with the use and wearing of personal protective equipment;
▶ working excessive hours;
▶ night working; and
▶ stress and violence to staff.

Detailed guidance is available on the HSE website at www.hse.gov.uk/mothers.

Workers with a disability

Organisations have been encouraged for many years to employ workers with disabilities and to ensure that their premises provide suitable access. From a health and safety point of view, it is important that workers with a disability are covered by special risk assessments so that appropriate controls are in place to protect them. Workers with disabilities may be at greater risk from particular hazards depending on the nature and extent of their disability. For example, employees with a hearing problem will need to be warned when the fire alarm sounds or a fork-lift truck approaches. Special vibrating signals or flashing lights may be used. Similarly workers in wheelchairs will require a clear, wheelchair-friendly route to a fire exit and onwards to the assembly point. Safe systems of work and welfare facilities need to be suitable for any workers with disabilities.

During the special risk assessment process it may be necessary to identify:

▶ the jobs with particular health and fitness requirements;
▶ the types of disability that would make certain jobs unsuitable;
▶ the staff whose disabilities would exclude them from undertaking those jobs safely;
▶ and screen staff against these criteria – this may have the effect of excluding those with a certain disability from doing these jobs (e.g. fork-lift truck driving).

It is important to recognise that there is other employment and anti-discrimination legislation that should be considered before the findings of the risk assessment are finalised. The Equality Act places duties on service providers where physical features make access to their services impossible or unreasonable for people with a disability. The Act requires equal opportunities for employment and access to workplaces to be extended to all people with disabilities.

4

Lone workers

People who work alone, like those in small workshops, remote areas of a large site, social workers, sales personnel or mobile maintenance staff, should not be at more risk than other employees. Lone workers are a group of workers who are especially vulnerable in certain situations. If, for example, the worker is in contact with members of the public, they may be at risk of violence from people who could be violent. People who work alone in confined spaces could also be at risk if they have an accident or become ill. There are, however, no absolute restrictions on working alone; it will depend on the findings of a risk assessment.

When risk assessment shows that it is not possible for the work to be done safely by a lone worker, arrangements for providing help or back-up should be put in place. Where a lone worker is working at the premises of another employer, that employer should inform the employer of the lone worker of any risks and the precautions that should be taken. A risk assessment is, therefore, essential for all instances of lone working.

Initially, it is important to consider whether the risks of the job can be properly controlled by one person. Other considerations in the risk assessment include:

▶ does the particular workplace present a special risk to someone working alone?
▶ is there safe access into and exit from the workplace?
▶ can all the equipment and substances be safely handled by one person?
▶ is violence from others a risk?
▶ would women and young persons be specially at risk?
▶ is the worker medically fit and suitable for working alone?
▶ are special training and supervision required?
▶ has the worker access to first aid?

The risk assessment should help decide the right level of supervision.

The controls resulting from the risk assessment for lone working are covered in Chapter 9 (9.4).

Figure 4.22 A lone worker – special arrangements required

4.2 Management of change

4.2.1 Types of changes

Effective management of change should ensure that safety, health, and environmental hazards and risks are properly controlled when an organisation makes changes to their facilities, operations or personnel. Whether the change is small or large, it needs to be managed effectively. Typical types of change management likely to be encountered in the workplace include:

▶ a change in the management structure following a takeover or merger of the organisation;
▶ construction work possibly of a temporary nature;
▶ a change of process;
▶ a change of equipment; and
▶ a change in working practices such as in the supervisory arrangements.

One of the major factors which hinder the change management process is the tendency for people to resist change in organisations because it can be uncomfortable.

4.2.2 Managing the impact of change – control measures

The organisation should plan for change and ensure sufficient resources are available to make sure that changes do not introduce new and unforeseen hazards or increase health and safety risks. The objective of the control measures is to mitigate the impact of change, and includes the following issues:

▶ communication and cooperation – between those implementing the change and those affected by it before, during and after the change takes place;
▶ risk assessment – to examine the health, safety and environmental risks associated with the change. Suitable controls should be defined and implemented;
▶ appointment of competent people to oversee the change and advise senior managers on progress and any problems;
▶ segregation of work areas – this may be necessary if only a particular work area is affected by the change;
▶ amendment of emergency procedures – this will be necessary if escape routes, for example, are compromised by the change;
▶ welfare provision – in particular counselling arrangements for those who have concerns about the change.

Effective management of change should ensure that those affected by the changes are:

▶ provided with timely information to enable them to understand the reasons for proposed changes;
▶ adequately consulted on the proposed changes and provided with opportunities to influence the proposals;

- aware of the probable impact of any changes to their jobs;
- given, if necessary, training to support them during the changes in their jobs;
- aware of the timetable for changes; and
- given access to relevant support during the changes.

4.2.3 Review of changes

It is important that changes are reviewed both during and after their introduction. Plans will be required to monitor the performance indicators beyond the end of the introduction of the changes. All relevant risk assessments will need to be checked and possibly modified to take into account the effect of the changes.

4.3 Safe system of work for general work activities

4.3.1 What is a safe system of work?

A safe system of work has been defined as:

'The integration of personnel, articles and substances in a laid out and considered method of working which takes proper account of the risks to employees and others who may be affected, such as visitors and contractors, and provides a formal framework to ensure that all of the steps necessary for safe working have been anticipated and implemented.'

In simple terms, a safe system of work is a defined method for doing a job in a safe way. It takes account of all foreseeable hazards to health and safety and seeks to eliminate or minimise these. Safe systems of work are normally formal and documented, for example in written operating procedures but, in some cases, they may be verbal.

The particular importance of safe systems of work stems from the recognition that most accidents are caused by a combination of factors (plant, substances, lack of training and/or supervision, etc.). Hence prevention must be based on an integral approach and not one which only deals with each factor in isolation. The adoption of a safe system of work provides this integral approach because an effective safe system:

- is based on looking at the job as a whole;
- starts from an analysis of all foreseeable hazards, for example physical, chemical, health;
- brings together all the necessary precautions, including design, physical precautions, training, monitoring, procedures and PPE.

It follows from this that the use of safe systems of work is in no way a replacement for other precautions, such as good equipment design, safe construction and the use of physical safeguards. However, there are many situations where these will not give adequate protection in themselves, and then a carefully thought-out and properly implemented safe system of work is

especially important. The best example is maintenance and repair work, which will often involve, as a first stage, dismantling the guard or breaking through the containment, which exists for the protection of the ordinary process operator. In some of these operations, a permit-to-work procedure will be the most appropriate type of safe system of work.

The operations covered may be simple or complex, routine or unusual.

Whether the system is verbal or written, and whether the operation it covers is simple or complex, routine or unusual, the essential features are forethought and planning – to ensure that all foreseeable hazards are identified and controlled. In particular, this will involve scrutiny of:

- the sequence of operations to be carried out;
- the equipment, plant, machinery and tools involved;
- chemicals and other substances to which people might be exposed in the course of the work;
- the people doing the work – their skill and experience;
- foreseeable hazards (health, safety, environment), whether to the people doing the work or to others who might be affected by it;
- practical precautions which, when adopted, will eliminate or minimise these hazards (Figure 4.23);
- the training needs of those who will manage and operate under the procedure;
- monitoring systems to ensure that the defined precautions are implemented effectively.

Figure 4.23 Multipadlocked hasp for locking off an isolation valve – each worker puts on their own padlock

4.3.2 Legal requirements

The HSW Act Section 2 requires employers to provide safe plant and systems of work. In addition, many Regulations made under the Act, such as the Provision and Use of Work Equipment Regulations 1998, require information and instruction to be provided to employees and others. In effect, this is also a more specific requirement to provide safe systems of work. Many of these safe systems, information and instructions will need to be in writing. There is also a need for employers to provide a safe system of work to fulfil their common law duty of care.

4.3.3 Assessment of what safe systems of work are required

Requirement

It is the responsibility of the management in each organisation to ensure that its operations are assessed to determine where safe systems of work need to be developed.

This assessment must, at the same time, decide the most appropriate form for the safe system; that is:

▶ Is a written procedure required?
▶ Should the operation only be carried out under permit to work?
▶ Is an informal system sufficient?

Factors to be considered

It is recognised that each organisation must have the freedom to devise systems that match the risk potential of their operations and which are practicable in their situation. However, they should take account of the following factors in making their decision:

▶ types of risk involved in the operation;
▶ magnitude of the risk, including consideration of the worst foreseeable loss;
▶ complexity of the operation;
▶ past accident and loss experience;
▶ requirements and recommendations of the relevant health and safety authorities;
▶ the type of documentation needed;
▶ resources required to implement the safe system of work (including training and monitoring).

4.3.4 Development of safe systems

Role of competent person

The competent person appointed under the MHSW Regulations and/or appointed under the CDM Regulations should assist managers to draw up guidelines for safe systems of work. This will include, where necessary, particularly in construction work, method statements. The competent person should prepare suitable forms and should advise management on the adequacy of the safe systems produced.

Role of managers

Primarily management is responsible for providing safe systems of work, as they will know the detailed way in which the task should be carried out.

Management is responsible to ensure that employees are adequately trained in a specific safe system of work and are competent to carry out the work safely. Managers need to provide sufficient supervision to ensure that the system of work is followed and the work is carried out safely. The level of supervision will depend on the experience of the particular employees concerned and the complexity and risks of the task.

When construction work is involved, principal contractors will need to monitor sub-contractors to check that they are providing suitable safe systems of work, have trained their employees and are carrying out the tasks in accordance with the safe systems.

Role of employees and consultation

Many people operating a piece of machinery or a manufacturing process are in the best position to help with the preparation of safe systems of work. Consultation with those employees who will be exposed to the risks, either directly or through their representatives, is also a legal requirement. The importance of discussing the proposed system with those who will have to work under it, and those who will have to supervise its operation, cannot be emphasised enough.

Employees have a responsibility to follow the safe system of work.

Analysis

The safe system of work should be based on a thorough analysis of the job or operation to be covered by the system. The way this analysis is done will depend on the nature of the job/operation.

If the operation being considered is a new one involving high loss potential, the use of formal hazard analysis techniques such as a hazard and operability (HAZOP) study, fault tree analysis (FTA) or failure modes and effects analysis should be considered.

However, where the potential for loss is lower, a more simple approach, such as job safety analysis (JSA), will be sufficient. This will involve three key stages:

▶ identification of the key steps in the job/operation – what activities will the work involve?
▶ analysis and assessment of the risks associated with each stage – what could go wrong?
▶ definition of the precautions or controls to be taken – what steps need to be taken to ensure the operation proceeds without danger, either to the people doing the work, or to anyone else?

The results of this analysis are then used to draw up the safe operating procedure or method statement. (See Appendix 4.4 for a suitable form.)

Introducing controls

There are a variety of controls that can be adopted in safe systems of work. They can be split into the following three basic categories:

(i)   **Technical** – these are engineering or process type controls which engineer out or contain the hazard so that the risks are acceptable. For example exhaust ventilation, a machine guard, dust respirator.

(ii) **Procedural** – these are ways of doing things to ensure that the work is done according to the procedure, legislation or cultural requirements of the organisation. For example a supervisor must be involved, the induction course must be taken before the work commences, a particular type of form or a person's signature must be obtained before proceeding, the names of the workforce must be recorded.

(iii) **Behavioural** – these are controls which require a certain standard of behaviour from individuals or groups of individuals. For example no smoking is permitted during the task, hard hats must be worn, all lifts are to be in tandem between two workers.

4.3.5 Preparation of safe systems

A checklist for use in the preparation of safe systems of work is set out as follows:

▶ What is the work to be done?

▶ What are the potential hazards?

▶ Is the work covered by any existing instructions or procedures? If so, to what extent (if any) do these need to be modified?

▶ Who is to do the work?

▶ What are their skills and abilities – is any special training needed?

▶ Under whose control and supervision will the work be done?

▶ Will any special tools, protective clothing or equipment (e.g. breathing apparatus) be needed? Are they ready and available for use?

▶ Are the people who are to do the work adequately trained to use the above?

▶ What isolations and locking-off will be needed for the work to be done safely?

▶ Is a permit to work required for any aspect of the work?

▶ Will the work interfere with other activities? Will other activities create a hazard to the people doing the work?

▶ Have other departments been informed about the work to be done, where appropriate?

▶ How will the people doing the work communicate with each other?

▶ Have possible emergencies and the action to be taken been considered?

▶ Should the emergency services be notified?

▶ What are the arrangements for handover of the plant/equipment at the end of the work? (For maintenance/project work, etc.)

▶ Do the planned precautions take account of all foreseeable hazards?

▶ Who needs to be informed about or receive copies of the safe system of work?

▶ What arrangements will there be to see that the agreed system is followed and that it works in practice?

▶ What mechanism is there to ensure that the safe system of work stays relevant and up to date?

4.3.6 Documentation

Safe systems of work should be properly documented.

Wherever possible, they should be incorporated into normal process operating procedures. This is so that:

▶ health and safety are seen as an integral part of, and not an add-on to, normal production procedures;

▶ the need for operators and supervisors to refer to separate manuals is minimised.

Whatever method is used, all written systems of work should be signed by the relevant managers to indicate approval or authorisation. Version numbers should be included so that it can quickly be verified that the most up-to-date version is in use. Records should be kept of copies of the documentation, so that all sets are amended when updates and other revisions are issued.

As far as possible, systems should be written in a non-technical style and should specifically be designed to be as intelligible and user-friendly as possible. It may be necessary to produce simple summary sheets which contain all the key points in an easy-to-read format.

4.3.7 Communication and training

People doing work or supervising work must be made fully aware of the laid-down safe systems that apply. The preparation of safe systems will often identify a training need that must be met before the system can be implemented effectively.

In addition, people should receive training in how the system is to operate. This applies not only to those directly involved in doing the work but also to supervisors/managers who are to oversee it.

In particular, the training might include:

▶ why a safe system is needed;

▶ what is involved in the work;

▶ the hazards which have been identified;

▶ the precautions which have been decided and, in particular:

▷ the isolations and locking-off required, and how this is to be done;

▷ details of the permit-to-work system, if applicable;

▷ any monitoring (e.g. air testing) which is to be done during the work, or before it starts;

▷ how to use any necessary PPE;

▷ emergency procedures.

4.3.8 Monitoring safe systems

Safe systems of work should be monitored to ensure that they are effective in practice. This will involve:

▶ reviewing and revising the systems themselves, to ensure they stay up to date;

▶ inspecting to identify how fully they are being implemented.

In practice, these two things go together, as it is likely that a system that is out of date will not be fully implemented by the people who are intended to operate it.

All organisations are responsible for ensuring that their safe systems of work are reviewed and revised as appropriate. Monitoring of implementation is part of all line managers' normal operating responsibilities, and should also take place during health and safety audits.

4.4 Permit-to-work system

4.4.1 Introduction

Safe systems of work are crucial in work such as the maintenance of chemical plant where the potential risks are high and the careful coordination of activities and precautions is essential to safe working. In this situation and others of similar risk potential, the safe system of work is likely to take the form of a permit-to-work procedure.

The role of the permit-to-work procedure is to provide a specialised type of safe system of work for ensuring that potentially very dangerous work (e.g. entry into process plant and other confined spaces) is done safely.

Although this procedure has been developed and refined by the chemical industry, the principles of the permit-to-work procedure are equally applicable to the management of complex risks in other industries.

The function of the permit-to-work procedure is to ensure that certain defined operations are prohibited without the specific permission of a responsible manager, this permission being only granted once stringent checks have been made to ensure that all necessary precautions have been taken and that it is safe for work to go ahead.

The people doing the work take on responsibility for following and maintaining the safeguards set out in the permit, which will define the work to be done (no other work being permitted) and the timescale in which it must be carried out.

To be effective, the permit system requires the training needs of those involved to be identified and met, and the monitoring procedures must ensure that the system is operating as intended.

4.4.2 The principles that apply to permits to work

Permit systems must adhere to the following eight principles:

1. Wherever possible, and especially with routine jobs, hazards should be eliminated so that the work can be done safely without requiring a permit to work.

2. Although the Site Manager may delegate the responsibility for the operation of the permit system, the overall responsibility for ensuring safe operation rests with him/her.

3. The permit must be recognised as the master instruction which, until it is cancelled, overrides all other instructions.

4. The permit applies to everyone on site, including contractors.

5. Information given in a permit must be detailed and accurate. It must state:
   (a) which plant/equipment has been made safe and the steps by which this has been achieved;
   (b) the work that may be done;
   (c) the time at which the permit comes into effect.

6. The permit remains in force until the work has been completed and the permit is cancelled by the person who issued it or by the person nominated by management to take over the responsibility (e.g. at the end of a shift or during absence).

7. No work other than that specified is authorised. If it is found that the planned work has to be changed, the existing permit should be cancelled and a new one issued.

8. Responsibility for the plant must be clearly defined at all stages.

4.4.3 Permit-to-work procedures

The permit-to-work procedure is a specialised type of safe system of work, under which certain categories of high risk-potential work may only be done with the specific permission of an authorised manager. This permission (in the form of the permit to work) will be given only if the laid-down precautions are in force and have been checked.

The permit document should specify the following key items of information:

▶ the date, time and duration of the permit;

▶ a description and assessment of the task to be performed and its location;

▶ the plant/equipment involved, and how it is identified;

▶ the authorised persons to do the work;

▶ the steps which have already been taken to make the plant safe;

▶ potential hazards which remain, or which may arise as the work proceeds;

▶ the precautions to be taken against these hazards;

- the action to be taken prior to the task being started, such as:
  - ▷ the isolation of sources of energy and inlets;
  - ▷ emergency procedures and equipment;
  - ▷ ensuring the competency of those involved;
  - ▷ communication arrangements; and
  - ▷ reference to any other relevant documents;
- the length of time that the permit is valid;
- the equipment to be released to those who are to carry out the work.

In accepting the permit, the person in charge of doing the authorised work normally undertakes to take/maintain whatever precautions are outlined in the permit, such as:

- isolation of the area;
- carrying out atmospheric monitoring;
- the provision and use of personal protective equipment;
- the provision of suitable equipment including lighting and tools;
- ensuring an adequate level of supervision; and
- arrangements for any extension to or handover of the permit.

The permit will also include spaces for:

- signature certifying that the work is complete; and
- signature confirming re-acceptance of the plant/equipment.

See Chapter 17 for an example of a general permit to work.

4.4.4 Work requiring a permit

The nature of permit-to-work procedures will vary in their scope depending on the job and the risks involved. However, a permit-to-work system is unlikely to be needed where, for example:

(a) the assessed risks are low and can be controlled easily;
(b) system of work is very simple;
(c) other work being done nearby cannot affect the work concerned in say a confined space entry, or a welding operation.

However, where there are high risks and the system of work is complex and other operations may interfere, a formal permit to work should be used.

The main types of permit and the work covered by each are identified below.

General permit

The general permit should be used for work such as:

- alterations to or overhaul of plant or machinery where mechanical, toxic or electrical hazards may arise. This is particularly important for:
  - ▷ large machines where visual contact between workers is difficult;
  - ▷ where work has to be done near dangerous parts of the machine;
  - ▷ where there are multiple isolations for energy; sources and/or dangerous substances; and
  - ▷ where dangerous substances are being used in confined areas with poor ventilation;
- work on or near overhead crane tracks;
- work on pipelines with hazardous contents;
- repairs to railway tracks, tippers, conveyors;
- work with asbestos-based materials;
- work involving ionising radiation; and
- excavations to avoid underground services.

Typical work tasks that might require a permit to work

Hot work

Hot work is potentially hazardous because:

- it may act as a source of ignition in any plant in which flammable materials are handled;
- it may act as a cause of fires in all processes, regardless of whether flammable materials are present.

Hot work includes cutting, welding, brazing, soldering and any process involving the application of a naked flame. Drilling and grinding should also be included where a flammable atmosphere is potentially present (Figure 4.24).

Figure 4.24 A hot work permit is usually essential for welding, cutting and burning except in designated areas like a welding shop

Hot work should therefore be done under the terms of a hot work permit, the only exception being where hot work is done in a designated maintenance area suitable for the purpose. Typical controls include:

- a suitable fire extinguisher nearby;
- prompt removal of flammable waste material; and
- the damping down of nearby wooden structures such as floors.

Work on high-voltage apparatus (including testing)

Work on high-voltage apparatus (over about 600V) is potentially high risk. Hazards include:

▶ possibly fatal electric shock/burns to the people doing the work;

▶ electrical fires/explosions;

▶ consequential danger from disruption of power supply to safety-critical plant and equipment.

In view of the risk, this work must only be done by suitably trained and competent people acting under the terms of a high-voltage permit. The most important control is to ensure that the necessary isolation is provided. Other controls for this type of work are covered in Chapter 12.

Confined spaces

A confined space can be any space of an enclosed nature where there is a risk of death or serious injury from hazardous substances or dangerous conditions (e.g. lack of oxygen).

Examples of confined spaces include:

▶ storage tanks;

▶ silos;

▶ reaction vessels;

▶ enclosed drains;

▶ sewers;

▶ combustion chambers in furnaces; and

▶ ductwork.

If the work and/or confined space is very hazardous, then a permit to work will be required. This will ensure that a formal check is undertaken to make sure all the elements of a safe system of work are in place before people are allowed to enter or work in the confined space. It will also be a means of communication between site management, supervisors and those carrying out the hazardous work. The essential features of such a permit to work are:

▶ the clear identification of who may authorise particular jobs (and any limits to their authority) and who, including any contractors, are responsible for specifying the necessary precautions (e.g. isolation, air testing and emergency arrangements);

▶ any particular training and instruction that may be required; and

▶ monitoring and auditing of the permit-to-work system.

Many fatal accidents have occurred where inadequate precautions were taken before and during work involving entry into confined spaces (Figure 4.25). Two hazards are the potential presence of toxic or other dangerous substances, and the absence of adequate oxygen. In addition, there may be mechanical hazards (entanglement on agitators), ingress of fluids, risk of engulfment in a free-flowing solid like grain or sugar, and raised temperatures. The work to be carried out

may itself be especially hazardous when done in a confined space, for example cleaning using solvents, or cutting/welding work. Should the person working in a confined space get into difficulties for whatever reason, getting help in and getting the individual out may prove difficult and dangerous.

Figure 4.25 Entering a confined space with breathing apparatus, rescue tripod and rescue watcher

Stringent preparation, isolation, air testing and other precautions are therefore essential, and experience shows that the use of a confined space entry permit is essential to confirm that all the appropriate precautions have been taken.

The Confined Spaces Regulations are summarised in Chapter 13. They detail the specific controls that are necessary when people enter confined spaces. The risks and associated controls for confined spaces are discussed in Chapter 9 (9.3).

Machinery maintenance

The Provision and Use of Work Equipment Regulations require that all work equipment should be maintained in an efficient state and in good repair. Any maintenance operations on the equipment should be carried out safely. The frequency and nature of maintenance should be determined through risk assessment, taking full account of:

▶ the manufacturer's recommendations;

▶ the intensity of use;

▶ operating environment (e.g. the effect of temperature, corrosion, weathering);

▶ user knowledge and experience; and

▶ the risk to health and safety from any foreseeable failure or malfunction.

Maintenance work should only be undertaken by those who are competent to do the work, who have been provided with sufficient information, instruction and training. For high-risk equipment, positive means of disconnecting the equipment from the energy source may be required (e.g. isolation), along with means to prevent inadvertent reconnection (e.g. by locking off). Such procedures are best controlled using permits to work. Where possible, equipment should normally be shut down and any residual/stored energy safely released (e.g. pneumatic pressure dumped, parts with gravitational/rotational energy stopped or brought to a safe position). In some cases, it may not be possible to avoid particular significant hazards during the maintenance of work equipment so appropriate measures should be taken to protect people and minimise the risk. These may include:

▶ physical measures, such as temporary guarding, slow speed hold-to-run control devices, safe means of access and personal protective equipment;
▶ management issues, including safe systems of work, supervision and monitoring; and
▶ personnel competence (training, skill, awareness and knowledge of risk).

Work at height

A permit to work may be required for some hazardous work at height such as roof work to ensure that a fall arrest strategy is in place. This is particularly important where there are:

▶ no permanent work platforms with fixed handrails on flat roofs;
▶ on sloping or fragile roofs;
▶ where specialist access equipment, like rope-hung cradles, is required; and
▶ where access is difficult.

4.4.5 **Responsibilities**

The effective operation of the permit system requires the involvement of many people. The following specific responsibilities can be identified:

(Note: all appointments, definitions of work requiring a permit, etc. must be in writing. All the categories of people identified below should receive training in the operation of the permit system as it affects them.)

Site Manager

▶ has overall responsibility for the operation and management of the permit system;
▶ appoints a senior manager (normally the Chief Engineer) to act as a senior authorised person.

Senior authorised person

▶ is responsible to the Site Manager for the operation of the permit system;
▶ defines the work on the site which requires a permit;
▶ ensures that people responsible for this work are aware that it must only be done under the terms of a valid permit;

▶ appoints all necessary authorised persons;
▶ appoints a deputy to act in his/her absence.

Authorised persons

▶ issue permits to competent persons and retain copies;
▶ personally inspect the site to ensure that the conditions and proposed precautions are adequate and that it is safe for the work to proceed;
▶ accompany the competent person to the site to ensure that the plant/equipment is correctly identified and that the competent person understands the permit;
▶ cancel the permit on satisfactory completion of the work.

Competent persons

▶ receive permits from authorised persons;
▶ read the permit and make sure they fully understand the work to be done and the precautions to be taken;
▶ signify their acceptance of the permit by signing both copies;
▶ comply with the permit and make sure those under their supervision similarly understand and implement the required precautions;
▶ on completion of the work, return the permit to the authorised person who issued it.

Operatives

▶ read the permit and comply with its requirements, under the supervision of the competent person.

Specialists

A number of permits require the advice/skills of specialists in order to operate effectively. Such specialists may include chemists, electrical engineers, health and safety advisers and fire officers. Their role may involve:

▶ isolations within his/her discipline – for example electrical work;
▶ using suitable techniques and equipment to monitor the working environment for toxic or flammable materials, or for lack of oxygen;
▶ giving advice to managers on safe methods of working.

Specialists must not assume responsibility for the permit system. This lies with the Site Manager and the senior authorised person.

Engineers (and others responsible for work covered by permits)

▶ ensure that permits are raised as required.

Contractors

The permit system should be applied to contractors in the same way as to direct employees.

The contractor must be given adequate information and training on the permit system, the restrictions it imposes and the precautions it requires.

4.5 Emergency procedures

4.5.1 Introduction

Emergency procedures are a critically important part of the health and safety organisation. They are concerned about the control procedures and equipment required to limit the damage to people and property caused by an incident. Local fire and rescue authorities will often be involved and are normally prepared to give advice to employers.

Under Regulation 8 of the Management of Health and Safety at Work Regulations, procedures must be established and set in motion when necessary to deal with serious and imminent danger to persons at work. Necessary links must be maintained with local authorities, particularly with regard to first aid, emergency medical care and rescue work (Figure 4.26).

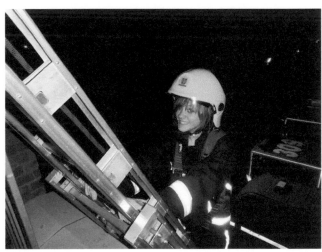

Figure 4.26 Emergency services at work

Although fire is the most common emergency likely to be faced, there are many other possibilities, which should be considered including:

- gas explosion;
- electrical burn or electrocution;
- escape of toxic gases or fumes;
- discovery of dangerous dusts like asbestos in the atmosphere;
- terrorist threat;
- large vehicle crashing into the premises;
- aircraft crash if near a flight path;
- spread of highly infectious disease;
- severe weather with high winds and flooding.

For fire emergencies, see Chapter 11.

4.5.2 Points to include in emergency procedures

Emergency procedures form an essential part of any health and safety management system. A quick and effective response to an emergency may well help to ease the situation and reduce the consequences. In any emergency, people are more likely to respond quickly and effectively if they:

- are well trained and competent;
- take part in regular and realistic practice exercises; and
- have clearly agreed, recorded and rehearsed plans, actions and responsibilities.

The following points should be considered when developing an emergency procedure:

- decide on the possible nature of any emergency that might occur and the potential consequences to the organisation;
- consider how the alarm will be raised. This may be particularly important when the organisation is closed at weekends or holiday times;
- decide on how the emergency services will be contacted – again including times when the organisation is closed (if there are 25 tonnes or more of dangerous substances on the premises, the fire and rescue service must be notified and warning signs erected);
- decide on an emergency assembly point for staff;
- ensure there are enough emergency exits for everyone to escape quickly, and keep emergency doors and escape routes unobstructed and clearly marked;
- nominate competent people to take control in the event of an emergency;
- decide which other key people are needed, such as activators of emergency equipment, a nominated incident controller and first-aiders;
- plan essential actions such as emergency plant shutdown, isolation or making processes safe. Clearly identify important items such as shut-off valves and electrical isolators;
- ensure that everyone is included in emergency procedures including people with disabilities and vulnerable workers; and finally
- decide on the arrangements for liaising with the media during and after the event.

4.5.3 Supervisory duties

A member of the site staff should be nominated to supervise and coordinate all emergency arrangements. This person should be in a senior position or at least have direct access to a senior manager. Senior members of the staff should be appointed as departmental fire/emergency procedure wardens, with deputies for every occasion of absence, however brief. They should ensure that the following precautions are taken:

- Everyone on site can be alerted to an emergency.
- Everyone on site knows what signal will be given for an emergency and knows what to do.
- Someone who has been trained in what to do is on site and ready to coordinate activities.
- Emergency routes are kept clear, signed and adequately lit.
- There are arrangements for calling the fire and rescue services and to give them special information about high-hazard work, for example in tunnels or confined spaces.
- There is adequate access to the site for the emergency services and this is always kept clear.
- Suitable arrangements for treating and recovering injured people are set up.
- Someone is posted to the site entrance to receive and direct the emergency services.

4.5.4 Assembly and roll call

Assembly points should be established for use in the event of evacuation. It should be at a position, preferably undercover, which is unlikely to be affected at the time of emergency. In some cases, it may be necessary to make mutual arrangements with the client or occupiers of nearby premises.

In the case of small sites, a complete list of the names of all staff should be maintained so that a roll call can be made if evacuation becomes necessary.

In those premises where the number of staff would make a single roll call difficult, each area warden should maintain a list of the names of employees and contractors in their area. Roll call lists must be updated regularly.

4.5.5 Contacting the emergency services

Emergency services should be called as soon as possible after the start of an emergency. The emergency services prefer to be called early as this gives them the best chance to contain the emergency and save lives. It is important that the organisation has a well-defined policy to call the emergency services in good time and a person nominated to make the call. Many fires have become established because everyone assumed that somebody else had called the fire service. A rota may need to be organised so that there is a nominated person available at all times even when the premises of the organisation are unoccupied. Some businesses use a specialist company to monitor the premises when it is unoccupied.

An emergency telephone number call may be answered by either a telephone operator or an emergency service dispatcher. The nature of the emergency (police, fire or medical) is determined first. If the call has been answered by a telephone operator, the call is transferred to the appropriate emergency service, which dispatches the appropriate help. If multiple services are required, the most urgent need must be determined and other services called in as needed.

Emergency dispatchers are trained to control the calling order to provide help in an appropriate manner. The dispatcher may find it necessary to give urgent advice in life-threatening situations. Some dispatchers have special training in telling people how to perform first aid or **cardiopulmonary resuscitation (CPR)** in an effort to return life to a person in cardiac arrest. The emergency dispatcher will need the following information:

- the caller's telephone number;
- the exact location of the incident (e.g. the road name and any important details about approaching and accessing the site);
- the type and seriousness of the incident; and
- details of any hazards (e.g. gas leak or fire).

The emergency service may be able to identify the telephone number of the caller. This is normally done using the system that the telephone company uses to bill calls, making the number visible even for users who have unlisted numbers or who block caller identity. For an individual fixed landline telephone, the caller's number is often associated with the caller's address and therefore their location. However, with mobile phones and business telephones, the address may be a mailing address rather than the caller's location.

4.5.6 Testing and training for emergencies

It is important that the emergency procedures are covered during the induction training. Separate training courses should also be given to staff on a regular basis on the following topics:

- use of firefighting equipment;
- regular refresher first-aid training;
- suitable training (and competency assessment) for all those allocated particular roles in an emergency; and
- regular timed fire drills.

Emergency lighting should be regularly tested and firefighting equipment examined. In high hazard plants, there should be an occasional full-scale rehearsal of the emergency procedure in action.

4.5.7 Requirements for, and effective provision of, first aid in the workplace

Introduction

People at work can suffer injuries or fall ill. It does not matter whether the injury or the illness is caused by the work they do. What is important is that they receive immediate attention and that an ambulance is called in serious cases. First Aid at Work (FAW) covers the arrangements employers must make to ensure this happens. It can save lives and prevent minor injuries becoming major ones.

The Health and Safety (First-Aid) Regulations 1981 require employers to provide adequate and appropriate equipment, facilities and personnel to enable first aid to be given to employees if they are injured or become ill at work.

What is adequate and appropriate will depend on the circumstances in a particular workplace.

The minimum first-aid provision on any work site is:

▶ a suitably stocked first-aid box;
▶ an appointed person to take charge of first-aid arrangements.

It is also important to remember that accidents can happen at any time. First-aid provision needs to be available at all times to people who are at work.

Many small firms will only need to make the minimum first-aid provision. However, there are factors which might make greater provision necessary. The following checklist covers the points that should be considered.

Aspects to consider

The risk assessments carried out under the Management of Health and Safety at Work and COSHH Regulations should show whether there are any specific risks in the workplace. The following should be considered:

▶ Are there hazardous substances, dangerous tools and equipment; dangerous manual handling tasks, electrical shock risks, dangers from neighbours or animals?
▶ Are there different levels of risk in parts of the premises or site?
▶ What is the history of accident and ill health, and type and location of incidents?
▶ What is the total number of persons likely to be on site?
▶ Are there young people, pregnant or nursing mothers on site and employees with disabilities or special health problems?
▶ Are the facilities widely dispersed with several buildings or compact in a multi-storey building?
▶ What is the pattern of working hours? Does it involve night work?
▶ Is the site remote from emergency medical services?
▶ Do employees travel or work alone for a considerable amount of time?
▶ Do any employees work at sites occupied by other employers?
▶ Are members of the public regularly on site?

First-aid needs assessment

This should consider the following:

▶ the type of the work being undertaken;
▶ hazards and the likely risk of causing harm from them;
▶ the size of the workforce;
▶ work patterns of the workers;
▶ holiday and other absences of those who will be first-aiders and appointed persons;
▶ the history of accidents within the organisation;
▶ the needs of travelling, remote and lone workers;
▶ the proximity of the sites of the organisation to emergency medical services;
▶ the involvement of the organisation in shared or multi-occupancy sites; and
▶ first aid for non-employees including members of the public.

Employers are responsible for meeting the first-aid needs of their employees working away from the main site. The assessment of first-aid needs should determine whether:

▶ those who travel long distances or are continuously mobile should carry a personal first-aid box; and
▶ employees should be issued with personal communicators/mobile phones.

Where an employer introduces an automated external defibrillator (AED) into the organisation, relevant information and written instructions, probably from the AED's manufacturer, on its use should be provided. However, fuller training is likely to make the user more confident and is now an integral part of the syllabus for FAW and Emergency First Aid at Work (EFAW) courses.

Impact on first-aid provision if risks are significant

Qualified first-aiders may need to be appointed if risks are significant. This will involve a number of factors which must be considered, including:

▶ training for first-aiders;
▶ additional first-aid equipment and the contents of the first-aid box;
▶ siting of first-aid equipment to meet the various demands in the premises. For example provision of equipment in each building or on several floors. There needs to be first-aid provision at all times during working hours;
▶ informing local medical services of the site and its risks;
▶ any special arrangements that may be needed with the local emergency services.

Any first-aid room provided under these Regulations must be easily accessible to stretchers and to other equipment needed to convey patients to and from the room. They must be signposted according to the Safety Signs and Signals Regulations (Figure 4.27).

Figure 4.27 (*left*) First-aid and stretcher sign; (*right*) First-aid sign

If employees travel away from the site, the employer needs to consider:

- issuing personal first-aid kits and providing training;
- issuing mobile phones to employees;
- making arrangements with employers on other sites.

Although there are no legal responsibilities for non-employees, the HSE strongly recommends that they are included in any first-aid provision.

Contents of the first-aid box

There is no standard list of items to put in a first-aid box. It depends on what the employer assesses the needs to be. Where there is no special risk in the workplace, a minimum stock of first-aid items is listed (see Table 4.1).

Tablets or medicines should not be kept in the first-aid box. Table 4.1 shows a suggested contents list only; equivalent but different items will be considered acceptable.

Table 4.1 Typical contents of first-aid box – low risk

Stock for up to 50 persons:	
A leaflet giving general guidance on first-aid, for example HSE leaflet *Basic advice on first-aid at work*.	
Medical adhesive plaster	40
Sterile eye pads	4
Individually wrapped triangular bandages	6
Safety pins	6
Individually wrapped medium sterile unmedicated wound dressings	8
Individually wrapped large sterile unmedicated wound dressings	4
Individually wrapped wipes	10
Paramedic shears	1
Pairs of latex gloves	2
Sterile eyewash if no clean running water	2

Appointed persons

An appointed person is someone who is appointed by management to:

- take charge when someone is injured or falls ill. This includes calling an ambulance if required;

- look after the first-aid equipment, for example keeping the first-aid box replenished;
- keeping records of treatment given.

Appointed persons should never attempt to give first aid for which they are not competent. Short emergency first-aid training courses are available. An appointed person should be available at all times when people are working at the work site – this may mean appointing more than one. The training should be repeated every three years to keep up to date.

A first-aider

A first-aider is someone who has undergone an HSE-approved training course in administering FAW and holds a current FAW certificate. Lists of local training organisations are available from the local environmental officer or HSE offices. The training should be repeated every three years to maintain a valid certificate and keep the first-aider up to date.

It is not possible to give definitive rules on when or how many first-aiders or appointed persons might be needed. This will depend on the circumstances of each particular organisation or work site. Table 4.2 offers suggestions on how many first-aiders or appointed persons might be needed in relation to categories of risk

Table 4.2 Number of first-aid personnel

Category of risk	Numbers employed at any location	Suggested number of first-aid personnel
Lower risk		
For example shops and offices, libraries	Fewer than 50	At least one appointed person
	50–100	At least one first-aider
	More than 100	One additional first-aider for every 100 employees
Medium risk		
For example light engineering and assembly work, food processing, warehousing	Fewer than 20	At least one appointed person
	20–100	At least one first-aider for every 50 employed (or part thereof)
	More than 100	One additional first-aider for every 100 employees
Higher risk		
For example most construction, slaughterhouses, chemical manufacture, extensive work with dangerous machinery or sharp instruments	Fewer than 5	At least one appointed person
	5–50	At least one first-aider
	More than 50	One additional first-aider for every 50 employees

Source: HSE.

and number of employees. The details in the table are suggestions only; they are not definitive, nor are they a legal requirement.

Employees must be informed of the first-aid arrangements. Putting up notices telling staff who and where the first-aiders or appointed persons are and where the first-aid box is will usually be enough. Special arrangements will be needed for employees with reading or language difficulties.

To ensure cover at all times when people are at work and where there are special circumstances, such as remoteness from emergency medical services, shift work or sites with several separate buildings, there may need to be more first-aid personnel than set out in Table 4.2.

Implementation of changes to first-aid training and approval arrangements

The HSE published a report in 2003 following a consultation of the Health and Safety (First-aid) at Work Regulations. Since these Regulations were introduced in 1981, there has been a significant transformation of the economy from manufacturing to service-based industries. The report identified the following points:

▶ small organisations found it difficult to release employees to attend a four-day first-aid training course;

▶ confusion over the role and training requirement of an appointed person;

▶ a significant depletion in the retained information of a trained first-aider after a short period of time after the training course.

Training guidelines have been issued by the HSE to address these issues. These guidelines are:

▶ a short first-aid course – Emergency First Aid at Work (EFAW). This lasts for one day and will enable the first-aider to give emergency first aid following an accident or ill-health problem. This is suitable for an organisation where the risk assessment has indicated that it is suitable for the level of risks present;

▶ a reduction to three days for the First Aid at Work (FAW) course;

▶ all first-aiders should undertake a half-day refresher course annually during the three-year certification period of the FAW and EFAW. This should address the depletion of information problem.

The first-aid training requirements have been amended so that the four-day course has been replaced by a First Aid at Work (FAW) course taught over three days. The syllabuses of the two courses are very similar with some topics being brought up to date. For example, the protocols used for resuscitation have become much easier to understand. Many medical professionals recommend that there should be an annual refresher course to accompany the course. The one-day

Emergency First Aid at Work (EFAW) course is designed for low-risk organisations and appointed persons.

Employers are able to send suitable employees on either a 6-hour (minimum) EFAW or an 18-hour (minimum) FAW course, based on the findings of their first-aid needs assessment (see Figure 4.28). After three years, first-aiders need to complete another course (either a 6-hour EFAW or a 12-hour FAW requalification course, as appropriate) to obtain a new certificate. Within any 3-year certification period, first-aiders should complete two annual refresher courses, covering basic life support/skills updates, which will each last for at least three hours.

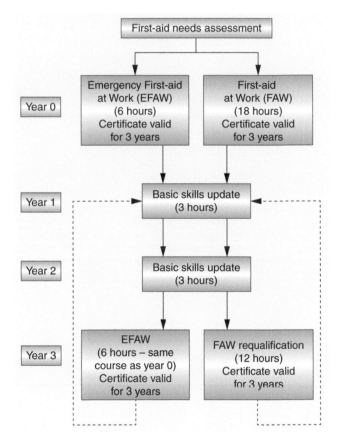

Figure 4.28 Flow chart showing courses to be completed over a 3-year certification period for EFAW and FAW. The dotted line indicates the route to be taken in subsequent years after completion of the relevant course at year 3

Detailed guidance for employers is available as a revision to the document produced by HSE: *First-aid at Work – The Health and Safety (First-Aid) Regulations 1981 – Approved Code of Practice and Guidance (L74)*. Within this, it is only the guidance that has been revised – the Regulations and Approved Code of Practice remain the same. HSE has revised its guidance for small and medium-sized enterprises in *First-aid at Work – Your Questions Answered* (INDG214).

Detailed guidance for first-aid training organisations is available on the HSE website: First-aid training and qualifications for the purposes of the Health and Safety (First-Aid) Regulations 1981 www.hse.gov.uk/pubns/web41.pdf.

4.6 Further information

The Health and Safety at Work etc. Act 1974

The Management of Health and Safety at Work Regulations 1999 (as amended)

The Health and Safety (First-Aid) Regulations 1981 as amended

Confined Spaces Regulations 1997

Emergency Procedures, HSE website guidance, www.hse.gov.uk/toolbox/managing/emergency.htm

Managing for health and safety, HSG65 third edition 2013, HSE Books, ISBN: 978-0-7176-6456-6 www.hse.gov.uk/pubns/priced/hsg65.pdf

Occupational Health and Safety Assessment Series (OHSAS 18000): Occupational Health and Safety Management Systems OHSAS 18001:2007 ISBN: 978-0-5805-9404-5; OHSAS18002:2008 ISBN: 978-0-5806-2686-9

HSE, 'Managing for health and safety', Parts 3 and 4 www.hse.gov.uk/managing/

Five Steps to Risk Assessment (INDG163, rev 3), HSE Books, ISBN: 978-0-7176-6440-5 www.hse.gov.uk/pubns/indg163.pdf

Young people and work experience, A brief guide to health and safety for employers, INDG364 (rev 1), HSE books, www.hse.gov.uk/pubns/indg364.pdf

New and expectant mothers www.hse.gov.uk/mothers/

Safe work in confined spaces; Confined Spaces Regulations 1997, ACoP regulations and guidance, (L101), second edition, HSE Books, ISBN: 978-0-7176-6233-3 www.hse.gov.uk/pubns/priced/l101.pdf

Guidance on permit-to-work systems. A guide for the petroleum, chemical and allied industries (HSG250) HSE Books, ISBN: 978-0-7176-2943-5 www.hse.gov.uk/pubns/priced/hsg250.pdf

Human factors: permit to work systems www.hse.gov.uk/humanfactors/topics/ptw.htm

Bird, F E, 1974. *Management guide to Loss Control*, Institute Press, Atlanta, Georgia, USA

First aid at work, The Health and Safety (First-Aid) Regulations 1981 (L74, third edition 2013), HSE Books, ISBN: 978-0-7176-6560-0 www.hse.gov.uk/pubns/priced/l74.pdf

4.7 Practice questions

1. General risk assessments on activities within a workplace are required under the Management of Health and Safety at Work Regulations:
   (a) **Explain** the criteria that must be met for the assessment to be deemed 'suitable and sufficient'.
   (b) **Outline** the **FIVE** key stages of the risk assessment process, identifying the issues that would need to be considered at **EACH** stage.
   (c) **State** the legal requirements for recording workplace risk assessments.

2. A plastics manufacturing company has moved to a new larger site and a programme of general risk assessments needs to be undertaken.
   (a) **Outline** the factors that should be considered when planning the risk assessment programme.
   (b) **Identify** the important characteristics required of the risk assessment team to ensure that it has the necessary competence.
   (c) **Outline** the content of a training course for the risk assessment team.

3. (a) **Outline** the hazards that might be encountered in a busy hotel kitchen.
   (b) **Identify SIX** hazards that might be considered when assessing the risk to

health and safety of a multi-storey car park attendant.
   (c) **Outline** the hazards that might be encountered by a gardener employed by a local authority parks department.

4. (a) **Define** the meaning of the term 'young person' as used in health and safety legislation.
   (b) **Identify FOUR** 'personal' factors that may place young persons at a greater risk from workplace hazards.
   (c) **Outline** the factors to be taken into account when undertaking a risk assessment on young persons who are to be employed in the workplace.

5. (a) **Identify** work activities that may present a particular risk to pregnant women at work giving an example of **EACH** type of activity.
   (b) **Outline** the factors that the employer should consider when undertaking a specific risk assessment in relation to a pregnant employee.

6. **Outline** the issues to be considered to ensure the health and safety of disabled workers in the workplace.

7. **Outline** the issues that should be considered to ensure the health and safety of cleaners employed in a school out of normal working hours.

8. (a) **Identify** the personal factors that may increase the risks to an individual who is required to work alone away from their workplace.
    (b) **Describe** the controls that employers could implement to help minimise the risk to lone workers.

9. (a) **Identify SIX** of the General Principles of Prevention given in Schedule 1 of the Management of Health and Safety at Work Regulations that could be used to reduce the risk of injury from hazards found in a workplace.
    (b) **Outline**, with examples, the general hierarchy of control that should be applied when controlling health and safety risks in the workplace.

10. **Outline** the possible consequences of poor standards of housekeeping on health and safety in the workplace.

11. **Identify** the shape and colour, and give a relevant example, of each of the following types of safety sign:
    (a) prohibition
    (b) warning
    (c) mandatory
    (d) safe condition.

12. **Explain** why personal protective equipment (PPE) should be considered as a last resort in the control of occupational health hazards.

13. **Outline FOUR** sources of internal and **FOUR** sources of external information that may be consulted when dealing with a health and safety issue in the workplace.

14. A company is about to introduce a significant change in the working processes in the company.

Outline the issues that will need to be managed so that the impact of the change on the workforce will be reduced.

15. (a) **Explain** the meaning of the term 'safe system of work'.
    (b) **State** the legal requirements for an employer to provide a safe system of work to an employee.
    (c) **Outline** the factors that should be considered when developing a safe system of work.
    (d) **Outline SIX** issues that should be addressed during the preparation of a safe system of work.

16. (a) **Define** the term 'permit-to-work system'.
    (b) **Outline FIVE** types of work situation that may require a permit-to-work system, giving reasons in **EACH** case for the requirement.
    (c) **Outline** the details that a permit to work should specify.

17. (a) **Identify FOUR** types of emergency procedure that an organisation might need to have in place.
    (b) **Outline** reasons why workplace emergency procedures should be developed and practised.
    (c) **Explain** why visitors to a workplace should be informed of the emergency procedures.

18. (a) **Identify** the two main functions of first-aid treatment.
    (b) **Outline** the factors to consider when making a risk assessment of first-aid provision in a workplace.

APPENDIX 4.1 Hazard checklist

The following checklist may be helpful.

1. Equipment/mechanical

entanglement
friction/abrasion
cutting
shearing
stabbing/puncturing
impact
crushing
drawing-in
air or high-pressure fluid injection
ejection of parts
pressure/vacuum
display screen equipment
hand tools

2. Transport

works vehicles
mechanical handling
people/vehicle interface

3. Access

slips, trips and falls
falling or moving objects
obstruction or projection
working at height
confined spaces
excavations

4. Handling/lifting

manual handling
mechanical handling

5. Electricity

fixed installation
portable tools and equipment

6. Chemicals

dust/fume/gas
toxic
irritant
sensitising
corrosive
carcinogenic
nuisance

7. Fire and Explosion

flammable materials/gases/liquids
explosion
means of escape/alarms/detection

8. Particles and dust

inhalation
ingestion
abrasion of skin or eye

9. Radiation

ionising
non-ionising

10. Biological

bacterial
viral
fungal

11. Environmental

noise
vibration
light
humidity
ventilation
temperature
overcrowding

12. The individual

individual not suited to work
long hours
high work rate
violence to staff
unsafe behaviour of individual
stress
pregnant/nursing women
young people

13. Other factors to consider

people with a disability
lone workers
poor maintenance
lack of supervision
lack of training
lack of information
inadequate instruction
unsafe systems

4

131

APPENDIX 4.2 Risk assessment example 1: Hairdressing salon

Name of Company:	His and Hers Hairdressers	Date of Assessment:	11 January 2020
Name of Assessor:	A. R. Smith	Date of Review:	12 July 2020

Hazards	Persons affected	Risks	Initial risk level	Existing controls	Additional controls	Action by whom?	Action by when?	Done
Hairdressing products and chemicals Various bleaching and cleansing products in particular	Staff and customers	Eye and/or skin irritation Possible allergic reaction	Medium	• COSHH assessment completed • Non-latex gloves are provided for staff when using products • Customers are protected with single use towels	• Needs to be reviewed • Eye baths to be purchased for treatment of eye splashes	Manager Owner	12/2/20 12/2/20	8/2/20 24/1/20
Lightening (bleach) product Hydrogen peroxide Oxidative colourants				• Only non-dusty bleaches used • Staff report any allergies at induction	• Repeat allergy checks every 3 months and records kept	Manager	14/2/20 then every 3 months	12/2/20
				• Storeroom and salon well ventilated	• Storage in salon kept to 1-day requirement	Manager	18/1/20	16/01/20
				• Products stored as per manufacturer's recommendations				
				• Staff are specifically trained in the correct use of products	• Refresher training every 3 months and records kept	Manager	11/4/20	8/4/20
				• Staff check whether customers have allergies to any products	• Records kept of customer allergies and simple patch tests introduced	Staff Manager	25/4/20	7/5/20
Sharp instruments	Staff and customers	Cuts, grazes and blood-borne infections	Medium	• All sharp instruments sterilized after use • Sterilizing liquid changed daily • Sharps box available for disposable blades	• Regular recorded checks that sterilization procedures correctly followed	Manager	18/1/20 then monthly	18/1/20
				• First-aid box available	• Contents of first-aid box checked weekly	Manager	18/1/20 then weekly	18/1/20
Fire	Staff and customers	Smoke inhalation and burns	Low	• Fire risk assessment completed	• No flammable products will be displayed in the windows or near heater	Manager	18/1/20	18/1/20

Hazards	Persons affected	Risks	Initial risk level	Existing controls	Additional controls	Action by whom?	Action by when?	Done
Electricity	Staff and customers	Electrical shocks and burns. Also a risk of fire	Low	• Any damaged cables, plugs or electrical equipment is reported to the manager • All portable electrical equipment and thermostats checked every 6 months by a competent person • All electrical equipment purchased from a reliable supplier • Staff shown how to use and store hairdryers, etc. safely and isolate electrical supply	• An appropriate fire extinguisher (carbon dioxide) should be available • Make a visual check of electrical equipment, cables and sockets every month and record findings • All internal wiring should be checked by a qualified electrician	Owner Manager Owner	25/1/20 25/1/20 then every month 25/4/20	23/1/20 25/1/20 3/4/20
Standing for long periods	Staff	Back pain and pain in neck, shoulders, legs and feet – musculoskeletal injuries	Medium	• Staff given regular breaks • Stools available for staff for use while trimming hair • Customer chairs are adjustable in height	• Develop a formal rota system for breaks	Manager	25/1/20	21/1/20
Wet hand work	Staff	Skin sensitization, dry skin, dermatitis	Medium	• Staff are trained to wash and dry hands thoroughly between hair washes • Non-latex gloves are provided for staff • Moisturizing hand cream is provided for staff	• Ensure that a range of glove sizes are available for staff • Staff will always wear gloves for all wet work	Manager Manager	25/1/20 25/1/20	31/1/20 1/2/20
Slips and trips	Staff and customers	Bruising, lacerations and possible fractures	Low	• Cut hair is swept up regularly • Staff must wear slip-resistant footwear • No trailing leads on floor • Any spills are cleaned immediately • Door mat provided at shop entrance	• Organize repair of worn floor covering	Manager	8/2/20	5/2/20

4

133

APPENDIX 4.3 Risk assessment example 2: Office cleaning

Name of Company: Apex Cleaning Company   Date of Assessment: 14 May 2020

Name of Assessor: TW James   Date of Review: 14 November 2020

Hazards	Persons affected	Risks	Initial risk level	Existing controls	Additional controls	Action by whom?	Action by when?	Done
Machine cleaning of floors	Staff and others	Injury to ankles due to incorrect use of machinery	Low	• Machine supplied for the job is suitable • Machine is maintained regularly and examined by a competent person • Cleaners trained in the safe use of the machine	• Maintenance and inspections to be documented • Training to be documented	Manager Manager	30/5/20 30/5/20	21/5/20 28/5/20
Electrical	Staff	Electric shock and burns	Low	• Staff trained in visual inspection of plugs, cables and switches before use on each shift • Staff inform manager of any defects • All portable electrical equipment is regularly tested by a competent person • Staff trained not to splash water near machines or wall sockets	• Training to be documented • Defect forms and a procedure to be developed and communicated to staff • All electrical equipment to be listed and results of portable appliance testing (PAT) tests recorded • Training to be documented	Manager Manager Manager Manager	30/5/20 13/6/20 20/6/20 30/5/20	28/5/20 18/6/20 18/6/20 28/5/20
Lone working	Staff	Accident, illness or attack by intruder	Medium	• Staff sign in and out at either end of shift with security staff • Security staff visit staff regularly • All staff issued with mobile telephone number of manager for emergency use	• Check that staff are aware that security staff are trained first-aiders • Check on regularity of these visits • If manager cannot respond, staff to be told to ring emergency services	Manager Manager Manager	22/5/20 22/5/20 22/5/20	20/5/20 20/5/20 20/5/20
Work at height	Staff	Bruising, sprains, lacerations and fractures	Low	• Staff issued with long-handled equipment for work at high level • Staff told not to stand on chairs or stepladders	• Staff to be trained in safe system of work for cleaning stairways, escalators and external windows	Manager	11/6/2020	3/6/20

Hazards	Persons affected	Risks	Initial risk level	Existing controls	Additional controls	Action by whom?	Action by when?	Done
Contact with bleach and other cleaning chemicals	Staff	Skin irritation. Possible allergic reaction and eye injuries from splashes	Low	• COSHH assessment completed • Staff trained in safe use and storage of cleaning chemicals • Staff induction questionnaire includes details of any allergies and skin problems • Impervious rubber gloves are issued for use with any chemical cleaners • Equipment used with chemicals is regularly serviced and/or cleaned	• COSHH assessment to be reviewed • Records of the staff training to be kept and updated • Check whether products marked 'harmful' or 'irritant' can be substituted with milder alternatives • Staff to report any health problems	Manager Manager Manager Staff	27/6/20 30/5/20 20/6/20 Ongoing	18/6/20 28/5/20 25/6/20
Musculoskeletal disorders and injuries	Staff	Musculoskeletal injuries to the back, neck, shoulders, legs and feet	Medium	• Staff trained in correct lifting technique • Long-handled equipment used to prevent stooping • Each floor is provided with all necessary cleaning equipment • Staff trained not to overfill buckets	• Check that staff are not lifting heavy objects or furniture or unduly stretching while cleaning • Check whether more up-to-date equipment is available (long-handled wringers and buckets on wheels)	Manager Manager	28/5/20 28/5/20	26/5/20 28/5/20
Slips, trips and falls	Staff and others	Bruising, sprains, lacerations and possible fractures	Medium	• Warning signs placed on wet floors • Staff use nearest electrical socket to reduce trip hazards • Staff told to wear slip-resistant footwear • Wet floor work restricted to less busy times at client offices • Client encouraged in a good housekeeping policy	• Introduce a wet and dry mops cleaning system for floors	Manager	28/5/20	26/5/20
Fire	Staff and others	Smoke inhalation and burns	Low	• Staff trained in client's emergency fire procedures including assembly point	• On each floor, a carbon dioxide fire extinguisher is available for cleaning staff	Manager	6/6/20	2/6/20

4

APPENDIX 4.4 Asbestos examples of safe systems of work

Standards required

All work involving asbestos in any form will be carried out in accordance with the current Control of Asbestos Regulations and Approved Code of Practice. Any asbestos removal must be done in accordance with the Asbestos Licensing Regulations. The use of new asbestos-containing materials is prohibited.

Planning procedures

All work will be tendered for or negotiated in accordance with the approved standards.

The Contracts Manager will ascertain at an early stage whether asbestos in any form is likely to be present or used on the site. If details provided by the client are inconclusive, then an occupational hygiene specialist will be asked to take and analyse samples.

Method statements will be prepared by the Contracts Manager in conjunction with an occupational hygiene specialist, and, where necessary, a licensed asbestos removal contractor will be selected to carry out the work.

The Contracts Manager will ensure that any requirement to give notice of the work to the Health and Safety Executive is complied with.

Where any work involving asbestos materials not subject to the licensing requirements is to be carried out by employees, the working methods, precautions, safety equipment, protective clothing, special tools, etc. will be arranged by the Contracts Manager.

Supervision

Before work starts, all information on working methods and precautions agreed will be issued to site supervision by the Contracts Manager in conjunction with the Safety Adviser/Officer.

The Site Supervisor in conjunction with management will ensure that the licensed contractor contracted to carry out the removal work has set up operations in accordance with the agreed method statement and that the precautions required are fully maintained throughout the operation so that others not involved are not exposed to risk.

Where necessary, smoke testing of the enclosure and monitoring of airborne asbestos fibre concentrations outside the removal enclosure will be carried out by an occupational hygiene specialist.

The Site Supervisor will ensure that when removal operations have been completed, no unauthorised person enters the asbestos removal area until clearance samples have been taken by an occupational hygiene specialist and confirmation received that the results are satisfactory.

Where employees are required to use or handle materials containing asbestos not subject to the Licensing Regulations, the Site Supervisor will ensure that the appropriate safety equipment and protective clothing is provided and that the agreed safe working procedures are understood by employees and complied with.

All warning labels will be left in place on any asbestos materials used on site.

Examples

1. Painting undamaged asbestos insulating boards

Description

This task guidance sheet can be used where undamaged asbestos insulating boards need to be painted. This may be to protect them, or for aesthetic reasons.

It is not appropriate where the material is damaged. Use a specialist contractor licensed by HSE.

Carry out this work only if you are properly trained.

PPE

▶ disposable overalls fitted with a hood;
▶ boots without laces (laced boots can be difficult to decontaminate);
▶ disposable particulate respirator (FF P3).

Equipment

▶ 500-gauge polythene sheeting and duct tape;
▶ warning tape and notices;
▶ Type H vacuum cleaner to BS5415 (if dust needs to be removed from the asbestos insulating board);
▶ paint conforming to the original specification, for example fire resistant. Select one low in hazardous constituents, for example solvents;
▶ low-pressure spray or roller/brush;
▶ bucket of water and rags;
▶ suitable asbestos waste container, for example a labelled polythene sack;
▶ appropriate lighting.

Preparing the work area

This work may be carried out at height; if so, the appropriate precautions MUST be taken.

▶ Carry out the work with the minimum number of people present.
▶ Restrict access, for example close the door and/or use warning tape and notices.
▶ Use polythene sheeting, secured with duct tape, to cover surfaces within the segregated area, which could become contaminated.
▶ Ensure adequate lighting.

Painting

- Never prepare surfaces by sanding.
- Before starting, check there is no damage.
- Repair any minor damage.
- If dust needs to be removed, use a Type H vacuum cleaner or rags.
- Preferably use the spray to apply the paint.
- Spray using a sweeping motion.
- Do not concentrate on one area as this could cause damage.
- Alternatively, apply the brush/roller lightly to avoid abrasion/damage.

Cleaning

- Use wet rags to clean the equipment.
- Use wet rags to clean the segregated area.
- Place debris, used rags, polythene sheeting and other waste in the waste container.

Personal decontamination

- Use suitable personal decontamination procedure.

Clearance procedure

- Visually inspect the area to make sure that it has been properly cleaned.
- Clearance air sampling is not normally required.

2. Removal of asbestos cement sheets, gutters, etc.

Description

This task guidance sheet can be used where asbestos cement sheets, gutters, drains and ridge caps, etc. need to be removed.

For the large-scale removal of asbestos cement, for example demolition, read *Working with Asbestos Cement* HSG189/2 HSE Books 1999 ISBN 978 0 7176 1667 1.

It is not appropriate for the removal of asbestos insulating board.

Carry out this work only if you are properly trained.

PPE

- Use disposable overalls fitted with a hood.
- Waterproof clothing may be required outside.
- Boots without laces (laced boots can be difficult to decontaminate).
- Use disposable particulate respirator (FIF P3).

Equipment

- 500- and 1,000-gauge polythene sheeting and duct tape;
- warning tape and notices;
- bolt cutters;
- bucket of water, garden type spray and rags;
- suitable asbestos waste container, for example a labelled polythene sack;
- lockable skip for larger quantities of asbestos cement;

- asbestos warning stickers;
- appropriate lighting.

Preparing the work area

This work may be carried out at height; if so, the appropriate precautions to prevent the risk of fails MUST be taken.

- Carry out the work with the minimum number of people present.
- Restrict access, for example close the door and/ or use warning tape and notices.
- Use 500-gauge polythene sheeting, secured with duct tape, to cover any surface within the segregated area, which could become contaminated.
- It is dangerous to seal over exhaust vents from heating units in use.
- Ensure adequate lighting.

Overlaying

- Instead of removing asbestos cement roofs, consider overlaying with a non-asbestos material.
- Attach sheets to existing purlings but avoid drilling through the asbestos cement.
- Note the presence of the asbestos cement so that it can be managed.

Removal

- Avoid breaking the asbestos cement products.
- If the sheets are held in place with fasteners, dampen and remove – take care not to create a risk of slips.
- If the sheets are bolted in place, use bolt cutters avoiding contact with the asbestos cement. Remove bolts carefully.
- Unbolt or use bolt cutters to release gutters, drain pipes and ridge caps, avoiding contact with the asbestos cement.
- Lower the asbestos cement to the ground. Do not use rubble chutes.
- Check for debris in fasteners or bolt holes. Clean with wet rags.
- Single asbestos cement products can be double wrapped in 1,000-gauge polythene sheeting (or placed in waste containers if small enough). Attach asbestos warning stickers.
- Where there are several asbestos cement sheets and other large items, place in a lockable skip.

Cleaning

- Use wet rags to clean the equipment.
- Use wet rags to clean segregated area.
- Place debris, used rags, polythene sheeting and other waste in the waste container.

Personal decontamination

- Use a suitable personal decontamination system.

Clearance procedure

▶ Visually inspect the area to make sure that it has been properly cleaned.

▶ Clearance air sampling is not normally required.

3. Personal decontamination system

Description

This guidance sheet explains how you should decontaminate yourself after working with asbestos materials.

If you do not decontaminate yourself properly, you may take asbestos fibres home on your clothing. You or your family and friends could be exposed to them if they were disturbed and became airborne.

It is important that you follow the procedures given in the task guidance sheets and wear PPE such as overalls correctly; this will make cleaning easier.

Removing and decontaminating PPE

▶ Remove your respirator last.

▶ Clean your boots with wet rags.

▶ Where available, use a Type H vacuum cleaner to clean your overalls.

▶ Otherwise use a wet rag – use a 'patting' action – rubbing can disturb fibres.

▶ Where two or more workers are involved they can help each other by 'buddy' cleaning.

▶ Remove overalls by turning inside out – place in suitable asbestos waste container.

▶ Use wet rags to clean waterproof clothing.

▶ Disposable respirators can then be removed and placed in a suitable asbestos waste container.

Personal decontamination

▶ Site-washing facilities can be used but restrict access during asbestos work.

▶ Wash each time you leave the work area.

▶ Use wet rags to clean washing facilities at the end of the job.

▶ Clean facilities daily if the job lasts more than a day.

▶ Visually inspect the facilities once the job is finished.

▶ Clearance air sampling is not normally required.

Further information

These examples are taken from Asbestos Essentials Task Manual HSG210 (now revised) HSE Books 2008 ISBN 978 0 7176 6263 0. Many more examples are contained in the publication including equipment and method guidance sheets. Obtainable from HSE Books.

CHAPTER 5

Health and safety monitoring and measuring

5.1 Active and reactive monitoring

5.1.1 Introduction

Why measure performance?

You can't manage what you can't measure
– Drucker

Measurement is an accepted part of the 'plan–do–check–act' management process discussed in Chapter 2. Monitoring performance is as much part of a health and safety management system as financial, production or service delivery management. The HSG65 framework for managing health and safety, illustrated in Figure 5.1, shows where monitoring performance standards fits within the overall health and safety management system shown in Figure 2.1.

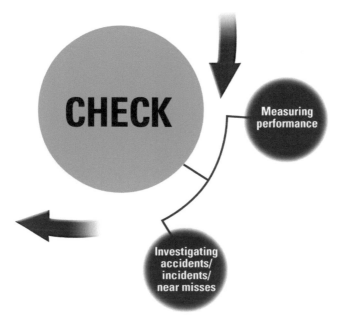

Figure 5.1 CHECK involves measuring performance and investigating incidents

The main purpose of monitoring health and safety performance is to provide information on the progress and current status of the strategies, processes and activities employed to control health and safety risks. Effective measurement not only provides information on what the levels are but also why they are at this level, so that corrective action can be taken.

Senior managers often measure company performance by using, for example, percentage profit, or return on investment or market share. A common feature of these measures would be that they are generally positive in nature, which demonstrates achievement, rather than negative, which demonstrates failure.

Yet, if senior managers are asked how they measure their companies' health and safety performance, it is likely that the only measure would be accident or injury statistics. Although the general business performance of an organisation is subject to a range of positive measures, for health and safety it too often comes down to one negative measure of failure.

Health and safety differs from many areas measured by managers because:

▶ improvement in performance means **fewer** outcomes from the measure (injuries or ill health) rather than more;
▶ a low injury or ill-health rate trend over years is still no guarantee that risks are being controlled and that incidents will not happen in the future. This is particularly true in organisations where major hazards are present but there is a low probability of accidents. This is particularly true in organisations when major hazards are present, where fortunately, there is a low probability of an incident occurring but it would be catastrophic if it happened.

There is no single reliable measure of health and safety performance. What is required is a 'basket' of measures, providing information on a range of health and safety issues.

There are some significant problems with the use of injury/ill-health statistics in isolation:

▶ there may be under-reporting – focusing on injury and ill-health rates as a measure, especially if a reward system is involved, can lead to non-reporting to keep up performance;

▶ it is often a matter of chance whether a particular incident causes an injury, and they may not show whether or not a hazard is under control. Luck, or a reduction in the number of people exposed, may produce a low injury/accident rate rather than good health and safety management;

▶ an injury is the particular consequence of an incident and often does not reflect the potential severity. For example an unguarded machine could result in a cut finger or an amputation;

▶ people can be absent from work for reasons which are not related to the severity of the incident;

▶ there is evidence to show that there is little relationship between 'occupational' injury statistics (e.g. slips, trips and falls) and the reasons for the lack of control of major accident hazards (e.g. loss of containment of flammable or toxic material);

▶ a small number of accidents may lead to complacency;

▶ injury statistics demonstrate outcomes not causes.

Because of the potential shortcomings related to the use of accident/injury and ill-health data as a single measure of performance, more active measures are required. These require a systematic approach to deriving positive measures and how they link to the overall risk control process, rather than a quick-fix based on things that can easily be counted, such as the numbers of training courses or numbers of inspections, which has limited value. The resultant data provide no information on how the figure was arrived at, whether it is 'acceptable' (i.e. good/bad) or the quality and effectiveness of the activity. A more disciplined approach to health and safety performance standards measurement is required. This needs to develop as the health and safety management system develops.

Managers should check by asking key questions to ensure that arrangements for health and safety risk control are in place, comply with the law as a minimum, and operate effectively.

There are two basic types of checking or monitoring:

Active monitoring (a 'leading indicator' see 5.1.3), by taking the initiative before things go wrong, involves routine inspections and checks to make sure that standards and policies are being implemented and that controls are working.

Reactive monitoring (a 'lagging indicator' see 5.1.2), after things go wrong, involves looking at historical events to learn from mistakes and see what can be put right to prevent a recurrence.

The UK Health and Safety Executive's (HSE) experience is that organisations find health and safety performance measurement a difficult subject. They struggle to develop health and safety performance measures which are not based solely on injury and ill-health statistics.

5.1.2 Lagging indicators of safety performance

What is a lagging indicator?

Lagging indicators measure a company's incidents in the form of past accident statistics.

Examples include:

▶ Injury frequency and severity
▶ RIDDOR reportable injuries
▶ Lost workdays
▶ Employers' liability compensation costs.

Why use lagging indicators?

Lagging indicators are the traditional safety metrics used to indicate progress toward compliance with safety rules. These are the bottom-line numbers that evaluate the overall effectiveness of safety at your facility. They tell you how many people got hurt and how badly.

The drawbacks of lagging indicators

The major drawback to only using lagging indicators of safety performance is that they tell you how many people got hurt and how badly, but not how well your company is doing at preventing incidents and accidents.

The reactionary nature of lagging indicators makes them a poor gauge of prevention. For example, when managers see a low injury rate, they may become complacent and put safety on the bottom of their to-do list, when, in fact, there are numerous risk factors present in the workplace that will contribute to future injuries.

5.1.3 Leading indicators of safety performance

What is a leading indicator?

A leading indicator is a measure preceding or indicating a future event used to drive and measure activities carried out to prevent and control injury.

Examples include:

▶ Safety training
▶ Ergonomic opportunities identified and corrected
▶ Reduction of musculoskeletal disorder (MSD) risk factors
▶ Employee perception surveys
▶ Safety audits.

5

Figure 5.2 Conceptual spectrum of leading to lagging indicators

Why use leading indicators?

Leading indicators are focused on future safety performance and continuous improvement. These measures are proactive in nature and report what employees are doing on a regular basis to prevent injuries.

5.1.4 Active monitoring – leading indicators

Active monitoring is a means of verifying the adequacy of, and the degree of compliance with, the risk control systems that have been established. It is intended to identify deficiencies in the safety management system for subsequent remedial action, and thus prevent accidents by eradicating their potential causes. It also enables a company to gain a picture of its health and safety performance and chart the progress of improvement.

Active monitoring employs several complementary methods which address differing aspects and areas of the organisation. These methods may be usefully categorised as follows:

(i) The **Safety Management System (SMS) Audit** see 5.3 for more information on auditing.

(ii) The **Safety Survey** is usually a detailed assessment of one aspect of an organisation's SMS, e.g. the organisation's training arrangements.

(iii) The **Safety Inspection** consists of a formal assessment of workplace safety, and the identification of hazardous conditions or practices, for subsequent remedial action. It is normally carried out by first line managers referring to, and completing, a checklist.

(iv) The **Safety Tour** addresses the 'people' aspects of workplace safety, and by discussions with a range of staff, establishes their familiarity with safety procedures and requirements. It is normally carried out by middle and senior management, as one means of demonstrating their commitment to safety. A questionnaire is frequently used.

(v) **Safety sampling** is a helpful technique that helps organisations to concentrate on one particular area or subject at a time. A specific area is chosen which can be inspected in about 30 minutes. A checklist is drawn up to facilitate the inspection looking at specific issues. These may be different types of hazard: they may be unsafe acts or conditions noted; they may

be proactive, good behaviour or practices noted. The inspection team or person then carries out the sampling at the same time each day or week in the specified period. The results are recorded and analysed to see if the changes are good or bad over time. Of course, defects noted must be brought to the notice of the appropriate person for action on each occasion.

Management roles

Performance should be measured at each management level from directors downwards. It is not sufficient to monitor by exception, where, unless problems are raised, it is assumed to be satisfactory. Senior managers must satisfy themselves that the correct arrangements are in place and working properly. Responsibilities for both active and reactive monitoring must be laid down and managers need to be personally involved in making sure that plans and objectives are met and compliance with standards is achieved. Although systems may be set up with the guidance of safety professionals, managers should be personally involved and given sufficient training to be competent to make informed judgements about monitoring performance.

Other people, such as safety representatives, will also have the right to inspect the workplace. Each employee should be encouraged to inspect their own workplace frequently to check for obvious problems and rectify them if possible or report hazards to their supervisors.

Specific statutory (or thorough) examinations of, for example, lifting equipment or pressure vessels, have to be carried out at intervals laid down in written schemes by competent persons – usually specially trained and experienced inspection/insurance company personnel.

This frequency of monitoring and inspections will depend on the level of risk and any statutory inspection requirement. Directors may be expected to examine the premises formally at an annual audit, whereas departmental supervisors may be expected to carry out inspection every week. Senior managers should regularly monitor the health and safety plan to ensure that objectives are being met and to make any changes to the plan as necessary.

5.1.5 Safety inspection programme

Within any active monitoring programme there should be a system for workplace safety inspections. It often forms part of the preventative maintenance scheme for plant and equipment. However, they may also be covered by legal examination and inspection requirements. Equipment in this category includes steam boilers, other pressure vessels, hoists, lifts, cranes, chains, ropes, lifting tackle, power presses, scaffolds, trench supports and local exhaust ventilation. Workplace inspections should include other precautions, such as those covering the use of premises, welfare facilities, behaviour of people and systems of work.

(a)

(b)

(c)

Figure 5.3 (a) Poor conditions – inspection needed; (b) inspection in progress; (c) poor conditions in offices can cause accidents

Inspections should be done by people who are competent to identify the relevant hazards and risks and who can assess the conditions found. Preferably they should mainly be local staff who understand the problems and can immediately present solutions. It is essential that they do not, in any way, put themselves or anyone else at risk during the inspection. Particular care must be taken with regard to safe access. In carrying out inspections, the safety of people's actions should be considered, in addition to the safety of the conditions they are working in – a ladder might be in perfect condition but it has to be used properly too.

A properly thought-out approach to inspection will include:

▶ well-designed inspection forms to help plan and initiate remedial action by requiring those doing the inspection to rank any deficiencies in order of importance;
▶ summary lists of remedial action with names and deadlines to track progress on implementing improvements;
▶ periodic analysis of inspection forms to identify common features or trends which might reveal underlying weaknesses in the system;
▶ information to aid judgements about any changes required in the frequency or nature of the inspection programme.

The inspection programme should be properly targeted whilst considering all risks in the premises. For example, low risks could be covered by monthly or bi-monthly general inspections involving a wide range of workplace safety issues such as the general condition of premises, floors, passages, stairs, heating, ventilation, lighting, welfare facilities and first aid. Higher risks need more frequent and detailed inspections; perhaps weekly or even, in extreme cases, daily or before using a piece of equipment like an injection moulding machine or mobile plant.

The daily, weekly or monthly inspections will be aimed at checking conditions in a specified area against a fixed checklist drawn up by local management. It will cover specific items, such as the guards at particular machines, whether access/agreed routes are clear, whether fire extinguishers are in place, etc. The checks should be carried out primarily by competent staff of the department who should sign off the checklist. Checks should not last more than half an hour, perhaps less. This is not a specific hazard-spotting operation, but there should be a space on the checklist for the inspectors to note down any particular problems encountered.

In order to get maximum value from inspection checklists, they should be designed so that they require objective rather than subjective judgements of conditions. For example, asking the people undertaking a general inspection of the workplace to rate housekeeping as good or bad, begs questions as to what does good and bad mean, and what criteria should be used to judge this. If good housekeeping means there is no rubbish left on the floor, all waste bins are regularly emptied and not overflowing, floors are swept each day and cleaned once a week, decorations should be in good condition with no peeling paint, then this should be stated. Adequate expected standards should be provided in separate notes so that those inspecting know the standards that are required.

The checklist or inspection form should facilitate:

▶ the planning and initiation of remedial action, by requiring those doing the inspection to rank deficiencies in priority order (those actions which are most important rather than those which can easily be done quickly);
▶ identifying those responsible for taking remedial actions, with sensible timescales to track progress on implementation;
▶ periodic monitoring to identify common themes which might reveal underlying problems in the system;
▶ management information on the frequency or nature of the monitoring arrangements.

The checklist could be structured using the 'four Ps' (note that the examples are not a definitive list); see Chapter 17 for an example:

Premises, including:
  ▶ work at height
  ▶ access
  ▶ working environment
  ▶ welfare
  ▶ services
  ▶ fire precautions
Plant and substances, including:
  ▶ work equipment
  ▶ manual and mechanical handling
  ▶ vehicles
  ▶ dangerous/flammable substance
  ▶ hazardous substance
Procedures, including:
  ▶ risk assessments
  ▶ safe systems of work
  ▶ permits to work
  ▶ personal protective equipment
  ▶ contractors
  ▶ notices, signs and posters
People, including:
  ▶ health surveillance
  ▶ people's behaviour
  ▶ training and supervision
  ▶ appropriate authorised persons
  ▶ violence
  ▶ those especially at risk.

Reports from inspections

Some of the items arising from safety inspections will have been dealt with immediately; other items will require action by specified people. Where there is some doubt about the problem, and what exactly is required, advice should be sought from the safety adviser or external expert. A brief report of the inspection and any resulting action list should be submitted to the safety committee. While the committee may not have the time available to consider all reports in detail, it will want to be satisfied that appropriate action is taken to resolve all matters; it will be necessary for the committee to follow up the reports until all matters are resolved.

Essential elements of a report are:

▶ identification of the organisation, workplace, inspector and date of inspection;
▶ list of observations;
▶ priority or risk level;
▶ actions to be taken;
▶ timescale for completion of the actions.

Appendix 5.1 gives examples of poor workplaces which can be used for practice exercises. In Chapter 17 a specimen workplace inspection form can be adapted for use at many workplaces. Chapter 17 also provides a workplace inspection checklist which can be used to assist in drawing up a specific checklist for any particular workplace or as an aide-memoire for the workplace inspection report form. It is important to check with the organisation requesting the report in case their in-house format differs. Large contracts often use a common digital platform to circulate and capture incidents and information.

Advantages of using a checklist

The strengths of using a checklist for inspection of a workplace are:

▶ the opportunity given for prior preparation and planning in order to ensure that the inspection is both structured and systematic;
▶ the degree of consistency obtained first by those carrying out the inspection and second in the areas and issues to be covered;
▶ it can be easily adapted or customised for different areas;
▶ it results in an immediate record of finding;
▶ it provides an easy method of comparison and audit.

Checklists can be used to improve safety performance by:

▶ checking adherence to standards;
▶ allowing an analysis of trends to be made;
▶ identifying problems which if quickly rectified can boost the morale of the workforce;
▶ demonstrating the commitment of management; and
▶ the involvement of employees which will increase the ownership of health and safety within the organisation.

There are a few downsides with checklists including:

▶ the lack of comments so that reasons for passing or failing are not given;

▶ the tick-box mentality when it is all too easy to merely tick a box than provide any original thinking;

▶ people may shut their eyes to items not on the list.

These can be overcome by requiring inspectors to fill in the comments section with each item and providing space for other issues with perhaps a few prompts of additional hazards which could be present.

5.1.6 Measuring failure – reactive monitoring

So far, this chapter has concentrated on monitoring activities designed to prevent the occurrence of injuries and work-related ill health (active monitoring using leading indicators). Failures in risk control also need to be monitored (reactive monitoring using lagging indicators), to provide opportunities to check performance, learn from failures and improve the health and safety management system.

Reactive monitoring arrangements include systems to identify lagging indicators and report:

▶ injuries and work-related ill health;

▶ other losses such as damage to property;

▶ incidents, including those with the potential to cause injury, ill health or loss (near misses);

▶ hazards and faults;

▶ weaknesses or omissions in performance standards and systems, including complaints from employees and enforcement action by the authorities.

Each of the above provides opportunities for an organisation to check performance, learn from mistakes, and improve the health and safety management system and risk control systems. In some cases, the organisation must send a report of the circumstances and causes of incidents such as major injuries to the appropriate enforcing authority. Statutorily appointed safety representatives and representatives of employee safety are entitled to investigate.

The 'corporate memory' is enhanced by these events and information obtained by investigations can be used to reinforce essential health and safety messages. Safety committees and safety representatives should be involved in discussing common features or trends. This will help the workforce to identify jobs or activities which cause the most serious or the greatest number of injuries where remedial action may be most beneficial. Investigations may also provide valuable information if an insurance claim is made or legal action taken.

Most organisations should not find it difficult to collect data on serious injuries and ill health. However, learning about minor injuries, other losses, incidents and hazards can prove more challenging. There is value in collecting

information on all actual and potential losses to learn how to prevent more serious events. Accurate reporting can be promoted by:

▶ training which clarifies the underlying objectives and reasons for identifying such events;

▶ a culture which emphasises an observant and responsible approach and the importance of having systems of control in place before harm occurs;

▶ open, honest communication in a just environment, rather than a tendency merely to allocate 'blame';

▶ cross-referencing and checking first-aid treatments, health records, maintenance or fire reports and insurance claims to identify any otherwise unreported events.

Guidance on recording, investigating and analysing these incidents is given later in this chapter.

5.2 Investigating incidents

5.2.1 Function of incident investigation

Incidents and accidents rarely result from a single cause and many turn out to be complex. Most incidents involve multiple, interrelated causal factors. They can occur whenever significant deficiencies, oversights, errors, omissions or unexpected changes occur. Any one of these can be the precursor of an accident or incident. There is a value in collecting data on all incidents and potential losses as it helps to prevent more serious events.

Incidents and accidents, whether they cause damage to property or more serious injury and/or ill health to people, should be properly and thoroughly investigated to allow an organisation to take the appropriate action to prevent a recurrence (Figure 5.4). Good investigation is a key element to making improvements in health and safety performance.

Figure 5.4 A dangerous occurrence – fire

Incident investigation is considered to be part of a reactive monitoring system because it is triggered after an event.

Incident/accident investigation is based on the logic that:

▶ all incidents/accidents have causes – eliminate the cause and eliminate future incidents;

▶ the direct and indirect causes of an incident/accident can be discovered through investigation;

▶ corrective action indicated by the causation can be taken to eliminate future incidents/accidents.

Investigation is not intended to be a mechanism for apportioning blame. There are often strong emotions associated with injury or significant losses. It is all too easy to look for someone to blame without considering the reasons why a person behaved in a particular way. Often short cuts to working procedures that may have contributed to the accident give no personal advantage to the person injured. The short cut may have been taken out of loyalty to the organisation or ignorance of a safer method.

There are many **benefits** from investigating accidents/incidents. These include:

▶ the prevention of similar events occurring again. Where the outcomes are serious injuries the enforcing authorities are likely to take a tough stance if previous warnings have been ignored;

▶ the prevention of business losses due to disruption immediately after the event, loss of production, loss of business through a lowering of reputation or inability to deliver, and the costs of criminal and legal actions;

▶ improvement in employee morale and general attitudes to health and safety, particularly if they have been involved in the investigations;

▶ improving management skills to improve health and safety performance throughout the organisation.

The case for investigating near misses and undesired circumstances may not be so obvious but it is just as useful and much easier as there are no injured people to deal with. There are no demoralised people at work or distressed families and seldom a legal action to answer. Witnesses will be more willing to speak the truth and help with the investigation.

Managers need to:

▶ communicate the type of accident and incident that needs to be reported;

▶ provide a system for reporting and recording;

▶ check that proper reports are being made;

▶ make appropriate records of accidents and incidents;

▶ investigate incidents and accidents reported;

▶ analyse the events routinely to check for trends in performance and the prevalence of types of incident or injury;

▶ monitor the system to make sure that it is working satisfactorily.

5.2.2 Types of incident or adverse events

In HSG245 Investigating Injury and Accidents at Work 2004, the HSE define different types of incident as follows:

Figure 5.5 Accident at work – reconstruction of a ladder accident showing where the deceased person was found under the ladder which had toppled over while he was attempting to adjust the height

'Adverse event' includes:

1. **Accident:** is defined by the HSE as *'any unplanned event that results in injury or ill health of people, or damage or loss to property, plant, materials or the environment or a loss of a business opportunity'*. Other authorities define an accident more narrowly by excluding events that do not involve injury or ill health. This book will always use the HSE definition.

2. **Incident:**
   – *near miss*: an event that, while not causing harm, has the potential to cause injury or ill health. (In this guidance, the term near miss will be taken to include dangerous occurrences);
   – *undesired circumstance*: a set of conditions or circumstances that have the potential to cause injury or ill health, e.g. untrained nurses handling heavy patients.

3. **Dangerous occurrence:** one of a number of specific, reportable adverse events, as defined in the Reporting of Injuries, Diseases and Dangerous Occurrences Regulations 2013 (RIDDOR).

4. **Immediate cause:** the most obvious reason why an adverse event happens, e.g. the guard is missing; the employee slips, etc. There may be several immediate causes identified in any one adverse event.

5. **Root cause:** an initiating event or failing from which all other causes or failings spring. Root causes are generally management, planning or organisational failings.

6. **Underlying cause:** the less obvious 'system' or 'organisational' reason for an adverse event happening, e.g. pre-start-up machinery checks are not carried out by supervisors; the hazard has not been adequately considered via a suitable and sufficient risk assessment; production pressures are too great, etc.

Consequence:

1. **Fatal:** *work-related death;*
2. **Specified serious injury/ill health:** (as defined in RIDDOR 2013, Regulation 4 see Chapter 13), including a
   ▶ fracture, other than to fingers, thumbs and toes;
   ▶ amputation of an arm, hand, finger, thumb, leg, foot or toe;
   ▶ permanent loss of sight or reduction of sight;
   ▶ crush injuries leading to internal organ damage;
   ▶ serious burns (covering more than 10% of the body, or damaging the eyes, respiratory system or other vital organs);
   ▶ scalpings (separation of skin from the head) which require hospital treatment;

▶ unconsciousness caused by head injury or asphyxia;
▶ any other injury arising from working in an enclosed space, which leads to hypothermia, heat-induced illness or requires resuscitation or admittance to hospital for more than 24 hours.

3. **Over 7-day injury/ill health:** where the person affected is unfit to carry out his or her normal work for more than seven consecutive days;
4. **Minor injury:** all other injuries, where the injured person is unfit for his or her normal work for less than seven days;
5. **Damage only:** damage to property, equipment, the environment or production losses.

Figure 5.6 demonstrates the difference between an accident, near miss and undesired circumstances.

Each type of event/incident gives the opportunity to:

▶ check performance;
▶ identify underlying deficiencies in management systems and procedures;

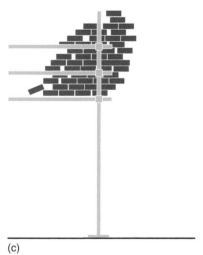

Figure 5.6 (a) Accident; (b) near miss (includes dangerous occurrence) damage only; (c) undesired circumstances

▶ learn from mistakes and add to the corporate memory;

▶ reinforce key health and safety messages;

▶ identify trends and priorities for prevention;

▶ provide valuable information if there is a claim for compensation;

▶ help meet legal requirements for reporting certain incidents to the authorities.

5.2.3 The different levels of investigations

The types of incident which may need to be investigated and the depth of the investigation are usually determined by their outcomes or consequences. Should every accident be investigated or only those that lead to serious injury? In fact, the main determinant is the potential of the accident to cause harm rather than the actual harm resulting. For example, a slip can result in an embarrassing flailing of arms or, just as easily, a broken leg. The frequency of occurrence of the accident type is also important – a stream of minor cuts from paper needs looking into.

As it is not possible to determine the potential for harm simply from the resulting injury, the only really sensible solution is to investigate all accidents. The amount of time and effort spent on the investigation should, however, vary depending on the level of risk (severity of potential harm, frequency of occurrence). The most effort should be focused on significant events involving serious injury, ill health or losses and events which have the potential for multiple or serious harm to people or substantial losses. These factors should become clear during the accident

investigation and be used to guide how much time should be taken.

Figure 5.7 has been developed by the Health and Safety Executive (HSE) to help to determine the level of investigation which is appropriate. The potential worst injury consequences in any particular situation should be considered when using the table. A particular incident like a scaffold collapse may not have caused an injury but had the potential to cause major or fatal injuries.

In a **minimal-level** investigation, the relevant supervisor will look into the circumstances of the accident/incident and try to learn any lessons which will prevent future incidents.

A **low-level** investigation will involve a short investigation by the relevant supervisor or line manager into the circumstances and immediate underlying and root causes of the accident/incident, to try to prevent a recurrence and to learn any general lessons.

A **medium-level** investigation will involve a more detailed investigation by the relevant supervisor or line manager, the health and safety adviser and employee representatives and will look for the immediate, underlying and root causes.

A **high-level** investigation will involve a team-based investigation, involving supervisors or line managers, health and safety advisers and employee representatives. It will be carried out under the supervision of senior management or directors and will look for the immediate, underlying and root causes.

Likelihood of recurrence	Potential worst injury consequences of accident/incident			
	Minor	**Serious**	**Major**	**Fatal**
Certain				
Likely				
Possible				
Unlikely				
Rare				

Risk	**Minimal**	**Low**	**Medium**	**High**
Investigation level	Minimal Level	Low Level	Medium Level	High Level

Figure 5.7 Appropriate levels of investigation

5.2.4 Basic incident investigation steps

Investigations should be led by supervisors, line managers or other people with sufficient status and knowledge to make recommendations that will be respected by the organisation.

The investigation should be carried out as soon as possible after the incident to allow the maximum amount of information to be obtained. There may be difficulties which should be considered in setting up the investigation quickly – if, for example, the victim is removed from the site of the accident, or if there is a lack of a particular expert. An immediate investigation is advantageous because:

▶ factors are fresh in the minds of witnesses;
▶ witnesses have had less time to talk (there is an almost automatic tendency for people to adjust their story of the events to bring it into line with a consensus view);
▶ physical conditions have had less time to change;
▶ more people are likely to be available, for example delivery drivers, contractors and visitors, who will quickly disperse following an incident, making contact very difficult;
▶ there will probably be the opportunity to take immediate action to prevent a recurrence and to demonstrate management commitment to improvement;
▶ immediate information from the person suffering the accident often proves to be most useful.

Investigation method

There are four basic steps to a sound investigation:

▶ Step one: gathering the information
▶ Step two: analysing the information
▶ Step three: identify risk control measures
▶ Step four: the action plan and its implementation.

Step one: information should be gathered from all available sources, for example, witnesses, supervisors, physical conditions, hazard data sheets, written systems of work, training records. Photographs are invaluable aids to investigation but of course with digital photography they can be easily altered. It is a good idea to print on the time and date of the picture and be prepared to verify that it is accurate. Printing out the picture as soon as possible captures it for the records. Plans and simple sketches of the incident site are also valuable.

The amount of time spent should not, however, be disproportionate to the risk. The aim of the investigation should be to explore the situation for possible underlying factors, in addition to the immediately obvious causes of the accident. For example, in a machinery accident it would not be sufficient to

conclude that an accident occurred because a machine was inadequately guarded. It is necessary to look into the possible underlying system failure that may have occurred.

Investigations have three facets, which are particularly valuable and can be used to check against each other:

▶ direct observation of the scene, premises, workplace, relationship of components, materials and substances being used, possible reconstruction of events and injuries or condition of the person concerned;
▶ documents including written instructions, training records, procedures, safe operating systems, risk assessments, policies, records of inspections or tests and examinations carried out;
▶ interviews (including written statements) with persons injured, witnesses, people who have carried out similar functions or examinations and tests on the equipment involved and people with specialist knowledge.

Investigation interview techniques

It must be made clear at the outset and during the course of the interview that the aim is not to apportion blame but to discover the facts and use them to prevent similar accidents or incidents in the future.

A witness should be given the opportunity to explain what happened in their own way without too much interruption and suggestion. Questions should then be asked to elicit more information. These should be of the open type, which do not suggest the answer. Questions starting with the words in Figure 5.8 are useful.

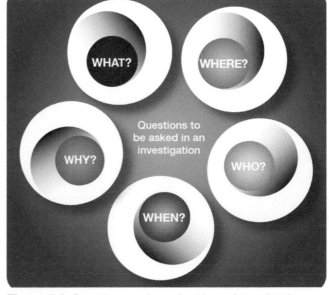

Figure 5.8 Questions to be asked in an investigation

'Why' should not be used at this stage. The facts should be gathered first, with notes being taken at the end of the explanation. The investigator should then read them or give a summary back to the witness, indicating clearly that they are prepared to alter the notes if the witness is not content with them.

If possible, indication should be given to the witness about immediate actions that will be taken to prevent a similar occurrence and that there could be further improvements depending on the outcome of the investigation.

Seeing people injured can often be very upsetting for witnesses, which should be borne in mind. This does not mean they will not be prepared to talk about what has happened. They may in fact wish to help, but questions should be sensitive; upsetting the witness further should be avoided.

5.2.5 Incident causes and analysis

Immediate causes – unsafe acts and conditions

An investigation should look at the following factors as they can provide useful information about **immediate causes** that have been manifested in the incident/accident and result directly from unsafe acts and conditions in the workplace.

Personal factors:

▶ behaviour of the people involved (unsafe acts)
▶ suitability of people doing the work
▶ training and competence.

Task factors:

▶ workplace conditions and precautions or controls (unsafe conditions)
▶ actual method of work adopted at the time
▶ ergonomic factor
▶ normal working practice either written or customary.

Underlying and root causes – management system failures

A thorough investigation should also look at the following factors as they can provide useful information about **underlying and root causes** that have been manifested in the incident/accident:

Underlying causes are the less obvious system or organisational reasons for an accident or incident such as:

▶ pre-start-up machinery checks were not made by supervisors
▶ the hazard had not been considered in the risk assessment
▶ there was no suitable method statement
▶ pressures of production had been more important
▶ the employee was under a lot of personal pressure at the time
▶ have there been previous similar incidents?
▶ was there adequate supervision, control and coordination of the work involved?

Root causes involve an initiating event or failing from which all other causes or failings arise. Root causes are generally management, planning or organisational failings including:

▶ quality of the health and safety policy and procedures;
▶ quality of consultation and cooperation of employees;
▶ the adequacy and quality of communications and information;
▶ deficiencies in risk assessments, plans and control systems;
▶ deficiencies in monitoring and measurement of work activities;
▶ quality and frequency of reviews and audits.

Comparison with relevant standards

There are usually suitable and relevant standards which may come from the HSE, industry or the organisation itself. These should be carefully considered to see if:

▶ suitable standards are available to cover legal standards and the controls required by the risk assessments;
▶ the standards are sufficient and available to the organisation;
▶ the standards were implemented in practice;
▶ the standards were implemented, why there was a failure;
▶ changes should be made to the standards.

Simple root cause analysis – the 'five whys'

This type of simple root cause analysis is ideal for a minimal or low level of investigation. More complex methods like the Tree of Causes is suitable for a higher level of investigation and is not covered here. Now is the time after obtaining all the initial information to ask 'why'.

▶ Basic question – keeping asking 'What caused or allowed this condition/practice to occur?' until you get to root causes.
▶ The 'five whys' is one of the simplest of the root cause analysis methods. It is a question-asking method used to explore the cause/effect relationships underlying a particular problem. Ultimately, the goal of applying the five whys method is to determine a **root cause** of a defect or problem.

The following example demonstrates the basic process

Car will not start (the problem)

1. *Why?* – The battery is dead (first 'why')
2. *Why?* – The alternator is not functioning (second 'why')
3. *Why?* – The alternator belt has broken (third 'why')
4. *Why?* – The alternator belt was well beyond its useful service life and has never been replaced (fourth 'why')
5. *Why?* – I have not been maintaining my car according to the recommended service schedule (fifth 'why' and the root cause).

Benefits of asking the five whys

▶ **Simplicity.** It is easy to use and requires no advanced mathematics or tools.

▶ **Effectiveness.** It truly helps to quickly separate symptoms from causes and identify the root cause of a problem.

▶ **Comprehensiveness.** It aids in determining the relationships between various problem causes.

▶ **Flexibility.** It works well alone and when combined with other quality improvement and troubleshooting techniques.

▶ **Engaging.** By its very nature, it fosters and produces teamwork and teaming within and without the organisation.

▶ **Inexpensive.** It is a guided team-focused exercise. There are no additional costs.

Often the answer to the one 'why' uncovers another reason and generates another 'why'. It often takes 'five whys' to arrive at the root cause of the problem. Investigators will probably find that they ask more or less than 'five whys' in practice.

5.2.6 Remedial actions

The investigation should have highlighted both immediate causes and underlying causes. Recommendations with priorities, both for immediate action and for longer-term improvements, should come out of this. It may be necessary to ensure that the report goes further up the management chain if the improvements recommended require authorisation which cannot be given by the investigating team.

It is essential that a follow-up is made to check on the implementation of the recommendations. It is also necessary to review the effect of the recommendations to check whether they have achieved the desired result and whether they have had unforeseen 'knock-on' effects, creating additional risks and problems.

The accident or incident investigation should not only be used to generate recommendations but should also be used to generate safety awareness. The investigation report or a summary should therefore be circulated locally to relevant people and, when appropriate, summaries circulated throughout the organisation. The accident or incident does not need to have resulted in a 7-day lost time injury for this system to be used.

Investigation form

Headings which could be used to compile an accident/incident investigation form are given below:

▶ date and location of accident/incident;
▶ circumstances of accident/incident;
▶ immediate cause of accident/incident;
▶ underlying cause of accident/incident;
▶ immediate action taken;
▶ recommendation for further improvement;

▶ report circulation list;
▶ date of investigation;
▶ signature of investigating team leader;
▶ names of investigating team.

Follow-up

▶ were the recommendations implemented?
▶ were the recommendations effective?

Key data for medium level of investigation

The HSE has suggested that the key data included in Box 5.1 should be covered in an investigation report. This level of data is more appropriate to a medium level of investigation.

Box 5.1 Key data for medium level of investigation

Key data to be covered in accident, ill-health and incident reports.

The event

▶ Details of any injured person, including age, gender, experience, training, etc.
▶ A description of the circumstances, including the place, time of day and conditions.
▶ Details of the event, including:
  ▷ any actions which led directly to the event;
  ▷ the direct causes of any injuries, ill health or other loss;
  ▷ the immediate causes of the event;
  ▷ the underlying causes – for example failures in workplace precautions, risk control systems or management arrangements.
▶ Details of the outcomes, including in particular:
  ▷ the nature of the outcome – for example injuries or ill health to employees or members of the public; damage to property, process disruption, emissions to the environment, creation of hazards;
  ▷ the severity of the harm caused, including injuries, ill health and losses.
▶ The immediate management response to the situation and its adequacy:
  ▷ Was it dealt with promptly?
  ▷ Were continuing risks dealt with promptly and adequately?
  ▷ Was the first-aid response adequate?
  ▷ Were emergency procedures followed?
  ▷ Whether the event was preventable and if so how?

The potential consequences

▶ What was the worst that could have happened?
▶ What prevented the worst from happening?

> ▶ How often could such an event occur (the 'recurrence potential')?
> ▶ What was the worst injury or damage which could have resulted (the 'severity potential')?
> ▶ How many people could the event have affected (the 'population potential')?
>
> **Recommendations**
> ▶ Prioritised actions with responsibilities and targets for completion.

5.2.7 Recording and reporting incidents

Typical examples

This section covers the legal and organisational requirements for recording and reporting accidents and other incidents. It also covers simple analysis of the data to help employers and employees benefit from the investigation and recording process.

Typical examples covered by this section are:

▶ a work-related death;
▶ a Specified Injury – such as a fractured leg, amputation of an arm or hand, loss of an eye or a serious burn;
▶ a Reportable Disease such as asbestosis from contact with asbestos-containing material, anthrax from contact with an infected animal or lead poisoning from breathing lead fumes while soldering or burning off old paint and occupational dermatitis from the unprotected use of industrial solvents or frequent use of shampoos in hairdressing saloons; and
▶ a Dangerous Occurrence such as the failure of a pressure vessel, collapse of a crane or a scaffold or a serious fire.

More detail is given later in this chapter and a full summary of RIDDOR is contained in Chapter 13.

Statutory requirements for recording and reporting incidents

Accident book

Under the Social Security (Claims and Payments) Regulations 1979, Regulation 25, employers must keep a record of accidents at premises where more than ten people are employed. Anyone injured at work is required to inform the employer and record information on the accident in an accident book, including a statement on how the accident happened.

The employer is required to investigate the cause and enter this in the accident book if they discover anything that differs from the entry made by the employee. The purpose of this record is to ensure that information is available if a claim is made for compensation.

The HSE Accident Book BI 510 (see Figure 5.9) 2018 Edition has notes on these Regulations, and the Reporting of Injuries, Diseases and Dangerous Occurrences Regulations 2013 (RIDDOR). It also complies with the Data Protection Act 1998 by requiring injured people to give their consent for the information to be passed on to Safety Representatives or Representatives of Employee Safety. The Accident/incident or First aid treatment and accident record report forms (see Chapter 17 Form M10 or M11) can be used as an initial record and as a substitute for form BI 510.

RIDDOR 2013

RIDDOR 2013 requires employers, the self-employed and those in control of premises, to report certain more serious accidents and incidents to the HSE or other enforcing authority and to keep a record. There are no exemptions for small organisations. A full summary of the Regulations is given in Chapter 13. The reporting and recording requirements are as follows:

Death or serious injury

If an accident occurs at work and:

▶ an employee or self-employed person working on the premise is killed or suffers a specified injury

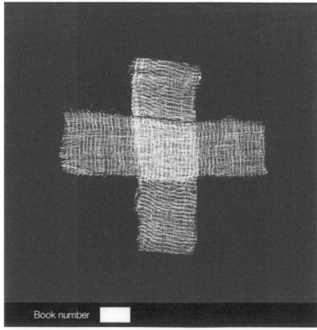

Figure 5.9 (a) The Accident Book BI 510 (2018 Edition)

Report Number

Accident record

1 About the person who had the accident

Name
....

Address
....

Postcode
....

Occupation
....

2 About you, the person filling in this record

▼ If you did not have the accident write your address and occupation.

Name
....

Address
....

Postcode
....

Occupation
....

3 About the accident   *Continue on the back of this form if you need to*

▼ Say when it happened.   Date   /   /   Time
....

▼ Say where it happened. State which room or place.
....

▼ Say how the accident happened. Give the cause if you can.
....
....
....
....

▼ If the person who had the accident suffered an injury, say what it was.
....

▼ Please sign the record and date it.

Signature   Date   /   /
....

4 For the employee only

▼ By ticking this box I give my consent to my employer to disclose my personal information and details of the accident which appear on this form to safety representatives and representatives of employee safety for them to carry out the health and safety functions given to them by law.   ☐

Signature   Date   /   /

5 For the employer only

▼ Complete this box if the accident is reportable under the Reporting of Injuries, Diseases and Dangerous Occurrences Regulations (RIDDOR). To report, go to page 4 of this book or go to www.hse.gov.uk/riddor/report.htm2.

How was it reported?
....

Date reported   /   /   Signature

Figure 5.9 (b) Record form from BI 510

(including the effects of physical violence) (see Chapter 13 under RIDDOR for definition of 'specified injury'). Specified injuries include:

▷ a fracture, other than to fingers, thumbs and toes;
▷ amputation of an arm, hand, finger, thumb, leg, foot or toe;
▷ permanent loss of sight or reduction of sight;
▷ crush injuries leading to internal organ damage;
▷ serious burns (covering more than 10% of the body, or damaging the eyes, respiratory system or other vital organs);

153

▷ scalpings (separation of skin from the head) which require hospital treatment;

▷ unconsciousness caused by head injury or asphyxia;

▷ any other injury arising from working in an enclosed space, which leads to hypothermia, heat-induced illness or requires resuscitation or admittance to hospital for more than 24 hours;

▶ a member of the public or person not at work is killed or taken to hospital from the scene of the accident. If the accident occurred at a hospital, the report only needs to be made if the injury is a 'specified injury'.

the responsible person must notify the enforcing authority without delay by the quickest practicable means, for example telephone (0345 300 9923) or online to the HSE Incident Contact Centre (ICC).

Over 7-day lost time injury

If there is an accident connected with work (including physical violence), and an employee or self-employed person working on the premises suffers an injury and is away from work or not doing his/her normal duties for more than 7 days (including weekends, rest days or holidays but not counting the day of the accident), the responsible person must report to the enforcing authority within 15 days. The best and quickest way is via telephone or online to the ICC.

Reportable occupational disease

If a doctor notifies the responsible person that an employee suffers from a reportable work-related disease they must report to the enforcing authority within 10 days and a completed disease report form, F2508A, must be sent to the enforcing authority. A summary is included in Chapter 13 and a full list is included with a pad of report forms. The best and quickest way is via telephone or online to the ICC. These diseases include:

▶ carpal tunnel syndrome;
▶ severe cramp of the hand or forearm;
▶ occupational dermatitis;
▶ hand-arm vibration syndrome;
▶ occupational asthma;
▶ tendonitis or tenosynovitis of the hand or forearm;
▶ any occupational cancer;
▶ any disease attributed to an occupational exposure to a biological agent.

The HSE ICC can be contacted to check whether a particular disease is reportable.

Reportable dangerous occurrences

Dangerous occurrences are certain, specified near-miss events. Not all such events require reporting. There are 27 categories of dangerous occurrence that are relevant to most workplaces. For example:

▶ the collapse, overturning or failure of load-bearing parts of lifts and lifting equipment;
▶ plant or equipment coming into contact with overhead power lines;
▶ the accidental release of any substance which could cause injury to any person.

See Chapter 13 for a summary of dangerous occurrences. All dangerous occurrences must be reported immediately to the authorities. The best and quickest way is via telephone or online to the ICC. The HSE Incident Contact Centre can also be contacted to check whether a dangerous occurrence is reportable.

Reportable gas incidents

If a distributor, filler, importer or supplier of flammable gas learns, either directly or indirectly, that someone has died, lost consciousness or been taken to hospital for treatment to an injury arising in connection with the gas they distributed, filled, imported or supplied, this can be reported online.

A gas engineer registered with the Gas Safe Register must provide details of any gas appliances or fittings that they consider to be dangerous to the extent that people could die, lose consciousness or require hospital treatment. This may be due to the design, construction, installation, modification or servicing, and could result in:

▶ an accidental leakage of gas;
▶ inadequate combustion of gas; or
▶ inadequate removal of products of the combustion of gas.

Records

Records of incidents covered by RIDDOR are also important. They ensure that sufficient information to allow properly managed health and safety risks is collected. This information is a valuable management tool that can be used as an aid to risk assessment, helping to develop solutions to potential risks. In this way, records also help to prevent injuries and ill health, and control costs from accidental loss.

Records must be kept of:

▶ any accident, occupational disease or dangerous occurrence which requires reporting under RIDDOR; and
▶ any other occupational accident causing injuries that result in a worker being away from work or incapacitated for more than three consecutive days (not counting the day of the accident but including any weekends or other rest days). Over-three-day injuries do not have to be reported, unless the incapacitation period goes on to exceed seven days.

Organisations can keep the record in any form they wish, for example, they may choose to keep records by:

▶ keeping copies of report forms in a file;
▶ recording the details on a computer;
▶ using the Accident Book entry;
▶ maintaining a written log.

RIDDOR records must be produced when asked by HSE, Local Authority or Office of Rail Regulator inspectors.

Whom to report to

Online

Go to www.hse.gov.uk/riddor and complete the appropriate online report form. The form will then be submitted directly to the RIDDOR database. A copy will be received for records.

Telephone

All incidents can be reported online but a telephone service remains for reporting **fatal and specified injuries only**. The HSE Incident Contact Centre can be called on 0345 300 9923 (opening hours Monday to Friday 8.30 a.m. to 5 p.m.).

Reporting out of hours

HSE has an out-of-hours duty officer. Circumstances where HSE may need to respond out of hours include:

▶ a work-related death or situation where there is a strong likelihood of death following an incident at, or connected with, work;
▶ a serious accident at a workplace so that HSE can gather details of physical evidence that would be lost with time;
▶ following a major incident at a workplace where the severity of the incident, or the degree of public concern, requires an immediate public statement from either HSE or government ministers.

Information supplied to HSE in a RIDDOR report is **not** passed on to the organisation's insurance company. If the relevant insurer needs to know about a work-related accident, injury or case of ill health they must be contacted separately – insurers have told the HSE that reporting injuries and illnesses at work to them quickly could save time and money.

Internal systems for collecting and analysing incident data

Managers need effective internal systems to know whether the organisation is getting better or worse, to know what is happening and why, and to assess whether objectives are being achieved. Earlier in the chapter, Section 5.1 deals with monitoring generally, but here the basic requirements of a collection and analysis system for incidents are discussed.

The incident report form (discussed earlier) is the basic starting point for any internal system. Each organisation needs to lay down what the system involves and who

is responsible to do each part of the procedure. This will involve:

▶ what type of incidents should be reported;
▶ who completes the incident report form – normally the manager responsible for the investigation;
▶ how copies should be circulated in the organisation;
▶ who is responsible to provide management measurement data;
▶ how the incident data should be analysed and at what intervals;
▶ the arrangements to ensure that action is taken on the data provided.

The data should seek to answer the following questions:

▶ are failure incidents occurring, including injuries, ill health and other loss incidents?
▶ where are they occurring?
▶ what is the nature of the failures?
▶ how serious are they?
▶ what are the potential consequences?
▶ what are the reasons for the failures?
▶ how much has it cost?
▶ what improvements in controls and the management system are required?
▶ how do these issues vary with time?
▶ is the organisation getting better or worse?

Type of accident/incident

Most organisations will want to collect data on:

▶ all injury accidents;
▶ cases of ill health;
▶ sickness absence;
▶ dangerous occurrences;
▶ damage to property, the environment, personal effects and work in progress;
▶ incidents with the potential to cause serious injury, ill health or damage (undesired circumstances).

Not all of these are required by law, but this should not deter the organisation that wishes to control risks effectively.

Analysis

All the information, whether in accident books or report forms, will need to be analysed so that useful management data can be prepared. Many organisations look at the analysis every month and annually. However, where there are very few accidents/incidents, quarterly may be sufficient. The health and safety information should be used alongside other business measures and should receive equal status.

There are several ways in which data can be analysed and presented. The most common ways are:

▶ by causation using the classification used on the RIDDOR form F2508. This has been used on the example accident/incident report form (see Form M10 in Chapter 17);

155

- ▶ by the nature of the injury, such as cuts, abrasions, asphyxiation and amputations;
- ▶ by the part of the body affected, such as hands, arms, feet, lower leg, upper leg, head, eyes, back and so on. Sub-divisions of these categories could be useful if there were sufficient incidents;
- ▶ by age and experience at the job;
- ▶ by time of day;
- ▶ by occupation or location of the job;
- ▶ by type of equipment used.

There are a number of up-to-date computer programs which can be used to manipulate the data if significant numbers are involved. The trends can be shown against monthly, quarterly and annual past performance of, preferably, the same organisation. If indices are calculated, such as Incident Rate, comparisons can be made nationally with HSE figures and with other similar organisations or businesses in the same industrial group. This is really of major value only to larger organisations with significant numbers of events. (See Chapter 3 for more information on definitions and calculation of incident rates.)

Although the number of accidents may give a general indication of the health and safety performance, a more detailed examination of accidents and accident statistics is normally required. A calculation of the rate of accidents enables health and safety performance to be compared between years and organisations.

The simplest measure of accident rate is called the incidence rate.

HSE's formula for calculating an annual injury incidence rate is:

$$\frac{\text{Number of reportable injuries in financial year}}{\text{Average number employed during the year}} \times 100,000$$

This gives the rate per 100,000 employees. The formula makes no allowances for variations in part-time employment or overtime. It is an annual calculation and the figures need to be adjusted pro rata if they cover a shorter period. Such shorter-term rates should be compared only with rates for exactly similar periods – not the national annual rates.

While HSE and industry calculates injury **incidence rates** per 100,000 or 1,000 employees, some parts of industry prefer to calculate injury **frequency rates**, usually per million hours worked. This method, by counting hours worked rather than the number of employees, avoids distortions which may be caused in the incidence rate calculations by part- and full-time employees and by overtime working. Frequency rates can be calculated for any time period.

The calculation is:

$$\frac{\text{Number of injuries in period}}{\text{Total hours worked during the period}} \times 1,000,000$$

The HSE produces annual bulletins of national performance along with a detailed statistical report, which can be used for comparisons. See: www.hse.gov.uk/statistics/index.htm. There are difficulties in comparisons across Europe and, say, with the USA, where the definitions of accidents or time lost vary.

Reports should be prepared with simple tables and graphs showing trends and comparisons. Line graphs, bar charts and pie charts are all used quite extensively with good effect. All analysis reports should be made available to employees as well as managers. This can often be done through the Health and Safety Committee and safety representatives, where they exist, or directly to all employees in small organisations. Other routine meetings, team briefings and noticeboards can all be used to communicate the message.

It is particularly important to make sure that any actions recommended or highlighted by the reports are taken quickly and employees kept informed.

The report form shown in Chapter 17 uses these immediate causes which can be used for analysis purposes. The categories can easily be changed to suit local needs.

> **Box 5.2 The following categories of immediate causes of accident are used in F2508:**
>
> 1. contact with moving machinery or material being machined;
> 2. struck by moving, including flying or falling, object;
> 3. struck by moving vehicle;
> 4. struck against something fixed or stationary;
> 5. injured whilst handling, lifting or carrying;
> 6. slip, trip or fall on same level;
> 7. fall from height; indicate approximate distance of fall in metres;
> 8. trapped by something collapsing or overturning;
> 9. drowning or asphyxiation;
> 10. exposure to or contact with harmful substances;
> 11. exposure to fire;
> 12. exposure to an explosion;
> 13. contact with electricity or an electrical discharge;
> 14. injured by an animal;
> 15. violence;
> 16. other kind of accident.

Collection of information, compensation and insurance issues

Accidents/incidents arising out of the organisation's activities resulting in injuries to people and incidents resulting in damage to property can lead to compensation claims. The second objective of an investigation should be to collect and record relevant information for the purposes of dealing with any claim. It must be remembered that, in the longer term, prevention is the best way to reduce claims and must be the first objective in the investigation. An overzealous approach to gathering information concentrating on the compensation aspect can, in fact, prompt a claim from the injured party where there was no particular intention to take this route before the investigation. Nevertheless, relevant information should be collected. Sticking to the collection of facts is usually the best approach.

As mentioned earlier the legal system in England and Wales changed dramatically with the introduction of the Woolf reforms in 1999. These reforms apply to injury claims. This date was feared by many because of the uncertainty and the fact that the pre-action protocols were very demanding.

The essence of the pre-legal action protocols is as follows:

1. 'letter of claim' to be acknowledged within 21 days;
2. ninety days from date of acknowledgement to either accept liability or deny. If liability is denied then full reasons must be given;
3. agreement to be reached on using a single expert.

The overriding message is that to comply with the protocols quick action is necessary. It is also vitally important that records are accurately kept and accessible.

Lord Woolf made it clear in his instructions to the judiciary that there should be very little leeway given to claimants and defendants who did not comply.

What has been the effect of these reforms on day-to-day activity? Initially there was a 25% drop in the number of cases moving to litigation. Whilst this was the primary objective of the reforms, the actual drop was far more significant than anyone anticipated. By 2007 cases began to rise significantly again and in some cases judges have been very reluctant to strike out cases because they are out of time, except in extremes circumstances.

However, some of the positive effects have been:

▶ the elimination of speculative actions due to the requirement to fully outline the claimant's case in the letter of claim;
▶ earlier and more comprehensive details of the claim allowing a more focused investigation and response;
▶ 'Part 36 offers' (payments into court) seem to be having greater effect in deterring claimants from pursuing litigation;
▶ overall faster settlement.

The negative effects have principally arisen from failure to comply with the timescales, particularly relating to the gathering of evidence and records and having no time therefore to construct a proper defence.

Section 47 of the Health and Safety at Work Act 1974, by virtue of Section 6 of the Enterprise and Regulatory Reform Act 2013, removes the standard of strict liability from certain health and safety regulations. No civil claim may be brought for breach of statutory duty unless a regulation expressly provides for it; this effectively reverses the previous position.

In almost all cases it will be for the injured employee to rely on common law negligence and prove that their injuries were caused by the employer's negligence. Negligence is generally a higher hurdle for employees than a breach of statutory duty – particularly those imposing strict liability – as the standard required of employers is higher; greater emphasis is placed on the 'reasonable practicability' defence which balances the expense of potential preventative measures against the scale of the risk.

The removal of strict liability for health and safety breaches is one of many changes designed to combat the perceived 'compensation culture' and reflects a growing trend against punishing employers for injuries that they took all reasonable steps to prevent. Under the previous regime businesses have paid out significant sums in compensation with around 78,000 civil liability claims brought each year for injuries sustained at work.

Overburdened employers are likely to welcome this cutting of red tape and the freedom to fulfil their obligations under health and safety law without the worry of potential, unforeseen claims against them. This in turn will increase employers' confidence to not only protect their employees but also to develop and expand their business.

Employees on the other hand might not welcome the changes so readily. They are now faced with the prospect of discharging a heavier legal and evidential burden to establish liability. In fact the Trades Union Congress argues that the changes do nothing to remove unfairness but merely shift it to another place.

Appendix 5.2 provides a checklist of headings, which may assist in the collection of information. It is not expected that all accidents and incidents will be investigated in depth and a dossier with full information prepared. Judgement has to be applied as to which incidents might give rise to a claim and when a full record of information is required. All accident/incident report forms should include the names of all witnesses as a minimum. Where the injury is likely to give rise to lost time, a photograph(s) of the situation should be taken.

Lessons learnt from an incident

After a thorough investigation there should be an action plan for the implementation of additional risk control measures The action plan should have SMART objectives, i.e. Specific, Measurable, Agreed, and Realistic, with Timescales.

A good knowledge of the organisation and the way it carries out its work is essential to know where improvements are needed. Management, safety professionals, employees and their representatives should all contribute to a constructive discussion on what should be in the action plan in order to make the proposals SMART.

Not every recommendation for further risk controls will be implemented, but the ones accorded the highest priority should be implemented immediately. Organisations need to ask 'What is essential to securing the health and safety of the workforce today? What cannot be left until another day? How high is the risk to employees if this risk control measure is not implemented immediately?'

Despite financial constraints, failing to put in place measures to control serious and imminent risks is totally unacceptable – either reduce the risks to an acceptable level, or stop the work. Each risk control measure should be assigned a priority and a timescale, with a designated person to carry out the recommendation. It is crucial that the action plan as a whole is properly monitored with a specific person, preferably a director, partner or senior manager, made responsible for its implementation.

Progress on the action plan should be regularly monitored and significant departures should be explained and risk control measures rescheduled, if necessary. There should be regular consultation with employees and their representatives to keep them fully informed of progress with implementation of the action plan.

Relevant safety instructions, safe working procedures and risk assessments should be reviewed after an incident. It is important to ask what the findings of the investigation indicate about risk assessments and procedures in general, to see if they really are suitable and sufficient.

It is also useful to estimate the cost of incidents to fully appreciate the true cost of accidents and ill health to the organisation. To find out more about the costs of accidents and incidents visit HSE's website cost calculator at: www.hse.gov.uk/costs.

5.3 Health and safety auditing

5.3.1 Audits – definition, scope and purpose

The final ACT steps in the health and safety management control cycle are auditing, performance review and taking action on lessons learnt. Organisations need to be able to reinforce, maintain and develop the ability to reduce risks and control hazards in the workplace. This is not a once for all procedure and should form part of a continual improvement programme.

Figure 5.10 ACT part of the health and safety management system

Audit is a business discipline which is frequently used, for example, in finance, environmental matters and quality. It can equally well be applied to health and safety.

The term is often used to mean inspection or other monitoring activity. Here, the following definition is used, which follows HSG65:

'The structured process of collecting independent information on the efficiency, effectiveness and reliability of the total health and safety management system and drawing up plans for corrective action.'

The **Safety Management System (SMS) Audit** is a major exercise, typically carried out every two to four years, as a means of assessing the adequacy of the whole organisation's SMS. It addresses all aspects of safety management in a structured manner, using written questions. Answers will be confirmed by a review of records, staff interviews, and observation of workplaces and operations. The SMS Audit may benefit from being undertaken by a person or persons independent of the organisation and should, where practicable, be carried out in real time.

Over time, it is inevitable that control systems will decay and may even become obsolete as things change. Auditing is a way of supporting monitoring by providing managers with information. It will show how effectively plans and the components of health and safety management systems are being implemented. In addition, it will provide a check on the adequacy and effectiveness of the management arrangements and risk control systems (RCSs).

Auditing is critical to a health and safety management system, but it is not a substitute for other essential parts of the system like workplace inspections (see 5.1). Companies need systems in place to manage cash flow and pay the bills – this cannot be managed through an annual or less frequent audit. In the same way, health and safety needs to be managed on a day-to-day basis and for this organisations need to have systems in place. A periodic audit will not achieve this.

The main difference between audits and other forms of monitoring (safety surveys, inspections, tours and sampling) is primarily the breadth and depth of an **audit**. **Surveys** look at only one aspect of the safety management system, inspections are frequent regular local monitoring normally carried out by line managers, **tours** concentrate on management commitment, **sampling** looks at only one area or subject over a short limited time.

The aims of auditing should be to establish that the three major components of a safety management system are in place and operating effectively. It should show that:

▶ appropriate management arrangements are in place;
▶ adequate risk control systems (RCSs) exist, are implemented and are consistent with the hazard profile of the organisation;
▶ appropriate workplace precautions are in place.

Where the organisation is spread over a number of sites, the management arrangements linking the centre with the business units and sites should be covered by the audit.

There are a number of ways in which this can be achieved and some parts of the system do not need auditing as often as others. For example an audit to verify the implementation of RCSs would be made more frequently than a more overall audit of the capability of the organisation or of the management arrangements for health and safety. Critical RCSs, which control the principal hazards of the business, would need to be audited more frequently. Where there are complex workplace precautions, it may be necessary to undertake technical audits. An example would be chemical process plant integrity and control systems.

A well-structured auditing programme will give a comprehensive picture of the effectiveness of the health and safety management system in controlling risks. Such a programme will indicate when and how each component part will be audited. Managers, safety representatives and employees, working as a team, will effectively widen involvement and cooperation needed to put together the programme and implement it.

The process of auditing involves:

▶ gathering information from all levels of an organisation about the health and safety management system;
▶ making informed judgements about its adequacy and performance.

5.3.2 Pre-audit preparations

Decisions will need to be made about the level and detail of the audit before starting to gather information

about the health and safety management of an organisation. Auditing involves sampling; so initially it is necessary to decide how much sampling is needed for the assessment to be reliable. The type of audit and its complexity will relate to its objectives and scope, to the size and complexity of the organisation and to the length of time that the existing health and safety management system has been in operation.

Figure 5.11 The Audit Process

Information sources including interviewing people, looking at documents and checking physical conditions are usually approached in the following order:

(a) Preparatory work
▶ meet with relevant managers and employee representatives to discuss and agree the objectives and scope of the audit;
▶ prepare and agree the audit procedure with managers;
▶ gather and consider documentation.

(b) On site
▶ interviewing;
▶ review and assessment of additional documents;
▶ observation of physical conditions and work activities.

(c) Conclusion
▶ assemble the evidence;
▶ evaluate the evidence;
▶ write an audit report;
▶ presentation of findings to management and workforce representatives where appropriate.

It is essential to start with a relevant standard or benchmark against which the adequacy of a health and safety management system can be judged. If standards are not clear, assessment cannot be reliable. Audit judgements should be informed by legal standards, HSE guidance and applicable industry standards. HSG65 sets out benchmarks for management arrangements and for the design of RCSs. This book follows the same concepts.

Auditing should not be seen as a fault-finding activity. It should make a valuable contribution to the health and safety management system and to learning. It should recognise achievement as well as highlight areas where more needs to be done.

Scoring systems can be used in auditing along with judgements and recommendations. This can be seen as a useful way to compare sites or monitor progress over time. However, there is no evidence that quantified results produce a more effective response than the use of qualitative evidence. Indeed, the introduction of a scoring system can, the HSE believes, have a negative effect, as it may encourage managers to place more emphasis on high-scoring questions which may not be as relevant to the development of an effective health and safety management system.

To achieve the best results, auditors should be competent people who are independent of the area and of the activities being audited. External consultants or staff from other areas of the organisation can be used. An organisation can use its own auditing system or one of the proprietary systems on the market or, as it is unlikely that any ready-made system will provide a perfect fit, a combination of both. With any scheme, cost and benefits have to be taken into account. Common problems include:

▶ systems can be too general in their approach. These may need considerable work to make them fit the needs and risks of the organisation;

▶ systems can be too cumbersome for the size and culture of the organisation;

▶ scoring systems may conceal problems in underlying detail;

▶ organisations may design their management system to gain maximum points rather than using one which suits the needs and hazard profile of the business.

Figure 5.12 Using the audit questions for interviews and collecting information

HSE encourages organisations to assess their health and safety management systems using in-house or proprietary schemes but without endorsing any particular one.

People selected for interviews need reasonable notice and should agree to be interviewed. They should be assured that the audit is not to apportion blame but to improve health and safety management in the organisation.

To achieve the best results, auditors should be competent people who are independent of the area and of the activities being audited. External consultants or staff from other areas of the organisation can be used. It is essential that auditors are experienced in meeting and interviewing people as well as knowledgeable in the subject being audited. In limited audits and smaller companies one person may carry out the audit, but in larger sites and organisations it is often necessary to have a team approach with a variety of disciplines. Sufficient time must be allocated to achieve the objectives of the audit and it is essential that local management organise themselves to give adequate time to the audit process including pre-audit preparation, interviews, accompanying auditors on plant inspection and post-audit remedial work.

5.3.3 Responsibility for audits – external v. internal audits

Directors and senior managers have a primary duty to ensure that they establish adequate systems for managing health and safety issues. Equally (as set out in 'Leading health and safety at work', HSE UK INDG417(rev1) and the ILO-OSH 2001) they have a duty to ensure that:

▶ appropriate weight is given to reporting both preventative information (such as progress of training and maintenance programmes) and incident data (such as accident and sickness absence rates);

▶ periodic audits of the effectiveness of management structures and risk controls for health and safety are carried out;

▶ the impact of changes such as the introduction of new procedures, work processes or products, or any major health and safety failure, is reported as soon as possible to the Board;

▶ there are procedures to implement new and changed legal requirements and to consider other external developments and events.

The summary of audit results and the fact that they are carried out should be part of directors' annual reporting procedures just as they are required to have financial audited reports. This gives confidence to workers, investors and customers that the company is operating proper health and safety management systems and considers the welfare of all those affected by its business operations. A failure to do this adequately has severely damaged the reputation of many large organisations.

Auditing may be carried out by internal or external consultants or, as is often the case, a combination of the two. Internal auditors:

- are familiar with the workplace, systems, processes and the organisation;
- are likely to be aware of what is practicable for the industry;
- have the ability to see improvements or a deterioration from the last audit;
- are familiar with the workforce and an individual's qualities and attitude; and
- may be less costly and easier to arrange than an external auditor.

But on the other hand internal auditors:

- may miss or gloss over some issues because of their familiarity;
- may not get honest views from the workforce for fear of the consequences;
- may not be in possession of recognised auditing skills;
- may not be up to date with legal requirements;
- may be less likely to be aware of best practice in other organisations;
- may be subject to pressure from management and the workforce; and
- may have time constraints imposed upon them.

External auditors:

- come with a new perspective, a fresh pair of eyes;
- need to ask more questions to understand the systems in operation which can elicit underlying problems;
- may have solutions learnt elsewhere that would benefit the organisation;
- can be more impartial in their presentation of the audit results;
- are more likely to have the necessary auditing skills;
- will not be inhibited from criticising members of management or the workforce;
- are more likely to be up to date with legal requirements and best practice in other companies.

The disadvantage of external auditors is that they:

- are unlikely to be familiar with the workplace, tasks and processes;
- will not be familiar with the workforce and their individual attitudes to health and safety and will have difficulty in obtaining their full cooperation;
- may be unfamiliar with the industry and seek unrealistic standards; and
- may well be more costly than an internal member of staff.

Multi-site organisations often use a hybrid arrangement with cross-site audits where people from other locations inside the organisation will audit a particular site, perhaps with one independent external expert on the team.

5.3.4 Actions taken after the audit

It is essential for the organisation including and especially the Board of Directors to take the appropriate action after the audit report is received. It is often a very good idea for the lead auditor to make a presentation to senior managers on the findings of the audit and the recommendations. It may be necessary to ensure that some or all board members have some health and safety training to promote their understanding of the importance and recommendations from audit and other health and safety monitoring or investigation reports.

An implementation programme will probably be needed to ensure that adequate capital finance is made available. This will need to be closely monitored by a responsible manager allocated to the task. This should be periodically reviewed by senior managers to ensure that implementation is on track. A follow-up independent audit to check progress might be useful.

Figure 5.13 The audit report should be reviewed by senior managers with an action plan and follow-up

The audit report and recommendations should be a live document which helps to improve the health and safety management system and produce a new impetus to improve performance and enhance risk control systems. The results should be fed into formal internal reviews, covered in the next section, to see whether policy or other changes are required.

5.4 Review of health and safety performance

5.4.1 Purpose of reviewing performance

When performance is reviewed, judgements are made about its adequacy and decisions are taken about how and when to rectify problems. The ACT part in the health and safety management system is needed by organisations so that they can see whether the health and safety management system is working as intended. This feedback on both successes and failures is an essential element to keep employees at all levels motivated to improve health and safety performance. Many successful organisations encourage positive reviews and concentrate on those indicators that demonstrate risk control improvements.

The information for review of performance comes from audits of risk control systems (RCSs) and workplace precautions, and from the measurement of activities. There may be other influences, both internal and external, such as reorganisation, new legislation or changes in current good practice. These may result in the necessity to redesign or change parts of the health and safety management system or to alter its direction or objectives.

SMART (see Figure 2.4) performance standards need to be established which will identify the systems requiring change, responsibilities and completion dates. It is essential to feed back the information about success and failure so that employees are motivated to maintain and improve performance.

In a review, the following areas will need to be examined:

▶ the operation and maintenance of the existing system;
▶ how the safety management system is designed, developed and installed to accommodate changing circumstances.

5.4.2 People involved and planned intervals

Reviewing is a continuous process. It should be undertaken at different levels within the organisation. Responses will be needed as follows:

▶ by first line supervisors or other managers to remedy failures to implement workplace precautions which they observe in the course of routine activities;
▶ to remedy sub-standard performance identified by active and reactive monitoring;
▶ to the assessment of plans at individual, departmental, site, group or organisational level;
▶ to the results of audits.

Senior managers, departmental managers and health and safety professionals will be involved in many reviews where these are relevant to their responsibilities.

Review plans may include:

▶ monthly reviews of individuals, supervisors or sections;
▶ three-monthly reviews of departments;
▶ annual reviews of sites or of the organisation as a whole.

Figure 5.14 Review of performance

The frequency of review at each level should be decided upon by the organisation and reviewing activities should be devised which will suit the measuring and auditing activities. The review will identify specific remedial actions which establish who is responsible for implementation and set deadlines for completion. Reviews should always be looking for opportunities to improve performance and be prepared to recommend changes to policies and the organisation to achieve the necessary change. The output from reviews must be consistent with the organisation's policy, performance, resources and objectives. They must be properly documented since they will be the base line from which continual improvement will be measured.

5.4.3 Items to be considered in reviews

Reviews will be wide ranging and may cover one specific subject or a range of subjects for an area of the organisation. They should aim to include:

▶ evaluation of compliance with legal and organisational requirements;
▶ incident data, recommendations and action plans from investigations;
▶ inspections, surveys, tours and sampling;
▶ absences and sickness records and their analysis;
▶ any reports on quality assurance or environmental protection;
▶ audit results and implementation;
▶ monitoring of data, reports and records;
▶ communications from enforcing authorities and insurers;
▶ any developments in legal requirements or best practice within the industry;
▶ changed circumstances or processes;
▶ benchmarking with other similar organisations;

- complaints from neighbours, customers and the public;
- effectiveness of consultation and internal communications;
- whether health and safety objectives have been met;
- whether actions from previous reviews have been completed.

5.4.4 Role of directors and senior managers

a) Reporting on performance

The management systems must allow the Board and senior managers to receive both specific (e.g. incident-led) and routine reports on the performance of health and safety policy. Much day-to-day health and safety information need be reported only at the time of a formal review. But only a strong system of monitoring can ensure that the formal review can proceed as planned – and that relevant events in the interim are brought to the Board's attention.

The Board should ensure that:

- appropriate weight is given to reporting both preventative information (results of active monitoring), such as progress of training and maintenance programmes, and incident data (results of reactive monitoring), such as accident and sickness absence rate;
- periodic audits of the effectiveness of management structures and risk controls for health and safety are carried out;
- the impact of changes such as the introduction of new procedures, work processes or products, or any major health and safety failure, is reported as soon as possible to the Board;
- there are procedures to implement new and changed legal requirements and to consider other external developments and events.

Good practice involves:

- effective monitoring of sickness absence and workplace health. This can alert the Board to underlying problems that could seriously damage performance or result in accidents and long-term illness;
- the collection of workplace health and safety data. This can allow the Board to benchmark the organisation's performance against others in its sector;
- appraisals of senior managers which include an assessment of their contribution to health and safety performance;
- boards receiving regular reports on the health and safety performance and actions of contractors;
- winning greater support for health and safety by involving workers in monitoring.

b) Reviewing health and safety

Reviewing performance should be supported at the highest level in the organisation and built into the safety management system. A formal boardroom review of health and safety performance is essential. It allows the Board to establish whether the essential health and safety principles – strong and active leadership, worker involvement, and assessment and review – have been embedded in the organisation. It tells senior managers whether their system is effective in managing risk and protecting people. Directors will need to know whether they are provided with sufficient information to make sound health and safety judgements about the organisation.

Performance on health and safety and well-being is increasingly being recorded in organisations' annual reports to investors and stakeholders. Board members can make extra 'shop floor' visits to gather information for the formal review. Good health and safety performance can be celebrated at central and local level both inside and outside the organisation (for example at RoSPA or IOSH annual awards).

The Board should review health and safety performance at least once a year. The review process should:

- examine whether the health and safety policy reflects the organisation's current priorities, plans and targets;
- examine whether risk management and other health and safety systems have been effectively reported to the Board;
- report health and safety shortcomings, and the effect of all relevant Board and management decisions;
- decide actions to address any weaknesses and a system to monitor their implementation;
- consider immediate reviews in the light of major shortcomings or events.

Setting up a separate risk management or health and safety committee as a subset of the Board, chaired by a senior executive, can make sure the key issues are addressed and guard against time and effort being wasted on trivial risks and unnecessary bureaucracy. This is referred to in Chapter 3 in more detail. The results of these reviews need to be properly recorded and fed into action and development plans for the coming year or so. This committee can ensure ongoing monitoring and review of performance and provide an annual report for the formal board review.

5.4.5 Continual improvement

Many organisations have a policy on continual improvement which should be applied to health and safety management in the same way as other management issues. The health and safety commitment on continuous improvement might include statements like: 'The organisation will continuously seek to improve its safety performance.' Continual improvement of safety performance will be achieved through:

► active and reactive evaluations of facilities, equipment, documentation and procedures through safety audits and surveys;
► active evaluation of each individual's performance to verify the fulfilment of their safety responsibilities; and
► a reactive evaluation in order to verify the effectiveness of the system for control and mitigation of risk.

The organisation will also continuously seek to improve its safety management processes.

Figure 5.15 Continual improvement part of the health and safety management process

Measures that could be used to improve safety management include:

► more succinct procedures;
► improved safety reviews, studies and audits;
► improved reporting and analysis tools;
► improved hazards identification and risk assessment processes and improved awareness of risks in the organisation;
► improved relations with the sub-contractors, suppliers and customers regarding safety;
► improved communication processes, including feedback from the personnel.

5.5 Additional useful background information that is not part of the NEBOSH NGC syllabus

5.5.1 Digital technology and EHS strategy

Digital technology is changing our economies, our societies and our way of life. The management of environment, health and safety (EHS) stands to be significantly impacted by digital technology. Technological advancements are changing the nature of how EHS professionals mitigate risk, make decisions and allocate resources. From the automation of processes, through the proliferation of in-field sensors and the evolution toward predictive analytics, workers and organisations stand to experience a step change in EHS performance, if these opportunities can be grasped.

As the business environment moves further into the digital age, leading organisations are leveraging existing and emerging technologies to improve their EHS outcomes. However, the majority of organisations are still reconciling how to translate these technological changes into improved performance.

Current state of play (February 2020)

Compared with other functions such as marketing and finance, the EHS profession has been slow to embrace and adopt digital technology. This is because organisations do not always fully understand the problem they are trying to solve, resulting in systemic, people- or technology-related gaps in the solutions implemented. And where digital technologies are employed, they are often poorly designed, disconnected and implemented to solve isolated problems.

Other barriers that could explain the lack of digital technology progress in this field include:

Focus on incidents

Traditionally, the management of EHS places a strong focus on measuring incidents. In a broader organisational context, incident datasets are limited and restrictive. Often, digital traces of incident causal factors are not identified. Even for those rare instances where data related to causal factors are identified, the profession grapples with the integrity and statistical rigour of incident causation models.

Small datasets

Despite the growing digitisation of traditional EHS activities (e.g. inspections and incident reports), EHS datasets are still comparatively small. Safety activities alone are unlikely to ever provide big data at the organisational level. Rather, to generate higher value insights, EHS systems and professionals increase the number of end-users and sensors providing data. Through thoughtfully designed user interfaces, greater, high-quality data can be used to reduce EHS risks.

Non-digital backgrounds

EHS is a multidisciplinary profession. Therefore, EHS professionals come from diverse and varied backgrounds; these backgrounds rarely include digital, data or statistics-intensive occupations, such as finance or IT.

Failing to demonstrate the business case

Quantifying the benefits of digital initiatives, decisions or operations that lead to healthy and safe workplaces can be challenging. By failing to identify 'what success looks like', ongoing support for initiatives is more difficult to secure. In other words, comprehensive integration can remain elusive.

The trust deficit

Big data analytics and the use of other digital technologies raise several key areas of concern, including privacy, discrimination, security and quality. These risks can hinder initiatives, particularly in low-trust workplaces.

Box 5.3 Examples of successful digital technology EHS initiatives are beginning to emerge

These include:

Technology	EHS initiative
Manual handling risk mitigation technology	This is used to reduce the incidence of musculoskeletal injuries. By measuring 'at risk' body movements, users can receive real-time alerts to mobile applications via ongoing data feeds. Tutorials are then provided to assist workers in their manual handling techniques.
Virtual reality (VR) technology	This is used to significantly improve training effectiveness. Technologies can provide organisations with a high-impact, scalable and efficient method to rapidly build the capabilities of workers, particularly those with less experience in high-risk environments.
Drones and robotics	These are used to perform jobs of the 'three Ds': dangerous, dirty and difficult. Such technology can access areas that are difficult to reach, such as those in emergency response scenarios. They can also be used to save significant periods of time, such as large-site scanning in agriculture and mining applications.
Artificial intelligence (AI) bots	These bots are increasingly being used to automate complex tasks, and being embedded into many applications to guide users – including manual handling where an AI bot can coach workers through customised manual handling training.
Workplace analytics	This involves installing sensors into workstations in order to measure elements of performance. The data streams generated can be used to coach workers, and improve workstation and interface design. 'Ergonomic analytics' is being performed in transport, power and utilities, and manufacturing settings.
Biometrics	This involves the application of sensors onto workers themselves, in order to gather physical health, mental health and other metrics. For example, fatigue levels and body temperature can be measured in real time, with alerts designed to intervene and warn workers and their organisations. Manual handling sensors have been developed to measure 'at risk' body movements, and provide workers with targeted coaching tutorials.
Hearables	These are ear-worn devices that assist users with hearing or audio-related tasks. An example of this is real-time language translation devices. This technology can be used to improve cross-cultural and language communications. With authentic, effective communication being a vital aspect of positive health and safety performance, this emerging technology could be used to significantly improve relationships.

5.5.2 Key points in becoming a good observer

To improve health and safety performance, managers and supervisors must eliminate unsafe acts by observing them, taking immediate corrective action and following up to prevent recurrence. To become a good observer, they must improve their observation skills and must learn how to observe effectively. Effective observation includes the following key points:

▶ be selective
▶ know what to look for
▶ practise
▶ keep an open mind
▶ guard against habit and familiarity
▶ do not be satisfied with general impressions
▶ record observations systematically.

Observation techniques

In addition, to become a good observer, a person must:

▶ stop for 10–30 seconds before entering a new area to ascertain where employees are working;
▶ be alert for unsafe practices that are corrected as soon as you enter an area;
▶ observe activity – do not avoid the action;
▶ remember ABBI – look above, below, behind, inside;
▶ develop a questioning attitude to determine what injuries might occur if the unexpected happened and how the job might be accomplished more safely. Ask 'why?' and 'what could happen if …?';
▶ use all senses: sight, hearing, smell, touch;
▶ maintain a balanced approach. Observe all phases of the job;
▶ be inquisitive;
▶ observe for ideas – not just to determine problems;
▶ recognise good performance.

Some commonly observed poor work practices include:

▶ using machinery or tools without authority
▶ operating at unsafe speeds or in other violation of safe work practice
▶ removing guards or other safety devices, or rendering them ineffective
▶ using defective tools or equipment or using tools or equipment in unsafe ways
▶ using hands or body instead of tools or push sticks
▶ overloading, crowding or failing to balance materials or handling materials in other unsafe ways, including improper lifting
▶ repairing or adjusting equipment that is in motion, under pressure or electrically charged
▶ failing to use or maintain, or improperly using, personal protective equipment or safety devices
▶ creating unsafe, unsanitary or unhealthy conditions by improper personal hygiene, by using compressed

air for cleaning clothes, by poor housekeeping, or by smoking in unauthorised areas

▶ standing or working under suspended loads, scaffolds, shafts or open hatches.

Additional principles for conducting inspections:

▶ Draw attention to the presence of any immediate danger – other items can await the final report.

▶ Shut down and 'lock out' any hazardous items that cannot be brought to a safe operating standard until repaired.

▶ Do not operate equipment. Ask the operator for a demonstration. If the operator of any piece of equipment does not know what dangers may be present, this is cause for concern. Items should never be ignored because the observer did not have the knowledge to make an accurate assessment of safety.

▶ Clearly describe each hazard and its exact location in rough notes. Allow 'on-the-spot' recording of all findings before they are forgotten. Record what have or have not been examined in case the inspection is interrupted.

▶ Ask questions, but do not unnecessarily disrupt work activities. This may interfere with efficient assessment of the job function and may also create a potentially hazardous situation.

▶ Consider the static (stop position) and dynamic (in motion) conditions of the item you are inspecting. If a machine is shut down, consider postponing the inspection until it is functioning again.

▶ Discuss as a group, 'Can any problem, hazard or accident generate from this situation when looking at the equipment, the process or the environment?' Determine what corrections or controls are appropriate.

▶ Do not try to detect all hazards simply by relying on human senses or by looking at them during the inspection. Monitoring equipment may have to be used to measure the levels of exposure to chemicals, noise, radiation or biological agents.

▶ It is increasingly possible to take photographs, particularly on tablet computers or smartphones, to help with any description of a particular situation, and transfer them directly to those who can take management actions. In fact many large contracts insist on inspections being recorded on special apps/platforms to obtain rapid dissemination of information in a common format. Make sure it is safe to operate an electronic device in the workplace concerned.

5.6 Further information

Investigating Incidents and Accidents at Work HSG245 HSE Books 2004 ISBN 978 0 7176 2827 8. www.hse.gov.uk/pubns/books/hsg245.htm

Reporting accidents and incidents at work, A brief guide to the Reporting of Injuries, Diseases and Dangerous Occurrences Regulations 2013 (RIDDOR), INDG453 (rev 1), October 2013 www.hse.gov.uk/pubns/indg453.pdf

The Management of Health and Safety at Work Regulations 1999 see HSE management site at: www.hse.gov.uk/managing/index.htm

The Reporting of Injuries, Disease and Dangerous Occurrences Regulations 2013 see HSE reporting website: www.hse.gov.uk/riddor/index.htm

HSE, 'Managing for health and safety' www.hse.gov.uk/managing/

HSE, The Health and Safety Toolbox: how to control risks at work www.hse.gov.uk/toolbox/

Managing for health and safety, HSG65 third edition 2013, HSE Books, ISBN: 978-0-7176-6456-6 www.hse.gov.uk/pubns/priced/hsg65.pdf

5.7 Practice questions

1. **Identify FIVE** active (or proactive) and **FIVE** reactive measures that can be used to monitor an organisation's health and safety performance.

2. A large company is planning to introduce a programme of regular inspections of the workplace.
   (a) **Outline** the factors that should be considered when planning such inspections.
   (b) **Outline** factors that determine the frequency with which health and safety inspections should be undertaken.

3. (a) **Define** the terms 'safety survey', 'safety tour' and 'safety sampling'.
   (b) **Outline** the issues which should be considered when a safety survey of a workplace is to be undertaken.

4. (a) **Outline** the strengths **AND** weaknesses of using a checklist to undertake a health and safety inspection of a workplace.
   (b) **Identify** the questions that might be included on a checklist to gather information following an accident involving slips, trips and falls.

5. An employee of a company is to be given duties to undertake a safety inspection.
   (a) **Outline** the competencies expected to carry out the duties.
   (b) **Identify** the principal issues that should be included in a safety inspection report so that managers can make decisions on any required remedial actions.
   (c) **Explain** how the report should be structured and presented so as to increase the likelihood of action being taken by managers.

6. (a) **Give FOUR** reasons why an organisation should have a system for the internal reporting of accidents.
   (b) **Identify** the issues that should be included in a typical workplace accident reporting procedure.
   (c) **Outline** factors that may discourage employees from reporting accidents at work.

7. An employee has been seriously injured after being struck by a fork-lift truck in a warehouse.
   (a) **Give FOUR** reasons why the accident should be investigated by the person's employer.
   (b) **Outline** the information that should be included in the investigation report.
   (c) **Outline FOUR** possible immediate causes and **FOUR** possible underlying (root) causes of the accident.
   (d) Giving reasons in **EACH** case, **identify FOUR** people who may be considered useful members of the accident investigation team.

8. **Outline** the issues to be considered to ensure an effective witness interview following a workplace accident.

9. (a) **Outline**, using a workplace example, the meaning of the terms:
      (i) near miss;
      (ii) dangerous occurrence.
   (b) **Explain** the purpose and benefits of collecting 'near miss' incident data.
   (c) **Outline** how an 'accident ratio study' can contribute to an understanding of accident prevention.

10. An employee sustained a serious injury while using an unguarded drilling machine and was admitted to hospital where he remained for several days. The machine had been unguarded for several days before the accident.
    (a) **Outline** the legal requirements for reporting the accident to the enforcing authority.
    (b) **Identify** the possible immediate **AND** root causes of the accident.
    (c) **Outline** the actions that should be taken by the management to improve health and safety standards in the workplace following the accident.

11. With reference to the Reporting of Injuries, Diseases and Dangerous Occurrences Regulations 2013:
    (a) **Identify** the legal requirements for reporting a fatality resulting from an accident at work to an enforcing authority.
    (b) **Identify FOUR** types of specified (serious) injury and **THREE** reportable work-related illnesses.

12. **Outline** reasons why employers should keep records of occupational ill health amongst employees.

13. An employee is claiming compensation for injuries received during a fall down a flight of stairs at the place of work. **Identify** the documented information required when preparing a possible defence against the claim.

14. **Outline** how the following techniques may be used to improve health and safety performance within an organisation and the differences between them.
    (a) Safety inspections.
    (b) Externally led health and safety audits.
    (c) Analysis of accident statistics.

15. (a) **Explain** the meaning of the term health and safety *'audit'*.
    (b) **Outline** the issues that need to be considered at the planning stage of the audit.
    (c) **Identify TWO** methods of gathering information during an audit.

16. (a) **Outline** key areas that may be covered within a health and safety audit.
    (b) **Identify** the documents that are likely to be examined during a health and safety audit.
    (c) **Explain** how the findings of a health and safety audit can be used to improve health and safety performance.

17. **Outline FOUR** advantages **AND FOUR** disadvantages of undertaking a health and safety audit of an organisation's activities by:
    (a) an internal auditor; or
    (b) an external auditor.

18. A health and safety audit of a manufacturing company has identified a lack of compliance with many of its health and safety procedures.
    (a) **Describe** the possible reasons for procedures not being followed.
    (b) the practical measures that could be taken to motivate employees to comply with health and safety procedures.

19. (a) **Outline** ways in which an organisation can monitor its health and safety performance.
    (b) **Identify EIGHT** measures that could be used by an organisation in order to monitor its health and safety performance.
    (c) **Outline** the reasons why an organisation should review and monitor its health and safety performance.

20. The Board of Directors of a large company decides to review its health and safety performance over the past year.
    (a) **Outline** the role of the Board in such a review.
    (b) **Identify** the topics that should be covered in the review.
    (c) **Outline** the possible review periods for the various parts of the company.

5

APPENDIX 5.1 Workplace inspection exercises

Figures 5.16–5.19 show workplaces with numerous inadequately controlled hazards. They can be used to practise workplace inspections and risk assessments.

Figure 5.16 Office

Figure 5.17 Road repair

Figure 5.18 Workshop

Figure 5.19 Roof repair/unloading

APPENDIX 5.2 Information for insurance/compensation claims

Factual information needs to be collected where there is the likelihood of some form of claim either against the organisation or by the organisation (e.g. damage to equipment). This aspect should be considered as a second objective in accident/incident investigation, the first being to learn from the accident/incident to reduce the possibility of accidents/incidents occurring in the future.

Workplace claims

▶ accident book entry;
▶ first-aider report;
▶ surgery record;
▶ foreman/supervisor accident report;
▶ safety representative accident report;
▶ RIDDOR report to HSE;
▶ other communications between defendants and HSE;
▶ minutes of Health and Safety Committee meeting(s) where accident/incident considered;
▶ report to Department for Work and Pensions;
▶ documents listed above relative to any previous accident/incident identified by the claimant and relied upon as proof of negligence.

Documents produced to comply with requirements of the Management of Health and Safety at Work Regulations 1999:

▶ pre-accident/incident risk assessment;
▶ post-accident/incident re-assessment;
▶ Accident/incident Investigation Report prepared in implementing the requirements; Health Surveillance Records in appropriate cases; information provided to employees;
▶ documents relating to the employee's health and safety training.

Workplace claims – examples of disclosure where specific Regulations apply

Section A – Workplace (Health Safety and Welfare) Regulations 1992

▶ repair and maintenance records;
▶ housekeeping records;
▶ hazard warning signs or notices (traffic routes).

Section B – Provision and Use of Work Equipment Regulations 1998

▶ manufacturers' specifications and instructions in respect of relevant work equipment establishing its suitability to comply with Regulation 5;
▶ maintenance log/maintenance records required; documents providing information and instructions to employees; documents provided to the employee in respect of training for use;

▶ any notice, sign or document relied upon as a defence against alleged breaches dealing with controls and control systems.

Section C – Personal Protective Equipment at Work Regulations 1992

▶ documents relating to the assessment of the PPE;
▶ documents relating to the maintenance and replacement of PPE;
▶ record of maintenance procedures for PPE;
▶ records of tests and examinations of PPE;
▶ documents providing information, instruction and training in relation to the PPE;
▶ instructions for use of PPE to include the manufacturers' instructions.

Section D – Manual Handling Operations Regulations 1992

▶ manual handling risk assessment carried out;
▶ re-assessment carried out post-accident;
▶ documents showing the information provided to the employee to give general indications related to the load and precise indications on the weight of the load and the heaviest side of the load if the centre of gravity was not positioned centrally;
▶ documents relating to training in respect of manual handling operations and training records.

Section E – Health and Safety (Display Screen Equipment) Regulations 1992

▶ analysis of workstations to assess and reduce risks;
▶ re-assessment of analysis of workstations to assess and reduce risks following development of symptoms by the claimant;
▶ documents detailing the provision of training including training records;
▶ documents providing information to employees.

Section F – Control of Substances Hazardous to Health Regulations 2002

▶ risk assessments and any reviews;
▶ copy labels from containers used for storage handling and disposal of carcinogens;
▶ warning signs identifying designation of areas and installations which may be contaminated by carcinogens;
▶ documents relating to the assessment of the PPE;
▶ documents relating to the maintenance and replacement of PPE;
▶ record of maintenance procedures for PPE;
▶ records of tests and examinations of PPE;
▶ documents providing information, instruction and training in relation to the PPE;
▶ instructions for use of PPE to include the manufacturers' instructions;

- air monitoring records for substances assigned a workplace exposure limit;
- maintenance examination and test of control measures records;
- monitoring records;
- health surveillance records;
- documents detailing information, instruction and training including training records for employees;
- labels and health and safety data sheets supplied to the employers.

Section G Construction Design and Management Regulations 2015

- documents relating to the appointment of principal designer and/or contractor;

- documents relating to the pre-construction phase health and safety information;
- documents relating to the construction phase health and safety plan;
- documents relating to the health and safety file;
- documents covering appropriate risk assessments method statements;
- information on sub-contractors working on site who are relevant to the incident concerned;
- any relevant inspection records of the work site, scaffolds, excavations or pieces of plant.

5

Physical and psychological health

> **Chapter summary**
>
> **This chapter covers the following topics in Element 5 of the NEBOSH NGC syllabus:**
> Noise
> Vibration
> Radiation
> Mental ill health
> Violence at work
> Substance abuse at work
>
> **Learning outcomes**
>
> **These are the learning outcomes covered by this chapter:**
> **The learner will be able to:**
> 1. Outline the health effects associated with exposure to noise and appropriate control measures
> 2. Outline the health effects associated with exposure to vibration and appropriate control measures
> 3. Outline the health effects associated with ionising and non-ionising radiation and appropriate control measures
> 4. Outline the meaning, causes and effects of mental health problems at work and appropriate control measures
> 5. Explain the risk factors and appropriate controls for violence at work
> 6. Explain the effects of substance misuse on health and safety at work and control measures to reduce such risks

Introduction

Occupational health is concerned with physical and psychological hazards as well as chemical and biological hazards. The physical occupational hazards have been well known for many years, and recent emphasis has been on the development of lower-risk workplace environments. Physical hazards include topics such as electricity, display screen equipment (DSE) and manual handling, which were covered in earlier chapters; and noise, vibration and radiation, which are discussed in this chapter.

However, it is only really in the last 20 years that psychological hazards have been included among the occupational health hazards faced by many workers. This is now the most rapidly expanding area of occupational health, and includes topics such as mental health, violence to staff and substance abuse.

Occupational ill health accounts for about 99% of work-related deaths each year, according to health and safety statistics from the HSE.

The biggest causes of work-related diseases are cancer, including mesothelioma and lung cancer due to asbestos exposure, and chronic obstructive pulmonary disease (COPD) with the main causal agents being dust, gases, vapours and fumes.

Following asbestos-induced cancer and COPD, the next four biggest diseases are:

- Lung cancer – silica
- Lung cancer – diesel engine exhaust
- Breast cancer – shiftwork
- Lung cancer – mineral oils.

6.1 Noise

6.1.1 Introduction

There was considerable concern for many years over the increasing cases of occupational deafness, and this led to the introduction of the Noise at Work Regulations in 1989 and the revised Noise at Work Regulations in 2005. HSE estimated that an additional 1.1 million workers will be covered by the revised Regulations. These Regulations, which are summarised in Chapter 13, require the employer to:

- assess noise levels and keep records;
- roduco tho risks from noise exposure by using engineering controls in the first instance and the provision and maintenance of hearing protection as a last resort;
- provide employees with information and training;
- if a manufacturer or supplier of equipment, provide relevant noise data on that equipment (particularly if any of the three action levels is likely to be reached).

Many occupations have potential noise problems including construction, manufacturing, entertainment, the uniformed services and call centres.

The main purpose of the Noise Regulations is to control noise levels rather than measuring them. This involves the better design of machines, equipment and work processes, and ensuring that personal protective equipment is correctly worn and employees are given adequate training and health surveillance.

Sound is transmitted through the air by sound waves which are produced by vibrating objects. The vibrations cause a pressure wave which can be detected by a receiver, such as a microphone or the human ear. The ear may detect vibrations which vary from 20 to 20,000 (typically 50–16,000) cycles each second (or Hertz – Hz). Sound travels through air at a finite speed (342 m/s at 20°C and sea level). The existence of this speed is shown by the time lag between lightning and thunder during a thunderstorm. Noise normally describes loud, sudden, harsh or irritating sounds although noise is defined as any audible sound.

Noise may be transmitted directly through the air, by reflection from surrounding walls or buildings or through the structure of a floor or building. In construction work, the noise and vibrations from a pneumatic drill will be transmitted from the drill itself, from the ground being drilled and from the walls of surrounding buildings

Figure 6.1 Better to control noise at source than wear ear protection

internal) ear. The sound pressure wave passes into and through the outer ear and strikes the eardrum causing it to vibrate. The eardrum is situated approximately 25mm inside the head. The vibration of the eardrum causes the proportional movement of three interconnected small bones in the middle ear, thus passing the sound to the cochlea situated in the inner ear.

6.1.2 The physical and psychological effects of exposure to noise

The human ear

There are three sections of the ear – the outer (or external) ear, the middle ear and the inner (or

(a)

Outer ear	Middle ear	Inner ear		
Sound waves in air	Tympanic membrane	Hammer→Anvil→Stirrup Mechanical movement of ossicles	Oval window	Perilymph Fluid wave	Endolymph Fluid wave stimulates basilar membrane and spiral organ	Perilymph Fluid wave	Round window

(b)

Figure 6.2 Passage of sound waves: (a) the ear with cochlea uncoiled; (b) summary of transmission

Within the cochlea the sound is transmitted to a fluid causing it to vibrate. The motion of the fluid induces a membrane to vibrate which, in turn, causes hair cells attached to the membrane to bend. The movement of the hair cells causes a minute electrical impulse to be transmitted to the brain along the auditory nerve. Those hairs nearest to the middle ear respond to high frequency, while those at the tip of the cochlea respond to lower frequencies. (**Frequency** is the number of occurrences of a repeating event over a given period of time and is measured in Hertz (Hz).)

There are about 30,000 hair cells within the ear and noise-induced hearing loss causes irreversible damage to these hair cells.

Ill-health effects of noise

Noise can lead to ear damage on a temporary (acute) or permanent (chronic) basis.

There are three principal acute effects:

▶ **temporary threshold shift** – caused by short excessive noise exposures and affects the cochlea by reducing the flow of nerve impulses to the brain. The result is a slight deafness, which is reversible when the noise is removed;

▶ **tinnitus** – a ringing in the ears caused by an intense and sustained high noise level. It is caused by the over-stimulation of the hair cells. The ringing sensation continues for up to 24 hours after the noise has ceased;

▶ **acute acoustic trauma** – caused by a very loud noise such as an explosion. It affects either the eardrum or the bones in the middle ear and is usually reversible. Severe explosive sounds can permanently damage the eardrum.

Occupational noise can also lead to one of the following three **chronic** hearing effects:

▶ **noise-induced hearing loss** – results from permanent damage to the cochlear hair cells. It affects the ability to hear speech clearly but the ability to hear is not lost completely;

▶ **permanent threshold shift** – this results from prolonged exposure to loud noise and is irreversible due to the permanent reduction in nerve impulses to the brain. This shift is most marked at the 4,000 Hz frequency, which can lead to difficulty in hearing certain consonants and some female voices;

▶ **tinnitus** – is the same as the acute form but becomes permanent. It is a very unpleasant condition, which can develop without warning.

It is important to note that, if the level of noise exposure remains unchanged, noise-induced hearing loss will lead to a permanent threshold shift affecting an increasing number of frequencies. The hearing loss caused can be temporary or permanent.

▶ **temporary deafness** can occur after leaving a noisy place. Hearing usually recovers within a couple of hours. This is a sign that continued exposure to loud noise could permanently damage your hearing;

▶ **sudden extremely loud noise** can cause instant damage; and

▶ **repeated exposure** causes gradual hearing loss due to repeated exposure. This is more common and it can take years for a worker to realise just how deaf they have become.

Presbycusis is the term used for hearing loss in older people which may have been exacerbated by occupational noise earlier in their lives.

6.1.3 Noise assessments including commonly used terms

The Control of Noise at Work Regulations specify action levels at which the hearing of employees must be protected. The conclusion as to whether any of those levels has been breached is reached after an assessment of noise levels has been made. However, before noise assessment can be discussed, noise measurement and the statutory action levels must be described.

Noise measurement

Sound intensity is measured by a unit known as a Pascal (Pa – N/m^2), which is a unit of pressure similar to that used when inflating a tyre. If noise was measured in this way, a large scale of numbers would be required ranging from 1 at one end to 1 million at the other. The sound pressure level (SPL) is a more convenient scale because:

▶ it compresses the size of the scale by using a logarithmic scale to the base 10;

▶ it measures the ratio of the measured pressure, p, to a reference standard pressure, p_0, which is the pressure at the threshold of hearing (2×10^{-5} Pa).

The unit is called a decibel (dB) and is defined as:

$$SPL = 20 \log_{10}(p/p_0) \text{ dB}$$

It is important to note that since a logarithmic scale to the base 10 is used, each increase of 3 dB is a doubling in the sound intensity. Thus, if a sound reading changes from 75 dB to 81 dB, the sound intensity or loudness has increased by four times.

Finally, as the human ear tends to distort its sensitivity to the sound it receives by being less sensitive to lower frequencies, the scale used by sound meters is weighted so that readings mimic the ear. This scale is known as the A scale and the readings known as dB(A). There are also three other scales known as B, C and D.

Originally the A scale was used for sound pressure levels (SPLs) up to 55 dB, the B scale for levels between 55 dB and 85 dB and the C scale for values above 85 dB. However, today, the A scale is used for nearly all levels except for very high SPLs when the C scale is used. The B scale is rarely used and the D scale is mainly used to monitor jet aircraft engine noise.

Table 6.1 gives some typical decibel readings for common activities and Table 6.2 gives some values for woodworking machines published by the HSE.

Table 6.1 Some typical sound pressure levels (SPLs) (dB(A) values)

Activity or environment	SPL [dB(A)]
Threshold of pain	140
Pneumatic drill	125
Pop group or disco	110
Heavy lorry	93
Street traffic	85
Conversational speech	65
Business office	60
Living room	40
Bedroom	25
Threshold of hearing	0

Table 6.2 Typical noise levels at woodworking machines

Machine	Noise level (dB)
Beam panel saws and sanding machines	97
Boring machines	98
Band re-saws, panel planers and vertical spindle moulders	100
Portable woodworking tools	101
Bench saws and multiple ripsaws	102
High-speed routers and moulders	103
Thicknessers	104
Edge banders and multi-cutter moulding machines	105
Double-end tenoners	107

Noise is measured using a sound level meter which reads SPLs in dB(A) and the **peak sound pressure** in Pascal (Pa), which is the highest noise level reached by the sound. There are two basic types of sound meter – integrated and direct reading meters. Meters which integrate the reading provide an average over a particular time period, which is an essential technique when there are large variations in sound levels. This value is known as the **continuous equivalent noise level (L_{Eq})**, which is normally measured over an 8-hour period.

Direct reading devices, which tend to be much cheaper, can be used successfully when the noise levels are continuous at a near constant value.

Another important noise measurement is the **daily personal exposure level** of the worker, $L_{EP,d}$, which is measured over an 8-hour working day. Hence, if a

person was exposed to 87 dB(A) over 4 hours, this would equate to an $L_{EP,d}$ of 84 dB(A) since a reduction of 3 dB(A) represents one half of the noise dose.

The HSE Guidance on the Control of Noise at Work Regulations, L108, offers some very useful advice on the implementation of the Regulations and should be read by anyone who suspects that they have a noise problem at work. The guidance covers 'equipment and procedures for noise surveys' and contains a noise exposure ready reckoner which can be used to evaluate $L_{EP,d}$ when the noise occurs during several short intervals and/or at several different levels during the 8-hour period.

A survey by The Chartered Institute of Environmental Health (CIEH) has found that up to 50% of pub and club owners are failing to protect their staff from excessive noise at work. The findings also show that a majority of responding venue owners had not taken steps to reduce noise exposure since the elements of the Control of Noise at Work Regulations regarding the entertainment sector came into force.

Noise action levels

Regulation 6 of the Control of Noise at Work Regulations places a duty on employers to reduce the risk of damage to the hearing of their employees from exposure to noise to the lowest level reasonably practicable.

The Regulations introduce **exposure action** level values and **exposure limit** values. An **exposure action** value is a level of noise at which certain action must be taken. An **exposure limit** value is a level of noise at the ear above which an employee must not be exposed. Therefore if the workplace noise levels are above this value, any ear protection provided to the employee must reduce the noise level to the limit value at the ear.

These exposure action and limit values are as follows.

1. The lower exposure action levels are:
   (a) a daily or weekly personal noise exposure of **80 dB(A)**;
   (b) a peak sound pressure of **135 dB(C)**.
2. The upper exposure action levels are:
   (a) a daily or weekly personal noise exposure of **85 dB(A)**;
   (b) a peak sound pressure of **137 dB(C)**.
3. The exposure limit values are:
   (a) a daily or weekly personal noise exposure of **87 dB(A)**;
   (b) a peak sound pressure of **140 dB(C)**.

The peak exposure action and limit values are defined because high-level peak noise can lead to short-term and long-term hearing loss. Explosives, guns (including nail guns), cartridge tools, hammers and stone chisels can all produce high peak sound pressures.

If the daily noise exposure exceeds the lower exposure action level, then a noise assessment should be carried out and recorded by a competent person. There is a very simple test which can be done in any workplace to determine the need for an assessment. Table 6.3 gives information on the simple test to determine the need for a noise risk assessment.

Table 6.3 Simple observations to determine the need for a noise risk assessment

Observation at the workplace	Likely noise level [dB(A)]	A noise risk assessment must be made if this noise level persists for
The noise is noticeable but does not interfere with normal conversation – equivalent to a domestic vacuum cleaner	80	6 hours
People have to shout to be heard if they are more than 2 m apart	85	2 hours
People have to shout to be heard if they are more than 1 m apart	90	45 minutes

If there is a marked variation in noise exposure levels during the working week, then the Regulations allow a weekly rather than daily personal exposure level, $L_{EP,w}$, to be used. It is only likely to be significantly different to the daily exposure level if exposure on one or two days in the working week is 5 dB(A) higher than on the other days, or the working week has three or fewer days of exposure. The weekly exposure rate is not a simple arithmetic average of the daily rates. If an organisation is considering the use of a weekly exposure level, then the following provisions must be made:

▶ hearing protection must be provided if there are very high noise levels on any one day;
▶ the employees and their representatives must be consulted on whether weekly averaging is appropriate;
▶ an explanation must be given to the employees on the purpose and possible effects of weekly averaging.

Finally, if the working day is 12 hours, then the action levels must be reduced by 3 dB(A) because the action levels assume an 8-hour working day.

The HSE Guidance document L108 gives detailed advice on noise assessments and surveys. The most important points are that the measurements should be taken at the working stations of the employees closest to the source of the noise and over as long a period as possible, particularly if there is a variation in noise levels during the working day. Other points to be included in a noise assessment are:

▶ details of the noise meter used and the date of its last calibration;
▶ the number of employees using the machine, time period of usage and other work activities;

▶ an indication of the condition of the machine and its maintenance schedule;
▶ the work being done on the machine at the time of the assessment;
▶ a schematic plan of the workplace showing the position of the machine being assessed;
▶ other noise sources, such as ventilation systems, that should be considered in the assessment. The control of these sources may help to reduce overall noise levels;
▶ recommendations for future actions, if any.

Other actions which the employer must undertake when the lower exposure action level is exceeded are to:

▶ inform, instruct and train employees on the hearing risks;
▶ supply hearing protection to those employees requesting it;
▶ ensure that any equipment or arrangements provided under the Regulations are correctly used or implemented.

The additional measures which the employer must take if the upper exposure action level is reached are:

▶ reduce and control exposure to noise by means other than hearing protection;
▶ establish hearing protection zones, marked by notices and ensure that anybody entering the zone is wearing hearing protection (Figure 6.3);
▶ supply hearing protection and ensure that it is worn.

Figure 6.3 Hearing protection zone notice

6.1.4 Basic noise control methods

In addition to **reduced time exposure** of employees to the noise source, there is a simple hierarchy of control techniques:

▶ reduction of noise at source;
▶ reduction of noise levels received by the employee (known as attenuation);
▶ personal protective equipment, which should only be used when the above two remedies are insufficient.

Figure 6.4 Noise paths found in a workplace: (a) the quiet area is subjected to reflected noise from a machine elsewhere in the building; (b) the correct use of roof absorption will reduce the reflected noise; (c) segregation of the noisy operation will benefit the whole workplace

Reduction of noise at source

There are several means by which noise could be reduced at source:

▶ change the process or equipment (e.g. replace solid tyres with rubber tyres or replace diesel engines with electric motors);
▶ change the speed of the machine;
▶ improve the maintenance regime by regular lubrication of bearings, tightening of belt drives.

Attenuation of noise levels

There are many methods of attenuating or reducing noise levels and these are covered in detail in the guide to the Regulations. The more common ones will be summarised here.

Orientation or relocation of the equipment – turn the noisy equipment away from the workforce or locate it away in separate and isolated areas.

Enclosure – surrounding the equipment with a good sound-insulating material can reduce sound levels by up to 30 dB(A). Care will need to be taken to ensure that the machine does not become overheated.

Screens or absorption walls – can be used effectively in areas where the sound is reflected from walls. The walls of the rooms housing the noisy equipment are lined with sound absorbent material, such as foam or mineral wool, or sound absorbent (acoustic) screens are placed around the equipment.

Damping – the use of insulating floor mountings to remove or reduce the transmission of noise and vibrations through the structure of the building such as girders, wall panels and flooring.

Lagging – the insulation of pipes and other fluid containers to reduce sound transmission (and, incidentally, heat loss).

Silencers – normally fitted to engines which are exhausting gases to atmosphere. Silencers consist of absorbent material or baffles.

Isolation of the workers – the provision of soundproofed workrooms or enclosures isolated away from noisy equipment (a power station control room is an example of worker isolation).

Immediate noise reduction benefits are often achieved for noise control equipment (e.g. noise enclosures) that is in good repair.

6.1.5 Personal ear protection

The provision of personal hearing protection should only be considered as a last resort. There is usually resistance from the workforce to use them and they are costly to maintain and replace. They interfere with communications, particularly alarm systems, and they can present hygiene problems.

The following factors should be considered when selecting personal ear protection:

▶ suitability for the range of sound spectrum of frequencies to be encountered;
▶ noise reduction (attenuation) offered by the ear protection;
▶ pattern of the noise exposure;
▶ acceptability and comfort of the wearer, particularly if there are medical problems;
▶ durability;
▶ hygiene considerations;
▶ compatibility with other personal protective equipment;

6

▶ ease of communication and ability to hear warning alarms;

▶ maintenance and storage arrangements;

▶ cost.

There are two main types of ear protection – earplugs and ear defenders (earmuffs).

Earplugs are made of sound absorbent material and fit into the ear. They can be reusable or disposable, and are able to fit most people and can easily be used with safety glasses and other personal protective equipment. Their effectiveness depends on the quality of the fit in the ear which, in turn, depends on the level of training given to the wearer. Permanent earplugs come in a range of sizes so that a good fit is obtained. The effectiveness of earplugs decreases with age and they should be replaced at the intervals specified by the supplier. A useful simple rule to ensure that the selected ear plug reduces the noise level at the ear to 87 dB(A) is to choose one with a manufacturer's rating of 83 dB(A). This should compensate for any fitting problems. The main disadvantage of ear plugs is that they do not reduce the sound transmitted through the bone structure which surrounds the ear and they often work loose with time.

Ear defenders (earmuffs) offer a far better reduction of all sound frequencies. They are generally more acceptable to workers because they are more comfortable to wear and they are easy to monitor as they are clearly visible. They also reduce the sound intensity transmitted through the bone structure surrounding the ear. A communication system can be built into earmuffs. However, they may be less effective if the user has long hair or is wearing spectacles or large earrings. They may also be less effective if worn with helmets or face shields and uncomfortable in warm conditions. Maintenance is an important factor with earmuffs and should include checks for wear and tear and general cleanliness.

Selection of suitable ear protection is very important as it should not just reduce sound intensities below the statutory action levels but also reduce those intensities at particular frequencies. Normally advice should be sought from a competent supplier who will be able to advise on ear protection to suit a given spectrum of noise using 'octave band analysis'.

Finally, it is important to stress that the use of ear protection must be well supervised to ensure that not only is it being worn correctly, but that it is, in fact, actually being worn. This is a particular problem where lone workers are required to wear personal protective equipment, such as ear defenders. The requirement can be emphasised by the use of signs and warning notices on equipment and in company vehicles. Unannounced monitoring visits to the work site also encourage the use of personal protective equipment. If there is a persistent refusal to wear personal protective equipment, then disciplinary procedures must be initiated.

Noise-induced hearing loss (NIHL) will be prevented by fit-testing hearing protection.

The World Health Organization (WHO) has identified noise-induced hearing loss as being the most common permanent and preventable injury in the world and tinnitus as the third most serious non-fatal medical condition.

The issue of ear defenders to the workforce is a last resort after measures outlined in 6.1.4 have been considered.

6.1.6 Role of health surveillance

The Control of Noise at Work Regulations place a duty on the employer to undertake health surveillance for employees whose exposure regularly exceeds the upper action level irrespective of whether ear protection is worn. The recommended **health surveillance** is a hearing test at induction, followed by an annual check and review of hearing levels. The checks may be extended to every three years if no adverse effects are found during earlier tests. If exposure continues over a long period, health surveillance of employees is recommended using a more substantial audiometric test. This will indicate whether there has been any deterioration in hearing ability. Where exposure is between the lower and upper exposure action levels or occasionally above the upper action values, health surveillance will only be required if information becomes available, perhaps from medical records, that the employee is particularly sensitive to noise-induced hearing loss.

The Regulations also place the following statutory duties on employees:

▶ for noise levels above the lower exposure action level, they must use any control equipment (other than hearing protection), such as silencers, supplied by the employer and report any defects;

▶ for noise levels above the upper exposure action level, they fulfil the obligations given above and wear the hearing protection provided;

▶ they must take care of any equipment provided under these Regulations and report any defects;

▶ they must see their doctor if they feel that their hearing has become damaged;

▶ present themselves for health surveillance.

The Management of Health and Safety Regulations prohibits the employment of anyone under the age of 18 years where there is a risk to health from noise.

6.2 Vibration

6.2.1 The effects on the body of exposure to hand-arm vibration and whole-body vibration

Hand-held vibrating machinery (such as pneumatic drills, sanders and grinders, powered lawn mowers and strimmers and chainsaws) can produce health risks from hand-arm or whole-body vibration (WBV).

Hand-arm vibration syndrome

Hand-arm vibration syndrome (HAVS) describes a group of diseases caused by the exposure of the hand and arm to external vibration. Some of these have been described under work-related upper limb disorders (WRULDs), such as carpal tunnel syndrome.

However, the best known disease is **vibration white finger (VWF)** in which the circulation of the blood, particularly in the hands, is adversely affected by the vibration. The early symptoms are tingling and numbness felt in the fingers, usually sometime after the end of the working shift. As exposure continues, the tips of the fingers go white and then the whole hand may become affected. This results in a loss of grip strength and manual dexterity. Attacks can be triggered by damp and/or cold conditions and, on warming, 'pins and needles' are experienced (Figure 6.5). If the condition is allowed to persist, more serious symptoms become apparent including discoloration and enlargement of the fingers. In very advanced cases, gangrene can develop leading to the amputation of the affected hand or finger. VWF was first detailed as an industrial disease in 1911.

The risk of developing HAVS depends on the frequency of vibration, the length of exposure and the tightness of the grip on the machine or tool. Some typical values of vibration measurements for common items of equipment used in industry are given in Table 6.4.

Table 6.4 Examples of vibration exposure values measured by HSE on work equipment

Equipment	Condition	Vibration reading (m/s²)
Road breakers	Typical	12
	Modern design and trained operator	5
	Worst tools and operating conditions	20
Demolition hammers	Modern tools	8
	Typical	15
	Worst tools	25
Hammer drills	Typical	9
	Best tools and operating conditions	6
	Worst tools and operating conditions	25
Large-angle grinders	Modern vibration-reduced designs	4
	Other types	8
Small-angle grinders	Typical	2–6
Chainsaws	Typical	6
Brush cutter or strimmer	Typical	4
	Best	2
Sanders (orbital)	Typical	7–10

Regular exposure to HAV can cause a range of permanent injuries to hands and arms, collectively known as hand–arm vibration syndrome (HAVS). The injuries can include damage to the:

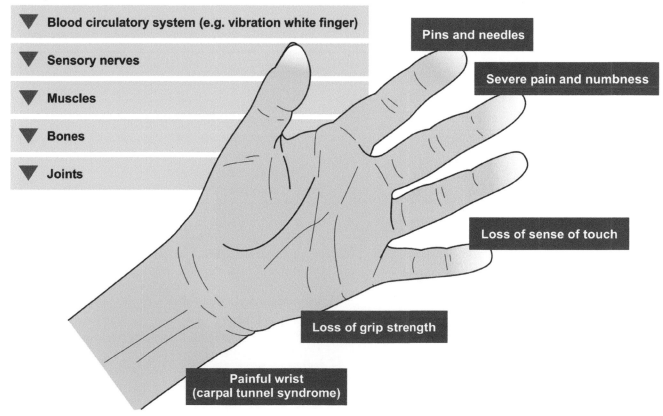

▼ **Blood circulatory system (e.g. vibration white finger)**

▼ **Sensory nerves**

▼ **Muscles**

▼ **Bones**

▼ **Joints**

Pins and needles

Severe pain and numbness

Loss of sense of touch

Loss of grip strength

Painful wrist (carpal tunnel syndrome)

Figure 6.5 Injuries which can be caused by hand-arm vibration

When assessing the risk of HAVS developing among employees, the source of the vibration, such as reciprocating, rotating and vibrating tools and equipment, needs to be considered first together with the age of the equipment, its maintenance record, its suitability for the job and any information or guidance available from the manufacturer. The number of employees using the tooling or equipment, the duration and frequency of their use and any relevant personal factors, such as a pre-existing circulatory problem, all form part of the assessment. Environmental factors, particularly exposure to cold and/or wet weather and the nature of the job itself, are also important factors to be considered during such a risk assessment. Finally, an examination of the existing controls and their effectiveness and the frequency, magnitude and direction of the vibration are important elements of the evaluation. Other issues could include the effectiveness of any personal protective equipment provided and any instruction or training given.

Whole-body vibration

Whole-body vibration (WBV) is caused by vibration from machinery passing into the body either through the feet of standing workers or the buttocks of sitting workers. It is the shaking or jolting of the human body through a supporting surface (usually a seat or the floor), for example when driving or riding on a vehicle along an unmade road, operating earth-moving machines or standing on a structure attached to a large, powerful, fixed machine, which is impacting or vibrating.

The most common ill-health effect is severe back pain which, in severe cases, may result in permanent injury. Reasons for back pain in drivers can include:

▶ poor design of controls, making it difficult for the driver to operate the machine or vehicle easily or to see properly without twisting or stretching;
▶ incorrect adjustment by the driver of the seat position and hand and foot controls, so that it is necessary to continually twist, bend, lean and stretch to operate the machine;
▶ sitting for long periods without being able to change position;
▶ poor driver posture;
▶ repeated manual handling and lifting of loads by the driver;
▶ excessive exposure to whole-body vibration, particularly to shocks and jolts;
▶ repeatedly climbing into or jumping down from a high cab or one which is difficult to get in and out of.

The risk increases where the driver or operator is exposed to two or more of these factors together.

Other acute effects include reduced visual and manual control, increased heart rate and blood pressure. Chronic or long-term effects include permanent spinal damage, damage to the central nervous system, hearing loss and circulatory and digestive problems.

The most common occupations which generate WBV are driving fork-lift trucks, construction vehicles and agricultural or horticultural machinery and vehicles. There is growing concern throughout the European Union about this problem. Control measures include the proper use of the equipment including correct adjustments of air or hydraulic pressures, seating and, in the case of vehicles, correct suspension, tyre pressures and appropriate speeds to suit the terrain. Other control measures include the selection of suitable equipment with low-vibration characteristics, work rotation, good maintenance and fault reporting procedures.

The HSE commissioned measurements of WBV on several machines and some of the results are shown in Table 6.5.

Table 6.5 Machines which could produce significant whole-body vibration

Machine	Activity	Vibration reading (m/s^2)
3-tonne articulated site	Removal of spoil	0.78
Dumper truck	Transport of spoil	1.13
25-tonne articulated dumper truck	Transport of spoil	0.91
Bulldozer	Dozing	1.16
4-tonne twin-drum	Finishing tarmac	0.86
80-tonne rigid dumper truck	Transport of spoil	1.03

Preventative and precautionary measures

The control strategy outlined in Chapter 4 can certainly be applied to ergonomic risks. The common measures used to control ergonomic ill-health effects are to:

▶ implement results of task analysis and identification of repetitive actions;
▶ eliminate vibration-related or hazardous tasks by performing the job in a different way;
▶ ensure that the correct equipment (properly adjusted) is always used;
▶ introduce job rotation so that workers have a reduced time exposure to the hazard;
▶ during the design of the job ensure that poor posture is avoided;
▶ undertake a risk assessment;
▶ reports from employees and safety representatives;
▶ ill-health reports and absence records;
▶ introduce a programme of health surveillance;
▶ ensure that employees are given adequate information on the hazards and develop a suitable training programme;
▶ ensure that a programme of preventative maintenance is introduced and include the regular inspection of items such as vibration isolation mountings;

- ▶ keep up to date with advice from equipment manufacturers, trade associations and health and safety sources (more and more low-vibration equipment is becoming available).

Figure 6.6 Breaker mounted on an excavator's jib to reduce vibrations

A very useful and extensive checklist for the identification and reduction of WRULDs is given in Appendix 2 of the HSE *Guide to work-related upper limb disorders* HSG60. HSE has also produced two very useful guides – *Hand-arm vibration, The Control of Vibration at Work Regulations 2005 Guidance on Regulations* L140 and *Whole-body vibration, The Control of Vibration at Work Regulations 2005 Guidance on Regulations* L141. These guidance documents offer detailed advice on the implementation of the Regulations.

6.2.2 The Control of Vibration at Work Regulations and basic control measures

The Control of Vibration at Work Regulations introduced, for both HAV and WBV, a daily exposure limit and action values. These values are as follows:

1. For HAV:
   (a) the daily exposure limit value normalised to an 8-hour reference period is 5 m/s²;
   (b) the daily exposure action value normalised to an 8-hour reference period is 2.5 m/s².
2. For WBV:
   (a) the daily exposure limit value normalised to an 8-hour reference period is 1.15 m/s²;
   (b) the daily exposure action value normalised to an 8-hour reference period is 0.5 m/s².

An exposure limit value must not be exceeded. If an exposure action value is exceeded, then action must be taken to reduce the value. However, operators of some off-road machines and vehicles may exceed the limit value but this will depend on the task, vehicle

speed, ground conditions, driver skill and duration of the operation (see the HSE website).

The term A(8) is added to the exposure limit or action value to denote that it is an average value spread over an 8-hour working day. Thus the daily exposure limit value for HAV is 5 m/s² A(8).

Hand-arm vibration

Many machines and processes used in industry produce HAV. Typical high-risk processes include:

- ▶ grinding, sanding and polishing wood and stone;
- ▶ cutting stone, metal and wood;
- ▶ riveting, caulking and hammering;
- ▶ compacting sand, concrete and aggregate;
- ▶ drilling and breaking rock, concrete and road surfaces; and
- ▶ surface preparation, including de-scaling and paint removal.

There are several ways to ascertain the size of the vibration generated by equipment and machines. Manufacturers must declare vibration emission values for portable hand-held and hand-guided machines and provide information on risks. Other important sources for vibration information include scientific and technical journals, trade associations and online databases. HSE experience has shown that the vibration level is higher in practice than that quoted by many manufacturers. The reasons for this discrepancy may be that:

- ▶ the equipment is not well maintained;
- ▶ the equipment is not suitable for the material being worked;
- ▶ the tool has not been purchased from a reputable supplier;
- ▶ the accessories are not appropriate or are badly fitted;
- ▶ the operative is not using the tool properly.

In view of these problems, it is recommended that the declared value should be doubled when comparisons are made with exposure limits. As the exposure limit or action value is averaged over 8 hours, it is possible to work with higher values for a reduced exposure time. Table 6.6 shows the reduction in exposure time as the size of the vibration increases.

Table 6.6 The change in exposure times as vibration increases

The change in exposure times as vibration increases							
Value of vibration (m/s²)	2.5	3.5	5	7	10	14	20
Exposure time to reach action value (hours)	8	4	2	1	30 min	15 min	8 min
Exposure time to reach limit value (hours)	over 24	16	8	4	2	1	30 min

The Guidance (L140) gives very useful advice on the measurement of vibration, undertaking a suitable and sufficient risk assessment, control measures, health surveillance and training of employees. The following points summarise the important measures which should be taken to reduce the risks associated with HAV:

1. avoid, whenever possible, the need for vibration equipment;
2. undertake a risk assessment which includes a soundly based estimate of the employees' exposure to vibration;
3. develop a good maintenance regime for tools and machinery. This may involve ensuring that tools are regularly sharpened, worn components are replaced or engines are regularly tuned and adjusted;
4. introduce a work pattern that reduces the time exposure to vibration;
5. issue employees with gloves and warm clothing. There is a debate as to whether anti-vibration gloves are really effective but it is agreed that warm clothing helps with blood circulation which reduces the risk of VWF. Care must be taken so that the tool does not cool the hand of the operator;
6. introduce a reporting system for employees to use so that concerns and any symptoms can be recorded and investigated; and
7. encourage employees to check regularly for any symptoms of vibration-related ill health and to participate in any health surveillance measures provided.

It is important that drill bits and tools are kept sharp and used intact – an angle grinder with a chipped cutting disc will lead to a large increase in vibration as well as being dangerous.

A hand-arm vibration exposure calculator, developed by the HSE, helps estimate and record workers' exposures to HAV and compare them with the action values in the Regulations as part of a HAV risk assessment. It includes dropdown vibration magnitudes for common tools that can be used to make cautious estimates of exposure before representative in-use data for specific tools is available.

Whole-body vibration

The Regulations require the control of the risks from whole-body vibration. This should be based on an assessment of the risk and exposure. In most cases it is simpler to make a broad assessment of the risk rather than try to assess exposure in detail, concentrating the main effort on introducing controls. The requirements of the Regulations are to:

▶ assess the vibration risk to employees;
▶ decide if they are likely to be exposed above the daily exposure action value (EAV) and if they are:
  ▷ introduce a programme of controls to eliminate or reduce their daily exposure so far as is reasonably practicable;

  ▷ decide if they are likely to be exposed above the daily exposure limit value (ELV) and if they are:
  ▷ take immediate action to reduce their exposure below the limit value;
▶ provide information and training on health risks and controls to employees at risk;
▶ consult trade union safety representative or employee representative about the risks and the plans to address these risks;
▶ keep a record of the risk assessment and control actions;
▶ review and update the risk assessment regularly.

If a broad risk assessment has been made and the appropriate and reasonable control actions taken, there is no need to measure the exposure to vibration of the employee. Most machine and vehicle activities in normal use produce daily exposures below the limit value. But some off-road machinery operated for long periods in conditions that generate high levels of vibration or jolting may exceed the exposure limit value.

The measurement of WBV is very difficult and can only be done accurately by a specialist competent person. If the risk assessment has been made and the recommended control actions are in place, there is no need to measure the exposure of employees to vibration. However, HSE has suggested that employers can use the following checklist to estimate whether exposure to WBV is high:

1. there is a warning in the machine manufacturer's handbook that there is a risk of WBV;
2. the task is not suitable for the machine or vehicle being used;
3. operators or drivers are using excessive speeds or operating the machine too aggressively;
4. operators or drivers are working too many hours on machines or vehicles that are prone to WBV;
5. road surfaces are too rough and potholed or floors uneven;
6. drivers are being continually jolted or when going over bumps rising from their seats;
7. vehicles designed to operate on normal roads are used on rough or poorly repaired roads;
8. operators or drivers have reported back problems.

If one or more of the above applies, then exposure to WBV may be high.

If checks on exposure levels are required, information in the vehicle manufacturer's handbook or data published by HSE may be used. An exposure calculator is available on the HSE website at www.hse.gov.uk/vibration for a range of machines and vehicles in different working conditions. However, it is more effective for most employers to direct their efforts towards controlling the risks rather than trying to assess vibration exposures precisely. WBV in industry arises from driving vehicles, such as tractors or fork-lift trucks, over rough terrain or uneven floors. It is highly unlikely that driving vehicles on smooth roads will produce WBV problems. The HSE Guidance document, L141, gives detailed advice to help with the risk assessment and on estimating daily exposure levels. WBV risks are low for exposures around the action value and usually only simple control measures are necessary.

As explained earlier, the most common health problem associated with WBV is back pain. This pain may well have been caused by other activities but WBV will aggravate it.

The actions for controlling the risks from WBV need to ensure that:

▶ the driver's seat is correctly adjusted so that all controls can be reached easily and that the driver weight setting on their suspension seat, if available, is correctly adjusted. The seat should have a back rest with lumbar support;
▶ anti-fatigue mats are used if the operator has to stand for long periods;
▶ the speed of the vehicle is such that excessive jolting is avoided. Speeding is one of the main causes of excessive WBV;
▶ all vehicle controls and attached equipment are operated smoothly;
▶ only established site roadways are used;
▶ only suitable vehicles and equipment are selected to undertake the work and cope with the ground conditions;
▶ the site roadway system is regularly maintained;
▶ all vehicles are regularly maintained with particular attention being paid to tyre condition and pressures, vehicle suspension systems and the driver's seat;
▶ work schedules are regularly reviewed so that long periods of exposure on a given day are avoided and drivers have regular breaks away from the vehicle;
▶ prolonged exposure to WBV is avoided for at-risk groups (older people, young people, people with a history of back problems and pregnant women);
▶ employees are aware of the health risks from WBV, the results of the risk assessment and the ill-health reporting system. They should also be trained to drive in such a way that excessive vibration is reduced.

A simple health monitoring system that includes a questionnaire checklist should be agreed with employees or their representatives (available on the HSE website) to be completed once a year by employees at risk. Employees with back

Figure 6.7 Remote control vibrating plate weighing 1.2 tons with compaction in excess of a 7 ton roller which eliminates the risk of whole-body vibration. The operator is protected from vibrations, noise and dust. The machine can only be operated if line of sight is intact. In case of loss of control the proximity sensor keeps the operator safe

problems should be referred to an occupational health specialist. Personal information about the health of individual employees must be treated confidentially.

6.2.3 Role of health surveillance

The Control of Vibration at Work Regulations require that the employer ensures that employees should be given suitable health surveillance if:

▶ the risk assessment indicates that there is a risk to the health of any employees who are, or are liable to be, exposed to vibration; or
▶ employees are exposed to vibration at or above an exposure action value.

Records of the health surveillance must be kept, employees given access to their own records and the enforcing authorities provided with copies, as required by them. A range of specified action must be taken if problems are found with the health surveillance results.

More information on the Control of Vibration at Work Regulations is given in Chapter 13.

6.3 Radiation

6.3.1 Ionising radiation

Ionising radiation is emitted from radioactive materials, either in the form of directly ionising alpha and beta particles or indirectly ionising X-rays and gamma rays or neutrons. It has a high energy potential and an ability to penetrate, ionise and damage body tissue and organs.

Figure 6.8 Typical ionisation warning sign

All matter consists of atoms within which are a nucleus containing protons and neutrons, and orbiting electrons. The number of protons within the atom defines the element – for example, hydrogen has 1 proton and lead has 82 protons. Some atoms are unstable and will change into atoms of another element, emitting ionising radiation in the process. The change is called radioactive decay, and the ionising radiations most commonly emitted are alpha and beta particles and gamma rays. X-rays are produced by bombarding a metal target with electrons at very high speeds using very high-voltage electrical discharge equipment. Neutrons are released by nuclear fission and are not normally found in manufacturing processes.

Alpha particles consist of two protons and two neutrons and have a positive charge. They have little power to penetrate the skin and can be stopped using very flimsy material, such as paper. Their main route into the body is by ingestion.

Beta particles are high-speed electrons whose power of penetration depends on their speed, but penetration is usually restricted to 2 cm of skin and tissue. They can be stopped using aluminium foil. There are normally two routes of entry into the body – inhalation and ingestion.

Gamma rays, which are similar to **X-rays,** are electromagnetic radiations and have far greater penetrating power than alpha or beta particles. They are produced from nuclear reactions and can pass through the body.

The two most common measures of radiation are the Becquerel (Bq), which measures the activity of a radioactive substance per second, and the Sievert (Sv), which measures the biological effects of the radiation – normally measured in milliSieverts (mSv).

Thus, the Becquerel (Bq) measures the amount of radiation in a given environment and the milliSievert (mSv) measures the ionising radiation dose received by a person.

Ionising radiation occurs naturally from man-made processes and about 87% of all radiation exposure is from natural sources. The Ionising Radiations Regulations specify a range of dose limits, some of which are given in Table 6.7.

Table 6.7 Typical radiation dose limits

	Dose (mSv)	Area of body
Employees aged 18 years	20	Whole body per year
Trainees 16–18 years	6	Whole body per year
Any other person	1	Whole body per year
Women employees of child-bearing age	13	The abdomen in any consecutive 3-month period
Pregnant employees	1	During the declared term of the pregnancy

Harmful effects of ionising radiation

Ionising radiation attacks the cells of the body by producing chemical changes in the cell DNA by ionising it (thus producing free radicals), which leads to abnormal cell growth. The effects of these ionising attacks depend on the following factors:

▶ the size of the dose – the higher the dose then the more serious will be the effect;

▶ the area or extent of the exposure of the body – the effects may be far less severe if only a part of the body (e.g. an arm) receives the dose;

▶ the duration of the exposure – a long exposure to a low dose is likely to be more harmful than a short exposure to the same quantity of radiation.

Acute exposure can cause, dependent on the size of the dose, blood cell changes, nausea and vomiting, skin burns and blistering, collapse and death. **Chronic exposure** can lead to anaemia, leukaemia and other forms of cancer. It is also known that ionising radiation can have an adverse effect on the function of human reproductive organs and processes. Increases in the cases of sterility, stillbirths and malformed foetuses have also been observed.

The health effects of ionising radiation may be summarised into two groups – **somatic effects,** which refer to cell damage in the person exposed to the radiation dose and **genetic effects,** which refer to the damage done to any future children of the irradiated person.

Sources of ionising radiation

The principal workplaces which could have ionising radiation present are the nuclear industry, medical centres (hospitals and research centres) and educational centres. Radioactive processes are used for the treatment of cancers, and radioactive isotopes are used for many different types of scientific research. X-rays are used extensively in hospitals, but they are also used in industry for non-destructive testing (e.g. crack detection in welds). Smoke detectors, used in most workplaces, also use ionising radiations.

Ionising radiations can also occur naturally – the best example being radon, which is a radioactive gas that occurs mainly at or near granite outcrops where there is a presence of uranium. It is particularly prevalent in Devon and Cornwall. The gas enters buildings normally from the substructure through cracks in flooring or around service inlets.

The Ionising Regulations have set two action levels above which remedial action, such as fitting sumps and extraction fans, has to be taken to lower the radon level in the building. The first action level is 400 Bq/m^3 in workplaces and 200 Bq/m^3 in domestic properties. At levels above 1,000 Bq/m^3, remedial action should be taken within 1 year. The average background radiation in the UK is 2.4 mSv per year. Background radiation levels are much higher elsewhere in the world – 260 mSv have been recorded in Iran.

About half of the background radiation in the UK is caused by radon. Radon gas is responsible for 5% of lung cancer cases. Radon is not generally a problem unless it is able to accumulate in confined spaces such as basements, cellars or lift shafts. An impermeable membrane beneath the floor of the building can be laid to prevent the gas entering the building through the floor. This latter treatment is normally only suitable for new buildings but does cost less than 10% of the cost of a sump and has no running costs. There are now government-defined 'radon-affected areas' throughout the UK, so that checks should be made if underground spaces in buildings are to be used.

About 14% of exposures to ionising radiation are due to medical exposures during diagnostic or treatment processes.

The hazards are categorised as either stochastic or non-stochastic. The principal stochastic hazard is cancer in various forms. The main non-stochastic effects include radiation burns, radiation sickness, cataracts and damage to unborn children.

Figure 6.9 X-ray generator cabinet

Figure 6.10 Radon monitoring equipment

Personal radiation exposure can be measured using a film badge, which is worn by the employee over a fixed time interval. The badge contains a photographic film which, after the time interval, is developed and an estimate of radiation exposure is made. A similar device, known as a radiation dose meter or detector, can be positioned on a shelf in the workplace for

6

3 months, so that a mean value of radiation levels may be measured. Instantaneous radiation values can be obtained from portable hand-held instruments, known as Geiger counters, which continuously sample the air for radiation levels. Similar devices are available to measure radon levels.

6.3.2 Non-ionising radiation

Non-ionising radiation includes ultraviolet, visible light (this includes lasers which focus or concentrate visible light), infrared and microwave radiations. As the wavelength is relatively long, the energy present is too low to ionise atoms which make up matter. The action of non-ionising radiation is to heat cells rather than change their chemical composition.

Other than the Health and Safety at Work Act there is no specific set of Regulations governing non-ionising radiation. However, the Personal Protective Equipment at Work Regulations are particularly relevant, since the greatest hazard is tissue burning of the skin or the eyes.

Ultraviolet radiation (UV) occurs with sunlight and with electric arc welding. In both cases, the skin and the eyes are at risk from the effect of burning. The skin will burn (as in sunburn) and repeated exposure can lead to skin cancer. Skin which is exposed to strong sunlight should be protected either by clothing or sun creams. This problem has become more common with the reduction in the ozone layer (which filters out much ultraviolet light). The eyes can be affected by a form of conjunctivitis which feels like grit in the eye, and is called by a variety of names dependent on the activity causing the problem. Arc welders call it 'arc eye' or 'welder's eye' and skiers 'snow blindness'. Cataracts caused by the action of ultraviolet radiation on the eye lens are another possible outcome of exposure.

The most dangerous form of skin cancer, malignant melanoma, has increased by over 40% over the last 10 years making it the cancer with the fastest rising number of cases in the UK. Outdoor workers that could be at risk include farm or construction workers, market gardeners, outdoor activity workers and some public service workers. Those who are particularly at risk have:

▶ fair or freckled skin that does not tan, or goes red or burns before it tans;
▶ red or fair hair and light coloured eyes;
▶ a large number of moles.

Employers need to be aware of the risks their employees are taking when they work outside without adequate protection from the sun. With growing concern following the rise in skin cancers, the HSE has suggested the following hierarchy of controls for outdoor working:

▶ relocate some jobs inside a building or to a shady location;
▶ undertake some outdoor work earlier or later in the day;
▶ provide personal protection such as:
  ▷ wearing long sleeve shirts or loose clothing with a close weave;
  ▷ wearing hats with a wide brim;
  ▷ using a high factor sunscreen of at least SPF15 on any exposed skin;
▶ provide suitable education and training for outdoor workers;
▶ provide suitable information and supervision to instigate safe systems of work that protect workers from the sun.

In the UK each year there are almost 3,000 cases of non-melanoma skin cancer caused by exposure at work to UV radiation from the sun and around 250 cases of malignant melanoma. The following advice is suggested for outdoor workers who are exposed to the sun:

▶ Protect the skin by wearing suitable clothing.
▶ Wear a hat with a brim or a flap that covers the ears and the back of the neck.
▶ Stay in the shade whenever possible, during your breaks and especially at lunch time.
▶ Use a high factor sunscreen of at least SPF15 on any exposed skin.
▶ Drink plenty of water to avoid dehydration.
▶ Check your skin regularly for any unusual moles or spots. See a doctor promptly if there is anything that is changing in shape, size or colour, itching or bleeding.

UV radiation has some beneficial effects such as the accumulation of vitamin D which strengthens bones.

Where UV insectocutors are used, normally in kitchens, to control flying insects, it is important that the correct emitting labels are fitted and not ones that, for example, are used to sterilise surgical instruments as has happened on one occasion.

Lasers use visible light and light from the invisible wavelength spectrum (infrared and ultraviolet). As the word laser implies, they produce 'light amplification by stimulated emission of radiation'. This light is highly concentrated and does not diverge or weaken with distance and the output is directly related to the chemical composition of the medium used within the particular laser. The output beam may be pulsating or continuous – the choice being dependent on the task of the laser. Lasers have a large range of applications including bar code reading at a supermarket checkout, the cutting and welding of metals and accurate measurement of distances and elevations required in land and mine surveying. They are also extensively used in surgery for cataract treatment and the sealing of blood vessels.

The International Electrotechnical Commission (IEC) and the American National Standards Institute (ANSI) have defined seven classes of laser (1, 1M, 2, 2M, 3R, 3B and 4) in ascending size of power output. Classes 1, 2 and 2M are relatively low hazard and only emit light in

Figure 6.11 A proper Class 4 laser area set up for defeatable access control. Curtain design and layout varies with the environment. Class 4 laser areas in a research or university environment usually run long-term experiments that require unattended operation. In such cases defeatable entryway controls are appropriate. By design persons who have been properly trained and given the keypad access code may momentarily 'defeat' the interlock to enter and exit the laser-controlled area protected by a magnetically locked door (MPE is Maximum Permissible Exposure)

the visible band. Direct eye contact must be avoided with classes 3a, 3b and 4. These are more hazardous than classes 1 and 2 and the appointment of a laser safety officer is recommended. A Class 4 laser can burn the skin and cause permanent eye damage as a result of direct, diffuse or indirect exposure to the beam. Such Class 4 lasers can ignite combustible materials, and are a fire risk. All lasers should carry information stating their class and any precautions required during use.

The main hazards associated with lasers are eye and skin burns, toxic fumes, electricity and fire. The vast majority of accidents with lasers affect the eyes. Retinal damage is the most common and is irreversible. Cataract development and various forms of conjunctivitis can also result from laser accidents. Skin burning and reddening (erythema) are less common and are reversible.

Infrared radiation is generated by fires and hot substances and can cause eye and skin damage similar to that produced by ultraviolet radiation. It is a particular problem for firefighters and those who work in foundries or near furnaces. Eye and skin protection are essential.

Microwaves are used extensively in cookers and mobile telephones and there are ongoing concerns about associated health hazards (and several inquiries are currently under way). The severity of any hazard is proportional to the power of the microwaves. The principal hazard is the heating of body cells, particularly those with little or no blood supply to dissipate the heat. This means that tissues such as the eye lens are most at risk from injury. However, it must be stressed that any risks are higher for items such as cookers than for low-powered devices such as mobile phones.

The measurement of non-ionising radiation normally involves the determination of the power output being received by the worker. Such surveys are best performed by specialists in the field, as the interpretation of the survey results requires considerable technical knowledge.

Figure 6.12 Metal furnace – source of infrared heat

6.3.3 Radiation protection strategies

Ionising radiation

Protection is obtained by the application of shielding, time and distance either individually or, more commonly, using a mixture of all three.

Shielding is the best method because it is an 'engineered' solution. It involves the placing of a physical shield, such as a layer of lead, steel and concrete, between the worker and the radioactive source. The thicker the shield the more effective it is.

Time involves the use of the reduced time exposure principle and thus reduces the accumulated dose.

Distance works on the principle that the effect of radiation reduces as the distance between the worker and the source increases.

Other measures include the following:

▶ effective emergency arrangements;
▶ training of employees;
▶ the prohibition of eating, drinking and smoking adjacent to exposed areas;
▶ a high standard of personal cleanliness and first-aid arrangements;
▶ strict adherence to personal protective equipment arrangements, which may include full body protection and respiratory protection equipment;
▶ procedures to deal with spillages and other accidents;

▶ prominent signs and information regarding the radiation hazards;
▶ medical surveillance of employees.

The Ionising Radiations Regulations specify a range of precautions which must be taken, including the appointment of a Radiation Protection Supervisor as the competent person and a Radiation Protection Adviser.

The **Radiation Protection Supervisor** must be appointed by the employer to advise on the necessary measures for compliance with the Regulations and its Approved Code of Practice. The person appointed, who is normally an employee, must be competent to supervise the arrangements in place and have received relevant training.

The **Radiation Protection Adviser** is appointed by the employer to give advice to the Radiation Protection Supervisor and employer on any aspect of working with ionising radiation including the appointment of the Radiation Protection Supervisor. The Radiation Protection Adviser is often an employee of a national organisation with expertise in ionising radiation. A Radiation Protection Adviser is not needed when the only work with ionising radiation is as specified in Schedule 1 of the Regulations.

The Regulations were amended in 2017 and the main changes were:

▶ The dose limit for exposure to the lens of the eye has been reduced from 150 mSv to 20 mSv in a year.
▶ There is a new system of authorisation for work with ionising radiation – the higher the radiation protection risk, the greater the requirements. A three-tier system of regulatory control, notification, registration and consent has replaced the previous requirement for notification and prior authorisation.
▶ The requirement for notification has been changed to a lower level of activity.
▶ The definition of an outside worker has been broadened so that it includes both classified and non-classified workers.
▶ There is now a requirement to put procedures in place to estimate doses to members of the public.

Non-ionising radiation

For ultraviolet and infrared radiation, eye protection in the form of goggles or a visor is most important, particularly when undertaking arc welding or furnace work. Skin protection is also likely to be necessary for the hands, arms and neck in the form of gloves, sleeves and a collar. For construction and other outdoor workers, protection from sunlight is important, particularly for the head and nose. Sun creams should also be used.

For laser operations, engineering controls such as fixed shielding and the use of non-reflecting surfaces around the workstation are recommended. For lasers in the higher class numbers, special eye protection is recommended. A risk assessment should be undertaken before a laser is used.

Engineering controls are primarily used for protection against microwaves. Typical controls include the enclosure of the whole microwave system in a metal surround and the use of an interlocking device that will not allow the system to operate unless the door is closed.

Intense sources of artificial light in the workplace, particularly from UV radiation and powerful lasers, can harm the eyes and skin of workers and need to be properly managed. The Control of Artificial Optical Radiation at Work Regulations came into force in 2010 and implements the Physical Agents (Artificial Optical Radiation) Directive (2006/25/EC). The aim of the Regulations is to ensure that standards are set so that all workers are protected from harm arising from exposure to hazardous sources of artificial light. As with the previous Noise and Vibration Regulations, it contains provisions on risk assessment, control of exposure, health surveillance and information, instruction and training. The Regulations are based on the limit values incorporated in the guidelines issued by the International Commission on Non-Ionising Radiation Protection (ICNIRP).

Workers should be protected from hazardous sources of light in the workplace, to ensure that the eyes and skin of workers are properly protected from intense sources of light at work that can damage the eyes and skin. Such light sources include ultraviolet, visible and infrared radiation produced by artificial sources, such as lasers and welding arcs. Examples of such hazardous light sources include:

▶ welding work (both arc and oxy-fuel) and plasma cutting causing mainly eye damage;
▶ furnaces and foundries causing eye and skin damage;
▶ printing involving the UV curing of inks causing mainly skin damage;
▶ motor vehicle repairs involving the UV curing of paints causing mainly skin damage;
▶ medical and cosmetic treatments involving laser surgery, blue light and UV therapies causing both eye and skin damage;
▶ all use of Class 3B and Class 4 lasers potentially causing permanent eye and skin damage.

More information is given on these Regulations in Chapter 13.

The EMF Directive

Electric, magnetic, and electromagnetic fields (EMFs) are a form of non-ionising radiation that arises whenever electrical energy is used. An electric field is generated by electric charges, while a magnetic field occurs around an electric current. Electromagnetic fields are mutually produced by time-varying magnetic and electric fields. When electric fields act on conductive materials (such as the human body), they influence the distribution of electric charges at their surface causing current to flow through the body to the ground. Magnetic fields induce circulating currents within the human body. The strength of these currents depends on the intensity of the outside magnetic field. If sufficiently large, these currents can stimulate nerves and muscles or cause other biological effects such as nausea.

Common sources of electromagnetic fields include work processes such as radiofrequency heating and welding; household electrical wiring and appliances; electrical motors; computer screens; telecommunications; transport and distribution of electricity; broadcasting; and security detection devices.

The EMF Directive covers a frequency spectrum from 0 to 300 GHz, which includes static magnetic fields and low-frequency electric and magnetic fields, through to radio frequency and microwave frequencies.

Welding operations

Over 1,000 accidents involving welding work are reported to the HSE each year. There are several different types of welding operation. The most common are:

▶ manual metal arc welding;
▶ metal inert-gas welding (MIG);
▶ tungsten inert-gas welding (TIG);
▶ oxy-acetylene welding.

The non-ionising radiation hazards caused by arc welding are not the only hazards associated with welding operations. The hazards from fume inhalation, trailing cables and pipes and the manual handling of cylinders are also present. There have also been serious injuries resulting from explosions and fires during welding processes. Accidents are often caused by lack of training and faulty equipment and they are often made worse by the lack of complete personal protective equipment. Many of these accidents occur on farms where welding equipment is used to make on-the-spot repairs to agricultural machinery.

6.3.4 The role of monitoring and health surveillance

The monitoring and health surveillance arrangements for exposure to ionising radiation are detailed under Part 5 of the Ionising Radiation Regulations. The Regulations specify that the employer of any individual exposed to ionising radiation as a result of work activities has the responsibility for deciding when such an employee needs to be designated as a classified person in accordance with Regulation 20. Before that person is classified, the employer will need to ensure that the employee has been certified as fit for the intended type of work within the last 12 months. This may require a medical examination. The employer then needs to make arrangements with the appointed doctor (or employment medical adviser) for continuing medical surveillance. Also, the employer will need to arrange for adequate medical surveillance for any employee who has received an overexposure, whether or not that employee has been designated as a classified person.

The ACoP further states that the main purpose of medical surveillance is to determine an individual's fitness or continuing fitness for the intended work with ionising radiation. The fitness of the person is not restricted to possible health effects from exposure to ionising radiation. The appointed doctor (or employment medical adviser) will need to take account of specific features of the work with ionising radiation, such as the fitness of the individual:

▶ to wear any personal protective equipment (including respiratory protective equipment) required to restrict exposure;
▶ with a skin disease, to undertake work involving unsealed radioactive materials; and
▶ with serious psychological disorders, to undertake work with radiation sources that involve a special level of responsibility for safety.

The Regulations also require that the employer maintains a health record of any employee receiving health surveillance, and that record (or a copy) is kept until the person to whom the record relates has or would have attained the age of 75 years but in any event for at least 50 years from the date of the last entry made in it.

6.4 Mental ill health in the workplace

The focus of business is changing rapidly and now, in the 21st century we find that we have to take mental health seriously in order to be successful. In the past there was emphasis on intellect, strength and dexterity but now emotional intelligence, innovation and communication have come to the fore and with this there is a new awareness of the importance of keeping the workforce mentally as well as physically healthy.

– (Dr Paul Litchfield, BT)

Recent statistics and their implications are a warning to everyone who is managing a business (HSE work-related stress, depression or anxiety GB stats 2019). In recent years there has been an increase in the incidence of self-reported work-related stress, depression and anxiety. When asked, people say that their problems stem from workload pressures such as tight deadlines and too much responsibility, and from a lack of support from managers.

It has been estimated by the Work Foundation (https://www.lancaster.ac.uk/work-foundation/) that one in five UK employees with a physical health condition also has a mental health condition. The most common cause of mental health illness in the workplace is work-related stress.

6.4.1 The frequency and extent of mental ill health at work

In 2018/19 there were 602,000 reports of stress, depression or anxiety and the total number of working days lost due to this was 15.4 million. This accounts for 44% of all work-related ill health cases and 54% of all the working days lost due to ill health. The HSE has estimated that work-related stress, anxiety or depression costs society around £5.2bn a year (2018/19 Labour Force Survey – LFS & HSE WBK01 2019).

Proportionately, cases of work-related stress occur most commonly in larger organisations, particularly among managerial staff and in professional occupations notably public administration, health and social care, and education. Levels are highest among the 35–54 age group with women suffering higher rates of stress disorders than men. The numbers are increasing annually as are the resulting civil claims. Stress and the related issues of workplace bullying and harassment are issues that simply cannot be ignored.

There are many studies and reviews to back up the view that mental ill health is a heavy financial burden on business. Not only that, it is an enormous social and emotional cost to families and friends, as well as colleagues in the workplace. Additionally, there are risks associating stress with heart disease. Musculoskeletal disorders, especially back pain, can also be linked to the effects of stress.

Clearly we need to pay attention. Financially, socially and emotionally, mental ill health is costing us dear.

Work-related stress

In 2001, the HSE defined work-related stress as the reaction people have to excessive pressure or other types of demand placed on them. An important distinction is made in this definition between pressure, which can be a positive state if managed appropriately and a normal reaction to reasonable demands, and stress which can arise in response to intense, continuous or prolonged exposure to excessive pressures and can be detrimental to health. However, a report commissioned by the HSE has concluded that there is no simple and universal definition of work-related stress, largely because of the complex nature of work-related stress.

However, the HSE report distinguished work-related stress as different from occupational stress. The former includes cases where work may have aggravated the experience of stress and associated symptoms of ill health regardless of original cause. Here work may be

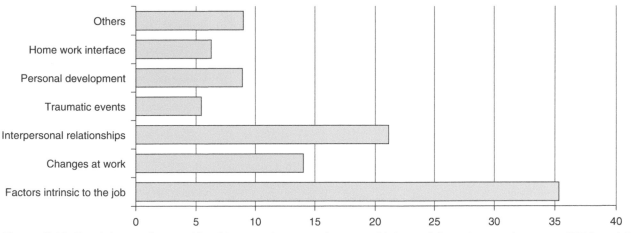

Figure 6.13 Breakdown of mental health cases by type of event which precipitated stress between 2010 and 2012

a contributory factor but not necessarily the sole cause. The latter refers to cases where work is the sole cause of the experience of stress and associated symptoms of ill health.

Stress is not a disease – it is the natural reaction to excessive pressure. It can be defined as the reaction that people have when they are unable to cope with excessive pressures and demands. Stress can lead to an improved performance but is not generally a good thing as it is likely to lead to both physical and mental ill health, such as high blood pressure, peptic ulcers, skin disorders and depression.

6.4.2 Common symptoms of workers with mental ill health: depression, anxiety, panic attacks, post-traumatic stress disorder (PTSD)

Depression and anxiety

Mental ill health affects people in many different ways. Some of the more challenging forms of mental illness such as schizophrenia and bi-polar disorder, while serious, can be managed using care planning, medication and complementary therapies. Depression and anxiety, while perhaps not so clearly identifiable, can impact on a person's life in equally severe and debilitating ways. The important point here is, don't focus on the diagnosis, focus on the person and what they are able to achieve.

People are often reluctant to admit to a mental health problem because of the fear of stigma. Encouraging a culture of openness, focusing on mental and physical well-being, allowing people to talk about their feelings and difficulties, provides a safe and nurturing environment. Staff who are aware of these things are more likely to be able to help themselves as well as being sensitive to the needs of others.

Early recognition of our own and our colleagues' problems will enable us all to give appropriate support

and to ask for help when it is needed, before a situation has escalated.

Some early signs might be:

▶ poor concentration;
▶ being easily distracted;
▶ worrying more;
▶ finding it hard to make decisions;
▶ feeling less interested in day-to-day activities;
▶ low mood;
▶ feeling overwhelmed by things;
▶ tearfulness;
▶ tiredness and lack of energy;
▶ headaches;
▶ sleeping more or less;
▶ talking less and avoiding social activities;
▶ talking more or talking very fast, jumping between topics and ideas;
▶ finding it difficult to control emotions;
▶ increased alcohol consumption;
▶ irritability and short temper;
▶ aggression.

As the level of mental distress increases people are likely to become less reliable, to be late for work and appointments and to be absent from work.

Panic attacks

The exact cause of panic disorder is not fully understood, but it is likely to be a combination of things. It is often associated with stress and in particular a traumatic or stressful life experience. Panic attacks can occur suddenly and for no apparent reason and this makes it particularly difficult for people who experience them as it can create a cycle of living 'in fear of fear'.

For this reason it is important to understand the problem. These attacks are frightening but they are not dangerous. A person experiencing one will need to be helped not to fight the attack, to slow their breathing and to breathe deeply. It helps if the person closes their eyes, particularly if the environment is busy. Remind them that the attack will pass and that they should focus on peaceful and relaxing images. Relaxation techniques will help to further relax them.

Complementary therapies are particularly helpful in dealing with panic attacks as are yoga, pilates and breathing exercises.

Post-traumatic stress disorder (PTSD)

This type of mental illness may occur when someone has had a frightening, stressful or overwhelming experience, especially if it has made them feel powerless to stop or change the event.

Witnessing a workplace accident resulting in severe injury or death of a colleague is an example of the type of event that may result in PTSD. An incident such as a robbery or other event involving violence can be equally traumatic.

PTSD does not always develop immediately and it sometimes lies dormant for many years, so that the connection between symptoms and disorder are not always made. The illness manifests itself in flashbacks to the event(s), nightmares and persistent feelings of fear, but can also be felt as anger, numbness or 'being on edge', premonitions of disaster, problems with concentration, a general dissatisfaction with life and drug and alcohol used to help the person through their despair.

As with other forms of mental ill health already discussed, the problem must be taken seriously and ways found to make the workplace an emotionally safe space where people can talk about their problems and expect to experience openness, sensitivity and support.

Things that are useful to know:

▶ Anger and withdrawal may be signs of PTSD.
▶ Even when a person does not want to talk about the problem, they will need reassurance that the offer is there and continues to be available. It may take a long time.

There are support services and people need:

a) to be made aware of them
b) to be given time to attend them.

There may be things in the workplace that produce environmental triggers to the traumatic event and where possible it will help if they can be removed.

After an incident colleagues will need patience and understanding when the person returns to work.

6.4.3 The causes of, and controls for, work-related mental ill health

Causes of work-related stress

Most people experience stress at some time during their lives, e.g. during an illness or death of a close relative or friend. However, recovery normally occurs after the particular crisis has passed. The position is,

however, often different in the workplace because the underlying causes of the stress, known as work-related stressors, are not relieved but continue to build up the stress levels until the employee can no longer cope.

The basic workplace stressors are:

▶ the job itself – boring or repetitive, unrealistic performance targets or insufficient training, job insecurity or fear of redundancy;
▶ individual responsibility – ill-defined roles and too much responsibility with too little power to influence the job outputs;
▶ working conditions – cramped, dirty and untidy workplace; unsafe practices; lack of privacy or security; inadequate welfare facilities; threat of violence; excessive noise, vibration or heat; poor lighting; lack of flexibility in working hours to meet domestic requirements and adverse weather conditions for those working outside;
▶ management attitudes – poor communication, consultation or supervision, negative health and safety culture, lack of support in a crisis;
▶ relationships – unhappy relationship between workers, bullying, sexual and racial harassment.

Control measures for work-related stress

Possible solutions to all these stressors have been addressed throughout this book and involve the creation of a positive health and safety culture, effective training and consultation procedures and a set of health and safety arrangements which work on a day-to-day basis. The HSE has produced its own generic stress audit survey tool which is available free of charge on its website and advises the following action plan:

▶ identify the problem;
▶ identify the background to the problem and how it was discovered;
▶ identify the remedial action required and give reasons for that action;
▶ identify targets and reasonable target dates;
▶ agree a review date with employees to check that the remedial action is working.

The following additional measures have also been found to be effective by some large employers:

▶ take a positive attitude to stress issues by becoming familiar with its causes and controls;
▶ take employees' concerns seriously and develop a counselling system (or better still give access to independent counselling) which will allow a frank, honest and confidential discussion of stress-related problems;
▶ develop an effective system of communication and consultation and ensure that periods of uncertainty are kept to a minimum;
▶ set out a simple policy on work-related stress and include stressors in risk assessments;

- ensure that employees are given adequate and relevant training and realistic performance targets;
- develop an effective employee appraisal system which includes mutually agreed objectives;
- discourage employees from working excessive hours and/or missing break periods (this may involve a detailed job evaluation);
- introduce job rotation and increase job variety;
- develop clear job descriptions and ensure that the individual is matched to them;
- encourage employees to improve their lifestyle (e.g. many local health authorities provide smoking cessation advice sessions);
- monitor incidents of bullying, sexual and racial harassment and, where necessary, take disciplinary action;
- train supervisors to recognise stress symptoms among the workforce;
- avoid a blame culture over accidents and incidents of ill health.

Workplace stress has no specific health and safety regulations but is covered by the duties imposed by the Health and Safety at Work Act and the Management of Health and Safety at Work Regulations to:

- ensure so far as is reasonably practicable that workplaces are safe and without risk to health;
- carry out a risk assessment relating to the risks to health;
- introduce and maintain appropriate control measures.

The risk assessment will need to relate to the individual, such as absentee record, production performance and which of the stressors are applicable, and to the organisation, such as training programmes, communication and appraisal procedures. Many stressful conditions can be reduced by an effective 'whistle-blowing' policy that enables individuals to highlight any concerns to senior managers or directors in a confidential manner.

6.4.4 HSE management standards

The HSE has produced a step-by-step workbook Tackling work-related stress using the management standards, WBK01, 2019, HSE https://www.hse.gov.uk/pubns/wbk01.pdf.

The standards cover six key areas of work design that, if not properly managed, are associated with poor health, lower productivity and increased absence and sickness rates. The following is a brief summary of what is contained in the standards but it is recommended that you take advantage of this extremely useful free aid and download the workbook. Useful links are provided with each one.

Demands:

- employees indicate that they are able to cope with the demands of their jobs;

- that systems are in place locally to respond to any individual concerns.

Control:

- employees indicate that they are able to have a say about the way they do their work;
- systems are in place locally to respond to any individual concerns.

Support:

- employees indicate that they receive adequate information and support from their colleagues and superiors;
- systems are in place locally to respond to any individual concerns.

Relationships:

- employees indicate that they are not subjected to unacceptable behaviours such as bullying at work;
- systems are in place locally to respond to any individual concerns.

Role:

- employees indicate that they understand their role and responsibilities;
- systems are in place locally to respond to any individual concerns.

Change:

- employees indicate that the organisation engages them frequently when undergoing an organisational change;
- systems are in place locally to respond to any individual concerns.

This is just a brief summary so make sure to download the workbook.

6.4.5 Home–work interface

Typical issues that can occur outside the workplace but can contribute to mental health problems at the workplace include:

- commuting over long distances particularly in overcrowded circumstances;
- childcare issues including sickness;
- relocation;
- care of frail (vulnerable) relatives; and
- the death of a close relative.

The individual can also take action if he/she feels that he/she is becoming over-stressed.

There are a number of simple ways to look after our mental and physical health, and if we are all aware of these we can create a positive culture. Good work/life balance may need, for instance, flexible working hours. Talking to somebody, preferably a trained counsellor, is a good possibility. Here are some tips, but bear in mind that the culture needs to exist to make these possible:

- keep physically active;
- eat well;

▶ only drink in moderation;
▶ value yourself and others;
▶ talk about your feelings;
▶ keep in touch with friends and family;
▶ care for others;
▶ get involved, make a contribution;
▶ learn a new skill;
▶ do something creative;
▶ take a break;
▶ ask for help.

How managers can help

▶ have a conversation in a quiet and private place;
▶ allow plenty of time and make sure there are no interruptions;
▶ focus on getting information that will help achieve the goal of supporting the member of staff;
▶ ask open questions, that is, questions that cannot be answered 'Yes' or 'No';
▶ give plenty of time for the person to answer;
▶ try to see things from the other person's point of view, stand in their shoes;
▶ make sure that there is a firm arrangement for follow-up.

To sum up:

▶ focus on the person rather than the diagnosis;
▶ encourage a supportive culture of openness and sensitivity;
▶ enable a healthy lifestyle in the workplace.

6.4.6 Recognition that most people with mental ill health can continue to work effectively

Stress usually occurs when people feel that they are losing control of a situation. In the workplace, this means that the individual no longer feels that they can cope with the demands made on them. Many such problems can be partly solved by listening to rather than talking at people. Many organisations have found that the active participation of union and employee representatives is a critical component in delivering improvements in employee health and well-being.

6.4.7 Additional information

There have been several successful civil actions for compensatory claims resulting from the effects of workplace stress. The Court of Appeal in the case of *Sutherland v. Hatton and others* in 2002 set out a number of practical propositions for future claims concerning workplace stress:

▶ Employers are entitled to take what they are told by employees at face value unless they have good reason to think otherwise. They do not have a duty to make searching enquiries about employee's mental health.

▶ An employer will not be in breach of duty in allowing a willing employee to continue in a stressful job if the only alternative is to dismiss or demote them. The employee must decide whether to risk a breakdown in their health by staying in the job.
▶ Indications of impending harm to health at work must be clear enough to show an employer that action should be taken, in order for a duty on an employer to take action to arise.
▶ The employer is in breach of duty only if he/she fails to take steps which are reasonable, bearing in mind the size of the risk, gravity of harm, the cost of preventing it and any justification for taking the risk.
▶ No type of work may be regarded as intrinsically dangerous to mental health.
▶ Employers who offer a confidential counselling advice service, with access to treatment, are unlikely to be found in breach of duty.
▶ Employees must show that illness has been caused by a breach of duty, not merely occupational stress.
▶ Compensation will be reduced to take account of pre-existing conditions or the chance that the claimant would have fallen ill in any event.

Their full guidelines should be consulted and consist of 16 points. However, more recent Court of Appeal decisions have suggested that the provision of a confidential counselling advice service alone may not be enough to discharge an employer's duty where an employee's problems could only be addressed by the intervention of management.

6.5 Violence at work

6.5.1 Types of violence at work

It has been estimated that one in eight people has had a violent encounter in the workplace.

Violence at work, particularly from dissatisfied customers, clients, claimants or patients, can cause stress and in some cases injury. This is not only physical violence, as people may face verbal and mental abuse, discrimination, harassment and bullying. Fortunately, physical violence is still rare, but violence of all types has risen significantly in recent years. Violence at work is known to cause pain, suffering, anxiety and stress, leading to financial costs due to absenteeism and higher insurance premiums to cover increased civil claims. It can be very costly to ignore the problem.

Many people resort to violence due to frustration. Common causes of such frustration are the following:

▶ dissatisfaction with a product or service, including the cost;
▶ a perception of being unreasonably penalised over an incident such as car parking; and
▶ a general lack of information following a problem, such as aircraft delays or long delays at hospitals.

Physical assaults include the offences of common assault, wounding, robbery and snatch theft. Threats include both verbal threats made to or against the victim and non-verbal intimidation. These are mainly threats to assault the victim and, in some cases, to damage property.

In over a third of the incidents, the victim believed the offender was under the influence of alcohol, and in nearly a fifth of incidents under the influence of drugs.

Almost half of the assaults and a third of the threats happen after 1800 hrs, which suggests that the risks are higher if people work at night or in the late evening.

6.5.2 Jobs and activities which increase the risk of violence

Violence from members of the public is a higher risk with several occupations including:

▶ health and social workers;
▶ police and firefighters;
▶ various types of enforcement officers, education, benefit services, various service industries and debt collectors;
▶ shop staff
▶ lone workers;
▶ those working with people under the influence of drugs or alcohol; and
▶ those who handle money or valuables.

Such incidents, whilst they are still high, have reduced over the last decade. Over this period many of the protections outlined in this chapter and advocated by the HSE have been in place. For example several hospitals now employ their own police officers to police accident and emergency departments and there has been much more use of closed circuit television (CCTV).

6.5.3 Control measures to reduce the risks from violence at work

Violence at work is defined by the HSE as: *'any incident in which an employee is abused, threatened or assaulted in circumstances relating to their work.'*

In recognition of this, the HSE has produced a useful guide for employers which includes a four-stage action plan and some advice on precautionary measures (*'Violence at Work: a Guide for Employers', INDG69 (rev)*). The employer is just as responsible, under health and safety legislation, for protecting employees from violence as they are for any other aspects of their safety.

The Health and Safety at Work Act puts broad general duties on employers and others to protect the health and safety of staff. In particular, Section 2 of the HSW Act gives employers a duty to safeguard, so far as is reasonably practicable, the health, safety and welfare at work of their employees and any visitors.

Employers also have a common law general duty of care towards their staff, which extends to the risk of violence at work. Legal precedents (see *West Bromwich Building Society v. Townsend* [1983] IRLR 147 and *Charlton v. Forrest Printing Ink Company Limited* [1980] IRLR 331) show that employers have a duty to take reasonable care to see that their staff are not exposed to unnecessary risks at work, including the risk of injury by criminals. In carrying out their duty to provide a safe system of work and a safe working place, employers should therefore have regard to, and safeguard their staff against, the risk of injury from violent criminals.

Figure 6.14 Security access and surveillance CCTV camera

The HSE recommends the following four-point action plan:

1. find out if there is a problem;
2. decide on what action to take;
3. take the appropriate action;
4. check that the action is effective.

Find out if there is a problem

This involves a risk assessment to determine what the real hazards are. It is essential to ask people at the workplace and, in some cases, a short questionnaire may be useful. Record all incidents to get a picture of what is happening over time, making sure that all relevant detail is recorded. The records should include:

▶ a description of what happened;
▶ details of who was attacked, the attacker and any witnesses;
▶ the outcome, including how people were affected and how much time was lost;
▶ information on the location of the event.

Owing to the sensitive nature of some aggressive or violent actions, employees may need to be encouraged to report incidents and be protected from future aggression.

All incidents should be classified so that an analysis of the trends can be examined.

Consider the following:

▶ fatalities;
▶ major injury;
▶ less severe injury or shock which requires first-aid treatment, outpatient treatment, time off work or expert counselling;
▶ threat or feeling of being at risk or in a worried or distressed state.

Decide on what action to take

It is important to evaluate the risks and decide who may be harmed and how this is likely to occur. The threats may be from the public or co-workers at the workplace or may be as a result of visiting the homes of customers. Consultation with employees or other people at risk will improve their commitment to control measures and will make the precautions much more effective. The level of training and information provided, together with the general working environment and the design of the job, all have a significant influence on the level of risk.

Consider the following issues:

▶ quality of service provided;
▶ design of the operating environment;
▶ type of equipment used;
▶ designing the job.

Some violence may be deterred if measures are taken which suggest that any violence may be recorded. Many public bodies use the following measures:

▶ informing telephone callers that their calls will be recorded;
▶ displaying prominent notices that violent behaviour may lead to the withdrawal of services and prosecution;
▶ using CCTV or security personnel.

Four in ten of all employment enquiries to one employment law firm in 2009 were related to bullying. Of those who reported suffering bullying, 80% said it had affected their physical and mental health and a third had taken time off work or left their jobs as a result.

Quality of service provided

The type and quality of service provision has a significant effect on the likelihood of violence occurring in the workplace. Frustrated people whose expectations have not been met and who are treated in an unprofessional way may believe they have the justification to cause trouble.

Sometimes circumstances are beyond the control of the staff member and potentially violent situations need to be defused. The use of correct skills can turn a dissatisfied customer into a confirmed supporter simply by careful response to their concerns. The perceived lack of, or incorrect, information can cause significant frustrations.

Design of the operating environment

Personal safety and service delivery are very closely connected and have been widely researched in recent years. This has resulted in many organisations altering their facilities to reduce customer frustration and enhance sales. It is interesting that most service points experience less violence when they remove barriers or screens, but the transition needs to be carefully planned in consultation with staff, and other measures adopted to reduce the risks and improve their protection.

The layout, ambience, colours, lighting, type of background music, furnishings including their comfort, information, things to do while waiting and even smell all have a major impact. Queue-jumping causes a lot of anger and frustration and needs effective signs and proper queue management, which can help to reduce the potential for conflict.

Wider desks, raised floors and access for special needs, escape arrangements for staff, carefully arranged furniture and screening for staff areas can all be utilised.

Type of security equipment used

There is a large amount of equipment available and expert advice is necessary to ensure that it is suitable and sufficient for the task. Some measures that could be considered include the following:

▶ **Access control** to protect people and property. There are many variations from staffed and friendly receptions, barriers with swipe-cards and simple coded security locks. The building layout and design may well partly dictate what is chosen. People inside the premises need access passes so they can be identified easily.
▶ **Closed circuit television** is one of the most effective security arrangements to deter crime and violence. Because of the high cost of the equipment, it is essential to ensure that proper independent advice is obtained on the type and the extent of the system required.

- **Alarms** – there are three main types:
  - ▷ intruder alarms fitted in buildings to protect against unlawful entry, particularly after working hours
  - ▷ panic alarms used in areas such as receptions and interview rooms covertly located so that they can be operated by the staff member threatened
  - ▷ personal alarms carried by an individual to attract attention and to temporarily distract the attacker.
- **Radios and pagers** can be a great asset to lone workers in particular, but special training is necessary, as good radio discipline with a special language and codes are required.
- **Mobile phones** are an effective means of communicating and keeping colleagues informed of people's movements and problems such as travel delays. Key numbers should be inserted for rapid use in an emergency.

Job design

Many things can be done to improve the way in which the job is carried out to improve security and avoid violence. These include:

- using cashless payment methods;
- keeping money on the premises to a minimum;
- careful check of customer or client's credentials;
- careful planning of meetings away from the workplace;
- teamwork where suspected aggressors may be involved;
- regular contact with workers away from their base. There are special services available to provide contact arrangements;
- avoidance of lone working as far as is reasonably practicable;
- thinking about how staff who have to work shifts or late hours will get home. Safe transport and/or parking areas may be required;
- setting up support services to help victims of violence and, if necessary, other staff who could be affected. They may need debriefing, legal assistance, time off work to recover or counselling by experts.

A busy accident and emergency department of a general hospital, for example, has to balance the protection of staff from violent attack with the need to offer patients a calm and open environment. Protection could be given to staff by the installation of wide counters, coded locks on doors, CCTV systems, panic buttons and alarm systems. The employment of security staff and strict security procedures for the storage and issuing of drugs are two further precautions taken by such departments. Awareness training for staff so that they can recognise early signs of aggressive behaviour and an effective counselling service for those who have suffered from violent behaviour should be provided.

Take the appropriate action

The arrangements for dealing with violence should be included in the safety policy and managed like any other aspect of the health and safety procedures. Action plans should be drawn up and followed through using the consultation arrangements as appropriate. The police should also be consulted to ensure that they are happy with the plan and are prepared to play their part in providing back-up and the like.

Check that the action is effective

Ensure that the records are being maintained and any reported incidents are investigated and suitable action taken. The procedures should be regularly audited and changes made if they are not working properly.

Victims should be provided with help and assistance to overcome their distress, through debriefing, counselling, time off to recover, legal advice and support from colleagues.

6.6 Substance abuse at work

6.6.1 Risks to health and safety from substance abuse at work

Alcohol and drug abuse damages health and causes absenteeism and reduced productivity. The HSE is keen to see employers address the problem and offers advice in two separate booklets.

Alcohol abuse is a considerable problem when vehicle driving is part of the job, especially if driving is required on public roads. Misuse of alcohol can:

- reduce productivity;
- increase absenteeism;
- increase accidents and near misses at work; and
- in some cases, endanger the public.

The HSE has estimated that between 3% and 5% of all absences from work are due to alcohol and result in approximately 14 million working days lost each year.

Drug and solvent abuse presents similar problems to those found with alcohol abuse – absenteeism, reduced productivity and an increase in the risk of accidents. The drugs that are involved in drug abuse may be legal drugs, such as painkillers of various types or illegal drugs such as heroin.

The types of work where drug or alcohol misuse could have a serious outcome include:

- using machinery
- using electrical equipment or ladders
- driving or operating heavy lifting equipment.

Where employees in safety-critical jobs seek help for alcohol or drug misuse, it may be necessary to transfer them to other work, at least temporarily.

6.6.2 Control measures to reduce risks from substance abuse at work

Employers need to adopt an alcohol policy following employee consultation. The following matters need to be considered:

▶ how the organisation expects employees to restrict their level of drinking;

▶ how drinking problems can be recognised and help offered; and

▶ at what point and under what circumstances will an employee's drinking be treated as a disciplinary rather than a health problem.

Prevention of the problem is better than remedial action after a problem has occurred. During working hours, there should be no drinking, and drinking during break periods should be discouraged. Induction training should stress this policy and managers should set it. Posters can also help to communicate the message. However, it is important to recognise possible symptoms of an alcohol problem, such as lateness and absenteeism, poor work standards, impaired concentration, memory and judgement, mood swings and slurring of speech and deteriorating relations with colleagues. It is always better to offer counselling rather than dismissal. The policy should be monitored to check on its effectiveness.

A successful drug misuse policy will benefit the organisation and workers by reducing absenteeism, poor productivity and the risk of accidents. There is no simple guide to the detection of drug abuse, but the HSE has suggested the following signals:

▶ sudden mood changes;
▶ unusual irritability or aggression;
▶ a tendency to become confused;
▶ abnormal fluctuations in concentration and energy;
▶ impaired job performance;
▶ poor time-keeping;
▶ increased short-term sickness leave;
▶ a deterioration in relationships with colleagues, customers or management;
▶ dishonesty and theft (arising from the need to maintain an expensive habit).

A policy on drug abuse can be established by:

1. **Investigation of the size of the problem**
   Examination of sicknesses, behavioural and productivity changes and accident and disciplinary records is a good starting point.
2. **Planning actions**
   Develop an awareness programme for all staff and a special training programme for managers and supervisors. Employees with a drug problem should be encouraged to seek help in a confidential setting.
3. **Taking action**
   Produce a written policy that includes everyone in the organisation and names the person responsible for implementing the policy. It should include details of the safeguards to employees and the confidentiality given to anyone with a drug problem. It should also clearly outline the circumstances in which disciplinary and/or reported action will be taken (the refusal of help, gross misconduct and possession/dealing in drugs).

4. **Monitoring the policy**
   The policy can be monitored by checking for positive changes in the measures made during the initial investigation (improvements in the rates of sickness and accidents). Drug screening and testing is a sensitive issue and should only be considered with the agreement of the workforce (except in the case of pre-employment testing). Screening will only be acceptable if it is seen as part of the health policy of the organisation and its purpose is to reduce risks to the misusers and others. Laboratories are accredited by the United Kingdom Accreditation Service (UKAS) and are able to undertake reliable testing.

It is important to stress that some drugs are prescribed and controlled. The side effects of these can also affect performance and pose risks to colleagues. Employers should encourage employees to inform them of any possible side effects from prescribed medication and be prepared to alter work programmes accordingly.

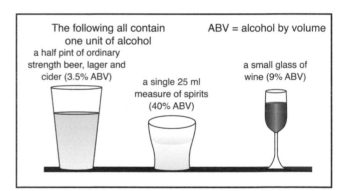

The following all contain one unit of alcohol

a half pint of ordinary strength beer, lager and cider (3.5% ABV)

a single 25 ml measure of spirits (40% ABV)

a small glass of wine (9% ABV)

ABV = alcohol by volume

Figure 6.15 It takes a healthy liver about one hour to break down and remove one unit of alcohol. A unit is equivalent to 8 mg or 10 ml of pure alcohol.

Managers and supervisors should receive training so that they are able to:

▶ recognise the signs of drug or alcohol misuse;
▶ follow the organisation's rules on drug and alcohol misuse;
▶ deal with an employee that they suspect is misusing drugs or alcohol; and
▶ respond to an employee who tells them about a drug or alcohol problem.

6.7 Further information

Statutory provisions

The Control of Noise at Work Regulations 2005

The Control of Vibration at Work Regulations 2005

The Ionising Radiations Regulations 2017

The Personal Protective Equipment at Work Regulations 1992 (as amended)

Other references

Controlling Noise at Work. The Control of Noise at Work Regulations 2005, L108 (second edition), 2005, HSE, ISBN 9780717661640 www.hse.gov.uk/pubns/books/l108.htm

Health and Safety Toolbox, online resource, HSE, www.hse.gov.uk/toolbox/index.htm

Hand-Arm Vibration: Control of Vibration at Work Regulations L140, HSE, 20019, ISBN 9780717665655 www.hse.gov.uk/pubns/books/l140.htm

How to tackle work-related stress INDG430, 2009, HSE, ISBN 9780717663781 https://www.hse.gov.uk/pubns/indg430.pdf

HSE Stress Management Standards www.hse.gov.uk/stress/standards

Tackling work-related stress using the management standards. A step-by-step workbook, WBK01, 2019, HSE ISBN 9780717667154 https://www.hse.gov.uk/pubns/wbk01.pdf

Personal Protective Equipment at Work (second edition), Personal Protective Equipment at Work Regulations 1992 (as amended), Guidance on Regulations, L25 (third edition) 2015, HSE, ISBN: 9780717665976 www.hse.gov.uk/pubns/priced/l25.pdf

Radon in the workplace www.hse.gov.uk/radiation/ionising/radon.htm#testingradon

Upper Limb Disorders in the Workplace – A Guide, second edition 2002, HSG60, HSE Books, ISBN: 9780717619788 www.hse.gov.uk/pubns/priced/hsg60.pdf

Whole-body vibration; The Control of Vibration at Work Regulations 2005, Guidance on ISBN 9780717661268 https://www.hse.gov.uk/pubns/books/l141.htm

Violence at work, INDG69 (rev), 1996, HSE Books www.hse.gov.uk/pubns/indg69.pdf

Managing drug and alcohol misuse at work, HSE https://www.hse.gov.uk/alcoholdrugs/

6

6.8 Practice questions

1. (a) **Outline** the mechanism by which noise is transmitted by the human ear to the brain.
   (b) **Explain** the following terms in relation to noise exposure at work in both the short and long term:
   (i) 'noise-induced hearing loss'
   (ii) 'tinnitus'
   (iii) temporary and permanent threshold shift.
   (c) **Outline** the options that might be considered to reduce the risk of hearing damage to workers who are exposed to high levels of noise.

2. (a) **Give** the meaning of the following terms used in noise measurement:
   (i) frequency;
   (ii) daily personal noise exposure' ($L_{EP'd}$).
   (iii) decibel (dB);
   (iv) 'A' weighting.
   (b) Noise action levels are used to limit the exposure of workers to excessive noise levels. **Explain** the difference between a noise exposure value and a noise exposure limit value
   (c) **Identify TWO** types of noise measurement techniques.

3. (a) **Outline** appropriate control measures to reduce the levels of noise to which the workers are exposed.
   (b) **Outline** the criteria that should be used when selecting suitable hearing protection

   and **identify** the limitations of such protective equipment.
   (c) **Outline** the contents of a health surveillance programme required to protect workers who are regularly exposed to high noise levels.

4. **Explain** the meaning of the following terms in relation to noise control in the workplace:
   (i) silencing
   (ii) absorption
   (iii) damping
   (iv) isolation
   (v) lagging.

5. (a) **Describe** hand-arm vibration (HAV) and whole-body vibration (WBV).
   (b) **Outline** the ill-health effects and the associated symptoms caused by the exposure of workers to HAV and WBV.

6. A worker operates a hammer drill for long periods of the working day.
   (a) **Outline** the issues to be considered when assessing the risk of hand-arm vibration syndrome (HAVS) being developed by the worker.
   (b) **Outline** the control measures that should be taken to minimise the risk of the worker developing HAVS.

7. A dumper truck driver, working on a construction site, spends much of the working day driving over very rough ground. After several weeks, the driver complains of back pain.
   (a) **Identify** possible reasons for the back pain associated with the truck driving.

(b) **Outline** actions that could be taken by the employer to control the risks from whole-body vibration.

8. (a) **Identify TWO** types of ionising radiation
   (b) **Outline** possible means of ensuring that workers are not exposed to unacceptable levels of ionising radiation.

9. (a) For each of the following types of non-ionising radiation, **identify** a source and the possible ill-health effects on exposed individuals:
   (i) infrared radiation
   (ii) ultraviolet radiation
   (iii) lasers
   (iv) microwaves.
   (b) **Identify** the controls available to protect people against exposure to these non-ionising radiations.

10. Construction workers are often required to work outdoors in a variety of weather conditions. **Outline** the hazards and the associated control measures that should be taken to protect these workers when they are working outdoors:
    (a) in strong sunlight
    (b) in cold temperatures.

11. (a) **Explain** the meaning of 'work-related stress'.
    (b) **Identify** the ill-health effects of excessive stress to workers in the workplace.
    (c) **Identify FOUR** issues that could lead to workplace stress.
    (d) **Outline** the control measures that could be taken to minimise the risk of occupational stress.

12. (a) **Identify FOUR** types of work where employees may be at an increased risk of violence when dealing with members of the public.
    (b) **Outline** an action plan that can be developed to protect employees from the risk of violence in the workplace.
    (c) **Identify FOUR** types of security equipment that could be employed to protect vulnerable workers from the risk of violence when dealing with members of the public.

13. (a) **Outline FIVE** possible indications that an employee may be suffering from alcohol, solvent or drug abuse.
    (b) **Identify** the risks to health from the misuse of these substances.
    (c) **Outline** the control measures that could reduce these risks to health.

CHAPTER 7

Musculoskeletal health

> **Chapter summary**
>
> **This chapter covers the following topics in Element 6 of the NEBOSH NGC syllabus:**
> Work-related upper limb disorders
> Manual handling
> Load handling equipment
>
> **Learning outcomes**
>
> **These are the learning outcomes covered by this chapter:**
> **The learner will be able to:**
> 1. Describe work processes and practices that may give rise to work-related upper limb disorders and appropriate control measures
> 2. Outline the hazards and control measures which should be considered when assessing risks from manual handling activities
> 3. Outline the hazards and controls associated with load handling equipment and the requirements for lifting operations.

7.1 Work-related upper limb disorders

7.1.1 Introduction

Until a few years ago, accidents caused by the manual handling of loads were the largest single cause of minor accidents reported to the Health and Safety Executive (HSE). The Manual Handling Operations Regulations recognised this fact and helped to reduce the number of these accidents. However, accidents due to poor manual handling technique still account for over 25% of all reported accidents and in some occupational sectors, such as the health service, the figure rises above 50%. An understanding of the factors causing some of these accidents is essential if they are to be further reduced. Mechanical handling methods should always be used whenever possible, but they are not without their hazards, many of which are outlined later in this chapter. Much mechanical handling involves the use of lifting equipment, such as cranes and lifts, which present specific hazards to both the users and bystanders. The risks from these hazards are reduced by thorough examinations and inspections as required by the Lifting Operations and Lifting Equipment Regulations (LOLER).

7.1.2 Principles and scope of ergonomics

Ergonomics is the study of the interaction between workers and their work in the broadest sense, in that it encompasses the whole system surrounding the work process. It is, therefore, as concerned with the work organisation, process and design of the workplace and work methods as it is with work equipment. The common definitions of ergonomics, the 'person machine interface' or 'fitting the person to the machine rather than vice versa' are far too narrow.

It is concerned with the physical and mental capabilities of an individual as well as their understanding of the job under consideration. Ergonomics includes the limitations of the worker in terms of skill level, perception and other personal factors in the overall design of the job and the system supporting and surrounding it. It is the study of the relationship between the worker, the machine and the environment in which it operates and attempts to optimise the whole work system, including the job, to the capabilities of the worker so that maximum output is achieved for minimum effort and discomfort by the worker. Cars, buses and lorries are all ergonomically designed so that all the important controls, such as the steering wheel, brakes, gear stick and instrument panel, are easily accessed by most drivers within a wide range of sizes. Ergonomics is sometimes described as human engineering and as working practices become more and more automated, the need for good ergonomic design becomes essential.

Figure 7.1 Handling roof slates onto a roof using a teleporter lift truck

The scope of ergonomics and an ergonomic assessment is very wide, incorporating the following areas of study:

▶ personal factors of the worker, in particular physical, mental and intellectual abilities, body dimensions and competence in the task required;

▶ the machine and associated equipment under examination;

▶ the interface between the worker and the machine – controls, instrument panel or gauges and any aids including seating arrangements and hand tools;

▶ environmental issues affecting the work process such as lighting, temperature, humidity, noise and atmospheric pollutants;

▶ the interaction between the worker and the task, such as the production rate, posture and system of working;

▶ the task or job itself – the design of a safe system of work, checking that the job is not too strenuous or repetitive and the development of suitable training packages;

▶ the organisation of the work, such as shift work, breaks and supervision.

The reduction of the possibility of human error is one of the major aims of ergonomics and an ergonomic assessment. An important part of an ergonomic study is to design the workstation or equipment to fit the worker. For this to be successful, the physical measurement of the human body and an understanding of the variations in these measurements between people are essential. Such a study is known as **anthropometry,** which is defined as the scientific measurement of the human body and its movement. Since there are considerable variations in, for example, the heights of people, it is common for some part of the workstation to be variable (e.g. an adjustable seat).

7.1.3 The ill-health effects of poor ergonomics

Ergonomic hazards are those hazards to health resulting from poor ergonomic design. They generally fall within the physical hazard category and include the manual handling and lifting of loads, pulling and pushing loads, prolonged periods of repetitive activities and work with vibrating

tools. The condition of the working environment, such as low lighting levels, can present health hazards to the eyes. It is also possible for psychological conditions, such as occupational stress, to result from ergonomic hazards.

The common ill-health effects of ergonomic hazards are musculoskeletal disorders (MSDs), work-related upper limb disorders (WRULDs), including repetitive strain injury (RSI) and deteriorating eyesight.

7.1.4 Work-related upper limb disorders

WRULDs describe a group of conditions which can affect the neck, shoulders, arms, elbows, wrists, hands and fingers. **Tenosynovitis** (affecting the tendons), **carpal tunnel syndrome** (affecting the tendons which pass through the carpal bone in the hand) and **frozen shoulder** are all examples of WRULDs which differ in the manifestation and site of the illness. The term **RSI** is commonly used to describe WRULDs.

WRULDs are caused by repetitive movements of the fingers, hands or arms which involve pulling, pushing, reaching, twisting, lifting, squeezing or hammering. These disorders can occur to workers in offices as well as in factories or on construction sites. Typical occupational groups at risk include painters and decorators, riveters and pneumatic drill operators and desktop computer users.

The main symptoms of WRULDs are aching pain to the back, neck and shoulders, swollen joints and muscle fatigue accompanied by tingling, soft tissue swelling, similar to bruising, and a restriction in joint movement. The sense of touch and movement of fingers may be affected. The condition is normally a chronic one in that it gets worse with time and may lead eventually to permanent damage. The injury occurs to muscle,

Figure 7.2 A tilted worktable. The distance between the operator and the work can be reduced by putting the table at a more vertical angle. The table is adjustable in height and angle to suit the particular job

tendons and/or nerves. If the injury is allowed to heal before being exposed to the repetitive work again, no long-term damage should result. However, if the work is repeated again and again, healing cannot take place and permanent damage can result leading to a restricted blood flow to the arms, hands and fingers.

The risk factors, which can lead to the onset of WRULDs, are repetitive actions of lengthy duration, the application of significant force and unnatural postures, possibly involving twisting and over-reaching and the use of vibrating tools. Cold working environments, work organisation and worker perception of the work organisation have all been shown in studies to be risk factors, as is the involvement of vulnerable workers such as those with pre-existing ill-health conditions and pregnant women.

7.1.5 Display screen equipment

Display screen equipment (DSE), which includes visual display units, is a good example of a common work activity which relies on an understanding of ergonomics and the ill-health conditions which can be associated with poor ergonomic design.

A survey of safety representatives by the Trades Union Congress (TUC) found that injuries/illnesses caused by the poor use of DSE and repetitive strain injuries together with stress or overwork were among their major concerns.

Legislation governing DSE is covered by the Health and Safety (Display Screen Equipment) Regulations and a detailed summary of them is given in Chapter 13. The Regulations apply to a user or operator of DSE and define a user or operator. The definition is not as tight as many employers would like it to be, but usage in excess of approximately one hour continuously each day would define a user. The definition is important as users are entitled to free eye tests and, if required, a

pair of spectacles specifically for DSE use. The basic requirements of the Regulations are:

▶ a suitable and sufficient risk assessment of the workstation, including the software in use, trip and electrical hazards from trailing cables and the surrounding environment;
▶ workstation compliance with the minimum specifications laid down in the schedules appended to the Regulations;
▶ a plan of the work programme to ensure that there are adequate breaks in the work pattern of workers;
▶ the provision of free eyesight tests and, if required, spectacles to users of DSE;
▶ a suitable programme of training and sufficient information given to all users.

The risk assessment of a DSE workstation needs to consider the following factors:

▶ the height and adjustability of the monitor;
▶ the adjustability of the keyboard, the suitability of the mouse and the provision of wrist support;
▶ the stability and adjustability of the DSE user's chair;
▶ the provision of ample foot room and suitable foot support;
▶ the effect of any lighting and window glare at the workstation;
▶ the storage of materials around the workstation;
▶ the safety of trailing cables, plugs and sockets;
▶ environmental issues – noise, heating, humidity and draughts.

See Chapter 17 Form H3, which shows an example of a checklist that can be used for a DSE workstation assessment by the user.

There are three basic ill-health hazards associated with DSE. These are:

▶ musculoskeletal problems;
▶ visual problems;
▶ psychological problems.

Figure 7.3 Pump juice from a bulk container to save awkward handling

Forearms approximately horizontal

Minimal extension, flexion or deviation of wrist

Seat back adjustability

Screen height and angle should allow comfortable head position

Good lumbar support

Seat height adjustability

Space in front of the keyboard to support hands or wrists during pauses in keying

Space for postural change – no obstacles under desk

No excess pressure on underside of thighs and backs of knees

Foot support if required

Figure 7.4 Workstation design

A fourth hazard, of radiation, has been shown from several studies to be very small and is now no longer normally considered in the risk assessment.

Similarly, in the past, there have been suggestions that DSE could cause epilepsy, and there were concerns about adverse health effects on pregnant women and their unborn children. All these risks have been shown by various studies to be very low.

The provision of DSE training and risk assessments online has become more common. The risk assessment, however, still has to be managed, made appropriate to the particular workplace setting and reviewed from time to time. Various studies have shown that users of any e-learning package lose concentration after 30–40 minutes.

Musculoskeletal problems

Tenosynovitis is the most common and well-known problem which affects the wrist of the user. The symptoms and effects of this condition have already been covered. Suffice it to say that if the condition is ignored, then the tendon and tendon sheath around the wrist will become permanently injured. Tenosynovitis is caused by the continual use of a keyboard and can be relieved by the use of wrist supports. Other WRULDs are caused by poor posture and can produce pains in the back, shoulders, neck or arms. Less commonly, pain may also be experienced in the thighs, calves and ankles. These problems can be mitigated by the application of ergonomic principles in the selection of working desks, chairs, footrests and document holders. It is also important to ensure that the desk is at the correct height and the computer screen is tilted at the correct angle to avoid putting too much strain on the neck. (Ideally the user's eyes should be at the same height as the top of the screen.)

The keyboard should be detachable so that it can be positioned anywhere on the desktop and a correct posture adopted while working at the keyboard. The chair should be adjustable in height, stable and have an adjustable backrest. If the knees of the user are lower than the hips when seated, then a footrest should be provided. The surface of the desk should be non-reflecting and uncluttered but ancillary equipment (e.g. telephone and printer) should be easily accessible.

Visual problems

There does not appear to be much medical evidence that DSE causes deterioration in eyesight, but users may suffer from visual fatigue which results in eye strain, sore eyes and headaches. Less common ailments are skin rashes and nausea.

The use of DSE may indicate that reading spectacles are needed and the Regulations make provision for this. It is possible that any prescribed lenses may only be suitable for DSE work as they will be designed to give optimum clarity at the normal distance at which screens are viewed (50–60 cm).

Eye strain is a particular problem for people who spend a large proportion of their working day using DSE. A survey has indicated that up to 70% of DSE users complain of eye fatigue. Eye strain can be reduced by the following steps additional to those already identified in this section:

▶ Train staff in the correct use of the equipment.
▶ Ensure that a font size of at least 12 is used on the screen.
▶ Ensure that users take regular breaks away from the screen (up to 10 minutes every hour).

Sore eyes are a common complaint associated with low humidity. Various professional building research organisations have agreed that an appropriate range of relative humidity in an office environment is 40–60%. Air conditioning systems should be properly maintained to ensure that they maintain a reasonable temperature and humidity level in the workplace. Sore eyes are often a product of dry eyes which may be controlled by:

▶ restraining from rubbing the eyes;
▶ focusing on an object in the near distance (preferably green in colour);
▶ blinking can encourage the moistening of the eye;
▶ drinking water regularly;
▶ looking away from the screen periodically to rest the eyes.

The screen should be adjustable in tilt angle and screen brightness and contrast. Finally, the lighting around the workstation is important. It should be bright enough to allow documents to be read easily but not too bright such that either headaches are caused or there are reflective glares on the computer screen.

Psychological problems

These are generally stress-related problems. They may have environmental causes, such as noise, heat, humidity or poor lighting, but they are usually due to high-speed working, lack of breaks, poor training and poor workstation design. One of the most common problems is the lack of understanding of all or some of the software packages being used.

There are several other processes and activities where ergonomic considerations are important. These include the assembly of small components (microelectronics assembly lines) and continually moving assembly lines (car assembly plants).

Many of these processes including some of the chemical hazards (such as soldering fumes) are described in Chapter 8.

7.2 Manual handling hazards and control measures

The term 'manual handling' is defined as the movement of a load by human effort alone. This effort may be applied directly or indirectly using a rope or a lever.

Manual handling may involve the transportation of the load or the direct support of the load including pushing, pulling, carrying, moving using bodily force and, of course, straightforward lifting. Back injury due to the lifting of heavy loads is very common and several million working days are lost each year as a result of such injuries.

Typical hazards of manual handling include:

▶ lifting a load which is too heavy or too cumbersome, resulting in back injury;
▶ poor posture during lifting or poor lifting technique, resulting in back injury;
▶ dropping a load, resulting in foot injury;
▶ lifting sharp-edged or hot loads resulting in hand injuries.

7.2.1 Injuries caused by manual handling

Manual handling operations can cause a wide range of acute and chronic injuries to workers. Acute injuries normally lead to sickness leave from work and a period of rest during which time the damage heals. Chronic injuries build up over a long period of time and are usually irreversible, producing illnesses such as arthritic and spinal disorders. There is considerable evidence to suggest that modern lifestyles, such as a lack of exercise and regular physical effort, have contributed to the long-term serious effects of these injuries.

Figure 7.5 Manual handling: there are many potential hazards

The most common injuries associated with poor manual handling techniques are all musculoskeletal in nature and are:

▶ muscular sprains and strains – caused when a muscular tissue (or ligament or tendon) is stretched beyond its normal capability leading to a weakening, bruising and painful inflammation of the area affected. Such injuries normally occur in the back or in the arms and wrists;

▶ back injuries – include injuries to the discs situated between the spinal vertebrae (i.e. bones) and can lead to a very painful prolapsed disc lesion (commonly known as a slipped disc). This type of injury can lead to other conditions known as lumbago and sciatica (where pain travels down the leg);

▶ trapped nerve – usually occurring in the back as a result of another injury but aggravated by manual handling;

▶ hernia – this is a rupture of the body cavity wall in the lower abdomen, causing a protrusion of part of the intestine. This condition eventually requires surgery to repair the damage;

▶ cuts, bruising and abrasions – caused by handling loads with unprotected sharp corners or edges;

▶ fractures – normally of the feet due to the dropping of a load. Fractures of the hand also occur but are less common;

▶ work-related upper limb disorders (WRULDs);

▶ rheumatism – this is a chronic disorder involving severe pain in the joints. It has many causes, one of which is believed to be the muscular strains induced by poor manual handling lifting technique.

The sites on the body of injuries caused by manual handling accidents are shown in Figure 7.6. Nearly 11 million working days are lost each year due to musculoskeletal disorders, which are thought to cost the UK more than £7 billion each year and affect 6.5 million people.

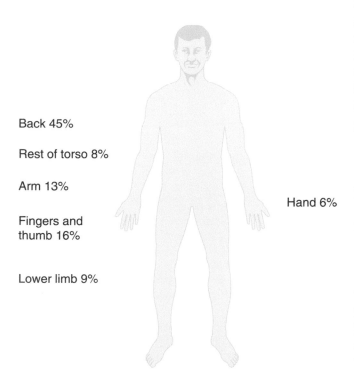

Back 45%

Rest of torso 8%

Arm 13%

Fingers and thumb 16%

Lower limb 9%

Hand 6%

Figure 7.6 Main injury sites caused by manual handling accidents

Musculoskeletal problems are the most common cause of absence, followed by viral infections and stress-related illnesses. These findings are based on an analysis of sickness management records for 11,000 individual employees across a range of private sector organisations. A study has found that musculoskeletal disorders (MSDs) account for nearly half (47%) of all absences from work and 60% of permanent work incapacity in the European Union. It is estimated that this costs the UK economy £7bn every year and costs Europe £240bn each year. The study, conducted across 25 European countries, found that 100 million Europeans suffer from chronic musculoskeletal pain – over 40 million of whom are workers – with up to 40% having to give up work due to their condition. In the UK alone, 7.5 million working days were lost in one year due to musculoskeletal problems.

In general, pulling a load is much easier for the body than pushing one. If a load can only be pushed, then pushing backwards using the back is less stressful on body muscles. Lifting a load from a surface at waist level is easier than lifting from floor level and most injuries during lifting are caused by lifting and twisting at the same time. If a load has to be carried, it is easier to carry it at waist level and close to the body trunk. A firm grip is essential when moving any type of load.

7.2.2 Hierarchy of measures for manual handling operations

With the introduction of the Manual Handling Operations Regulations, the emphasis during the assessment of lifting operations changed from a simple reliance on safe lifting techniques to an analysis, using risk assessment, of the need for manual handling. The Regulations established a clear hierarchy of measures to be taken when an employer is confronted with a manual handling operation:

▶ Avoid manual handling operations so far as is reasonably practicable either by redesigning the task to avoid moving the load or by automating or mechanising the operations.

▶ If manual handling cannot be avoided, then a suitable and sufficient risk assessment should be made.

▶ Reduce the risk of injury from those operations so far as is reasonably practicable, by either the use of mechanical handling or making improvements to the task, the load and the working environment.

The guidance given to the Manual Handling Operations Regulations (available in the HSE Legal series – L23) is a very useful document. It gives very helpful advice on manual handling assessments and manual handling training. The advice is applicable to all occupational sectors. In Chapter 17 Forms H5 and H6 show examples of manual handling assessment forms.

7.2.3 Manual handling assessments

The Regulations specify four main factors which must be taken into account during the assessment. These are the task, the load, the working environment and the capability of the individual who is expected to do the lifting.

The **task** should be analysed in detail so that all aspects of manual handling are covered including the use of mechanical assistance. The number of people involved and the cost of the task should also be considered. Some or all of the following questions are relevant to most manual handling tasks:

▶ Is the load held or manipulated at a distance from the trunk? The further from the trunk, the more difficult it is to control the load and the stress imposed on the back is greater.

▶ Is a satisfactory body posture being adopted? Feet should be firmly on the ground and slightly apart and there should be no stooping or twisting of the trunk. It should not be necessary to reach upwards, as this will place additional stresses on the arms, back and shoulders. The effect of these risk factors is significantly increased if several are present while the task is being performed.

▶ Are there excessive distances to carry or lift the load? Over distances greater than 10 m, the physical demands of carrying the load will dominate the operation. The frequency of lifting and the vertical and horizontal distances the load needs to be carried (particularly if it has to be lifted from the ground and/or placed on a high shelf) are very important considerations.

▶ Is there excessive pulling and pushing of the load? The state of floor surfaces and the footwear of the

individual should be noted so that slips and trips may be avoided.

▶ Is there a risk of a sudden movement of the load? The load may be restricted or jammed in some way.

▶ Is frequent or prolonged physical effort required? Frequent and prolonged tasks can lead to fatigue and a greater risk of injury.

▶ Are there sufficient rest or recovery periods? Breaks and/or the changing of tasks enables the body to recover more easily from strenuous activity.

▶ Is there an imposed rate of work on the task? This is a particular problem with some automated production lines and can be addressed by spells on other operations away from the line.

▶ Are the loads being handled while the individual is seated? In these cases, the legs are not used during the lifting processes and stress is placed on the arms and back.

▶ Does the handling involve two or more people? The handling capability of an individual reduces when he/she becomes a member of a team (e.g. for a three-person team, the capability is half the sum of the individual capabilities). Visibility, obstructions and the roughness of the ground must all be considered when team handling takes place.

The **load** must be carefully considered during the assessment and the following questions asked:

▶ Is the load too heavy? The maximum load that an individual can lift will depend on the capability of the individual and the position of the load relative to the body. There is therefore no safe load. Figure 7.7 is reproduced from the HSE guidance, which does give some advice on loading levels. It recommends that loads in excess of 25 kg should not be lifted or carried by a man (and this

Figure 7.7 HSE guidance for manual lifting – recommended weights

is only permissible when the load is at the level of and adjacent to the thighs). For women, the guideline figures should be reduced by about one-third.

▶ Is the load too bulky or unwieldy? In general, if any dimension of the load exceeds 0.75 m (c. 2 ft.), its handling is likely to pose a risk of injury. Visibility around the load is important. It may hit obstructions or become unstable in windy conditions. The position of the centre of gravity is very important for stable lifting – it should be as close to the body as possible.

▶ Is the load difficult to grasp? Grip difficulties will be caused by slippery surfaces, rounded corners or a lack of foot room.

▶ Are the contents of the load likely to shift? This is a particular problem when the load is a container full of smaller items, such as a sack full of nuts and bolts. The movements of people (in a nursing home) or animals (in a veterinary surgery) are loads which fall into this category.

▶ Is the load sharp, hot or cold? Personal protective equipment may be required.

The **working environment** in which the manual handling operation is to take place, must be considered during the assessment. The following areas will need to be assessed:

▶ any space constraints which might inhibit good posture. Such constraints include lack of headroom, narrow walkways and items of furniture;

▶ slippery, uneven or unstable floors;

▶ variations in levels of floors or work surfaces, possibly requiring the use of ladders;

▶ extremes of temperature and humidity. These effects are discussed in detail in Chapter 6;

▶ ventilation problems or gusts of wind;

▶ poor lighting conditions.

Finally, the **capability of the individual** to lift or carry the load must be assessed. The following questions will need to be asked:

▶ Does the task require unusual characteristics of the individual (e.g. strength or height)? It is important to remember that strength and general manual handling ability depend on age, gender, state of health and fitness.

▶ Are employees who might reasonably be considered to be pregnant or to have a health problem put at risk by the task? Particular care should be taken to protect pregnant women or those who have recently given birth from handling loads. Allowance should also be given to any employee who has a health problem which could be exacerbated by manual handling.

The assessment must be reviewed if there is reason to suspect that it is no longer valid or there has been a significant change to the manual handling operations to which it relates.

7.2.4 Reducing the risk of injury

This involves the introduction of control measures resulting from the manual handling risk assessment. The guidance to the Regulations (L23) and the HSE publication *Manual Handling – Solutions You Can Handle* (HSG115) contain many ideas to reduce the risk of injury from manual handling operations. An ergonomic approach is generally required to design and develop the manual handling operation as a whole. The control measures can be grouped under five headings. However, the first consideration, when it is reasonably practicable, is **mechanical assistance** (see Section 7.3).

The **task** can be improved by changing the layout of the workstation by, for example, storing frequently used loads at waist level. The removal of obstacles and the use of a better lifting technique that relies on the leg rather than back muscles should be encouraged. When pushing, the hands should be positioned correctly. The work routine should also be examined to see whether job rotation is being used as effectively as it could be. Special attention should be paid to seated manual handlers to ensure that loads are not lifted from the floor while they are seated. Employees should be encouraged to seek help if a difficult load is to be moved, so that a team of people can move the load. Adequate and suitable personal protective equipment should be provided where there is a risk of loss of grip or injury. Care must be taken to ensure that the clothing does not become a hazard in itself (e.g. the snagging of fasteners and pockets).

The **load** should be examined to see whether it could be made lighter, smaller or easier to grasp or manage. This could be achieved by splitting the load, the positioning of handholds or a sling, or ensuring that the centre of gravity is brought closer to the handler's body. Attempts should be made to make the load more stable and any surface hazards, such as slippery deposits or sharp edges, should be removed. It is very important to ensure that any improvements do not, inadvertently, lead to the creation of additional hazards.

The **working environment** can be improved in many ways. Space constraints should be removed or reduced. Floors should be regularly cleaned and repaired when damaged. Adequate lighting is essential and working at more than one level should be minimised so that hazardous ladder work is avoided. Attention should be given to the need for suitable temperatures and ventilation in the working area.

The **capability of the individual** is the fifth area where control measures can be applied to reduce the risk of injury. The state of health of the employee and his/her medical record will provide the first indication as to whether the individual is capable of undertaking

7

the task. A period of sick leave or a change of job can make an individual vulnerable to manual handling injury. The Regulations require that the employee be given information and training. The information includes the provision, where it is reasonably practicable to do so, of precise information on the weight of each load and the heaviest side of any load whose centre of gravity is not centrally positioned. In a more detailed risk assessment, other factors will need to be considered such as the effect of personal protective equipment and psychosocial factors in the work organisation. The following points may need to be assessed:

▶ Does protective clothing hinder movement or posture?
▶ Is the correct personal protective equipment being worn?
▶ Is proper consideration given to the planning and scheduling of rest breaks?
▶ Is there good communication between managers and employees during risk assessment or workstation design?
▶ Is there a mechanism in place to deal with sudden changes in the volume of workload?
▶ Have employees been given sufficient training and information?
▶ Does the worker have any learning disabilities and, if so, has this been taken into account in the assessment?

Amendments to the Manual Handling Operations Regulations have emphasised that a worker may be at risk if he/she:

▶ is physically unsuited to carry out the tasks in question;
▶ is wearing unsuitable clothing, footwear or other personal effects;
▶ does not have adequate or appropriate knowledge or training.

The HSE have developed a Manual Handling Assessment Chart (MAC) tool to help the user identify high-risk workplace manual handling activities and in the assessment of common risks associated with lifting, carrying and handling. The tool can be used to assess the risks posed by lifting, carrying and team manual handling activities. It incorporates a numerical and a colour coding score system to highlight high-risk manual handling tasks. There are three types of assessment that can be carried out with the MAC:

1. lifting operations;
2. carrying operations;
3. team handling operations.

The MAC is available on the HSE website.

The training requirements are given in the following section.

7.2.5 Manual handling training

Training alone will not reduce manual handling injuries – there still need to be safe systems of work in place and the full implementation of the control measures highlighted in the manual handling assessment. The following topics should be addressed in a manual handling training session:

▶ types of injuries associated with manual handling activities;
▶ the findings of the manual handling assessment;
▶ the recognition of potentially hazardous manual handling operations;
▶ the correct use of mechanical handling aids;
▶ the correct use of personal protective equipment;
▶ features of the working environment which aid safety in manual handling operations;
▶ good housekeeping issues;
▶ factors which affect the capability of the individual;
▶ good lifting or manual handling technique as shown in Figure 7.8

1. Check suitable clothing and assess load. Heaviest side to body.

2. Place feet apart – bend knees.

3. Firm grip – close to body. Slight bending of back, hips and knees at start.

4. Lift smoothly to knee level and then waist level. No further bending of back.

5. With clear visibility move forward without twisting. Keep load close to the waist. Turn by moving feet. Keep head up. Do not look at load.

6. Set load down at waist level or to knee level and then on the floor.

Figure 7.8 The main elements of a good lifting technique

The 'good handling technique' given in the Guidance on the Manual Handling Operations Regulations (L23) advises:

- think before you lift;
- keep the load close to your waist;
- adopt a stable position;
- ensure a good hold on the load;
- at the start of the lift, moderate flexion (slight bending) of the back, hips and knees is preferable to fully flexing the back (stooping) or the hips and knees (squatting);
- don't flex your spine any further as you lift;
- avoid twisting the trunk or leaning sideways, especially while the back is bent;
- keep your head up when handling;
- move smoothly;
- don't lift more than you can easily manage; and
- put down then adjust.

Finally, it needs to be stressed that if injuries involving manual handling operations are to be avoided, planning, control and effective supervision are essential.

7.3 Load handling equipment

7.3.1 Manually operated load handling equipment

Types of manually operated load handling equipment

As mentioned earlier (7.2.4), the first consideration to reduce the risk of injury due to manual handling is, when it is reasonably practicable, to use mechanical assistance in the form of:

- simple tools;
- wheelbarrows;
- trucks and trolleys;
- roller tracks and chutes;
- pallet trucks;
- conveyors; and
- various types of hoists that can be used to lift people as well as other loads.

The aids vary from simple, manually operated tools to power-assisted trucks and lifting devices. All of them 'lighten the load' and reduce the risk of injury.

All kinds of **simple tools** can aid the manual handling of loads. **Lifting hooks** can be used to lift sheets of steel or glass, timber boards and large awkward loads. **Log tongs** will help lift logs, and other devices can be used for cylindrical loads. These tools help to grip the load and reduce the need for bending.

Trucks and trolleys allow one person to transport loads between different locations. They can be inexpensive and are available in various sizes to suit the load and type of workplace. **Sack trucks,** like **wheelbarrows,** move loads by balancing them on the truck axle. Some trucks have lifting mechanisms

that allow loads to be raised and lowered and others are fitted with special wheels for climbing and descending stairs.

General purpose trucks can be used to support larger loads than with sack trucks. They can be flat-topped or fitted with a variety of sides and wheels to suit different uses. The wheels can be fitted with swivels to improve manoeuvrability. Some **platform trucks** are fitted with detachable tug units. This helps to reduce congestion and obstacles in busy production areas. The platform can be designed to be raised or lowered – this further reduces the manual handling because it reduces the need for bending when loading and unloading. Trucks can have big wheels for use on rough ground. When fitted with removable sides they can be used for bulk loads like sand and gravel. **Balance trucks** have a central axle with swivel wheels at each end so they can spin around their centres. This makes them highly manoeuvrable so they can be used in restricted spaces.

There is no clear distinction between general purpose trucks and trolleys. **Trolleys** tend to be of lighter weight construction and designed for more specific applications. Container trolleys allow mixed loads to be carried. A supermarket trolley is a simple example of this. Shelf trolleys can have fixed or removable shelves. Drum trolleys are useful for transporting drums. Brakes can be fitted to trucks and trolleys to keep them stationary and may be necessary where the ground slopes. Garment rails can be used in clothes factories, shops and theatres.

Roller tracks and chutes allow heavy and bulky loads to be moved manually or by gravity under their own weight. They can be portable or set into the floor. Carefully designed work areas with appropriate tracks and chutes can reduce the number of manual handling tasks. When using tracks, the potential for creating a tripping hazard must be kept as low as possible. Using gentle gradients can ensure that the load is moved with very little effort. Chutes are normally used instead of tracks when there are significant changes of level, for example movement between floors. Spiral chutes are typically used for sacks but almost any kind of load can be handled on a straight chute.

Pallet trucks (Figure 7.9) are moved by pedestrians. Manual effort is required to transfer the load but hydraulic power is normally used to raise and lower the load. They can be used in fairly congested and confined areas and are designed to move different types of loads.

Portable conveyors are used to transport loads between places at the same level or different heights. Different types can transport a wide variety of loads including bulk materials like sand and grain. Using portable conveyors can significantly reduce (and often avoid) manual handling. The height of the conveyor at loading and unloading points is important. Recommended heights to reduce manual handling efforts are:

▶ around 0.7 m above floor level when light loads which can normally be lifted with one hand are to be lifted;

▶ around 0.75 m above floor level when heavier loads, normally lifted with two hands, are loaded; and

▶ around floor level when heavy loads, like drums, are loaded onto conveyors, either by rolling or by lifting devices or lift trucks.

Figure 7.9 A pallet truck

Figure 7.10 Mechanical aids to lift patients in hospital

Portable conveyors for occasional or semi-permanent occasions can be used in different work areas. Some conveyors are fitted with wheels to increase their mobility. They can be used to move loads at the same level or at different levels. Some types can be inclined by hydraulic rams, to adjust the height of transfer to suit a particular elevation. Non-powered roller conveyors on a slight incline allow loads to move under gravity. Powered conveyers and elevators are covered in Section 7.3.2.

Mechanical assistance, therefore, involves the use of mechanical aids to assist the manual handling operation such as wheelbarrows, hand-powered hydraulic hoists, specially adapted trolleys, hoists for lifting patients, roller conveyors and automated systems using robots.

Some examples of manually operated load handling equipment are shown in Appendix 7.2.

Hazards associated with manually operated load handling equipment

The principal hazards with the use of load handling equipment are due to its incorrect use, such as overloading or attempting to carry unstable loads. Pushing or pulling a truck or trolley is still a manual handling operation and can create different kinds of risks. It may be necessary, for example, to segregate pedestrians from vehicles.

Lack of maintenance can also create additional hazards. This is particularly important for certain lifting machines and tackle where there may be a legal requirement for a regular examination by a competent person (see 7.3.3).

Precautions with the use of manually operated load handling equipment

When handling aids are being selected, the subsequent user should be consulted whenever it is possible. Moving and handling tasks are often made easier by good design. For example, in the health and social care sector, workers are often required to care for service users in bed and the provision of adjustable height beds can prevent the risks of back injuries. Raising the height of laundry equipment, such as washing machines or driers, by placing it on a platform can also reduce bending and stooping. Further measures to reduce the amount of handling by care staff are:

▶ raising the height of beds and chairs using wooden blocks;

▶ handrails at strategic heights adjacent to the bath and toilet;

▶ the use of bath hoists and overhead hoists;

▶ walk-in showers with seats; and

▶ cutting slots in bath panels to allow for mobile hoist wheels to fit underneath a bath fixed against a wall.

All operators of mechanical handling equipment must be properly trained in its use and supervised while they are using it. These precautions are particularly important when people are being lifted or aided using a manual hoist. Other precautions include:

▶ a safe system of work in place;
▶ sufficient room available to easily manoeuvre the equipment;
▶ adequate visibility and lighting available;
▶ floors in a stable condition;
▶ regular safety checks to identify any faults with the equipment; and
▶ a regular maintenance schedule for the equipment.

Other important considerations are that:

▶ the proposed use will be within the safe working load of the equipment;
▶ it is suitable for the area in which it will operate (e.g. is there enough room to manoeuvre and enough headroom?);
▶ it suits the terrain in terms of stability and ground surface;
▶ the lifting equipment is CE-marked; and
▶ advice is sought from the suppliers/hirers on its suitability for the proposed task and any maintenance requirements.

A useful precaution is to request the equipment on a trial basis, if possible, to check that it is suitable for the required task and to involve the employees who will be expected to use it in this appraisal exercise. Consideration should also be given to other risks associated with the introduction of the lifting aid, such as site safety and the lack of communication with fork-lift truck drivers.

7.3.2 Mechanically operated load handling equipment

1. Safety in the use of lifting and moving equipment

An amendment to the Lifting Operations and Lifting Equipment Regulations (LOLER) included the positioning and installation of lifting equipment and the organisation of lifting operations.

Positioning and installation of lifting equipment

Lifting equipment must be positioned and installed so as to reduce the risks, so far as is reasonably practicable, from:

▶ equipment or a load striking a person;
▶ a load drifting, falling freely or being released unintentionally.

Lifting equipment should be positioned and installed to minimise the need to lift loads over people and to prevent crushing in extreme positions. It should be designed to stop safely in the event of a power failure and not release its load. Lifting equipment, which follows a fixed path, should be enclosed with suitable and substantial interlocked gates and any necessary protection in the event of power failure.

The organisation of lifting operations

Every lifting operation, that is lifting or lowering of a load, shall be:

▶ properly planned by a competent person;
▶ appropriately supervised;
▶ carried out in a safe manner.

The person planning the operation should have adequate practical and theoretical knowledge and experience of planning lifting operations. The plan needs to address the risks identified by the risk assessment and identify the resources, the procedures and the responsibilities required so that any lifting operation is carried out safely. For routine simple lifts, a plan will normally be the responsibility of the people using the lifting equipment. For complex lifting operations, for example where two cranes are used to lift one load, a written plan may need to be produced each time.

The planning should include: the need to avoid suspending loads over occupied areas, visibility, the attaching/detaching and securing of loads, the environment, the location, the possibility of overturning, the proximity to other objects, any lifting of people and the pre-use checks required for the equipment.

In 2010, the Arthritis and Musculoskeletal Alliance launched a charter to help people with MSDs to remain in work. The charter calls for early diagnosis and treatment of MSDs and training of line managers so that they can identify potential problem areas.

Summary of the requirements for lifting operations

There are four general requirements for all lifting operations:

▶ Use strong, stable and suitable lifting equipment.
▶ The equipment should be positioned and installed correctly.
▶ The equipment should be visibly marked with the safe working load (SWL).
▶ Lifting operations must be planned, supervised and performed in a safe manner by competent people.

2. Types of mechanical handling and lifting equipment

There are four elements to mechanical handling, each of which can present hazards. These are handling equipment, the load, the workplace and the employees involved.

The **mechanical handling equipment** must be capable of lifting and/or moving the load. It must be fault-free, well maintained and inspected on a regular basis. The hazards related to such equipment include collisions between people and the equipment and personal injury from being trapped in moving parts of the equipment (such as belt and screw conveyors).

The **load** should be prepared for transportation in such a way as to minimise the possibility of accidents. The hazards will be related to the nature of the load (e.g. substances which are flammable or hazardous to health) or the security and stability of the load (e.g. collapse of bales or incorrectly stacked pallets).

The **workplace** should be designed so that, whenever possible, workers and the load are kept apart. If, for example, an overhead crane is to be used, then people should be segregated away or barred from the path of the load.

The **employees** and any other people who are to use the equipment must be properly trained and competent in its safe use.

Conveyors and elevators

Conveyors transport loads along a given level which may not be completely horizontal, whereas elevators move loads from one level or floor to another.

(a)

(b)

(c)

Figure 7.11 Conveyor systems: (a) roller conveyor (these may have powered and free running rollers); (b) an overhead conveyor moving wheels. Other designs of overhead conveyor are useful for transferring components and garments between workstations, in, for example, manufacture of machines or clothing; (c) a slat conveyor in use in a food factory

There are three common forms of **conveyor** – belt, roller and screw conveyors. The most common hazards and preventative measures are:

▶ the in-running nip, where a hand is trapped between the rotating rollers and the belt. Protection from this hazard can be provided by nip guards and trip devices;

▶ entanglement with the power drive requiring the fitting of fixed guards and the restriction of loose clothing which could become caught in the drive;

▶ loads falling from the conveyor. This can be avoided by edge guards and barriers;

▶ impact against overhead systems. Protection against this hazard may be given by the use of bump caps, warning signs and restricted access;

▶ contact hazards prevented by the removal of sharp edges, conveyor edge protection and restricted access;

▶ manual handling hazards;

▶ noise and vibration hazards.

Screw conveyors, often used to move very viscous substances, must be provided with either fixed guards or covers to prevent accidental access. People should be prohibited from riding on belt conveyors, and emergency trip wires or stop buttons must be fitted and be operational at all times.

Elevators are used to transport goods between floors, such as the transportation of building bricks to upper storeys during the construction of a building (Figure 7.12) or the transportation of grain sacks into the loft of a barn. Guards should be fitted at either end of the elevator and around the power drive. The most common hazard is injury due to loads falling from elevators. There are also potential manual handling problems at both the feed and discharge ends of the elevator.

Fork-lift trucks

The most common form of mobile handling equipment is the fork-lift truck (shown in Figure 7.13).

It comes from the group of vehicles, known as lift trucks, and can be used in factories, on construction sites and on farms. The term fork-lift truck is normally applied to the counterbalanced lift truck, where the load on the forks is counterbalanced by the weight of the vehicle over the rear wheels. The reach truck is designed to operate in narrower aisles in warehouses and enables the load to be retracted within the wheelbase. The very narrow aisle (VNA) truck does not turn within the aisle to deposit or retrieve a load. It is often guided by guides or rails on the floor. Other forms of lift truck include the pallet truck and the pallet stacker truck, both of which may be pedestrian or rider controlled. Around 400 workers are seriously injured in fork-lift truck accidents every year and some of these accidents cause about 10 fatalities annually. Accidents frequently occur when fork-lift truck drivers or pedestrians in warehouses are distracted, or simply assume that they have been seen by the other party.

Figure 7.12 A brick elevator

Figure 7.13 Reach truck, designed so that the load retracts within the wheelbase to save space

There are many hazards associated with the use of fork-lift trucks. These include:

▶ overturning – manoeuvring at too high a speed (particularly cornering); wheels hitting an obstruction such as a kerb; sudden braking; poor tyre condition leading to skidding; driving forwards down a ramp; movement of the load; insecure, excessive or uneven loading; incorrect tilt or driving along a ramp;

▶ overloading – exceeding the rated capacity of the machine;

▶ collisions – particularly with warehouse racking which can lead to a collapse of the whole racking system;

▶ silent operation of the electrically powered fork-lift truck – can make pedestrians unaware of its presence;

▶ uneven road surface – can cause the vehicle to overturn and/or cause musculoskeletal problems for the driver;

▶ overhead obstructions – a particular problem for inexperienced drivers;

▶ loss of load – shrink wrapping or sheeting will reduce this hazard;

▶ inadequate maintenance leading to mechanical failure;

▶ use as a work platform;

▶ speeding – strict enforcement of speed limits is essential;

▶ poor vision around the load;

▶ pedestrians – particularly when pedestrians and vehicles use the same roadways. Warning signs, indicating the presence of fork-lift trucks, should be posted at regular intervals;

▶ dangerous stacking or de-stacking technique – this can destabilise a complete racking column;

▶ carrying passengers – this should be a disciplinary offence;

▶ battery charging – presents an explosion and fire risk;

▶ fire – often caused by poor maintenance resulting in fuel leakages or engine/motor burnout, or through using an unsuitable fork-lift truck in areas where flammable liquids or gases are used and stored;

▶ lack of driver training.

If fork-lift trucks are to be used outside, visibility and lighting, weather conditions and the movement of other vehicles become additional hazards.

There are also the following physical hazards:

▶ noise – caused by poor silencing of the power unit;

▶ exhaust fumes – should only be a problem when the maintenance regime is poor;

▶ vibrations – often caused by a rough road surface or wide expansion joints. Badly inflated tyres will exacerbate this problem;

▶ manual handling – resulting from manoeuvring the load by hand or lifting batteries or gas cylinders;

▶ ergonomic – musculoskeletal injuries caused by soft tyres and/or undulating road surface or holes or cracks in the road surface (e.g. expansion joints).

Regular and documented maintenance by competent mechanics is essential. However, the driver should undertake the following checks at the beginning of each shift:

▶ condition of tyres and correct tyre pressures;

▶ effectiveness of all brakes;

▶ audible reversing horn and light working properly;

▶ lights, if fitted, working correctly;

▶ mirrors, if fitted, in good working order and properly set;

▶ secure and properly adjusted seat;

▶ correct fluid levels, when appropriate;

▶ fully charged batteries, when appropriate;

▶ correct working of all lifting and tilting systems.

A more detailed inspection should be undertaken by a competent person within the organisation on a weekly basis to include the mast and the steering gear. Driver training is essential and should be given by a competent trainer. The training session must include the site rules covering items such as the fork-lift truck driver code of practice for the organisation, speed limits, stacking procedures and reversing rules. Refresher training should be provided at regular intervals and a detailed record kept of all training received. Table 7.1 illustrates some key requirements of fork-lift truck drivers and the points listed should be included in most codes of practice. Appendix 7.3 gives advice from the HSE on the safe use of fork-lift trucks.

Finally, care must be taken with the selection of drivers, including relevant health checks and previous experience. Drivers should be at least 18 years of age and their fitness to drive should be re-assessed regularly (every five years after the age of 40 and every year after 65 (HSG6)).

Table 7.1 Safe driving of lift trucks

Drivers must:
▶ drive at a suitable speed to suit road conditions and visibility
▶ use the horn when necessary (at blind corners and doorways)
▶ always be aware of pedestrians and other vehicles
▶ take special care when reversing (do not rely on mirrors)
▶ take special care when handling loads which restrict visibility
▶ travel with the forks (or other equipment fitted to the mast) lowered
▶ use the prescribed lanes
▶ obey the speed limits
▶ take special care on wet and uneven surfaces
▶ use the handbrake, tilt and other controls correctly
▶ take special care on ramps
▶ always leave the truck in a state which is safe and discourages/prevents unauthorised use (brake on, motor off, forks down, key out).
Drivers must not:
▶ operate in conditions in which it is not possible to drive and handle loads safely (e.g. partially blocked aisles)
▶ travel with the forks raised
▶ use the forks to raise or lower persons unless a purpose-built working cage is used
▶ carry passengers
▶ park in an unsafe place (e.g. obstructing emergency exits)
▶ turn round on ramps
▶ drive into areas where the truck would cause a hazard (flammable substance store)
▶ allow unauthorised use.

Other forms of lifting equipment

The other types of lifting equipment to be considered are cranes (mobile overhead and jib), lifts and hoists and lifting tackle. A sample risk assessment for the use of lifting equipment is given in Appendix 7.3.

The lifting operation should be properly prepared and planned. This involves the selection of a suitable crane having up-to-date test certificates and examination reports that have been checked. A risk assessment of the task will be needed which would ascertain the weight, size and shape of the load and its final resting place. A written plan for completing the lift should be drafted and a competent person appointed to supervise the operation.

Lift plans

A lift plan identifies the ways in which the risks involved in a lifting operation can be eliminated or controlled. The degree of planning for a lifting operation should be proportional to the risk and will vary considerably depending upon the complexity of the lifting operation. The complexity will depend on the load to be lifted, the equipment to be used and the environment in which the operation is to be undertaken.

Well-planned lifting operations are a combination of two parts:

1. initial planning to ensure that the lifting equipment provided is suitable for the range of tasks that it will have to undertake; and
2. the planning of individual lifting operations so that they can be performed safely with the lifting equipment provided.

The balance between the two parts of the planning process will vary depending upon the lifting equipment and the particular lifting operation.

The person planning the operation should have adequate practical and theoretical knowledge and experience of planning lifting operations. The plan will need to address the risks identified by the risk assessment and identify the resources, the procedures and the responsibilities required so that any lifting operation is carried out safely. For routine simple lifts a plan will normally be left to the people using the lifting equipment. For complex lifting operations, for example where two cranes are used to lift one load, a written plan may need to be produced each time.

The planning should take account of avoiding suspending loads over occupied areas, visibility, attaching/detaching and securing loads, the environment, location, overturning, proximity to other objects, lifting of people and pre-use checks of the equipment.

Cranes

Cranes may be either a **jib crane** or an **overhead gantry travelling crane**. The safety requirements are similar for each type. All cranes need to be properly designed, constructed, installed and maintained. They must also be operated in accordance with a safe system of work. They should only be driven by authorised persons who are fit and trained. Each crane is issued with a certificate by its manufacturer giving details of the **safe working load (SWL)**. The SWL must **never** be exceeded and should be marked on the crane structure. If the SWL is variable, as with a jib crane (the SWL decreases as the operating radius increases), an SWL indicator should be fitted. Care should be taken to avoid sudden shock loading, as this will impose very high stresses on the crane structure. It is also very important that the load is properly shackled and all eyebolts tightened. Safe slinging should be included in any training programme. All controls should be clearly marked and be of the 'hold-to-run' type.

Figure 7.14 Manoeuvring a yacht using a large overhead travelling gantry and slings in a marina

Large cranes, which incorporate a driving cab, often work in conjunction with a banksman, who will direct the lifting operation from the ground. Banksmen are operatives trained to direct vehicle movement on or around site. They are often called traffic marshals. Banksmen should only be used in circumstances where other control measures are not possible. It is important that banksmen are trained so that they understand recognised crane signals. Signallers are operatives who are trained to direct crane drivers during lifting operations. It is a hazardous job and measures will help to keep the signaller safe:

▶ Ensure that they are trained and competent to direct lifting operations.
▶ Provide a protected position from which they can work in safety.
▶ Provide distinctive 'hi vis' clothing for identification.
▶ Tell drivers that if they cannot see the signaller they should stop immediately.
▶ Agree on the use of standard signals.

The lifting operation should be properly prepared and planned. This involves the selection of a suitable crane having up-to-date test certificates and examination reports that have been checked. A risk assessment of the task will be needed which would ascertain the weight, size and shape of the load and its final resting place. A written plan for completing the lift should be drafted and a competent person appointed to supervise the operation. During the planning of a lifting operation close to a public highway, there should be liaison with the police and local authority over any road closures and the lift should be timed to take place when there will be a minimum number of people about. Arrangements may be needed to give advanced warning to motorists and other road users and appropriate signs and traffic controls used.

The basic principles for the safe operation of cranes are as follows. **For all cranes,** the driver must:

▶ undertake a brief inspection of the crane and associated lifting tackle each time before it is used;
▶ check that all lifting accessory statutory inspections are in place and up to date;
▶ check that tyre pressures, where appropriate, are correct;
▶ ensure that loads are not left suspended when the crane is not in use;
▶ before a lift is made, ensure that nobody can be struck by the crane or the load;
▶ ensure that loads are never carried over people;
▶ ensure good visibility and communications;
▶ lift loads vertically – cranes must not be used to drag a load;
▶ travel with the load as close to the ground as possible;
▶ switch off power to the crane when it is left unattended.

For **mobile jib cranes**, the following points should be considered:

▶ Each lift must be properly planned, with the maximum load and radius of operation known.
▶ Overhead obstructions or hazards must be identified; it may be necessary to protect the crane from overhead power lines by using goal posts and bunting to mark the safe headroom.
▶ The ground on which the crane is to stand should be assessed for its load-bearing capacity.
▶ If fitted, outriggers should be used.

The principal reasons for crane failure, including loss of load, are:

▶ overloading;
▶ poor slinging of load;
▶ insecure or unbalanced load;
▶ overturning;
▶ collision with another structure or overhead power lines;
▶ foundation failure;
▶ structural failure of the crane;
▶ operator error;

▶ lack of maintenance and/or regular inspections;
▶ no signaller used when driver's view is obscured;
▶ incorrect signals given.

Tower cranes

Typical causes of recent serious incidents with tower cranes include:

▶ mechanical failure of the brake or lifting ram;
▶ overturn of the crane;
▶ jib collapse;
▶ a load or dropped load striking a worker;
▶ sling failure.

The reasons for some of these incidents are:

▶ poor site induction training – not dealing with site-specific risks and lasting too long (20–30 minutes maximum is sufficient time);
▶ problems with crane maintenance and thorough examinations;
▶ operators working long hours without a break;
▶ poor operator cabin design and too high a climbing distance;
▶ operator health problems;
▶ problems in communicating health and safety issues by crane operators on site.

During lifting operations using cranes, it must be ensured that:

▶ the driver has good visibility;
▶ there are no pedestrians below the load by using barriers, if necessary;
▶ there is adequate space for the installation, manoeuvring and operation of the crane;
▶ all pedestrians are re-routed during the lifting operation;
▶ everyone within the lifting area is wearing a safety helmet; and
▶ an audible warning is given prior to the lifting operation.

If lifting takes place in windy conditions, tag lines may need to be attached to the load to control its movement.

Tower cranes are being used more and more often on construction sites across Europe. Approximately 1,800 tower cranes have operated in the UK. Since 2001 in the UK, there have been 61 accidents involving tower cranes – 7 people have died, and 25 have been seriously injured (e.g. in Canary Wharf, Worthing, Battersea and Liverpool). These accidents led to public concern over tower crane safety and to the introduction by the HSE of the Notification of Conventional Tower Cranes Regulations in 2010. The Regulations required employers who had primary responsibility for the safety of cranes to notify certain information to HSE. This included name and address of the crane owner and the site address, as well as sufficient information to identify the crane and the date of its last thorough examination.

The Löfstedt report into health and safety in 2011 noted that,

'when these Regulations were put into place, the Impact Assessment (IA) carried out then did not predict that they would have "direct health and safety benefits, (i.e. reductions in injury or ill-health"), but that the main benefit of a tower crane register would be "an increase in public assurance". It was not clear that a statutory requirement to register tower cranes is the most appropriate way to provide public assurance and suggested that non-regulatory methods should be explored.'

The findings of the IA suggested that the revocation of these Regulations would:

▶ have little effect on public perception of risk to their safety arising from the use of tower cranes;
▶ not reduce safety standards required when erecting, using or dismantling tower cranes on construction sites; and
▶ remove a source of cost and burden on the industry.

The key issues for the safe operation of all cranes including tower cranes are:

1. The planning of safe lifting operations by a competent person following a detailed risk assessment;
2. Safe systems of work for the installation, operation and dismantling stages. The main elements of the safe system or method statement are:
   ▶ planning – including site preparation, crane erection and dismantling;
   ▶ selection, provision and use of a suitable crane and work equipment; including safe slinging and signalling arrangements;
   ▶ maintenance and examination of the crane and equipment;
   ▶ provision of properly trained and competent personnel;
   ▶ supervision of operations by personnel having the necessary authority;
   ▶ thorough examinations, reports and other documents;
   ▶ preventing unauthorised movement or use of the crane; and
   ▶ measures to secure safety of persons not involved in the lifting;
3. Supervision of all lifting operations; and
4. The thorough examination of the crane and, when required, the notification of the HSE.

Recommendations were made by a Construction Industry Forum that site-specific induction training should be given to everybody involved with tower crane erection, operation and dismantling. A specific risk assessment and a regular inspection of the crane should be undertaken and only competent and trained operators should be involved with tower crane activities.

A **lift or hoist** incorporates a platform or cage and is restricted in its movement by guides. Hoists are generally used in industrial settings (e.g. construction sites and garages), whereas lifts are normally used inside buildings. Lifts and hoists may be designed to carry passengers and/or goods alone. They should be of sound mechanical construction and have interlocking doors or gates, which must be completely closed before the lift or hoist moves. The hoist-way should be properly enclosed so that the moving parts of the hoist are guarded. The loads should be secured on the hoist platform so that they cannot fall and the operator should have a clear view of the landing levels. There should be no unauthorised use of the hoist by untrained personnel. Passenger-carrying lifts must be fitted with an automatic braking system to prevent overrunning, at least two suspension ropes, each capable alone of supporting the maximum working load, and a safety device which could support the lift in the event of suspension rope failure. Maintenance procedures must be rigorous, recorded and only undertaken by competent persons. It is very important that a safe system of work is employed during maintenance operations to protect others, such as members of the public, from falling down the lift shaft and other hazards.

Figure 7.15 Hoist for lifting a car

Other **items of lifting tackle,** usually used with cranes, include chain slings and hooks, wire and fibre rope slings, eyebolts and shackles. Special care should be taken when slings are used, to ensure that the load is properly secured and balanced. Lifting hooks should be checked for signs of wear and any distortion of the hook. Shackles and eyebolts must be correctly tightened. Slings should always be checked for any damage before they are used and only competent people should use them. Training and instruction in the use of lifting tackle is essential and should include regular inspections of the tackle, in addition to the mandatory thorough examinations. Finally, care should be taken when these items are being stored between uses.

A hook designed to prevent displacement of the load.

A hook with a safety catch.

Figure 7.16 Specially designed safety hooks

Passenger lifts

The Lifting Operations and Lifting Equipment Regulations (LOLER) and the ACoP L113 set out precise requirements for inspection and use of passenger lifts. These are:

▶ Regulation 7 requires passenger lifts to be thoroughly examined for defects before first use and every six months thereafter (12 months for goods-only lifts). Further examinations may be required in accordance with a written scheme, or if an incident occurs that may affect the safe operation of the equipment. This regulation also requires equipment to be inspected at suitable intervals between thorough examinations.

▶ Regulation 10 requires thorough-examiners to notify defects that are a danger to persons to both the employer (duty-holder) and the enforcing authority, and make a report of their inspection in writing as soon as is practicable. This same regulation requires the employer to ensure equipment is, as identified in the report, either not used until a defect is repaired, or is repaired within a specified time frame.

▶ Regulation 11 requires thorough examination reports to be retained for at least two years.

The following issues are relevant to the safe operation of passenger lifts;

▶ An emergency mushroom stop button should be fitted in the lift.

▶ A direct telephone line is required if the internal alarm sounder/siren cannot be heard by rescue staff in the building.

▶ A risk assessment for possible injuries to the lift passengers should be undertaken on the design, sensitivity and performance of the door edges. The lift inspector should have the relevant standard to help with this.

▶ A lift stop switch should be available in the pit as well as the motor room at the top of the lift. Provision and Use of Work Equipment Regulations (PUWER) requires that this switch must be 'readily available'.

▶ The lift beam should be inspected by the lift inspector – the safe working load should be stated on it to prevent overload.

▶ If there are any moving or revolving mechanisms accessible by lift technicians performing their maintenance tasks, then they should be protected from inadvertent contact by suitable guards.

▶ It is advisable to locate a fire extinguisher inside the access door of the motor room although this is not a statutory requirement.

▶ Emergency door release devices must not be routinely operated during non-emergency situations. The emergency keys are intended to allow emergency access to the lifting platform in the event of people becoming trapped and should be under strict control.

▶ Emergency unlocking should be undertaken only in exceptional circumstances and by suitably trained and authorised people.

▶ Safe working procedures and arrangements should be in place setting out what to do in the event of an emergency or failure. For example, how to deal with trapped people and the arrangements for repairing faults.

▶ All lifts and lifting platforms must be inspected, serviced and maintained. Where the lift is used for work purposes, it must be thoroughly examined by a competent person.

7.3.3 Requirements for the statutory examination of lifting equipment

LOLER specify examinations required for lifting equipment by using two terms – an inspection and a thorough examination. Both terms are defined by the HSE in guidance accompanying various Regulations.

Lifting equipment includes any equipment used at work to lift or lower loads including any anchoring, fixing or supporting attachments.

An **inspection** is used to identify whether the equipment can be operated, adjusted and maintained safely so that any defect, damage or wear can be detected before it results in unacceptable risks. It is normally performed by a competent person appointed by the employer (often an employee).

Figure 7.17 Truck-mounted lifting equipment

A **thorough examination** is a detailed examination, which may involve a visual check, a disassembly and testing of components and/or an equipment test under operating conditions. Such an examination must normally be carried out by a competent person who is independent of the employer. The examination is usually carried out according to a written scheme and a written report is submitted to the employer.

The Code of Practice for the Safe Use of Lifting Equipment defines a thorough examination as a visual examination carried out by a competent person carefully and critically and, where appropriate, supplemented by other means, such as measurement and testing, in order to check whether the equipment is safe to use.

A detailed summary of the thorough examination and inspection requirements of the LOLER is given in Chapter 13.

A thorough examination of lifting equipment should be undertaken at the following times:

▶ before the equipment is used for the first time;
▶ after it has been assembled at a new location;
▶ at least every 6 months for equipment used for lifting persons or as a lifting accessory;
▶ at least every 12 months for all other lifting equipment including the lifting of loads over people;
▶ in accordance with a particular examination scheme drawn up by an independent competent person;
▶ each time that exceptional circumstances, which are likely to jeopardise the safety of the lifting equipment, have occurred (such as severe weather).

The person making the thorough examination of lifting equipment must:

▶ notify the employer forthwith of any defect which, in their opinion, is or could become dangerous;

▶ as soon as practicable (within 28 days) write an authenticated report to the employer and any person who leased or hired the equipment.

The Regulations specify the information that should be included in the report.

The initial report should be kept for as long as the lifting equipment is used (except for a lifting accessory which need only be kept for two years). For all other examinations, a copy of the report should be kept until the next thorough examination is made or for two years (whichever is the longer). If the report shows that a defect exists which could lead to an existing or imminent risk of serious personal injury, a copy of the report must be sent, by the person making the thorough examination, to the appropriate enforcing authority.

The equipment should be inspected at suitable intervals between thorough examinations. The frequency and the extent of the inspection are determined by the level of risk presented by the lifting equipment. A report or record should be made of the inspection which should be kept until the next inspection. Unless stated otherwise, lifts and hoists should be inspected every week.

It is important to stress that thorough examinations must be accompanied by meticulous in-service inspection that will detect any damage every time the equipment is used. For most workplaces, it is better to store every item of portable lifting equipment in a central store where these inspections can take place and records of the equipment be kept. Users must also be encouraged to report any defects in equipment that they have used.

7.4 Further information

Statutory provisions

The Health and Safety (Display Screen Equipment) Regulations 1992 (as amended)

The Lifting Operations and Lifting Equipment Regulations 1998

The Manual Handling Operations Regulations 1992 (as amended)

The Provision and Use of Work Equipment Regulations 1998

Other references

Manual Handling, Manual Handling Operations Regulations 1992 (as amended), Guidance on Regulations, L23, third edition 2016, HSE Books, ISBN: 778-0-7176-66539 www.hse.gov.uk/pubns/priced/l23.pdf

Rider-operated lift trucks L117 (third edition) 2013 ISBN 9780717664412 https://www.hse.gov.uk/pUbns/priced/l117.pdf

Safe use of lifting equipment, Lifting Operations and Lifting Equipment Regulations 1998, ACoP and Guidance, L113, HSE Books, 2014 ISBN: 9780717665860 www.hse.gov.uk/pubns/priced/l113.pdf

Safe use of work equipment, Provision and Use of Work Equipment Regulations, ACoP and Guidance, L22, fourth edition, 2014 HSE Books ISBN: 9780717666195 www.hse.gov.uk/pubns/priced/l22.pdf

The law on VDUs – An Easy Guide, HSG90, second edition 2003, HSE Books, ISBN: 9780717626021 www.hse.gov.uk/pubns/priced/hsg90.pdf

Ergonomics and human factors at work, INDG90(rev3), HSE Books https://www.hse.gov.uk/pubns/indg90.pdf

Work Related Upper Limb Disorders – A Guide, HSG60, second edition 2002, HSE Books, ISBN: 778-0-7176-1778-8 www.hse.gov.uk/pubns/priced/hsg60.pdf

Work with display screen equipment: Health and Safety (Display Screen Equipment) Regulations 1992 as amended by the Health and Safety (Miscellaneous Amendments) Regulations 2002: Guidance on Regulations, L26, 2003, HSE Books ISBN: 9780717625826 www.hse.gov.uk/pubns/priced/l26.pdf

7.5 Practice questions

1. (a) **Define** the term 'ergonomics'.
   (b) **Identify SIX** factors that need to be considered during an ergonomic assessment.

2. Work-related upper limb disorders (WRULDs) are responsible for many cases of work-related ill health:
   (a) **Identify TWO** examples of WRULDs.
   (b) **Identify THREE** work activities that may cause WRULDs and the typical symptoms that might be experienced by affected individuals.
   (c) **Describe** the measures that should be taken to minimise the risk to these individuals.

3. Employees who work for a publishing company are required to use display screen equipment (computers) for eight hours a day.
   (a) **Identify** the risks to the health and safety of these employees.
   (b) **Outline** the factors to consider when making an assessment of a display screen equipment (DSE) workstation.
   (c) **Outline** the control measures that may be taken to minimise the risks to these employees.

4. (a) **Identify FOUR** types of injury that may be caused by the incorrect manual handling of loads.
   (b) **Outline** a good lifting technique that should be adopted by a person required to lift a load from the ground to a work bench.
   (c) **Give TWO** examples of how a manual handling task might be avoided.

5. **Outline** the factors that may affect the risk from manual handling activities in relation to:
   (a) The task
   (b) The individual
   (c) The load
   (d) The environment.

6. Metal boxes, each weighing 5 kg, are lifted by workers from the ground to a conveyor belt for onward transmission to a paint shop.
   (a) **Outline** the issues that should be considered when undertaking a manual handling assessment of this task.
   (b) **Identify FOUR** factors associated with the worker that may affect the risk of injury when lifting the boxes.
   (c) **Outline** the measures that may be needed in order to reduce the risk of injury to the workers undertaking this task.

7. In order to minimise the risk of injury when undertaking a manual handling operation:
   (a) **Identify FOUR** types of manually operated load handling equipment that can be used to assist the manual handling operation;
   (b) **Outline FOUR** hazards and the corresponding precautions to be taken when using conveyor systems for moving materials within a workplace.

8. Fork-lift trucks are used extensively in busy warehouses.
   (a) **Identify EIGHT** general and **FOUR** physical hazards associated with their use.
   (b) **Outline** the checks that the driver should undertake at the beginning of a shift.
   (c) **Identify FOUR** rules that the driver should follow when a fork-lift truck is left unattended.
   (d) **Outline** the principal requirements for lifting operations.

9. A pedestrian has been injured after a collision with a fork-lift truck in a warehouse. **Outline** the possible **immediate** causes of the accident associated with:
   (a) the pedestrian
   (b) the way in which the vehicle was driven

(c) the workplace

(d) the vehicle.

10. (a) **Identify THREE** types of crane used for lifting operations.

(b) **Outline** the key issues for the safe management of all crane operations.

(c) **Identify EIGHT** reasons for accidents during the operation of cranes.

(d) **Identify EIGHT** requirements of crane drivers to ensure the safe operation of the crane.

(e) **Outline** the issues that could be included in a lifting plan.

11. An overhead gantry crane is used in a plastic manufacturing factory to transport heavy equipment between machines

(a) **Identify FOUR** reasons why loads may fall from the crane.

(b) **Outline** the precautions to be taken so that accidents to those working at ground level are prevented when gantry cranes are in use.

12. A mechanical hoist with specially designed lifting tackle is used to remove engines from vehicles in a motor repair shop.

(a) **Outline** the precautions to be taken to reduce the risk of injury to employees and others during the lifting operation.

(b) **Identify** the two types of safety inspection required for this equipment.

(c) **Identify** the factors relating to the condition of lifting accessories that could form part of a pre-use checklist.

13. The Lifting Operations and Lifting Equipment Regulations 1998 require the statutory inspection of lifting equipment

(a) **Identify** the occasions when a thorough examination of a fork-lift truck is required under these regulations.

(b) **Outline** the differences between an inspection and a thorough examination of lifting equipment when applied to these regulations.

(c) **Identify** the duties of the person making a thorough examination of lifting equipment.

7

APPENDIX 7.1 A typical risk assessment for the use of lifting equipment

INITIAL RISK ASSESSMENT	Use of Lifting Equipment		
SIGNIFICANT HAZARDS	Low	Medium	High
1. Unintentional release of load			✓
2. Unplanned movement of load	✓		
3. Damage to equipment	✓		
4. Crush injuries to personnel		✓	
5.			

ACTION ALREADY TAKEN TO REDUCE THE RISKS:

Compliance with:

Lifting Operations and Lifting Equipment Regulations (LOLER)

Safety Signs and Signals Regulations (SSSR). Provision and Use of Work Equipment Regulations (PUWER)

British Standard – Specification for flat woven webbing slings

BS – Guide to selection and use of lifting slings for multi-purposes

Planning:

Copies of statutory thorough examinations of lifting equipment will be kept on site. Before selection of lifting equipment, the above standards will be considered as well as the weight, size, shape and centre of gravity of the load. Lifting equipment is subject to the planned maintenance programme.

Physical:

All items of lifting equipment will be identified individually and stored so as to prevent physical damage or deterioration. Safe working loads of lifting equipment will be established before use. Packing will be used to protect slings from sharp edges on the load. All items of lifting equipment will be visually examined for signs of damage before use. Ensuring the eyes of strops are directly below the appliance hook and that tail ropes are fitted to larger loads will check swinging of the load. Banksmen will be used where the lifting equipment operator's vision is obstructed. Approved hand signals will be used.

Managerial/Supervisory:

Only lifting equipment that is in date for statutory examination will be used. Manufacturer's instructions will be checked to ensure that methods of sling attachment and slinging arrangements generally are correct.

Training:

Personnel involved in the slinging of loads and use of lifting equipment will be required to be trained to CITB or equivalent standard. Supervisors will be trained in the supervision of lifting operations.

Date of Assessment..........................Assessment made by........................

Risk Re-Assessment Date...................Site Manager's Comments:

APPENDIX 7.2 Examples of manually operated load handling equipment

(Source: HSE)

Lifting hooks

Paving slab and general purpose handler

A hydraulically lifted mobile belt conveyor

A chute being used to transfer sacks from a high to low level

A sack truck with star truck wheels for climbing and descending stairs

Platform truck

APPENDIX 7.3 Safe use of fork-lift trucks (based on an HSE document)

1. Operating

Always:

▶ carry out a pre-shift check of the lift truck;
▶ wear operator restraints, where fitted;
▶ look all around before moving off;
▶ look in the direction of travel;
▶ travel at a speed suitable for the location and the load you are carrying;
▶ travel with the forks lowered, but clear of the ground;
▶ watch out for pedestrians;
▶ watch out for obstructions;
▶ avoid sudden stops and violent braking;
▶ slow down at corners, doorways, and at every danger spot and sound the horn, where necessary;
▶ apply the parking brake when leaving the lift truck;
▶ face the lift truck and use steps and handholds when getting in or out.

Never:

▶ operate a lift truck unless you are trained and authorised to do so;
▶ use a lift truck or equipment you know is not working properly;
▶ operate controls from outside the cab, unless it is designed so you can do this;
▶ stand on or near the controls to reach the load or anything outside the cab;
▶ travel on uneven ground unless the lift truck is suitable for this;
▶ run over unprotected cables or flexible pipes;
▶ try to carry out repairs – leave this to a qualified maintenance engineer;
▶ operate a lift truck when under the influence of alcohol or drugs;
▶ use mobile phones or other hand-held devices while operating or travelling.

2. Loads

Always:

▶ observe floor loading limits – find out the weight of the laden lift truck;
▶ ensure there is adequate clearance for the lift truck and load, including overhead;
▶ lower loads at a safe speed;
▶ use suitable attachments for lifting unusual or wide loads and follow the manufacturer's instructions;
▶ ensure you are properly trained, competent and authorised to operate the lift truck with the attachment being used.

Never:

▶ lift loads greater than the capacity of the lift truck;
▶ move a load that appears unsuitable or unstable (including on a damaged pallet);
▶ travel with a bulky load which blocks your view;
▶ travel with a raised load, unless the lift truck is designed for this;
▶ use an attachment unless a competent person, an authorised dealer or manufacturer has derated the lift truck (reduced the actual capacity).

3. Slopes

Always:

▶ travel slowly when going down slopes;
▶ when carrying a load, ensure the forks face uphill when travelling up or down slopes;
▶ without a load, ensure the forks face downhill when travelling up or down slopes;
▶ adjust the tilt (where fitted) to suit the gradient and raise the forks to clear the ground.

Never:

▶ attempt to turn on or travel across a ramp or a slope;
▶ leave a lift truck on a gradient except in an emergency, in which case always chock the wheels.

4. People

Always:

▶ use a safe work method when using working platforms.

Never:

▶ lift a person on the forks, or on a pallet, or similar, balanced on the forks;
▶ carry passengers, unless the lift truck is designed for this;
▶ allow people to walk under raised forks or loads;
▶ pick up a load if someone is standing close to it.

5. When you have finished working

Always:

▶ park the lift truck on level ground, never on a slope;
▶ leave the lift truck with the mast tilted forwards and the forks fully lowered, with the tips on the floor;
▶ apply the parking brake, select neutral, switch off the engine and remove the key;
▶ return keys or other activating devices to their place of safe-keeping.

CHAPTER 8

Chemical and biological agents

The contents of this chapter are:

Chapter summary

This chapter covers the following topics in Element 7 of the NEBOSH NGC syllabus:

Hazardous substances

Assessment of health risks

Occupational exposure limits

Control measures

Specific agents

Learning outcomes

These are the learning outcomes covered by this chapter:

The learner will be able to:

1. Outline the forms of, the classification of, and the health risks from exposure to, hazardous substances
2. Describe what should be considered when undertaking an assessment of the health risks from substances commonly encountered in the workplace
3. Outline the use and limitations of occupational exposure limits including the purpose of long-term and short-term exposure limits
4. Outline control measures that should be used to reduce the risk of ill health from exposure to hazardous substances
5. Outline the hazards, risks and controls associated with specific agents

8.1 Forms of, classification of, and health risks from hazardous substances

8.1.1 Introduction

Occupational health is as important as occupational safety, but generally receives less attention from managers. Every year twice as many people suffer ill health that is caused or exacerbated by the workplace than suffer workplace injury. Although these illnesses do not usually kill people, they can lead to many years of discomfort and pain. Such illnesses include respiratory disease, hearing problems, asthmatic conditions and back pain. Furthermore, it has been estimated in the UK that 30% of all cancers probably have an occupational link – that linkage is known for certain in 8% of cancer cases. Occupational exposures to fumes, dusts and chemicals in various forms account for around 4,000 deaths a year and some 38,000 individuals suffer breathing or lung problems possibly caused by their work.

Work in the field of occupational health has been taking place for the last four centuries and possibly longer. The main reason for the relatively low profile for occupational health over the years has been the difficulty in linking the ill-health effect with the workplace cause. Many illnesses, such as asthma or back pain, can have a workplace cause but can also have other causes. Many of the advances in occupational health have been as a result of statistical and epidemiological studies (one well-known such study linked the incidence of lung cancer to cigarette smoking). While such studies are invaluable in the assessment of health risk, there is always an element of doubt when trying to link cause and effect. The measurement of gas and dust concentrations is also subject to doubt when a correlation is made between a measured sample and the workplace environment from which it was taken. Occupational health, unlike occupational safety, is generally more concerned with probabilities than certainties.

In this chapter, chemical and biological health hazards are considered – other forms of health hazard are covered in Chapters 6 and 7.

The chemical and biological health hazards described in this chapter are covered by the following health and safety regulations:

▶ Control of Substances Hazardous to Health Regulations;
▶ Control of Lead at Work Regulations;
▶ Control of Asbestos at Work Regulations.

Risk assessment is not only concerned with injuries in the workplace but also needs to consider the possibility of occupational ill health. Health risks fall into the following four categories:

▶ chemical (e.g. paint solvents, exhaust fumes);
▶ biological (e.g. bacteria, pathogens);
▶ physical (e.g. noise, vibrations);
▶ psychological (e.g. occupational stress).

8.1.2 Forms of chemical agent

Chemicals can be transported by a variety of agents and in a variety of forms. They are normally defined in the following ways.

Dusts are solid particles slightly heavier than air but often suspended in it for a period of time. The size of the particles ranges from about 0.4 μm (fine) to 10 μm (coarse). Dusts are created either by mechanical processes (e.g. grinding or pulverising), construction processes (e.g. concrete laying, demolition or sanding) or by specific tasks (e.g. furnace ash removal). The fine dust is much more hazardous because it penetrates deep into the lungs and remains there – known as **respirable dust**. In rare cases, respirable dust enters the bloodstream directly causing damage to other organs. Examples of such fine dust are cement, granulated plastic materials and silica dust produced from stone or concrete dust. Repeated exposure may lead to permanent lung disease. Any dusts which are capable of entering the nose and mouth during breathing are known as **inhalable dust**.

Fibres are threads or filaments that can occur naturally (e.g. asbestos) or be man-made such as glass-fibre, nylon and polyester. Man-made fibres are commonly used in insulation boards, blankets for the purpose of heat treatment, electrical insulation and in the reinforcement of plastic and cement. Fibres have a very high length to width ratio of at least 100 and many fibres are in the respirable range causing concern about the effects of exposure to many fibres – fibrosis of the lung and various cancers. Fibres with diameters in excess of 4 μm can cause irritation of the skin and eyes. Exposure to large concentration of such fibres may also cause irritation in the upper respiratory tract. Although the hazards and risks of asbestos and glass-fibre have been well documented, there are many other types of fibre in use. These include synthetic and semi-synthetic fibres (e.g. cellulosic fibres), a range of other non-organic fibres and whiskers and specialist technical fibres including various types of carbon fibres and nanofibres.

Gases are any substances present at a temperature above their boiling point. Steam is the gaseous form of water. Common gases include carbon monoxide, carbon dioxide, nitrogen and oxygen. Gases are absorbed into the bloodstream where they may be beneficial (oxygen) or harmful (carbon monoxide).

Vapours are substances which are at or very close to their boiling temperatures. They are gaseous in form. Many solvents, such as cleaning fluids, fall into this category. The vapours, if inhaled, enter the bloodstream and some can cause short-term effects (dizziness) and long-term effects (brain damage).

Liquids are substances which normally exist at a temperature between freezing (solid) and boiling (vapours and gases) points. They are sometimes referred to as fluids in health and safety legislation.

Mists are similar to vapours in that they exist at or near their boiling temperature but are closer to the liquid phase. This means that very small liquid droplets are suspended in the vapour. A mist is produced during a spraying process (such as paint spraying). Many industrially produced mists can be very damaging if inhaled, producing similar effects to vapours. It is possible for some mists to enter the body through the skin or by ingestion with food.

Fume is a collection of very small metallic particles (less than 1 μm) which have condensed from the gaseous state. They are most commonly generated by the welding process. The particles tend to be within the respirable range (approximately 0.4–1.0 μm) and can lead to long-term permanent lung damage. The exact nature of any harm depends on the metals used in the welding process and the duration of the exposure.

8.1.3 Forms of biological agent

As with chemicals, biological hazards may be transported by any of the following forms of agent.

Fungi are very small organisms, sometimes consisting of a single cell, and can appear as plantlike (e.g. mushrooms and yeast). Unlike plants, they cannot produce their own food but either live on dead organic matter, or on living animals or plants as parasites. Fungi reproduce by producing spores, which can cause allergic reactions when inhaled. The infections produced by fungi in humans may be mild, such as athlete's foot, or severe, such as ringworm. Many fungal infections can be treated with antibiotics.

Moulds are a particular group of very small fungi which, under damp conditions, will grow on surfaces such as walls, bread, cheese, leather and canvas. They can be beneficial (penicillin) or cause allergic reactions (asthma). Asthma attacks, athlete's foot and farmer's lung are all examples of fungal infections.

Bacteria are very small single-celled organisms which are much smaller than cells within the human body. They can live outside the body and be controlled and destroyed by antibiotic drugs. There is evidence that bacteria are becoming resistant to most antibiotics. This has been caused by the widespread misuse of antibiotics. It is important to note that not all bacteria are harmful to humans. Bacteria aid the digestion of food, and babies would not survive without their aid to break down the milk in their digestive system. Legionellosis, tuberculosis and tetanus are all bacterial diseases.

Viruses are minute, non-cellular organisms which can only reproduce within a host cell. They are very much smaller than bacteria and cannot be controlled by antibiotics. They appear in various shapes and are continually developing new strains. They are usually only defeated by the defence and healing mechanisms of the body. Drugs can be used to relieve

8

the symptoms of a viral attack but cannot cure it. The common cold is a viral infection as are hepatitis, AIDS (HIV) and influenza.

A number of diseases including bovine spongiform encephalopathy (BSE) in cattle and Creutzfeldt–Jakob disease (CJD) in humans are caused by another biological agent known as a prion. A **prion** is an infectious agent that is composed primarily of protein. Such agents induce existing substances, called polypeptides, in the host organism to take on a rogue form. All known prion diseases affect the structure of the brain or other neural tissue, are currently untreatable and are always fatal.

8.1.4 Difference between acute and chronic health effects

The effects on health of hazardous substances may be either acute or chronic.

Acute effects are of short duration and appear fairly rapidly, usually during or after a single or short-term exposure to a hazardous substance. Such effects may be severe and require hospital treatment but are usually reversible. Examples include asthma-type attacks, nausea and fainting.

Chronic effects develop over a period of time which may extend to many years. The word 'chronic' means 'with time' and should not be confused with 'severe' as its use in everyday speech often implies. Chronic health effects are produced from prolonged or repeated exposures to hazardous substances resulting in a gradual, latent and often irreversible illness, which may remain undiagnosed for many years. Many cancers and mental diseases fall into the chronic category. During the development stage of a chronic disease, the individual may experience no symptoms.

8.1.5 Classification of hazardous substances and their associated health risks

A hazardous substance is one which can cause ill health to people at work. Such substances may include those used directly in the work processes (glues and paints), those produced by work activities (welding fumes) or those which occur naturally (dust). Hazardous substances are classified according to the severity and type of hazard which they may present to people who may come into contact with them. The contact may occur while working or transporting the substances or might occur during a fire or accidental spillage. There are several classifications but here only the most common will be described.

Irritant is a non-corrosive substance which can cause skin (dermatitic) or lung (bronchial) inflammation after repeated contact. People who react in this way

to a particular substance are **sensitised** or **allergic** to that substance. In most cases, it is likely that the concentration of the irritant may be more significant than the exposure time. Many household substances, such as wood preservatives, bleaches and glues are irritants. Many chemicals used as solvents are also irritants (white spirit, toluene and acetone). Formaldehyde and ozone are other examples of irritants.

Corrosive substances are ones which will attack, normally by burning, living tissue. Usually strong acids or alkalis, examples include sulphuric acid and caustic soda. Many tough cleaning substances, such as kitchen oven cleaners, are corrosive as are many dishwasher crystals.

Harmful is the most commonly used classification and describes a substance which, if swallowed, inhaled or penetrates the skin, **may** pose limited health risks. These risks can usually be minimised or removed by following the instruction provided with the substance (e.g. by using personal protective equipment). There are many household substances which fall into this category including bitumen-based paints and paint brush restorers. Many chemical cleansers are categorised as harmful. It is very common for substances which are labelled as harmful also to be categorised as an irritant.

Toxic substances are ones which impede or prevent the function of one or more organs within the body, such as the kidney, liver and heart. A toxic substance is, therefore, a poisonous one. Lead, mercury, pesticides and the gas carbon monoxide are toxic substances. The effect on the health of a person exposed to a toxic substance depends on the concentration and toxicity of the substance, the frequency of the exposure and the effectiveness of the control measures in place. The state of health and age of the person and the route of entry into the body have influence on the effect of the toxic substance.

Very toxic substances are very hazardous to health when inhaled, swallowed or when they come in contact with the skin and could be fatal. Examples of such substances include prussic acid, nicotine and carbon monoxide.

Carcinogenic substances are ones which are known or suspected of promoting abnormal development of body cells to become cancers. Asbestos, hardwood dust, creosote and some mineral oils are carcinogenic. It is very important that the health and safety rules accompanying the substance are strictly followed.

Mutagenic substances are those which damage genetic material within cells, causing abnormal changes that can be passed from one generation to another. A **reproductive toxin** is a chemical which affects adversely the reproductive process including mutations and other effects on the unborn child. One group of such toxins are teratogens that can cause congenital abnormalities, which are also called birth defects.

Each of the classifications may be identified by a symbol and a symbolic letter – the most common of these are shown in Figure 8.1.

8.1.6 Hazard warnings and precautionary statements

The European Regulation on classification, labelling and packaging of substances and mixtures became law across the EU in 2009 and is known as either the CLP Regulation or simply as 'CLP'. It adopted the United Nations Globally Harmonised System on the classification and labelling of chemicals (GHS) across all EU states.

Worldwide, there are many different laws to identify hazardous chemicals (classification) and to communicate this information to users. This is often confusing as the same chemical can have different hazard descriptions in different countries. The UN brought together experts from different countries to create the GHS.

The aim of GHS is to have, throughout the world, the same:

▶ criteria for classifying chemicals according to their health, environmental and physical hazards;
▶ hazard communication requirements for labelling and safety data sheets.

(a)

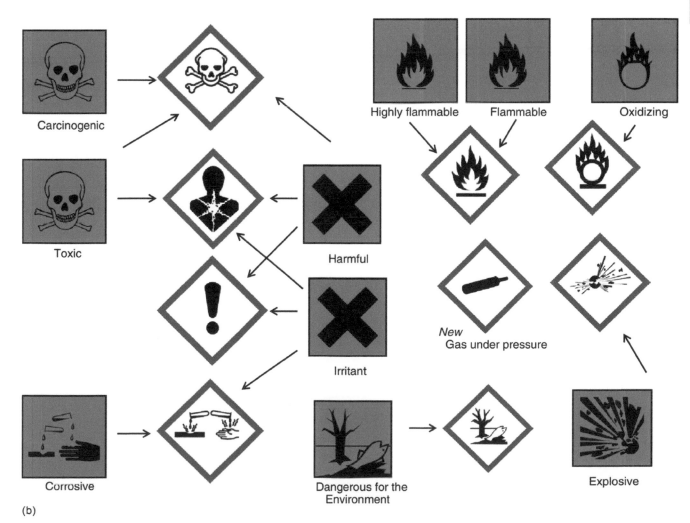

(b)

Figure 8.1 (a) GHS symbols in use and (b) how the European packaging symbols relate to the new GHS labels

GHS provides a basis for globally uniform physical, environmental, health and safety information on hazardous chemicals through the harmonisation of the criteria for their classification and labelling. It was developed at UN level with the aim of overcoming differing labelling information requirements on physical, health and environmental hazards for the same chemicals around the world. Moreover, it also aims to lower barriers to trade caused by the fact that every time a product was exported, it mostly had to be classified and labelled differently because of differing criteria.

Since GHS is a voluntary agreement, it has to be adopted through a suitable national or regional legal mechanism to ensure it becomes legally binding. That is achieved by the CLP Regulation. As GHS was heavily influenced by the old EU system, the CLP Regulation is very similar in many ways. The duties on suppliers are broadly the same: classification, labelling and packaging. The rules for the classification of hazardous substances have changes and a new set of hazard pictograms (quite similar to the old ones) are used (see Figure 8.1).

The EU legislation on classification, labelling and packaging consists of three acts: The Dangerous Substances Directive (Directive 67/548/EEC, DSD), the Dangerous Preparations Directive (Directive 1999/45/EC, DPD) and the Regulation on Classification, Labelling and Packaging of Substances and Mixtures, Regulation (EC) No. 1272/2008 (CLP Regulation or CLP) which applies directly in all member states.

Further information on the stage of implementation of the UN GHS in different countries is available on the UN ECE website, see: http://unece.org/trans/danger/publi/ghs/implementation_e.html.

Under the CLP Regulations there are:

▶ new scientific criteria to assess hazardous properties of chemicals;
▶ two new harmonised hazard warning symbols for labels (known as 'pictograms');
▶ a new design for existing symbols;
▶ new harmonised **hazard warning (H)** and **precautionary statements (P)** for labels; and
▶ two new **signal words** – 'Danger' and 'Warning'.

A **hazard statement** is a phrase that describes the nature of the hazard in the substance or mixture. A hazard statement will be determined by the application of the classification criteria. It replaces the 'risk or R-phrase' used previously. A list of some of these is given in Appendix 8.1.

A **precautionary statement** is a phrase that describes recommended measure(s) to minimise or prevent adverse effects resulting from exposure to a hazardous substance or mixture due to its use or disposal.

Suppliers determine the appropriate precautionary statements (usually no more than six) based on the required hazard statements. It replaces the 'safety or S-phrase' used previously.

The CLP Regulation also introduced two **signal words**: 'Danger' and 'Warning'. If the chemical has a more severe hazard, the label includes the signal word 'Danger'; in case of less severe hazards, the signal word is 'Warning'.

Examples of the new hazard statements and the new precautionary statements for labels are shown in Table 8.1.

Table 8.1 Examples of the new hazard warning (H) and precautionary statements (P)

Hazard warning statement (H)	Precautionary statement (P)
H240 – Heating may cause an explosion	P102 – Keep out of reach of children
H320 – Causes eye irritation	P271 – Use only outdoors or in well-ventilated area
H401 – Toxic to aquatic life	P410 – Protect from sunlight

8.2 Assessment of health risks

8.2.1 Types of health risk

The principles of control for health risks are the same as those for safety. However, the nature of health risks can make the link between work activities and employee ill health less obvious than in the case of injury from an accident.

Figure 8.2 Paint spraying – risk of sensitising particularly if isocyanate-based paint used and inadequate local exhaust ventilation

The COSHH Amendment Regulations 2004 set out the principles of good practice for the control of exposure. Figure 8.3 shows a route map for achieving adequate control.

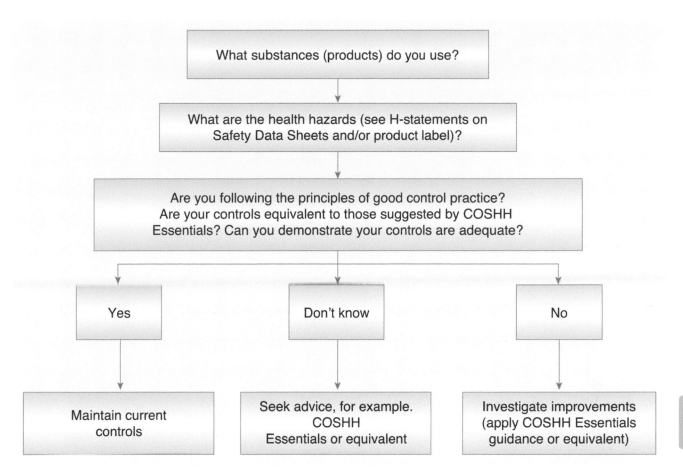

Figure 8.3 Route map for adequate control for SMEs non-experts

Unlike safety risks, which can lead to immediate injury, the result of daily exposure to health risks may not manifest itself for months, years and, in some cases, decades. Irreversible health damage may occur before any symptoms are apparent. It is, therefore, essential to develop a preventative strategy to identify and control risks before anyone is exposed to them.

Risks to health from work activities include:

▶ skin contact with irritant substances, leading to dermatitis, etc.;

▶ inhalation of respiratory sensitisers, triggering immune responses such as asthma;

▶ badly designed workstations requiring awkward body postures or repetitive movements, resulting in upper limb disorders, repetitive strain injury and other musculoskeletal conditions;

▶ noise levels which are too high, causing deafness and conditions such as tinnitus;

▶ too much vibration, for example from hand-held tools leading to hand-arm vibration syndrome and circulatory problems;

▶ exposure to ionising and non-ionising radiation including ultraviolet in the sun's rays, causing burns, sickness and skin cancer;

▶ infections ranging from minor sickness to life-threatening conditions, caused by inhaling or being contaminated with microbiological organisms;

▶ stress causing mental and physical disorders.

Some illnesses or conditions, such as asthma and back pain, have both occupational and non-occupational causes and it may be difficult to establish a definite causal link with a person's work activity or their exposure to particular agents or substances. But, if there is evidence that shows the illness or condition is prevalent among the type of workers to which the person belongs or among workers exposed to similar agents or substances, it is likely that their work and exposure has contributed in some way.

8.2.2 Routes of entry of hazardous substances into the body

There are three principal routes of entry of hazardous substances into the human body:

▶ **inhalation** – breathing in the substance with normal air intake. This is the main route of contaminants into the body. These contaminants may be chemical (e.g. solvents or welding fume) or biological (e.g. bacteria or fungi) and become airborne by a variety of modes, such as sweeping, spraying, grinding and bagging. They enter the lungs where they have access to the bloodstream and many other organs;

▶ **absorption through the skin** – the substance comes into contact with the skin and enters through either the pores or a wound. Tetanus can enter in this way as can toluene, benzene and various phenols;

▶ **ingestion** – through the mouth and swallowed into the stomach and the digestive system. This is not a significant route of entry to the body. The most common occurrences are due to airborne dust or poor personal hygiene (not washing hands before eating food).

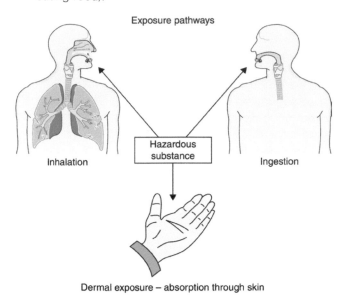

Figure 8.4 Hazardous substances – principal routes of entry into the human body

Another very rare entry route is by **injection** or **skin puncture**. The abuse of compressed air lines by shooting high pressure air at the skin can lead to air bubbles entering the bloodstream. Accidents involving hypodermic syringes in a health or veterinary service setting are rare but illustrate this form of entry route.

Body cells have a natural defence system against illness caused by harmful invasive foreign particles and bacteria. White blood cells, known as phagocytes, give protection by destroying these harmful foreign particles, bacteria and dead cells. This protection is basically a scavenging action at the level of cells. Other defence mechanisms include the secretion of defensive substances; the prevention of excessive blood loss and the repair of damaged tissue.

The most effective control measures which can reduce the risk of infection from biological organisms are disinfection, proper disposal of clinical waste (including syringes), good personal hygiene and, where appropriate, personal protective equipment. Other measures include vermin control, water treatment and immunisation.

There are five major functional systems within the human body – respiratory, nervous, cardiovascular (blood), urinary and the skin.

The respiratory system

This comprises the lungs and associated organs (e.g. the nose). Air is breathed in through the nose, and passes through the trachea (windpipe) and the bronchi into the two lungs. Within the lungs, the air enters many smaller passageways (bronchioles) and thence to one of 300,000 terminal sacs called alveoli. The alveoli are approximately 0.1 mm across, although the entrance is much smaller. On arrival in the alveoli, there is a diffusion of oxygen into the bloodstream through blood capillaries and an effusion of carbon dioxide from the bloodstream. While soluble dust which enters the alveoli will be absorbed into the bloodstream, insoluble dust (respirable dust) will remain permanently, leading to possible chronic illness (Figure 8.5).

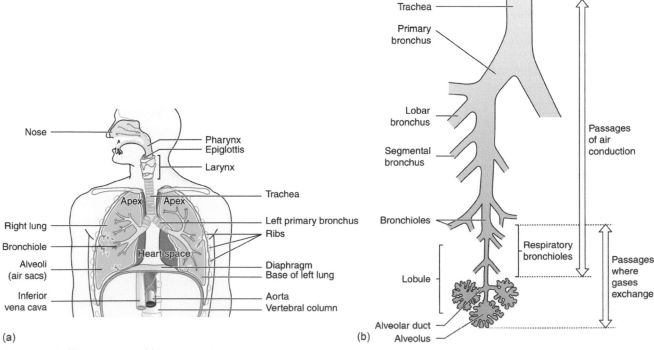

Figure 8.5 The upper and lower respiratory system

The whole of the bronchial system is lined with hairs, known as cilia. The cilia offer some protection against insoluble dusts. These hairs will arrest all non-respirable dust (above 5 μm) and, with the aid of mucus, pass the dust from one hair to a higher one and thus bring the dust back to the throat (this is known as the ciliary escalator). It has been shown that smoking damages this action. The nose will normally trap large particles (greater than 20 μm) before they enter the trachea. There are over 40 conditions that can affect the lungs and/or airways and impinge on the ability of a person to breathe normally.

Respirable dust tends to be long thin particles with sharp edges which puncture the alveoli walls. The puncture heals producing scar tissues which are less flexible than the original walls – this can lead to fibrosis. Such dusts include asbestos, coal, silica, some plastics and talc. The possible indicators of a dust problem in the workplace are fine deposits on surfaces, people and products or blocked filters on extraction equipment. Ill-health reports or complaints from the workforce could also indicate a dust problem.

Acute effects on the respiratory system include bronchitis and asthma and chronic effects include fibrosis and cancer. Hardwood dust, for example, can produce asthma attacks and nasal cancer.

Finally, asphyxiation, due to a lack of oxygen, is a problem in confined spaces particularly when MIG (metal inert gas) welding is taking place.

The nervous system

The nervous system consists primarily of the brain, the spinal cord and nerves extending throughout the body. Any muscle movement or sensation (e.g. hot and cold) is controlled or sensed by the brain through small electrical impulses transmitted through the spinal cord and nervous system.

The effectiveness of the nervous system can be reduced by **neurotoxins** and lead to changes in mental ability (loss of memory and anxiety), epilepsy and narcosis (dizziness and loss of consciousness). Organic solvents (trichloroethylene) and heavy metals (mercury) are well-known neurotoxins. The expression 'mad hatters' originated from the mental deterioration of top hat polishers in the 19th century who used mercury to produce a shiny finish on the top hats.

The cardiovascular system

The blood system uses the heart to pump blood around the body through arteries, veins and capillaries. Blood is produced in the bone marrow and consists of a plasma within which are red cells, white cells and platelets. The system has three basic objectives:

▶ to transport oxygen to vital organs, tissues and the brain and carbon dioxide back to the lungs (red cell function);
▶ to attack foreign organisms and build up a defence system (white cell function);
▶ to aid the healing of damaged tissue and prevent excessive bleeding by clotting (platelets).

There are several ways in which hazardous substances can interfere with the cardiovascular system. Benzene can affect the bone marrow by reducing the number of blood cells produced. Carbon monoxide prevents the red cells from absorbing sufficient oxygen and the effects depend on its concentration. Symptoms begin with headaches and end with unconsciousness and possibly death.

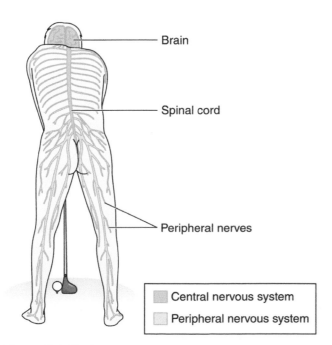

Figure 8.6 The nervous system

Figure 8.7 The cardiovascular system

237

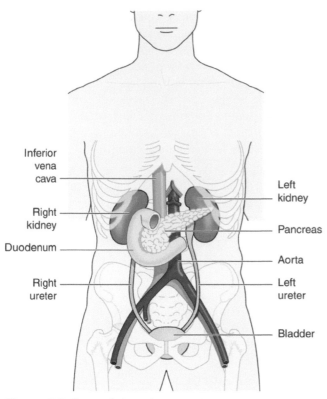

and kidney, both of which can be affected by hazardous substances within the bloodstream.

The liver removes toxins from the blood, maintains the levels of blood sugars and produces protein for the blood plasma. Hazardous substances can cause the liver to be too active or inactive (e.g. xylene), lead to liver enlargement (e.g. cirrhosis caused by alcohol) or liver cancer (e.g. vinyl chloride).

The kidneys filter waste products from the blood as urine, regulate blood pressure and liquid volume in the body and produce hormones for making red blood cells. Heavy metals (e.g. cadmium and lead) and organic solvents (e.g. glycol ethers used in screen printing) can restrict the operation of the kidneys possibly leading to failure.

The skin

The skin holds the body together and is the first line of defence against infection. It regulates body temperature, is a sensing mechanism, provides an emergency food store (in the form of fat) and helps to conserve water. There are two layers – an outer layer called the epidermis (0.2 mm) and an inner layer called the dermis (4 mm). The epidermis is a tough protective layer and the dermis contains the sweat glands, nerve endings and hairs.

The most common industrial disease of the skin is **dermatitis** (non-infective dermatitis). It begins with a mild irritation on the skin and develops into blisters which can peel and weep becoming septic. It can be

Figure 8.8 Parts of the urinary system

The urinary system

The urinary system extracts waste and other products from the blood. The two most important organs are the liver (normally considered part of the digestive system)

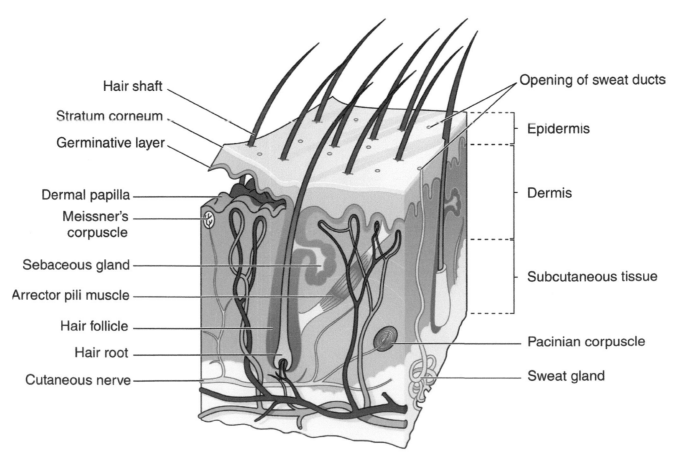

Figure 8.9 The skin – main structures of the dermis

caused by various chemicals, mineral oils and solvents. There are two types:

▶ **irritant contact dermatitis** – occurs soon after contact with the substance and the condition reverses after contact ceases (detergents and weak acids);

▶ **allergic contact dermatitis** – caused by a skin sensitiser such as turpentine, epoxy resin, solder flux and formaldehyde that exerts its effects via the immune system. Once sensitised to a substance, a severe dermatitis may occur following a small exposure to the same substance at a later date.

Dermatitis is on the rise and is costing business more than £20m a year, even though the cost of control measures to prevent the disease is minimal. Workers in the hotel and catering industry are particularly vulnerable to this debilitating disease as are print workers who have contracted dermatitis after coming into contact with UV curable lacquers used to produce a glossy finish to magazine covers.

For many years, dermatitis was seen as a 'nervous' disease which was psychological in nature. Nowadays, it is recognised as an industrial disease which can be controlled by good personal hygiene, personal protective equipment, use of barrier creams and health screening of employees. Dermatitis can appear anywhere on the body but it is normally found on the hands. Therefore, gloves should always be worn when there is a risk of dermatitis.

The risk of dermatitis occurring increases with the presence of cuts or abrasions on the skin, which allow chemicals to be more easily absorbed. The risk also depends on the type, sensitivity and existing condition of the skin.

8.2.3 The role of COSHH

The Control of Substances Hazardous to Health Regulations (COSHH) 1988 were the most comprehensive and significant piece of health and safety legislation to be introduced since the Health and Safety at Work Act 1974. They were enlarged to cover biological agents in 1994 and further amended in 1997, 1999, 2002 and 2004. A detailed summary of these Regulations appears in Chapter 13. The Regulations impose duties on employers to protect employees and others who may be exposed to substances hazardous to their health and requires employers to control exposure to such substances.

The COSHH Regulations offer a framework for employers to build a management system to assess health risks and to implement and monitor effective controls. Adherence to these Regulations will provide the following benefits to the employer and employee:

▶ improved productivity due to lower levels of ill health and more effective use of materials;

▶ improved employee morale;

▶ lower numbers of civil court claims;

▶ better understanding of health and safety legal requirements.

The Approved Code of Practice (ACoP) to the COSHH Regulations covers all substances to which the Regulations apply. It outlines the preferred or recommended methods that can be used to comply with the Regulations and the accompanying guidance also provides advice on achieving compliance, such as the control of carcinogenic substances or those causing occupational asthma, monitoring control measures and conducting health surveillance. The latest edition of this ACoP and guidance takes account of regulatory changes following the introduction of the EU Regulations for REACH (Registration, Evaluation, Authorisation and restriction of Chemicals) and CLP on classification, labelling and packaging of substances and mixtures.

Organisations which ignore COSHH requirements will be liable for enforcement action, including prosecution, under the Regulations.

8.2.4 Requirements of the COSHH Regulations

When hazardous substances are considered for use (or in use) at a place of work, the COSHH Regulations impose certain duties on employers and require employees to cooperate with the employer by following any measures taken to fulfil those duties. The principal requirements are as follows.

1. Employers must undertake a suitable and sufficient assessment of the health risks created by work which is liable to expose their employees to substances hazardous to health and of the steps that need to be taken by employers to meet the requirements of these Regulations (Regulation 6).

2. Employers must prevent, or where this is not reasonably practicable, adequately control the exposure of their employees to substances hazardous to health. Workspace Exposure Limits (WELs), which should not be exceeded, are specified by the HSE for certain substances. As far as inhalation is concerned, control should be achieved by means other than personal protective equipment. If, however, respiratory protective equipment, for example, is used, then the equipment must conform to HSE standards (Regulation 7). The control of exposure can only be treated as adequate if the principles of good practice (discussed under Section 8.4.1) are applied (Schedule 2A).

3. Employers and employees must make proper use of any control measures provided (Regulation 8).

4. Employers must maintain any installed control measures on a regular basis, keep suitable records (Regulation 9) and review systems of work.

5. Monitoring must be undertaken of any employee exposed to items listed in Schedule 5 of the

Regulations or in any other case where monitoring is required for the maintenance of adequate control or the protection of employees. Records of this monitoring must be kept for at least 5 years, or 40 years where employees can be identified (Regulation 10).

6. Health surveillance must be provided to any employees who are exposed to any substances listed in Schedule 6. Records of such surveillance must be kept for at least 40 years after the last entry (Regulation 11).

7. Emergency plans and procedures must be prepared to deal with accidents or incidents involving exposure to hazardous substances beyond normal day-to-day risks. This will involve warnings and communication systems to give appropriate response immediately after any incident occurs.

8. Employees who may be exposed to substances hazardous to their health must be given information, instruction and training sufficient for them to know the health risks created by the exposure and the precautions which should be taken (Regulation 12).

8.2.5 **Details of a COSHH assessment**

Not all hazardous substances are covered by the COSHH Regulations. If there is no warning symbol on the substance container or it is a biological agent which is not directly used in the workplace (such as an influenza virus), then no COSHH assessment is required. The COSHH Regulations do not apply to those hazardous substances which are subject to their own individual Regulations (asbestos, lead or radioactive substances). The COSHH Regulations do apply to the following substances:

▶ substances having occupational exposure limits as listed in the HSE publication EH40 (Occupational Exposure Limits);
▶ substances or combinations of substances listed in the CLP Regulations;
▶ biological agents connected with the workplace;
▶ substantial quantities of airborne dust (more than 10 mg/m^3 of total inhalable dust or 4 mg/m^3 of respirable dust, both 8-hour time-weighted average (TWA), when there is no indication of a lower value);
▶ any substance creating a comparable hazard which for technical reasons may not be documented.

The following factors should be considered when assessing health risks:

▶ the health risk assessment should be suitable and sufficient (see Chapter 4);
▶ the exposure of employees to hazardous substances is adequately controlled;
▶ the control measures provided are properly used;
▶ the control measures and any associated equipment is properly maintained;
▶ monitor employees exposed to hazardous substances;

▶ there is adequate health surveillance; and
▶ information, instruction and training.

A COSHH assessment is very similar to a risk assessment but is applied specifically to hazardous substances. The HSE has suggested five steps to COSHH assessments but within these steps there are a number of sub-sections. The steps are as follows.

Step 1

Gather information about the substances, the work and working practices. Assessors should:

▶ identify the hazardous substances present or likely to be present in the workplace, the categories and numbers of persons (e.g. employees and visitors);
▶ gather information about the hazardous substances including the quantity of the substances used;
▶ identify the hazards from these substances by reviewing labels, material safety data sheets, HSE guidance and published literature;
▶ decide who could be affected by the hazardous substances and the possible routes of entry to people exposed (i.e. inhalation, ingestion or absorption). There is a need to look at both the substances and the activities where people could be exposed to hazardous substances.

Step 2

Evaluate the risks to health either individually or collectively. Assessors should:

▶ evaluate the risks to health including the duration and frequency of the exposure of those persons to the substances;
▶ evaluate the level of exposure, for example the concentration and length of exposure to any airborne dusts, gases, fumes or vapours;
▶ consider any WELs (see Section 8.3);
▶ decide if existing and potential exposure presents any insignificant risks to health or they pose a significant risk to health.

Step 3

Decide what needs to be done to control the exposure to hazardous substances. Assessors should:

▶ evaluate the existing control measures, including any PPE and respiratory protective equipment (RPE), for their effectiveness (using any available records of environmental monitoring) and compliance with relevant legislation;
▶ decide on additional control measures, if any are required (see Section 8.4);
▶ decide what maintenance and supervision of the use of the control measures are needed;
▶ plan what to do in an emergency;
▶ set out how exposure should be monitored;

- decide what, if any, health surveillance is necessary;
- decide what information, instruction and training is required.

Step 4

Record the assessment. Assessors should:

- decide if a record is required (five or more employed, and significant findings);
- decide on the format of the record;
- decide on storage and how to make records available to employees, safety representatives, etc.

Step 5

Review the assessment. Assessors should:

- decide when a review is necessary (e.g. changes in substances used, processes or people exposed);
- decide what needs to be reviewed.

An example of a typical form that can be used for a COSHH assessment is given in Chapter 17.

It is important that the assessment is conducted by somebody who is competent to undertake it. Such competence will require some training, the extent of which will depend on the complexities of the workplace. For large organisations with many high-risk operations, a team of competent assessors will be needed. If the assessment is simple and easily repeated, a written record is not necessary. In other cases, a concise and dated record of the assessment together with recommended control measures should be made available to all those likely to be affected by the hazardous substances. The assessment should be reviewed on a regular basis, particularly when there are changes in work process or substances or when adverse ill health is reported.

8.2.6 Assessing exposure and health surveillance

Some aspects of health exposure will need input from specialist or professional advisers, such as occupational health hygienists, nurses and doctors. However, considerable progress can be made by taking straightforward measures such as:

- consulting the workforce on the design of workplaces;
- talking to manufacturers and suppliers of substances and work equipment about minimising exposure;
- enclosing machinery to cut down dust, fumes and noise;
- researching the use of less hazardous substances;
- ensuring that employees are given appropriate information and are trained in the safe handling of all the substances and materials to which they may be exposed.

To assess health risks and to make sure that control measures are working properly, it may be necessary,

for example, to measure the concentration of substances in the air to make sure that exposures remain within the assigned WELs. Sometimes health surveillance of workers who may be exposed will be needed. This will enable data to be collected to check control measures and for early detection of any adverse changes to health. Health surveillance procedures available include biological monitoring for bodily uptake of substances, examination for symptoms and medical surveillance – which may entail clinical examinations and physiological or psychological measurements by occupationally qualified registered medical practitioners. The procedure chosen should be suitable for the case concerned. Sometimes a method of surveillance is specified for a particular substance, for example, in the COSHH ACoP. Whenever surveillance is undertaken, a health record has to be kept for the person concerned.

Health surveillance should be supervised by a registered medical practitioner or, where appropriate, it should be done by a suitably qualified person (e.g. an occupational nurse). In the case of inspections for easily detectable symptoms like chrome ulceration or early signs of dermatitis, health surveillance should be done by a suitably trained responsible person. If workers could be exposed to substances listed in Schedule 6 of the COSHH Regulations, medical surveillance, under the supervision of an HSE employment medical adviser or a doctor appointed by HSE, is required. There is more on health surveillance later in this chapter.

8.2.7 Sources of information

There are other important sources of information available for a COSHH assessment in addition to the HSE Guidance Note EH40.

Product labels include details of the hazards associated with the substances contained in the product and any precautions recommended. They may also bear one or more of the CHIP hazard classification symbols.

Material safety data sheets are another very useful source of information for hazard identification and associated advice. Manufacturers of hazardous substances are obliged to supply such sheets to users, giving details of the name, chemical composition and properties of the substance. Information on the nature of the health hazards and any relevant exposure standard (WEL) should also be given together with recommended exposure control measures and personal protective equipment. The sheets contain useful additional information on first aid and firefighting measures and handling, storage, transport and disposal information. The data sheets should be stored in a readily accessible and known place for use in the event of an emergency, such as an accidental release.

(a)

Product J

(abc chemical)

Danger

Fatal if swallowed
Causes skin irritation

Precautions:
Wear protective gloves.
Take off contaminated clothing and wash before reuse.
Wash hands thoroughly after handling.
Do not eat, drink or smoke when using this product.

Store locked up.
Dispose of contents/containers in accordance with local regulations.

IF ON SKIN: Rinse skin with water/shower.
IF IN EYES: Rinse cautiously with water.
IF SWALLOWED: Immediately call a Poison Center or
doctor/physician. Do not induce vomiting.

ABC Chemical Co., 123 Anywhere St., (123) 456-7890
See the SDS for more information

(b)

Figure 8.10 (a) Typical symbols and (b) product label on containers

Other sources of information include trade association publications, industrial codes of practice and specialist reference manuals.

8.2.8 Survey techniques for health risks

An essential part of the COSHH assessment is the measurement of the quantity of the hazardous substance in the atmosphere surrounding the workplace. This is known as air sampling. There are four common types of air-sampling technique used for the measurement of air quality:

1. **Stain tube detectors** use direct reading glass indicator tubes filled with chemical crystals

which change colour when a particular hazardous substance passes through them. The method of operation is very similar to the breathalyser used by the police to check alcohol levels in motorists. The glass tube is opened at each end and fitted into a pumping device (either hand or electrically operated). A specific quantity of contaminated air containing the hazardous substance is drawn through the tube and the crystals in the tube change colour in the direction of the air flow. The tube is calibrated such that the extent of the colour change along the tube indicates the concentration of the hazardous substance within the air sample.

This method can only be effective if there are no leakages within the instrument and the correct volume of sampled air is used. The instrument should be held within 30 cm of the nose of the person whose atmosphere is being tested. A large range of different tubes is available. This technique of sampling is known as **grab** or **spot sampling** since it is taken at one point.

The advantages of the technique are that it is quick, relatively simple to use and inexpensive. There are, however, several disadvantages:

▶ The instrument cannot be used to measure concentrations of dust or fume.
▶ The accuracy of the reading is approximately 25%; it will yield a false reading if other contaminants present react with the crystals.
▶ The instrument can only give an instantaneous reading, not an average reading over the working period (TWA).
▶ The tubes are very fragile with a limited shelf life.
▶ Some disposal issues.

Figure 8.11 Hand pump and stain detector

2. **Passive sampling** is measured over a full working period by the worker wearing a badge containing absorbent material. The material will absorb the contaminant gas and, at the end of the measuring period, the sample is sent to a laboratory for analysis. The advantages of this method over

the stain tube are that there is less possibility of instrument errors and it gives a TWA reading.

3. **Sampling pumps and heads** can be used to measure gases and dusts. The worker, whose breathing zone is being monitored, wears a collection head as a badge and a battery-operated pump on his/her back at waist level. The pump draws air continuously through a filter, fitted in the head, which will either absorb the contaminant gas or trap hazardous dust particles. If this filter is used for dust measurement, it is sometimes called a static dust sampler. Before sampling takes place, the filter is weighed and the pump flow rate preset. After the designated testing period, it is sent to a specialist laboratory for analysis. The quantity of dust present would be determined either by measuring the weight change of the filter or by chemical analysis (e.g. for lead) or by using microscopes to count the number of fibres (e.g. for asbestos). This system is more accurate than stain tubes and gives a TWA result but can be uncomfortable to wear over long periods. Such equipment can only be used by trained personnel.

4. **Direct reading instruments** are available in the form of sophisticated analysers which can only be used by trained and experienced operatives. Infrared gas analysers are the most common but other types of analysers are also available. They are very accurate and give continuous or TWA readings. They tend to be very expensive and are normally hired or used by specialist consultants.

Other common monitoring instruments include **vane anemometers,** used for measuring air flow speeds and hygrometers, which are used for measuring air humidity.

Qualitative monitoring techniques include smoke tubes and the dust observation lamp. **Smoke tubes** generate a white smoke which may be used to indicate the direction of flow of air – this is particularly useful when the air speed is very low or when testing the effectiveness of ventilation ducting. A **dust observation lamp** enables dust particles which are normally invisible to the human eye to be observed in the light beam. This dust is usually in the respirable range and, although the lamp does not enable any measurements of the dust to be made, it will illustrate the operation of a ventilation system and the presence of such dust.

Regulation 10 requires routine sampling or monitoring of exposure where there could be serious health effects if the controls fail, exposure limits might be exceeded or the control measures may not be working properly. Air monitoring should also be undertaken for any hazardous substances listed in Schedule 5 of the COSHH Regulations. Records of this monitoring should be kept for 5 years unless an employee is identifiable in the records, in which case they should be kept for 40 years.

8.3 Occupational exposure limits

An occupational exposure limit (OEL) is the acceptable upper limit on the concentration of a hazardous substance in the air within a workplace for a particular or group of hazardous substances. OELs are normally established by competent national authorities and enforced by appropriate legislation.

One of the main purposes of a COSHH assessment is to adequately control the exposure of employees and others to hazardous substances. This means that such substances should be reduced to levels which do not pose a health threat to those exposed to them day after day at work. Under the 2004 amendments to the COSHH Regulations 2002, the HSE assigned WELs (as its OEL) to a large number of hazardous substances and publishes any updates in a publication called 'Occupational Exposure Limits' EH40. The WEL is related to the concentration of airborne hazardous substances that people breathe over a specified period of time – known as 'time-weighted average'. Before the introduction of WELs, there were two types of exposure limit published: the maximum exposure limit (MEL) and the occupational exposure standard (OES). Under the amended COSHH the OES/MEL system was replaced with a single WEL. This removed the concern of HSE that the OES was seen as a 'safe' limit rather than a 'likely safe' limit. Hence, the WEL must not be exceeded. Hazardous substances which have been assigned a WEL fall into two groups.

1. Substances which are carcinogenic or mutagenic or could cause occupational asthma (or listed in section C of the HSE publication '*Asthmagen? Critical assessment for the agents implicated in occupational asthma*' as updated from time to time) or are listed in Schedule 1 of the COSHH Regulations. These are substances which were assigned a MEL before 2005. The level of exposure to these substances should be reduced as far as is reasonably practicable.

2. All other hazardous substances which have been assigned a WEL. Exposure to these substances by inhalation must be controlled adequately to ensure that the WEL is not exceeded. These substances were previously assigned an OES before 2005. For these substances, employers should achieve adequate control of exposure by inhalation by applying the principles of good practice outlined in the ACoP and listed in Section 8.4.1. The implication of these principles is discussed later in this chapter.

The WELs are subject to time-weighted averaging. There are two such time-weighted averages (TWA): the long-term exposure limit (LTEL) or 8-hour reference period and the short-term exposure limit (STEL) or 15-minute reference period. The 8-hour TWA is the maximum exposure allowed over an 8-hour period so that if the exposure period was less than 8 hours the

8

WEL is increased accordingly with the proviso that exposure above the LTEL value continues for no longer than 1 hour. Table 8.2 shows some typical WELs for various hazardous substances.

For example, if a person was exposed to a hazardous substance with a WEL of 100 mg/m^3 (8-hour TWA) for 4 hours, no action would be required until an exposure level of 200 mg/m^3 was reached (further exposure at levels between 100 and 200 mg/m^3 should be restricted to 1 hour).

Table 8.2 Examples of workplace exposure limits (WELs)

Group 1 WELs	LTEL (8 h TWA)	STEL (15 min)
All isocyanates	0.02 mg/m^3	0.07 mg/m^3
Styrene	430 mg/m^3	1080 mg/m^3
Group 2 WELs		
Ammonia	18 mg/m^3	25 mg/m^3
Toluene	191 mg/m^3	574 mg/m^3

If, however, the substance has a STEL of 150 mg/m^3, then action would be required when the exposure level rose above 150 mg/m^3 for more than 15 minutes.

The STEL always takes precedence over the LTEL. When a STEL is not given, it should be assumed that it is three times the LTEL value.

The publication EH40 is a valuable document for the health and safety professional as it contains much additional advice on hazardous substances for use during the assessment of health risks, particularly where new medical information has been made public. The HSE is constantly revising WELs and introducing new ones and it is important to refer to the latest publication of EH40.

It is important to stress that if a WEL from Group 1 is exceeded, the process and use of the substance should cease immediately and employees should be removed from the immediate area until it can be made safe. In the longer term the process and the control and monitoring measures should be reviewed and health surveillance of the affected employees considered.

The overriding requirement for any hazardous substance which has a WEL from Group 1, is to reduce exposure to as low as is reasonably practicable.

Finally, there are certain limitations on the use of the published WELs:

▶ They are specifically quoted for an 8-hour period (with an additional STEL for many hazardous substances). Adjustments must be made when exposure occurs over a continuous period longer than 8 hours.

▶ They can only be used for exposure in a workplace and not to evaluate or control non-occupational exposure (e.g. to evaluate exposure levels in a neighbourhood close to the workplace, such as a playground).

▶ WELs are only approved where the atmospheric pressure varies from 900 to 1100 millibars. This could exclude their use in mining and tunnelling operations.

▶ They should also not be used when there is a rapid build-up of a hazardous substance due to a serious accident or other emergency. Emergency arrangements should cover these eventualities.

The fact that a substance has not been allocated a WEL does not mean that it is safe. The exposure to these substances should be controlled to a level which nearly all of the working population could experience all the time without any adverse effects to their health. Detailed guidance and references are given for such substances in the HSE ACoP and guidance to the COSHH Regulations (L5).

HSE publishes a revised ACoP and EH40 to include these changes. More information on the latest COSHH Regulations is given in Chapter 13.

8.4 Control measures

Exposure to hazardous substances needs to be prevented or, where this is not reasonably practicable, adequately controlled.

8.4.1 The principles of good practice for the control of exposure to substances hazardous to health

The objective of the COSHH Regulations is to prevent ill health due to the exposure to hazardous substances. Employers are expected to develop suitable and sufficient control measures by:

1. identifying hazards and potentially significant risks;
2. taking action to reduce and control risks;
3. keeping control measures under regular review.

In order to assist employers with these duties, the HSE has produced the following eight principles of good practice:

(a) Design and operate processes and activities to minimise the emission, release and spread of substances hazardous to health.

(b) Take into account all relevant routes of exposure – inhalation, skin absorption and ingestion – when developing control measures.

(c) Control exposure by measures that are proportionate to the health risk.

(d) Choose the most effective and reliable control options which minimise the escape and spread of substances hazardous to health.

(e) When adequate control of exposure cannot be achieved by other means, provide, in combination with other control measures, suitable personal protective equipment.

(f) Check and review regularly all elements of control measures for their continuing effectiveness.

(g) Inform and train all employees on the hazards and risks from the substances with which they work and the use of control measures developed to minimise the risks.

(h) Ensure that the introduction of control measures does not increase the overall risk to health and safety.

All these principles are embodied in the following sections on COSHH control measures.

The frequency and type of future monitoring of exposure levels will depend on the exposure found in relation to recognised exposure limits. If the exposure level is very much lower than the limit and there is no change in process or other reason, then repeat measurement may only be needed occasionally.

If the exposure level is relatively high, then measurement may be needed several times between assessment reviews, to ensure that these levels have not been altered by some unidentified factor.

8.4.2 Common methods used to implement the principles of good practice

Hierarchy of control measures

The COSHH Regulations require the prevention or adequate control of exposure by measures other than personal protective equipment, so far as is reasonably practicable, taking into account the degree of exposure and current knowledge of the health risks and associated technical remedies. The hierarchy of control measures is as follows:

▶ elimination;
▶ substitution;
▶ provision of engineering controls;
▶ provision of supervisory (people) controls;
▶ provision of personal protective equipment.

Examples where engineering controls are not reasonably practicable include emergency and maintenance work, short-term and infrequent exposure and where such controls are not technically feasible.

Measures for preventing or controlling exposure to hazardous substances include one or a combination of the following:

▶ elimination of the substance;
▶ substitution of the substance (or the reduction in the quantity used);
▶ total or partial enclosure of the process;
▶ local exhaust ventilation;
▶ dilution or general ventilation;
▶ reduction of the number of employees exposed to a strict minimum;
▶ reduced time exposure by task rotation and the provision of adequate breaks;

▶ good housekeeping;
▶ training and information on the risks involved;
▶ effective supervision to ensure that the control measures are being followed;
▶ personal protective equipment (such as clothing, gloves and masks);
▶ welfare (including first aid);
▶ medical records;
▶ health surveillance.

Preventative control measures

Prevention is the safest and most effective of the control measures and is achieved either by changing the process completely or by substituting with a less hazardous substance (the change from oil-based to water-based paints is an example of this). It may be possible to use a substance in a safer form, such as a brush paint rather than a spray.

The EU introduced chemical safety regime REACH (Registration, Evaluation, Authorisation and restriction of Chemicals) Regulations, which restrict the use of high-risk substances or substances of very high concern and require that safer substitutes must be used. Manufacturers and importers of chemicals are responsible for understanding and managing the risks associated with their products. Authorisation seeks to ensure that risks from these substances of very high concern (SVHCs) are properly controlled. These are substances that are carcinogenic, mutagenic or toxic for reproduction or substances that are persistent, bio-accumulative and toxic. The objective is that all such substances should be replaced where possible with less dangerous alternatives.

REACH does not replace the COSHH Regulations which place duties on employers to control the exposure of employees to hazardous substances in the workplace.

The Regulations apply to many common items, such as glues, paints, solvents, detergents, plastics, additives, polishes, pens and computers. The three main types of REACH duty-holder are:

▶ **Manufacturers/importers** – businesses that manufacture or import (from outside the EU) 1 ton or more of any given substance each year are responsible for registering a dossier of information about that substance with the European Chemicals Agency. If substances are not registered, then the data on them will not be available and it will no longer be legal to manufacture or supply them within the EU. Suppliers will be obliged to carry out an inventory and identify where in the supply chain the chemicals come from. Under the REACH system, industry will also have to prepare risk assessments and provide control measures for safe use of the substance by downstream users and get community-wide authorisations for the use of any substances considered to be of high concern.

▶ **Downstream users** – downstream users include any businesses using chemicals, which probably includes most businesses in some way. Companies that use chemicals have a duty to use them in a safe way, and according to the information on risk management measures that should be passed down the supply chain.

▶ **Other users in the supply chain** – in order for suppliers to be able to assess these risks they need information from the downstream users about how they are used. REACH provides a framework in which information can be passed both up and down supply chains by using the safety data sheet. This should accompany materials down through the supply chain so that users are provided with the information that they need to ensure chemicals are safely managed. It is envisaged that, in the future, the safety data sheets will include information on safe handling and use. Some users may also be importers and have a duty to register.

An important aspect of chemical safety is the need for clear information about any hazardous chemical properties. The classification of different chemicals according to their characteristics (for example, those that are corrosive, or toxic to fish) currently follows an established system, which is reflected in REACH. (Chapter 13 contains more information on REACH.)

Engineering controls

Occupational exposures to fumes, dusts and chemicals together account for around 4,000 deaths a year: while some 38,000 individuals suffer breathing or lung problems that they believe are caused by their work.

The simplest and most efficient engineering control is the segregation of people from the process; a chemical

(b)

Figure 8.12 (a) A LEV system; (b) welding with an adjustable LEV system to remove dust and fumes

fume cupboard is an example of this as is the handling of toxic substances in a glove box. Modification of the process is another effective control to reduce human contact with hazardous substances.

More common methods, however, involve the use of forced ventilation – local exhaust ventilation and dilution ventilation.

Local exhaust ventilation

Local exhaust ventilation removes the hazardous gas, vapour or fume at its source before it can contaminate the surrounding atmosphere and harm people working in the vicinity. Such systems are commonly used for the extraction of welding fumes and dust from woodworking machines. All exhaust ventilation systems have the following five basic components (Figure 8.12).

1. A **collection hood and intake** – sometimes this is a nozzle-shaped point which is nearest to the work-

piece, while at other times it is simply a hood placed over the workstation. The speed of the air entering the intake nozzle is important; if it is too low then hazardous fume may not be removed (air speeds of up to 1 m/s are normally required). A suitable airflow indicator should make it easy to see whether airflow is adequate. New LEV systems are likely to be fitted with airflow indicators as standard.

LEV hood design is very important and poor design is the main reason that LEV fails to control exposure. It should be as close as possible to the work-piece and ergonomic principles used in the designing of LEV hoods. LEV suppliers tend to concentrate on the fan and filter system and assume that hoods are standard for all applications.

The most commonly used hood – the capturing hood has a particular limitation as the process to be controlled takes place outside the hood and the LEV system needs to generate sufficient air flow at and around the process to 'capture' and draw in the airborne contaminant cloud. One company replaced a capturing hood with an enclosing hood fitted with a glass panel so that soldering fumes were separated from the operative's breathing zone.

2. **Ventilation ducting** – this normally acts as a conduit for the contaminated air and transports it to a filter and settling section. It is very important that this section is inspected regularly and any dust deposits removed. It has been known for ventilation ducting attached to a workshop ceiling to collapse under the added weight of metal dust deposits. It has also been known for them to catch fire.

3. **Filter** or **other air cleaning device** – normally located between the hood and the fan, the filter removes the contaminant from the air stream. The filter requires regular attention to remove contaminant and to ensure that it continues to work effectively.

4. **Fan** – this moves the air through the system. It is crucial that the correct type and size of fan is fitted to a given system and it should only be selected by a competent person. It should also be positioned so that it can easily be maintained but does not create a noise hazard to nearby workers.

5. **Exhaust duct** – this exhausts the air to the outside of the building. The duct should be positioned on the outside wall of the building such that the air is not discharged into a public area or close to an air inlet for an air conditioning system of any building. It should be checked regularly to ensure that the correct volume of air is leaving the system and that there are no leakages. The exhaust duct should also be checked to ensure that there is no corrosion due to adverse weather conditions.

Controlling Airborne Contaminants at Work, HSG258, HSE Books, is a very useful document on ventilation systems.

The COSHH Regulations require that such ventilation systems should undergo a periodic thorough examination and test at least every 14 months by a competent person to ensure that they are still operating effectively. A record of the examination should be kept for at least 5 years.

The effectiveness of a ventilation system will be reduced by damaged ducting, blocked or defective filters and poor fan performance. More common problems include the unauthorised extension of the system, poor initial design, poor maintenance, incorrect adjustments and a lack of inspection or testing.

Routine maintenance should include repair of any damaged ducting, checking filters, examination of the fan blades to ensure that there has been no dust accumulation, tightening all drive belts and a general lubrication of moving parts. LEV systems will degrade if they are not regularly checked and maintained. Dirty ventilation systems can seriously affect staff well-being and working environments, and both the bacteria MRSA and Clostridium difficile have been found in duct systems. A 'responsible person' should be appointed who follows the instructions in the user manual, arranges monitoring and maintenance, and keeps records. The employer will also need to check that operators are using the LEV correctly.

The local exhaust ventilation system will have an effect on the outside environment in the form of noise and odour. Both these problems can be reduced by regular routine maintenance of the fan and filter. The waste material from the filter may be hazardous and require the special disposal arrangements described later in this chapter.

Dilution (or general) ventilation

Dilution (or general) ventilation uses either natural ventilation (doors and windows) or fan-assisted forced ventilation system to ventilate the whole working room by inducing a flow of clean air, using extraction fans fitted into the walls and the roof, sometimes assisted by inlet fans. It operates by either removing the contaminant or reducing its concentration to an acceptable level. It is used when airborne contaminants are of low toxicity, low concentration and low vapour density or contamination occurs uniformly across the workroom.

Paint-spraying operations often use this form of ventilation as does the glass reinforced plastics (GRP) boat-building industry – these being instances where there are no discrete points of release of the hazardous substances. It is also widely used in kitchens and bathrooms. It is not suitable for dust extraction or where it is reasonably practicable to reduce levels by other means.

There are limitations to the use of dilution ventilation. Certain areas of the workroom (e.g. corners and beside cupboards) will not receive the ventilated air and a build-up of hazardous substances occurs. These areas are known as 'dead areas'. The flow patterns are also significantly affected by doors and windows being opened or the rearrangement of furniture or equipment.

Figure 8.13 (a) Local exhaust ventilation systems can quickly corrode and need regular maintenance of fans and filters; (b) large roof-level system with access platform for maintenance of fans and filters; (c) paint-spraying inside a special room with general ventilation to remove fumes

Supervisory or people controls

Many of the supervisory controls required for COSHH purposes are part of a good safety culture and were discussed in detail in Chapter 3. These include items such as systems of work, arrangements and procedures, effective communications and training. Additional controls when hazardous substances are involved are as follows:

▶ Reduced time exposure – thus ensuring that workers have breaks in their exposure periods. The use of this method of control depends very much on the nature of the hazardous substance and its STEL.

▶ Reduced number of workers exposed – only persons essential to the process should be allowed in the vicinity of the hazardous substance. Walkways and other traffic routes should avoid any area where hazardous substances are in use.

▶ Eating, drinking and smoking must be prohibited in areas where hazardous substances are in use.

▶ Any special rules, such as the use of personal protective equipment, must be strictly enforced.

Personal protective equipment

Personal protective equipment (PPE) is to be used as a control measure only as a last resort. It does not eliminate the hazard and will present the wearer with the maximum health risk if the equipment fails. Successful use of PPE relies on good user training, the availability of the correct equipment at all times and good supervision and enforcement.

The 'last resort' rule applies in particular to respiratory protective equipment (RPE) within the context of hazardous substances. There are some working conditions when RPE may be necessary:

▶ during maintenance operations;

▶ as a result of a new assessment, perhaps following the introduction of a new substance;

▶ during emergency situations, such as fire or plant breakdown;

▶ where alternatives are not technically feasible.

The Personal Protective Equipment at Work Regulations state in Regulation 4 *'Every employer shall ensure that suitable personal protective equipment is provided to his employees who may be exposed to a risk to their health or safety while at work except where and to the extent that such risk has been adequately controlled by other means which are equally or more effective'*. The accompanying guidance recommends that employers should, therefore, provide appropriate PPE and training in its usage to their employees wherever there is a risk to health and safety that cannot be adequately controlled by other means. Employers must do more than simply have the equipment on the premises – it must be readily available. These

Figure 8.14 Personal protective equipment

Regulations cover all PPE except respiratory protective equipment (which is covered by specific Regulations such as COSHH and Lead at Work).

The principal requirements of these Regulations are as follows:

▶ PPE which is suitable for the wearer and the task;
▶ compatibility and effectiveness of the use of multiple personal protective equipment;
▶ a risk assessment to determine the need and suitability of proposed PPE;
▶ a suitable maintenance programme for the personal protective equipment;
▶ suitable accommodation for the storage of the PPE when not in use;
▶ information, instruction and training for the user of PPE including a demonstration on how to use the equipment properly;
▶ the supervision of the use of PPE by employees and a reporting system for defects.

A summary of these Regulations is given in Chapter 13.

For complete protection, personal protective equipment must be put on and taken off properly and in the right order. Overalls should not be passed over footwear because hazardous residues on the soles of shoes or boots from the floor of the changing area are likely to rub off on the inside of coveralls or safety trousers.

Respiratory protection should not be removed before the rest of the PPE. This will guard against any hazardous material released from clothing being inhaled.

Types of personal protective equipment

There are several types of PPE such as footwear, hearing protectors and hard hats which are not primarily concerned with protection from hazardous substances; those which are used for such protection include:

▶ respiratory protection PPE;
▶ hand and skin protection PPE;
▶ eye protection PPE;
▶ protective clothing.

Table 8.3 shows the types of PPE recommended by the HSE for various parts of the body.

Table 8.3 The hazards and types of PPE for various parts of the body

	Hazards	PPE
Eyes	chemical or metal splash, dust, projectiles, gas and vapour, radiation.	safety spectacles, goggles, face shields, visors.
Head	impact from falling or flying objects, risk of head bumping, hair.	a range of helmets and bump caps.
Respiratory system	dust, vapour, gas, oxygen-deficient atmospheres.	disposable filtering face piece or respirator, half- or full-face respirators, air-fed helmets, breathing apparatus.
Hand and arms	abrasion, temperature extremes, cuts and punctures, impact, chemicals, electric shock, skin infection, disease or contamination.	gloves, gauntlets, mitts, wrist-cuffs, armlets.
Feet and legs	wet, electrostatic build-up, slipping, cuts and punctures, falling objects, metal and chemical splash, abrasion.	safety boots and shoes with protective toe caps and penetration-resistant mid-sole, gaiters, leggings, spats.
Whole body	temperature extremes, adverse weather, chemical or metal splash, spray from pressure leaks or spray guns, impact or penetration, contaminated dust, excessive wear or entanglement of own clothing.	conventional or disposable overalls, boiler suits, specialist protective clothing, e.g. chainmail aprons, high-visibility clothing.

Source: HSE.

For all types of PPE, there are some basic standards that should be reached. The PPE should fit well, be comfortable to wear and not interfere with other equipment being worn or present the user with additional hazards (e.g. impaired vision due to scratched eye goggles). Training in the use of particular PPE is essential, so that it is not only used correctly, but the user knows when to change an air filter or to change a type of glove. Supervision is essential, with disciplinary procedures invoked for non-compliance with PPE rules.

It is also essential that everyone who enters the proscribed area, particularly senior managers, wear the specified PPE.

Respiratory protective equipment

Respiratory protective equipment (RPE) can be subdivided into two categories – respirators (or face masks), which filter and clean the air, and breathing apparatus which supplies breathable air.

Respirators should not be worn in air which is dangerous to health, including oxygen deficient atmospheres. They are available in several different forms but the common ones are:

▶ a **filtering half-mask** often called disposable respirator – made of the filtering material. It covers the nose and mouth and removes respirable size dust particles. It is normally replaced after 8–10 hours of use. It offers protection against some vapours and gases;

▶ a **half-mask respirator** – made of rubber or plastic and covering the nose and mouth. Air is drawn through a replaceable filter cartridge. It can be used for vapours, gases or dusts but it is very important that the correct filter is used (a dust filter will not filter vapours);

▶ a **full-face mask respirator** – similar to the half-mask type but covers the eyes with a visor;

▶ a **powered respirator** – a battery-operated fan delivers air through a filter to the face mask, hood, helmet or visor.

Breathing apparatus is used in one of three forms (Figure 8.15):

▶ **self-contained breathing apparatus** – where air is supplied from compressed air in a cylinder and forms a completely sealed system;

▶ **fresh air hose apparatus** – fresh air is delivered through a hose to a sealed face mask from an uncontaminated source. The air may be delivered by the wearer, by natural breathing or mechanically by a fan;

▶ **compressed air-line apparatus** – air is delivered through a hose from a compressed air line. This can be either continuous flow or on demand. The air must be properly filtered to remove oil, excess water and other contaminants and the air pressure must be reduced. Special compressors are normally used.

The selection of appropriate RPE and correct filters for particular hazardous substances is best done by a competent specialist person.

There are several important technical standards which must be considered during the selection process. RPE must either be CE marked or HSE approved

(a)

(b)

(c)

Figure 8.15 Types of respiratory protective equipment: (a) filtering half-mask; (b) half-mask – reusable with filters; (c) compressed air-line breathing apparatus with full face fitted with demand value

(HSE approval ceased in 1995 but such approved equipment may still be used). Other standards include the minimum protection required (MPR) and the assigned protection factor (APF). The CE mark does not indicate that the equipment is suitable for a particular hazard. The following information will be needed before a selection of suitable RPE can be made:

- details of the hazardous substance, in particular whether it is a gas, vapour or dust or a combination of all three;
- presence of a beard or other facial hair which could prevent a good leak-free fit (a simple test to see whether the fit is tight or not is to close off the air supply, breathe in and hold the breath. The respirator should collapse onto the face. It should then be possible to check to see if there is a leak);
- the size and shape of the face of the wearer and physical fitness;
- compatibility with other personal protective equipment, such as ear defenders;
- the nature of the work and agility and mobility required.

Filters and masks should be replaced at the intervals recommended by the supplier or when taste or smell is detected by the wearer.

The performance of RPE with a tight-fitting face piece depends on a good contact between the wearer's skin and the face seal of the mask. However, research by the British Safety Industry Federation indicated that up to 50% of all RPE used does not offer the wearer the level of protection assumed, usually because it was not fitted correctly.

The COSHH Approved Code of Practice states that *'Employers should ensure that the selected face piece is of the right size and can correctly fit each wearer. For a tight-fitting face piece, the initial selection should include fit testing to ensure the wearer has the correct device. Also, employers must ensure that whoever carries out the fit testing is competent to do so.'*

The British Safety Industry Federation (working with the HSE) has developed 'Fit2Fit', an accreditation scheme for people performing face piece fit testing. An HSE document 'Fit testing of RPE face pieces' provides more detail on fit testing methods. The HSE uses the term 'face piece' and defines this to mean a full-face mask, a half-face mask or a filtering face piece/disposable mask. Fit2Fit standards are based on this HSE document.

Fit testing is needed when RPE is used as a control measure, not when it is worn by choice for comfort. Fit testing should be repeated if the shape of the face of the wearer changes for any reason, such as weight loss or gain. Re-testing is recommended to check that the face piece remains suitable and that the wearer is taking care to wear and remove the mask correctly. A two-year cycle has been suggested, and a one-year re-test is suggested for work with asbestos.

Fit testing can be qualitative or quantitative. Qualitative testing involves a simple pass or fail based on whether the user can taste or smell a harmless aerosol or odour through the RPE. Qualitative testing is only appropriate for disposable or half-face masks.

A quantitative fit test may use a laboratory test chamber or a portable fit test device, and produces a numerical 'fit factor' measure. The British Safety Industry Federation 'Fit2Fit' website provides details of accredited Fit2Fit training providers. The results of fit tests should be recorded. Fit test records must be made available to the HSE on request, and to the employees who wear RPE. RPE should be checked every time that it is used to make sure it has not been damaged and is being worn correctly.

A filtering face piece (FFP3) device is a mask which is certified to the PPE Directive. It provides a high level of filtering capability and face fit. It can be supplied with an exhale valve so that it can be worn comfortably over a fairly long period of time. It will provide an effective barrier to both droplets and fine aerosols and is the type recommended particularly for people in the healthcare sector dealing with symptomatic patients undergoing treatment where aerosols are likely to be generated.

All RPE should be examined at least once a month except for disposable respirators. A record of the inspection should be kept for at least 5 years. There should be a routine cleaning system in place and proper storage arrangements.

Respiratory Protective Equipment – A Practical Guide for Users, HSG53, HSE Books, contains comprehensive advice and guidance on RPE selection, use, storage, maintenance and training, and should be consulted for more information.

Hand and skin protection

Hand and skin protection is mainly provided by gloves (arm shields are also available). A wide range of safety gloves is available for protection from chemicals, sharp objects, rough working and temperature extremes. Many health and safety catalogues give helpful guidance for the selection of gloves. For protection from chemicals, including paints and solvents, impervious gloves are recommended. These may be made of PVC, nitrile or neoprene. For sharp objects, such as trimming knives, a Kevlar-based glove is the most effective. Gloves should be regularly inspected for tears or holes since this will obviously allow skin contact to take place.

Common issues with safety gloves include:

- dexterity problems when users cannot feel what they are handling;
- poor grip when handling wet/oily substances;
- entanglement, where gloves (particularly loose fitting or longer gloves/gauntlets) become caught in moving machinery; and
- gloves filling with substances from the open-end during immersion due to short length.

Another effective form of skin protection is the use of barrier creams and these come in two forms – pre-work and after-work. Pre-work creams are designed to provide a barrier between the hazardous substance and the skin. After-work creams are general purpose moisteners which replace the natural skin oils removed either by solvents or by washing.

Eye protection

Eye protection comes in three forms – spectacles (safety glasses), goggles and face visors. Eyes may be damaged by chemical and solvent splashes or vapours, flying particles, molten metals or plastics, non-ionising radiation (arc welding and lasers) and dust. Spectacles are suitable for low-risk hazards (low-speed particles such as machine swarf). Some protection against scratching of the lenses can be provided but this is the most common reason for replacement. Prescription lenses are also available for people who normally wear spectacles.

Figure 8.16 Variety of eye protection goggles

Goggles are best to protect the eyes from dust or solvent vapours because they fit tightly around the eyes. Visors offer protection to the face as well as the eyes and do not steam up so readily in hot and humid environments. For protection against very bright lights, special light filtering lenses are used (e.g. in arc welding). Maintenance and regular cleaning are essential for the efficient operation of eye protection.

When selecting eye protection, several factors need to be considered. These include the nature of the hazard (the severity of the hazard and its associated risk will determine the quality of protection required), comfort and user acceptability, compatibility with other personal protective equipment, training and maintenance requirements and costs.

Protective clothing

Protective clothing includes aprons, boots and headgear (hard hats and bump caps). Aprons are normally made of PVC and protect against spillages but can become uncomfortable to wear in hot environments. Other lighter fabrics are available for use in these circumstances. Safety footwear protects against falling

objects, collision with hard or sharp objects, hot or molten materials, slippery surfaces and chemical spills. It has metal toecaps and comes in the form of shoes, ankle boots or knee-length boots and is made of a variety of materials dependent on the particular hazard (e.g. thermally insulated against cold environments). It must be used with care near live, unprotected-against electricity. Specialist advice is needed for use with flammable liquids.

Appropriate selection of safety footwear involves the matching of the workplace hazards to the performance requirements of the footwear. The key issues are:

▶ the type of hazard (e.g. physical, chemical or thermal);
▶ the type of environment (e.g. indoors or out of doors);
▶ the ergonomics of the job (e.g. standing up, constant movement).

The footwear must have the correct grip for the environment, a hard-wearing sole unit and, possibly, a good shock-absorbing capability.

It is important to note that appropriate personal protective equipment should be made available to work-related visitors and other members of the public visiting workplaces where hazardous substances are being used. It is also important to stress that managers and supervisors must lead by example, particularly if there is a legal requirement to wear particular personal protective equipment. Refusals by employees to wear mandatory personal protective equipment must lead to some form of disciplinary action.

In 2002, some extra requirements were added to the Personal Protective Equipment Regulations.

▶ The personal protective equipment must satisfy the basic health and safety requirements which are applicable to that class or type of personal protective equipment.
▶ The appropriate conformity assessment procedures must be carried out.
▶ CE marking must be correctly affixed.
▶ The personal protective equipment must not compromise the safety of individuals, domestic animals or property when properly maintained and used.

Additional controls are needed for substances that can cause cancer, asthma or genetic damage (see Chapter 13).

Health surveillance and personal hygiene

The need for health surveillance in assessing exposure to hazardous substances has already been introduced in Section 8.2.6. It is required when there appears to be a reasonable chance that ill-health effects are occurring in a particular workplace as a result of reviewing sickness records or when a substance listed in Schedule 6 under Regulation 11 of the COSHH Regulations is being used.

Schedule 6 lists the substances and the processes in which they are used. There are a limited number of such substances and includes medical surveillance by an employment medical adviser or appointed doctor at intervals not exceeding 12 months. The need for health surveillance is not common, and further advice on the necessary procedures is available from the Employment Medical Advisory Service.

Health surveillance enables the identification of those employees most at risk from occupational ill health. It should not be confused with health monitoring procedures such as pre-employment health checks or drugs and alcohol testing, but it covers a wide range of situations, from a responsible person looking for skin damage on hands to medical surveillance by a medical doctor. Health surveillance allows for early identification of ill health and helps identify any corrective action needed. It may be required by law if employees are exposed to noise or vibration, solvents, isocyanates or other respiratory sensitisers, fumes, dusts, biological agents and asbestos, lead, work in compressed air or with ionising radiation. Health surveillance is a system of ongoing health checks and is important for:

▶ detecting ill-health effects at an early stage, so employers can introduce better controls to prevent them getting worse;

▶ providing data to help employers evaluate health risks;

▶ enabling employees to raise concerns about how work affects their health;

▶ highlighting lapses in workplace control measures, therefore providing invaluable feedback to the risk assessment;

▶ providing an opportunity to reinforce training and education of employees (e.g. on the impact of health effects and the use of protective equipment).

When initiating a health surveillance programme, it is important to avoid blanket coverage for all employees as this can produce misleading results and waste money.

The risk assessment for the organisation should indicate when and where health surveillance is required. In its simplest form, health surveillance could involve employees checking themselves for signs or symptoms of ill health following a training session – for example, soreness, redness and itching on their hands and arms if they work with substances that can irritate or damage the skin. A responsible person, who has been trained by a competent medical practitioner, can also make routine basic checks, such as skin inspections or signs of rashes.

For more complicated assessments, an occupational health nurse or an occupational health doctor can ask about symptoms or carry out periodic examinations. Statutory medical surveillance involves a medical examination and possibly tests by a doctor with appropriate training and experience.

A health record must be kept for all employees under health surveillance. Health records, or a copy, should be kept in a suitable form for at least 40 years from the date of last entry because often there is a long period between exposure and onset of ill health. Recorded details of each health surveillance check should include:

▶ the date they were carried out and by whom;

▶ the outcome of the test/check;

▶ the decision made by the occupational health professional in terms of fitness for task and any restrictions required. This should be factual and only relate to the employee's functional ability and fitness for specific work, with any advised restrictions.

Health records are different to medical records in that they should not contain confidential medical information.

Personal hygiene has already been covered under supervisory controls. It is very important for workers exposed to hazardous substances to wash their hands thoroughly before eating, drinking or smoking. Protection against biological hazards can be increased significantly by vaccination (e.g. tetanus). Finally, contaminated clothing and overalls need to be removed and cleaned on a regular basis.

Maintenance and emergency controls

Engineering control measures will only remain effective if there is a programme of preventative maintenance available. Indeed the COSHH Regulations require that systems of adequate control are in place and the term 'adequate control' spans normal operations, emergencies and maintenance. Maintenance will involve the cleaning, testing and, possibly, the dismantling of equipment. It could involve the changing of filters in extraction plant or entering confined spaces. It will almost certainly require hazardous substances to be handled and waste material to be safely disposed of. It may also require a permit-to-work procedure to be in place since the control equipment will be inoperative during the maintenance operations. Records of maintenance should be kept for at least 5 years.

Emergencies can range from fairly trivial spillages to major fires involving serious air pollution incidents. The following points should be considered when emergency procedures are being developed:

▶ the possible results of a loss of control (e.g. lack of ventilation);

▶ dealing with spillages and leakages (availability of effective absorbent materials);

▶ raising the alarm for more serious emergencies;

▶ evacuation procedures, including the alerting of neighbours;

▶ firefighting procedures and organisation;

▶ availability of respiratory protective equipment;

▶ information and training.

The Emergency Services should be informed of the final emergency procedures and, in the case of the Fire Service, consulted for advice during the planning of the procedures. (See Chapter 4 for more details on emergency procedures.)

The transport of hazardous substances by road

Although this topic is not in the NEBOSH National General Certificate syllabus, a brief mention of the main precautions required to safeguard the health and safety of those directly involved in the transport of hazardous substances and of general members of the public is important.

Data sheets from the manufacturer of the hazardous substance should indicate the safest method of handling it and will give information on emergency procedures (e.g. for spillages and fire). These sheets should be available to all concerned with the transportation of the substance, in particular those responsible for loading/unloading, as well as the driver. The hazardous substance should be loaded correctly on the vehicle in suitable containers and segregated from incompatible materials. There must be adequate emergency information with the substance containers and attached to the vehicle.

Drivers of the vehicles must receive special training which covers issues such as emergency procedures and route planning. There should also be emergency provisions for first aid and personal protective equipment on the vehicle. Finally, the Transport of Dangerous Goods (Safety Adviser) Regulations require the appointment of a trained and competent safety adviser to ensure the safe loading, transportation and unloading of the hazardous substance. HSE has produced several guidance publications which offer more detailed advice on this topic.

An illustrative example using COSHH controls

Organic solvents are widely used throughout industry and commerce in paints, inks, glues and adhesives. The COSHH hierarchy discussed under 8.4.2 should be applied to minimise the health risks from the use of these solvents. The top of this hierarchy is to eliminate or substitute the use of the organic solvent by using a less volatile or water-based alternative. If this is not possible, then some form of engineering control should be applied, such as dilution or local exhaust ventilation. Alternatively, the workplace, where the solvents are being used, could be enclosed or isolated from the main work activities. Other engineering type controls could include the use of properly labelled anti-spill containers, the use of covered disposal units for any used cloths and the transfer of large quantities using pumping/pipe arrangements rather than simply pouring the solvent.

Supervisory controls include the reduction in the length of time any employee is exposed to the solvent and the provision of good housekeeping, such as ensuring that containers are kept closed when not in use and any spills are quickly removed. The provision of barrier creams and personal protective equipment (eye protection, gloves and aprons) and RPE may also be required. Welfare issues will include first-aid provision, washing facilities and the encouragement of high levels of personal hygiene. Smoking and the consumption of food and drink should be prohibited where there might be contamination from organic solvents. All these supervisory items should be reinforced in training sessions and employees given appropriate information on the risks associated with the solvents. Finally, some form of health surveillance will be needed so that employees who show allergies to the solvents can be treated and, possibly, assigned other duties.

8.5 Specific agents

8.5.1 Health risks and controls associated with asbestos

It is estimated that 1.8 million workers remain at risk every day from the possible disturbance of this carcinogenic material. Those most at risk are workers such as electricians, plumbers, joiners and IT installers. Asbestos is responsible for over 4,500 deaths every year and 4 plumbers, 20 tradesmen, 6 electricians and 8 joiners die every week. Younger people, if routinely exposed to asbestos fibres over time, are at greater risk of developing asbestos-related disease than older workers. This is due to the time it takes for the body to develop symptoms after exposure to asbestos (latency). Exposure to asbestos can cause four main diseases:

▶ Mesothelioma (a cancer of the lining of the lungs; it is always fatal and is almost exclusively caused by exposure to asbestos)
▶ Asbestos-related lung cancer (which is almost always fatal)
▶ Asbestosis (a scarring of the lungs which is not always fatal but can be a very debilitating disease, greatly affecting quality of life).
▶ Diffuse pleural thickening (a thickening of the membrane surrounding the lungs which can restrict lung expansion leading to breathlessness.)

The number of mesothelioma deaths is more than 2,000 each year and increasing year on year. Most mesothelioma deaths occurring now are a legacy of past occupational exposures to asbestos when it was widely used in the building industry.

Asbestos appears in three main forms – crocidolite (blue), amosite (brown) and chrysotile (white). The blue and brown asbestos are considered to be the most dangerous and may be found in older buildings where they were used as heat insulators around boilers and hot water pipes and as fire protection of structures. White asbestos has been used in asbestos cement products and brake linings. It is difficult to identify an asbestos product by its colour alone – laboratory identification is usually required. Many asbestos-containing materials (ACMs) are difficult to distinguish from other materials. It is easy to drill or cut ACMs unwittingly and release large quantities of airborne fibres that could cause long-term health problems to the operator. Asbestos produces a fine fibrous dust of respirable dust size which can become lodged in the lungs. The fibres can be very sharp and hard causing damage to the lining of the lungs over a period of many years.

The condition of the ACM will need to be monitored regularly and a record kept. The time between these monitoring checks should not exceed 12 months and may need to be more frequent. Monitoring involves visual inspection to see whether there has been any deterioration at the surface of the ACM. This may be remedied by resealing the surface or removing a section of the ACM.

If asbestos is discovered during the performance of a contract, work should cease immediately and the employer informed. Typical sites of asbestos include ceiling tiles, asbestos cement roof and wall sheets, sprayed asbestos coatings on structural members, loft insulation and asbestos gaskets. Asbestos has its own Regulations (Control of Asbestos Regulations) and a summary of these is given in Chapter 13. These cover the need for a risk assessment, a method statement covering removal and disposal, air-monitoring procedures and the control measures (including personal protective equipment and training) to be used.

Training is required for the majority of workers involved in maintenance, refurbishment and demolition. The Health and Safety Executive (HSE) has estimated that approximately 50% of buildings still contain some form of asbestos and about 1.5 million workers require asbestos training.

The most recent Asbestos Regulations have removed textured coatings (decorative products such as Artex, Wondertex and Pebblecoat) from the asbestos licensing regime. Until 1992, these products contained white asbestos. This amendment by the HSE followed research work by the Health and Safety Laboratory. The new Regulations introduce an additional training requirement for asbestos awareness training. Such training should include:

▶ the health risks caused by exposure to asbestos;
▶ the materials that are likely to contain asbestos and where they are likely to be found;
▶ the methods to reduce asbestos risks during work;
▶ the action to take in an emergency, such as an uncontrolled release of asbestos dust.

Asbestos is the single biggest workplace killer. According to HSE statistics, there are 15 times as many deaths from asbestos as there are deaths from workplace accidents. Asbestos is responsible for at least 4,000 deaths in the UK each year, and the HSE felt that there was a need to increase awareness amongst the workforce of the risks associated with this material.

8.5.2 Managing asbestos in buildings

A study published by the *British Medical Journal* found that there are 1,800 mesothelioma deaths each year in Britain. Since this disease can take between 15 and 60 years to develop, the peak of the epidemic has still to be reached. In the construction industry, those at risk are asbestos removal workers and those, such as electricians, plumbers and carpenters, who are involved in refurbishment, maintenance or repair of buildings.

It was used widely as a building material until the mid-1980s. Although much asbestos has been removed from buildings, it has been estimated that over half a million non-domestic buildings still have asbestos in them amounting to many thousands of tons.

The strategy of the HSE is to ensure that those involved in the repair, removal or disturbance of asbestos-containing materials (ACMs), such as insulation, coatings or insulation boards, are licensed, competent and working to the strict requirements of the Control of Asbestos Regulations. This requires the identification of ACMs and the planning of any subsequent work. This should prevent inadvertent exposure to asbestos and minimise the risks to those who have to work with it.

The Regulations bring together the three previous sets of Regulations covering the prohibition of asbestos, the control of asbestos at work and asbestos licensing. The duty to manage asbestos is a legal requirement under the Control of Asbestos Regulations (Regulation 4). It applies to the owners and occupiers of commercial premises (such as shops, offices, industrial units, etc.) who have responsibility for maintenance and repair activities. In addition to these responsibilities, they also have a duty to assess the presence and condition of any asbestos-containing materials. If asbestos is present, or is presumed to be present, then it must be managed appropriately.

The Regulations also prohibit the importation, supply and use of all forms of asbestos. They continue the ban introduced for blue and brown asbestos in 1985 and for white asbestos in 1999. They also continue the ban on the second-hand use of asbestos products such as asbestos cement sheets and asbestos boards and tiles; including panels which have been covered with paint or textured plaster containing asbestos. The ban applies to new uses of asbestos. If existing ACMs are in good condition, they may be left in place and their condition monitored and managed to ensure that they are not disturbed. The Regulations only cover the safe management of asbestos in industrial and commercial

8

premises. Domestic premises are covered by the Defective Premises Act and the Civic Government (Scotland) Act although the Control of Asbestos Regulations apply to the shared parts of some domestic premises.

The Regulations also include the duty to manage asbestos in non-domestic premises. Guidance on the duty to manage asbestos can be found in the HSE ACoP – *Managing and working with asbestos, L143*.

The Regulations require mandatory training for anyone liable to be exposed to asbestos fibres at work, including maintenance workers and others who may come into contact with or who may disturb asbestos (e.g. cable installers) as well as those involved in asbestos removal work.

When work with asbestos or which may disturb asbestos is being carried out, the Regulations require employers and the self-employed to prevent exposure to asbestos fibres. Where this is not reasonably practicable, they must make sure that exposure is kept as low as reasonably practicable by measures other than the use of respiratory protective equipment. The spread of asbestos must be prevented. The Regulations specify the work methods and controls that should be used to prevent exposure and spread.

Worker exposure must be below the airborne exposure limit (the Control Limit) of 0.1 fibres per cm^3 for all types of asbestos. The Control Limit is the maximum concentration of asbestos fibres in the air (averaged over any continuous 4-hour period) and must not be exceeded. Short-term exposures must be strictly controlled and worker exposure should not exceed 0.6 fibres per cm^3 of air averaged over any continuous 10-minute period using respiratory protective equipment if exposure cannot be reduced sufficiently using other means. Respiratory protective equipment is an important part of the control regime but it must not be the sole measure used to reduce exposure and should only be used to supplement other measures.

Most asbestos removal work must be undertaken by a licensed contractor but any decision on whether particular work needs to be licensed is based on the level of risk. Details are given in Chapter 13 of the six possible reasons for work with asbestos being exempt from licensing.

The Control of Asbestos Regulations requires those in control of premises and the duty-holders to:

▷ take reasonable steps to determine the location and condition of materials likely to contain asbestos;
▷ presume materials contain asbestos unless there is strong evidence that they do not;
▷ make and keep an up-to-date record of the location and condition of the ACMs or presumed ACMs in the premises;

▷ assess the risk of the likelihood of anyone being exposed to fibres from these materials;
▷ prepare a plan setting out how the risks from the materials are to be managed;
▷ take the necessary steps to put the plan into action;
▷ review and monitor the plan periodically;
▷ provide such information and asbestos awareness training to anyone who is liable to work on these materials or otherwise disturb them.

In addition, the Regulations include the following main provisions:

1. a single tighter control limit for all types of asbestos;
2. specific training requirements for those working with asbestos;
3. a clear hierarchy of controls to be used to reduce exposure.

The HSE has produced a document – *HSG264 Asbestos: The survey guide* – that gives useful guidance on asbestos surveys. There are **two** types of survey, known in the UK as the **management survey** and the **refurbishment and demolition survey.**

A **management survey** is the standard survey. Its purpose is to locate, as far as reasonably practicable, the presence and extent of any suspect ACMs in the building which could be damaged or disturbed during normal occupancy, including foreseeable maintenance and installation, and to assess their condition. A management survey is required during the normal use of the building to ensure continued management of the ACMs. It should include an assessment of the condition of the various ACMs and their ability to release fibres into the air should they be disturbed in some way. It will often involve minor intrusive work and some disturbance.

Management surveys can involve a combination of sampling to confirm asbestos is present or presuming asbestos to be present. All ACMs should be identified as far as is reasonably practicable. The areas inspected should include under floor coverings, above false ceilings (ceiling voids), cladding and partitions, under carpets, tiles and floors, lofts, inside risers, service ducts and lift shafts and basements, cellars or underground rooms. In these situations, controls should be put in place to prevent the spread of debris, which may include asbestos.

A **refurbishment or demolition survey** is used to locate and describe, as far as reasonably practicable, all ACMs in the area where the refurbishment work will take place or in the whole building if demolition is planned. It is necessary when the building (or part of it) is to be upgraded, refurbished or demolished. It is required for all work which disturbs the fabric of the building in areas particularly where the management survey has not been intrusive. A refurbishment and demolition survey is needed before any refurbishment or demolition work is carried out.

Refurbishment and demolition surveys should only be conducted in unoccupied areas to minimise risks to the public or employees on the premises. Ideally, the building should not be in service and all furnishings removed.

For minor refurbishment, this would only apply to the room involved or even part of the room where the work is small and the room large. In these situations, there should be effective isolation of the survey area (e.g. full floor to ceiling partition), and furnishings should be removed as far as possible or protected using sheeting.

The person who undertakes any of these surveys must be suitably trained and experienced in such work.

Some types of work, of an intermittent and low intensity nature, will not have to be done by a licensed contractor – artex work is an example of this.

There are many issues which need to be addressed when asbestos is possibly present in a workplace. Several publications are available from HSE Books which cover all these issues in some detail and the reader should refer to them for more information. Here a brief summary of the principal issues will be given.

Identification of the presence of asbestos is the first action. Asbestos is commonly found as boiler and pipe lagging, insulation panels around pillars and ducting for fire protection and heat insulation, ceiling tiles and asbestos cement products, including asbestos cement sheets. The main duty-holder is required to ensure that a written plan is prepared that shows where the ACM is located and how it will be managed to prevent exposure to asbestos, including to contractors and other workers who may undertake work on the fabric of the building that could disturb the ACM. This plan then needs to be put into action and communicated to those affected. The duty-holder should ensure that the plan is reviewed regularly and updated as circumstances change, in consultation with all those who may be affected. The plan, including drawings, should be available on site for the entire life of the premises and should be kept up to date.

Initial investigations will involve the examination of building plans, the determination of the age of the building and a thorough examination of the building. Advice is available from a number of reputable specialist consultants and details may be obtained from the local authority, who often offers such a service. If the specialist is in any doubt, a sample of the suspect material will be sent to a specialist laboratory for analysis. It is important for a specialist to take the sample because the operation is likely to expose loose fibres.

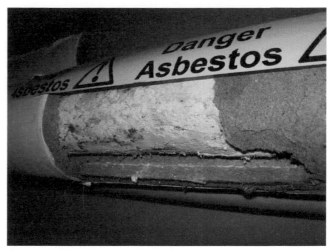

Figure 8.17 Damaged asbestos lagging on pipework

The asbestos risk register is a key component of the asbestos management. The management plan must contain current information about the presence and condition of any asbestos in the building. The asbestos risk register will therefore need to be updated on a regular basis (at least once a year). To do this you should make:

▶ regular inspections to check the current condition of asbestos materials;
▶ deletions to the register when any asbestos is removed;
▶ additions to the register when new areas are surveyed and asbestos is located;
▶ changes to the register (at any time asbestos-containing materials are found to have deteriorated).

The register can be kept as a paper or electronic record and it is very important that this is kept up to date and easily accessible should any future maintenance be necessary. Paper copies may be easier to pass on to visiting maintenance workers, who will need them to know the location and condition of any asbestos before they start work. Electronic copies are easier to update and are probably better suited for people responsible for large numbers of properties or bigger premises.

An example of an asbestos register is available on the HSE website.

Assessment is an evaluation or asbestos survey to determine whether the location or the condition could lead to the asbestos being disturbed. If it is in good condition, undamaged and not likely to be disturbed, then it is usually safer to leave it in place and monitor it. However, if it is in a poor condition, it may need to be repaired, sealed, enclosed or removed. If there is doubt, then specialist advice should be sought. The condition of the ACM will need to be monitored regularly and a record kept. The time between these monitoring checks should not exceed 12 months and may need to be more frequent. Monitoring involves visual inspection to see whether there has been any deterioration at the surface of the ACM. This may be remedied by resealing the surface or removing a section of the ACM.

The asbestos survey should provide enough information so that an asbestos register, a risk assessment and a management plan can then be prepared. The survey will usually involve sampling and analysis to determine the presence of asbestos, so asbestos surveys should only be carried out by competent surveyors who can clearly demonstrate they have the necessary skills, experience and qualifications.

Removal must only be done by a licensed contractor. A detailed plan of work is essential before work begins. The plan should give details of any equipment to be used for the protection and decontamination of employees and others. This process will also require an assessment to be made to ensure that people within the building and neighbours are properly protected. At the planning stage, generally, the HSE or the local authority must be notified of the intention to remove asbestos (at least 14 days' notice) and again when the work begins. This is particularly important when the exposure to asbestos is likely to exceed the action level. The assessment should include details of the type and location of the asbestos, the number of people who could be affected, the controls to be used to prevent or control exposure, the nature of the work, the removal methods, the procedures for the provision of personal and respiratory protective equipment and details of emergency procedures. If asbestos cement sheeting is to be removed the following procedure is recommended:

▶ Where reasonably practicable, remove the asbestos sheets before any other operation, such as demolition.
▶ Avoid any breaking of the sheets.
▶ Dampen the sheets while working on them.
▶ Lower the sheets on to a clean hard surface.
▶ Remove all waste and debris from the site as soon as possible to prevent its spread around the site.
▶ Do not bulldoze broken asbestos cement or sheets into piles.
▶ Do not dry sweep asbestos cement debris.
▶ Dispose of the waste and debris safely, separate from general waste as hazardous waste.

Control measures during the removal of asbestos include the provision of personal and RPE including overalls, good ventilation arrangements in and the segregation or sealing of the working area, suitable method statements and air monitoring procedures. A decontamination unit should also be provided. The sealed area will need to be tested for leaks. Good supervision and induction of the workforce are also essential. A high level of personal hygiene must be expected for all workers and the provision of welfare amenity arrangements, particularly washing and catering facilities and the separation of working and personal

Figure 8.18 High hazard vacuum cleaner to clear up asbestos material

clothing. Suitable warning signs must be displayed and extra controls provided if the work is taking place at height. After the work is completed, the area must be thoroughly cleaned and a clean air certificate provided after a successful air test.

Work with high-risk asbestos products, such as asbestos insulating board, sprayed fire or insulation coatings or lagging, must only be carried out by an asbestos licence holder who works within the conditions of their licence.

If asbestos is disturbed, the first response should be to stop the disturbance and the second should be to isolate the location.

The HSE advise that *'exposure to asbestos is reportable under RIDDOR when a work activity causes the accidental release or escape of asbestos fibres into the air in a quantity sufficient to cause damage to the health of any person'*.

Medical surveillance in the form of a regular medical examination by an appointed doctor should be given to any employee who has been exposed to asbestos at levels above the action level. The first medical examination should take place within 2 years of exposure and further examinations at intervals of not more than 2 years. A health record of such surveillance should be kept for a period of at least 40 years after the last entry.

Awareness training is an important feature of the Control of Asbestos Regulations and they require that adequate information, instruction and training are given to those of his/her employees:

(a) who are or who are liable to be exposed to asbestos, or who supervise such employees; and/or

(b) who carry out work in connection with the employer's duties under these Regulations, so that they can carry out that work effectively.

Such training should cover the following topics:

▷ properties of asbestos and its effects on health, including its interaction with smoking;

▷ the types of products or materials likely to contain asbestos;

▷ the uses and location of ACMs in buildings and plant;

▷ the operations which could result in asbestos exposure and the importance of preventative controls to minimise exposure;

▷ the presence of other hazards such as working at height;

▷ the requirements of the Control of Asbestos Regulations;

▷ safe work practices, control measures, and protective equipment;

▷ the purpose, choice, limitations, proper use and maintenance of RPE;

▷ emergency procedures;

▷ hygiene requirements;

▷ decontamination procedures;

▷ waste handling procedures;

▷ medical examination requirements; and

▷ the control limit and the need for air monitoring.

Refresher training should be given at regular intervals (at least annually) and adapted to take account of significant changes in the type of work carried out or methods of work used by the employer. It should be provided in a manner appropriate to the nature and degree of exposure identified by the risk assessment, and so that the employees are aware of:

▷ the significant findings of the risk assessment; and

▷ the results of any air monitoring carried out with an explanation of the findings.

Disposal of asbestos waste is subject to the Hazardous Waste Regulations which require it to be consigned to an authorised asbestos waste site only. Asbestos waste describes any asbestos products or materials that are ready to be disposed. This includes any contaminated building materials, dust, rubble, used tools that cannot be decontaminated, disposable PPE and damp rags that have been used for cleaning. Asbestos waste must be placed in suitable packaging to prevent any fibres being released. This should be double wrapped and appropriately labelled. Standard practice is to use a red inner bag marked up with asbestos warning labels and a clear outer bag with appropriate hazard markings. Intact asbestos cement sheets and textured coatings that are firmly attached to a board should not be broken up into smaller pieces. These should instead be carefully double wrapped in suitable polythene sheeting (1,000 gauge) and labelled.

The waste container must be strong enough to securely contain the waste and not become punctured; it must be easily decontaminated, kept securely on the site until required. The waste must only be carried by a licensed carrier.

Accidental exposure to ACMs can occur even when all reasonable precautions have been taken. If an ACM has been worked on by a worker who did not realise that it was an ACM or has accidentally damaged an ACM, then all work must stop immediately and nobody should be allowed to enter the area in question. The work supervisor must be informed and a sample of the material sent for analysis. If the result is positive, then a specialist licensed contractor should be employed unless the work is exempt from the need for a licence. Any contaminated clothing must be removed and placed in a plastic bag. The affected worker should shower as soon as possible or wash thoroughly.

More information on the legislative requirements is given in Chapter 13.

8.5.3 Health risks and controls associated with other specific agents

The health hazards associated with hazardous substances can vary from very mild (momentary dizziness or a skin irritation) to very serious, such as a cancer.

Cancer is a serious body cell disorder in which the cells develop into tumours. There are two types of tumour – benign and malignant. Benign tumours do not spread but remain localised within the body and grow slowly. Malignant tumours are called cancers and often grow rapidly, spreading to other organs using the bloodstream and lymphatic system. Survival rates have improved dramatically in recent years as detection methods have improved and the tumours can be found in their early stages of development.

Workplace exposures leading to occupational illness and work-related cancers account for 8,000, of which half are caused by asbestos. Five of the key causes of occupational cancer registrations and deaths are:

▷ Diesel engine exhaust emissions – related deaths in the UK are estimated at 650 a year to lung or bladder cancer.

▷ Solar radiation – at over 1,500 a year.

▷ Asbestos – as covered earlier.

▷ Shift work – causes an estimated 2,000 cases of breast cancer a year.

▷ Silica – over 600 cancer registrations.

A minority of cancers are believed to be occupational in origin.

Occupational asthma has approximately 4 million sufferers in the UK and it is estimated that 8 million working days are lost each year as a result of it. It is mainly caused by breathing in respiratory sensitisers, such as wood dusts, organic solvents, solder flux fumes or animal hair. The symptoms are coughing, wheezing, tightness of the chest and breathlessness due to a constriction of the airways. It can be a mild attack or a serious one that requires hospitalisation. There is some evidence that stress can trigger an attack.

Chronic obstructive pulmonary disease (COPD) is a lung disease causing chronic obstruction of lung airflow that interferes with normal breathing. It has been estimated that it affects 1.2 million people in the UK.

The following common agents of health hazards will be described together with the circumstances in which they may be found (those marked* are not in the NEBOSH National Certificate syllabus).

Ammonia* is a colourless gas with a distinctive odour which, even in small concentrations, causes the eyes to smart and run and a tightening of the chest. It is a corrosive substance which can burn the skin, burn and seriously damage the eye, cause soreness and ulceration of the throat and severe bronchitis and oedema (excess of fluid) of the lungs. Good eye and respiratory protective equipment (RPE) is essential when maintaining equipment containing ammonia. Any such equipment should be tested regularly for leaks and repaired promptly if required. Ammonia is also used in the production of fertilisers and synthetic fibres. Most work on ammonia plant should require a permit-to-work procedure.

Chlorine* is a greenish, toxic gas with a pungent smell which is highly irritant to the respiratory system, producing severe bronchitis and oedema of the lungs and may also cause abdominal pain, nausea and vomiting. It is used as a disinfectant for drinking water and swimming pool water and in the manufacture of chemicals.

Organic solvents* are used widely in industry as cleansing and degreasing agents. There are two main groups – the hydrocarbons (includes the aromatic and aliphatic hydrocarbons, such as toluene and white spirit) and the non-hydrocarbons (such as trichloroethylene and carbon tetrachloride). All organic solvents are heavier than air and most are sensitisers and irritants. Some are narcotics, while others can cause dermatitis and after long exposure periods liver and kidney failure. It is very important that the safety data sheet accompanying the particular solvent is read and the recommended personal protective equipment is worn at all times.

Solvents are used extensively in a wide variety of industries as varnishes, paints, adhesives, glue strippers, printing inks and thinners. They are highest risk when used as sprays. One of the most hazardous is dichloromethane (DCM), also known as methylene chloride. It is used as a paint stripper, normally as a gel. It can produce narcotic effects and has been classified as a Category 3 carcinogen in the European Union. The minimum personal protective equipment requirements are impermeable overalls, apron, footwear, long gloves and gauntlet and chemically resistant goggles or visor. Respiratory protective equipment is also required if it cannot be demonstrated that exposure is below the appropriate workplace exposure limit (WEL).

Carbon dioxide* is a colourless and odourless gas which is heavier than air. It represses the respiratory system, eventually causing death by asphyxiation. At low concentrations it will cause headaches and sweating followed by a loss of consciousness. The greatest hazard occurs in confined spaces, particularly where the gas is produced as a by-product.

Carbon monoxide is a colourless, tasteless and odourless gas which makes it impossible to detect without special measuring equipment. As explained earlier, carbon monoxide enters the blood (red cells) more readily than oxygen and restricts the supply of oxygen to vital organs. At low concentrations in the blood (less than 5%), headaches and breathlessness will occur, while at higher concentrations unconsciousness and death will result. The most common occurrence of carbon monoxide is as an exhaust gas either from a vehicle or a heating system. In either case, it results from inefficient combustion and, possibly, poor maintenance.

Carbon monoxide exposure may be prevented by ensuring that any work carried out in relation to gas appliances in domestic or commercial premises is undertaken by a Gas Safe Registered engineer, competent in that area of work and that there is enough fresh air in the room containing a gas appliance. In addition the HSE recommends the use of audible carbon monoxide (CO) alarms as a useful back-up precaution.

All gas workers must be signed up to the **Gas Safe Register**. This register replaces the requirement for all gas workers to be registered with CORGI and the scheme is overseen by the HSE. Consumers and landlords should check whether gas workers are Gas Safe registered since the failure to check that an installer is registered under the **Gas Safe Scheme** may result in a breach of the Building Regulations.

Employers and landlords should ensure their appliances are serviced regularly by an approved and competent engineer. Registered engineers can be identified by the display of a new yellow Gas Safe Register logo, and are required to carry a Gas Safe Register identification card with their own unique licence number.

There have been cases where gas leaks, electricity and water faults have developed in buildings constructed with timber frames as a result of the frame shrinking, thereby placing excessive weight on these services. Shrinkage usually only takes place with new wood, and within its first year of use.

Diesel-engine exhaust emissions* have been classified as carcinogenic by experts from the World Health Organization (WHO). The International Agency for Research on Cancer (IARC) found sufficient evidence that exposure to diesel exhausts is associated with an increased risk of lung cancer. They also noted limited evidence that exposure could lead to an increased risk of bladder cancer. The classification follows the findings in a US study of occupational exposure to such emissions in underground miners, which showed an increased risk of death from lung cancer in exposed workers. Exposure to these particulates may also aggravate other respiratory diseases such as asthma.

Isocyanates* are volatile organic compounds widely used in industry for products such as printing inks, adhesives, and two-pack paints (particularly in vehicle body shops) and in the manufacture of plastics (polyurethane products). They are irritants and sensitisers. Inflammation of the nasal passages and the throat and bronchitis are typical reactions to many isocyanates. When a person becomes sensitised to an isocyanate, very small amounts of the substance often provoke a serious reaction similar to an extreme asthma attack. Isocyanates also present a health hazard to firefighters. They are subject to a workplace exposure limit (WEL) and respiratory protective equipment (RPE) should be worn.

Lead* is a heavy, soft and easily worked metal. It is used in many industries but is most commonly associated with plumbing and roofing work. Over 300,000 tonnes of lead are used annually in the UK, of which 80% is used in the manufacture of lead-acid batteries and 6% is extruded into sheets for use in construction. The toxicity of lead affects nerve conduction and red blood cell production. Symptoms include abdominal pains, muscular weakness and tiredness. Inhalation is the most common means for the entry of lead into the body. From the lungs, lead enters the bloodstream and is distributed throughout the body.

Lead enters the body normally by inhalation but can also enter by ingestion and skin contact. The main targets for lead are the spinal cord and the brain, the blood and blood production. The effects are normally chronic and develop as the quantity of lead builds up. Headaches and nausea are the early symptoms followed by anaemia, muscle weakening and (eventually) coma. Regular blood tests are a legal and sensible requirement as are good ventilation and the use of appropriate personal protective equipment. High personal hygiene standards and adequate welfare (washing) facilities are essential and must be used before smoking or food is consumed. The reduction in the use of leaded petrol was an acknowledgement of the health hazard represented by lead in the air. Lead is covered by

its own set of regulations – the Control of Lead at Work Regulations (summarised in Chapter 13). These Regulations require risk assessments to be undertaken and engineering controls to be in place. They also recognise that lead can be transferred to an unborn child through the placenta and, therefore, offer additional protection to women of reproductive capacity. Medical surveillance in the form of a blood test of all employees who come into contact with lead operations, is required by the Regulations. Such tests should take place at least once a year. Regular health surveillance of lead workers by a medical practitioner is required by these Regulations.

Silica is the main component of most rocks and is a crystalline substance made of silicon and oxygen. It occurs in quartz (found in granite), sand and flint. Harm is caused by the inhalation of silica dust, which can lead to silicosis (acute and chronic), fibrosis and pneumoconiosis. The dust which causes the most harm is respirable dust which becomes trapped in the alveoli. This type of dust is sharp and very hard and, probably, causes wounding and scarring of lung tissue. As silicosis develops, breathing becomes more and more difficult and eventually as it reaches its advanced stage, lung and heart failure occur. It has also been noted that silicosis can result in the development of tuberculosis as a further complication. Hard rock miners, quarrymen, stone and pottery workers are most at risk. Health surveillance is recommended for workers in these occupations at initial employment and at subsequent regular intervals. Prevention is best achieved by the use of good dust extraction systems and respiratory protective equipment. Over 100 enforcement notices were served in 2009/10 on silica-dust related issues compared to 8 in 2008/9.

Silica is commonly produced during construction activities. Such activities which can expose workers and members of the public to silica dust include:

▷ cutting building blocks and other stone masonry work;
▷ cutting and/or drilling paving slabs and concrete paths;
▷ demolition work;
▷ sand blasting of buildings;
▷ tunnelling.

In general, the use of power tools to cut or dress stone and other silica-containing materials will lead to very high exposure levels while the work is occurring. In most cases, exposure levels are in excess of workplace exposure limits (WELs) by factors greater than 2 and in some cases as high as 8. In response to the growing evidence of ill-health effects of silica inhalation, the HSE revised the WEL down from 0.3 mg/m^3 to 0.1 mg/m^3.

The use of compressed air to remove dust is not recommended. There is a risk of compressed air entering the operator's bloodstream, which can result

261

in death. Eye injury including blindness can also occur if dust particles bounce back at the operator. The inhalation of very fine silica dust can lead to the development of silicosis. HSE has produced a detailed information sheet on silica – CIS No. 36 (Rev 1).

In addition to silica, there are three hazardous substances that are particularly relevant to construction activities – cement dust and wet cement, wood dust and the biological hazard tetanus. Respiratory protective equipment (RPE), usually in the form of a mask, will normally be needed for high-risk tasks such as:

▶ using a cut-off saw, grinder or wall chaser on material containing silica;
▶ using powered cross-cut saws and sanders on hardwood, red cedar or medium-density fibreboard (MDF);
▶ sanding softwood in an enclosed space.

A filtering face piece FFP3 is the most advisable type of mask to use when high dust levels, silica or wood dust is produced. There are also filtering face pieces FFP1 and FFP2 available but these provide less respiratory protection than a properly fitting FFP3 device.

Cement dust and wet cement is important in construction and is also a hazardous substance. Contact with wet cement can cause serious burns or ulcers which will take several months to heal and may need a skin graft. Dermatitis, both irritant and allergic, can be caused by skin contact with either wet cement or cement powder. Allergic dermatitis is caused by an allergic reaction to hexavalent chromium (chromate) which is present in cement. Cement powder can also cause inflammation and irritation of the eye, irritation of the nose and throat, and, possibly, chronic lung problems. Research has shown that between 5% and 10% of construction workers are probably allergic to cement. And plasterers, concreters and bricklayers or masons are particularly at risk. A plasterer who knelt in wet cement for 5 hours while working required skin grafts to his legs.

Figure 8.19 Wet concrete or cement can cause serious dermatitis

Manual handling of wet cement or cement bags can lead to musculoskeletal health problems and cement bags weighing more than 25 kg should not be carried by a single worker. PPE in the form of gloves, overalls with long sleeves and full-length trousers and waterproof boots must be worn on all occasions. If the atmosphere is dusty, goggles and respiratory protection equipment must be worn. An important factor in the possibility of dermatitis occurring is the sensitivity of the worker to the chromate in the cement and the existing condition of the skin including cuts and abrasions. Finally, adequate welfare facilities are essential so that workers can wash their hands at the end of the job and before eating, drinking or using the toilet. If cement is left on the skin for long periods without being washed off, the risk of an allergic reaction to hexavalent chromium will increase.

A 2005 amendment to the COSHH Regulations prohibits the supply of cement which has a concentration of more than 2 parts per million of chromium VI. This measure is designed to prevent allergic contact dermatitis when wet cement comes into contact with the skin. However, since the strong alkalinity of cement will remain, there is still the potential for skin burns.

The Approved Code of Practice (ACoP) gives useful advice on possible control measures by offering two routes for employers to comply with the amended COSHH Regulations. They can either use the generic advice given in COSHH Essentials HSG193 or design a solution themselves with the help of competent advice. In any event, the controls should be proportionate to the health risk. The Regulations stress the need to provide adequate washing, changing, eating and drinking facilities.

Wood dust can be hazardous, particularly when it is hard wood dust which is known, in rare cases, to lead to nasal cancer. Composite boards, such as **medium-density fibreboard (MDF),** are hazardous due to the resin bonding material used, which can also be carcinogenic. There are three types of wood-based boards available: laminated board, particle board and fibreboard. The resins used to bond the fibreboard together contain formaldehyde (usually urea formaldehyde). It is generally recognised that formaldehyde is 'probably carcinogenic to humans' and is subject to a WEL. At low exposure levels, it can cause irritation to the eyes, nose and throat and can lead to dermatitis, asthma and rhinitis. The main problems are most likely to occur when the MDF is being machined and dust is produced. A suitable risk assessment should be made and gloves and appropriate masks should always be worn when machining MDF. However, it is important to stress that safer materials are available which do not contain formaldehyde and these should be considered for use in the first instance.

Wood dust is produced whenever wood materials are machined, particularly sawed, sanded, bagged as dust from dust extraction units or during cleaning operations, especially if compressed air is used. The main hazards associated with all wood dusts are skin disorders, nasal problems, such as rhinitis, and asthma. There is also a hazard from fire and explosion. A COSHH assessment is essential to show whether the particular wood dust is hazardous. When the wood dust is created inside a woodworking shop, a well-designed extraction system is essential. PPE in the form of gloves, suitable respiratory protective equipment, overalls and eye protection may also be necessary as a result of the assessment. Finally, good washing and welfare facilities are also essential.

Tetanus* is a serious, sometimes fatal, disease caused by a bacterium that lives in the soil. It usually enters the human body through a wound from an infected object, such as a nail, wood splinter or thorn. On entering the wound, it produces a powerful toxin which attacks the nerves that supply muscle tissue. It is commonly known as lockjaw because after an incubation period of approximately a week, stiffness around the jaw area occurs. Later the disease spreads to other muscles including the breathing system and this can be fatal. The disease has been well controlled with anti-tetanus immunisation and it is important that all construction workers are so immunised. Booster shots should be obtained every few years. Any flesh wound should be thoroughly cleaned immediately and an antiseptic cream applied.

Leptospirosis and Weil's disease are caused by a bacterium found in the urine of rats. In humans, the kidneys and liver are attacked causing high temperatures and headaches followed by jaundice and, in up to 20% of cases, it can be fatal. It enters the body either through the skin or by ingestion. The most common source is contaminated water in a river, sewer or ditch and workers, such as canal or sewer workers, are most at risk. Leptospirosis is always a risk where rats are present, particularly if the associated environment is damp. Good, impervious protective clothing, particularly wellington boots, is essential in these situations and the covering of any skin wounds. For workers who are frequently in high-risk environments (sewer workers), immunisation with a vaccine may be the best protection. Weil's disease is, strictly, a severe form of leptospirosis. The symptoms of leptospirosis are similar to influenza but those for Weil's disease are anaemia, nose bleeds and jaundice. While the most common source of infection is from the urine of rats, Weil's disease has been found in other animals, such as cattle; therefore, farm and veterinary workers may also be at risk.

Legionella is an airborne bacterium and is found in a variety of water sources. It produces a form of pneumonia caused by the bacteria penetrating to the alveoli in the lungs. This disease is known as Legionnaires' disease, named after the first

documented outbreak at a State Convention of the American Legion held at Pennsylvania in 1976. During this outbreak, 200 men were affected, of whom 29 died. That outbreak and many subsequent ones were attributed to air-conditioning systems. Legionnaires' disease is a potentially fatal form of pneumonia and everyone is susceptible to infection. However, some people are at higher risk, including:

▶ people over 45 years of age;
▶ smokers and heavy drinkers;
▶ people suffering from chronic respiratory or kidney disease;
▶ anyone with an impaired immune system.

The symptoms are a high temperature, fever and chills, cough, muscle pains and headache. In a severe case, there may also be pneumonia, and occasionally diarrhoea, as well as signs of mental confusion. Legionnaires' disease is not known to spread from person to person.

The legionella bacterium cannot survive at temperatures above 60°C but grows between 20°C and 45°C, being most virulent at 37°C. It also requires food in the form of algae and other bacteria. Control of the bacteria involves the avoidance of water temperatures between 20°C and 45°C, avoidance of water stagnation and the build-up of algae and sediments and the use of suitable water-treatment chemicals. This work is often done by a specialist contractor.

The most common systems at risk from the bacterium are:

▶ water systems incorporating a cooling tower;
▶ water systems incorporating an evaporative condenser;
▶ hot and cold water systems and other plant where the water temperature may exceed 20°C.

An Approved Code of Practice (*Legionnaires' disease – The control of legionella bacteria in water systems* – L8) is available from the HSE. The latest edition of the ACoP and guidance contains revisions to simplify and clarify the text. The main changes are removing Part 2, the technical guidance, which is published separately in HSG274, and giving the following issues ACoP status:

▶ risk assessment;
▶ the specific role of an appointed competent person, known as the 'responsible person';
▶ the control scheme;
▶ review of control measures;
▶ duties and responsibilities of those involved in the supply of water systems.

Where plant at risk of the development of legionella exists, the following is required:

▶ a written 'suitable and sufficient' risk assessment;
▶ the preparation and implementation of a written control scheme involving the treatment, cleaning and maintenance of the system;

▶ appointment of a named person with responsibility for the management of the control scheme;

▶ the monitoring of the system by a competent person;

▶ record keeping and the review of procedures developed within the control scheme.

The code of practice also covers the design and construction of hot and cold water systems and cleaning and disinfection guidance. There have been several cases of members of the public becoming infected from a contaminated cooling tower situated on the roof of a building. It is required that all cooling towers are registered with the local authority. People are more susceptible to the disease if they are older or weakened by some other illness. It is, therefore, important that residential and nursing homes and hospitals are particularly vigilant. The most common source of isolated outbreaks of legionella is showerheads, particularly when they remain unused for a period of time. Showerheads should be cleaned and descaled at least every three months.

All warm water systems should be checked annually by a competent person. However, there are some simple checks that can be made as follows:

▶ each month, check that the temperature of water from hot water taps reaches 50°C after the water has run for 1 minute;

▶ each month, check that the temperature of water from cold water taps is below 20°C after the water has run for 2 minutes;

▶ each quarter, clean and descale shower heads and hoses;

▶ each week, purge little used water outlets.

Hepatitis* is a disease of the liver and can cause high temperatures, nausea and jaundice. It can be caused by hazardous substances (some organic solvents) or by a virus. The virus can be transmitted from infected faeces (hepatitis A) or by infected blood (hepatitis B and C). The normal precautions include good personal hygiene, particularly when handling food and in the use of blood products. Hospital workers who come into contact with blood products are at risk of hepatitis as are drug addicts who share needles. It is also important that workers at risk regularly wash their hands and wear protective disposable gloves.

Hepatitis is an example of a **blood-borne virus**. Another example is human immunodeficiency virus (HIV) which causes acquired immune deficiency syndrome (AIDS), affecting the immune system of the body. Some people carry a blood-borne virus in their blood which may cause disease in some people and few or no symptoms in others. The virus can spread to another person, whether the carrier of the virus is ill or not.

In the workplace, direct exposure can happen through accidental contamination by a sharp instrument or through contamination of open wounds, skin abrasions

or through splashes to the eyes, nose or mouth. In occupations where there is a risk of exposure to blood-borne viruses, the following measures to prevent or control risks are recommended by the HSE:

▶ prohibit eating, drinking, smoking and the application of cosmetics in working areas where there is a risk of contamination;

▶ prevent puncture wounds, cuts and abrasions, especially in the presence of blood and body fluids;

▶ when possible avoid use of, or exposure to, sharps such as needles, glass or metal or if unavoidable take care in handling and disposal;

▶ consider the use of devices incorporating safety features, such as safer needle devices and blunt-ended scissors;

▶ cover all breaks in exposed skin by using waterproof dressings and suitable gloves;

▶ protect the eyes and mouth by using a visor/goggles/safety spectacles and a mask, where splashing is possible;

▶ avoid contamination by using water-resistant protective clothing;

▶ wear rubber boots or plastic disposable overshoes when the floor or ground is likely to be contaminated;

▶ use good basic hygiene practices, such as hand washing;

▶ control contamination of surfaces by containment and using appropriate decontamination procedures; and

▶ dispose of contaminated waste safely.

The risk of first aiders being infected with a blood-borne virus while carrying out their duties is small. There has been no recorded case of HIV or hepatitis B being passed on during mouth-to-mouth resuscitation. The following precautions are recommended by the HSE to reduce the risk of infection:

▶ cover any cuts or grazes on your skin with a waterproof dressing;

▶ wear suitable disposable gloves when dealing with blood or any other body fluids;

▶ use suitable eye protection and a disposable plastic apron where splashing is possible;

▶ use devices such as face shields when giving mouth-to-mouth resuscitation, but only if training in their use has been received; and

▶ wash hands after each procedure.

8.6 Additional useful background information that is not part of the NEBOSH NGC syllabus

8.6.1 Safe handling and storage of waste

Work-related deaths in the waste management industry are ten times the national average. Accident rates in the industry are four times the national average. The statutory duty of care for the management of waste derives from Part 2 of the Environmental Protection Act (EPA). The principal requirements are as follows:

▶ to handle waste so as to prevent any unauthorised escape into the environment;
▶ to pass waste only to an authorised person as defined by EPA;
▶ to ensure that a written description accompanies all waste. The Environmental Protection (Duty of Care) Regulations 1991 requires holders or producers of waste to complete a 'Transfer Note' giving full details of the type and quantity of waste for collection and disposal. Copies of the note should be kept for at least 2 years;
▶ to ensure that no person commits an offence under the Act.

Figure 8.20 Commercial waste collection

The EPA is concerned with controlled waste. **Controlled waste** comprises household, industrial or commercial waste. It is a criminal offence to deposit controlled waste without an environmental permit and/or in a manner likely to cause environmental pollution or harm to human health.

The EPA also covers **hazardous waste and spillage.** Hazardous waste can only be disposed of using special arrangements. There are two categories of hazardous waste – absolute hazardous waste, such as hydrochloric acid, fuel, oil, or diesel, or mirror hazardous waste, such as sawdust, shavings, cuttings, wood particle board and veneer containing dangerous substances. If waste does fall into these categories, it needs to be handled, stored, transported and disposed of appropriately. The EPA imposes a legal duty of care to ensure that waste is not illegally disposed of, or dealt with without an Environmental Permit or Registered Exemption (see Chapter 13 for more information on these topics).

The waste must be stored securely and safely in sealed waterproof containers with clear labelling and written instructions for storage and disposal to ensure that the environment is protected. It must be regularly checked

for leaks, deteriorating containers, or other potential risks. It is essential to carry out a risk assessment to identify the risks associated with handling and storing hazardous waste.

These substances are sometimes life threatening (toxic, corrosive or carcinogenic) or highly flammable. Clinical waste falls within this category. A consignment note system accompanies this waste at all the stages to its final destination. Before hazardous waste is removed from the originating premises, a contract should be in place with a licensed carrier.

The Hazardous Waste Regulations, with the exclusion of Scotland, replace the Special Waste Regulations and cover many more substances; for example, computer monitors, fluorescent tubes, end-of-life vehicles and television sets. Hazardous waste is waste which can cause damage to the environment or to human health. The hazardous properties of waste covered by these Regulations are listed in Appendix 8.2 and producers of such waste may need to notify the Environment Agency. The Regulations seek to ensure that hazardous waste is safely managed and its movement is documented. The following points are important for construction sites:

▶ Sites that produce more than 200 kg of hazardous waste each year for removal, treatment or disposal must register with the Environment Agency.
▶ Different types of hazardous waste must not be mixed.
▶ Producers must maintain registers of their hazardous wastes.

8.6.2 Waste disposal

The collection and removal of waste from a workplace is normally accomplished using a skip. Every year, activities involving the movement of skips and containers cause death and serious injury. The hazards include:

▶ being struck by vehicles;
▶ falling and slipping;
▶ failures of lifting equipment;
▶ striking overhead cables/obstructions;
▶ vehicle overturns;
▶ runaway vehicles.

The skip should be located on firm, level ground away from the main construction work, particularly excavation work. This will allow clear access to the skip for filling and removal from site. On arrival at site, the integrity of the skip should be checked. It should be filled either by chute or by mechanical means unless items can be placed in by hand. Skips should not be overfilled and be netted or sheeted over when it is full. Any hazardous waste should be segregated from controlled waste.

Waste skip selection should be made during the site planning process. The selected skip must be suitable for the particular job. The following points should be considered:

- ▶ sufficient strength to cope with its load;
- ▶ stability while being filled;
- ▶ a reasonable uniform load distribution within the skip at all times;
- ▶ the immediate removal of any damaged skip from service and the skip inspected after repair before it is used again;
- ▶ sufficient space around the skip to work safely at all times;
- ▶ the skip should be resting on firm level ground;
- ▶ the skip should never be overloaded or overfilled;
- ▶ there must be sufficient headroom for the safe removal of the skip when it is filled.

Figure 8.21 A designated waste collection area with two types of skip commonly used for waste collection. Heavy materials would be transported in the smaller skip. Sizes of skip range from about 4 cu metres (small skip shown) to about 35 cu metres (large skip shown)

There are hazards present during the movement of a loaded skip from the ground to the back of a skip loader vehicle. Entanglement with the vehicle lifting mechanisms, such as the hydraulic arms and lifting chains, is a major hazard. Other hazards include contact by the skip with overhead obstructions, movement of the skip contents and skip overload leading to mechanical or structural failure. Slip hazards may be present due to spillages from the skip and the skip contents could be contaminated with biological material, asbestos or syringes. Passing traffic during the loading operation may also present a hazard.

The control measures for these hazards include the use of outriggers to increase the stability of the loader vehicle and the provision of steps for the driver to alight from the cab or the vehicle flatbed. The contents of the skip should be secured using netting or tarpaulin. Adhering to the safe working loads of the skip and lifting equipment and the use of a banksman during the lifting process are additional controls. The area around the vehicle may need to be cordoned off to protect passing pedestrians and road traffic. All workers concerned with the operation should wear suitable PPE, such as high-visibility jackets, gloves and suitable footwear. Finally, all lifting equipment, including chains and shackles, must be subject to a periodic statutory examination.

8.6.3 Waste management

Some form of training may be required to ensure that employees segregate hazardous and non-hazardous wastes on site and fully understand the risks and necessary safety precautions which must be taken. Personal protective equipment, including overalls, gloves and eye protection, must be provided and used. The storage site should be protected against trespassers, fire and adverse weather conditions. If flammable or combustible wastes are being stored, adequate fire protection systems must be in place. Finally, in the case of liquid wastes, any drains must be protected and bunds used to restrict spreading of the substance as a result of spills.

A hierarchy for the management of waste streams has been recommended by the Environment Agency.

1. Prevention – by changing the process so that the waste is not produced (e.g. substitution of a particular material).
2. Reduction – by improving the efficiency of the process (e.g. better machine maintenance).
3. Reuse – by recycling the waste back into the process (e.g. using reground waste plastic products as a feed for new products).
4. Recovery – by releasing energy through the combustion, recycling or composting of waste (e.g. the incineration of combustible waste to heat a building).
5. Responsible disposal – by disposal in accordance with regulatory requirements.

Each year in the UK, commercial and industrial sectors generate over 30 million tonnes of waste.

8.6.4 Other waste issues

In 1998, land disposal accounted for approximately 58% of waste disposal, 26% was recycled and the remainder was incinerated with some of the energy recovered as heat. The Producer Responsibility Obligations (Packaging Waste) Regulations 1997 placed legal obligations on employers to reduce their packaging waste by either recycling or recovery as energy (normally as heat from an incinerator attached to a district heating system). A series of targets have been stipulated which will reduce the amount of waste progressively over the years. These Regulations are enforced by the Environment Agency which has powers of prosecution in the event of non-compliance. The UK recycled almost two-thirds of all packaging produced in 2008.

The Landfill (England and Wales) Regulations 2002 are part of the Government's drive to encourage recycling and reduce the amount of rubbish sent to landfill sites, and this means that companies now have to either recycle or treat their waste before it is taken to landfill. The rules are also in force in Scotland.

The main changes are that liquid wastes are banned from landfill and other waste must be treated before it can be passed into landfill. Businesses need to demonstrate that their waste has been treated in either a physical, thermal, chemical or biological process. One major environmental problem is that of contaminated land. Contaminated land is defined in the Environmental Protection Act and is produced by leakage, accidental spillage and uncontrolled waste disposal. The contaminator of the land has the primary responsibility and liability to clean up the contamination. If the contaminator is unknown, the current owner of the land has the responsibility. If no current owner can be found then the local authority becomes responsible for the remediation.

Figure 8.22 Electronic waste under WEEE

The Environment Agency has similar powers following the introduction of the Waste Electrical and Electronic Equipment (WEEE) Directive, the Restrictions of the use of certain Hazardous Substances in electrical and electronic equipment (RoHS) and the End of Life Vehicle (ELV) Directive. The aim of the WEEE Directive is to minimise the environmental impact of electrical and electronic equipment both during their lifetime and when they are discarded. All electrical and electronic equipment must be returned to the retailer from its end user and reused or reprocessed by the manufacturer. Manufacturers must register with the Environment Agency, who will advise on these obligations. The WEEE Regulations have been amended so that producer-compliance schemes report their activities in a more precise way (by providing evidence of recycling in kilograms instead of tonnes). It is hoped that this will reduce the delays in recycling and lead to the recycling of smaller amounts at more frequent intervals. The European Commission has proposed changes to both WEEE and RoHS. On WEEE, a major proposal is to increase the amounts of electrical and electronic waste that are separately collected and recycled, while the proposals on RoHS aim for a higher level of environmental protection

by revising the scope of the restrictions and the applicable substances.

There is a useful HSE Waste management and recycling website www.hse.gov.uk/waste/index.htm

8.7 Further information

Statutory provisions

The Chemicals (Hazard Information and Packaging for Supply) Regulations 2009

The Control of Asbestos Regulations 2012

The Control of Substances Hazardous to Health Regulations 2002

The Hazardous Waste (England and Wales) Regulations 2005

The Personal Protective Equipment at Work Regulations 1992 (as amended)

The Special Waste Amendment (Scotland) Regulations 2004

The Waste (England and Wales) Regulations 2011 as amended

Other references

Asbestos: The Survey Guide HSG264, HSE Books, ISBN: 978-0-7176-6502-0 http://books.hse.gov.uk/hse/public/saleproduct.jsf?catalogueCode=9780717665020

Control of Substances Hazardous to Health (fifth edition), ACoP and guidance, L5, HSE Books ISBN: 978-0-7176-2981-7 www.hse.gov.uk/pubns/priced/l5.pdf

Controlling Airborne Contaminants at Work: A Guide to Local Exhaust Ventilation, HSG258, second edition 2011, HSE Books, ISBN: 978-0-7176-6415-3 www.hse.gov.uk/pubns/priced/hsg258.pdf

Managing Asbestos in Buildings: A brief guide, INDG223(rev5) 2012, HSE Books www.hse.gov.uk/pubns/indg223.pdf

Occupational Exposure Limits, EH40/2005, HSE Books, ISBN: 978-0-7176-6446-7 www.hse.gov.uk/pubns/priced/eh40.pdf

Personal Protective Equipment at Work (second edition), Personal Protective Equipment at Work Regulations 1992 (as amended), Guidance on Regulations, HSE Books, ISBN: 978-0-7176-689-3 www.hse.gov.uk/pubns/priced/l25.pdf

Step by Step Guide to COSHH Assessment, HSG97, second edition 2004, HSE Books, ISBN: 978-0-7176-2785-1 www.hse.gov.uk/pubns/priced/hsg97.pdf

The management of asbestos in non-domestic premises, Regulation 4 of the Control of Asbestos Regulations 2006, Approved Code of Practice and guidance (L127), ISBN: 978-0-7176-6209-8 www.hse.gov.uk/pubns/priced/l127.pdf

Information on the Control of Asbestos Regulations 2012 www.hse.gov.uk/asbestos/regulations.htm

8.8 Practice questions

1. Health hazards may be present in the workplace through various chemical and biological agents.
   (a) **Describe FOUR** chemical and **THREE** biological hazardous agents and **give** an example of each agent.
   (b) **Explain** the term 'respirable dust' and **outline** its health effect on the human body.

2. For each of the following types of hazardous substance, **give** a typical example and **outline** its primary effect on the human body:
   (a) irritant
   (b) corrosive
   (c) toxic
   (d) carcinogenic
   (e) mutagenic.

3. (a) **Describe** the differences between acute and chronic health effects giving an example in each effect.
   (b) **Identify TWO** types of cellular defence mechanisms that the body has as a natural defence system.
   (c) **Identify** the factors that could affect the level of harm experienced by a worker who has been exposed to a toxic substance.

4. (a) **Identify FOUR** possible routes of entry of a hazardous substance into the body **AND, in EACH** case, **give** an example of how an employee might be at risk of such exposure.
   (b) **Define** the term 'target organ' within the context of occupational health.
   (c) **Outline** the personal hygiene practices that should be followed to reduce the risk of ingestion of a hazardous substance.

5. (a) **Identify** possible routes of entry of biological organisms into the body.
   (b) **Outline** control measures that could be used to reduce the risk of infection from biological organisms.

6. (a) **Identify FOUR** respiratory diseases that could be caused by exposure to dust at work.
   (b) **Describe** the respiratory defence mechanisms of the body against atmospheric dust.

7. (a) **Describe** the typical symptoms of occupational dermatitis.
   (b) **Identify TWO** common causative agents.
   (c) **Outline** control measures which may be taken to prevent the occurrence of occupational dermatitis.

8. (a) **Outline** the issues to be considered when undertaking an assessment of health risks from hazardous substances that may be used in a workplace.
   (b) **Identify** the information that should be included on a manufacturer's safety data sheet supplied with a hazardous substance.
   (c) **Outline FOUR** sources of information that might be consulted when assessing the risk of a hazardous substance in a workplace.

9. (a) **Identify** the main advantages and limitations of chemical indicator (stain detector) tubes.
   (b) **Explain** how a dust lamp can give an indication of airborne dust levels in a workplace.
   (b) **Outline** how a personal dust sampler is used to measure the levels of airborne dust in a workplace.

10. (a) **Outline** the main purposes of occupational exposure limits (OELs).
    (b) **Explain** the meaning of the following terms:
       (i) the short-term exposure limit (STEL)
       (ii) the long-term exposure limit (LTEL)
       (iii) the maximum permissible concentration.
    (c) **Outline** the limitations of occupational exposure limits.

11. (a) **Explain** the meaning of the term 'workplace exposure limit' (WEL).
    (b) **Identify** the two classifications of hazardous substances that require exposure to be reduced to as low a level as is reasonably practicable below the WEL.
    (c) **Outline FOUR** actions that could be taken when a WEL has been exceeded.

12. (a) **Outline** the Principles of Good Practice that will help to prevent ill health due to exposure of workers to hazardous substances.
    (b) **Outline** a hierarchy of control measures that could be used to implement the Principles of Good Practice for the control of exposure to hazardous substances.

13. A local exhaust ventilation (LEV) system is to be installed to extract welding fume from an engineering workshop.
    (a) **Identify** the five basic components of the LEV system.
    (b) **Outline** the issues that might reduce the effectiveness of the LEV system.
    (c) **Outline** the routine maintenance that should be carried out between statutory examinations so that the ventilation system continues to operate effectively.

14. (a) **Explain** the meaning of the term 'dilution ventilation'.

(b) **Identify THREE** instances when dilution ventilation may be used in a workplace.

(c) **Identify** the limitations to the use of dilute ventilation.

15. **Identify** the possible health effects on workers and **identify FOUR** control measures to minimise those health effects for the following hazardous substances:

(a) solvent-based adhesives and paints

(b) wood dust

(c) glass-fibres

(d) isocyanates.

16. (a) **Outline** the issues that should be considered when selecting and using personal protective equipment.

(b) **Identify** the hazards against which the following items of personal protective equipment should offer protection and **outline** the practical limitations of each item:

(i) safety gloves

(ii) safety footwear

(iii) safety goggles

(iv) safety spectacles.

17. (a) **Outline** the health and safety hazards associated with welding operations.

(b) **Outline** the factors to be considered in the selection of respiratory protective equipment for persons carrying out welding activities.

(c) **Identify** the difference between breathing apparatus and respirators.

18. (a) **Outline** the purpose of health surveillance of workers who may have come into contact with hazardous substances.

(b) **Give THREE** examples of health surveillance in the workplace.

(c) **Outline** the importance of personal hygiene in the workplace.

19. (a) **Identify TWO** respiratory diseases that may be caused by exposure to asbestos.

(b) **Identify** the **THREE** types of asbestos commonly found in buildings and **identify** the common sources of asbestos in buildings.

(c) **Outline** the control measures required before and during the removal of asbestos from a building.

20. For **EACH** of the following agents, **outline** the principal health effects **AND identify** a typical workplace situation in which a person might be exposed:

(a) carbon monoxide

(b) legionella bacteria

(c) leptospira

(d) blood-borne viruses.

21. **Identify** the possible health hazards to which construction workers may be exposed from the following substances. In each case **give** an example of a likely source and suitable control measures:

(a) silica dust

(b) wet cement.

22. A toxic substance has been transported by lorry to a warehouse in a number of drums. At the warehouse, one of the drums is ruptured during unloading and some of the contents are spilt on the floor.

(a) **Outline** the main controls that should be in place to ensure the safe transport of the hazardous substance by road.

(b) **Outline** a procedure that would deal safely with the spillage.

(c) **Identify THREE** ways in which persons working near the spillage might be harmed.

23. A company produces a range of solid and liquid wastes, both hazardous and non-hazardous.

(a) **Identify** the main hazards associated with the storage of wastes.

(b) **Outline** the required arrangements to ensure the safe storage of the wastes prior to their collection and disposal.

(c) **Outline** the issues that should be addressed by the company when developing a system for the safe collection and disposal of waste.

24. Skips are used to store solid waste materials prior to collection and disposal.

(a) **Identify** the hazards a skip collector could be exposed to when moving a full skip from the ground onto the back of a skip loader vehicle.

(b) **Identify EIGHT** safe practices to be followed when using a skip for the collection and removal of waste from a construction site.

8

269

APPENDIX 8.1 GHS Hazard (H) Statements (Health only)

H-statement	Phrase
300	**Fatal** if **swallowed**
301	**Toxic** if **swallowed**
302	**Harmful** if **swallowed**
303	**May be harmful** if **swallowed**
304	**May be fatal** if **swallowed** and **enters airways**
305	**May be harmful** if **swallowed** and **enters airways**
310	**Fatal** in contact with skin
311	**Toxic** in contact with skin
312	**Harmful** in contact with skin
313	**May be harmful** in contact with skin
314	Causes **severe burns** and **eye damage**
315	Causes skin irritation
316	Causes mild skin irritation
317	May cause an allergic skin reaction
318	Causes serious **eye damage**
319	Causes serious **eye irritation**
320	Causes **eye irritation**
330	**Fatal** if **inhaled**
331	**Toxic** if **inhaled**
332	**Harmful** if **inhaled**
333	**May be harmful** if **inhaled**
334	May cause allergy or asthma symptoms or breathing difficulties if inhaled
335	May cause **respiratory irritation**
336	May cause **dizziness** or **drowsiness**
340	May cause **genetic defects** *(route if relevant)*
341	Suspected of causing **genetic defects** *(route if relevant)*
350	May **cause cancer** *(route if relevant)*
351	Suspected of **causing cancer** *(route if relevant)*
360	May **damage fertility** or the **unborn child** *(effect if known, route if relevant)*
361	Suspected of **damaging fertility** or the unborn child *(effect if known, route if relevant)*
362	May cause harm to breast-fed children
370	Causes **damage to organs** *(organ if known, route if relevant)*
371	May cause **damage to organs** *(organ if known, route if relevant)*
372	Causes **damage to organs** through prolonged or repeated exposure *(organ if known, route if relevant)*
373	May cause **damage to organs** through prolonged or repeated exposure *(organ if known, route if relevant)*
EU66	Repeated exposure may cause skin dryness or cracking
EU70	**Toxic** by **eye contact**
EU71	**Corrosive** to the **respiratory tract**

APPENDIX 8.2 Hazardous properties of waste (as listed in the Hazardous Waste (England and Wales) Regulations 2005)

	Hazard	Description.
H1	Explosive	Substances and preparations which may explode under the effect of flame or which are more sensitive to shocks or friction than dinitrobenzene.
H2	Oxidising	Substances and preparations which exhibit highly exothermic reactions when in contact with other substances, particularly flammable substances.
H3-A	Highly flammable	▶ liquid substances and preparations having a flash point below 21°C (including extremely flammable liquids), or ▶ substances and preparations which may become hot and finally catch fire in contact with air at ambient temperature without any application of energy, or ▶ solid substances and preparations which may readily catch fire after brief contact with a source of ignition and which continue to burn or to be consumed after removal of the source of ignition, or ▶ gaseous substances and preparations which are flammable in air at normal pressure, or ▶ substances and preparations which, in contact with water or damp air, evolve highly flammable gases in dangerous quantities.
H3-B	Flammable	Liquid substances and preparations having a flash point equal to or greater than 21°C and less than or equal to 55°C.
H4	Irritant	Non-corrosive substances and preparations which, through immediate, prolonged or repeated contact with the skin or mucous membrane, can cause inflammation.
H5	Harmful	Substances and preparations which, if they are inhaled or ingested or if they penetrate the skin, may involve limited health risks.
H6	Toxic	Substances and preparations (including very toxic substances and preparations) which, if they are inhaled or ingested or if they penetrate the skin, may involve serious, acute or chronic health risks and even death.
H7	Carcinogenic	Substances and preparations which, if they are inhaled or ingested or if they penetrate the skin, may induce cancer or increase its incidence.
H8	Corrosive	Substances and preparations which may destroy living tissue on contact.
H9	Infectious	Substances containing viable micro-organisms or their toxins which are known or reliably believed to cause disease in humans or other living organisms.
H10	Teratogenic or Toxic for reproduction	Substances and preparations which have adverse effects on sexual function and fertility in adult males and females, as well as developmental toxicity in the offspring.
H11	Mutagenic	Substances and preparations which may cause a mutation, that is a permanent change in the amount or structure of the genetic material in a cell.
H12	Release of an acute toxic gas	Substances and preparations which release acute toxic gases (Acute Tox. 1, 2 or 3) in contact with water or an acid.
H13	Sensitising	Substances and preparations which contain one or more substances known to cause sensitising effects to the skin or the respiratory organs.
H14	Ecotoxic	Substances and preparations which present or may present immediate or delayed risks for one or more sectors of the environment.

8

APPENDIX 8.3 Different types of protective glove

Remember: Select your gloves according to the Hazard and Risk Assessment. Keep your gloves in good condition. Seek advice if you are in any doubt.

GENERAL USE

TYPE 1 Heavy work, General Protection (Rigger's type replacement)
✓ Some Abrasion & Good Tear Resistance
✗ Low Cut & Puncture Protection. No Resistance to Fluids
2142

TYPE 2 Fine, Dextrous Work
✓ Abrasion & Tear Resistant
✗ Low Cut & Puncture Protection. No Resistance to Fluids
3131

TYPE 3 Work involving Water & Oil
✓ Water & Oil Resistant, Good Abrasion Protection
✗ Low Cut Resistance
3121

TYPE 4 Handling Acid and other Chemicals
✓ Chemical Resistant, Long to Protect Arm
✗ Minimal Cut & Abrasion Protection
N/A

TYPE 5 Skin Protection from Oils and other Low Hazard Substances
✓ Light and Thin.
✗ No Mechanical Protection at all
N/A

SPECIALIST USE

TYPE 6 Welders Leather Gauntlets. For protection against heat and burns
✓ Very Durable for Heat Protection
✗ Poor Chemical & Cut Resistance
N/A

TYPE 7 High Pressure Water Jetting. Slip Resistant Grip
✓ Waterproof with Good Wet Grip. Abrasion & Tear Resistant
✗ Low Cut & Puncture Resistance
4131

TYPE 8 Kevlar, with High Cut Resistance. Use with Sharps, e.g. cent screens, glass
✓ Strong & Dextrous. High Cut Resistance
✗ No Resistance to Fluids
3444

TYPE 9 Heavy Engineering Work e.g. on Pellet Mills
✓ Good Resistance to all Mechanical Hazards, Very Strong
✓ Some Heat and Fluid Protection
4532

Explanation of Protection Ratings

MECHANICAL HAZARDS EN 388

PERFORMANCE LEVELS			
0 to 4	0 to 5	0 to 4	0 to 4
Abrasion resistance	Blade cut resistance	Tear resistance	Puncture resistance

CHEMICAL HAZARDS EN 374
Liquidproof test. Permeation test
Liquidproof according to the air leak test

COLD HAZARDS EN 511

PERFORMANCE LEVELS		
0 to 4	0 to 4	0 to 1
Convective cold resistance	Contact cold resistance	Water permeability

MICRO ORGANISMS

RADIOACTIVE CONTAMINATION EN 421

HEAT AND FIRE EN 407

PERFORMANCE LEVELS				
0 to 4	0 to 4	0 to 3	0 to 4	0 to 4
Burning behaviour	Contact heat resistance	Convective heat resistance	Radiant heat resistance	Resistance to small drops of molten metal	Resistance to large quantities of molten metal

* Level X : The test is not applicable, or the glove is not tested
* Level 0 : The minimum level is not attained
For additional advice regarding CE categories on listed gloves contact the SAFETY DEPARTMENT
ref: CAN-DOC-01 rev.1

Note that types provided may vary from those illustrated above, but will give similar protection levels

General workplace issues

The contents of this chapter are:

Chapter summary

This chapter covers the following topics in Element 8 of the NEBOSH NGC syllabus:

Health, welfare and work environment

Working at height

Safe working in confined spaces

Lone working

Slips and trips

Safe movement of people and vehicles in the workplace

Work-related driving

Learning outcomes

These are the learning outcomes covered by this chapter:

The learner will be able to:

1. Outline common health, welfare and work environment requirements in the workplace
2. Outline the hazards and control measures for safe working at height
3. Outline safe work practices for working within confined spaces
4. Outline the hazards and controls associated with lone working
5. Outline the main causes of and controls for slips and trips in the workplace
6. Outline the hazards and control measures for the safe movement of people and vehicles in the workplace
7. Outline the advantages and disadvantages of road traffic safety management systems, factors to consider when assessing the risk from work-related driving and the hazards associated with electric and hybrid vehicles

9.1 Health, welfare and work environment requirements

9.1.1 Introduction

People are most often involved in accidents as they walk around the workplace, or when they come into contact with vehicles in or around the workplace. It is therefore important to understand the various common causes of accidents, and the control strategies that can be employed to reduce them. Slips, trips and falls account for the majority of accidents in the workplace. However, the workplace may be a construction site where working at height or excavation work will present risks to health and safety. There are also health hazards and risks present in some workplaces due to violence and substance abuse. Many of the risks associated with these hazards can be significantly reduced by an effective management system.

Finally, there are welfare and work environment requirements to ensure a healthy and safe workplace. This is where this chapter begins.

9.1.2 Health and welfare

Welfare and work environment issues are covered by the Workplace (Health, Safety and Welfare) Regulations together with an Approved Code of Practice and additional guidance. Welfare arrangements include the provision of sanitary conveniences and washing facilities, drinking water, accommodation for clothing, facilities for changing clothing and facilities for rest and eating meals. First-aid provision is also a welfare issue, but is covered in Chapter 4.

Sanitary conveniences and washing facilities must be provided together and in proportion to the size of the

workforce. The Approved Code of Practice provides two tables offering guidance on the requisite number of water closets, wash stations and urinals for varying sizes of workforce (approximately one of each for every 25 employees). Special provision should be made for disabled workers and there should normally be separate facilities for men and women. A single convenience would only be acceptable if it were situated in a separate room whose door could be locked from the inside. There should be adequate protection from the weather and only as a last resort should public conveniences be used. A good supply of warm water, soap and towels must be provided as close to the sanitary facilities as possible. The facilities should be well lit and ventilated and their walls and floors easy to clean. It may be necessary to install a shower for certain types of work. Hand dryers are permitted but there are concerns about their effectiveness in drying hands completely and thus removing all bacteria. In the case of temporary or remote work sites, sufficient chemical closets and sufficient washing water in containers must be provided.

All such facilities should be well ventilated and lit and cleaned regularly.

Drinking water must be readily accessible to the entire workforce. The supply of drinking water must be adequate and wholesome. Normally mains water is provided and should be marked as 'drinking water' if water unfit for drinking is also available.

Accommodation for clothing and facilities for changing clothing must be provided. This should be clean, warm, dry, well ventilated and secure. Such accommodation is only necessary when the work activity requires employees to change into specialist

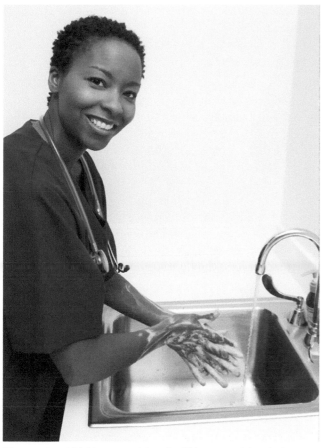

Figure 9.1 Welfare washing facilities: washbasin large enough for people to wash their forearms

clothing. Where workers are required to wear special or protective clothing, arrangements should be such that the workers' own clothing is not contaminated by any hazardous substances.

Facilities for rest and eating meals must be provided so that workers may sit down during break times in areas where they do not need to wear personal protective equipment. Facilities should also be provided for pregnant women and nursing mothers to rest. Arrangements must be in place to ensure that food is not contaminated by hazardous substances.

Workplace environment

The issues governing the workplace environment are ventilation, heating and temperature, lighting, workstations and seating.

Ventilation

The ventilation of the workplace should be effective and sufficient and free of any impurity, and air inlets should be sited clear of any potential contaminant (e.g. a chimney flue). Care needs to be taken to ensure that workers are not subject to uncomfortable draughts. The ventilation plant should have an effective visual or audible warning device fitted to indicate any failure of the plant. The plant should be properly maintained and records kept. The supply of fresh air should not normally fall below 5–8 litres per second per occupant (see Figure 9.2).

The Workplace (Health, Safety and Welfare) Regulations under Regulation 6 states that 'effective and suitable provision shall be made to ensure that every enclosed workplace is ventilated by a sufficient quantity of fresh or purified air'. The fresh air should be free of contaminants such as vehicle exhaust fumes or boiler chimney emissions. If, following a risk assessment, there is a problem of particulate pollution of the incoming air to a workplace, then the air should be filtered.

9

Figure 9.2 Natural ventilation in a building

Air containing contaminants escapes to atmosphere via roof vents

Air enters and escapes through gaps around building

Open window

Fresh air enters via windows and/or openings or vents

Heating and temperature

During working hours, the temperature in all workplaces inside buildings shall be reasonable (not uncomfortably high or low). 'Reasonable' is defined in the Approved Code of Practice as at least 16°C, unless much of the work involves severe physical effort in which case the temperature should be at least 13°C. These temperatures refer to readings taken close to the workstation at working height and away from windows. The Approved Code of Practice recognises that these minimum temperatures cannot be maintained where rooms open to the outside or where food or other products have to be kept cold. A heating or cooling method must not be used in the workplace which produces fumes, or is injurious or offensive to any person. Such equipment needs to be regularly maintained to prevent this.

There is no maximum workplace temperature given in the Workplace (Health, Safety and Welfare) Regulations, which place a legal obligation on employers to provide a 'reasonable' temperature in the workplace. The Approved Code of Practice suggests a minimum temperature in workrooms should normally be at least 16°C or 13°C if much of the work indoors involves severe physical effort.

HSE guidance on workplace temperature states that 'a meaningful figure cannot be given at the upper end of the scale. This is because the factors, other than air temperature which determine thermal comfort, (radiant temperature, humidity and air velocity), become more significant and the interplay between them more complex as temperatures rise.' The HSE recommends that employers should consult with employees or their representatives, to establish sensible means to cope with high temperatures.

Advice is available from professional organisations. The Chartered Institution of Building Services Engineers recommends, for air-conditioned buildings in the UK, a temperature range of between 21°C and 23°C during winter and between 22°C and 24°C in summer for continuous sedentary occupancy. Room temperatures in buildings without artificial cooling will exceed the summer values for some of the time but should not exceed 25°C for more than 5% of the annual occupied period (typically 125 hours).

A sufficient number of thermometers should be provided and maintained to enable workers to determine the temperature in any workplace inside a building (but need not be provided in every workroom).

Where, despite the provision of local heating or cooling, the temperatures are still unreasonable, suitable protective clothing and rest facilities should be provided.

Lighting

Every workplace shall have suitable and sufficient lighting and this shall be natural lighting so far as is reasonably practicable. Suitable and sufficient emergency lighting must also be provided and maintained in any room where workers are particularly exposed to danger in the event of a failure of artificial lighting (normally due to a power cut and/or a fire). Windows and skylights should be kept clean and free from obstruction so far as is reasonably practicable unless it would prevent the shading of windows or skylights or prevent excessive heat or glare.

Figure 9.3 A well-lit workplace

When deciding on the suitability of a lighting system, the general lighting requirements will be affected by the following factors:

▶ the availability of natural light;
▶ the specific areas and processes, in particular any colour rendition aspects or concerns over stroboscopic effects (associated with fluorescent lights);
▶ the type of equipment to be used and the need for specific local lighting;
▶ the lighting characteristics required (type of lighting, its colour, intensity and local adjustability);
▶ the location of visual display units and any problems of glare;
▶ structural aspects of the workroom, such as the use of screens in open office layout and the reduction of shadows;
▶ the presence of atmospheric dust;
▶ the heating effects of the lighting;

- lamp and window cleaning and repair (and disposal issues); and
- the need and required quantity of emergency lighting.

Light levels are measured in illuminance, having units of lux (lx), using a light meter. A general guide to lighting levels in different workplaces is given in Table 9.1.

Poor lighting levels will increase the risk of accidents such as slips, trips and falls. More information is available on lighting from *Lighting at Work,* HSG38, HSE Books.

Table 9.1 Typical workplace lighting levels

Workplace or type of work	Illuminance (lx)
Warehouses and stores	150
General factories or workshops	300
Offices	500
Drawing offices (detailed work)	700
Fine working (ceramics or textiles)	1,000
Very fine work (watch repairs or engraving)	1,400

Workstations and seating

Workstations should be arranged so that work may be done safely and comfortably. The worker should be at a suitable height relative to the work surface and there should be no need for undue bending and stretching. Workers must not be expected to stand for long periods of time, particularly on solid floors. A suitable seat should be provided when a substantial part of the task can or must be done sitting. The seat should, where possible, provide adequate support for the lower back and a footrest be provided for any worker whose feet cannot be placed flat on the floor. It should be made of materials suitable for the environment, be stable and, possibly, have arm rests.

It is also worth noting that sitting for prolonged periods can present health risks, such as blood circulation and pressure problems, and vertebral and muscular damage.

Seating at Work, HSG57, HSE Books, provides useful guidance on how to ensure that seating in the workplace is safe and suitable.

Other factors

The Workplace (Health, Safety and Welfare) Regulations cover other important work environment issues, and details of these are given in Chapter 13. The condition of floors, stairways and traffic routes should be suitable for the purpose and well maintained and undue space constraints anywhere in the workplace should be avoided. Translucent or transparent doors should be constructed with safety glass and properly marked to warn pedestrians of their presence. Windows and skylights should be designed so that, when they are opened, they do not present an obstruction to passing pedestrians. There must be adequate arrangements in place to ensure the safe cleaning of windows and skylights. Finally, there need to be adequate provisions for the needs of disabled workers.

9.1.3 The effects of extremes of temperature

The human body is very sensitive to relatively small changes in external temperatures. Food not only provides energy and the build-up of fat reserves, but also generates heat, which needs to be dissipated to the surrounding environment. The body also receives heat from its surroundings. The temperature of the body is normally around 37°C, and it will attempt to maintain this temperature irrespective of the temperature of the surroundings. Therefore, if the surroundings are hot, sweating will allow heat loss to take place by evaporation caused by air movement over the skin. On the other hand, if the surroundings are cold, shivering causes internal muscular activity, which generates body heat.

At high temperatures, the body has more and more difficulty in maintaining its natural temperature unless sweating can take place and therefore water must be replaced by drinking. If the surrounding air has high humidity, evaporation of the sweat cannot take place and the body begins to overheat. This leads to heart strain and, in extreme cases, heat stroke. It follows that when working is required at high temperatures, a good supply of drinking water should be available and, further, if the humidity is high, a good supply of ventilation air is also needed. Heat exhaustion is a particular hazard in confined spaces so more breaks and air-moving plant are needed.

Over time people adapt to hot conditions by sweating more, and by changing their behaviour to try and cool down, e.g. removing clothing, taking cool drinks, fanning themselves, sitting in the shade or a cool area, and/or reducing their work rate. However, in many work situations such behavioural changes may not be possible, e.g. during asbestos removal. The effects of climate change mean that work in extreme temperatures is likely to occur more frequently.

When carrying out a risk assessment, the major factors that need to be considered are:

- work rate – the harder someone works the greater the amount of body heat generated;
- working climate – this includes air temperature, humidity, air movement and effects of working near a heat source;
- employee clothing and respiratory protective equipment – may impair the efficiency of sweating and other means of temperature regulation;
- employee's age, build and medical factors – may affect an individual's tolerance.

9

At low temperatures, the body will lose heat too rapidly and the extremities of the body will become very cold leading to frostbite and possibly the loss of limbs. Under these conditions, thick, warm (thermal) clothing, the provision of hot drinks and external heating will be required. For those who work in sub-zero temperatures, such as cold store workers, additional precautions will be needed. The store doors must be capable of being unlocked from the inside and an emergency alarm system should be installed. Appropriate equipment selection and a good preventative maintenance system are very important as well as a regular health surveillance programme for the workers, who should be provided with information and training on the hazards associated with working in very low temperatures.

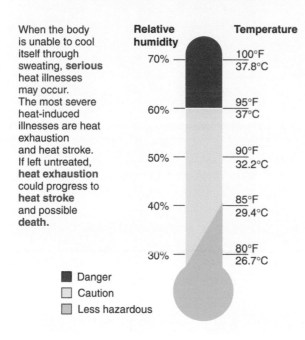

Figure 9.4 The heat equation

When working in cold conditions the HSE advises that to ensure thermal comfort the following actions could be taken:

▷ Provide adequate workplace heating, e.g. portable heaters.
▷ Reduce cold exposure by designing processes that minimise exposure to cold areas and cold products where possible.
▷ Reduce draughts.
▷ Provide insulating floor coverings or special footwear when employees have to stand for long periods on cold floors.

▷ Provide appropriate protective clothing for cold environments.
▷ Introduce formal systems of work to limit exposure, e.g. flexible working patterns, job rotation.
▷ Provide sufficient breaks to enable employees to get hot drinks or to warm up in heated areas.

In summary, extremes of temperature require special measures, particularly if accompanied by extremes of humidity. Frequent rest periods will be necessary to allow the body to acclimatise to the conditions. An index called WBGT (Wet Bulb Globe Temperature) is normally used.

9.2 Working at height

Work at height accounts for about 50–60 deaths – more than any other workplace activity – and 4,000 injuries each year. The Work at Height Regulations attempts to address these problems. They apply to all operations carried out at height, not just construction work, so they are also relevant to, for example, window cleaning, tree surgery, maintenance work at height and the changing of street lamps. The Work at Height Regulations affect approximately 3 million workers where working at height is essential to their work. During the inspection of 2,400 refurbishment projects in construction, the HSE found that one in five sites failed to address work at height risks on the site.

A report from a UK parliamentary group makes a number of recommendations to reduce the overall number of falls related to work at height, the cause of one in four workplace fatalities.

The recommendations were:

1. The introduction of an enhanced reporting system through the Reporting of Injuries, Diseases and Dangerous Occurrences Regulations (RIDDOR) 2013, that as a minimum would record the scale of a fall and the method and circumstances surrounding the activity.
2. The appointment of an independent body, which would allow for the confidential, enhanced and digital reporting of all near misses and accidents that do not qualify under RIDDOR.
3. The extension of existing work at height safety campaigns to industries outside the construction sector.
4. The extension to the rest of the UK of a system equivalent to Scotland's, under which ministers must report on workplace deaths in line with Section 29 of the Inquiries into Fatal Accidents and Sudden Deaths (Scotland) Act 2016. A system such as this would ensure that the circumstances of fatal injuries relating to falls from height are properly investigated and reported on.

9.2.1 The Work at Height Regulations

The Work at Height Regulations have no minimum height requirement for work at height. They include all work activities where there is a need to control a risk of falling a distance liable to cause personal injury. This is regardless of the work equipment being used, the duration of the work at height involved or the height at which the work is performed. They include access to and egress from a place of work. They would therefore include:

▷ working on a scaffold or from a mobile elevating work platform (MEWP);
▷ sheeting a lorry or dipping a road tanker;
▷ working on the top of a container in docks or on a ship or storage area;
▷ tree surgery and other forestry work at height;
▷ using cradles or rope for access to a building or other structure like a ship under repair;
▷ climbing permanent structures like a gantry or telephone pole;
▷ working near an excavation area or cellar opening if a person could fall into it and be injured;
▷ painting or pasting and erecting bill posters at height;
▷ work on staging or trestles for example, for filming or events;
▷ using a ladder/stepladder or kick stool for shelf filling, window cleaning and the like;
▷ using manriding harnesses in ship repair, or offshore or steeple jack work;
▷ working in a mine shaft or chimney;
▷ work carried out at a private house by a person employed for the purpose, for example, a painter and decorator (but not if the private individual carries out work on their own home).

However, it would not include:

▷ slips, trips and falls on the same level;
▷ falls on permanent stairs if there is no structural or maintenance work being done;
▷ work in the upper floor of a multi-storey building where there is no risk of falling (except separate activities like using a stepladder).

At the centre of the Regulations (Regulation 6), the employer is expected to apply a three-stage hierarchy to all work which is to be carried out at height. The three steps are the avoidance of work at height, the prevention of workers from falling and the mitigation of the effect on workers of falls should they occur.

The Regulations require that:

▷ work is not carried out at height when it is reasonably practicable to carry the work out safely other than at height (e.g. the assembly of components should be done at ground level);

▷ when work is carried out at height, the employer shall take suitable and sufficient measures to prevent, so far as is reasonably practicable, any person falling a distance liable to cause injury (e.g. the use of guard rails);
▷ the employer shall take suitable and sufficient measures to minimise the distance and consequences of a fall (collective measures, for example airbags or safety nets, must take precedence over individual measures, for example safety harnesses).

9.2.2 Hazards from working at height

The risks from working at height are affected by the following hazards:

▷ The vertical distance of the height;
▷ fragile roofs;
▷ unprotected edges;
▷ unstable and/or poorly maintained access equipment;
▷ adverse weather conditions; and
▷ falling materials.

The risk assessment and action required to control risks from using a kick stool to collect books from a shelf should be simple (not overloading, not overstretching, etc.). However, the action required for a complex construction project would involve significantly greater consideration and assessment of risk.

The Work at Height Regulations were amended in 2007 and apply to those who work at height providing instruction or leadership to one or more people engaged in caving or climbing by way of sport, recreation, team building or similar activities in Great Britain.

A risk assessment for working at height should first consider whether the work could be avoided. If this is not possible, then the risk assessment should consider the following issues:

▷ the nature and duration of the work;
▷ the competence level of all those involved with the work and any additional training requirements;
▷ the required level of supervision;
▷ use of guard rails, toe boards, working platforms and means of access and egress;
▷ required personal protective equipment (PPE), such as helmets and harnesses;
▷ the presence of fall arrest systems, such as netting or soft landing systems;
▷ the health status of the workers;
▷ the possible weather conditions;
▷ compliance with the relevant legislation, in particular the Work at Height Regulations and the Management of Health and Safety at Work Regulations.

A summary of the Regulations is given in Chapter 13.

9

9.2.3 Protection against falls from work at height

The Work at Height Regulations require all duty-holders to ensure that:

- all work at height is properly planned and organised;
- those involved in work at height are competent;
- the risks from work at height are assessed, and appropriate work equipment is selected and used;
- the risks of working on or near fragile surfaces are properly managed; and
- the equipment used for work at height is properly inspected and maintained.

The regulations require that guard rails on scaffolds are at a minimum of 950 mm and the maximum unprotected gap between the toe and guard rail of a scaffold is 470 mm. This implies the use of an intermediate guard rail although other means, such as additional toe boards or screening, may be used. They also specify requirements for personal suspension equipment and means of arresting falls (such as safety nets).

9.2.4 Hierarchy for selection of equipment for working safely at height

When working at height, a hierarchy of measures given by the Work at Height Regulations should then be followed, to prevent falls from occurring and selecting appropriate access equipment. These measures are:

- Eliminate or avoid working at height.
- Work from an existing safe workplace provided, such as a properly constructed working platform, complete with toe boards and guard rails.
- Ensure that there is sufficient work equipment or other measures to minimise the distance and consequence of a fall by the use of:
  - collective measures rather than individual measures (e.g. a handrail instead of a harness); or where this is not practicable
  - collective fall arrest equipment (airbags or safety nets); or where this is not practicable
  - individual fall restrainers (safety harnesses) should be used.
- Provide supervision, training and instruction.
- Only when none of the above measures is practicable, should ladders or stepladders be considered.

Much of the work which is done at height could often be done, or partly done, at ground level – thus avoiding the hazards of working at height. The partial erection of scaffolding or edge protection at ground level and the use of cranes to lift it into place at height are examples of this. The manufacture of complete window frames in a workshop and then the final installation of the frame into the building is another. By the use of suitable extension equipment high windows can be cleaned from the ground and high walls can be painted from the ground. However, in most construction work at height, the work cannot be done at ground level and suitable control measures to address the hazards of working at height will be required.

9.2.5 Fragile roofs and surfaces

Work on or near fragile surfaces is also covered by the Work at Height Regulations. Roof work, particularly work on pitched roofs, is hazardous and requires a specific risk assessment and method statement (see Chapter 1 for a definition of a method statement) prior to the commencement of work. Particular hazards are fragile roofing materials, including those materials which deteriorate and become more brittle with age and exposure to sunlight, exposed edges, unsafe access equipment and falls from girders, ridges or purlins. There must be suitable means of access such as scaffolding, ladders and crawling boards; suitable barriers, guard rails or covers where people work near to fragile materials and roof lights; and suitable warning signs indicating that a roof is fragile should be on display at ground level.

Figure 9.5 Proper precautions must always be taken when working on or near fragile roofs or surfaces

Where possible, work on a fragile roof should be avoided by doing the following:

- work from underneath the roof using a suitable work platform or
- where this is not possible, use a mobile elevating work platform that allows people to work from within the basket without having to stand on the roof.

If access onto the fragile roof cannot be avoided, perimeter edge protection should be installed and staging used to spread the load. Unless all the work and access is on staging or platforms that are fitted with guard rails then safety nets should be installed underneath the roof or a harness system used. Where a harness is used, adequate anchorage points will be required.

A roof should always be treated as fragile until a competent person has ruled otherwise. Fragility can be caused by:

▷ general deterioration of the roof through ageing, neglect and lack of maintenance;
▷ corrosion of cladding and fixings;
▷ quality of the original installation and selection of materials;
▷ thermal and impact damage;
▷ deterioration of the supporting structure; and
▷ weather damage.

Asbestos cement sheets and old roof lights should always be treated as fragile.

There are other hazards associated with roof work – overhead services and obstructions, the presence of asbestos or other hazardous substances, the use of equipment such as gas cylinders and bitumen boilers and manual handling hazards.

All roofs, once fixed, should be treated as fragile until a competent person has confirmed that they are non-fragile. In particular, the following are likely to be fragile:

▷ fibre-cement sheets – non-reinforced sheets irrespective of profile type;
▷ liner panels – on built-up sheeted roofs;
▷ metal sheets – where corroded;
▷ glass – including wired glass;
▷ chipboard – or similar material where rotted; and
▷ roof lights.

It is essential that only trained and competent persons are allowed to work on roofs and that they wear footwear having a good grip. It is a good practice to ensure that a person does not work alone on a roof.

Guidance from the Advisory Committee for Roof Work states that any person undertaking roof work needs to be 'both mentally and physically fit, competent to do the work, and be fully aware of all the dangers that exist and the actions necessary to overcome those dangers'. The HSE has published the third edition of the guidance booklet on safety during roof work – HSG33, *Health and safety in roof work*.

A roof ladder should only be used if more suitable equipment cannot be used and should only be used for low-risk, short-duration work. Roof ladders should be of an industrial grade, in good condition and

secured to prevent movement. The anchorage at the top of the roof ladder should be by some method which does not depend on the ridge capping, as this is liable to break away from the ridge. The anchorage should bear on the opposite slope by a properly designed and manufactured ridge hook or be secured by other means.

Roof work should not be attempted in poor weather conditions such as rain, ice, frost or strong winds (particularly gusting) or if slippery conditions exist on the roof. Winds in excess of 23 mph (Force 5) will affect the balance of a roof worker.

9.2.6 Protection against falling objects

This is also covered by the Work at Height Regulations (see summary in Chapter 13). Both construction workers and members of the public need to be protected from the hazards associated with falling objects. Both groups should be protected by the use of covered walkways or suitable netting to catch falling debris. Waste material should be brought to ground level by the use of chutes or hoists. Waste should not be thrown and only minimal quantities of building materials should be stored on working platforms.

The Construction (Head Protection) Regulations 1989 virtually mandated employers to supply head protection (hard hats) to employees whenever there was a risk of head injury from falling objects. (Sikhs wearing turbans can be exempted from this requirement.) The employer is also responsible for ensuring that hard hats are properly maintained and replaced when they are damaged in any way. Self-employed workers had to supply and maintain their own head protection. Visitors to construction sites had to be supplied with head protection and mandatory head protection signs displayed around the site.

Following the recommendations of the Löfstedt report in 2011, the Construction (Head Protection) Regulations were revoked since it was felt that they largely replicate the Personal Protective Equipment at Work Regulations which could be relied on to regulate the use of head protection on construction sites. Sikhs wearing turbans are still exempted from the use of head protection.

Under the Personal Protective Equipment at Work Regulations there will still be a requirement for employers to provide, and for workers to wear, head protection where there is a risk of head injury. HSE will review and update existing guidance (either published or on its website) on compliance with the Regulations to ensure it adequately covers the provision and use of head protection on construction sites.

9

Adequate and suitable head protection should be provided and employers should ensure it is worn if there is a risk of injury to the head, which will be the case on many work sites. They must also comply with the requirements of the Personal Protective Equipment (PPE) Regulations 1992, which were amended in April 2013 to cover the provision and use of head protection on construction sites, when the Construction (Head Protection) Regulations 1989 were revoked.

The different types of helmet used in construction and industry, and the environments in which they are most suitable, are as follows:

▷ EN 812 bump cap: A close-fitting baseball cap style offering the lowest level of protection.

▷ EN 397 standard safety helmet: This will be suitable if a falling object could hit the head, and is the most common type of helmet used in the workplace.

▷ EN 397 lightweight helmet: They are useful when there is a very low risk of anything falling on the head.

▷ EN 397 industrial safety helmet: This is ideal for applications such as workers excavating foundations where there is a chance that something may fall on them from the side as well as the top and hit them on the head.

▷ High-performance EN 14052 industrial safety helmet: This is intended for use on sites where objects and debris are likely to fall, such as in mining and quarrying.

9.2.7 Fall arrest equipment

The three most common types of fall arrest equipment are safety harnesses, safety nets and air bags.

Safety harnesses should only be used alone when conventional protection, using guard rails, is no longer practicable. Such conditions occur when it is possible to fall 2 m or more from an open edge. It is important that the following points are considered when safety harnesses are to be used:

1. The length of fall only is reduced by a safety harness. The worker may still be injured due to the shock load applied to him when the fall is arrested. A free fall limit of about 2 m is maintained to reduce this shock loading. Lanyards are often fitted with shock absorbers to reduce the effect of the shock loading.

2. The worker must be attached to a secure anchorage point before they move into an unsafe position. The lanyard should always be attached above the worker, whenever possible.

3. Only specifically trained and competent workers should attach lanyards to anchorage points and work in safety harnesses. Those who wear safety harnesses must be able to undertake safety checks and adjust the harness before it is used.

Figure 9.6 Fall arrest harness and device

Three levels of inspection are recommended for safety harnesses as follows:

1. A pre-use check undertaken by the user at the beginning of each shift to check there are no visible or surface defects.

2. A detailed inspection undertaken at least every 6 months. However, for frequently used equipment, this should be increased to at least every three months, particularly when the equipment is used in arduous environments (e.g. demolition, steel erection, scaffolding, steel masts/towers with edges). The results of the inspection should be formally recorded.

3. An interim inspection is also an in-depth, recorded inspection and may be needed between detailed inspections because the associated risk assessment has identified a risk that could result in significant deterioration, affecting the safety of the lanyard before the next detailed inspection is due. The need for and frequency of interim inspections will depend on use. Examples of situations where they may be appropriate include risks from arduous working environments involving paints, chemicals or grit blasting.

Any defects discovered by any of these inspections must be reported to the employer as soon as possible.

Safety nets are widely used to arrest falls of people, tools and materials from height but competent installation is essential. The correct tensioning of the net is important and normally specialist companies are available to fit nets. The popularity of nets has grown since the Construction Regulations came into force in 1996 and the subsequent advocacy of their use by the HSE. Nets are used for roofing work and for some refurbishment work. Nets, however, have a limited application since they are not suitable for use in low-level construction where there is insufficient clearance below the net to allow it to deflect the required distance after impact. Nets should be positioned so that workers will not fall more than 2 m, in case they hit the ground or other obstructions.

Air bags are used when it is either not possible or practical to use safety nets. Therefore they are used extensively in domestic house building or when it is difficult to position anchorage points for safety harnesses. When air bags are used, it is important to ensure that the bags are of sufficient strength and the air pressure high enough to ensure that any falling person does not make contact with the ground. Only reputable suppliers should be employed for the provision of air bags. Air bags or bean bags are known as soft landing systems and are used to protect workers from the effects of inward falls. Other possible solutions to this problem of the inward fall are the use of internal scaffolding or lightweight 'crash deck systems'.

Air bags may be linked together to form an inflated crash deck system. Such a system is suitable when a safety net is not viable. It consists of a series of inter-linked air mattresses that are positioned beneath the working area and is suitable for working at height inside a building where safety harnesses would not be practicable. The mattresses are interconnected by secure couplings and inflated in position on site using an air pump. The mattresses are made in various sizes so that any floor area configuration can be covered. A similar form of crash deck can be made from bean bags that are clipped together.

After the completion of the construction project, cable-based fall arrest systems are normally suitable for ongoing building maintenance work. They may also be used for the construction of complex roof structures, such as parabolic or dome structures, where a safety net may be more than 2 m below the highest point of the structure.

9.2.8 Emergency procedures (including rescue)

A suitable emergency and rescue procedure needs to be in place for situations that could be expected to occur on the construction site. There is a legal requirement to plan for emergencies and rescue when there is working at height. Such foreseeable situations could include a crane driver trapped in his cab due to a power failure or a serious ill-health problem, or the rescue of a person who has fallen into a safety net. The method of rescue may be simple and straightforward, such as putting a ladder up to a net and allowing the fallen person to descend or the use of a MEWP to lower the person to the ground.

The UK Work at Height Safety Association (WAHSA) recommends that an appropriate rescue plan should consider:

▶ the safety of the persons undertaking or assisting with the rescue;

▶ the anchor points to be used for the rescue equipment;

▶ the suitability of equipment (anchors, harnesses, attachments and connectors) that has already arrested the fall of the person for use during the rescue;

▶ the method that will be used to attach the person to the rescue system;

▶ the direction in which the person needs to be moved to get them to the point of safety; and

▶ the possible needs of the person after the rescue, including first aid.

The UK HSE guidance states that the key is to release the person from the suspended-harness position to the nearest point of safety, in the shortest possible time, and as soon as is safely possible. Otherwise, the person can suffer from pre-syncope (loss of consciousness). The symptoms for this include light-headedness, nausea, sensations of flushing, tingling or numbness of the arms or legs, anxiety, visual disturbance, or a feeling they are about to faint. Pre-syncope can occur within one hour (or sooner for some people) and if the rescuer is unable to immediately release the conscious person, elevation of the legs by the person or rescuer may prolong the tolerance of suspension.

Proprietary rescue systems are available and may well be suitable for many more complex rescues. The system selected needs to be proportionate to the risk and there should not be undue reliance on the emergency services. However, arrangements should be in place to inform the emergency services of any serious incident. It is important that rescue teams are available at all times with a designated leader. Instruction and training are essential for the teams, preferably by simulated rescue exercises.

9.2.9 Summary of the hazards and controls associated with working above ground level

Before any work at height starts, a risk assessment should be undertaken that considers the simple hierarchy given under 9.2.4.

The significance of injuries resulting from falls from height, such as fatalities and other major injuries, has been dealt with earlier in the chapter as has the importance and legal requirements for head protection. Also covered were the many hazards involved in working at height, including fragile roofs and the deterioration of materials, unprotected edges and falling materials. Additional hazards include the weather and unstable or poorly maintained access equipment, such as ladders and various types of scaffold.

The principal means of preventing falls of people or materials includes the use of fencing, guard rails, toe boards, working platforms, access boards, ladder hoops, safety nets and safety harnesses. Safety harnesses arrest the fall by restricting the fall to a given distance due to the fixing of the harness to a point on an adjacent rigid structure. They should only be used when all other possibilities are not practical.

The work must also be properly planned and supervised. Training and instruction are needed for workers both initially and before each new workplace. Such training should include safe working methods (including the mandatory wearing of hard hats), emergency procedures and accident procedures.

9.2.10 Safe working practices for common types of access equipment

There are many different types of access equipment, but only the following four categories will be considered here:

- ladders
- fixed scaffold
- pre-fabricated mobile scaffold towers
- mobile elevating work platforms (MEWPs).

Ladders

The main cause of accidents involving ladders is ladder movement while in use. This occurs when they have not been secured to a fixed point, particularly at the foot. Other causes include over-reaching by the worker, slipping on a rung, ladder defects and, in the case of metal ladders, contact with electricity. The main types of accident are falls from ladders.

There are three common materials used in the construction of ladders – aluminium, timber and glass fibre. Aluminium ladders have the advantage of being light but should not be used in high winds or near live electricity. Timber ladders need regular inspection for damage and should not be painted, as this could hide cracks. Glass fibre ladders can be used near electrical equipment and in food processing areas.

Every time a ladder is used, a pre-use check should be made. Such a check should be undertaken by the user to check the condition of the feet and rungs at the beginning of the working day or whenever the ladder is dropped or moved from a dirty area to a clean area.

The following factors should be considered when using ladders:

- Undertake as much work as possible from the ground.
- Ensure that the equipment is suitable, stable and strong enough for the job, maintained and checked regularly.
- Ensure that the use of a ladder can be justified and is the safest means of access given the work to be done and the height to be climbed.
- The location itself needs to be checked. The supporting wall and supporting ground surface should be dry, firm, level and slip free. Extra care will be needed if the area is busy with pedestrians or vehicles or the ladder is rested against weak upper surfaces, such as glazing or plastic gutters.
- The ladder needs to be stable in use. This means that the inclination should be as near the optimum as possible (1:4 ratio of distance from the wall to distance up the wall).
- Wherever possible, a ladder should be tied to prevent it from slipping. This can either be at the top, the bottom or both, making sure both stiles are tied. Never tie a ladder by its rungs.
- If the ladder cannot be tied, use an 'effective ladder' or one with an 'effective ladder-stability device' that the suppliers or manufacturers can confirm is stable enough to use unsecured in the worst-case scenario e.g. a ladder stay with anti-slip device.
- If the above three precautions are not possible, then the ladder stiles can be wedged against a wall or other similar heavy object, or, as a last resort, have a second person 'foot' the ladder.
- Weather conditions must be suitable (no high winds or heavy rain).
- The proximity of live electricity should also be considered, particularly when ladders are carried near or under power lines.
- There should be at least 1 m of ladder above the stepping off point.
- The work activity must be considered in some detail. Over-reaching must be eliminated and consideration given to the storage of paints or tools which are to be used from the ladder and any loads to be carried up the ladder. The ladder must be matched to work required. take precautions when working on or near fragile surfaces
- There should be protection provided from falling objects.
- Workers who are to use ladders must be trained in the correct method of use and selection. Such

training should include the use of both hands during climbing, clean non-slippery footwear, clean rungs and an undamaged ladder.

▷ Ladders should be inspected (particularly for damaged or missing rungs) and maintained on a regular basis and they should only be repaired by competent persons.

▷ The transportation and storage of ladders is important as much damage can occur at these times. They need to be handled carefully and stored in a dry place.

▷ When a ladder is left secured to a structure during non-working hours, a plank should be tied to the rungs to prevent unauthorised access to the structure.

▷ Emergency evacuation and rescue procedures should be considered.

Figure 9.7 Ladder showing correct 1 in 4 angle (means of securing omitted for clarity of illustration)

More information on the safe use of ladders and stepladders is given in INDG402 *Safe Use of Ladders and Stepladders – An Employers' Guide,* HSE Books.

Certain work should not be attempted using ladders. This includes work where:

▷ a secure handhold is not available;
▷ the work is at an excessive height or reach;
▷ the ladder cannot be secured or made stable;
▷ the work is of long duration or is strenuous;
▷ the work area is very large;
▷ the equipment or materials to be used are heavy or bulky;
▷ the weather conditions are adverse;
▷ there is no protection from passing vehicles.

There have been several rumours that the Work at Height Regulations have banned the use of ladders. This is not true. Ladders may be used for access and it is legal to work from ladders. Ladders may be used when a risk assessment shows that the risk of injury is low, and that the task is of short duration or there are unalterable features of the work site and that it is not reasonably practicable to use potentially safer alternative means of access, such as a MEWP or a mobile tower scaffold. There is no maximum height for using a ladder. However, where a ladder rises 9 m or more above its base, landing areas or rest platforms should be provided at suitable intervals. Ladder guidance from HSE recommends that users should maintain three points of contact when climbing a ladder and wherever possible at the work position. The three points of contact are a hand and two feet.

More information on ladders and their use within the requirements of the Work at Height Regulations is available from the British Ladder Manufacturers' Association. Ladders for industrial work in the UK should be marked to:

> Timber BS1129: Kite marked Class 1 Industrial
> Aluminium BS2037: 1994 Kite Marked Class 1 Industrial
> Glass fibre BSEN131: 1993 Kite marked Industrial
> Step stools BS7377: 1994.

Combination ladders

Following a recent fatal accident investigation, HSE is strongly advising all duty-holders and users of combination ladders to ensure that they:

▷ carry out pre-use checks;
▷ use them in accordance with instructions;
▷ check the locking mechanism(s).

Failure to do so could result in serious accidents.

Telescopic ladders

The use of telescopic ladders is growing more popular due to their ease of storage and convenience; however, there are numerous issues with many of these products due to the number of components involved and their construction. Namely:

▶ they are often rated for a lower load (person, tools and materials);

▶ the stiles are prone to greater bending;

▶ they are prone to greater bending of the frame.

Stepladders, trestles and staging

Many of the points discussed for ladders apply to stepladders and trestles, where stability and over-reaching are the main hazards.

All equipment must be checked by the supervisor before use to ensure that there are no defects and must be checked at least weekly by the user(s) whilst in use on site. If a defect is noted, or the equipment is damaged, it must be taken out of use immediately. Any repairs must only be carried out by competent persons.

Supervisors must also check that the equipment is being used correctly and not being used where a safer method should be provided.

Where staging, such as a 'Youngmans' staging platform, is being used in roof areas, supervisors must ensure that only experienced operatives are permitted to carry out this work and that all necessary safety harnesses and anchorage points are provided and used.

The main hazards associated with stepladders, trestles and staging are:

▶ unsuitable base (uneven or loose materials);

▶ unsafe and incorrect use of equipment (e.g. the use of staging for barrow ramps);

▶ overloading;

▶ use of equipment where a safer method should be provided;

▶ overhang of boards or staging at supports ('trap ends');

▶ use of defective equipment.

(a)

(b)

(c)

(d)

Figure 9.8 (a) Ladder tied at top style (correct for working on, not for access); (b) tying part-way down; (c) tying near base; (d) securing at base

Stepladders and trestles must be:

▶ manufactured to a recognised industrial specification;

▶ stored and handled with care to prevent damage and deterioration;

▶ subject to a programme of regular inspection (there should be a marking, coding or tagging system to confirm that the inspection has taken place);

▶ checked by the user before use;

▶ taken out of use if damaged – and destroyed or repaired;

▶ used on a secure surface, and with due regard to ensuring stability at all times;

▶ kept away from overhead cables and similar hazards.

The small platform fitted at the top of many stepladders is designed to support tools, paint pots and other working materials. It should not be used as a working place unless the stepladder has been constructed with a suitable handhold above the platform. Stepladders must not be used if they are too short for the work being undertaken, or if there is not enough space to open them out fully. If, when on a stepladder, two hands need to be free for a brief period of light work (e.g. to change a light bulb), keep two feet on the same step and the body supported by the stepladder to maintain three points of contact.

Figure 9.9 Working with stepladders

Platforms based on trestles should be fully boarded, adequately supported (at least one support for each 1.5 m of board for standard scaffold boards) and provided with edge protection when there is a risk of falling a distance liable to cause injury.

Fixed scaffolds

It is quicker and easier to use a ladder as a means of access, but it is not always the safest. Jobs, such as painting, gutter repair, demolition work or window

replacement, are often easier done using a scaffold. If the work can be completed comfortably using ladders, a scaffold need not be considered. Scaffolds must be capable of supporting building workers, equipment, materials, tools and any accumulated waste. A common cause of scaffold collapse is the 'borrowing' of boards and tubes from the scaffold, thus weakening it. Falls from scaffolds are often caused by badly constructed working platforms, inadequate guard rails or climbing up the outside of a scaffold. Falls also occur during the assembly or dismantling process.

There are two basic types of external scaffold:

Independent tied – These are scaffolding structures which are independent of the building but tied to it often using a window or window recess. This is the most common form of scaffolding.

Putlog – This form of scaffolding is usually used during the construction of a building. A putlog is a scaffold tube which spans horizontally from the scaffold into the building – the end of the tube is flattened and is usually positioned between two brick courses.

The important components of a scaffold have been defined in a guidance note issued by the HSE as follows:

Standard – An upright tube or pole used as a vertical support in a scaffold.

Ledger – A tube spanning horizontally and tying standards longitudinally.

Transom – A tube spanning across ledgers to tie a scaffold transversely. It may also support a working platform.

Bracing – Tubes which span diagonally to strengthen and prevent movement of the scaffold.

Guard rail – A horizontal tube fitted to standards along working platforms to prevent persons from falling; it should be at least 950 mm high and no more than 470 mm between the toe boards and the intermediate rail and top rail.

Toe boards – These are fitted at the base of working platforms to prevent persons, materials or tools falling from the scaffold.

Base plate – A square steel plate fitted to the bottom of a standard at ground level.

Sole board – *Normally* a timber plank positioned beneath at least two base plates to provide a more uniform distribution of the scaffold load over the ground.

Ties – Used to secure the scaffold by anchoring it to the building. The scaffold in Figure 9.10 (a) is tied to the building using a through-tie.

Working platform – An important part of the scaffold, as it is the platform on which the building workers operate and where building materials are stored prior to use. These are laid on the transoms in various quantities; usually for general purpose, they should be four boards wide. A working platform

can be almost any surface from which work can be undertaken, such as:
- a roof
- a floor
- a platform on a scaffold

- mobile elevating work platforms (MEWPs)
- the treads of a stepladder

Other components of a scaffold include access ladders, brick or block guards and chutes to dispose of waste materials.

(a)

(b)

Figure 9.10 (a) Typical independent tied scaffold; (b) fixed scaffold left in place to reach gutters

The following factors must be addressed if a scaffold is being considered for use for construction purposes:

- Scaffolding must only be erected by competent people who have attended recognised training courses. Any work carried out on the scaffold must be supervised by a competent person. Any changes to the scaffold must be done by a competent person.
- Adequate toe boards, guard rails and intermediate rails must be fitted to prevent people or materials from falling. The top guard rail should be a minimum of 950 mm above the working platform and any gap between the top rail and the intermediate rail should not exceed 470 mm. The toe boards to be suitable and sufficient to prevent people or materials from falling.
- The scaffold must rest on a stable surface; uprights should have base plates and timber sole plates if necessary.
- The scaffold must have safe access and egress.
- Work platforms should be fully boarded with no tipping or tripping hazards.

▶ The scaffold should be sited away from or protected from traffic routes so that it is not damaged by vehicles.

▶ Lower level uprights should be prominently marked in red and white stripes.

▶ The scaffold should be properly braced, secured to the building or structure.

▶ Overloading of the scaffold must be avoided.

▶ The public must be protected at all stages of the work. Barriers/signage must be displayed and what measures to take if unattended.

▶ Regular inspections of the scaffold must be made and recorded.

A checklist of typical scaffolding faults is given in Appendix 9.1.

Pre-fabricated mobile scaffold towers

Mobile scaffold towers are frequently used throughout industry. It is essential that the workers are trained in

their use as recent research has revealed that 75% of lightweight mobile pre-fabricated tower scaffolding is erected, used, moved or dismantled in an unsafe manner.

The following points must be considered when mobile scaffold towers are to be used:

▶ The selection, erection and dismantling of mobile scaffold towers must be undertaken by competent and trained persons with maximum height to base ratios not being exceeded.

▶ Diagonal bracing and stabilisers should always be used.

▶ Access ladders must be fitted to the narrowest side of the tower or inside the tower and persons should not climb up the frame of the tower.

▶ All wheels must be locked while work is in progress and all persons must vacate the tower before it is moved.

▶ The tower working platform must be boarded, fitted with guard rails and toe boards and not overloaded.

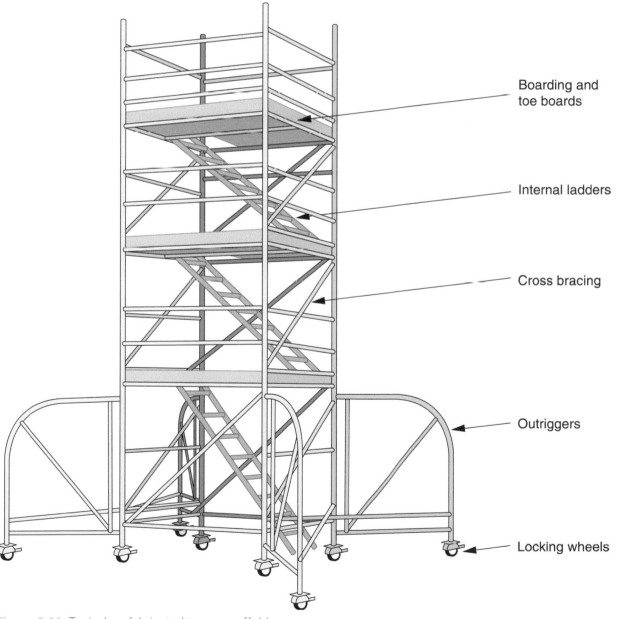

Boarding and toe boards

Internal ladders

Cross bracing

Outriggers

Locking wheels

Figure 9.11 Typical prefabricated tower scaffold

▶ Towers must be tied to a rigid structure if exposed to windy weather or to be used for work such as jet blasting.

▶ Persons working from a tower must not over-reach or use ladders from the work platform.

▶ Safe distances must be maintained between the tower and overhead power lines both during working operations and when the tower is moved.

▶ The tower should be inspected on a regular basis and a report made.

Anyone erecting a tower scaffold should be competent to do so and should have received training under an industry-recognised training scheme, such as the Prefabricated Access Suppliers' and Manufacturers' Association (PASMA) or under a recognised manufacturer or supplier scheme.

Mobile elevating work platforms (MEWPs)

Most fatal and serious injuries involving MEWPs arise from:

▶ entrapment of the operator;

▶ overturning;

▶ the operator falling from the basket;

▶ the vehicle may collide with pedestrians, overhead cables or nearby vehicles.

MEWPs are very suitable for high-level work such as changing light bulbs in a warehouse. The following factors must be considered when using a MEWP:

▶ It should be maintained regularly, have a current Certificate of Thorough Inspection within the last six months and be safe to use. Procedures should be in place in the event of machine failure.

▶ The MEWP must only be operated by trained and competent persons.

▶ It must never be moved in the elevated position.

▶ It must be operated on level and stable ground with consideration being given for the stability and loading of floors.

▶ The tyres must be properly inflated and the wheels immobilised.

▶ Outriggers should be fully extended and locked in position.

▶ Due care must be exercised with overhead power supplies, obstructions and adverse weather conditions.

▶ Warning signs should be displayed and barriers erected to avoid collisions.

▶ Drivers of MEWPs must be instructed in emergency procedures, particularly to cover instances of power failure.

▶ All workers on MEWPs should wear safety harnesses.

Workers using mobile elevating work platforms have been injured by falling from the platform due to a lack of handrails or inadvertent movement of the equipment because the brakes had not been applied before raising the platform. Injuries have also resulted from the mechanical failure of the lifting mechanism and by workers becoming trapped in the scissor mechanism.

When working on a MEWP, there is a danger that the operator may become trapped against an overhead or adjacent object, preventing him/her from releasing the controls. There have also been accidents caused when an MEWP is reversed into areas where there is poor pedestrian segregation and the driver has limited visibility. During any manoeuvring operation, a dedicated banksman should be used.

There is evidence of a growing number of serious and fatal accidents from entrapment accidents between guard rails on MEWPs and adjacent obstructions. There have been several serious accidents caused by workers becoming trapped by overhead obstructions while the MEWP was in motion. These accidents often occur during maintenance operations. It is important that risk assessments are undertaken and safe systems of work developed for such work. The Strategic Forum for Construction Plant Safety Group have published guidance on best practice when using MEWPs in confined overhead spaces. The guidance covers hazards, risk assessment, controls and responsibilities of managers.

If **a MEWP is to be used**, the HSE suggest that the following questions should be considered:

▶ Height – How high is the job from the ground?

▶ Application – Do you have the appropriate MEWP for the job? (If you're not sure, check with the hirer or manufacturer)

▶ Conditions – What are the ground conditions like? Is there a risk of the MEWP becoming unstable or overturning?

▶ Operators – Are the people using the MEWP trained, competent and fit to do so?

▶ Obstructions – Could the MEWP be caught on any protruding features or overhead hazards, such as steelwork, tree branches or power lines?

▶ Traffic – Is there passing traffic and, if so, what do you need to do to prevent collisions?

▶ Restraint – Do you need to use either work restraint (to prevent people climbing out of the MEWP) or a fall arrest system (which will stop a person hitting the ground if they fall out)? Allowing people to climb out of the basket is not normally recommended – do you need to do this as part of the job?

▶ Checks – Has the MEWP been examined, inspected and maintained as required by the manufacturer's instructions and have daily checks been carried out?

▷ the location of the work equipment inspected;
▷ a description of the work equipment inspected;
▷ the date and time of the inspection;
▷ details of any matter identified that could give rise to a risk to the health or safety of any person;
▷ details of any action taken as a result of any matter identified above;
▷ details of any further action considered necessary;
▷ the name and position of the person making the report.

All scaffolding inspection should be carried out by a competent person whose combination of knowledge, training and experience is appropriate for the type and complexity of the particular scaffold to be inspected. Competence may have been assessed under the Construction Industry Scaffolders Record Scheme (CISRS) or an individual may be suitably experienced in scaffolding work and have received additional training under a recognised manufacturer/supplier scheme for the specific configuration being inspected.

A non-scaffolder, such as a site manager, who has attended a suitable scaffold inspection course and has the necessary background experience would be considered competent to inspect a basic scaffold.

There is no a legal requirement to use scafftags, but using a visible tag system to supplement inspection records is one way of recording that the scaffold has been checked before use.

Appendix 9.2 shows the inspection format under the CDM Regulations, which still covers the essential items under the Work at Height Regulations and has been slightly adapted to suit.

9.2.12 Prevention of falling materials through safe stacking and storage

It is estimated that two million working days are lost each year on account of handling injuries and slip and trip incidents. Such incidents can occur in warehouses and storage facilities when palletised goods are stacked higher than two storeys and often weigh several tonnes. Implementing tried and tested methods on safe racking and storage is therefore essential to mitigate the risks of an incident arising.

It is often possible to remove high-level storage from offices and other general workplaces, such as construction sites, and provide storage in warehouses or similar storage facilities. If a storage facility is to be installed for the first time, then the following points should be considered:

▷ The racking must be erected on and fixed securely to a sound, level floor.
▷ The storage system must be installed in accordance with manufacturers' instructions.
▷ If the racking is to be secured to the wall of a building, has this been proved by structural

Figure 9.12 Mobile elevating work platforms: (a) cherry-picker; (b) scissor lift

9.2.11 Inspection

Equipment for work at height needs regular inspection to ensure that it is fit for use. A marking system is probably required to show when the next inspection is due. Formal inspections should not be a substitute for any pre-use checks or routine maintenance. Inspection does not necessarily cover the checks that are made during maintenance although there may be some common features. Inspections need to be recorded but checks do not.

Under the Work at Height Regulations weekly inspections are still required for scaffolding used in construction, as required by the Construction Design and Management (CDM) Regulations where a person could fall 2 m or more. The requirements for inspection are set out in the Regulations as follows:

▷ the name and address of the person for whom the inspection was carried out;

calculations which should be made to ensure that the walls can support the racking and its contents securely.

▶ Beam-connector locks must be fixed at both ends of the beam.

▶ Maximum-load notices must be displayed.

▶ Correct pallets must be used.

▶ Sufficient protective equipment must be used, such as column guards and rack-end protectors.

Management and employees should familiarise themselves with the racking systems used and ensure they understand the difference between general wear and tear and real damage, in order to help them identify potentially dangerous situations as early as possible.

Storage racking is particularly vulnerable and should be strong and stable enough for the loads it has to carry. Damage from vehicles in a warehouse can easily weaken the structure and cause collapse. Uprights need protection, particularly at corners.

The following action can be taken to keep racking serviceable:

▶ Inspect them regularly and encourage workers to report any problems/defects.

▶ Post notices with maximum permissible loads and never exceed the loading.

▶ Use good pallets and safe stacking methods.

▶ Band, box or wrap articles to prevent items falling.

▶ Set limits on the height of stacks and regularly inspect to make sure that limits are being followed.

▶ Provide instruction and training for staff and special procedures for difficult objects.

Regular visual inspections should be carried out and documented, so that any damage can be quickly resolved. In particular, staff should be trained to act if damage occurs to and affects:

▶ the cross-sectional profile of a main load beam;

▶ the straightness of beams, bracing, or uprights; and

▶ the welds and joints, or bolts and clips.

To guard against back strains and other injuries, shelving and storage solutions can be installed that allow access and retrieval of stock at a comfortable, ergonomic height. Products such as vertical storage machines or pallet pull-out units are possible solutions, as they are designed so that stock can be reached without unnecessary straining.

Generally, storage racks should be examined by a qualified inspector approved by the Storage Equipment Manufacturers' Association (SEMA) at least once or twice a year.

During an inspection, particular attention will be paid to beams, uprights, frame bracing, floor fixings and lock-in clips, as indicated in the SEMA Code of Practice, guideline no. 6. The following will also be subject to general observations:

▶ pallet locations on beams;

▶ conditions and types of pallets;

▶ positioning of loads and types of loads stored on pallets;

▶ general fork-lift operatives' use of the racking;

▶ the condition and type of floor on which the racking is fixed;

▶ general housekeeping of the installation; and possible changes from the original design requirements.

With warehouse managers relying increasingly on temporary and agency workers, who sometimes have little, or no prior experience of working in warehouse or storage environments, it is vital that safety training becomes part of the induction process.

All materials used in the construction process on site must be either stacked or stored safely. This will keep the site tidy and reduce slip and trip hazards. It will also help to reduce the risk of fire. Dangerous substances should be kept in a safe place in a separate building or the open air. Only small quantities of hazardous and/or flammable substances should be kept on the site. Any larger amounts, which cannot be kept outside in a safe area, should be kept in a special fire-resisting store which is well ventilated and free of ignition sources. Flammable gas cylinders also need to be stored and used safely. More information on this is given in Chapter 11.

9.3 Safe working in confined spaces

9.3.1 Definition of and specific examples of safe systems of work

A confined space is defined in the **Confined Spaces Regulations** as:

'any place, including any chamber, tank, vat, silo, pit, trench, pipe, sewer, flue, well or other similar space in which, by virtue of its enclosed nature, there arises a reasonably foreseeable specified risk.'

It is, therefore, a place which is substantially enclosed (though not always entirely), where serious injury can occur from hazardous substances or conditions within the space or nearby.

These specified risks mean a risk to a worker of:

▶ serious injury arising from a fire or explosion;

▶ loss of consciousness arising from an increase in body temperature;

▶ loss of consciousness or asphyxiation arising from gas, fume, vapour or the lack of oxygen;

▶ drowning arising from an increase in the level of liquid;

▶ asphyxiation arising from a free-flowing solid or because of entrapment by it.

Therefore, confined spaces include chambers, tanks (sealed and open-top), vessels, furnaces, ducts, sewers, manholes, pits, flues, excavations, boilers, reactors and ovens.

The principal hazards associated with a confined space are the difficult access and egress, which can make escape and rescue more difficult. Other hazards associated with confined spaces include:

- asphyxiation due to oxygen depletion;
- poisoning by toxic substance or fumes;
- explosions due to gases, vapours and dust;
- fire due to flammable liquids;
- fall of materials leading to possible head injuries;
- free-flowing solid such as grain in a silo;
- electrocution from unsuitable equipment;
- difficulties of rescuing injured personnel;
- drowning due to flooding; and
- fumes from plant or processes entering confined spaces.

Since work in confined spaces is a high-risk work activity, a risk assessment is essential. The following items should be included in the assessment:

- the task;
- the working environment;
- working materials and tools;
- the suitability of those carrying out the task; and
- arrangements for emergency rescue.

If the assessment identifies risks of serious injury from work in the confined space, then the following key duties specified in the Confined Spaces Regulations apply. These duties are:

- if possible, avoid entry to confined space by doing the work outside the space (e.g. remote cameras can be used for internal inspection of vessels);
- if entry to a confined space is unavoidable, follow a safe system of work; and
- ensure that adequate emergency arrangements are in place before the work starts.

If entry into a confined space cannot be avoided, then a safe system of work for working inside the space must be developed that uses the results of the risk assessment to identify the controls needed to reduce the risk of injury. These controls will depend on the nature of the confined space, the associated risk and the work involved. It is important that the agreed safe system of work is fully implemented and everyone who is to work in the confined space must be trained and instructed on the controls required.

The following topics need to be addressed in a safe system of work for a confined space:

- the appointment of a supervisor;
- the competence and experience required of the workers in the confined space;
- the isolation of mechanical and electrical isolation of equipment in an emergency;
- the size of the entrance must enable rapid access and exit in an emergency;
- the provision of adequate ventilation;
- the testing of the air inside the space to ensure that it is fit to breathe. If the air inside the space is not fit to breathe, then breathing apparatus will be essential;

- the provision of special tools and lighting, such as extra-low-voltage equipment, non-sparking tools and specially protected lighting;
- the emergency arrangements to cover the necessary equipment, training, practice drills and the raising of the alarm; and
- adequate communications arrangements to enable communication between people inside and outside the confined space and to summon help in an emergency.

The provision of suitable rescue and resuscitation equipment will depend on the likely emergencies identified. Where such equipment is provided for rescuers to use, training in the correct operation of the equipment is essential. Rescuers should also be trained in all aspects of the emergency procedures and relevant first-aid procedures.

For particularly hazardous confined space working a permit to work may be required as discussed in Chapter 4.

The HSE publication INDG258 gives detailed advice on the possible risks and control measures associated with confined spaces.

9.4 Lone working

People who work by themselves without close or direct supervision are found in many work situations. In some cases they are the sole occupant of small workshops or warehouses; they may work in remote sections of a large site; they may work out of normal hours, like cleaners or security personnel; they may be working away from their main base as installers, social workers, sales personnel, or maintenance people; they could be people giving a service, like domiciliary care workers, drivers and estate agents.

Chapter 4 (Section 4.1.11) covered the need and contents of a risk assessment for lone workers. In this section the controls required to protect lone workers are discussed.

The most serious hazards that lone workers face are:

- violence from the public; and
- lack of access to help in the event of an accident or an attack of ill health.

There is no general legal reason why people should not work alone, but there may be special risks which require two or more people to be present; for example, during entry into a confined space in order to effect a rescue. It is important to ensure that a lone worker is not put at any higher risk than other workers. This is achieved by carrying out a specific risk assessment and introducing special protection arrangements for their safety. The overall health and suitability of a person to work alone should be taken into account. It is important to ascertain whether the work should be performed alone, particularly where there is a possibility of a serious risk, such as violence, being confronted by the worker.

Typical control procedures may include:

▶ documented records of the location or itineraries of the lone workers;

▶ periodic visits from the supervisor to observe what is happening;

▶ regular voice contact, using mobile phones or radios, between the lone worker and the supervisor;

▶ automatic warning devices to alert others if a specific signal is not received from the lone worker;

▶ other devices to raise the alarm, which are activated by the absence of some specific action;

▶ checks that the lone worker has returned safely home or to their base;

▶ special arrangements for first aid to deal with minor injuries – this may include mobile first-aid kits;

▶ arrangements for emergencies – these should be established and employees trained.

One of the largest increases in lone working has been in work-related driving. Just under half of such workers face road-rage incidents at least once a year and 11% are assaulted.

All lone workers at risk of violence should receive training to help them recognise and anticipate violence and difficult situations. Such training should develop communication skills to help the workers remain diplomatic and non-confrontational when facing potentially violent circumstances.

Lone workers should be required to leave a voicemail before working alone remotely, saying where they will be and for how long. If they fail to call the contact centre at the end of the specified period, emergency procedures should be enacted.

In remote areas with poor mobile phone coverage, SPOT trackers using the global positioning system may be used to pinpoint a lone worker's location.

The Institution of Occupational Safety and Health (IOSH) have produced a very useful document on lone working that contains an audit checklist for remote working.

9.5 Slips and trips

9.5.1 Common causes of slips and trips

Slipping and tripping is the single most common cause of injury in the workplace. These are the most common of the hazards faced by pedestrians and account for 38% of all the major accidents every year and 24% of minor injuries. Every 25 minutes someone breaks or fractures a bone due to slipping, tripping or falling at work. In addition to 40 fatalities each year, there are over 15,000 major injuries to workers, as well as more than 30,000 workers having to take over three days off work. HSE figures show that slips and trips are the most common cause of major workplace injury in Britain.

It has been estimated that the annual cost of these accidents to the nation is over £650 million and a direct cost to employers of £300 million. Civil compensation claims are becoming more common and costly to employers. The highest reported injuries are reported in the food and related industries. Older workers, especially women, are the most severely injured group from falls resulting in fractures of the hips and/or femur. Civil compensation claims are becoming more common and costly to employers and such claims are now being made by members of the public who have tripped on uneven paving slabs on pavements or in shopping centres. A worker was awarded compensation after slipping on a wet wheelchair ramp. The accident could have been avoided if the ramp had been covered in an anti-slip material.

Figure 9.13 Tripping hazards

The HSE has been so concerned at the large number of such accidents that it has identified slips, trips and falls on the same level as a key risk area. The costs of slips, trips and falls on the same level are high to the injured employee (lost income and pain), the employer (direct and indirect costs including lost production) and to society as a whole in terms of health and social security costs.

Slip hazards are caused by:

▶ wet or dusty floors;

▶ the spillage of wet or dry substances – oil, water, flour dust and plastic pellets used in plastic manufacture;

▶ loose mats on slippery floors;

▶ wet and/or icy weather conditions;

▶ unsuitable footwear or floor coverings or sloping floors.

Trip hazards are caused by:
- loose floorboards or carpets;
- obstructions, low walls, low fixtures on the floor;
- cables or trailing leads across walkways or uneven surfaces; leads to portable electrical hand tools and other electrical appliances (vacuum cleaners and overhead projectors); raised telephone and electrical sockets – also a serious trip hazard (this can be a significant problem when the display screen workstations are re-oriented in an office);
- rugs and mats – particularly when worn or placed on a polished surface;
- poor housekeeping – obstacles left on walkways, rubbish not removed regularly;
- poor lighting levels – particularly near steps or other changes in level;
- sloping or uneven floors – particularly where there is poor lighting or no handrails;
- unsuitable footwear – shoes with a slippery sole or lack of ankle support.

Over half of all trip accidents are caused by poor housekeeping. Many of these housekeeping problems can easily be solved by:
- ensuring that all walkways are suitable for purpose and clear of obstructions (e.g. trailing leads);
- ensuring that workers wear suitable footwear;
- training workers in the maintenance of trip free working areas;
- regular inspections of work areas by supervisors.

Slip and trip accidents increase during the autumn and winter season for a number of reasons: there is less daylight, leaves fall onto paths and become wet and slippery and cold weather spells cause ice and snow to build up on paths.

The selection of suitable footwear for the workplace, should involve consideration of the types of hazard present, the general environment and ergonomic issues.

Figure 9.14 Cleaning must be done carefully to prevent slipping

The vast majority of major accidents involving slips, trips and falls on the same level result in dislocated or fractured bones.

9.5.2 **Main control measures**

Slips, trips and falls on the same level

These may be prevented or, at least, reduced by several control strategies. These and all the other pedestrian hazards discussed should be included in the workplace risk assessments required under the Management of Health and Safety at Work Regulations by identifying slip or trip hazards, such as poor or uneven floor/pavement surfaces, badly lit stairways and puddles from leaking roofs. There is also a legal requirement in the Workplace (Health, Safety and Welfare) Regulations for all floors to be suitable, in good condition and free from obstructions. Traffic routes must be organised to enable people to move around the workplace safely.

The key elements of a health and safety management system are as relevant to these as to any other hazards:

Planning – remove or minimise the risks by using appropriate control measures and defined working practices (e.g. covering all trailing leads).

Organisation – involve employees and supervisors in the planning process by defining responsibility for keeping given areas tidy and free from trip hazards.

Control – record all cleaning and maintenance work. Ensure that anti-slip covers and cappings are placed on stairs, ladders, catwalks, kitchen floors and smooth walkways. Use warning signs when floor surfaces have recently been washed.

Monitoring and review – carry out regular safety audits of cleaning and housekeeping procedures and include trip hazards in safety surveys. Check on accident records to see whether there has been an improvement or if an accident black spot can be identified.

Slip and trip accidents are a major problem for large retail stores for both customers and employees. The provision of non-slip flooring, a good standard of lighting and minimising the need to block aisles during the restocking of merchandise are typical measures that many stores use to reduce such accidents. Other measures include the wearing of suitable footwear by employees, adequate handrails on stairways, the highlighting of any floor level changes and procedures to ensure a quick and effective response to any reports of floor damage or spillages. Good housekeeping procedures are essential. The design of the store layout and any associated warehouse can also ensure a reduction

9

in all types of accident. Many of these measures are valid for a range of workplaces.

Cleaning arrangements should be chosen to suit both the type of floor and the users. Floors with surface roughness are not difficult to clean despite the popular belief that they are. Stairs are a particularly hazardous part of the building and become even more so when being cleaned.

The prevention of access during cleaning or drying needs to be effective. Signs that warn people that floors are wet are only partially effective. HSE make the following suggestions:

▶ Physically exclude people from wet cleaning areas by the use of barriers or locking off an area.
▶ Clean during quiet hours.
▶ Clean in sections so there is always a dry path through the area.

Floor cleaners must be provided with appropriate slip-resistant footwear that can help reduce the risk of slipping.

Stairs can also present significant potential for harm to their users, with falls on stairs often leading to serious injury or even death. Around 20% of all major injuries reported to HSE each year from slips and trips occurred on stairs.

9.6 Safe movement of people and vehicles in the workplace

9.6.1 Hazards to pedestrians in the workplace

The most common hazards to pedestrians at work are slips, trips and falls on the same level (covered in 9.5 in this chapter).

Other hazards include falls from height, collisions with moving vehicles, being struck by moving, falling or flying objects and striking against fixed or stationary objects. Each of these will be considered in turn, including the conditions and environment in which the particular hazard may arise.

Falls from work at height

More workplace deaths are triggered by falls from height than any other cause. These are the most common cause of serious injury or death in the construction industry. These accidents are often concerned with falls of greater than about 2 m and often result in fractured bones, serious head injuries, loss of consciousness and death. Some 25% of all deaths at work and 19% of all major accidents are due to falls from a height. Falls down staircases and stairways, through fragile surfaces, off landings and stepladders and from vehicles, all come into this category. Often

injuries, sometimes serious, can also result from falls below 2 m, for example using swivel chairs for access to high shelves.

Collisions with moving vehicles

These can occur within the workplace premises or on the access roads around the building. It is a particular problem where there is no separation between pedestrians and vehicles or where vehicles are speeding. Poor lighting, blind corners, the lack of warning signs and barriers at road crossing points also increase the risk of this type of accident. Some 18% of fatalities at work are caused by collisions between pedestrians and moving vehicles with the greatest number occurring in the service sector (primarily in retail and warehouse activities).

Being struck by moving, falling or flying objects

This causes 18% of fatalities at work and is the second highest cause of fatality in the construction industry. It also causes 15% of all major and 14% of over 3-day accidents. Moving objects include articles being moved, moving parts of machinery or conveyor belt systems, and flying objects are often generated by the disintegration of a moving part or a failure of a system under pressure. Falling objects are a major problem in construction (due to careless working at height) and in warehouse work (due to careless stacking of pallets on racking). The head is particularly vulnerable to these hazards. Items falling off high shelves and moving loads are also significant hazards in many sectors of industry.

Striking against fixed or stationary objects

This accounts for over 1,000 major accidents every year. Injuries are caused to a person either by colliding with a fixed part of the building structure, work in progress, a machine member or a stationary vehicle or by falling against such objects. The head appears to be the most vulnerable part of the body to this particular hazard and this is invariably caused by the misjudgement of the height of an obstacle. Concussion in a mild form is the most common outcome and a medical check-up is normally recommended. It is a very common injury during maintenance operations when there is, perhaps, less familiarity with particular space restrictions around a machine. Effective solutions to all these hazards need not be expensive, time-consuming or complicated. Employee awareness and common sense combined with a good housekeeping regime will solve many of the problems.

Maintenance activities

Maintenance-related accidents are a serious cause of concern. For example, analysis of data from recent years indicates that 25–30% of manufacturing industry fatalities in Great Britain were related to maintenance activity. Maintenance work is often more hazardous than other types of work because it is often performed under pressure – possibly due to a plant failure during a production run – or the areas requiring access are difficult to reach.

The majority of recent accidents to workers engaged in maintenance work occurred during the following activities:

▷ roof work;
▷ machinery and plant maintenance;
▷ building maintenance including painting work;
▷ maintenance of services (gas, electricity, telecommunications and water);
▷ vehicle repair work; and
▷ window and building cleaning.

The main hazards associated with maintenance work are due to:

▷ electricity – electric shock;
▷ mechanical, including entanglement, ejection and unexpected start-up;
▷ hazardous substances – gases, dusts and fume;
▷ physical – noise and vibration;
▷ obstructions of various types;
▷ access – confined spaces and work at height; and
▷ unexpected releases of pressure or other forms of energy.

Inspections and even thorough examinations are not substitutes for properly maintaining equipment. The information gained in the maintenance work, inspections and thorough technical examinations should inform one another. A maintenance log should be kept and be up to date. The whole maintenance system will require proper management systems. The frequency will depend on the equipment, the conditions in which it is used and the manufacturers' instructions.

9.6.2 Control measures for the safe movement of people in the workplace

Falls from work at height

The Work at Height Regulations gives a legal framework for the protection of workers from falls when working at height. The principal means of preventing falls of people or materials includes the use of fencing, guard rails, toe boards, working platforms, access boards and ladder hoops, safety nets and safety harnesses. Safety harnesses arrest

the fall by restricting the fall to a given distance due to the fixing of the harness to a point on an adjacent rigid structure. Safety nets and safety harnesses should only be used when all other possibilities are not reasonably practicable. The use of banisters on open sides of stairways and handrails fitted on adjacent walls will also help to prevent people from falling. Holes in floors and pits should always be fenced or adequately covered. Precautions should be taken when working on fragile surfaces. More detail on the requirements of the Work at Height Regulations is given in Section 9.2.1.

Figure 9.15 Falling from a height – tower scaffold has inadequate handrails (no centre rail), external ladder and should never be moved during use

Permanent staircases are also a source of accidents included within this category of falling from a height and the following design and safety features will help to reduce the risk of such accidents:

▷ adequate width of the stairway, depth of the tread and provision of landings and banisters or handrails and intermediate rails; the treads and risers should always be of uniform size throughout the staircase and designed to meet the Building Regulations requirements for angle of incline (i.e. steepness of staircase);
▷ provision of non-slip surfaces and reflective edging;
▷ adequate lighting;
▷ adequate maintenance;
▷ special or alternative provision for disabled people (e.g. personnel elevator at the side of the staircase).

Great care should be used when people are loading or unloading vehicles; as far as possible people should avoid climbing onto vehicles or their loads. For example, sheeting of lorries should be carried out in designated places using properly designed access equipment.

Collisions with moving vehicles

These are best prevented by completely separating pedestrians and vehicles, providing well-marked, protected and laid-out pedestrian walkways. People should cross roads by designated and clearly marked pedestrian crossings. Suitable guard rails and barriers should be erected at entrances and exits from buildings and at 'blind' corners at the end of racking in warehouses. Particular care must be taken in areas where lorries are being loaded or unloaded. It is important that separate doorways are provided for pedestrians and vehicles and all such doorways should be provided with a vision panel and an indication of the safe clearance height, if used by vehicles. Finally, the enforcement of a sensible speed limit, coupled, where practicable, with speed governing devices, is another effective control measure.

Being struck by moving, falling or flying objects

These hazards may be prevented by guarding or fencing the moving part (as discussed in Chapter 10) or by adopting the measures outlined later in this chapter. Both construction workers and members of the public need to be protected from the hazards associated with falling objects. Both groups should be protected by the use of covered walkways or suitable netting to catch falling debris where this is a significant hazard. Waste material should be brought to ground level by the use of chutes or hoists. Waste should not be thrown from a height and only minimal quantities of building materials should be stored on working platforms. Appropriate personal protective equipment, such as hard hats or safety glasses, should be worn at all times when construction operations are taking place.

Striking against fixed or stationary objects

This hazard can only be effectively controlled by:

- having good standards of lighting and housekeeping;
- defining walkways and making sure they are used;
- the use of awareness measures, such as training and information in the form of signs or distinctive colouring;
- the use of appropriate personal protective equipment, such as head protection, as discussed previously.

General preventative measures for pedestrian hazards

Minimising pedestrian hazards and promoting good work practices requires a mixture of sensible planning, good housekeeping and common sense. Few of the required measures are costly or difficult to introduce and, although they are mainly applicable to slips, trips and falls on the same level and collisions with moving vehicles, they can be adapted to all types of pedestrian hazard. Typical measures include the following:

- Develop a safe workplace as early as possible and ensure that suitable floor surfaces and lighting are selected and vehicle and pedestrian routes are carefully planned. Lighting should not dazzle approaching vehicles nor should pedestrians be obscured by stored products. Lighting is very important where there are changes of level or stairways. Any physical hazards, such as low beams, vehicular movements or pedestrian crossings, should be clearly marked. Staircases need particular attention to ensure that they are slip resistant and the edges of the stairs marked to indicate a trip hazard.
- Consider pedestrian safety when re-orientating the workplace layout (e.g. the need to reposition lighting and emergency lighting).
- Adopt and mark designated walkways.
- Apply good housekeeping principles by keeping all areas, particularly walkways, as tidy as possible and ensure that any spillages are quickly removed.
- Ensure that all workers are suitably trained in the correct use of any safety devices (such as machine guarding or personal protective equipment) or cleaning equipment provided by the employer.
- Only use cleaning materials and substances that are effective and compatible with the surfaces being cleaned, so that additional slip hazards are not created.
- Ensure that a suitable system of maintenance, cleaning, fault reporting and repair are in place and working effectively. Areas that are being cleaned must be fenced and warning signs erected. Care must also be taken with trailing electrical leads used with the cleaning equipment. Records of cleaning, repairs and maintenance should be kept.
- Ensure that all workers are wearing high visibility clothing and appropriate footwear with the correct type of slip-resistant soles suitable for the type of flooring.
- Consider whether there are significant pedestrian hazards present in the area when any workplace risk assessments are being undertaken.

It can be seen, therefore, that floors and traffic routes should be of sound construction. If there are frequent, possibly transient, slip hazards, the provision of slip-resistant coating and/or mats should be considered and warning notices posted. Any damaged areas must be cordoned off until repairs are completed. Risk assessments should review past accidents and near misses to enable relevant controls, such as suitable footwear, to be introduced. Employees can often indicate problem areas, so employee consultation is important.

Figure 9.16 Typical separation of pedestrian/vehicle crossing

Figure 9.17 Well laid out and lit fabrication workshop with marked walkways which are kept clear of obstructions

Electrically powered gates

The HSE has issued a warning on the risks to pedestrians from crushing zones on electrically powered gates, such as those commonly used at car parks. Two young children suffered fatal injuries after being trapped by automatic sliding gates between the closing edge of the gate and the gate post at the end of the gates' travel. They were trapped because:

▶ Their presence in the vicinity of the closing edge was not detected; and

▶ The closing force of the gate when they obstructed it was not limited to the values specified in Annex A of BS EN 12453:2001.

The HSE Safety Notice reinforced and updated previous information to organisations and individuals involved in the design, construction, installation and commissioning of electrically powered gates and organisations in control of their use and/or maintenance. It is also relevant to companies carrying out ongoing maintenance of these types of gates. The British Standard (BS EN 12453:2001) recommends a minimum level of safeguarding against the crushing hazard at the closing edge of the gate depending on the type of environment in which the gate is operating. For electrically powered gates that operate in an area accessible to the public, the advised level of safeguarding is to:

▶ Limit forces according to Annex A of the Standard using force limitation devices or sensitive protective equipment AND provide a means for the detection of the presence of a person or an obstacle standing on the floor at one side of the gate; or

▶ Provide a means for detection of the presence a person, which is designed in a way that in no circumstances can that person be touched by the moving gate leaf.

The HSE recommend the following actions:

▶ All designers and installers of electrically powered gates should ensure that the forces generated by a gate when meeting a person or an obstacle are limited and that they do not exceed the values specified in Annex A of BS EN 12453:2001.

▶ Forces should be periodically re-measured and checked as part of the planned preventative maintenance schedule for the gates.

▷ In addition to force limitation, additional safeguards, such as pressure sensitive strips on the closing edge and photoelectric sensing devices, should be fitted where the risk assessment identifies the gate as high risk, in that it is operating automatically in a public place where children and other members of the public may be present.

▷ Persons or organisations in control of powered gates should periodically review their risk assessments to ensure that they identify any changes to the environment or operating conditions and that they have taken appropriate steps to address them. This is particularly important when the responsibility for management of the gate passes from one person or organisation to another.

▷ All safety devices and features should be checked on a regular basis and in accordance with the manufacturer's instructions to ensure they continue to function as designed to ensure that safety is maintained. This should be specified in a planned preventative maintenance schedule agreed by persons responsible for the gate's management and their appointed maintenance company.

9.6.3 Safe movement of vehicles in the workplace

Introduction

The safe movement of vehicles in the workplace is essential if accidents are to be avoided. The more serious accidents between pedestrians and vehicles can often be traced back to excessive speed or other unsafe vehicle practices, such as lack of driver training. More and more workers drive vehicles on public highways as part of their jobs, and some are involved in road accidents that may be classed as work – many of the risks associated with these hazards can be significantly reduced by an effective management system.

Hazards from workplace transport operations

Many different kinds of vehicle are used in the workplace, including dumper trucks, heavy goods vehicles, all-terrain vehicles and, perhaps the most common, the fork-lift truck. Each year, approximately 45 workers are killed and over 5,000 injured in workplace transport accidents. The third highest cause of accidental death in the workplace is being struck by a moving vehicle. Approximately a quarter of all vehicle-related deaths in the workplace occur while a vehicle is reversing. There are also over 1,000 major accidents (involving serious fractures, head injuries and amputations) caused by:

▷ collisions between pedestrians and vehicles;
▷ people falling from vehicles;
▷ people being struck by objects falling from vehicles;

▷ people being struck by an overturning vehicle;
▷ people injured while inside a vehicle during a collision; and
▷ communication problems between vehicle drivers and employees or members of the public.

The term 'workplace transport' covers any vehicle that is used in a work setting, such as fork-lift trucks, compact dumpers, tractors and mobile cranes. It can also include cars, vans and large goods vehicles when these are operating at the works premises. A goods vehicle that is loading or unloading on the public highway outside a works premises is considered as workplace transport. The Regulations that cover workplace traffic movements are the Workplace (Health, Safety and Welfare) Regulations. The ACoP to the Workplace Regulations gives specific advice and guidance on how to comply with them.

The main types of accident associated with workplace transport are people being struck by moving vehicles, falling from vehicles (jumping out of cabs or from the tops of high-sided trailers) and being hit by objects falling from vehicles (usually part of a load).

A key cause of these accidents is the lack of competent and documented driver training. HSE investigations, for example, have shown that in over 30% of dumper truck accidents on construction sites, the drivers had little experience and no training. Common forms of these accidents include driving into excavations, overturning while driving up steep inclines and runaway vehicles which have been left unattended with the engine running. Each year, the driving of tractors and other agricultural vehicles, such as all-terrain vehicles causes several fatalities in farming and other agriculture-related activities.

Risks of injuries to employees and members of the public involving vehicles could arise due to the following occurrences:

▷ collision with pedestrians;
▷ collision with other vehicles;
▷ overloading of vehicles;
▷ overturning of vehicles;
▷ general vehicle movements and parking;
▷ dangerous occurrences or other emergency incidents (including fire); and
▷ access and egress from the buildings and the site.

There are several other more general hazardous situations involving pedestrians and vehicles. These include the following:

▷ reversing of vehicles, especially inside buildings;
▷ poor road surfaces and/or poorly drained road surfaces;
▷ roadways too narrow with insufficient safe parking areas;
▷ roadways poorly marked out and inappropriate or unfamiliar signs used;

- too few pedestrian crossing points;
- the non-separation of pedestrians and vehicles;
- lack of barriers along roadways;
- lack of directional and other signs;
- poor environmental factors, such as lighting, dust and noise;
- ill-defined speed limits and/or speed limits which are not enforced;
- poor or no regular maintenance checks;
- vehicles used by untrained and/or unauthorised personnel;
- poor training or lack of refresher training.

Vehicle operations need to be carefully planned so that the possibility of accidents is minimised.

Hazards from mobile work equipment

Mobile work equipment is used extensively throughout industry – in factories, warehouses and construction sites. As mentioned in the previous section, the most common is the fork-lift truck (Figure 9.18).

Figure 9.18 Industrial counterbalance lift truck

Accidents, possibly causing injuries to people, often arise from one or more of the following events:

- poor maintenance with defective brakes, tyres and steering;

- poor visibility because of dirty mirrors and windows or loads which obstruct the driver's view; good visibility is essential at all times for mobile plant operators. The HSE advise that operators of mobile construction plant must ensure they regularly clean their windows so they can safely see all around. This should be combined with constant use of mirrors and a banksman where appropriate. Lights on all vehicles should be cleaned regularly to ensure vehicles are visible at all times, and vehicle depots should be well lit to avoid slip and trip hazards;
- operating on rough ground or steep gradients which causes the mobile equipment to turn on its side 90° plus or roll over 180° or more;
- carrying of passengers without the proper accommodation for them;
- people being flung out as the vehicle overturns and being crushed by it;
- being crushed under wheels as the vehicle moves;
- being struck by a vehicle or an attachment;
- lack of driver training or experience;
- underlying causes of poor management procedures and controls, safe working practices, information, instruction, training and supervision;
- collision with other vehicles;
- overloading of vehicles;
- general vehicle movements and parking;
- dangerous occurrences or other emergency incidents (including fire);
- access and egress from the buildings and the site.

The machines most at risk of rollover according to the HSE are:

- compact dumpers frequently used in construction sites;
- agricultural tractors;
- variable reach rough terrain trucks (telehandlers) (Figure 9.19).

Figure 9.19 Telescopic materials handler

Control measures for safe workplace transport operations

Any control strategy involving vehicle operations will involve a risk assessment to ascertain where, on traffic routes, accidents are most likely to happen. It is important that the risk assessment examines both internal and external traffic routes, particularly when goods are loaded and unloaded from lorries. It should also assess whether designated traffic routes are suitable for the purpose and sufficient for the volume of traffic.

The following need to be addressed:

▷ Traffic routes, loading and storage areas need to be well designed with enforced speed limits, good visibility and the separation of vehicles and pedestrians whenever reasonably practicable.

▷ Environmental considerations, such as visibility, road surface conditions, road gradients and changes in road level, must also be taken into account.

▷ The use of one-way systems and separate site access gates for vehicles and pedestrians may be required.

▷ The safety of members of the public must be considered, particularly where vehicles cross public footpaths.

▷ All external roadways must be appropriately marked, particularly where there could be doubt on right of way, and suitable direction and speed limit signs erected along the roadways. While there may well be a difference between internal and external speed limits, it is important that all speed limits are observed.

▷ Induction training for all new employees must include the location and designation of pedestrian walkways and crossings and the location of areas in the factory where pedestrians and fork-lift trucks use the same roadways.

▷ The identification of recognised and prohibited parking areas around the site should also be given during these training sessions.

Many industries have vehicles designed and used for specific workplace activities. The safe system of work for those activities should include:

▷ details of the work area (e.g. vehicle routes, provision for pedestrians, signage);

▷ details of vehicles (e.g. type, safety features and checks, maintenance requirements);

▷ information and training for employees (e.g. driver training, traffic hazard briefing);

▷ type of vehicle activities (e.g. loading and unloading, refuelling or recharging, reversing, tipping).

To summarise, there are three aspects to a control strategy for safe workplace transport operations:

1. **The design of the site (safe site)** – involves managers in:
   ▷ planning routes to separate pedestrians from vehicles whenever possible;

▷ reducing the need to reverse by using one-way systems;

▷ avoiding steep gradients and overhead cables and provide traffic routes on firm ground, minimising sharp and blind corners;

▷ marking out parking areas for vehicles;

▷ providing speed limit signs and traffic warning signs;

▷ ensuring a well-lit environment; and

▷ maintaining good housekeeping and a tidy site.

2. **Vehicle selection and maintenance (safe vehicle)** requires the provision of:
   ▷ vehicles with suitable and effective headlights, brakes, bumpers and horns, sufficient mirrors to reduce blind spots and seat belts for drivers and passengers;

   ▷ some additional vehicle features such as rear lens or radar sensors to provide extra safety when reversing and speed governors; and

   ▷ a regular and documented inspection and maintenance regime.

3. **Systems of work for system operatives (safe driver)** to include:
   ▷ effective supervision of everyone who is in areas where vehicles operate, including the provision of banksmen, when required;

   ▷ adequate training and refresher training for all drivers;

   ▷ relevant information for all drivers, including speed limits and parking areas;

   ▷ regular health checks on the suitability of employees for driving roles;

   ▷ the provision of high visibility clothing, appropriate protective clothing (such as steel toe-capped boots and hard hats);

   ▷ the control of vehicle movements at times of day when there are more people moving around. Access to vehicle areas should be restricted to those that need to be there.

Mobile work equipment legislation

(a) General

The main purpose of the mobile work equipment Provision and Use of Work Equipment Regulations (PUWER) Part III, Regulations 25 to 30 is to require additional precautions relating to work equipment while it is travelling from one location to another, or where it does work while moving. All appropriate sections of PUWER 98 will also apply to mobile equipment as it does to all work equipment; for example dangerous moving parts of the engine would be covered by Part II Regulations 10, 11 and 12 of PUWER 98. If the equipment is designed primarily for travel on public roads, the Road Vehicles (Construction and Use) Regulations will normally be sufficient to comply with PUWER 98.

Mobile equipment would normally move on wheels, tracks, rollers, skids, etc. Mobile equipment may be self-propelled, towed or remote controlled and may incorporate attachments. Pedestrian-controlled work equipment, such as a lawn mower, is not covered by Part III.

(b) Employees carried on mobile work equipment – Regulation 25

No employee may be carried on mobile work equipment:

► unless it is suitable for carrying persons;
► unless it incorporates features to reduce risks as low as is reasonably practicable, including risks from wheels and tracks.

Where there is significant risk of falling materials, falling-object protective structures (FOPS) must be fitted.

(c) Rolling over of mobile work equipment

Where there is a risk of overturning it must be minimised by:

► stabilising the equipment;
► fitting a structure so that it only falls on its side;
► fitting a structure which gives sufficient clearance for anyone being carried if it turns over further – rollover protective structure (ROPS);
► a device giving comparable protection;
► fitting a suitable restraining system for people if there is a risk of being crushed by rolling over.

(d) Self-propelled work equipment

Where self-propelled work equipment may involve risks while in motion it shall have:

► facilities to prevent unauthorised starting;
► facilities to minimise the consequences of collision (with multiple rail-mounted equipment);
► a device for braking and stopping;
► emergency facilities for braking and stopping, in the event of failure of the main facility, which have readily accessible or automatic controls (where safety constraints so require);
► devices fitted to improve vision (where the driver's vision is inadequate);
► appropriate lighting fitted or otherwise it shall be made sufficiently safe for its use (if used at night or in dark places);
► if there is anything carried or towed that constitutes a fire hazard liable to endanger employees (particularly, if escape is difficult such as from a tower crane), appropriate firefighting equipment carried, unless it is sufficiently close by.

(e) Rollover and falling-object protection (ROPS and FOPS)

Rollover protective structures (ROPS) are now becoming much more affordable and available for most types of mobile equipment where there is a high risk of turning over. Their use is spreading across most developed countries and even some less well-developed areas. A ROPS is a cab or frame that provides a safe zone for the vehicle operator in the event of a rollover.

The ROPS frame must pass a series of static and dynamic crush tests. These tests examine the ability of the ROPS to withstand various loads to see if the protective zone around the operator remains intact in an overturn.

A home-made bar attached to a tractor axle or simple shelter from the sun or rain cannot protect the operator if the equipment overturns.

The ROPS must meet International Standards such as ISO 3471:1994. All mobile equipment safeguards should comply with the essential health and safety requirements of the Supply of Machinery (Safety) Regulations but need not carry a CE marking.

ROPS must also be correctly installed strictly following the manufacturers' instructions and using the correct strength bolts and fixings. They should never be modified by drilling, cutting, welding or other means as this may seriously weaken the structure.

Falling-object protective structures (FOPS) are required where there is a significant risk of objects falling on the equipment operator or other authorised person using the mobile equipment. Canopies that protect against falling objects (FOPS) must be properly designed and certified for that purpose. Front loaders work in woods or construction sites near scaffolding or buildings under construction and high bay storage areas, these all being locations where there is a risk of falling objects. Purchasers of equipment should check that any canopies fitted are FOPS. ROPS should never be modified by the user to fit a canopy without consultation with the manufacturers.

Figure 9.20 Various construction plant with driver protection

ROPS provide some safety during an overturn, but only when operatives are confined to the protective zone of the ROPS. So where ROPS are fitted, a suitable restraining system must be provided for all seats. The use of seat restraints could avoid accidents where drivers are thrown from machines, thrown through windows or doors or thrown around inside the cab. In agriculture and forestry, 50% of overturning accidents occur on slopes of less than 10° and 25% on slopes of 5° or less. This means that seat restraints should be used most of the time that the vehicle is being operated.

Safe driving of mobile work equipment

Drivers have an important role to play in the safe use of mobile equipment. They should include the following in their safe working practice checklist:

- Make sure they understand fully the operating procedures and controls on the equipment being used.
- Only operate equipment for which they are trained and authorised.
- Never drive if abilities are impaired by, for example, alcohol, poor vision or hearing, ill health or drugs whether prescribed or not.
- Use the seat restraints where provided.
- Know the site rules and signals.
- Know the safe operating limits relating to the terrain and loads being carried.
- Keep vehicles in a suitably clean and tidy condition with particular attention to mirrors and windows or loose items which could interfere with the controls.
- Drive within speed limits and follow site rules and routes at all times.
- Allow passengers only when there are safe seats provided on the equipment.
- Park vehicles on suitable flat ground with the engine switched off and the parking brakes applied; use wheel chocks if necessary.
- Make use of visibility aids or a signaller when vision is restricted.
- Get off the vehicle during loading operations unless adequate protection is provided.
- Ensure that the load is safe to move.
- Do not get off vehicle until it is stationary, engine stopped and parking brake applied.
- Where practicable, remove the operating key when getting off the vehicle.
- Take the correct precautions such as not smoking and switching off the engine when refuelling.
- Report any defects immediately.

The management of vehicle movements

The movement of vehicles should be properly managed, as should vehicle maintenance and driver

Figure 9.21 Operating a piece of mobile equipment, an excavator, to excavate a swimming pool

training. The development of an agreed code of practice for drivers, to which all drivers should sign up, and the enforcement of site rules covering all vehicular movements are essential for effective vehicle management.

All vehicles should be subject to appropriate regular preventative maintenance programmes with appropriate records kept and all vehicle maintenance procedures properly documented. Many vehicles, such as mobile cranes, require regular inspection by a competent person and test certificates.

Certain vehicle movements, such as reversing, are more hazardous than others and particular safe systems should be set up. The reversing of lorries, for example, must be kept to a minimum (and then restricted to particular areas). Vehicles should be fitted with reversing warning systems as well as being able to give warning of approach. Refuges, where pedestrians can stand to avoid reversing vehicles, are a useful safety measure. Banksmen, who direct reversing vehicles, should also be alert to the possibility of pedestrians crossing in the path of the vehicle. Where there are many vehicle movements, consideration should be given to the provision of high visibility clothing. Pedestrians must keep to designated walkways and crossing points, observe safety signs and use doors that are separate to those used by vehicles. Visitors who are unfamiliar with the site and access points should be escorted through the workplace.

The risk of a fire is increased by vehicular-associated activities such as battery charging, refuelling, pallet storage, etc. All batteries should be recharged in a separate well-ventilated area.

As mentioned earlier, driver training, given by competent people, is essential. Only trained drivers should be allowed to drive vehicles and the training should be relevant to the particular vehicle (fork-lift truck, dumper truck, lorry, etc.). All drivers must receive specific training and instruction before they are permitted to drive vehicles. They must also be given refresher training and medical examinations at regular intervals. This involves a management system for ensuring driver competence, which must include detailed records of all drivers with appropriate training dates and certification in the form of a driving licence or authorisation. Competence and its definition are discussed in Chapter 3.

Where large vehicles are routinely stopping to load or unload at loading bays, a certain amount of reversing is probably inevitable. When large vehicles need to reverse in the workplace, the following precautions should be taken:

- Undertake a risk assessment and develop a written safe system of work.
- Restrict reversing to places where it can be carried out safely.
- Keep people on foot or in wheelchairs away from the area.
- Provide suitable high-visibility clothing for those people who are permitted in the area.
- Fit reversing alarms to alert, or a detection device to warn, the driver of an obstruction, or automatically apply the brakes.
- Employ banksmen to supervise the safe movement of vehicles.

The design features that may need to be considered to minimise risks associated with movement of vehicles in the workplace include:

- providing traffic routes with smooth and stable surfaces and with the right width and headroom for the types of vehicles that will use them;
- eliminating sharp bends, blind corners and steep gradients, and siting convex mirrors on those corners that are unavoidably blind;
- installing a one-way system, to minimise the need for reversing;
- including passing places for vehicles;
- introducing speed limits and providing speed retarders;
- providing a good standard of lighting for the traffic routes, and particularly at the transition areas between the inside and outside of buildings;

- segregating vehicles and pedestrians, including separate access and egress, and providing clearly marked crossing places (zebra crossings).

The procedural arrangements that should accompany these design features are:

- selecting and training competent drivers;
- implementing a regular health screening programme for all drivers;
- providing information on site rules for visitors, such as delivery drivers;
- procedures for the regular maintenance of the traffic routes and the in-house vehicles, including a system for the reporting of defects and near-miss accidents;
- rigorously enforcing speed limits, with the possibility of a points system on drivers' licences or permits.

The HSE publications 'Workplace Transport Safety, Guidance for Employers' HSG136, and 'Managing Vehicle Safety at the Workplace' INDG199 (revision 2) provide useful checklists of relevant safety requirements that should be in place when vehicles are used in a workplace.

9.7 Work-related driving

9.7.1 Introduction

The first fatal car accident in the UK occurred in 1896 when Mrs Bridget Driscoll was run over by a Roger-Benz which had a maximum speed of 8 mph. Since then, the Royal Society for the Prevention of Accidents (RoSPA) has estimated that more than 550,000 people have been killed on UK roads. It has been estimated that up to one-third of all road traffic accidents involve somebody who is at work at the time – accounting for over 20 fatalities and 250 serious injuries every week. Based on recent annual statistics, this means around 800–1,060 deaths a year on the road, compared with 241 fatal injuries to workers in the 'traditional workplace'. Accident rates are 30–40% higher for business drivers than for private drivers.

Some employers believe, incorrectly, that if they comply with certain road traffic law requirements, so that company vehicles have a valid MOT test certificate, and drivers hold a valid licence, this is enough to ensure the safety of their employees, and others, when they are on the road. However, health and safety law applies to on-the-road work activities as it does to all work activities, and the risks should be managed effectively within a health and safety management system.

These requirements are in addition to the duties employers have under road traffic law, for example the Road Traffic Act and Road Vehicle (Construction and Use) Regulations, which are administered by the police and other agencies such as the Vehicle and Operator Services Agency.

9

Just under half of all those who drive while at work, suffer from a road-rage incident at least once a year, and 11% are assaulted. Furthermore, pain in the lower back, neck and shoulder is linked to prolonged driving, as is poor mental health.

Health and safety law does not apply to commuting, unless the employee is travelling from their home to a location which is not their usual place of work.

9.7.2 Benefits of managing work-related road safety

The true costs of accidents to organisations are nearly always higher than just the costs of repairs and insurance claims. The benefits of managing work-related road safety can be considerable, no matter what the size of the organisations. There will be benefits in the area of:

► control: costs, such as wear and tear and fuel, insurance premiums and claims can be better controlled;
► driver training and vehicle purchase: better informed decisions can be made;
► lost time: fewer days will be lost due to injury, ill-health and work rescheduling;
► vehicles: fewer will need to be off the road for repair;
► orders: fewer orders will be missed;
► key employees: there is likely to be a reduction in driving bans.

9.7.3 Managing occupational road risks

Introduction

The Health and Safety Executive (HSE) estimates that over a quarter of all road traffic accidents involve someone who is driving as part of their work.

Where employees are using their own privately owned vehicles, they should consider appropriate policies and procedures for managing that risk, including:

► the vehicles in question are insured for business use;
► they are serviced in accordance with manufacturer's service schedule; and
► they have an appropriate MOT (if over three years old).

Where work-related road safety is integrated into the arrangements for managing health and safety at work, it can be managed effectively. The main areas to be addressed are policy, responsibility, organisation, systems and monitoring. Employees should be encouraged to report all work-related road incidents and be assured that punitive action will not be taken against them.

Figure 9.22 Vehicle recovery driver – professional driver at work

Figure 9.23 Road accidents are a significant risk when driving for work. Poor weather conditions increase the risk

The risk assessment should:

► consider the use, for example, of air or rail transport as a partial alternative to driving;
► consider the factors that might increase the risk of becoming involved in a road traffic incident – distance, driving hours, work schedules, stress due to traffic and road conditions and weather conditions;
► attempt to avoid situations where employees feel under pressure;
► make sure that maintenance work is organised to reduce the risk of vehicle failure. This is particularly important when pool cars are used, because pool car users often assume another user is checking on maintenance and the MOT. The safety critical systems that need to be properly maintained are the brakes, steering and tyres. Similarly, if the car is leased and serviced by the leasing company, a system should be in place to confirm that servicing is being done to a reasonable standard;

insist that drivers and passengers are adequately protected in the event of an incident. Crash helmets and protective clothing for those who ride motorcycles and other two-wheeled vehicles should be of the appropriate colour and standard;

ensure that company policy covers the important aspects of the Highway Code.

9.7.4 Evaluating the risks

A study by RoSPA has found that a new driving for work qualification would be welcomed by employers, because the current system of learner training and testing does not adequately prepare young people to drive as part of their job. The European Parliament and Council have introduced the Driving CPC (Certificate of Professional Competence) – under EU Directive 2003/59. There are two types of qualification – one for goods vehicles and one for drivers of passenger vehicles – and the Driver CPC will only apply to operators of vehicles over three-and-a-half tonnes or with more than eight passenger seats.

The following actions should be considered by employers to reduce the risks to employees who drive as part of their work:

- Journeys should be planned to reduce driving time.
- Rest breaks should be included in journey times.
- Encourage drivers to remain physical fit to reduce chronic fatigue.
- Ensure that drivers have hands-free mobile equipment.
- Monitor related working procedures to ensure safety.
- Ensure that drivers have necessary communication equipment (mobile phones, GPS, personal alarms).
- Keep a record of the location of driver destinations.
- Encourage drivers to regularly report back to colleagues in the base office.

The following considerations can be used to check on work-related road safety management.

The driver

Competency

- Is the driver competent, experienced and capable of doing the work safely?
- Is his or her licence valid for the type of vehicle to be driven?
- Is the vehicle suitable for the task or is it restricted by the driver's licence?
- Does the recruitment procedure include appropriate pre-appointment checks?
- Is the driving licence checked for validity on recruitment and periodically thereafter?
- When the driver is at work, is he or she aware of company policy on work-related road safety?
- Are written instructions and guidance available?

- Has the company specified and monitored the standards of skill and expertise required for the circumstances for the job?

Training

- Are drivers properly trained?
- Do drivers need additional training to carry out their duties safely?
- Does the company provide induction training for drivers?
- Are those drivers whose work exposes them to the highest risk given priority in training?
- Do drivers need to know how to carry out routine safety checks such as those on lights, tyres and wheel fixings?
- Do drivers know how to adjust safety equipment correctly, for example seat belts and head restraints?
- Is the headrest 3.8 cm (1.5 in.) behind the driver's head?
- Is the front of the seat higher than the back and are the legs 45° to the floor?
- Is the steering wheel adjustable and set low to avoid shoulder stress?
- Are drivers able to use anti-lock brakes (ABS) properly?
- Do drivers have the expertise to ensure safe load distribution?
- If the vehicle breaks down, do drivers know what to do to ensure their own safety?
- Is there a handbook for drivers?
- Are drivers aware of the dangers of fatigue?
- Do drivers know the height of their vehicle, both laden and empty?

Figure 9.24 Must have a valid vehicle licence to drive each type of vehicle used

Fitness and health

- The driver's level of health and fitness should be sufficient for safe driving.
- Drivers of Heavy Goods Vehicles (HGVs) must have the appropriate medical certificate.

- Drivers who are most at risk should also undergo regular medicals. Staff should not drive, or undertake other duties, while taking a course of medicine that might impair their judgement.
- All drivers should have regular (every 2 years) eyesight tests. Research has shown that one in four motorists have a level of eyesight below the legal standard for driving which is to be able to read a car number plate from a distance of 20.5 m.
- Drivers should rest their eyes by taking a break of at least 15 minutes every 2 hours.

Offences under the Road Safety Act allow courts to imprison drivers who cause deaths by not paying due care and attention to the road or to other road users. Avoidable distractions which courts will consider when sentencing motorists who have killed include:

- using a mobile phone (for either calling or texting);
- drinking and eating;
- applying make-up;
- anything else which takes their attention away from the road and which a court judges to have been an avoidable distraction.

Every year, over 87,000 motorists are disqualified for drink-driving or driving while under the influence of drugs and up to 20% of drink-drivers are caught the morning after drinking. The Department for Transport have calculated that 5% of drivers who failed a breath test after a crash were driving for work at the time.

The wearing of lanyards with staff identity tags has caused serious injuries to motorist after airbag impact following incidents.

The accurate recording of employee crashes can identify areas that require improvement in the management of vehicle safety by identifying underlying rather than the immediate causes of the crash. Such a record should include:

- vehicle and driver details;
- maintenance records of the vehicle;
- date and time of the crash;
- crash location and road conditions;
- crash severity;
- length of service and training of the driver; and
- the degree of driver culpability for the crash.

The reason for the inclusion of an accurate assessment of culpability in the record is so that a clear understanding of whether the driver's actions were a factor in the crash may be determined.

There are three considerations that need to be made before deciding upon culpability:

1. The driver's actions in relation to traffic law.
2. Although the crash may have occurred due to a wrong, negligent or erroneous action, it may not have involved any deliberate intent.
3. The culpability for the crash may need to be shared between different road users in a reliable way.

The vehicle

Suitability

All vehicles should be fit for the purpose for which they are used. When purchasing new or replacement vehicles, management should select those that are suitable for both driving and the health and safety of the public. The fleet should be suitable for the job in hand. Where privately owned vehicles are used for work, they should be insured for business use and have an appropriate MOT certificate test (for vehicles over three years old, in the UK).

Condition and safety equipment

Are vehicles maintained in a safe and fit condition? There will need to be:

- maintenance arrangements to acceptable standards;
- basic safety checks for drivers;
- a method of ensuring that the vehicle does not exceed its maximum load weight;
- reliable methods to secure goods and equipment in transit;
- checks to make sure that safety equipment is in good working order;
- checks on seat belts and head restraints. (Are they fitted correctly and functioning properly?);
- a defect reporting system for drivers to use if they consider their vehicle is unsafe.

The use of speed limiters may become mandatory over the next few years.

Ergonomic considerations

Ergonomics plays a much wider role in car design than simply seat design and includes how drivers physically use their vehicles. Ergonomic design has a role to play in safety in the event of an accident – such as how far a person needs to reach over to operate the radio or Sat Nav.

The health of drivers, and possibly also their safety, may be put at risk from an inappropriate seating position or driving posture. Ergonomic considerations should therefore be considered before purchasing or leasing new vehicles. Information may need to be provided to drivers about good posture and, where appropriate, on how to set their seat correctly.

The load

For any lorry driving, most of the topics covered in this section are relevant. However, the load being carried is an additional issue. If the load is hazardous, emergency procedures (and possibly equipment) must be in place and the driver trained in those procedures. The load should be stacked safely in the lorry so that it cannot move during the journey. There must also be satisfactory arrangements for handling the load at either end of the journey.

The journey

Routes

Route planning is crucial. Safe routes should be chosen which are appropriate for the type of vehicle undertaking the journey wherever practicable. Motorways are the safest roads. Minor roads are suitable for cars, but they are less safe and could present difficulties for larger vehicles. Overhead restrictions, for example bridges, tunnels and other hazards such as level crossings, may present dangers for long and/or high vehicles, so route planning should take particular account of these.

Scheduling

There are danger periods during the day and night when people are most likely, on average, to feel sleepy. These are between 2 a.m. and 6 a.m. and between 2 p.m. and 4 p.m. Schedules need to take sufficient account of these periods. Where tachographs are carried, they should be checked regularly to make sure that drivers are not putting themselves and others at risk by driving for long periods without a break. Periods of peak traffic flow should be avoided if possible and new drivers should be given extra support while training.

Time

Has enough time been allowed to complete the driving job safely? A realistic schedule would take into account the type and condition of the road and allow the driver rest breaks. A non-vocational driver should not be expected to drive and work for longer than a professional driver. The recommendation of the Highway Code is for a 15-minute break every 2 hours.

- Are drivers put under pressure by the policy of the company? Are they encouraged to take unnecessary risks, for example exceeding safe speeds because of agreed arrival times?
- Is it possible for the driver to make an overnight stay? This may be preferable to having to complete a long road journey at the end of the working day.
- Are staff aware that working irregular hours can add to the dangers of driving? They need to be advised of the dangers of driving home from work when they are excessively tired. In such circumstances they may wish to consider an alternative, such as a taxi.

Distance

Managers need to satisfy themselves that drivers will not be put at risk from fatigue caused by driving excessive distances without appropriate breaks. Combining driving with other methods of transport may make it possible for long road journeys to be eliminated or reduced. Employees should not be asked to work an exceptionally long day.

Weather conditions

When planning journeys, sufficient consideration will need to be given to adverse weather conditions, such as snow, ice, heavy rain and high winds. Routes should be rescheduled and journey times adapted to take adverse weather conditions into consideration. Where poor weather conditions are likely to be encountered, vehicles should be properly equipped to operate, with, for example, ABS.

Where there are ways of reducing risk, for example when driving a high-sided vehicle in strong winds with a light load, drivers should have the expertise to deal with the situation. In addition, they should not feel pressurised to complete journeys where weather conditions are exceptionally difficult and this should be made clear by management.

9.7.5 Typical health and safety rules for drivers of cars on company business

At least 25% of all road accidents are work-related accidents involving people who are using the vehicle on company business. Drivers are expected to understand and comply with the relevant requirements of the current edition of the Highway Code. The following rules have been produced to reduce accidents at work. Any breach of these rules will be treated as a disciplinary offence.

The following example shows typical rules that have been prepared for use by car drivers.

- All drivers must have a current and valid driving licence.
- All vehicles must carry comprehensive insurance for use at work.
- Plan the journey in advance to avoid, where possible, dangerous roads or traffic delays.
- Use headlights in poor weather conditions and fog lights in foggy conditions (visibility 100 m).
- Use hazard warning lights if an accident or severe traffic congestion is approached (particularly on motorways).
- All speed limits must be observed but speeds should always be safe for the conditions encountered.
- Drivers must not drive continuously for more than 2 hours without a break of at least 15 minutes.
- Mobile phones, including hands-free equipment, must not be used whilst driving. They must be turned off during the journey and only used during the rest periods or when the vehicle is safely parked and the handbrake on.
- No alcohol must be consumed during the day of the journey until the journey is completed. Only minimal amounts of alcohol should be consumed on the day before a journey is to be made.
- No recreational drugs should be taken on the day of a journey. Some prescribed and over the counter drugs and medicines can also affect driver awareness and speed of reaction. Always check with a doctor or pharmacist to ensure that it is safe to drive.

9

309

9.7.6 Hazards associated with the use of electric and hybrid vehicles

With the challenges of climate change and increasing environmental awareness, the use of electric and hybrid vehicles is increasing. The UK Government have a target to replace diesel and petrol driven vehicles with electric vehicles by 2035 or perhaps even 2032. It is becoming increasingly important that people, particularly those in the motor vehicle repair and recovery industry, become aware of some of the hazards associated with these vehicles both during their use on the road and repair in a workshop.

Essentially there are only really two types of electric vehicles or cars – those that run via electric battery only (Battery Electric Vehicles, BEVs or EVs) and those that use both a combustion engine and an electric battery (Hybrid Electric Vehicles, HEVs; Plug-in Hybrid Electric Vehicles, PHEVs; or Range-Extended Electric Vehicles, REEVs or E-Rev).

Electric vehicles use a large capacity battery and electric motor(s) to drive the vehicle. The battery needs to be charged from the electricity supply network when the vehicle is not in use although some energy may be recovered during slowing down and/or braking.

Hybrid vehicles typically have two sources of energy, an internal combustion engine using either diesel or petrol for fuel and a battery. Hybrid vehicles will use the two sources of power automatically and may use them together. The internal combustion engine and energy recovered from the vehicle braking systems are used to charge the battery. Some hybrid vehicles can have their battery charged directly from the electrical supply network (PHEVs).

The hazards associated with the use of electric and hybrid vehicles include:

▷ Silent operation so that pedestrians are not aware of vehicles manoeuvring. A slow speed pedestrian alarm sounder should be switched on but it should never be assumed that an EV is switched off because it is silent.

▷ Availability and location of charging points.

▷ Electric shock from high-voltage components and cabling.

Voltages present in these vehicles are significantly higher (currently up to 650 Volts direct current (DC)) than those used in other vehicles (12/24 Volts DC). In dry conditions, accidental contact with parts that are live at voltages above 110 Volts DC can be fatal.

▷ Retained electrical charge in components even when the vehicle is switched off.

▷ Unexpected movement of the vehicle or engine components due to the motor's magnetic forces.

Remote operation keys that only need to be close to the vehicle for the vehicle to be powered up should be kept away from vehicles when they are not in use.

▷ Potential for the release of explosive gases and harmful liquids from damaged or incorrectly modified batteries.

Battery systems may contain chemicals that can be harmful if released. They also store significant amounts of energy that can give rise to explosion if not dealt with correctly.

Other possible hazards

▷ Manual handling issues associated with battery replacement. However as this is a specialist task and batteries are normally guaranteed for eight years or leased this will be undertaken almost exclusively by vehicle repair specialists.

▷ Potential for electrical systems affecting the driver's health, i.e. where a pacemaker may be fitted.

The two main factors to consider when charging an EV are charging speed and type of connector. Different EVs offer different charging speeds. The length of time EV batteries take to recharge is determined by how many kilowatts (kW) the charging station can provide and how many the car can accept – the higher the wattage, the faster the charge. The following different rates exist.

Slow charging rate: 3 kW (at home or office, 8–15 hours to charge)

Fast charging rate: 7–22 kW (most public charging points and possible with special chargers at home, 3–5 hours to charge)

Rapid charging rate: 43–50 kW (EV must have compatible socket, 30–45 minutes to charge)

Super charging rate: 150 kW (recently announced which will charge in ten minutes).

The voltages used and the current flows are very high and it is essential that only properly designed cables in good condition with the correct connector are used. Extension cables are not suitable for the high currents involved and have resulted in fires.

Figure 9.25 Electric vehicle and fast home charger 7 kW

A home charging point is a compact weatherproof unit that mounts to a wall, with a connected (tethered) charging cable or a socket for plugging in a portable charging cable. They should be purchased and installed by a reputable supplier and qualified, competent electrician.

There are numerous charging points for those on longer journeys at motorway services, public car parks, garage forecourts and the like. These are normally accessed by smartphone apps and there are several charging station networks. There are apps like zap-map and opencharge.com to find the closest charging point, compatible with your vehicle. It is a requirement (under EU requirements) for all public charging points to have a Mennekes Type 2 socket. These are normally between 7 and 22 kW capacity.

Specialist knowledge is needed to safely move an electric vehicle as it can produce high voltages by turning the wheels. Towing is not recommended.

People involved in vehicle repair and maintenance, however, are likely to need a much greater level of competence in order to work on these types of vehicle safely. Four categories of work have been identified. These are:

- Valeting, sales and other lower risk activities
  - Keep remote operation keys away from vehicles to prevent accidental power up and movement of the EV;
  - People should remember that others may not hear an EV being moved or they could move without warning;
  - High-voltage cables usually orange, can be damaged by high pressure washers.
- Incident response including emergency services and vehicle recovery
  - Watch for damage to high-voltage cables or components and whether the battery has been affected by an incident;
  - Isolate the standard accessory battery, if possible isolate the high-voltage traction battery;
  - Keep the remote key away from the vehicle;
  - Have access to specialist information on EVs;
  - Do not tow vehicle unless safe to do so.
- Maintenance and repair excluding high-voltage electrical systems
  - Have specialist information on vehicle to hand;
  - Keep keys locked away. Check that vehicle is safe before using a key;
  - Where possible isolate batteries and check the location of high-voltage cables and make sure they are protected.
- Working on high-voltage electrical systems in addition to those above
  - High-voltage system should be isolated so that it cannot be accidently switched on. It should be checked to ensure that it is dead by suitable test equipment, before starting work;

- Use specialist tools and test equipment;
- Always follow manufacturers' guidance data and procedures. This is particularly important if the vehicle is damaged;
- If it is necessary to work on live electrical equipment fully trained and competent personnel only should be involved strictly following manufacturers' precautions and procedures.

There are substantial differences in the designs of electric and hybrid vehicles from different manufacturers. Having information specific to the manufacturer and the vehicle being worked on is important in identifying what actions are necessary to work safely.

More information on the hazards associated with the repair and maintenance of electric and hybrid vehicles, roadside recovery and other vehicle-related activities is available on the HSE website.

9.8 Additional useful background information that is not part of the NEBOSH NGC syllabus

9.8.1 Hazards and control measures for works of a temporary nature

The impact on workplaces from hazards associated with works of a temporary nature

Works of a temporary nature include such projects as building maintenance, renovation and refurbishment. These works usually take place on a client's site by contractors and often include building demolition and excavation work required to supply new underground services. The projects must be properly planned, supervised and monitored, particularly when the project is to take place within or near an occupied building. There are many other possible hazards associated with work of a temporary nature. Some of them are:

- slips, trips and falls are a frequent source of injury to members of the host organisation and the public;
- unfenced excavations;
- inadequate protection of holes, uneven surfaces, poor reinstatement, trailing leads and cables, spillage of oils and gravel;
- trailing cables (especially on stairways);
- poor lighting and ventilation in and around the temporary works;
- noise and vibration;
- hazardous substances including dust;
- falling or flying objects;
- the accumulation of waste materials;
- falling or moving objects;
- poor storage of materials and equipment and other obstructions in public areas, including inadequate control of waste materials, are common causes of accidents;

- fire and/or explosions;
- lack of project management and supervision; and
- hazards created outside the site perimeter (for example, unloading materials from a delivery lorry outside the perimeter).

Two relatively common problems with building projects concern asbestos and demolition. An up-to-date asbestos survey together with relevant plans and drawings should, therefore, be available before any renovation work is commenced.

Most of the health and safety risks in demolition activities are related to an unplanned or premature collapse of structure or of a part of it. Various accident investigation reports following such incidents have found:

- the absence of temporary structures to support unstable elements;
- the lack of risk assessment at design stage, neglecting CDM requirements;
- the lack of any preliminary structural survey or site investigation;
- poor planning of demolition sequences;
- the lack of demolition method statements; and
- the lack of supervision while undertaking demolition activities.

Several of these conclusions are valid for a wide range of large and small building projects, namely the lack of good communications, supervision, planning and risk assessment and the neglect of legislative requirements.

The majority of temporary work projects involve some form of construction work. Details of some of the basic hazards and controls for construction work are given later in this section.

Main control measures relating to the management of works of a temporary nature

The important controls for work of a temporary nature are communication and the cooperation of both contractors and host employees, a comprehensive risk assessment for the project and the management and supervision of the project to ensure that all the agreed controls are implemented.

Communication and cooperation

The organisation of liaison and consultation arrangements is essential to ensure good communication with both the host employees and the contractors. This should provide the cooperation and coordination of all those responsible for the project to ensure the health and safety of everyone in the workplace and enable any concerns to be addressed. Good communication between clients, contractors and residents is an important means of controlling the risks and minimising discomfort. It can also be essential

when arranging alternative escape routes during work. Advice should be sought from the local fire authority where it is necessary to block or alter the permanent escape routes.

Risk assessment

A detailed risk assessment should be made to cover the temporary works. This should include the identification of:

- all aspects of the works including the health and safety implications to the host employees and the contractors;
- the people at risk from the work including, where appropriate, members of the public;
- any risks to the contractors from the business of the host organisation; and
- any additional controls and procedures required to protect everyone affected by the project including:
  - any equipment that should or should not be worked on or used;
  - the personal protective equipment that should be used and who will provide it;
  - all the working procedures, including any permits to work and method statements that are to be used;
  - details of any site access restrictions;
  - any amendment required to existing emergency procedures and access to the host building; and
  - details of any safety signs and/or barriers required.

Often work must be undertaken in areas which need to remain occupied. The risk assessment should indicate the nature of the perimeter of the project and how it will be maintained. It might be possible for the work to take place outside normal hours – if not then the work areas must be segregated or access controlled by physical barriers or warning signs.

Management and supervision of the project

The first two important tasks for the management and supervision of the project are the formulation of a detailed specification of the project and the appointment of competent managers and contractors to complete the project (see Chapter 1).

A Temporary Works Coordinator should be appointed for all temporary works to oversee the project and to sign it off on completion. Such an appointment is particularly important for refurbishment projects involving complex activities that require information related to the existing structure and interrelated health and safety issues.

The coordinator would be responsible for:

- the appointment of competent contractors to undertake the project;
- the provision of information, instruction and training to the employees of the host organisation on any hazards associated with the temporary works;

▶ the provision of information to the temporary works contractors on any hazards from the activities of the host organisation and the controls in place to address them;

▶ the management and supervision of the work of the contractors and agreement on the nature of any temporary controls required before work begins; and

▶ the formulation of a detailed job specification for the work required which should include:

   ▷ safe access to both the temporary work site and to the rest of the building;

   ▷ safe method statements to cover all aspects of the work; and

   ▷ the arrangements for the examination of any temporary access equipment.

The provision of welfare facilities including first-aid arrangements must be discussed and agreed prior to the commencement of the project. It may be that project contractors can have use of the same facilities as the host employees. If this is not possible, then separate arrangements should be included in the project specification.

Other control measures

Where construction work is carried out in, or near occupied premises, such as houses, hospitals, factories and shops, it may be necessary to evacuate part or all of these premises, either for the full duration of the work, or for a limited period during hazardous operations. Certain organisations such as housing associations and care homes often have an established policy which identifies the circumstances when it is appropriate to evacuate premises.

Other control measures that will be relevant to many projects include the:

▶ use of temporary flooring material, such as plywood or steel plates, to cover uneven ground or potholes;

▶ covering or fixing of any training cables which need to cross pedestrian areas;

▶ provision of lighting at night and in dark areas;

▶ prompt removal of all waste and rubbish as it arises;

▶ storage of all hazardous substances in suitable containers or in secure compounds when not in use;

▶ provision of site storage compounds to accommodate all the plant, equipment and materials outside working hours. Strict control over the amount and timing of deliveries will help keep storage to a minimum outside this compound area;

▶ safe storing and stacking of materials and a limit to the height of all material stacks;

▶ use of suitably protected electrical cordless tools or reduced voltage equipment,

▶ gas cylinders and similar appliances are fitted with valves which require special tools to turn on the supply;

▶ isolation of gas cylinders when not in use and locking them in a secure cage out of hours;

▶ protection against hazards that are created within the project site as the work develops, such as around deep excavations which will need to be covered or fenced;

▶ protection of vulnerable groups such as the elderly, children and people with certain disabilities – work in certain premises such as schools and hospitals require careful planning;

▶ protection of wheelchair users and partially sighted people where construction work affects pedestrian routes;

▶ provision of suitable protection:

   ▷ from dust, noise and vibration (see Chapters 8 and 6);

   ▷ to contractors, host employees and members of the public from being struck by falling or ejected materials; and

   ▷ from delivery and other moving site vehicles. Consideration should be given to ensuring that deliveries are scheduled at times outside of large movements of people such as rush hours or the journeys to and from school;

▶ posting of warning signs around the site and at all entrances to the main building.

Whenever possible, school projects which present high risks to staff and pupils should be carried out during school holidays, at weekends or out of school hours. If this is not possible, the work should be programmed to avoid busy periods such as school start, finish and break times. Deliveries should also be arranged to avoid these times.

Certain premises have their particular hazards that will need to be addressed. Residents in care homes may need to be evacuated when hazardous substances are being used (such as timber treatment or damp course work). Where work takes place within occupied factories, offices, shops and other premises, the host organisation and the contractor will need to manage the risks created by their own work, and cooperate closely to manage those risks created by sharing the workplace.

In healthcare premises, the vulnerability of those who are within the premises must be considered – children and outpatients with restricted mobility or with partial sight. Some patients, such as those who have had major surgery, will be more open to infection than healthy people. Refurbishment works may disturb fungal spores and other organisms which could be a serious hazard to patients. Site vehicles and pedestrians must be separated from the hospital traffic as much as possible.

9.8.2 Construction activities

Many works of a temporary nature are construction projects so it is worthwhile to describe the hazards and controls that should be in place for construction work.

It is important to stress that the following topics go beyond the requirements of the NEBOSH NGC syllabus. However, many health and safety managers will be faced with construction projects at some stage in their careers.

The scope and definition of construction

The scope of the construction industry is very wide. The most common activity is general building work, which is domestic, commercial or industrial in nature. This work may be new building work, such as a building extension or, more commonly, the refurbishment, renovation, alteration, maintenance or repair of existing buildings. The buildings may be occupied or unoccupied. Such projects may begin with a partial or total demolition of a structure which is a particularly hazardous operation. Larger civil engineering projects involving road and bridge building, water supply and sewage schemes and river and canal work all come within the scope of construction.

Most construction projects cover a range of activities such as site clearance, the demolition or dismantling of building structures or plant and equipment, the felling of trees and the safe disposal of waste materials. The work could involve hazardous operations, such as roof work or excavation, or contact with hazardous materials, such as asbestos or lead. The site activities will include the loading, unloading and storage of materials and site movements of vehicles and pedestrians. Finally, the construction processes themselves are often hazardous. These processes include fabrication, decoration, cleaning, installation and the removal and maintenance of services (electricity, water, gas and telecommunications). Construction also includes the use of woodworking workshops together with woodworking machines and their associated hazards, painting and decorating and the use of heavy machinery. It will often require work to take place in confined spaces, such as excavations and underground chambers. At the end of most projects, the site is landscaped, which will introduce a new set of hazards.

Figure 9.26 Groundwork clearance and foundations

Particular issues relating to the construction industry

Nearly all construction sites are temporary in nature and, during the construction process, are constantly changing. The workforce itself is transitory in nature and, at any given time, there are often many young people receiving training on site in the various construction trades. These trainees need supervision and structured training programmes. The level of numeracy and literacy among some of the workforce may make the communication of health and safety information difficult.

The progress of work on the site can be impeded by poor weather conditions leading to the possibility of contract penalties. Clients are usually keen to see the work finished as soon as possible and put additional pressure on the contractor. All these issues always lead to the temptation to compromise on health and safety issues, such as the provision of adequate welfare facilities or the safe re-routing of site traffic. In recent years, some additional management issues have arisen with the increasing employment of migrant workers.

Construction hazards and controls

The Construction Design and Management (CDM) Regulations deal with many of the hazards likely to be found on a building site. In addition to these specific hazards, there will be the more general hazards (working at height, excavations, manual handling, electricity, noise, etc.) which have been discussed in more detail in earlier chapters. Other hazards and controls identified in CDM 2015 and the Working at Height Regulations are as follows (hazards associated with work at height have been covered earlier in the chapter).

Safe place of work

Safe access to and egress from the site and the individual places of work on the site are fundamental to a good health and safety environment. This clearly requires that all ladders, scaffolds, gangways, stairways and passenger hoists are safe for use. It further requires that all excavations are fenced, the site is tidy and proper arrangements are in place for the storage of materials and the disposal of waste. The site needs to be adequately lit and secured against intruders, particularly children, when it is unoccupied. Such security will include:

▶ secure and locked gates with appropriate notices posted;
▶ a secure and undamaged perimeter fence with appropriate notices posted;
▶ all ladders either stored securely or boarded across their rungs;
▶ all excavations covered;
▶ all mobile plant immobilised and fuel removed, where practicable, and services isolated;

- secure storage of all flammable and hazardous substances;
- visits to local schools to explain the dangers present on a construction site. This has been shown to reduce the number of child trespassers;
- if unauthorised entry persists, then security patrols and closed circuit television may need to be considered.

Figure 9.27 Secure site access gate

Demolition

Demolition is one of the most hazardous construction operations and is responsible for more deaths and major injuries than any other activity. The management of demolition work is controlled by the CDM Regulations.

The principal hazards associated with demolition work are:

- falls from height or on the same level;
- falling debris;
- premature collapse of the structure being demolished;
- dust and fumes;
- the silting up of drainage systems by dust;
- the problems arising from spilt fuel oils;
- manual handling;
- presence of asbestos and other hazardous substances;
- noise and vibration from heavy plant and equipment;
- electric shock;
- fires and explosions from the use of flammable and explosive substances;
- smoke from burning waste timber;
- pneumatic drills and power tools;
- the existence of services, such as electricity, gas and water;
- collision with heavy plant;
- plant and vehicles overturning.

Before any work is started, a full site investigation must be made by a competent person to determine the hazards and associated risks which may affect the demolition workers and members of the public who may pass close to the demolition site. The investigation should cover the following topics:

- construction details of the structures or buildings to be demolished and those of neighbouring structures or buildings;
- the presence of asbestos, lead or other hazardous substances;
- the location of any underground or overhead services (water, electricity, gas, etc.);
- the location of any underground cellars, storage tanks or bunkers, particularly if flammable or explosive substances were previously stored; and
- the location of any public thoroughfares adjacent to the structure or building.

The CDM Client is responsible for notifying the HSE of the proposed demolition work, and the Principal Designer, who will also plan the demolition work, must ensure that a written risk assessment is made of the design of the structure to be demolished and the influence of that design on the demolition method proposed. A further risk assessment should then be made by the Principal Contractor undertaking the demolition – this risk assessment will be used to draw up a method statement for inclusion in the Health and Safety Plan. A written method statement will be required before demolition takes place. The contents of the method statement will include the following:

- details of the method of demolition to be used, including the means of preventing premature collapse or the collapse of adjacent buildings and the safe removal of debris from upper levels;
- details of equipment, including access equipment, required and any hazardous substances to be used;
- arrangements for the protection of the public and the construction workforce, particularly if hazardous substances, such as asbestos or other dust, are likely to be released;
- details of the isolation methods of any services which may have been supplied to the site and any temporary services required on the site;
- details of PPE which must be worn;
- first-aid, emergency and accident arrangements;
- training and welfare arrangements;
- arrangements for waste disposal;
- names of site foremen and those with responsibility for health and safety and the monitoring of the work;
- control of Substances Hazardous to Health (COSHH) and other risk assessments (PPE, manual handling, etc.) should be appended to the method statement.

There are two forms of demolition:

- piecemeal – where the demolition is done using hand and mechanical tools such as pneumatic drills and demolition balls; and

▷ deliberate controlled collapse – where explosives are used to demolish the structure. This technique should be used only by trained, specialist competent persons.

A very important element of demolition is the training required by all construction workers involved in the work. Specialist training courses are available for those concerned with the management of the process, from the initial survey to the final demolition. However, induction training, which outlines the hazards and the required control measures, should be given to all workers before the start of the demolition work. The site should be made secure with relevant signs posted to warn members of the public of the dangers.

Prevention of drowning

Where construction work takes place over water, steps should be taken to prevent people falling into the water and rescue equipment should be available at all times.

Figure 9.28 Working near or over water – large scaffold with protection screens and a small boat moored under the bridge in case rescue is needed

Vehicles and traffic routes

All vehicles used on site should be regularly maintained and records kept. Only trained drivers should be allowed to drive vehicles and the training should be relevant to the particular vehicle (fork-lift truck, dumper truck, etc.). Vehicles should be fitted with reversing warning systems. HSE investigations have shown that in over 30% of dumper truck accidents, the drivers had little experience and no training. Common forms of accident include driving into excavations, overturning while driving up steep inclines and runaway vehicles which have been left unattended with the engine running. Many vehicles such as mobile cranes require regular inspection and test certificates.

The small dumper truck is widely used on all sizes of construction site. Compact dumper trucks are involved

in about 30% of construction transport accidents. The three main causes of such accidents are:

▷ overturning on slopes and at the edges of excavations;
▷ poorly maintained braking systems;
▷ driver error due to lack of training and/or inexperience.

Some of the hazards associated with these vehicles are: collisions with pedestrians, other vehicles or structures such as scaffolding. They can be struck by falling materials and tools or be overloaded. The person driving the truck can be thrown from the vehicle, come into contact with moving parts on the truck, suffer the effects of whole-body vibrations due to driving over potholes in the roadway and suffer from the effects of noise and dust. The precautions that can be taken to address these hazards include the use of authorised, trained, competent and supervised drivers only. As with so many other construction operations, risks should be assessed, safe systems of work followed and drivers forbidden from taking short cuts. The following site controls should also be in place:

▷ designated traffic routes and signs;
▷ speed limits;
▷ stop blocks used when the vehicle is stationary;
▷ proper inspection and maintenance procedures;
▷ procedures for starting, loading and unloading the vehicle;
▷ provision of rollover protective structures (ROPS) and seat restraints;
▷ provision of falling-object protective structures (FOPS) when there is a risk of being hit by falling materials;
▷ visual and audible warning of approach;
▷ where necessary, hearing protection.

For other forms of mobile construction equipment, such as fork-lift trucks, the risk to people from the overturning of the equipment must always be safeguarded. This can usually be achieved by the avoidance of working on steep slopes, the provision of stabilisers and ensuring that the load carried does not affect the stability of the equipment/vehicle.

Traffic routes and loading and storage areas need to be well designed with enforced speed limits, good visibility and the separation of vehicles and pedestrians being considered. The use of one-way systems and separate site access gates for vehicles and pedestrians may be required. Finally, the safety of members of the public must be considered, particularly where vehicles cross public footpaths.

Fire and other emergencies

Emergency procedures relevant to the site should be in place to prevent or reduce injury arising from fire, explosions, flooding or structural collapse. These procedures should include the location of fire points and assembly points, extinguisher provision, site evacuation, contact with the emergency services, accident reporting and investigation, and rescue from excavations and confined spaces. There also needs to be training in these procedures at the induction of new workers and ongoing for all workers.

Welfare facilities

The HSE have been concerned for some time at the poor standard of welfare facilities on many construction sites. Welfare arrangements include the provision of sanitary conveniences and washing facilities, drinking water, accommodation for clothing, facilities for changing clothing and facilities for rest and eating meals. First-aid provision is also a welfare issue and is covered in Chapter 4. However, the scale of first-aid provision will be related to the size and complexity of the project, the number of workers on site at any given time and the proximity of emergency hospital facilities.

Sanitary and washing facilities (including showers if necessary) with an adequate supply of drinking water should be provided for everybody working on the site. Accommodation will be required for the changing and storage of clothes and rest facilities for break times. There should be adequate first-aid provision (an accident book) and protective clothing against adverse weather conditions.

Detailed guidance is also available in HSG150 *Health and Safety in Construction*. **Sanitary conveniences and washing facilities** must be provided together and in proportion to the size of the workforce. The Approved Code of Practice of the Workplace (Health, Safety and Welfare) Regulations provides two tables offering guidance on the requisite number of water closets, wash stations and urinals for varying sizes of workforce (approximately one of each for every 25 employees). Special provision should be made for disabled workers and there should normally be separate facilities for men and women. There should be adequate protection from the weather and only as a last resort should public conveniences be used. A good supply of warm water, soap and towels must be provided as close to the sanitary facilities as possible. Hand dryers are permitted but there are concerns about their effectiveness in drying hands completely and thus removing all bacteria. In the case of temporary or remote work sites, sufficient chemical closets and sufficient washing water in containers must be provided.

All such facilities should be well ventilated and lit and cleaned regularly.

Drinking water must be readily accessible to all the workforce. The supply of drinking water must be adequate and wholesome. Normally mains water is used and should be marked as 'drinking water' if water not fit for drinking is also available. On some sites, particularly during the early stages of construction, it may be necessary to provide bottled water or water in tanks. In this event, the water should be changed regularly and protected from contamination. Cups or mugs should be available near the tap unless a drinking fountain is provided.

Accommodation for clothing and facilities for changing clothing must be provided which is clean, warm, dry, well ventilated and secure. Where workers are required to wear special or protective clothing, arrangements should be such that the workers' own clothing is not contaminated by any hazardous substances. On smaller sites, the site office may be suitable for the storage of clothing, provided that it can be kept secure and used as a rest facility.

Facilities for rest and eating meals must be provided so that workers may sit down during break times in areas where they do not need to wear PPE. It should be possible for workers to make hot drinks and prepare food. Facilities should also be provided for pregnant women and nursing mothers to rest. Arrangements must be in place to ensure that food is not contaminated by hazardous substances. Many fires have been caused by placing clothes on an electric heater. Damp clothes should not be positioned in contact with the heater. There should be adequate ventilation around the heater. If possible, the heater should be fitted with a high temperature cut-out device.

The location of the welfare facilities on the site is important. They should be located adjacent to each other and as close to the main working area as possible. It is likely that as the work progresses, the welfare facilities will need to be moved. It may be appropriate to use facilities in neighbouring premises. However, all such arrangements must be agreed in advance with the relevant neighbour. Reliance on public conveniences should be avoided due to their unreliability and lack of hot water for washing (normally).

Welfare arrangements for transient sites will be different to those for fixed sites. Transient sites include sites where the construction work is of short duration (up to one week in many cases), such as road works and emergency work. In these cases, welfare facilities (toilets, hand-washing basins, rest and eating arrangements) may be provided by a suitably equipped vehicle or by siting fixed installations at intervals along the length of the job.

HSE have produced two useful information leaflets on the provision of welfare facilities at fixed construction sites (CIS No. 18 (rev 1)) and transient construction sites (CIS No. 46).

317

Figure 9.29 Mobile site toilets and welfare facilities

Electricity

Electrical hazard is covered in detail in Chapter 12, and all the control measures mentioned apply on a construction site. However, due to the possibility of wet conditions, it is recommended that 110 V equipment only is used on site. If mains electricity is used (perhaps during the final fitting-out of a building), residual current devices should be used with all electrical equipment. Where workers or tall vehicles are working near or under overhead power lines, either the power should be turned off or 'goal posts' or taped markers used to prevent contact with the lines. Similarly, underground supply lines should be located and marked before digging takes place.

Noise

Noisy machinery should be fitted with silencers. When machinery is used in a workshop (such as woodworking machines), a noise survey should be undertaken and, if the noise levels exceed the second action level, the use of ear defenders becomes mandatory.

A cement mixer can be a particularly noisy piece of machinery on a construction site. The levels of noise exposure to workers on the site can be reduced by fixing silencers to diesel-powered mixers, ensuring that the mixer is regularly maintained, minimising the exposure time of the workers by job rotation and providing ear defenders to those working near to the mixer. A better solution would be to use a mixer with lower noise emissions.

Excessive noise hazards may also be present during demolition work and by pneumatic drills, compressors and vehicles used on site.

Health hazards

Health hazards are present on a construction site. These hazards include vibration, dust (including asbestos), cement, solvents and paints and cleaners. A COSHH assessment is essential before work starts with regular updates as new substances are introduced. Copies of the assessment and the related safety data sheets should be kept in the Site Office for reference after accidents or fires. They will also be required to check that the correct personal protective equipment (PPE) is available. A manual handling assessment should also be made to ensure that the lifting and handling of heavy objects is kept to a minimum. The findings of all assessments must be communicated to the workforce.

Many of the health hazards (both chemical and biological) are covered in Chapter 8 including silica, which is commonly produced during construction activities, cement dust and wet cement, wood dust and the biological hazard tetanus.

Waste disposal

The collection, storage and removal of waste from a construction site are described in Chapter 8.

Site security

Site security is required to ensure that construction materials, equipment and accessories are protected and this is usually achieved by the erection of a strong perimeter fence and a lockable gate. However, it is also important to protect members of the public including trespass by children. This is particularly relevant for a refurbishment project when strong hoardings or secure fencing must be provided. There is also a need to ensure that the site itself is safe should children manage to trespass onto the site. This involves the removal of access ladders, providing trench supports, covering holes where possible, immobilising plant, safe stacking of materials, locking away dangerous chemicals, isolating electrical supplies and locking off fuel storage tanks. Many of these problems may be avoided by visiting local schools to inform children of construction site dangers. Accidents to children on construction sites are a national concern and a widespread problem everywhere. A brickwork company has been fined after being convicted of failing to provide and maintain a suitable barrier around the boundary of a quarry where a 13-year-old schoolgirl died.

The provision of warning signs at the perimeter fence and suitable viewing points for the public to observe the construction work safely have been shown to reduce the incidents of site trespass.

Whilst storage is an important factor for site security, it is also needed for health and safety reasons. Cement, which is a hazardous substance, must be securely stored in a dry store. Gas cylinders, fuel and other flammable substances must be stored in a separate flameproof store (more details of these requirements are given in Chapter 11).

Environmental considerations

The public must also be protected from environmental hazards resulting from the work, such as excessive dust, noise and mud on the public highway. Excessive dust may be controlled by damping down haul roads regularly, the sheeting of lorry loads and imposing speed limits on roads close to the site. During road repair operations, dust should be prevented from entering surface drains and precautions taken to prevent or restrict the effects of fuel spillage.

Noise levels may be reduced by the regular maintenance of equipment and the erection of noise barriers. Other controls include the use of quieter equipment and restricting the use of noisy machinery to late morning or early afternoon.

Regular road sweeping will reduce mud on surrounding roads. The problem will also be alleviated by washing vehicle wheels before they leave the site.

For any of these measures to be successful, the site workforce must be informed of the procedures at induction and by posters and signs around the site.

Arrangements with the client and/or occupier of the premises

Such arrangements will include site rules for the protection of members of the public, visitors and other employees. They should also cover issues such as shared facilities (e.g. first-aid arrangements and the use of catering facilities) and the need for full cooperation from the occupier and his staff in the observation of health and safety rules.

When the construction work involves a public highway, measures should be introduced to ensure the safe passage of vehicles and pedestrians. This will involve the erection of warning signs, the use of cones with lead-in and exit tapers and an indication of a safety zone between the cones and the work area. Pedestrian safety will require the use of barriers and lighting, and traffic may need to be controlled using traffic lights or stop/go boards. Certain neighbours, such as schools, hospitals or factories, may also need to be informed of the timing of some aspects of the work.

The management of construction activities is governed by the CDM Regulations which are covered in Chapter 1.

Excavations

Hazards associated with excavations

There are about seven deaths each year due to work in excavations. Many types of soil, such as clays, are self-supporting but others, such as sands and gravel, are not. Many excavations collapse without any warning, resulting in death or serious injury. Many such accidents occur in shallow workings. It is important to note that, although most of these accidents affect workers, members of the public can also be injured. The specific hazards associated with excavations are as follows:

- collapse of the sides;
- materials falling on workers in the excavation;
- falls of people and/or vehicles into the excavation;
- workers being struck by plant;
- specialist equipment such as pneumatic drills;
- hazardous substances, particularly near the site of current or former industrial processes;
- influx of ground or surface water and entrapment in silt or mud;
- proximity of stored materials, waste materials or plant;
- proximity of adjacent buildings or structures and their stability;

319

▷ contact with underground services;
▷ access and egress to the excavation;
▷ fumes, lack of oxygen and other health hazards (such as Weil's disease); and
▷ contaminated ground.

Clearly, alongside these specific hazards, more general hazards, such as manual handling, electricity, noise and vibrations, will also be present.

Wedges

Poling boards or sheeting

Waling

Puncheons

Screw props or struts

Figure 9.30 Timbered excavation with ladder access and supported services (guard removed on one side for clarity)

Precautions and controls required for excavations

The following precautions and controls should be adopted:

▷ At all stages of the excavation, a competent person must supervise the work and the workers must be given clear instructions on working safely in the excavation.
▷ The sides of the excavation must be prevented from collapsing either by digging them at a safe angle (between 5° and 45° dependent on soil and dryness) or by shoring them up with timber, sheeting or a proprietary support system. Falls of material into the workings can also be prevented by not storing spoil material near the top of the excavation.
▷ The workers should wear hard hats.
▷ If the excavation is more than 2 m deep, a substantial barrier consisting of guard rails and toe boards should be provided around the surface of the workings.

▷ Vehicles should be kept away as far as possible using warning signs and barriers. Where a vehicle is tipping materials into the excavation, stop blocks should be placed behind its wheels.
▷ It is very important that the excavation site is well lit at night.
▷ All plant and equipment operators must be competent, and non-operators should be kept away from moving plant.
▷ PPE must be worn by operators of noisy plant.
▷ Nearby structures and buildings may need to be shored up if the excavation may reduce their stability. Scaffolding could also be de-stabilised by adjacent excavation trenches.
▷ The influx of water can only be controlled by the use of pumps after the water has been channelled into sumps. The risk of flooding can be reduced by the isolation of the mains water supply.
▷ The presence of hazardous substances or health hazards should become apparent during the original survey work and, when possible, removed or suitable control measures adopted. Any such hazards found after work has started, must be reported and noted in the inspection report and remedial measures taken. Exhaust fumes can be dangerous and petrol or diesel plant should not be sited near the top of the excavation.
▷ The presence of buried services is one of the biggest hazards and the position of such services must be ascertained using all available service location drawings before work commences. As these will probably not be accurate, service location equipment should be used by specifically trained people. The area around the excavation should be checked for service boxes. If possible, the supply should be isolated. Only hand tools should be used in the vicinity of underground services. Overhead services may also present risks to cranes and other tall equipment. If the supply cannot be isolated then 'goal posts' beneath the overhead supply together with suitable bunting and signs must be used.
▷ Safe access by ladders is essential, as are crossing points for pedestrians and vehicles. Whenever possible, the workings should be completely covered outside working hours, particularly if there is a possibility of children entering the site.
▷ Finally, care is needed during the filling-in process. Wells and disused mine shafts are found during construction work and must be treated with caution, and in the same way as an excavation. The obvious hazards include falling in and/or drowning and those associated with confined spaces – oxygen deficiency, the presence of toxic gases and the possible collapse of the walls. Controls include fencing off the well and covering it until the situation has been

reviewed by specialists. Shallow wells would normally be drained and filled with hard core whereas deeper ones would be capped.

Inspection and reporting requirements

The duty to inspect and prepare a report only applies to excavations which need to be supported to prevent accidental fall of material. Only persons with a recognised and relevant competence should carry out the inspection and write the report. Inspections should take place at the following timing and frequency:

▶ after any event likely to affect the strength or stability of the excavation;

▶ before work at the start of every shift; and

▶ after an accidental fall of any material.

Although an inspection must be made at the start of every shift, only one report is required of such inspections every seven days. However, reports must be completed following all other inspections. The report should be completed before the end of the relevant working period and a copy given to the manager responsible for the excavation within 24 hours. The report must be kept on site until the work is completed and then retained for three months at an office of the organisation which carried out the work.

A suitable form is shown in Appendix 9.2.

Figure 9.31 Barriers around excavation by footpath

9.9 Further information

Statutory provisions

Personal Protective Equipment at Work Regulations 1992

The Construction (Design and Management) Regulations 2015

The Health and Safety (Safety Signs and Signals) Regulations 1996

The Provision and Use of Work Equipment Regulations 1998 (as amended)

The Work at Height Regulations 2005 (as amended)

The Workplace (Health, Safety and Welfare) Regulations 1992

Other references

Health and Safety Toolbox, online resource, HSE, www.hse.gov.uk/toolbox/index.htm

Health and Safety in Construction, HSG150, third edition, HSE Books, ISBN: 978-0-7176-6182-2 www.hse.gov.uk/pubns/priced/hsg150.pdf

Lighting at Work, HSG38, second edition 1997, HSE Books, ISBN: 978-0-7176-1232-1 www.hse.gov.uk/pubns/priced/hsg38.pdf

Managing Health and Safety in Construction, Construction (Design and Management) Regulations, Approved Code of Practice (ACoP), L144, HSE Books, ISBN: 978-0-7176-6223-4 www.hse.gov.uk/pubns/priced/l144.pdf

Personal Protective Equipment at Work (second edition), Personal Protective Equipment at Work Regulations 1992 (as amended), Guidance on Regulations, HSE Books, ISBN: 978-0-7176-6139-3 www.hse.gov.uk/pubns/priced/l125.pdf

Safe Use of Work Equipment, ACoP and guidance (part III in particular), L22, third edition 2008, HSE Books, ISBN: 978-0-7176-6295-1 www.hse.gov.uk/pubns/priced/l22.pdf

Seating at Work, HSG57, third edition 2002, HSE Books, ISBN: 978-0-7176-1231-4 www.hse.gov.uk/pubns/priced/hsg57.pdf

The Health and Safety (Safety Signs and Signals) Regulations 1996, Guidance on regulations, second edition 2009, L64, HSE Books, ISBN: 978-0-7176-6359-0 www.hse.gov.uk/pubns/priced/l164.pdf

Violence at work: A guide for employers, HSE Books, INDG69, www.hse.gov.uk/pubns/indg69.pdf

Work at Height Regulations 2005 (as amended) – A Brief Guide, INDG401(rev1), HSE Books www.hse.gov.uk/pubns/indg401.pdf

Workplace health, safety and welfare, Workplace (Health, Safety and Welfare) Regulations 1992, ACoP, L24, HSE Books, ISBN: 978-0-7176-0413-5 www.hse.gov.uk/pubns/priced/l124.pdf

Understanding ergonomics at work, INDG90(rev2), HSE Books www.hse.gov.uk/pubns/indg90.pdf

Workplace Transport Safety – Guidance for Employers, HSG136, HSE Books, ISBN: 978-0-7176-6154-1 www.hse.gov.uk/pubns/priced/hsg136.pdf

Driving at Work – Managing Work-Related Road Safety INDG382, HSE Books. www.hse.gov.uk/pubns/priced/indg382.pdf

9

9.10 Practice questions

1. A manufacturing company has to relocate its workforce and workshops while a new factory is to be built. There will be 24-hour working at the new premises.
   (a) **Outline** the sanitary and washing facilities, drinking water supply and other welfare facilities that will be required at the new location.
   (b) **Identify** other aspects of the work environment that the company must consider so as to comply with the Workplace (Health, Safety and Welfare) Regulations.

2. Extremes of temperature in the workplace may have an adverse effect on the health of workers.
   (a) **Identify** the possible effects on health that may be caused by working in a hot and cold environment.
   (b) **Outline** the controls required when workers are exposed to;
      (i) a hot environment
      (ii) a cold environment.

3. An office building is to be fitted out and equipped by a magazine publishing company.
   (a) **Identify** the issues that should be considered by the company when the lighting requirements are being specified for the building.
   (b) **Outline FOUR** other health and safety issues associated with the physical working environment that should also be considered by the company.

4. (a) **Identify EIGHT** types of hazard that may cause slips or trips at work.
   (b) **Outline** the measures that may be needed to reduce the risk of slip and trip accidents in a workplace.

5. Permanent staircases are a source of accidents involving a fall from height. **Identify FIVE** design features and/or safe practices intended to reduce the risk of such accidents.

6. Both pedestrians and vehicles use the same traffic routes in a busy warehouse. **Identify** control measures that should be in place to reduce the risks to pedestrians in the warehouse.

7. During the maintenance of a machine that is in a fixed position, a fitter suffers a head injury when he collides with an overhead beam on the machine. **Describe** possible control measures that could have prevented this accident.

8. (a) **Identify FOUR** hazards associated with work at height above ground level.
   (b) **Outline** a hierarchy of control measures to be considered when a construction worker is at risk of falling while working at height.
   (c) **Outline** the precautions to be taken to ensure the safety of construction workers required to undertake repair work on a fragile roof.

9. The three most common types of fall arrest equipment are safety harnesses, safety nets and airbags. **Outline** the application, advantages and disadvantages of each of these devices to protect those working at height.

10. **Describe** safe practices required when ladders are to be used in the workplace.

11. (a) **Outline THREE** causes of a scaffold collapse.
    (b) **Outline** the precautions that should be taken to reduce the risk of injury to members of the public during the erection and use of the scaffold.

12. Mobile tower scaffolds must be used on stable, level ground.
    (a) **Give** reasons why a mobile tower scaffold may become unstable.
    (b) **Identify SIX** additional points that should be considered to ensure the safe use of a mobile tower scaffold.

13. Ceiling repairs are to be undertaken in a busy warehouse by construction workers using a mobile elevating working platform (MEWP).
    (a) **Identify** the hazards associated with this work.
    (b) **Outline** the controls required to ensure the safety of the maintenance workers and others who may be affected by the work.

14. A deep trench needs to be excavated and supported next to a four-storey building.
    (a) **Identify** the hazards that may be present in these circumstances.
    (b) **Outline** the control measures required so that the excavation workers, members of the public and the adjacent building are adequately protected.
    (c) **Identify** when inspections of the supported excavation must be carried out by the competent person.

15. (a) **Define** the term 'workplace transport' and give **THREE** examples of such transport.
    (b) **Outline** the hazards associated with workplace transport operations.
    (c) **Outline** the factors to be considered when planning traffic routes within a busy warehouse.

(d) **Describe** the main aspects of a control strategy for safe workplace transport operations.

16. (a) **Identify SIX** ways in which people may be injured by vehicles that are operating in a workplace.

(b) **Identify** the important contents of a training programme for vehicle drivers so that the risk of accidents to themselves and other employees may be reduced.

(c) **Outline** the actions required by both drivers and pedestrians to improve the safety of pedestrians in vehicle manoeuvring areas.

(d) **Outline** the hazards and controls associated with reversing vehicles in warehouse.

17. The warehouse of a plastic products manufacturer is to be enlarged to enable the use of internal transport to transfer the goods between racking and lorry loading bays.

(a) **Outline** the design features of the traffic routes that should be addressed in order to minimise the risk of fork-lift truck related accidents.

(b) **Describe** *additional* measures that need to be taken to protect pedestrians from the risk of being struck by a fork-lift truck in the warehouse.

18. **Describe** the rollover (ROPS) and falling-object (FOPS) protection techniques that are used to protect drivers of mobile work equipment. **Outline** the types of unsafe incidents from which drivers may be protected by these techniques.

19. Managing work-related road risk is an important part of a health and safety management system. **Outline** the possible risks and controls associated with:

(a) The driver

(b) The vehicle

(c) The journey.

20. (a) **Identify SIX** health and safety rules for drivers on company business.

(b) **Identify** the particular risks associated with long distance lorry driving.

9

APPENDIX 9.1 Checklist of typical scaffolding faults

Footings	Standards	Ledgers	Bracing	Putlogs and transoms	Couplings	Bridles	Ties	Boarding	Guard rails and toe boards	Ladders
Soft and uneven	Not plumb	Not level	Some missing	Wrongly spaced	Wrong fitting	Wrong spacing	Some missing	Bad boards	Wrong height	Damaged
No base plates	Jointed at same height	Joints in same bay	Loose	Loose	Loose	Wrong couplings	Loose	Trap boards	Loose	Insufficient length
No sole plates	Wrong spacing	Loose	Wrong fittings	Wrongly supported	Damaged	No check couplers	Not enough	Incomplete	Some missing	Not tied
Undermined	Damaged	Damaged			No check couplers			Insufficient supports		

APPENDIX 9.2 Inspection recording form with timing and frequency chart

Timing and frequency chart (reproduced from HSG150).

Place of work or work equipment	Timing and frequency of checks, inspections and examinations equipment								
	Inspect before work at the start of every shift (see note 1)	Inspect after any event likely to have affected its strength or stability	Inspect after accidental fall of rock or other material	Inspect after installation or assembly in any position (see notes 2 and 3)	Inspect at suitable intervals	Inspect after exceptional circumstances which are liable to jeopardise the safety at work equipment	Inspect at intervals not exceeding 7 days (see note 3)	Check on each occasion before use (REPORT NOT REQUIRED)	LOLER Thorough Examination (if work equipment subject to LOLER) (see note 4)
Excavations which are supported to prevent any person being buried or trapped by an accidental collapse or a fall or dislodgement of material	✓	✓	✓						
Cofferdams and caissons	✓	✓							
The surface and every parapet or permanent rail of every existing place of work at height								✓	
Guard rails, toe boards, barriers and similar collective means of fall protection				✓	✓	✓			
Scaffolds and other working platforms (including tower scaffolds and MEWPs) used for construction work and from which a person could fall more than 2 m				✓		✓	✓		✓
All other working platforms				✓	✓	✓			✓
Collective safeguards for arresting falls (e.g. nets, airbags, soft landing systems)				✓	✓	✓			
Personal fall protection systems (including work positioning, rope access, work restraint and fall arrest systems)				✓	✓	✓			✓
Ladders and stepladders					✓	✓		✓	

Notes:

1. Although an excavation must be inspected at the start of every shift, only one report is needed in any seven-day period. However, if something happens to affect its strength or stability, and/or an additional inspection is carried out, a report must then be completed. A record of this inspection must be made and retained for three months.

2. 'Installation' means putting into position and 'assembly' means putting together. You are not required to inspect and provide a report every time a ladder, tower scaffold or mobile elevating work platform (MEWP) is moved on site or a personal fall protection system is clipped to a new location.

3. An inspection and a report is required for a tower scaffold or MEWP (used for construction work and from which a person could fall 2 metres) after installation or assembly and every seven days thereafter, provided the equipment is being used on the same site. A record of this inspection must be made and retained for three months. If a tower scaffold is reassembled rather than simply moved, then an additional, pre-use inspection and report is required. It is acceptable for this inspection to be carried out by the person responsible for erecting the tower scaffold, provided they are trained and competent. A visible tag system, which supplements inspection records as it is updated following each pre-use inspection, is a way of recording and keeping the results until the next inspection.

4. All work equipment subject to LOLER Regulation 9, thorough examination and inspection requirements, will continue to be subject to LOLER Regulation 9 requirements.

325

CHAPTER 10

Work equipment

The contents of this chapter are:

Chapter summary

The chapter covers the following topics in Element 9 of the NEBOSH NGC syllabus:

General requirements

Hand-held tools

Machinery hazards

Control measures for machinery

Learning outcomes

These are the learning outcomes covered by this chapter:

The learner will be able to:

1. Outline general requirements for work equipment
2. Explain the hazards and controls for hand-held tools
3. Describe the main mechanical and non-mechanical hazards of machinery
4. Explain the main control measures for reducing risk from machinery hazards

---

10.1 General requirements for work equipment

10.1.1 Introduction and types of equipment

This chapter covers the scope and main requirements for work equipment as covered by Parts II and III of the Provision and Use of Work Equipment Regulations (PUWER). The requirements for the supply of new machinery are also included. Summaries of PUWER and the Supply of Machinery (Safety) Regulations are given in Chapter 13. The safe use of hand-held tools, hand-held power tools and the proper safeguarding of a small range of machinery used in industry and commerce are also included.

Any equipment used by an employee at work is generally covered by the term 'work equipment'. The scope is extremely wide and includes hand tools, power tools, ladders, photocopiers, laboratory apparatus, lifting equipment, fork-lift trucks and motor vehicles (which are not privately owned). Virtually anything used to do a job of work, including employees' own equipment, is covered. The uses covered include starting or stopping the equipment, repairing, modifying, maintaining, servicing, cleaning and transporting.

Employers and the self-employed must ensure that work equipment is suitable, maintained, inspected if necessary, provided with adequate information, instruction and only used by people who have received sufficient training.

Many serious accidents at work involve machinery. Hair or clothing can become entangled in moving parts, people can be struck by moving parts of machinery, parts of the body can be drawn into or trapped in machinery, and/or parts of the machinery or work tool can be ejected.

Many circumstances can increase the risks, including:

▶ not using the right equipment for the task, e.g. ladders instead of access towers for an extended task at high level;

▶ not fitting adequate controls on machines, or fitting the wrong type of controls, so that equipment cannot be stopped quickly and safely, or it starts accidentally;

▶ not guarding machines properly, leading to accidents caused by entanglement, shearing, crushing, trapping or cutting;

▶ not properly maintaining guards and other safety devices;

▶ not providing the right information, instruction and training;

▶ not fitting rollover protective structures (ROPS) and seat belts on mobile work equipment where there is a risk of rollover (*excluding* quad bikes);

▶ not maintaining work equipment or doing the regular inspections and thorough examinations;

▶ not providing (free) adequate personal protective equipment to use.

When identifying the risks, think about:

▶ the work being done during normal use of the equipment and also during setting-up, maintenance, cleaning and clearing blockages;

▶ which workers will use the equipment, including those who are inexperienced, have changed jobs or those who may have particular difficulties, e.g. those with language problems or impaired hearing;

▶ people who may act stupidly or carelessly or make mistakes;

▶ guards or safety devices that may be badly designed and difficult to use or are easy to defeat;

▶ other features of the equipment which could cause risks like vibration, electricity, wet or cold conditions.

Consider the following:

▶ Is the equipment suitable for the task?

▶ Are all the necessary safety devices fitted and in working order?

▶ Are there proper instructions for the equipment?

▶ Is the area around the machine safe and level with no obstructions?

▶ Has suitable lighting been provided?

▶ Has extraction ventilation been provided where required, e.g. on grinding and woodworking machinery?

- Has a risk assessment been done to establish a person's competence or training requirements to control particular machinery, which is very important for everyone?
- Are machine operators trained and do they have enough information, instruction, training?
- Are people adequately supervised?
- Are safety instructions and procedures being used and followed?
- Are machine operators using appropriate work clothing without loose sleeves, open jackets, dangling jewellery or sandals?
- Has the employer supplied all necessary special personal protective equipment (PPE)?
- Are safety guards or devices being used properly?
- Is maintenance carried out correctly and in a safe way?
- Are hand tools being used correctly and properly maintained, and only used by people who have received sufficient training?

ISO 12100:2010 requires that protective measures to achieve the necessary risk reduction with work equipment is undertaken in the following three-step sequence:

1. Inherently safe design measures which eliminate hazards or reduce risks for people using the machinery;
2. The provision of safeguarding and/or complementary protective measures which take into account the use and reasonably foreseeable misuse of the equipment;
3. The provision of information for use which cover: operating procedures, recommended safe working practices, warning of residual risks and other information for the different phases of the life of the equipment; the description of any PPE required.

10.1.2 Suitability of work equipment and CE marking within the UK and EU

Standards and requirements

The UK has now left the EU. The Transition period is in place while the EU and the UK negotiate new arrangements for a trade deal. During the transition period there are no changes to responsibilities for the health and safety of people at work. It is anticipated that the UK will continue to adopt similar requirements to the CE marking as currently exist.

When work equipment is provided it has to be suitable for its purpose and conform to standards which cover its supply as a new or second-hand piece of equipment and its use in the workplace. This involves:

- its initial integrity
- the place where it will be used
- the purpose for which it will be used.

There are two groups of law that deal with the provision of work equipment:

- One deals with what manufacturers and suppliers have to do. This can be called the 'supply' law. One

of the most common is the Supply of Machinery (Safety) Regulations 2008 as amended, which requires manufacturers and suppliers to ensure that machinery is safe when supplied and has CE marking. Its **primary** purpose is to prevent barriers to trade across the EU by ensuring uniform standards, rather than protecting people at work.

- The other deals with what the users of machinery and other work equipment have to do. This can be called the 'user' law, PUWER 98, and applies to most pieces of work equipment. Its **primary** purpose is to protect people at work.

Under 'user' law, employers have to provide safe equipment of the correct type, ensure that it is correctly used and maintain it in a safe condition. When buying new equipment, the 'user' has to check that the equipment complies with all the 'supply' law that is relevant. Figure 10.1(b) shows the division of responsibility between designers, manufacturers and suppliers and those of employers. However, whatever hazards have or have not been controlled effectively, **the user must check that the machine is safe before it is used**.

Most new equipment, including machinery in particular, should have 'CE' marking when purchased (Figure 10.1(a)). 'CE' marking is only a claim by the manufacturer that the equipment is safe and that they have met relevant supply law. If this is done properly manufacturers will have to do the following:

- Find out about the health and safety hazards (trapping, noise, crushing, electrical shock, dust, vibration, etc.) that are likely to be present when the machine is used.
- Assess the likely risks.
- Design out the hazards that result in risks; or, if that is not possible
  - Provide safeguards (e.g. guarding dangerous parts of the machine, providing noise enclosures for noisy parts); or, if that is not possible
  - use warning signs on the machine to warn of hazards that cannot be designed out or safeguarded (e.g. 'noisy machine' signs).

Figure 10.1 (a) CE marking

Figure10.1 (b) Division of responsibility for the safety of machinery

Manufacturers also have to:

▶ keep information, explaining what they have done and why, in a technical file;

▶ fix CE marking to the machine where necessary, to show that they have complied with all the relevant supply laws;

▶ issue a 'Declaration of Conformity' for the machine (see Figure 10.2). This is a statement that the machine complies with the relevant essential health and safety requirements or with the example that underwent type-examination. A declaration of conformity must:

▷ state the name and address of the manufacturer or importer into the EU;

▷ contain a description of the machine, and its make, type and serial number;

EC-Declaration of Conformity

MACHINERY DIRECTIVE

DC720, DC721, DC722, DC725, DC727, DC730, DC731, DC732, DC735, DC737, DC742, DC743, DC745

DeWALT declares that these products described under "technical data" are in compliance with: 98/37/EC (until Dec. 28, 2009), 2006/42/EC (from Dec. 29, 2009), EN 60745-1, EN 60745-2-1, EN 60745-2-2.

These products also comply with Directive 2004/108/EC. For more information, please contact DeWALT at the following address or refer to the back of the manual.

Horst Grossmann
Vice President Engineering and Product Development
DeWALT, Richard-Klinger-Strase 11, D-65510, Idstein, Germany
30.07.2009

Figure 10.2 Typical Certificate of Conformity

▷ indicate all relevant European Directives with which the machinery complies;

▷ state details of any notified body that has been involved;

▷ specify which standards have been used in the manufacture (if any);

▷ be signed by a person with authority to do so;

▷ provide the buyer with instructions to explain how to install, use and maintain the machinery safely.

This can help in deciding which equipment may be suitable, particularly if buying a standard piece of equipment 'off the shelf'.

If buying a more complex or custom-built machine the buyer should discuss their requirements with potential suppliers. For a custom-built piece of equipment, there is the opportunity to work with the supplier to design out the causes of injury and ill health. Time spent now on agreeing the necessary safeguards, to control health and safety risks, could save time and money later.

Note: Sometimes equipment is supplied via another organisation, for example, an importer, rather than direct from the manufacturer, so this other organisation is referred to as the supplier. It is important to realise that the supplier may not be the manufacturer.

When the equipment has been supplied the buyer should look for CE marking, check for a copy of the Declaration of Conformity and that there is a set of instructions in English on how the machine should be used and most important of all, **check to see if they think that it is safe.**

Before buying new equipment the buyer will need to think about:

▶ where and how it will be used;

▶ what it will be used for;

▶ who will use it (skilled employees, trainees);

▶ what risks to health and safety might result;

▶ how well health and safety risks are controlled by different manufacturers.

Limitations of CE marking

CE marking is not a guarantee that the machine is safe. It is a claim by the manufacturer that the machinery complies with the law. CE marking has many advantages if done properly, for example:

▶ it allows a common standard across Europe;

▶ it provides a means of selling to all European Union member states without barriers to trade;

- ▶ it ensures that instructions and safety information are supplied in a fairly standard way in most languages in the EU;
- ▶ it has encouraged the use of diagrams and pictorials which are common to all languages;
- ▶ it allows for independent type-examination for some machinery like woodworking machinery which has not been made to an EU-harmonised standard – identified by an EN marking before the standard number (e.g. BS EN).

Clearly there are disadvantages as well, for example:

- ▶ instruction manuals have become very long. Sometimes two volumes are provided, because of the number of languages required;
- ▶ translations can be very poor and disguise the proper meaning of the instruction;
- ▶ manufacturers can fraudulently put on the CE marking;
- ▶ manufacturers might make mistakes in claiming conformity with safety laws.

10.1.3 Prevention of access to dangerous parts of machinery

Regulation 11(1) of PUWER requires employers to take effective measures to prevent access to dangerous parts of machinery or stop their movement before any part of a person enters a danger zone. This regulation also applies to contact with a rotating stockbar which projects beyond the headstock of a lathe.

The term 'dangerous part' has been established in health and safety law through judicial decisions. In practice, this means that if a piece of work equipment could cause injury and if it is being used in a foreseeable way, it can be considered a dangerous part.

There are many HSE publications which are specific to a machine or industry, for example *Safe Use of Woodworking Machinery Approved Code of Practice L114*. They describe the measures that can be taken to protect against risks associated with dangerous parts of machinery. Current national, European and international standards may also be used for guidance, such as ISO 12100:2010 'Safety of Machinery'.

The risk assessment carried out under Regulation 3 of the Management Regulations should identify hazards presented by machinery. The risk assessment should evaluate the nature of the injury, its severity and likelihood of occurrence for each hazard identified. This will enable employers to decide whether the level of risk is acceptable or if risk reduction measures are needed. In most cases the objective of risk reduction measures is to prevent contact of part of the body or clothing with any dangerous part of the machine, for example by providing guards.

Regulation 11(2) of PUWER specifies the measures which should be taken to prevent access to the dangerous parts of the machinery and achieve compliance with Regulation 11(1). The measures are ranked in the order they should be implemented, where practicable, to achieve an adequate level of protection. The levels of protection are:

(a) fixed enclosing guards;
(b) other guards or protection devices such as interlocked guards and pressure mats;
(c) protection appliances such as jigs, holders and push sticks, etc.; and
(d) the provision of information, instruction, training and supervision.

The purpose of the risk assessment is to identify measures that can be taken to reduce the risks that the hazards present. When selecting measures employers should consider each level of protection from the first level of the scale listed above, and they should use measures from that level so far as it is practicable to do so, provided that they contribute to the reduction of risk. It may be necessary to select a combination of measures. The selection process should continue down the scale until the combined measures are effective in reducing the risks to an acceptable level thus meeting the requirements of Regulation 11(1). In selecting the appropriate combination employers will need to take account of:

- ▶ the requirements of the work;
- ▶ evaluation of the risks; and
- ▶ the technical features of possible safeguarding solutions.

Most machinery will present more than one mechanical hazard, and employers will need to deal with the risks associated with all of these. For example, at belt conveyors there is a risk of entanglement with the rotating shafts and of being trapped by the intake between drum and moving belt – so appropriate safety measures should be adopted.

Any risk assessment carried out under Regulation 3 of the Management Regulations should not just deal with the machine when it is operating normally, but must also cover activities such as setting, maintenance, cleaning or repair. The assessment may indicate that these activities require a different combination of protective measures from those appropriate to the machine doing its normal work. In particular, parts of machinery that are not dangerous in normal use because they are not then accessible may become accessible and therefore dangerous while this type of work is being carried out.

Certain setting or adjustment operations which may have to be done with the machine running may require a greater reliance on the provision of information, instruction, training and supervision than for normal use.

Regulation 11(3) of PUWER sets out various requirements for guards and protection devices which are covered later in this chapter. These are largely

10

detailed in relevant national, European and international standards such as BS EN 12100 Safety of Machinery. Chapter 13 includes a full summary of PUWER.

10.1.4 Use and maintenance of equipment with specific risks

Some pieces of work equipment involve specific risks to health and safety where it is not possible to control adequately the hazards by physical measures alone, for example the use of a bench-mounted circular saw or an abrasive wheel (Figure 10.3) and their use or maintenance needs to be restricted. In all cases the hierarchy of controls should be adopted to reduce the risks by:

▶ eliminating the risks; or, if this is not possible –
▶ taking physical measures to control the risks such as guards; but if the risks cannot be adequately controlled –
▶ taking appropriate software measures, such as a safe system of work.

PUWER 98 Regulation 7 restricts the use of such equipment to the persons designated to use it. These people need to have received sufficient information, instruction and training so that they can carry out the work using the equipment safely.

Figure 10.3 Typical bench-mounted abrasive wheel

Repairs, modifications, maintenance or servicing is also restricted to designated persons. A designated person may be the operator if he/she has the necessary skills and has received specific instruction and training. Another person specifically trained to carry out a particular maintenance task, for example dressing an abrasive wheel, may not be the operator but may be designated to do this type of servicing task on a range of machines.

Protection against other hazards associated with machinery are dealt with in Regulations 12 and 13. However, the measures required by these regulations may also protect or help to protect against those other hazards such as ejected particles and heat.

10.1.5 Information, instruction and training for specific risks

People using and maintaining work equipment, where there are residual risks that cannot be sufficiently reduced by physical means, require enough information, instruction and training to operate safely. Managers and supervisors will also need to be provided with sufficient information about a piece of equipment to enable them to fulfil their responsibilities towards people using and maintaining equipment.

The information and instructions are likely to come from the manufacturer in the form of operating and maintenance manuals. It is up to the employer to ensure that what is provided is easily understood, and set out logically with illustrations and standard symbols where appropriate. The information should normally be in good plain English but other languages may be necessary in some cases.

The extent of the information and instructions will depend on the complexity of the equipment and the specific risks associated with its use. They should cover:

▶ all safety and health aspects;
▶ any limitations on the use of the equipment;
▶ any foreseeable problems that could occur;
▶ safe methods to deal with the problems;
▶ any relevant experience with the equipment that would reduce the risks or help others to work more safely, being recorded and circulated to everyone concerned.

Everyone who uses and maintains work equipment needs to be adequately trained. For general training requirements see Section 10.4.6. Examples of specific requirements for woodworking and abrasive wheels are given below.

Employers should ensure that information and instructions provided on the use of woodworking machinery includes, where relevant:

(a) the speed, range, type and dimensions of tools suitable for the machine;
(b) any limitation on the cutting speeds of the machine, particular operations or size and material of any work-piece;
(c) procedures relating to the repair or replacement of any guard or protection device;
(d) the availability, suitability and use of any additional protection device or protection appliance;

(e) the correct procedures to be followed for setting and adjusting operations;

(f) safe methods of handling tools;

(g) correct procedures for start-up and shutdown, isolation and how to discharge any residual energy;

(h) procedures for cleaning saw blades by hand (which should be carried out with the machine isolated and the blade stopped);

(i) procedures for adjusting any guard, tool, clamp or other part of a machine (which should not be carried out while any part of the machine is in motion, unless they can be done safely).

Training of someone to use a grinding machine should cover the proper methods of mounting and abrasive wheel, see *Safety in the use of abrasive wheels HSG17*, 3rd Edition, HSE 2000. They are as follows:

Summary of mounting procedures for an abrasive wheel:

1. Wheel mounting should be carried out only by an appropriately trained person. A wheel should be mounted only on the machine for which it was intended. Before mounting, all wheels should be closely inspected to ensure that they have not been damaged in storage or transit.

2. The speed marked on the machine should not exceed the speed marked on the wheel, blotter or identification label.

3. The bush, if any, should not project beyond the sides of the wheel and blotters. The wheel should fit freely but not loosely on the spindle.

4. Flanges should not be smaller than their specified minimum diameter, and their bearing surfaces should be true and free from burrs.

5. With the exception of the single flange used with threaded-hole wheels, all flanges should be properly recessed or undercut.

6. Flanges should be of equal diameter and have equal bearing surfaces. Protection flanges should have the same degree of taper as the wheel. Blotters, slightly larger than the flanges, should be used with all abrasive wheels except those listed below in a–g. Wrinkles in blotters should be avoided.

   Blotters should not be used with the following types of wheels:
   (a) mounted wheels and points;
   (b) abrasive discs (inserted nut discs and cylinders);
   (c) plate-mounted wheels;
   (d) cylinder wheels mounted in chucks;

(e) rubber-bonded cutting-off wheels 0.5 mm or less in thickness;

(f) taper-sided wheels;

(g) wheels with threaded inserts.

7. Wheels, blotters and flanges should be free from foreign matter. Clamping nuts should be tightened only sufficiently to hold the wheel firmly. When the flanges are clamped by a series of screws they should be tightened uniformly in a criss-cross sequence. Screws for inserted nut mounting of discs, cylinders and cones should be long enough to engage a sufficient length of thread, but not so long that they contact the abrasive.

8. When mounting the wheels and points, the overhang appropriate to the speed, diameter of the mandrel and size of the wheel should not be exceeded, and there should be sufficient length of mandrel in the collet or chuck.

The training and supervision of young persons is particularly important because of their relative immaturity, unfamiliarity with a working environment and lack of awareness of existing or potential risks. Some Approved Codes of Practice, for example on the Safe Use of Woodworking Machinery, L114, 2nd edition, HSE 2014, restricts the use of high-risk machinery. Only young persons with sufficient maturity and competence who have finished their training may use the equipment unsupervised.

10.1.6 Maintenance and inspection

Maintenance

Work equipment needs to be properly maintained so that it continues to operate safely and in the way it was designed to perform. The amount of maintenance will be stipulated in the manufacturer's instructions and will depend on the amount of use, the working environment and the type of equipment. High-speed, high-risk machines, which are heavily used in an adverse environment like salt water, may require very frequent maintenance, whereas a simple hand tool, like a shovel, may require very little.

Maintenance management schemes can be based around a number of techniques designed to focus on those parts which deteriorate and need to be maintained to prevent health and safety risks. These techniques include the following:

▶ **Preventative planned maintenance** – this involves replacing parts and consumables or making necessary adjustments at preset intervals normally set by the manufacturer, so that there are no hazards created by component deterioration or failure. Vehicles are normally maintained on this basis.

10

▶ **Condition-based maintenance** – this involves monitoring the condition of critical parts and carrying out maintenance whenever necessary to avoid hazards which could otherwise occur.

▶ **Breakdown-based maintenance** – here maintenance is only carried out when faults or failures have occurred. This is only acceptable if the failure does not present an immediate hazard and can be corrected before the risk is increased. If, for example, a bearing overheating can be detected by a monitoring device, it is acceptable to wait for the overheating to occur as long as the equipment can be stopped and repairs carried out before the fault becomes dangerous to persons employed.

In the context of health and safety, maintenance is not concerned with operational efficiency but only with avoiding risks to people. It is essential to ensure that maintenance work can be carried out safely. This will involve the following:

▶ competent well-trained maintenance people;
▶ the equipment being made safe for the maintenance work to be carried out. In many cases, the normal safeguards for operating the equipment may not be sufficient as maintenance sometimes involves going inside guards to observe and subsequently adjust, lubricate or repair the equipment. Careful design allowing adjustments, lubrication and observation from outside the guards, for example, can often eliminate the hazard. Making equipment safe will usually involve disconnecting the power supply and then preventing anything moving, falling or starting during the work. It may also involve waiting for equipment to cool or warm up to room temperature;
▶ a safe system of work being used to carry out the necessary procedures to make and keep the equipment safe and perform the maintenance tasks. This can often involve a formal 'permit-to-work' scheme to ensure that the correct sequence of safety critical tasks has been performed and all necessary precautions taken;
▶ correct tools and safety equipment being available to perform the maintenance work without risks to people. For example special lighting or ventilation may be required.

Inspection under PUWER

Complex equipment and/or high-risk equipment will probably need a maintenance log and may require a more rigid inspection regime to ensure continued safe operation. This is covered by PUWER 98, Regulation 6.

PUWER requires, where safety is dependent on the installation conditions and/or the work equipment is exposed to conditions causing deterioration, which may result in a significant risk and a dangerous situation developing, that the equipment is inspected by a

competent person. In this case the competent person would normally be an employee, but there might be circumstances where an outside competent person would be used.

The inspection must be done:

▶ after installation for the first time;
▶ after assembly at a new site or in a new location; and thereafter
▶ at suitable intervals;
▶ each time exceptional circumstances occur which could affect safety.

The inspection under PUWER will vary from a simple visual inspection to a detailed comprehensive inspection, which may include some dismantling and/or testing. The level of inspection required would normally be less rigorous and intrusive than thorough examinations under the Lifting Operations and Lifting Equipment Regulations 1998 (LOLER) for certain lifting equipment (see Chapters 9 and 13). In the case of boilers and air receivers the inspection is covered by the Pressure Systems Safety Regulations 2000, which involves a thorough examination (see the summary in Chapter 13). The inspection under PUWER would only be needed in these cases if the examinations did not cover all the significant health and safety risks that are likely to arise.

Examination of boilers and air receivers

Under the Pressure Systems Safety Regulations 2000, a wide range of pressure vessels and systems require thorough examination by a competent person to an agreed specifically written scheme. This includes steam boilers, pressurised hot water plants and air receivers although these are no longer covered in the Certificate syllabus, but often encountered in small and medium-sized workplaces (Figure 10.4).

Figure 10.4 Typical electrically powered compressor with air receiver

The Regulations place duties on designers and manufacturers but this section is only concerned with the duty on users to have the vessels examined. An employer who operates a steam boiler and/or a pressurised hot water plant and/or an air receiver must ensure:

▶ that it is supplied with the correct written information and markings;
▶ that the equipment is properly installed;
▶ that it is used within its operating limits;
▶ there is a written scheme for periodic examination of the equipment certified by a competent person (in the case of standard steam boilers and air receivers the scheme is likely to be provided by the manufacturer);
▶ that the equipment is examined in accordance with the written scheme by a competent person within the specified period;
▶ that a report of the periodic examination is held on file giving the required particulars;
▶ that the actions required by the report are carried out;
▶ that any other safety critical maintenance work is carried out, whether or not covered by the report.

In these cases the competent person is usually a specialist inspector from an external inspection organisation. Many of these organisations are linked to the insurance companies who cover the financial risks of use of the pressure vessel.

10.1.7 Operation and working environment

To operate work equipment safely it must be fitted with easily reached and operated controls, kept stable, properly lit, kept clear and provided with adequate markings and warning signs. These are covered by PUWER 98, which applies to all types of work equipment.

Controls

Equipment should be provided with efficient means of:

▶ starting or making a significant change in operating conditions;
▶ stopping in normal circumstances;
▶ emergency stopping as necessary to prevent danger.

All controls should be well positioned, clearly visible and identifiable, so that it is easy for the operator to know what each control does. Markings should be clearly visible and remain so under the conditions met at the workplace. See Figure 10.5 for more information on controls.

Equipment controls should:

➤ be easily reached from the operating positions
➤ not permit accidental starting of equipment
➤ move in the same direction as the motion being controlled
➤ vary in mode, shape and direction of movement to prevent inadvertent operation of the wrong control
➤ incorporate adequate red emergency stop buttons of the mushroom-headed type with lock-off
➤ have shrouded or sunken green start buttons to prevent accidental starting of the equipment
➤ be clearly marked to show what they do.

Figure 10.5 Equipment controls – design features

(a) Start controls

It should only be possible to start the work equipment by using the designed start control. Equipment may well have a start sequence which is electronically controlled to meet certain conditions before starting can be achieved, for example preheating a diesel engine, or purge cycle for gas-fed equipment. Restarting after a stoppage will require the same sequence to be performed.

Stoppage may have been deliberate or as a result of opening an interlocked guard or tripping a switch accidentally. In most cases it should not be possible to restart the equipment simply by shutting the guard or resetting the trip. Operation of the start control should be required.

Any other change to the operating conditions such as speed, pressure or temperature should only be done by using a control designed for the purpose.

(b) Stop controls

The action of normal stopping controls should bring the equipment to a safe condition in a safe manner. In some cases immediate stopping may cause other risks to occur. The stop controls do not have to be instantaneous and can bring the equipment to rest in a safe sequence or at the end of an operating cycle. It is only the parts necessary for safety, that is, accessible dangerous parts, that have to be stopped. So, for example, suitably guarded cooling fans may need to run continuously and be left on.

In some cases where there is, for example, stored energy in hydraulic systems, it may be necessary to insert physical scotches to prevent movement and/or to exhaust residual hydraulic pressure. These should

10

be incorporated into the stopping cycle, which should be designed to dissipate or isolate all stored energy to prevent danger.

It should not be possible to reach dangerous parts of the equipment until it has come to a safe condition, e.g. stopped, cooled, electrically safe.

(c) Emergency stop controls

Emergency stop must be provided where the other safeguards in place are not sufficient to prevent danger to operatives and any other persons who may be affected. Where appropriate, there should be an emergency stop at each control point and at other locations around the equipment so that action can be taken quickly. Emergency stops should bring the equipment to a halt rapidly but this should be controlled where necessary so as not to create any additional hazards. Crash shutdowns of complex systems have to be carefully designed to optimise safety without causing additional risks.

Emergency stops are not a substitute for effective guarding of dangerous parts of equipment and should not be used for normal stopping of the equipment.

Emergency stop buttons should be easily identified, reached and operated. Common types are mushroom-headed buttons, bars, levers, kick-plates or pressure-sensitive cables. They are normally red and should need to be reset after use. With stop buttons this is either by twisting or a security key (see Figure 10.6).

Mobile work equipment is normally provided with effective means of stopping the engine or power source. In some cases large equipment may need emergency stop controls away from the operator's position.

Figure 10.6 Emergency stop button

(d) Isolation of equipment

Equipment should be provided with efficient means of isolating it from all sources of energy.

The purpose is to make the equipment safe under particular conditions, for example when maintenance is to be carried out or where adverse weather conditions may make it unsafe to use. On static equipment isolation will usually be for mains electrical energy; however, in some cases there may be additional or alternative sources of energy. The isolation should cover all sources of energy such as diesel and petrol engines, LPG, steam, compressed air, hydraulics, batteries and heat. In some cases special consideration is necessary where, for example, hydraulic pumps are switched off, so as not to allow heavy pieces of equipment to fall under gravity. An example would be the loading shovel on an excavator.

(e) Stability

Stability is important and is normally achieved by bolting equipment in place or, if this is not possible, by using clamps. Some equipment can be tied down, counterbalanced or weighted, so that it remains stable under all operating conditions. If portable equipment is weighted or counterbalanced, it should be reappraised when the equipment is moved to another position. If outriggers are needed for stability in certain conditions, for example to stabilise mobile access towers, they should be employed whenever conditions warrant the additional support. In severe weather conditions it may be necessary to stop using the equipment or reappraise the situation to ensure stability is maintained.

The quality of general and local lighting will need to be considered to ensure the safe operation of the equipment. The level of lighting and its position relative to the working area are often critical to the safe use of work equipment. Poor levels of lighting, glare and shadows can be dangerous when operating equipment. Some types of lighting, for example sodium lights, can change the colour of equipment, which may increase the level of risk. This is particularly important if the colour coding of pipework or cables is essential for safety.

(f) Markings

Markings on equipment must be clearly visible and durable. They should follow international conventions for some hazards like radiation and lasers and, as far as possible, conform to the Health and Safety (Safety Signs and Signals) Regulations (see Chapter 13 for a summary). The contents, or the hazards of the contents, as well as controls, will need to be marked on some equipment. Warnings or warning devices are required in some cases to alert operators or people nearby to any dangers, for example 'wear hard hats', a flashing light on an airport vehicle or a reversing horn on a truck.

10.1.8 User responsibilities

The responsibilities for users of work equipment are covered in Section 7 of the Health and Safety at Work Act (HSW Act) and Regulation 14 of the Management Regulations. Section 7 requires employees to:

▸ take reasonable care for themselves and others who may be affected; and

▸ to cooperate with the employer.

In addition Section 8 of HSW Act requires employees not to misuse or interfere with safety provisions.

Regulation 14 of the MHSW Regulations requires employees to use equipment properly in accordance with instructions and training. They must also inform employers of dangerous situations and shortcomings in protection arrangements. The self-employed who are users have similar responsibilities to employers and employees combined.

Where employees provide their own tools the employer still has a responsibility to ensure that work equipment is suitable.

It is not always possible to eliminate every hazard or design safeguards that protect people against all machinery hazards, particularly during commissioning, setting, adjustment cleaning and maintenance. Safe working practices, in some cases even permits to work, need to be adopted and followed by machinery operators or maintenance staff. These should be considered at the machine design and installation stages as the production of special jigs, fixtures, controls and isolation facilities may be needed.

Where mechanical hazards cannot be avoided there are many precautions which should be observed by properly trained and supervised people, these include:

▸ keeping the area around the machine clear and free of obstructions,

▸ wearing suitable clothing and footwear that does not have loose ends which could become entangled;

▸ avoiding the use of neckties, rings, necklaces and other jewellery;

▸ appropriate eye protection where there is a risk of particles being ejected;

▸ using precautions against kickback of work-pieces such as riving knives on circular saws, work rests on abrasive wheels, cutting speeds are correct for the task and cutting tool;

▸ taking precautions against bursting of abrasive wheels by ensuring they are not damaged and are running within their design speed;

▸ limiting closeness of approach to machines like overhead cranes, or removing work from the rear of circular saws;

▸ using manual handling devices such as tongs for hot steel handling or push sticks for pushing timber through a circular saw;

▸ using jigs or holders for work-pieces.

10.2 Hazards and controls for hand-held tools

10.2.1 Introduction

Work equipment includes hand-held tools whether powered or not. These tools need to be correct for the task, well maintained and properly used by trained people.

Five basic safety rules can help prevent hazards associated with the use of hand-held tools:

▸ Keep all tools in good condition with regular maintenance.

▸ Use the right tool for the job.

▸ Examine each tool for damage before use and do not use damaged tools.

▸ Use tools according to the manufacturer's instructions.

▸ Provide and use properly the right personal protective equipment (PPE).

10.2.2 Hazards of non-powered hand tools

Hazards from the misuse or poor maintenance of non-powered hand tools (Figure 10.7) include:

▸ broken handles on files/chisels/screwdrivers/hammers which can cause cut hands or hammer heads to fly off;

▸ incorrect use of knives, saws and chisels with hands getting injured in the path of the cutting edge;

▸ tools that slip causing stab wounds;

▸ poor-quality uncomfortable handles that damage hands;

▸ splayed spanners that slip and damage hands or faces;

▸ chipped or loose hammer heads that fly off or slip;

▸ incorrectly sharpened or blunt chisels or scissors that slip and cut hands. Dull tools can cause more hazards than sharp ones. Cracked saw blades must be removed from service;

▸ flying particles that damage eyes from breaking up stone or concrete;

▸ electrocution or burns by using incorrect or damaged tools for electrical work;

▸ use of poorly insulated tools for hot work in the catering or food industry;

▸ use of pipes or similar equipment as extension handles for a spanner which is likely to slip causing hand or face injury;

▸ mushroomed headed chisels or drifts which can damage hands or cause hammers (not suitable for chisels) and mallets to slip;

▸ use of spark-producing or percussion tools in flammable atmospheres;

▸ painful wrists and arms (upper limb disorders) from the frequent twisting on using screwdrivers;

▸ when using saw blades, knives or other tools, they should be directed away from aisle areas and away from other people working in close proximity.

337

Figure 10.7 (a) Damaged wood chisel with dangerous handle; (b) range of non-powered hand tools

10.2.3 Non-powered hand tools safety considerations

Use of non-powered hand tools should be properly controlled including those tools owned by employees. The following controls are important:

(a) Suitability

All tools should be suitable for the purpose and location in which they are to be used. Using the correct tool for the job is the first step in safe hand tool use. Tools are designed for specific needs. That is why screwdrivers have various lengths and tip styles and pliers have different head shapes. Using any tool inappropriately is a step in the wrong direction. To avoid personal injury and tool damage, select the proper tool to do the job well and safely.

High-quality professional hand tools will last many years if they are taken care of and treated with respect. Manufacturers design tools for specific applications. Use tools only for their intended purpose.

Suitability will include:

▶ specially protected and insulated tools for electricians;
▶ non-sparking tools for flammable atmospheres and the use of non-percussion tools and cold cutting methods;
▶ tools made of suitable quality materials which will not chip or splay in normal use;
▶ the correct tools for the job, for example using the right-sized spanner and the use of mallets not hammers on chisel heads. The wooden handles of tools must not be splintered;
▶ safety knives with enclosed blades for regular cutting operations;
▶ impact tools such as drift pins, wedges and cold-chisels being kept free of mushroomed heads;
▶ spanners not being used when jaws are sprung to the point that slippage occurs.

(b) Inspection

All tools should be maintained in a safe and proper condition. This can be achieved through:

▶ the regular inspection of hand tools;
▶ discarding or prompt repair of defective tools;
▶ taking time to keep tools in the proper condition and ready for use;
▶ proper storage to prevent damage and corrosion;
▶ locking tools away when not in use to prevent them being used by unauthorised people.

(c) Training

All users of non-powered hand tools should be properly trained in their use. This may well have been done through apprenticeships and similar training. This will be particularly important with specialist working conditions or work involving young people.

Always wear approved eye protection when using hand tools, particularly when percussion tools are being used. Metal and wood particles will fly out when the material is cut or planed, so other workers in the vicinity should wear eye protection as well.

(d) Use well-designed, high-quality tools

Finally, investing in high-quality tools makes the professional's job safer and easier:

▶ If extra leverage is needed, use high-leverage pliers, which give more cutting and gripping power than standard pliers. This helps, in particular, when making repetitive cuts or twisting numerous wire pairs.
▶ Serrated jaws provide sure gripping action when pulling or twisting wires.
▶ Some side-cutting and diagonal-cutting pliers are designed for heavy-duty cutting. When cutting screws, nails and hardened wire, only use pliers that are recommended for that use.

▶ Pliers with hot riveting at the joint ensure smooth movement across the full action range of the pliers, which reduces handle wobble, resulting in a positive cut. The knives align perfectly every time.

▶ Induction hardening on the cutting knives adds to long life, so the pliers cut cleanly day after day.

▶ Sharp cutting knives and tempered handles also contribute to cutting ease.

▶ Some pliers are designed to perform special functions. For example some high-leverage pliers have features that allow crimping connectors and pulling fish tapes.

▶ Tool handles with dual-moulded material allow for a softer, more comfortable grip on the outer surface and a harder, more durable grip on the inner surface and handle ends.

▶ Well-designed tools often include a contoured thumb area for a firmer grip or colour-coded handles for easy tool identification.

▶ Insulated tools reduce the chance of injury where the tool may make contact with an energised source.

Well-designed tools are a pleasure to use. They save time, give professional results and help to do the job more safely.

10.2.4 Hand-held portable power tools

(a) Introduction

The electrical hazards of portable hand-held tools and portable appliance testing (PAT) are covered in more detail in Chapter 12. This section deals mainly with other physical hazards and safeguards relating to hand-held power tools (Figure 10.8).

The section covers, in particular, electric drills, sanders and chainsaws which are commonly used in the workplace.

Figure 10.8 Range of hand-held power tools

(b) General hazards of hand-held power tools

The general hazards involve:

▶ mechanical entanglement in rotating spindles or sanding discs;

▶ waste material flying out of the cutting area;

▶ coming into contact with the cutting blades or drill bits;

▶ risk of hitting electrical, gas or water services when drilling into building surfaces;

▶ electrocution/electric shock from poorly maintained equipment and cables or cutting the electrical cable;

▶ manual handling problem with a risk of injury if the tool is heavy or very powerful;

▶ hand-arm vibration, especially with pneumatic drills and chainsaws, disc cutters and petrol-driven units;

▶ tripping hazard from trailing cables, hoses or power supplies;

▶ eye hazard from flying particles;

▶ injury from poorly secured or clamped work-pieces;

▶ fire and explosion hazard with petrol-driven tools or when used near flammable liquids, explosive dusts or gases;

▶ high noise levels with pneumatic chisels, planes and saws in particular (see Chapter 6);

▶ levels of dust and fumes given off during the use of the tools (but see Chapter 8).

(c) Typical safety controls and instructions

Guarding

The exposed moving parts of hand-held power tools need to be safeguarded. Belts, gears, shafts, pulleys, sprockets, spindles, drums, flywheels, chains or other reciprocating, rotating or moving parts of equipment must be guarded. Machine guards, as appropriate, must be provided to protect the operator and others from the following:

▶ point of operation;

▶ in-running nip points;

▶ rotating parts;

▶ flying chips and sparks.

Safety guards must never be removed when a tool is being used. For example portable circular saws must be equipped at all times with guards. An upper guard must cover the entire blade of the saw. A retractable lower guard must cover the teeth of the saw, except where it makes contact with the work material. The lower guard must automatically return to the covering position when the tool is withdrawn from the work material.

Operating controls and switches

Most hand-held power tools should be equipped with a constant-pressure switch or control that shuts off the power when pressure is released. On/off switches should be easily accessible without removing hands from the equipment.

Handles should be designed to protect operators from excessive vibration and keep their hands away from danger areas. In some cases handles are also designed to activate a brake of the cutting chain or blade, for example in a chainsaw. Equipment should be designed to reduce lifting and manual handling problems, with special harnesses being used as necessary, for example when using large strimmers or brush cutters.

Means of starting engines and holding equipment should be designed to minimise any musculoskeletal problems.

Safe operations/instructions

When using power tools, the following basic safety measures should be observed to protect against electrical shock, personal injury, ill health and risk of fire. See also more detailed electrical precautions in Chapter 12. Operators should read these instructions before using the equipment and ensure that they are followed; that is, they must:

▶ maintain a clean and tidy working area that is well lit and clear of obstructions;

▶ never expose power tools to rain. Do not use power tools in damp or wet surroundings unless especially designed and protected for this purpose;

▶ not use power tools in the vicinity of combustible fluids, dusts or gases unless they are specially protected and certified for use in these areas;

▶ protect against electric shock (if tools are electrically powered) by avoiding body contact with grounded objects such as pipes, scaffolds and metal ladders (see Chapter 12 for more electrical safety precautions);

▶ keep children away;

▶ not let other persons handle the tool or the cable. Keep them away from the working area;

▶ store tools in a safe place when not in use where they are in a dry, locked area which is inaccessible to children;

▶ tools should not be overloaded as they operate better and safer in the performance range for which they were intended;

▶ use the right tool. Do not use small tools or attachments for heavy work. Do not use tools for purposes and tasks for which they were not intended; for example, do not use a hand-held circular saw to cut down trees or cut off branches;

▶ wear suitable work clothes. Do not wear loose-fitting clothing or jewellery. They can get entangled in moving parts. For outdoor work, rubber gloves and non-skid footwear are recommended. Long hair should be protected with a hair net;

▶ use safety glasses;

▶ also use a suitable filtering respirator mask or preferably carry out the work in a suitable work area fitted with exhaust ventilation for work that generates dust;

▶ not abuse the power cable;

▶ not carry the tool by the power cable and do not use the cable to pull the plug out of the power socket. Protect the cable from heat, oil and sharp edges;

▶ secure the work-piece. Use clamps or a vice to hold the work-piece. It is safer than using hands and it frees both hands for operating the tool;

▶ not over-reach the work area. Avoid abnormal body postures. Maintain a safe stance and maintain a proper balance at all times;

▶ maintain tools with care. Keep your tools clean and sharp for efficient and safe work. Follow the maintenance Regulations and instructions for the changing of tools. Check the plug and cable regularly and in the case of damage, have them repaired by a qualified service engineer. Also inspect extension cables regularly and replace if damaged;

▶ keep the handle dry and free of oil or grease;

▶ disconnect the power plug when not in use, before servicing and when changing the tool parts, that is blade, bits, cutter, sanding disc, etc.;

▶ not forget to remove the key. Check before switching on that the key and any tools for adjustment are removed;

▶ avoid unintentional switch-on. Do not carry tools that are connected to power with your finger on the power switch. Check that the switch is turned off before connecting the power cable;

▶ outdoors, use extension cables. When working outdoors, use only extension cables which are intended for such use and marked accordingly;

▶ stay alert, keep eyes on the work. Use common sense. Do not operate tools when there are significant distractions;

▶ check the equipment for damage. Before further use of a tool, check carefully the protection devices or lightly damaged parts for proper operation and performance of their intended functions. Check movable parts for proper function, for whether there is binding or for damaged parts. All parts must be correctly mounted and meet all conditions necessary to ensure proper operation of the equipment;

▶ ensure damaged protection devices and parts are repaired or replaced by a competent service centre unless otherwise stated in the operating instructions. Damaged switches must be replaced by a competent service centre. Do not use any tool which cannot be turned on and off with the switch;

▶ only use accessories and attachments that are described in the operating instructions or are provided or recommended by the tool manufacturer. The use of tools other than those described in the operating instructions or in the catalogue of recommended tool inserts or accessories can result in a risk of personal injury;

▶ use engine-driven power tools in well-ventilated areas. Store petrol in a safe place in approved storage cans. Stop and let engines cool before refuelling.

(d) Hazards and control measures for a range of hand-held power tools

Electric drills and sanders are both commonly used. The use of chainsaws is highly hazardous and they are also widely used. They have been included because the requirements are specific and detailed in the general machinery section. The hazards and safety control measures in addition to the general ones covered earlier are set out for each type of equipment.

Electric drills

Hazards are:

▶ entanglement, particularly of loose clothing or long hair in rotating drill bits;
▶ high noise levels from the drill or attachment;
▶ eye injury from flying particles and chips, particularly from chisels;
▶ injury from poorly secured work-pieces;
▶ electrocution/electric shock from poorly maintained equipment;
▶ electric shock from drilling into a live hidden cable;
▶ hand-arm vibration hazard in hammer mode;
▶ dust given off from material being worked on;
▶ tripping hazard from trailing cables;
▶ upper limb disorder from powerful machines with a strong torque, particularly if they jam and kick back;
▶ foot injury hazards from dropping heavy units onto unprotected feet;
▶ manual handling hazards. particularly with heavy machines and intensive use or using at awkward heights and/or reaches;
▶ fire and explosion hazard when used near flammable liquids, explosive dusts or gases;
▶ using equipment in poor weather conditions with wet, slippery surfaces, poor visibility and cold conditions.

Figure 10.9 Electric drill with percussion hammer action to drill holes in masonry

Specific control measures include the following:

▶ using double-insulated tools or earthed reduced voltage tools with a residual current device (see Chapter 12 for electrical safeguards in detail);
▶ using a pilot hole or punch to start holes whenever possible;
▶ selecting the correct drill bit for the material being drilled;
▶ securing small pieces to be drilled to prevent spinning;
▶ protecting against damage or injury on the far side if the bit is long enough to pass through the material or there are buried services in, say, plaster walls;
▶ taking care to prevent loose sleeves or long hair from being wound around the drill bit; for example wear short or close-fitting sleeves;
▶ wearing suitable eye protection.

Sanders

There is a large range of hand-held sanders on the market, from rotating discs (Figure 10.10), random orbital, rectangular orbital finishing (Figure 10.11(b)), belt sanders and heavy-duty floor sanders of rotating drum (Figure 10.11(a)) and, recently, orbital types. The high-speed rotating discs and drum types are the most hazardous but they all need using with care.

Hazards include:

▶ high noise levels from the sander in operation;
▶ injury from poorly secured work-pieces;
▶ electrocution/electric shock from poorly maintained electrical equipment;
▶ potential of entanglement with rotating disc and drum sanders;
▶ sanding attachments becoming loose in the chuck – these can fling off;
▶ injury from contact with abrasive surfaces, particularly with coarse abrasives and high-speed rotating sanding discs and drums;
▶ hand-arm vibration hazard, particularly from reciprocating equipment;
▶ health hazards from extensive dust given off from material being worked on;
▶ fire and health hazards from overheating of abraded surfaces, particularly if plastics are being sanded;
▶ tripping hazard from trailing cables;
▶ large powerful sanders suddenly gripping the surface and pulling the operative off their feet;
▶ foot injury hazards from dropping heavy units onto unprotected feet;
▶ manual handling hazards, particularly with heavy machines and intensive use or using at awkward heights and/or reaches;
▶ fire and explosion hazard when used near flammable liquids, explosive dusts or gases;
▶ using equipment in poor weather conditions with wet, slippery surfaces, poor visibility and cold conditions.

10

Figure 10.10 Disc sander with dual hand holds

(a)

(b)

Figure 10.11 (a) Rotary drum floor sander; (b) orbital finishing sander

Specific control measures include the following:

▶ The work-pieces must be securely clamped or held in position during sanding. In some cases a jig will be necessary. The direction of spin of disc sanders (normally anti-clockwise) is important to ensure that a small work-piece is pushed towards a stop or fence which is normally at the left side of the work-piece, particularly when clamping is impossible.

▶ Abrasive sanding belts, discs and sheets should be properly and firmly attached to the machine without any torn parts or debris underneath. The manufacturer's instructions and fixing accessories should be used to ensure correct attachment of the abrasive. Operators should be trained, competent and registered to fit abrasive discs.

▶ Old nails and fixings should be sunk below the surface or removed to prevent the sander snagging.

▶ Always hold the equipment by the proper handles and particularly on large disc and floor sanders always use both hands (Figure 10.11). Excessive pressure should not be used as the surface will be rutted and the machine may malfunction.

▶ Ensure that the dust extractor is working properly and has been emptied when about one-third full. Some extraction systems draw dust and air through the sanding sheet, which must have correctly pre-punched holes to allow the passage of air.

▶ Operators should wear suitable dust respirators, eye protection and, where necessary, hearing protection.

▶ Operators should wear suitable clothing avoiding loose garments, long hair and jewellery, which could catch in the equipment.

▶ Protective gloves and footwear are recommended.

10.3 Mechanical and non-mechanical hazards of machinery

10.3.1 Hazard identification

Most machinery has the potential to cause injury to people, and machinery accidents figure prominently in official accident statistics. These injuries may range in severity from a minor cut or bruise, through various degrees of wounding and disabling mutilation, to crushing, decapitation or other fatal injury. It is not solely powered machinery that is hazardous, for many manually operated machines (e.g. hand-operated guillotines and fly presses) can still cause injury if not properly safeguarded.

In the risk assessment of machinery (see ISO 12100:2010) it is essential to consider both permanent hazards and those which can appear unexpectedly. Hazardous situations and events during all phases of the machine life cycle need to be covered. These include: transport, assembly and installation; commissioning; use; maintenance; dismantling, disabling and scrapping.

The human interaction with the machine involving numerous tasks throughout the life cycle of the machine should be considered. They include:

Setting; testing; teaching/programming; process/tool changeover; start-up; all modes of operation; feeding the machine; removal of product from the machine; stopping the machine; stopping the machine in case of emergency; recovery of operation from jam or blockage; restart after unscheduled stop; fault-finding/trouble-shooting (operator intervention); cleaning and housekeeping; preventative maintenance; corrective maintenance.

The possible states of the machine should be taken into account which are:

1. the machine performs the intended function (the machine operates normally);
2. the machine does not perform the intended function (i.e. it malfunctions) due to a variety of reasons, including:

   ▶ variation of a property or of a dimension of the processed material or of the work-piece;
   ▶ failure of one or more of its component parts or services;
   ▶ external disturbances (for example, shocks, vibration, electromagnetic interference);
   ▶ design error or deficiency (for example, software errors);
   ▶ disturbance of its power supply; and
   ▶ surrounding conditions (for example, damaged floor surfaces).

Finally, the unintended behaviour of the operator or reasonably foreseeable misuse of the machine must be considered. Examples include:

▶ loss of control of the machine by the operator (especially for hand-held or mobile machines);
▶ reflex behaviour of a person in case of malfunction, incident or failure during the use of the machine;
▶ behaviour resulting from lack of concentration or carelessness;
▶ behaviour resulting from taking the 'line of least resistance' in carrying out a task;

▶ behaviour resulting from pressures to keep the machine running in all circumstances; and
▶ behaviour of certain persons (for example, children, disabled persons).

10.3.2 Mechanical machinery hazards

The hazards of machinery are set out in ISO 12100:2010, which covers the classification of machinery hazards and how harm may occur. The following mechanical machinery hazards follow this standard (Figure 10.12 shows a number of these hazards).

A person may be injured at machinery as a result of:

▶ **acceleration, deceleration** – potential for impact, being thrown, run over, slipping, tripping or falling for example from a piece of mobile plant or an overhead travelling crane
▶ **angular parts** – potential consequences are impact, crushing or shearing when a person hits a stationary part of the machine or a moving part impacts on an operator particularly in mobile plant;
▶ **approach of a moving element to a fixed part** – potential consequences are crushing or impact through being trapped between a moving part of a machine and a fixed structure, such as a wall or any material in a machine;
▶ **cutting parts** – potential consequences are cutting or severing through contact with a cutting edge, such as a band saw or rotating cutting disc;
▶ **elastic elements** – consequences are crushing or impact when an elastic part is deformed and suddenly straightens or returns to its original dimensions or explodes when an airbag used for lifting fails;
▶ **falling objects** – potential consequences are crushing or impact for example if a load fell from an overhead hoist or rail system;
▶ **gravity** – potential consequences are crushing or trapping for example if a vehicle lift failed while a person was underneath;
▶ **height from the ground** – potential consequences are crushing, impact, slipping, tripping or falling for example getting out of a truck or large piece of mobile plant or gaining access to a tower crane;
▶ **high pressure** – potential consequences injection, stabbing or puncture, impact for example, from a high pressure hydraulic system leak;
▶ **instability** – potential for being thrown, crushing or impact when a machine breaks its holding down bolts and falls over or a tower crane is not left free to slew to the wide direction and is blown over;
▶ **kinetic energy** – potential for impact, puncture or severing, for example when a kinetic recovery rope is used to tow a vehicle or ship and it snaps the whiplash can be very hazardous;

10

Figure 10.12 Range of mechanical hazards

▶ **machinery mobility** – potential consequences are being run over, impact, crushing when for example a large machine is being manoeuvred in a small space;

▶ **moving elements** – potential consequences crushing, friction, abrasion, impact, shearing, severing, drawing-in, for example on grinding wheels or sanding machines. Or when a moving part directly strikes a person, such as with the accidental movement of a robot's working arm when maintenance is taking place. Further example is when persons are drawn-in between in-running gear wheels or rollers or between belts and pulley drives;

▶ **rotating elements** – potential consequences severing or entanglement, for example an exposed rotating shaft which grips loose clothing, hair or working material, such as emery paper. The smaller the diameter of the revolving part the easier it is to get a wrap or entanglement. Further example is a shearing which occurs when part of the body, typically a hand or fingers, is trapped between rotating and fixed parts of the machine;

▶ **rough, slippery surface** – potential consequences are friction, abrasion, slipping, tripping and falling, impact when for example the operating area round a machine is very slippery from product spillages and the operator falls over and/or into the machinery hazard zone;

- **sharp edges** – potential consequences are severing, impact, shearing, stabbing or puncture;
- **stored energy** – potential consequences are crushing, impact, puncture, suffocation when the stored energy is released suddenly, for example the valve on a welding gas bottle (other hazards such as explosion or fire may also be present) is broken off or a high pressure air hose breaks;
- **vacuum** – potential consequences are crushing, drawing-in suffocation, for example the sudden failure of a large vacuum vessel.

10.3.3 Non-mechanical machinery hazards

Non-mechanical machinery hazards are also set out in ISO 12100:2010 and include:

Electrical hazards from:
- arc;
- electromagnetic phenomena;
- electrostatic phenomena;
- live parts;
- not enough distance to live parts under high voltage;
- overload;
- parts which have become live under fault conditions;
- short circuit;
- thermal radiation.

Thermal hazards from:
- explosion;
- flame;
- objects or materials with a high or low temperature;
- radiation from heat sources.

Noise hazards from:
- cavitation phenomena;
- exhausting system;
- gas leaking at high speed;
- manufacturing process;
- moving parts;
- scraping surfaces;
- unbalanced rotating parts;
- whistling pneumatics;
- worn parts.

Vibration hazards from:
- cavitation phenomena;
- misalignment of moving parts;
- mobile equipment;
- scraping surfaces;
- unbalanced rotating parts;
- vibrating equipment;
- worn parts.

Radiation hazards from:
- ionising radiation sources;
- low-frequency electromagnetic radiation;

- optical radiation (infrared, visible and ultraviolet), including laser;
- radio frequency electromagnetic radiation.

Material/substance hazard from:
- aerosol;
- biological and microbiological (viral and bacterial) agent;
- combustible;
- dust;
- explosive;
- fibre;
- flammable;
- fluid;
- fume;
- gas;
- mist;
- oxidiser.

Ergonomic hazard from:
- access;
- design or location of indicators and visual display units;
- design, location or identification of control devices;
- effort;
- flicker, dazzling, shadow, stroboscopic effect;
- local lighting;
- mental overload/underload;
- posture;
- repetitive activity;
- visibility.

Hazards associated with the environment in which the machine is used from:
- dust and fog;
- electromagnetic disturbance;
- lightning;
- moisture;
- pollution;
- snow;
- temperature;
- water;
- wind;
- lack of oxygen.

Combination of hazards from:
- for example, repetitive activity + effort + high environmental temperature (or repetitive hard work in a hot climate).

In many cases it will be practicable to install safeguards which protect the operator from both mechanical and non-mechanical hazards.

For example a guard may prevent access to hot or electrically live parts as well as to moving ones. The use of guards which reduce noise levels at the same time is also common.

As a matter of policy, machinery hazards should be dealt with in this integrated way instead of dealing with each hazard in isolation.

See Chapter 17 for a form which could be used to assess machinery risks.

10

10.3.4 Examples of machinery hazards

The following examples are given to demonstrate a small range of machines found in industry and commerce, which are included in the Certificate syllabus. Examples are shown in the application of safeguards section (10.4.8).

Office – photocopier

The hazards are:

▶ contact (cutting or abrasion) with moving parts when clearing a jam;
▶ electrical – when clearing a jam, maintaining the machine or through poorly maintained plug and wiring;
▶ heat through contact with hot parts when clearing a jam;
▶ health hazard from ozone or lack of ventilation in the area or very rarely an allergy to paper dust;
▶ inhalation of dust from emptying toner;
▶ getting dust from toners in the eyes.

Office – document shredder

The hazards are:

▶ drawing in between the rotating cutters when feeding paper into the shredder;
▶ contact (cutting or severing) with the rotating cutters when emptying the waste container or clearing a jam;
▶ electrical through faulty plug and wiring or during maintenance;
▶ possible noise from the cutting action of the machine;
▶ possible dust from the cutting action.

Manufacturing and maintenance – bench-top grinding machine

The hazards are:

▶ contact (friction or abrasion) with the rotating wheel causing abrasion;
▶ drawing in between the rotating wheel and a badly adjusted tool rest;
▶ bursting of the wheel, ejecting fragments which wound the operator;
▶ electrical through faulty wiring and/or earthing or during maintenance;
▶ fragments given off during the grinding process causing eye injury;
▶ hot fragments given off which could cause a fire or burns;
▶ noise produced during the grinding process;
▶ possible health hazard from dust/particles/fumes given off during grinding.

Manufacturing and maintenance – pedestal drill

The hazards are:

▶ entanglement around the rotating spindle and chuck;
▶ contact (cutting, puncturing, severing) with the cutting drill or work-piece;
▶ being struck (impact) by the work-piece if it rotates;
▶ being cut or punctured by fragments ejected from the rotating spindle and cutting device;
▶ drawing in to the rotating drive belt and pulley;
▶ contact or entanglement with the rotating motor;
▶ electrical from faulty wiring and/or earthing or during maintenance;
▶ possible health hazard from cutting fluids or dust given off during the process.

Agricultural/horticultural – cylinder mower

The hazards are:

▶ trapping, (cutting or severing) typically hands or fingers, in the shear caused by the rotating cutters;
▶ contact (impact, cutting) and entanglement with moving parts of the drive motor;
▶ drawing in between chain and sprocket drives;
▶ impact and cutting injuries from the machine starting accidentally;
▶ burns from hot parts of the engine;
▶ fire from the use of highly flammable petrol as a fuel;
▶ possible noise hazard from the drive motor;
▶ electrical if electrically powered, but this is unlikely;
▶ possible sensitisation health hazard from cutting grass, for example, hay fever;
▶ possible health hazard from exhaust fumes.

Agricultural/horticultural – brush cutter/strimmer

The hazards are:

▶ entanglement with rotating parts of motor and shaft;
▶ cutting or severing from contact with cutting head/line;
▶ electric shock if electrically powered, but this is unlikely;
▶ burns from hot parts of the engine;
▶ fire from the use of highly flammable petrol as a fuel;
▶ possible noise hazard from the drive motor and cutting action;
▶ eye and face puncture wounds from ejected particles;
▶ health hazard from hand-arm vibration causing white finger and other problems;
▶ back strain from carrying the machine while operating;
▶ health hazards from animal faeces.

Chainsaw

The hazards are:

- very serious cutting or severing by contact with the high-speed cutting chain;
- kickback (impact, cutting, severing) due to being caught on the wood being cut or contact with the top front corner of the chain in motion with the saw chain being kicked upwards towards the face in particular;
- pull in (drawing-in) when the chain is caught and the saw is pulled forward;
- push-back (impact)when the chain on the top of the saw bar is suddenly pinched and the saw is driven straight back towards the operator;
- burns from hot parts of the engine;
- high noise levels;
- hand-arm vibration causing white finger and other problems;
- fire from the use of highly flammable petrol as a fuel;
- eye and face puncture wounds from ejected particles;
- back strain while supporting the weight of the chainsaw while operating;
- electric shock if electrically powered;
- falls from height if using the chainsaw in trees and the like;
- lone working and the risk of serious injury;
- contact with overhead power lines if felling trees;
- being hit by falling branches or whole trees while felling;
- possible health hazards from cutting due to wood dust, particularly if wood has been seasoned;
- using chainsaw in poor weather conditions with slippery surfaces, poor visibility and cold conditions;
- health hazards from the engine fumes (carbon dioxide and carbon monoxide), particularly if used inside a shed or other building.

Retail – compactor

The hazards are:

- crushing hazard between the ram and the machine sides;
- trapping (severing, crushing) between the ram and machine frame (shear action);
- crushing when the waste unit is being changed if removed by truck;
- entanglement with moving parts of pump motor;
- electrical from faulty wiring and/or earthing or during maintenance;
- failure of hydraulic hoses with liquid released under pressure causing puncture to eyes or other parts of body;
- falling of vertical ram under gravity if the hydraulic system fails;
- handling hazards during loading and unloading.

Retail – checkout conveyor system

The hazards are:

- entanglement with belt fasteners if fitted;
- drawing in between belt and rollers if under tension;
- drawing in between drive belt and pulley;
- contact (impact, drawing-in, entanglement) with motor drive;
- electrical from faulty wiring and/or earthing or during maintenance.

Construction – cement/concrete mixer

The hazards are:

- contact (impact, drawing-in and entanglement) with moving parts of the drive motor;
- crushing between loading hopper (if fitted) and drum;
- drawing in between chain and sprocket drives;
- electrical, if electrically powered;
- burns from hot parts of engine;
- fire if highly flammable liquids used as fuel;
- possible noise hazards from the motor and dry mixing of aggregates;
- eye injury from splashing cement slurry;
- possible health hazard from cement dust and cement slurry while handling.

Construction – bench-mounted circular saw

The hazards are:

- contact (cutting, severing) with the cutting blade above and below the bench;
- ejection of the work-piece or timber as it closes after passing the cutting blade;
- drawing in between chain and sprocket or V-belt drives;
- contact (impact, drawing-in, entanglement) with moving parts of the drive motor;
- likely noise hazards from the cutting action and motor;
- health hazards from wood dust given off during cutting;
- electric shock from faulty wiring and/or earthing or during maintenance;
- climatic conditions such as wet/extreme heat or cold;
- possible fire and/or explosion from wood dust caused by overheated blades from excessive friction or an electrical fault.

Emerging technologies – including drones and driverless vehicles

Drones

Small unmanned aircraft are now widely available for commercial use. More popularly known as drones,

347

just like many other devices, they can cause injury or damage if they are not used responsibly and so are subject to specific safety rules relating to the way they are operated, which are underpinned by UK law.

The regulations are contained within the Air Navigation Order 2016 (ANO 2016) and there are some specific additional steps that must be taken if a drone is being flown for commercial operations.

Anyone using a small unmanned aircraft needs to be aware of the regulations shown here https://www.caa.co.uk/Commercial-industry/Aircraft/Unmanned-aircraft/Small-drones/Regulations-relating-to-the-commercial-use-of-small-drones/

Visual inspection of tanks and vessels forms a cornerstone of integrity assurance and is often the primary means of detection and sizing of flaws and corrosion damage. Recent advances in access technologies such as unmanned aerial vehicles (UAVs) or drones and remotely operated vehicles (ROVs) coupled with imaging technology have enabled increasing replacement of the human element of visual inspection. This can present significant benefits in avoidance of high-risk manned interventions such as in confined spaces, working at height or in hazardous environments.

The hazards of drones include:

▶ untrained, unauthorised pilots;
▶ flying out of sight of the pilot;
▶ hitting people or buildings through flying too close or stalling with battery failure or strong winds;
▶ collision with planes and other flying or moving objects;
▶ intrusion of privacy by over-flying private areas;
▶ dropping banners or loads if used to transport goods.

Driverless vehicles

The biggest safety advantage to an autonomous vehicle (AV) is that a robot is not a human – it is programmed to obey all the rules of the road, will not speed and cannot be distracted by a text message flickering onto a phone. And, hypothetically at least, AVs can also detect what humans can't – especially at night or in low-light conditions – and react more quickly to avoid a collision.

The real safety promise of autonomous vehicles is the fact that they can be summoned on-demand, routed more efficiently and easily shared – meaning not just the overall number of single-passenger cars on streets will decline, but the number of single-passenger trips will be reduced, meaning a reduction in overall miles travelled – unless of course people return them home empty or frequently request a pick-up or even let them cruise the streets rather than pay for parking.

Hazards include:

▶ trapping pedestrians in dual-use areas;
▶ travelling too close to buildings or pedestrians;
▶ not recognising the mistakes of human drivers;
▶ a mix of incompatible vehicles particularly where drivers wish to get out of the vehicle;

▶ passengers unaware of road hazards getting out suddenly when AV stops or arrives at destination;
▶ people being lost because they do not know where they are or how they got there;
▶ AVs running out of power and suddenly stopping in dangerous locations.

10.4 Control measures for reducing risks from machinery hazards

10.4.1 Introduction

As discussed in 10.1.1 and 10.1.3, PUWER requires that access to dangerous parts of machinery should be prevented in a preferred order or hierarchy of control methods. The standard required is a 'practicable' one, so that the only acceptable reason for non-compliance is that there is no technical solution. Cost is not a factor. (See Chapter 1 for more details on standards of compliance.)

As the mechanical hazard of machinery arises principally from someone coming into contact or entanglement with dangerous components, risk reduction is based on preventing this contact occurring.

This may be by means of:

▶ a physical barrier between the individual and the component (e.g. a fixed enclosing guard);
▶ a device which allows access only when the component is in a safe state (e.g. an interlocked guard which prevents the machine starting unless a guard is closed and acts to stop the machine if the guard is opened);
▶ a device which detects that the individual is entering a risk area and then stops the machine (e.g. certain photoelectric guards and pressure-sensitive mats).

The best method should, ideally, be chosen by the designer as early in the life of the machine as possible. It is often found that safeguards which are 'bolted on' instead of 'built in' are not only less effective in reducing risk, but are also more likely to inhibit the normal operation of the machine. In addition, they may in themselves create hazards and are likely to be difficult and hence expensive to maintain.

10.4.2 Fixed guards

Fixed guards have the advantage of being simple, always in position, difficult to remove and almost maintenance-free. Their disadvantage is that they do not always properly prevent access, they are often left off by maintenance staff and they can create difficulties for the operation of the machine.

A fixed guard has no moving parts and should, by its design, prevent access to the dangerous parts of the machinery. It must be of robust construction and sufficient to withstand the stresses of the process and environmental conditions. If visibility or free air flow (e.g. for cooling) is necessary, this must be allowed for in the design and construction of the guard. If the guard can be opened or removed, this must only be possible with the aid of a tool.

An alternative fixed guard is the distance fixed guard, which does not completely enclose a hazard, but which reduces access by virtue of its dimensions and its distance from the hazard. Where perimeter fence guards are used, the guard must follow the contours of the machinery as far as possible, thus minimising space between the guard and the machinery. With this type of guard, it is important that the safety devices and operating systems prevent the machinery being operated with the guards closed and someone inside the guard, that is in the danger area. Figure 10.13 shows a range of fixed guards for some of the examples shown in Figure 10.12.

10.4.3 Adjustable guards

(a) User-adjusted guard

These are fixed or movable guards, which are adjustable for a particular operation during which they remain fixed. They are particularly used with machine tools where some access to the dangerous part is required (e.g. drills, circular saws, milling machines Figure 10.14) and where the clearance required will vary (e.g. with the size of the cutter in use on a horizontal milling machine or with the size of the timber being sawn on a circular saw bench).

Adjustable guards may be the only option with cutting tools, which are otherwise very difficult to guard, but they have the disadvantage of requiring frequent readjustment. By the nature of the machines on which they are most frequently used, there will still be some access to the dangerous parts, so these machines must only be used by suitably trained operators. Jigs, push sticks and false tables must be used wherever possible to minimise hazards during the feeding of the work-piece. The working area should be well lit and kept free of anything which might cause the operator to slip or trip.

Shaft with projections

Chain and sprockets

Rack and pinion gears

Meshing gears

Counterrotating rolls

Pulley belts

Axial flow fans

Belt conveyor

Figure 10.13 Range of fixed guards

Figure 10.14 Adjustable guard for a rotating drill bit on a pedestal drill

(b) Self-adjusting guard

A self-adjusting guard is one which adjusts itself to accommodate, for example, the passage of material.

A good example is the spring-loaded guard fitted to many portable circular saws.

As with adjustable guards (see Figure 10.14) they only provide a partial solution in that they may well still allow access to the dangerous part of the machinery. They require careful maintenance to ensure they work to the best advantage (Figure 10.15).

Figure 10.15 Self-adjusting guard on a circular saw

10.4.4 Interlocking guard

The advantages of interlocked guards are that they allow safe access to operate and maintain the machine without dismantling the safety devices. Their disadvantage stems from the constant need to ensure that they are operating correctly and designed to be fail-safe. Maintenance and inspection procedures must be very strict.

This is a guard which is movable (or which has a movable part) whose movement is connected with the power or control system of the machine.

An interlocking guard must be connected to the machine controls such that:

▶ until the guard is closed the interlock prevents the machinery from operating by interrupting the power medium;

▶ either the guard remains locked until the risk of injury from the hazard has passed or opening the guard causes the hazard to be eliminated before access is possible.

A passenger lift or hoist is a good illustration of these principles: the lift will not move unless the doors are closed, and the doors remain closed and locked until the lift is stationary and in such a position that it is safe for the doors to open.

Special care is needed with systems which have stored energy. This might be the momentum of a heavy moving part, stored pressure in a hydraulic or pneumatic system, or even the simple fact of a part being able to move under gravity even though the power is disconnected. In these situations, dangerous movement may continue or be possible with the guard open, and these factors need to be considered in the overall design. Braking devices (to arrest movement when the guard is opened) or delay devices (to prevent the guard opening until the machinery is safe) may be needed. All interlocking systems must be designed to minimise the risk of failure-to-danger and should not be easy to defeat (Figure 10.16).

(a)　　　　(b)

Figure 10.16 Typical interlocking guards: (a) sliding and (b) hinged

10.4.5 Other protective devices

(a) Trip devices

A trip device does not physically keep people away but detects when a person approaches close to a danger point. It should be designed to stop the machine before injury occurs. A trip device depends on the ability of the machine to stop quickly and in some cases a brake may need to be fitted. Trip devices can be:

▶ mechanical in the form of a bar or barrier;
▶ electrical in the form of a trip switch on an actuator rod, wire or other mechanism;
▶ photoelectric or an other type of presence-sensing device;
▶ a pressure-sensitive mat.

They should be designed to be self-resetting so that the machine must be restarted using the normal procedure (Figure 10.17).

(b) Two-handed control devices

These are devices which require the operator to have both hands in a safe place (the location of the controls) before the machine can be operated. They are an option on machinery that is otherwise very difficult to guard but they have the drawback that they only protect the operator's hands. It is therefore essential that the design does not allow any other part of the operator's body to enter the danger zone during operation. More significantly, they give no protection to anyone other than the operator.

Where two-handed controls are used, the following principles must be followed:

▶ The controls should be so placed, separated and protected as to prevent spanning with one hand only, being operated with one hand and another part of the body, or being readily bridged.
▶ It should not be possible to set the dangerous parts in motion unless the controls are operated within approximately 0.5 seconds of each other. Having set the dangerous parts in motion, it should not be possible to do so again until both controls have been returned to their off position.
▶ Movement of the dangerous parts should be arrested immediately or, where appropriate, arrested and reversed if one or both controls are released while there is still danger from movement of the parts.

Figure 10.17 Schematic diagram of a telescopic trip device fitted to a radial drill

▶ The hand controls should be situated at such a distance from the danger point that, on releasing the controls, it is not possible for the operator to reach the danger point before the motion of the dangerous parts has been arrested or, where appropriate, arrested and reversed (Figure 10.18).

Figure 10.18 Two-handed control device

(c) Hold-to-run control

This is a control which allows movement of the machinery only as long as the control is held in a set position. The control must return automatically to the stop position when released. Where the machinery runs at crawl speed, this speed should be kept as low as practicable.

Hold-to-run controls give even less protection to the operator than two-handed controls and have the same main drawback in that they give no protection to anyone other than the operator.

However, along with limited movement devices (systems which permit only a limited amount of machine movement on each occasion that the control is operated and are often called 'inching devices'), they are extremely relevant to operations such as setting, where access may well be necessary and safeguarding by any other means is difficult to achieve.

(d) Jigs holders and push sticks

There are a number of devices to limit the approach of people to mechanical hazards while using machinery.

These include manual handling devices such as push sticks for circular saws and spindle moulding machines, or push blocks for wood planing machines. With certain machines jigs and other holding devices may be necessary to hold the work-piece near to the cutter of the machine without exposing the operator's hands to the dangerous part of the machine.

There is a logical progression from push sticks through jigs and holders; through more sophisticated clamping devices and moving work tables for circular saws, to power driven feeds and automatic feeding arrangements. Emergency stops must always be immediately available.

10.4.6 Information, instruction, training and supervision

(a) Information and instructions

PUWER Regulation 8 requires the provision of information and written instructions for work equipment. This should cover

▶ the conditions in which the work equipment can be used;
▶ the way in which the work equipment can be used;
▶ any foreseeable difficulties that could arise, and instructions on how to deal with them;
▶ using any conclusions drawn from experience using the work equipment, should be either recorded or steps taken to make sure that all appropriate members of the workforce are aware of them.

Written instructions include all the information provided by manufacturers or suppliers such as manuals, instruction sheets, instruction placards, warning labels and training manuals. In addition there may be in-house instructions and information from training sessions. Manufacturers have a duty to provide sufficient information, including drawings, to enable the correct installation, safe operation and maintenance of the equipment. These are often available to download from the manufacturer's website.

Employers must ensure that written instructions are available to everyone using the equipment. This includes temporary workers and others like maintenance staff who will need information like maintenance instructions and manuals.

Supervisors and managers should also have access to the information and written instructions. The amount of detailed health and safety information they will need to have immediately available for day-to-day operations will vary, but it is important that they know what information is available and where it can be found.

Information should be provided in writing or verbally where that is considered sufficient. Managers should make a decision based on the unusual nature or complexity of the equipment or task. They should consider:

► the skill levels of the workers involved;
► their experience and existing training;
► the level of supervision required;
► the complexity and length of the task.

Information and instructions whether verbal or written should be easy to understand and must be provided in an appropriate language for the user/purchaser. They should be logical, illustrated where appropriate and use standard symbols. Particular attention should be paid where people's first language differs from the local official language, or they have language problems or a disability which makes receiving or understanding information or instructions difficult.

(b) training and supervision

PUWER Regulation 9 requires that employers ensure that all persons, including supervisors and managers, who use or supervise work equipment have received adequate training for purposes of health and safety.

Employers should:

► consider what are the existing levels of competence and skills of all users of work equipment including temporary and agency workers;
► evaluate the level of management or supervision needed;
► carry out a gap analysis to determine the training required to enhance workers, level of competence taking into consideration for example whether they work alone or under close supervision;
► carry out the training particularly on recruitment but also when:
  ▷ risks change;
  ▷ tasks change;
  ▷ new technology or equipment is introduced;
  ▷ the system of work changes.

Refresher training needs to be provided as necessary as skills will decline particularly if not used regularly, such as after length absence. Supervisors and managers also require adequate training to carry out their function, particularly if they only supervise a particular task occasionally.

The training and supervision of young persons (under 18) is particularly important because of their relative immaturity, unfamiliarity with a working environment and lack of awareness of existing or potential risks. Some codes of practice, for example on the UK's 'Safe Use of Woodworking Machinery', restrict the use of high-risk machinery, so that only young persons with sufficient maturity and competence who have finished their training may use the equipment unsupervised. ISO 12100:2010 states that the unintended behaviour of the operator or foreseeable misuse of the machinery should be taken into consideration when assessing the risks.

10.4.7 Personal Protective Equipment

This is covered in detail in Chapter 9 and also for chainsaws in Section 10.4.8. The use of PPE for users of work equipment is the last issue to be considered in the hierarchy of protection (see 10.1) after inherently safe design, safeguards, instructions and training. However, there are many tasks and pieces of work equipment where PPE is necessary for general protection, such as the use of suitable gloves which fit properly and do not restrict the operation.

Some pieces of equipment such as chainsaws are so inherently dangerous that despite all the safeguards and training specialised PPE is absolutely essential. These include Kevlar-based footwear and overtrousers to prevent chain cuts on the feet and legs. Special face screen and helmets to prevent flying chips and saw kickback damaging the face, head and eyes. Harnesses are necessary when working at height in trees.

It is important that the risk assessments identify the equipment which should be used to protect against specific risks as well as general standards required for the use of work equipment. It is becoming increasingly common for all people in process plants to wear eye protection, gloves, high-visibility jackets or overalls and hard hats. These should not be seen as a substitute for well-designed and properly safeguarded equipment.

10.4.8 Application of safeguards to the range of work equipment

The application of the safeguards to the Certificate range of equipment is as follows.

Office – printer/photocopier

Application of safeguards (Figure 10.19):

► the machines are provided with an all-enclosing case which prevents access to the internal moving, hot or electrical parts;
► the access doors are interlocked so that the machine is automatically switched off when gaining access to clear jams or maintain the machine.
► It is good practice to switch off when opening the machine;
► internal electrics are insulated and protected to prevent contact;
► regular inspection and maintenance should be carried out;
► the machine should be on the PAT schedule;
► good ventilation in the machine room should be maintained;
► care must be taken while changing cartridges or toner – use gloves and place in a plastic bag immediately.

Figure 10.19 Typical multifunction printer/photocopier

Office – document shredder

Application of safeguards (Figure 10.20):

▶ enclosed fixed guards surround the cutters with restricted access for paper only, which prevents fingers reaching the dangerous parts;

▶ interlocks are fitted to the cutter head so that the machine is switched off when the waste bin is emptied;

▶ a trip device is used to start the machine automatically when paper is fed in;

▶ machine should be on PAT schedule and regularly checked;

▶ general ventilation will cover most dust problems except for very large machines where dust extraction may be necessary;

▶ noise levels should be checked and the equipment perhaps placed on a rubber mat if standing on a hard, reflective floor.

Manufacturing and maintenance – bench-top grinder

Application of safeguards (Figure 10.21):

▶ wheel should be enclosed as much as possible in a strong casing capable of containing a burst wheel;

▶ grinder should be bolted down to prevent movement;

▶ an adjustable tool rest should be adjusted as close as possible to the wheel;

▶ adjustable screen should be fitted over the wheel to protect the eyes of the operator. Goggles should also be worn;

▶ only properly trained competent and registered people should mount an abrasive wheel. See 10.1.5 for a suitable mounting procedure;

▶ the maximum speed should be marked on the machine so that the abrasive wheel can be matched to the machine speed to ensure that the wheel

Figure 10.20 Typical office shredder

Figure 10.21 Typical bench-mounted grinder

permitted speed exceeds or equals the machine max speed;

▶ noise levels should be checked and attenuating screens used if necessary;

▶ the machine should be on the PAT schedule and regularly checked;

▶ if necessary extract ventilation should be fitted to the wheel encasing to remove dust at source.

Manufacturing and maintenance – pedestal drill

Application of safeguards (Figure 10.22):

▶ motor and drive should be fitted with fixed guard;

▶ machine should be bolted down to prevent movement;

▶ the spindle should be guarded by an adjustable guard, which is fixed in position during the work;

▶ a clamp should be available on the pedestal base to secure work-pieces;

▶ the machine should be on the PAT schedule and regularly checked;

▶ cutting fluid, if used, should be contained and not allowed to get onto clothing or skin. A splash guard may be required but is unlikely;

▶ goggles should be worn by the operator.

Figure 10.22 Typical pedestal drill – note the guard is adjustable and adjusted to cover the rotating drill

Agricultural/horticultural – cylinder mower

Application of safeguards (Figure 10.23):

▶ machine should be designed to operate with the grass collection box in position to restrict access to the bottom blade trap. A warning sign should be fitted on the machine;

▶ on pedestrian-controlled machines the control handle should automatically stop the blade rotation when the operator's hands are removed. It should take two separate actions to restart;

▶ ride-on machines should be fitted with a device to automatically stop the blades when the operator leaves the operating position. This is normally a switch under the seat and it should be tested to ensure that it is functioning correctly and is not defective;

▶ drives and motor should be completely encased with a fixed guard;

▶ the machine should only be refuelled in the open air with a cool engine, using a proper highly flammable fuel container with pourer to restrict spillage. No smoking should be allowed;

▶ hot surfaces like the exhaust should be covered;

▶ engine must only be run in the open air to prevent a build up of fumes;

▶ noise levels should be checked and if necessary an improved silencer fitted to the engine and, where required, hearing protection used;

▶ hay-fever-like problems from grass cutting are difficult to control. A suitable dust mask may be required to protect the user.

Figure 10.23 Typical large cylinder mower

Agricultural/horticultural – brush cutter/strimmer

Application of safeguards (Figure 10.24):

▶ moving engine parts should be enclosed;

▶ rotating shafts should be encased in a fixed drive shaft cover;

▶ rotating cutting head should have a fixed top guard, which extends out on the user side of the machine;

▶ line changes must only be done either automatically or with the engine switched off;

355

- engine should only be run in the open air;
- refuelling should only be done in the open air using a proper highly flammable liquid container with pouring spout;
- boots with steel toe cap and good grip, stout trousers and non-snagging upper garments should be worn in addition to hard hat fitted with full face screen and safety glasses;
- if the noise levels are sufficiently high (normally they are with petrol-driven units) suitable hearing protection should be worn;
- low-vibration characteristics should be balanced with engine power and speed of work to achieve the minimum overall vibration exposure. Handles should be of an anti-vibration type. Engines should be mounted on flexible mountings. Work periods should be limited to allow recovery;
- washing arrangements and warm impervious gloves should be provided to guard against health risks;
- properly constructed harness should be worn which comfortably balances the weight of the machine.

Figure 10.24 Typical brush cutter

Agriculture/horticulture – chainsaw

Application of safeguards (Figure 10.25):

Competence assessment

Everyone working with chainsaws on or in trees should hold such a certificate or award unless:

- the work is being done as part of agricultural operations (e.g. hedging, clearing fallen branches, pruning trees to maintain clearance for machines);
- the work is being done by the occupier or their employees;
- they have used a chainsaw before 5 December 1998.

In any case, operators using chainsaws for any task in agriculture or any other industry must be competent under PUWER 98.

Training should be carried out by specialist instructors at organised training courses. Advice on suitable training courses (and competence assessment where appropriate) is available from the industry Sector Skills Council and

- Lantra Awards, Stoneleigh Park, Kenilworth, Warwickshire CV8 2LG, Tel: 02476 419703, Fax: 02476 411655.
- NPTC, see www.nptc.org.uk

Where training is being consolidated through workplace-based experience, the trainee should be supervised by a person competent in the use of a chainsaw for the work being done by the trainee and who holds the relevant competence certificate or award.

It is recommended that all chainsaw operators have regular refresher/update training to ensure they work to industry best practice and maintain their levels of competence. The suggested intervals for such training are:

- occasional users – every two to three years;
- full-time users – every five years.

Specific control measures include:

- chainsaws may only be operated by fully trained, fit and competent people;
- avoid working alone with a chainsaw. Where this is not possible, establish procedures to raise the alarm if something goes wrong;
- moving engine parts should be enclosed;
- electrical units should be double-insulated and cables fitted with residual current devices;
- the saw must be fitted with a top handle and effective brake mechanism;
- chainsaws expose operators to high levels of noise and hand-arm vibration which can lead to hearing loss and conditions such as vibration white finger. These risks can be controlled by good management practice including:
  - ▷ purchasing policies for low-noise/low-vibration chainsaws (e.g. with anti-vibration mounts and heated handles)
  - ▷ providing suitable hearing protection
  - ▷ proper maintenance schedules for chainsaws and protective equipment
  - ▷ giving information and training to operators on the health risks associated with chainsaws and use of PPE, etc.
- operators need to be trained in the correct chain-sharpening techniques and chain and guide bar maintenance to keep the saw in safe working condition;
- make sure petrol containers are in good condition and clearly labelled, with securely fitting caps. Use containers which are specially designed for chainsaw fuelling and lubrication;

Figure 10.25 Typical chainsaw with rear handle. The rear handle projects from the back of the saw. It is designed to always be gripped with both hands, with the right hand on the rear handle. It may be necessary to have a range of saws with different guide bar lengths available. As a general rule, choose a chainsaw with the shortest guide bar suitable for the work. 1 – hand guard with integral chain brake; 2 – exhaust outlet directed to the right-hand side away from the operator; 3 – chain breakage guard at bottom of rear handle; 4 – chain designed to have low-kickback tendency; 5 – rubber anti-vibration mountings; 6 – lockout for the throttle trigger; 7 – guide bar (should be protected when transporting chainsaw); 8 – bottom chain catcher; 9 – PPE hand/eye/ear defender signs; 10 – on/off switch

► do not allow operators to use discarded engine oil as a chain lubricant – it is a very poor lubricant and may cause cancer if it is in regular contact with an operator's skin;

► when starting the saw, operators should maintain a safe working distance from other people and ensure the saw chain is clear of obstructions;

► kickback is the sudden uncontrolled upward and backward movement of the chain and guide bar towards the operator. This can happen when the saw chain at the nose of the guide bar hits an object. Kickback is responsible for a significant proportion of chainsaw accidents, many of which are to the face and parts of the upper body where it is difficult to provide protection. A properly maintained chain brake and use of low-kickback chains (safety chains) reduce the effect, but cannot entirely prevent it;

► training in good manual handling techniques and using handling aids/tools should reduce the risk of back injuries;

► suitable PPE should always be worn, no matter how small the job. European standards for chainsaw PPE are published as part of EN 381 Protective Clothing for Users of Hand-Held Chainsaws (Figure 10.26). Protective clothing complying with this standard should provide a consistent level of resistance to chainsaw cut-through. Other clothing worn with the PPE should be close-fitting and non-snagging. **NB. No protective equipment can**

ensure 100% protection against cutting by a hand-held chainsaw;

► if conditions are dusty, suitable filtering face masks should be worn.

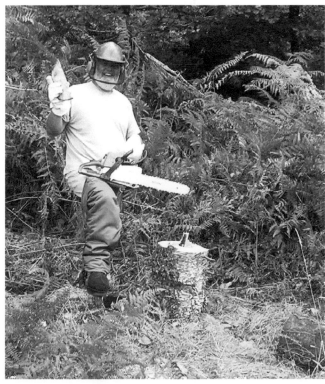

Figure 10.26 Kevlar gloves, overtrousers and overshoes providing protection against chainsaw cuts. Helmet, ear and face shield protect the head. Apprentice under training – first felling

357

Retail – compactor

Application of safeguards (Figure 10.27):

▶ access doors to the loading area, which gives access to the ram, should be positively interlocked with electrical and/or hydraulic mechanisms;
▶ the ram pressure should be dumped if it is hydraulic;
▶ drives of motors should be properly guarded;
▶ if the waste unit is removed by truck, the ram mechanism should be interlocked with the unit so that it cannot operate when the unit is changed for an empty one;
▶ the machine should be regularly inspected and tested by a competent person;
▶ if the hydraulic ram can fall under gravity, mechanical restraints should automatically move into position when the doors are opened;
▶ emergency stop buttons should be fitted on each side.

Figure 10.27 Typical retail compactor

Retail – checkout conveyor

Application of safeguards (Figures 10.28):

▶ all traps between belt and rollers are provided with either fixed guards or interlocked guards;
▶ motor and drive unit should be provided with a fixed guard and access to the underside of the conveyor is prevented by enclosure;

Figure 10.28 Typical retail checkout conveyor

▶ adequate emergency stop buttons must be provided to the checkout operator;
▶ machine should be on the PAT electrical inspection schedule and regularly checked;
▶ auto stop system should be fitted so that conveyor does not push products into the operator's working zone.

Construction – cement mixer

Application of safeguards (Figures 10.29(a) and (b)):

▶ operating position for the hopper hoist (Figure 10.29(b)) should be designed so that anyone in the trapping area is visible to the operator. The use of the machine should be restricted to designated operators only. As far as possible the trapping point should be designed out. The hoist operating location should be fenced off just allowing access for barrows, etc., to the unloading area;
▶ drives and rotating parts of engine should be enclosed;
▶ the drum gearing should be enclosed and persons kept away from the rotating drum, which is normally fairly high on large machines;

(a)

(b)

Figure 10.29 (a) Small concrete/cement mixer; (b) diesel concrete mixer with hopper

- no one should be allowed to stand on the machine while it is in motion;
- goggles should be worn to prevent cement splashes;
- if petrol-driven, care is required with flammable liquids and refuelling;
- engines must only be run in the open air;
- electric machines should be regularly checked and be on the PAT schedule;
- noise levels should be checked and noise attenuation, for example silencers and damping, fitted if necessary.

Construction – bench-mounted circular saw

Application of safeguards (Figure 10.30):

- a fixed guard should be fitted to the blade below the bench;
- fixed guards should be fitted to the motor and drives;
- an adjustable top guard should be fitted to the blade above the bench which encloses as much of the blade as possible. An adjustable front section should also be fitted;
- a riving knife should be fitted behind the blade to keep the cut timber apart and prevent ejection;
- a push stick should be used on short work-pieces (under 300 mm) or for the last 300 mm of longer cuts;
- blades should be kept properly sharpened and set with the diameter of the smallest blade marked on the machine;
- noise attenuation should be applied to the machine, for example damping and special quiet saw blades,

and, if necessary, fit in an enclosure. Hearing protection may have to be used;

- protection against wet weather should be provided;
- the electrical parts should be regularly checked in addition to all the mechanical guards;
- extraction ventilation will be required for the wood dust and shavings;
- suitable dust masks should be worn;
- suitable cool or warm clothing will be needed when used in hot or cold locations;
- space around machine should be kept clear.

Emerging technologies

Drones

Drones may only be flown by trained and authorised pilots who must pass the drone test and register with the CAA (Civil Aviation Authority) (at www.Register-drones.caa.co.uk) before flying.

Here are the rules from the UK Civil Aviation Authority 'The Drone Code' published in October 2019:

- Always keep your drone or model aircraft in direct sight.
- Never fly more than 400 ft (120 m) above the surface and stay well away from aircraft, airports and airfields.
- Never fly closer than 50 m to people. Even when your drone is more than 50 m away from people it is safer to avoid directly overflying them.
- Never fly closer than 50 m to buildings, cars, trains or boats.
- Never fly closer than 150 m to a crowd of 1,000 people or more. Never fly directly over a crowd.

10

Figure 10.30 Typical bench-mounted circular saw

Exhaust outlet
Extension table
Riving knife
Saw guard
Exhaust outlet
Cross-cut fence
Table
Rip fence

▶ Never fly closer than 150 m to built-up areas. Never fly directly over a built-up area.

▶ Never fly in an airport's flight restriction zone.

▶ It is illegal to fly a drone or model aircraft between 250 g and 20 kg that does not show a valid operator ID.

For further information see: www.dronesafe.uk/restrictions/

Figure 10.31 Typical drone

Driverless vehicles

Such vehicles, as of 2020, are best used on dedicated routes for mass transit of passengers. There are numerous examples of driverless railways and/or trams that operate very safely and a growing number of buses. They can also be used in industrial situations where electronic tracks can be laid in the ground/floor and vehicles fitted with obstacle detectors and rapid stopping devices.

However, when vehicles can move on standard roads in any direction at speed, the situation is very different and far from being perfected.

On routes used by both pedestrians and automatic (driverless) vehicles, it is important that vehicles are not able to trap pedestrians. Adequate clearance between the vehicles and pedestrians must be provided, and fixtures along the route must not create trapping hazards. The vehicles should be fitted with safeguards to keep the risk of injury low if they hit someone.

There should be audible warnings internally of EV starting, stopping and externally moving at slow speeds. Designated stopping places need to be provided with a ban on stopping in other places along the road. The issue of empty vehicles travelling randomly around or making frequent trips home to avoid parking fees will need to be addressed.

There's one safety debate that continues to divide the self-driving industry: some car manufacturers are still pushing for versions of vehicles which allow control to pass from human to computer, offering drivers the ability to toggle between semi-autonomous and fully autonomous modes. This dramatically changes the safety dynamics and must work flawlessly before being accepted for mass use.

At least for the near future, even full AVs will still have to contend with the mistakes of human drivers. To truly make self-driving technology the safest it can be, *all* the vehicles on the road should be fully autonomous – not just programmed to obey the rules of the road, but also to communicate with each other.

To plot the safest path forward for self-driving vehicles – and for urban areas to reap the many other potential environmental and social benefits of the technology – AVs should provide shared rides in regulated fleets, integrate with existing vehicles, and operate in a way that prioritises the most vulnerable humans above all users of the streets.

Figure 10.32 Driverless electric vehicle (AV)

10.4.9 Basic requirements for guards and safety devices

The design and construction of guards must be appropriate to the risks identified and the mode of operation of the machinery in question. ISO 12100:2010 gives the following general requirements for guards and protective devices which should protect against danger, including risks from moving parts. They should:

(a) be of robust construction;

(b) be securely held in place;

(c) not give rise to any additional hazard;

(d) not be easy to bypass or render non-operational, or be easily defeated;

(e) be located at an adequate distance from the danger zone;

(f) cause minimum obstruction of the view of the production process; and

(g) enable essential work to be carried out on the installation and replacement of tools and for maintenance purposes by restricting access exclusively to the area where the work has to be done, if possible without the guard having to be removed or the protective device having to be disabled.

In addition, guards should protect against the ejection or falling of materials or objects and against emissions generated by the machinery.

These general requirements involve consideration of the following factors:

▶ strength – guards and protection devices must be of good construction, sound material and adequate strength. They must be capable of doing the job they are intended to do. Several factors can be considered:

  ▷ material of construction (metal, plastic, laminated glass, etc.);

  ▷ form of the material (sheet, open mesh, bars, etc.);

  ▷ method of fixing.

They must be adequate for the purpose, able to resist the forces and vibration involved and able to withstand impact (where applicable). Good construction involves design and layout as well as the mechanical nature and quality of the construction. Foreseeable use and misuse should be taken into account. Guards and protection devices must be designed and installed so that they cannot be easily bypassed or disabled. This refers to accidental or deliberate action that removes the protection offered;

▶ weight and size – in relation to the need to remove and replace the guard during maintenance;

▶ compatibility with materials being processed and lubricants, etc.;

▶ hygiene and the need to comply with food safety regulations;

▶ visibility – may be necessary to see through the guard for both operational and safety reasons. Guards and protection devices must not unduly restrict the view of the operating cycle of the machinery, where such a view is necessary. It is not usually necessary to be able to see all the machine – the part that needs to be seen is normally that which is acting directly on material or a work-piece. Operations for which it is necessary to provide a view include those where the operator controls and feeds a loose work-piece at a machine. Examples include manually fed woodworking machines and food slicers. Many of these operations involve the use of protection appliances. Where an operation protected by guards needs to be seen, the guard should be provided with viewing slits or properly constructed panels, perhaps backed up by internal lighting, enabling the operator to see the operation;

▶ noise attenuation – guards are often utilised to reduce the noise levels produced by a machine. Conversely, the resonance of large undamped panels may exacerbate the noise problem;

▶ enabling a free flow of air – where necessary (e.g. for ventilation);

▶ avoidance of additional hazards – for example free of sharp edges. Guards must be constructed so that they are not themselves dangerous parts. If a guard is power operated or assisted, the closing or opening action might create a potentially dangerous trap which needs secondary protection, for example a leading-edge trip bar or pressure-sensitive strip;

▶ ease of maintenance and cleanliness – it should be necessary to use a tool such as a spanner or screwdriver to remove a fixed guard. Guards should be interlocked with the main drive if they need to be removed frequently for maintenance or cleaning. Movable panels in guards giving access to dangerous parts or movable guards themselves will often need to be fitted with an interlocking device. This device must be designed and installed so that it is difficult or impossible to bypass or defeat. Guards and protection devices must be maintained in an efficient state, in efficient working order and in good repair. This is an important requirement as many accidents have occurred when guards have not been maintained. It is a particular example of the general requirement under Regulation 5 of PUWER to maintain equipment. Compliance can be achieved by the use of an effective check procedure for guards and protection devices, together with any necessary follow-up action. In the case of protection devices or interlocks, some form of functional check or test is desirable;

▶ openings – the size of openings and their distance from the dangerous parts should not allow anyone to be able to reach into a danger zone. These values can be determined by experiment or by reference to standard tables. If doing so by experiment, it is essential that the machine is first stopped and made safe (e.g. by isolation). The detailed information on openings is contained in EN 294:1992, EN 349:1993 and EN 811:1997. Safeguarding is normally attached to the machine, but the regulation does not preclude the use of free-standing guards or protection devices. In such cases, the guards or protection devices must be fixed in an appropriate position relative to the machine.

10.5 Further information

Safe Use of Work Equipment Provision and Use of Work Equipment Regulations 1998 Approved Code of Practice and guidance L22 (fourth edition) HSE Books 2014 ISBN 978 0 7176 6619 5. https://www.hse.gov.uk/pubns/books/l22.htm

Using work equipment safely 2012 INDG229(rev2). https://www.hse.gov.uk/pubns/indg229.htm

Personal Protective Equipment at Work Regulations 1992 (as amended). Guidance on Regulations. 2005 L25 second edition, HSE Books ISBN 978 0 7176 6139 8. www.hse.gov.uk/pubns/books/l25.htm

Safety of machinery – General principles for design – Risk assessment and risk reduction, ISO 12100:2010 ISBN 9780 580 74262 0

The Supply of Machinery (Safety) Regulations 2008 Scope and application, and relationship to CE marking. www.hse.gov.uk/pubns/indg270.pdf

Also see HSE micro site on New Machinery: www. hse.gov.uk/work-equipment-machinery/new-machinery.htm

10.6 Practice questions

1. A manufacturing company has decided to purchase a new machine.
   (a) **Identify** the questions that must be answered before a purchase is made.
   (b) **Outline** the issues that will need to be considered before the machine is put into use.
   (c) **Identify** the main issues when assessing the suitability of controls (including emergency controls) of the new machine.

2. (a) **Describe** the contents of a 'Declaration of Conformity' that should be issued with a new machine.
   (b) **Outline** the advantages and limitations of CE marking on a new machine.

3. **Outline** a hierarchy of control measures that may be used to prevent contact with dangerous parts of machinery. Give an example of the level of protection required at each stage of the hierarchy.

4. An operator suffers a serious injury after coming into contact with a dangerous part of a machine, **describe**:
   (a) Possible immediate causes.
   (b) Possible root (underlying) causes.

5. Employers are required to provide adequate information, instruction and training to ensure the safe use of work equipment.
   (a) **Identify THREE** categories of employees that should receive information, instruction and training on the safe use of work equipment.
   (b) **Outline** the issues that could be included in such information, instruction and training.

6. (a) **Describe THREE** preventative maintenance schemes that may be used on machinery to prevent health and safety risks.
   (b) **Outline** the measures to be taken to reduce the risk of accidents associated with the routine maintenance of machinery.
   (c) **Identify FOUR** non-mechanical hazards that could lead to injury and/or ill-health when undertaking maintenance work on machinery in the workplace.

7. A bricklayer uses a hammer and chisel to split a building block.
   (a) **Identify FOUR** unsafe conditions, associated with the tools, which could affect the safety of the bricklayer.
   (b) **Outline** suitable control measures for minimising the risk of injury to the bricklayer when using the tools.

8. A carpenter regularly uses a hand-held electric sander for the preparation of wooden doors before painting them.
   (i) **Outline** the checks that should be made to ensure the electrical safety of the sander.
   (ii) Other than electricity, **identify EIGHT** hazards associated with the use of the sander.

9. (a) **Identify FOUR** non-mechanical hazards that may be encountered on woodworking machines and **outline** the possible health and safety effects from exposure in **EACH** case.
   (b) **Identify EIGHT** mechanical hazards associated with moving parts of machinery.

10. **Identify THREE** mechanical, **TWO** non-mechanical hazards and the associated control measures with the use of the following items of equipment:
    (a) a bench-top grinder
    (b) a document shredder
    (c) a photocopier
    (d) a bench-mounted circular saw
    (e) a strimmer.

11. A new pedestal (pillar) drill is to be used in a maintenance workshop.
    (a) **Identify** the issues that should be addressed before it is first used so that the risk of injury to operators of the machine is reduced.
    (b) **Identify FOUR** mechanical hazards presented by the machine and **outline** in each case how injury may occur and the appropriate control measures.

12. A bench-mounted circular saw is used in a carpentry workshop on a regular basis.
    (a) **Identify FOUR** risks to the health **AND FIVE** risks to the safety of the circular saw operators.
    (b) **Outline** the control measures that can be implemented to minimise the health and safety risks to these operators.

13. A petrol-driven chainsaw is to be used to remove several branches from a large tree.
    (a) **Identify EIGHT** hazards associated with the use of the chainsaw for this work.
    (b) **Outline EIGHT** safeguards that should be in place to ensure the safe use of chainsaws.
    (c) **Identify** the items of personal protective equipment that should be used by the chainsaw operative.

14. A local authority uses a rider-operated petrol-powered motor mower to cut the grass on roadside verges.
    (a) **Identify FIVE** mechanical and **FOUR** non-mechanical hazards that the operator could be exposed to while using the mower.
    (b) **Outline** the precautions that should be taken to address the hazards identified in part (a).

15. **Outline** the principles of operation, advantages and limitations of the following guards and safeguarding devices:
    (a) fixed guards
    (b) interlocked guards
    (c) adjustable guards
    (d) trip devices
    (e) two-handed control devices.

16. Other than contact with dangerous parts, **identify FOUR** types of hazard against which fixed guards on machines may provide protection.

10

CHAPTER 11

Fire

Chapter summary

This chapter covers the following topics in Element 10 of the NEBOSH NGC syllabus:
Fire principles
Preventing fire and fire spread
Fire alarms and firefighting arrangements
Fire evacuation

Learning outcomes

These are the learning outcomes covered by this chapter:
The learner will be able to:
1. Describe the principles of fire initiation, classification and spread
2. Describe the basic principles of fire prevention and the prevention of fire spread in buildings
3. Identify the appropriate fire alarm systems and firefighting equipment for a simple workplace
4. Outline the factors which should be considered when implementing a successful evacuation of a workplace in the event of a fire
5. Outline the legal requirements for fire prevention – not part of the NEBOSH NGC syllabus
6. Outline the principles of fire risk assessment – not part of the NEBOSH NGC syllabus

11.1 Fire principles

11.1.1 Introduction

This chapter covers fire prevention in the workplace, and how to ensure that people are properly protected if fire does occur. Each year, UK fire and rescue services attend over 35,000 fires at work in which about 30 people are killed and over 2,500 are injured. Fire and explosions at work account for about 2% of the major injuries reported under Reporting of Injuries, Diseases and Dangerous Occurrences Regulations (RIDDOR 2013).

The financial costs associated with serious fires are very high including, in many cases (believed to be over 40%), the failure to start up the business again. Never underestimate the potential of any fire. What may appear to be a small fire in a waste bin, if not dealt with, can quickly spread through a building or a structure. The Bradford City Football ground in 1985, or King's Cross Underground station in 1987 are examples of where small fires quickly became raging infernos, resulting in many deaths and serious injuries. The Buncefield fuel storage depot fire in December 2005 and the Deepwater Horizon oil drilling platform fire in 2010 are further examples of the destruction that a large fire and explosion can cause.

The Grenfell Tower fire in June 2017, where 72 people lost their lives in a 22-storey high-rise block of flats in Kensington, London, was the largest loss of life in a residential building since the Second World War. The Phase 1 enquiry has reported on the events of the fire itself in October 2019 and is currently (February 2020) continuing the enquiry into the prior refurbishment of the building and the materials used. The effects of the fire will have a lasting impact on materials used for cladding high-rise buildings and the response of the fire authorities.

Figure 11.1 The Grenfell Tower fire of June 2017 when 72 people lost their lives

11.1.2 Basic principles of fire

Fire triangle

Fire cannot take place unless three things are present. These are shown in Figure 11.2.

The absence of any one of these elements will prevent a fire from starting. Prevention depends on avoiding these three coming together. Fire extinguishing depends on removing one of the elements from an existing fire, and is particularly difficult if an oxidising substance is present.

Once a fire starts, it can spread very quickly from fuel to fuel as the heat increases.

Fuel
Flammable gases, liquids, solids

Ignition source
Hot surfaces
Electrical equipment
Static electricity
Smoking materials
Naked flame

Oxygen
From the air
Oxidizing substances

Figure 11.2 Fire triangle

Sources of ignition

Workplaces have numerous sources of ignition, some of which are obvious but others may be hidden inside machinery. Most of the sources may cause an accidental fire from sources inside but, in the case of arson (about 13% of industrial fires), the source of ignition may be brought from outside the workplace and will be deliberately used. The following are potential sources of ignition in the typical workplace:

Naked flames – from smoking materials, cooking appliances, heating appliances and process equipment.

External sparks – from grinding metals, welding, impact tools, electrical switch gear.

Internal sparking – from electrical equipment (faulty and normal), machinery, lighting.

Hot surfaces – from lighting, cooking, heating appliances, process equipment, poorly ventilated equipment, faulty and/or badly lubricated equipment, hot bearings and drive belts.

Static electricity – causing significant high-voltage sparks from the separation of materials such as unwinding plastic, pouring highly flammable liquids, walking across insulated floors or removing synthetic overalls.

Sources of fuel

If it will burn it can be fuel for a fire. The things which will burn easily are the most likely to be the initial fuel, which then burns quickly and spreads the fire to other fuels. The most common things that will burn in a typical workplace are:

Solids – these include: wood, paper, cardboard, wrapping materials, plastics, rubber, foam (e.g. polystyrene tiles and furniture upholstery), textiles (e.g. furnishings and clothing), wall paper, hardboard and chipboard used as building materials, waste materials (e.g. wood shavings, dust, paper), hair (see Figures 11.3(a) and (b)).

Liquids – these include: paint, varnish, thinners, adhesives, petrol, white spirit, methylated

Danger
Flammable solid

Figure 11.3 (a) Transport flammable solid sign; (b) GHS – packaging sign

spirits, paraffin, toluene, acetone and other chemicals.

Most flammable liquids give off vapours which are heavier than air so they will fall to the lowest levels. A flash flame or an explosion can occur if the vapour catches fire in the correct concentrations of vapour and air (see Figures 11.4(a) and (b)).

Danger
Highly Flammable Liquid

Figure 11.4 (a) Transport flammable liquid sign; (b) GHS – packaging sign

Gases – flammable gases include: LPG (liquefied petroleum gas in cylinders, usually butane or propane), acetylene (used for welding) and hydrogen. An explosion can occur if the air/gas mixture is within the explosive range (see Figures 11.5(a) and (b)).

Danger
Extremely Flammable Gas

Figure 11.5 (a) Transport flammable gas sign; (b) GHS – packaging sign

Oxygen

Oxygen is of course provided by the air all around but this can be enhanced by wind, or by natural or powered ventilation systems which will provide additional oxygen to continue burning.

Cylinders providing oxygen for medical purposes or welding can also provide an additional very rich source of oxygen.

In addition, some chemicals such as nitrates, chlorates, chromates and peroxides can release oxygen as they burn and therefore need no external source of air (see Figures 11.6(a) and (b)).

11.1.3 **Classification of fire**

Fires are classified in accordance with British Standard EN 2:1992 Classification of Fires. For all practical purposes there are FIVE main classes of fire – A, B, C, D and F, plus fires involving electrical equipment. BS 7937:2000 The Specification of Portable Fire Extinguishers for Use on Cooking Oil Fires introduced the new class F. The categories based on fuel and the means of extinguishing are as follows:

Class A – Fires which involve solid materials such as wood, paper, cardboard, textiles, furniture and

Danger
May cause or intensify fire
Oxidizer

Figure 11.6 (a) Transport oxidising agent sign; (b) GHS – packaging sign

plastics where there are normally glowing embers during combustion. Such fires are extinguished by cooling, which is achieved using water.

Class B – Fires which involve liquids or liquefied solids such as paints, oils or fats. These can be further subdivided into:

    Class B1 – Fires which involve liquids that are soluble in water such as methanol. They can be extinguished by carbon dioxide, dry powder, water spray, light water and vaporising liquids;

    Class B2 – Fires which involve liquids not soluble in water, such as petrol and oil. They can be extinguished by using foam, carbon dioxide, dry powder, light water and vaporising liquid.

Class C – Fires which involve gases such as natural gas, or liquefied gases such as butane or propane. They can be extinguished using foam or dry powder in conjunction with water to cool any containers involved or nearby.

Class D – Fires which involve metals such as aluminium or magnesium. Special dry powder extinguishers are required to extinguish these fires, which may contain powdered graphite or talc.

Class F – Fires which involve high-temperature cooking oils or fats in large catering establishments or restaurants.

Electrical fires – Fires involving electrical equipment or circuitry do not constitute a fire class on their own, as electricity is a source of ignition that will feed a fire until switched off or isolated. But there are some pieces of equipment that can store, within capacitors, lethal voltages even when isolated. Extinguishers specifically designed for electrical use like carbon dioxide or dry powder units should always be used for this type of fire hazard.

Fire extinguishers are usually designed to tackle one or more classes of fire. This is discussed in 11.3.

11.1.4 Principles of heat transmission and fire spread

Fire transmits heat in several ways, which need to be understood in order to prevent, plan escape from, and fight fires. Heat can be transmitted by convection, conduction, radiation and direct burning (Figure 11.7).

Convection

Hot air becomes less dense and rises, drawing in cold new air to fuel the fire with more oxygen. The heat is transmitted upwards at sufficient intensity to ignite combustible materials in the path of the very hot products of combustion and flames. This is particularly important inside buildings or other structures where the shape may effectively form a chimney for the fire.

Conduction

This is the transmission of heat through a material with sufficient intensity to melt or destroy the material and ignite combustible materials which come into contact or close to a hot section. Metals like copper, steel and aluminium are very effective or good conductors of heat. Other materials like concrete, brickwork and insulation materials are very ineffective or poor conductors of heat.

Poor conductors or good insulators are used in fire protection arrangements. When a poor conductor is also incombustible, it is ideal for fire protection. Care is necessary to ensure that there are no other issues, such as health risks, with these materials. Asbestos is a very poor conductor of heat and is incombustible. However, it can cause very severe health effects, which now outweighs its value as a fire protection material and it is banned in the United Kingdom. Although still found in many buildings where it was used extensively for fire protection, it now has to be managed under the Control of Asbestos Regulations 2006.

Radiation

Often in a fire, the direct transmission of heat through the emission of heat waves from a surface can be so intense that adjacent materials are heated sufficiently to ignite. A metal surface glowing red-hot would be typical of a severe radiation hazard in a fire.

Direct burning

This is the effect of combustible materials catching fire through direct contact with flames which causes fire to spread, in the same way that lighting an open fire, with a range of readily combustible fuels, results in its spread within a grate.

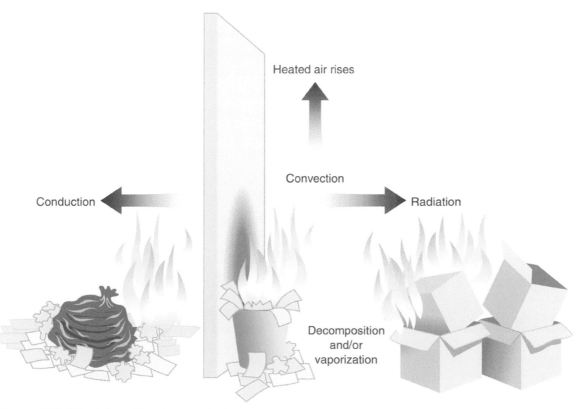

Heated air rises

Convection

Conduction

Radiation

Decomposition and/or vaporization

Figure 11.7 Principles of heat transmission

Fire and smoke spread in buildings

Where fire is not contained and people can move away to a safe location, there is little immediate risk to those people. However, where fire is confined inside buildings the fire behaves differently (Figure 11.8).

Figure 11.8 Smoke spread in buildings

The smoke rising from the fire gets trapped inside the space by the immediate ceiling, then spreads horizontally across the space deepening all the time until the entire space is filled. The smoke will also pass through any holes or gaps in the walls, ceiling or floor and get into other parts of the building. It moves rapidly up staircases or lift wells and into any areas that are left open, or rooms which have open doors connecting to the staircase corridors. The heat from the building gets trapped inside, raising the temperature very rapidly. The toxic smoke and gases are an added danger to people inside the building, who must be able to escape quickly to a safe location.

11.1.5 Common causes of fire and consequences

Causes

Government fire statistics for the causes of fires in buildings, excluding dwellings, in recent years are shown in Figure 11.9 The total shows a general trend downwards in the last ten years. The majority of fires occurred in:

- ► private garages and sheds (22%) – 6,700 fires;
- ► retail distribution (14%) – 4,200 fires;
- ► restaurants, cafes, public houses, etc. (9%) – 2,600 fires;
- ► industrial premises (other than construction) (8%) – 2,400 fires;
- ► recreational and other cultural services (6%) – 1,700 fires.

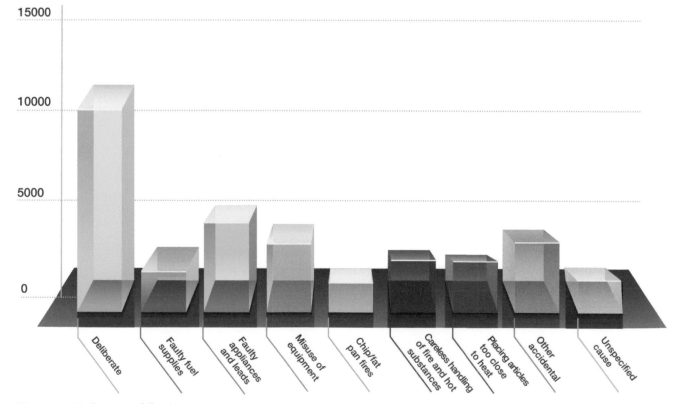

Figure 11.9 Causes of fire in recent years

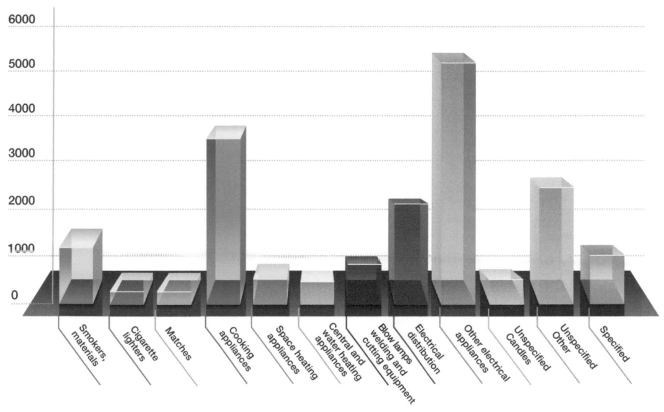

Figure 11.10 Accidental fires – sources of ignition in recent years

Further information on fire statistics are available at: https://assets.publishing.service.gov.uk/government/uploads/system/uploads/attachment_data/file/831136/detailed-analysis-fires-attended-fire-rescue-england-1819-hosb1919.pdf

About 35 people die in buildings other than dwellings each year. Also in fires in other buildings, about 1,300 injuries are sustained each year. These figures represent around 8% of all fire deaths and 10% of non-fatal casualties.

The sources of ignition are shown in Figure 11.10. Out of approximately 19,000 accidental fires in each year it shows that cooking appliances and electrical equipment account for over 60% of the total.

Consequences

The main consequences of fire are:

▶ Death – although this is a very real risk, relatively few people die in building fires that are not in dwellings. Each year about 35 (8%) people die out of a total of about 440 in all fires. The main causes of all deaths are:
  ▷ overcome by gas or smoke – 43%
  ▷ burns – 26%
  ▷ burns and overcome by gas or smoke – 20%
  ▷ other – 2%.
  ▷ Unspecified – 9%.

Clearly gas and smoke are the main risks;

▶ Personal injury – some 1,282 people were injured (10%) of total injuries in all fires;
▶ Building damage – can be very significant, particularly if the building materials have poor resistance to fire and there is little or no built-in fire protection;

▶ Flora and fauna damage – can be significant, particularly in a hot draught or forest fires, which have been massive in 2019 and 2020;
▶ Loss of business and jobs – it is estimated that about 40% of businesses do not start up again after a significant fire. Many are under- or not insured and small companies often cannot afford the time and expense of setting up again when they probably still have old debts to service;
▶ Transport disruption – rail routes, roads and even airports are sometimes closed because of a serious fire. The worst cases in recent years were 11 September 2001 in the USA and the Icelandic ash cloud over Europe in 2010, when airports around the world were disrupted;
▶ Environmental damage from the fire and/or fighting the fire – firefighting water, the products of combustion and exploding building materials, such as asbestos cement roofs, can contaminate significant areas around the fire site.

11.2 Preventing fire and fire spread

This section provides further information on evaluating the risk of a fire and its prevention in premises. The following should be considered:

▶ Good housekeeping;
▶ Control of flammable and combustible materials;
▶ Control of ignition sources;
▶ Systems of work.

11.2.1 **Control measures**

Housekeeping

Good housekeeping will lower the chances of a fire starting, so the accumulation of combustible materials in all premises should be monitored carefully. Good housekeeping is essential to reduce the chances of escape routes and fire doors being blocked or obstructed.

Waste material should be kept in suitable containers before it is removed from the premises. If bins, particularly wheeled bins, are used outside, secure them in a compound to prevent them being moved to a position next to the building and set on fire. Never place skips against a building – they should normally be a minimum of 6 m away from any part of the premises.

If considerable quantities of combustible waste material are generated then a formal plan to manage this effectively needs to be developed.

In higher risk areas special arrangements should be in place for close-down, e.g. checking all appliances are turned off and combustible waste has been removed.

Storage

Many of the materials found in premises will be combustible which will increase the fire risk. Combustible materials are not just those generally regarded as highly combustible, such as polystyrene, but all materials that will readily catch fire. Even non-combustible materials may present a fire hazard when packed in combustible materials.

The absence of adequate storage arrangements results in congestion on the factory floor, warehouse or shop. In offices, the retention of large quantities of paper records, especially if not filed away in proprietary cabinets, can increase the fire hazard. Such readily available flammable material makes the potential effect of arson more serious. Many shops or hotels will take great care to present an efficient and attractive image in the retail or public areas, while other areas are neglected and allowed to become over-stocked or dumping areas for unsold material, old furniture and the like (see Figure 11.11).

This may lead to a concealed fire, restriction of access to the fire; fire extinguishers; alarm points and escape routes. Discarded packaging materials, e.g. polystyrene and cardboard, and even piles of wooden pallets can introduce severe fire hazards. Poorly managed storage areas often become over-stocked or dumping areas for unwanted material. Do not pile combustible material against electrical equipment or heaters, even if turned off for the summer.

Figure 11.11 Partly blocked fire exit door

To reduce the risk, store excess materials and stock in a dedicated storage area, storeroom or cupboard. Do not store excess stock in areas where the public would normally have access.

Consider how stock is displayed in shops and evaluate any additional risk of fire that it generates. For example, rugs stacked on the floor on top of each other would not present a high fire risk, but rolls of carpet stored vertically up against a wall or hung on displays present a vertical surface for fire to spread rapidly upwards. The display of large quantities of clothing on vertical hangers is also likely to increase the risk of rapid fire development.

The fire risk assessment should also consider any additional risk generated by seasonal products such as fireworks and Christmas decorations.

Consider the following to reduce these risks:

▶ ensure storage and display areas are adequately controlled and monitored;
▶ use fire-resistant display materials wherever possible (suppliers should be able to provide evidence of this); and
▶ ensure electrical lighting used as part of the display does not become a potential source of ignition.

Voids (including roof voids) should not be used for the storage of combustible material. Such voids should be sealed off or kept entirely open to allow for easy access for inspection and the removal of combustible materials.

Equipment and machinery

Common causes of fire in equipment are:

▶ allowing ventilation points to become clogged or blocked, causing overheating;
▶ inadequate cleaning of heat-shrink packaging equipment, such as that used in in-store bakeries;

allowing extraction equipment in catering environments to build up excessive grease deposits;

misuse or lack of maintenance of cooking equipment and appliances; and

disabling or interfering with automatic or manual safety features and cut-outs.

All machinery, apparatus and office equipment should be properly maintained by a competent person. Appropriate signs and instructions on safe use may be necessary.

Fork-lift trucks and other vehicles

There are hazards associated with industrial vehicles, particularly during refuelling and maintenance operations; also when stored or in use. Battery charging of fork-lift trucks can give rise to sparks and hydrogen (a gas that is highly flammable, explosive and is lighter than air). Sparks can occur when connecting and disconnecting power supplies.

Fork-lift truck charging points should be carefully sited in a well-ventilated area (ideally direct to open air), clear of ignition sources and preferably in a separate dedicated non-combustible structure. However, if sited in the building, the charging point should be against a fire-resisting wall (e.g. 30-minute fire resistance).

Heating

Individual heating appliances require particular care if they are to be used safely, particularly those which are kept for emergency use during a power cut or as supplementary heating during severe weather. The greatest risks arise from lack of maintenance and staff unfamiliarity with them. Heaters should preferably be secured in position when in use and fitted with a fire guard if appropriate.

As a general rule, convector or fan heaters should be preferred to radiant heaters because they present a lower risk of fire and injury.

The following rules should be observed:

all heaters should be kept well clear of combustible materials and where they do not cause an obstruction;

heaters which burn a fuel should be sited away from draughts;

portable fuel burning heaters (including bottled gas (LPG)) should only be used in exceptional circumstances and if shown to be acceptable in your risk assessment.

All gas heating appliances should be used only in accordance with manufacturer's instructions and should be serviced annually by a competent person.

In general, staff should be discouraged from bringing in their own portable heaters and other electrical equipment (e.g. kettles) into the premises.

Cooking processes

Typical installations used in cooking processes include deep fat fryers, ovens, grills, surface cookers, ductwork, flues, filters, hoods, extract and ventilation ducts and dampers. These cooking processes can operate at high temperatures, involving large quantities of oil and combustible food stuffs. Heat sources used for cooking processes include: gas, electric and microwave. The main causes of fire are ignition of cooking oil, combustion of crumbs and sediment deposits, and ductwork fires from a build-up of fats and grease. The siting of cooking processes close to insulated wall panels with combustible insulation can lead to the likely ignition of the panels and consequent rapid fire spread to other parts of the building. This practice should therefore be avoided.

The following should be considered to reduce the risk from cooking processes:

regular cleaning to prevent build-up of crumbs and other combustible material;

fire resisting containers for waste products;

a fire suppression system capable of controlling an outbreak of fire;

monitored heat/oil levels, even after the cooking process is complete and installation of temperature control/cut-off/shut-off devices as appropriate;

duct, joints and supports able to withstand high cooking temperatures;

separation from wall and ceiling panels (with combustible insulation) e.g. 2.5 m for walls, 4 m for ceilings;

insulation of ducts to prevent heating/ignition of nearby combustible wall and ceiling materials;

a regular programme for inspection and cleaning;

a programme of electrical and mechanical maintenance; and

annual service of all gas heating appliances by a competent person.

Smoking

Carelessly discarded cigarettes and other smoking materials are still a major cause of fire. A cigarette can smoulder for several hours, especially when surrounded by combustible material.

Many fires are started several hours after the smoking materials have been emptied into waste bags and left for future disposal.

With the smoke-free legislation concerning enclosed premises this has greatly reduced the risk of fire inside premises. However the risks can still exist in outside areas and designated smoking shelters. Display the smoke-free signs throughout the premises.

In those areas where smoking is permitted, provide deep and substantial metal ashtrays to help prevent unsuitable containers being used. Empty all ashtrays

11

daily into a metal waste bin and keep it outside. It is dangerous to empty ashtrays into plastic waste sacks which are then left inside for disposal later.

Electrical safety

Electrical equipment can be a significant cause of accidental fires in premises. The main causes are:

▶ overheating cables and equipment, e.g. due to overloading circuits, bunched or coiled cables or impaired cooling fans;
▶ incorrect installation or use of equipment;
▶ little or no maintenance and testing of equipment;
▶ incorrect fuse ratings;
▶ damaged or inadequate insulation on cables or wiring;
▶ combustible materials being placed too close to electrical equipment which may give off heat even when operating normally or may become hot due to a fault;
▶ arcing or sparking by electrical equipment;
▶ overloading of local/home network through charging of electric vehicles which use high current flows;
▶ incorrectly installed or protected solar panels; and
▶ embrittlement and cracking of cable sheathing in cold environments.

All electrical equipment should be installed and maintained in a safe manner by a competent person. If portable electrical equipment is used, including items brought into a workplace by staff, then the fire risk assessment should ensure that it is visually inspected and undergoes portable appliance testing (PAT) at intervals suitable for the type of equipment and its frequency of use (see Chapter 12). If there is any doubt about the safety of your electrical installation then you should consult a competent electrician.

Issues to consider include:

▶ overloading of equipment;
▶ correct fuse ratings;
▶ PAT testing and testing of fixed installations;
▶ protection against overloading of installation;
▶ protection against short circuit;
▶ insulation, earthing and electrical isolation requirements;
▶ frequency of electrical inspection and test;
▶ temperature rating and mechanical strength of flexible cables;
▶ portable electrical equipment;
▶ physical environment in which the equipment is used (e.g. wet or dusty atmospheres); and
▶ suitable use and maintenance of personal protective equipment.

See Chapter 12 for more information.

Systems of work

Safe systems of work have an important part to play in preventing fires. This is particularly important in high hazard areas where dangerous substances are being used. Systems of work may involve a hot work permit where, for example, welding, cutting or burning are taking place and/or a method statement from a contractor stating how hot roof work using heated bitumen will take place. See Chapter 4 for more details.

11.2.2 Storage and use of flammable liquids and gases

Introduction

The NEBOSH General Certificate includes the storage of flammable liquids in workrooms and other locations. This is covered by the Dangerous Substances and Explosive Atmosphere Regulations (DSEAR) and the Approved Code of Practice L138 2nd edition, which are given as reference documents. Some of the information provided here goes beyond the storage of small quantities but is given to provide some basic information to managers in process industries.

DSEAR apply at most workplaces where a dangerous substance is present or could be present. This is a fire and explosion rather than a health risk which is covered by COSHH (see Chapter 8). The employer must:

▶ carry out a risk assessment of any work activities involving dangerous substances;
▶ provide a way of eliminating or reducing risks as far as is reasonably practicable;
▶ provide procedures and equipment to deal with accidents and emergencies;
▶ provide training and information for employees;
▶ classify places where explosive atmospheres may occur into zones and mark the zones where necessary.

The Regulations give a detailed definition of 'dangerous substance', which should be referred to for more information. They include any substance or preparation which, because of its properties or the way it is used, could be harmful because of fires and explosions. The list includes petrol, LPG, paints, varnishes, solvents and some dusts. These are dusts which, when mixed with air, can cause an explosive atmosphere. Dusts from milling and sanding operations are examples of this. Most workplaces contain a certain amount of dangerous substances.

An explosive atmosphere is an accumulation of gas, mist, dust or vapour, mixed with air, which has the potential to catch fire or explode. Although an explosive atmosphere does not always result in an explosion (detonation), if it catches fire, flames can quickly travel through the workplace. In a confined space (e.g. in plant or equipment) the rapid spread of the flame front or rise in pressure can itself cause an explosion and rupture of the plant and/or building. For a detailed summary of the Regulations, see Chapter 13.

Risk assessment

This is the process of identifying and carefully examining the dangerous substances present or likely to be present in the workplace, the work activities involving them and how they might fail and cause fire, explosion and similar events that could harm employees and the public. The purpose of a risk assessment is to enable the employer to decide what needs to be done to eliminate or reduce the safety risks from dangerous substances as far as is reasonably practicable. It should take account of the following:

▶ what hazardous properties the substances have;
▶ the way they are used and stored;
▶ the possibility of hazardous explosive atmospheres occurring;
▶ any potential ignition sources.

Regardless of the quantity of dangerous substance present, the employer must carry out a risk assessment. This will enable them to decide whether existing measures are sufficient or whether they need to make any additional controls or precautions. Non-routine activities need to be assessed as well as the normal activities within the workplace. For example in maintenance work, there is often a higher potential for fire and explosion incidents to occur.

Unless the employer has already carried out a detailed assessment under the Management Regulations, there must be an assessment of the risks from fire, explosion and other events arising from dangerous substances, including addressing requirements specified by DSEAR. This is separate from the Fire Risk assessment made under the Regulatory Reform (Fire Safety) Order 2005 (RRFSO) as Dangerous Substances are considered to be process risks covered by the HSW Act.

Employers are required to ensure that the safety risks from dangerous substances are eliminated or, when this is not reasonably practicable, to take measures to control risks and to reduce the harmful effects of any fire, explosion or similar events, so far as is reasonably practicable.

Substitution

Substitution is the best solution. It is much better to replace a dangerous substance with a substance or process that totally eliminates the risk. In practice this is difficult to achieve; so it is more likely that the dangerous substance will be replaced with one that is less hazardous (e.g. by replacing a low-flashpoint solvent with a high-flashpoint one).

Designing the process so that it is less dangerous is an alternative solution. For example a change could be made from a batch production to a continuous production process, or the manner or sequence in which the dangerous substance is added could be altered. However, care must be taken when carrying out these

steps to make sure that no other new safety or health risks are created or increased, as this would outweigh the improvements implemented as a result of DSEAR.

The fact is that where a dangerous substance is handled or stored for use as a fuel, there is often no scope to eliminate it and very little chance to reduce the quantities handled.

Where risk cannot be entirely eliminated, control and mitigation measures should be applied. These should reduce risk.

Control measures

Control measures should be applied in the following order of priority:

▶ reduce the amount of dangerous substances to a minimum;
▶ avoid or minimise releases;
▶ control releases at source;
▶ prevent the formation of an explosive atmosphere;
▶ use a method such as ventilation to collect, contain and remove any releases to a safe place;
▶ avoid ignition sources;
▶ avoid adverse conditions (e.g. exceeding the limits of temperature or other control settings) that could lead to danger;
▶ keep incompatible substances apart.

Mitigation measures

Choose mitigation measures which are consistent with the risk assessment and appropriate to the nature of the activity or operation. These can include:

▶ preventing fires and explosions from spreading to other plant and equipment or to other parts of the workplace;
▶ making sure that a minimum number of employees is exposed;
▶ in the case of a process plant, providing plant and equipment that can safely contain or suppress an explosion, or vent it to a safe place.

In workplaces where explosive atmospheres may occur, employers should ensure that:

▶ areas where hazardous explosive atmospheres may occur are classified into zones based on their likelihood and persistence;
▶ areas classified into zones are protected from sources of ignition by selecting suitable special equipment and protective systems;
▶ where necessary, areas classified into zones are marked with a specified 'EX' sign at their points of entry;
▶ employees working in zoned areas must be provided with appropriate clothing that does not create a risk of an electrostatic discharge igniting the explosive atmosphere;
▶ before coming into operation for the first time, areas where hazardous explosive atmosphere may be present are confirmed as being safe (verified)

375

by a person (or organisation) competent in the field of explosion protection. The person carrying out the verification must be competent to consider the particular risks at the workplace and the adequacy of control and other measures put in place.

Storage

Dangerous substances should be kept in a safe place in a separate building or the open air. Only small quantities of dangerous substances should be kept in a workroom or area as follows:

For flammable liquids that have a flashpoint above the maximum ambient temperature (normally taken as 32°C), the small quantity that may be stored in the workroom is considered to be an amount up to 250 litres.

For extremely and highly flammable liquids and those flammable liquids with a flashpoint below the maximum ambient temperature, the small quantity is considered to be up to 50 litres and this should be held in a special metal cupboard or container (Figure 11.12(a) and (b)).

Any larger amounts, which cannot be kept outside in a safe area, should be kept in a special fire-resisting store (Figure 11.12(c)), which should be:

- ▶ properly ventilated;
- ▶ provided with spillage retaining arrangements such as sills;
- ▶ free of sources of ignition, such as unprotected electrical equipment, sources of static electrical sparks, naked flames or smoking materials;
- ▶ arranged so that incompatible chemicals do not become mixed together either in normal use or in a fire situation;
- ▶ of fire-resisting construction;
- ▶ used for empty as well as full containers – all containers must be kept closed;
- ▶ kept clear of combustible materials such as cardboard or foam plastic packaging materials.

Flammable gases

Flammable gas cylinders also need to be stored and used safely. The following guidance should be adopted:

- ▶ Both full and empty cylinders should be stored outside. They should be kept in a separate secure compound at ground level with sufficient ventilation. Open mesh is preferable.
- ▶ Valves should be uppermost during storage to retain them in the vapour phase of the LPG.
- ▶ Cylinders must be protected from mechanical damage. Unstable cylinders should be together, for example, and cylinders must be protected from the heat of the summer sun.
- ▶ The correct fittings must be used. These include hoses couplers, clamps and regulators.
- ▶ Gas valves must be turned off after use at the end of the shift.

(a)

(b)

(c)

Figure 11.12 Various storage arrangements for highly flammable liquids

- ► Precautions must be taken to avoid welding flame 'flash back' into the hoses or cylinders. People need training in the proper lighting up and safe systems of work procedures; non-return valves and flame arrestors also need to be fitted.
- ► Cylinders must be changed in a well-ventilated area remote from any sources of ignition.
- ► Joints should be tested for gas leaks using soapy/detergent water – never use a flame.
- ► Flammable material must be removed or protected before welding or similar work.
- ► Cylinders should be positioned outside buildings with gas piped through in fixed metal piping.
- ► Both high and low ventilation must be maintained where LPG applications are being used.
- ► Flame failure devices are necessary to shut off the gas supply in the event of flame failure.

Aerosols

Some aerosols can contain flammable products stored at pressure and they can present a high level of hazard. When ignited they can explode, produce fireballs and rocket to distances of 40 m. Their presence in premises can make it unsafe for firefighters to enter a building and they have the potential for starting multiple fires.

The following should be considered to reduce these risks:

- ► All staff involved in the movement, storage and display of aerosol cans should be adequately instructed, trained and supervised.
- ► Damaged and leaking aerosol cans should be removed immediately to a safe, secure, well-ventilated place prior to disposal.
- ► Powered vehicles should not be used to move damaged stock, unless specially adapted for use in flammable atmospheres.
- ► Arrangements should be made for disposal at a licensed waste management facility.

Electrical and other equipment in potentially flammable atmospheres

Most electrical equipment either sparks in normal operation or is liable to spark under fault conditions. Some electrical appliances such as electric heaters, are specifically designed to produce high temperatures. These circumstances create fire and explosion hazards, which demand very careful assessment in locations where processes capable of producing flammable concentrations of gas or vapour are used, or where flammable liquids are stored.

It is likely that many fires are caused by static electrical discharges. Static electricity can, in general, be eliminated by the careful design and selection of materials used in equipment and plant, and the materials used in products being manufactured. When it is impractical to avoid the generation of static electricity, a means of control must be devised. Where flammable materials are present,

especially if they are gases or dusts, then there is a great danger of fire and explosion, even if there is only a small discharge of static electricity. The control and prevention of static electricity is considered in more detail later.

The use of electrical equipment in potentially flammable atmospheres should be avoided as far as possible. However, there will be many cases where electrical equipment must be used and, in these cases, the standards for the construction of the equipment should comply with the Equipment and Protective Systems Intended for Use in Potentially Explosive Atmospheres Regulations, known as ATEX. Details on the classification or zoning of areas are contained in the Dangerous Substances and Explosive Atmospheres Regulations and ACoPs.

Before electrical equipment is installed in any location where flammable vapours or gases may be present, the area must be zoned in accordance with the Dangerous Substances and Explosive Atmosphere Regulations, and records of the zoned areas must be marked on building drawings and revised when any zoned area is changed. The installation and maintenance of electrical equipment in potentially flammable atmospheres is a specialised task. It must only be undertaken by electricians or instrument mechanics who are trained to ATEX standards.

Static electricity is produced by the build-up of electrons on weak electrical conductors or insulating materials. These materials may be gaseous, liquid or solid and may include flammable liquids, powders, plastic films and granules. Plastics have a high resistance that enables them to retain static charges for long periods of time. The generation of static may be caused by the rapid separation of highly insulated materials by friction or by transfer from one highly charged material to another in an electric field by induction (see Figure 11.13).

Dispensing Flammable Liquids

Figure 11.13 Earth bonding of containers to prevent a static discharge – cotton outer clothing and conductive footwear should be used. Gloves and eye protection (not shown) should be worn.

A static electric shock, perhaps caused by closing a door with a metallic handle, can produce a voltage greater than 10,000 V. Since the current flows for a very short period of time, there is seldom any serious harm to an individual. However, discharges of static electricity may be sufficient to cause serious electric shock and are always a potential source of ignition when flammable liquid, dusts or powders are present. This is a particular problem in the parts of the printing industry where solvent-based inks are used on high-speed web presses. Flour dust in a mill has also been ignited by static electricity.

Static electricity may build up on both materials and people. When a charged person approaches flammable gases or vapours and a spark ignites the substance, the resulting explosion or fire often causes serious injury. In these situations, effective static control systems must be used which include:

▶ bonding or earthing continuity between pieces of equipment particularly portable equipment such as containers for carrying highly flammable substances, piping, filling funnels, drip trays, and the like;

▶ not wearing outer clothing which generates static charges. In practice this usually means avoiding man-made fibres and using cotton only;

▶ using conductive footwear to leak static charges to ground (these must not be used by electricians as they conduct electricity and will not protect them against electric shock);

▶ avoiding the free fall of highly flammable liquids from one container top another unless anti-static additives have been put into the liquid or its natural properties will not hold static charges. Filling funnels should reach down to near the bottom of the container being filled.

For more information see Chapter 12 on electrical hazards.

11.2.3 Principles of fire protection in buildings – structural measures to prevent the spread of fire and smoke

(a) General

The design of all new buildings and the design of extensions or modifications to existing buildings must be approved by the local planning authority.

Design data for new and modified buildings must be retained throughout the life of the structure.

Building legal standards are concerned mainly with safety of life. Therefore, it is necessary to consider the early stage of fire and how it affects the means of escape, and, also, aim to prevent eventual spread to other buildings.

Asset protection requires extra precautions that will have an effect at both early and later stages of the fire

growth by controlling fire spread through and between buildings and preventing structural collapse. However, this extra fire protection will also improve life safety not only for those escaping at the early stages of the fire but also for firefighters who will subsequently enter.

If a building is carefully designed and suitable materials are used to build it and maintain it, then the risk of injury or damage from fire can be substantially reduced. Three objectives must be met:

▶ it must be possible for everyone to leave the building quickly and safely;

▶ the building must remain standing for as long as possible;

▶ the spread of fire and smoke must be reduced.

These objectives can be met through the selection of materials and design of buildings.

(b) Fire loading

The fire load of a building is used to classify types of building use. It may be calculated simply by multiplying the weight of all combustible materials by their energy values and dividing by the floor area under consideration. The higher the fire load, the more the effort needed to offset this by building to higher standards of fire resistance.

(c) Surface spread of fire

Combustible materials, when present in a building as large continuous areas, such as for lining walls and ceilings, readily ignite and contribute to spread of fire over their surfaces. This can represent a risk to life in buildings, particularly where walls of fire-escape routes and stairways are lined with materials of this nature.

Materials are tested by insurance bodies and fire research establishments. The purpose of the test is to classify materials according to the tendency for flame to spread over their surfaces. As with all standardised test methods, care must be taken when applying test results to real applications.

In the UK, a material is classified as having a surface in one of the following categories:

Class 1 – Surface of very low flame spread;
Class 2 – Surface of low flame spread;
Class 3 – Surface of medium flame spread;
Class 4 – Surface of rapid flame spread.

The test shows how a material would behave in the initial stages of a fire.

As all materials tested are combustible, in a serious fire they would burn or be consumed. Therefore, there is an additional Class 0 of materials which are non-combustible throughout or, under specified conditions, non-combustible on one face and combustible on the other. The spread of flame rating of the combined Class 0 product must not be worse than that for Class 1.

The Building Regulations use three of these classes in their Approved Document for lining materials as follows:

Class 0: Materials suitable for circulation spaces and escape routes

Such materials include brickwork, blockwork, concrete, ceramic tiles, plaster finishes (including rendering on wood or metal lathes), wood-wool cement slabs and mineral fibre tiles or sheets with cement or resin binding.

Note: Additional finishes to these surfaces may be detrimental to the fire performance of the surface and if there is any doubt about this then consult the manufacturer of the finish.

Class 1: Materials suitable for use in all rooms but not on escape routes

Such materials include all the Class 0 materials referred to above. Additionally, timber, hardboard, block-board, particle board, heavy flock wallpapers and thermosetting plastics will be suitable if flame-retardant treated to achieve a Class 1 standard.

Class 3: Materials suitable for use in rooms of less than 30 m²

Such materials include all those referred to in Class 1, including those that have not been flame-retardant treated and certain dense timber or plywood and standard glass-reinforced polyesters.

The equivalent European classification standard will also be acceptable. Further details about internal linings and classifications are available in *The Building Regulations 2000 (as amended 2001): Approved document B Fire Safety*. The Stationery Office. ISBN 0 11 753911-2. Appropriate testing procedures are detailed in BS 476-7 and where appropriate BS EN 13501-1.

(d) Fire resistance of structural elements

If structural elements such as walls, floors, beams, columns and doors are to provide effective barriers to fire spread and to contribute to the stability of a building, they should be of a required standard of fire resistance.

In the UK, tests for fire resistance are made on elements of structure, full size if possible, or on a representative portion having minimum dimensions of 3 m long for columns and beams and 1 m² for walls and floors. All elements are exposed to the same standard fire provided by furnaces in which the temperature increases with time at a set rate. The conditions of exposure are appropriate to the element tested. Free-standing columns are subjected to heat all round, and walls and floors are exposed to heat on one side only. Elements of structure are graded by the length of time they continue to meet three criteria:

▷ the element must not collapse;
▷ the element must not develop cracks through which flames or hot gases can pass;
▷ the element must have enough resistance to the passage of heat so the temperature of the unexposed face does not rise by more than a prescribed amount.

The term fire resistance has a precise meaning. It should not be applied to such properties of materials as resistance to ignition or resistance to flame propagation. For example steel has a high resistance to ignition and flame propagation but will distort quickly in a fire and allow the structure to collapse – it therefore has poor 'fire resistance' (Figure 11.14). It must be insulated to provide good fire protection. This is normally done by encasing steel frames in concrete.

Figure 11.14 Steel structures can collapse in the heat of a fire

In the past, asbestos has been made into a paste and plastered onto steel frames, giving excellent fire protection, but it has caused major health problems and its use in new work is banned.

Building materials with high fire resistance are, for example, brick, stone, concrete, very heavy timbers (the outside chars and insulates the inside of the timber), and some specially made composite materials used for fire doors.

(e) Insulating materials

Building materials used for thermal or sound insulation could contribute to the spread of fire. Only approved fire-resisting materials should be used. Many buildings have insulated core panels as exterior cladding or for internal structures and partitions (Figure 11.15). The food industry, in particular, uses insulated core panels because they are easy to clean and facilitate consistent temperature control within the premises. The simple construction of these panels enables alterations and for additional internal partitions to be erected with minimum disruption to business.

Figure 11.15 Insulated core panels

They normally consist of a central insulated core, sandwiched between an inner and outer metal skin. There is no air gap. The external surface is then normally coated with a PVC covering to improve weather resistance or the aesthetic appeal of the panel. The central core can be made of various insulating materials, ranging from virtually non-combustible through to highly combustible. Differing fire hazards are associated with common types of insulation, when the panels are subjected to certain temperatures. Typical examples are:

▶ Mineral rock/modified phenolic will produce surface char and little smoke or gaseous combustion products, at temperatures above 230°C.

▶ Polyisocyanurate (PIR)/polyurethane (PUR) will char and will generate smoke and gaseous combustion products, at temperatures above 430°C PIR and 300°C PUR.

▶ Expanded polystyrene (EPS) will melt and will generate smoke and gaseous combustion products, at temperatures above 430°C.

Insulation charring can lead to panel delamination/collapse, and the gaseous combustion products can fill areas with the toxic gases carbon monoxide and styrene. A number of fires in buildings where insulated core panels have been used extensively in the fabric of the building have highlighted the particular dangers that may be associated with this form of construction, i.e. where the fabric of the building can contribute to the fire hazard. For more information see: www.communities.gov.uk/fire/firesafety/firesafetylaw/aboutguides/

Since the Grenfell fire in June 2017 there has been extensive attention focused on residential high-rise accommodation. The January 2020 report should be studied and acted upon by anybody responsible for or affected by such buildings (https://assets.publishing.service.gov.uk/government/uploads/system/uploads/attachment_data/file/860484/Building_safety_advice_for_building_owners__including_fire_doors_-_January_2020.pdf). The report states:

1.6 The government has announced its intention to introduce a Fire Safety Bill which will clarify that building owners and managers of multi-occupied residential premises of any height must fully consider and mitigate the risks of any external wall systems and fire doors in discharging their duties under the Fire Safety Order. **We strongly advise building owners to consider the risks of any external wall system and fire doors in their fire risk assessments, irrespective of the height of the building, ahead of the planned clarification.**

1.7 Building owners and residents should be aware that government has committed to reform the regulatory system and agreed with the principles outlined within Dame Judith Hackitt's review (Building a Safer Future, Independent Review of Building Regulations and Fire Safety). Under the proposed reforms, a new category of duty holder will be required to review the safety of their existing buildings – and remediate them where necessary. **We strongly advise that building owners should already be actively ensuring the safety of residents, and not wait for the regulatory system to be reformed.**

1.10 Existing residential buildings which have external wall systems that contain combustible materials may not meet an appropriate standard of safety and could pose a significant risk to the health and safety of residents, other building users, people in the proximity of the building or firefighters. External walls of residential buildings should not assist the spread of fire, irrespective of height. It is important therefore to understand both the materials used in the external wall construction and whether the entire system has been designed, installed, and, maintained appropriately.

1.11 While the use of combustible materials within or attached to external walls of residential buildings below 18m is not currently expressly prohibited, it has been a legal requirement since the 1980s to consider the risk from fire spread. This is in accordance with the functional requirements of the Building Regulations. It is also a requirement of the Fire Safety Order that any purpose-built block of flats – regardless of height – should have an up to date fire risk assessment and appropriate fire precautions in place.

(f) Fire compartmentation

A compartment is a part of a building that is separated from all other parts by walls and floors, and is designed to contain a fire for a specified time.

The principal object is to limit the effect of both direct fire damage and consequential business interruption caused by not only fire spread but also smoke and water damage in the same floor and other storeys.

Buildings are classified into purpose groups, according to their size. To control the spread of fire, any building whose size exceeds that specified for its purpose group must be divided into compartments that do not exceed the prescribed limits of volume and floor area. Otherwise, they must be provided with special fire protection. In the UK, the normal limit for the size of a compartment is 7,000 m³. Compartments must be separated by walls and floors of sufficient fire resistance. Any openings needed in those walls or floors must be protected by fire-resisting doors to ensure proper fire-tight separation.

Ventilation and heating ducts must be fitted with fire dampers where they pass through compartment walls and floors. Firebreak walls must extend completely across a building from inside wall to outside wall. They must be stable; they must be able to stand even when the part of the building on one side or the other is destroyed.

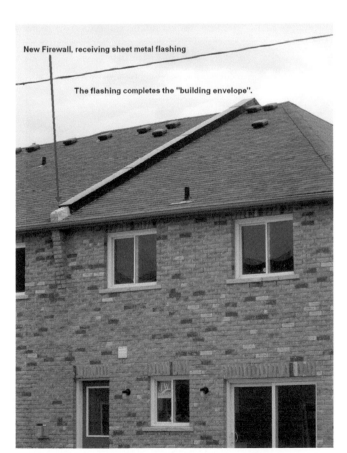

Figure 11.16 Firebreak wall between dwellings

No portion of the wall should be supported on unprotected steelwork nor should it have the ends of unprotected steel members embedded in it. The wall must extend up to the underside of a non-combustible roof surface, and sometimes above it. Any openings must be protected to the required minimum grade of fire resistance. If an external wall joins a firebreak wall and has an opening near the join, the firebreak wall may need to extend beyond the external wall.

An important function of external walls is to contain a fire within a building, or to prevent fire spreading from outside.

The fire resistance of external walls should be related to the:

▶ purpose for which the building is used;
▶ height, floor area and volume of the building;
▶ distance of the building from relevant boundaries and other buildings;
▶ extent of doors, windows and other openings in the wall.

A wall which separates properties from each other should have no doors or other openings.

11.3 Fire alarms and firefighting arrangements

11.3.1 Fire detection and alarm systems

In the event of fire, it is vital that everyone in the workplace is alerted as soon as possible. The earlier the fire is discovered, the more likely it is that people will be able to escape before the fire takes hold and before it blocks escape routes or makes escape difficult.

Every workplace should have detection and warning arrangements. Usually the people who work there will detect the fire and in many workplaces nothing further will be needed.

In small workplaces where occupancy is low, a shouted warning should be all that is needed, so long as the warning can be heard and understood everywhere on the premises. Where a simple shout of 'fire' or the operation of a manual device such as a gong, whistle or air horn is not adequate to warn everyone when operated from any single point within the building, it is likely that an electrical fire warning system may be required. This will typically include the following:

▶ manual call points (break-glass call points) at storey exit and final exit doors;
▶ electronic sirens or bells; and
▶ a simple control and indicator panel.

In some cases a heat or smoke fire detector will also be fitted in the system.

Figure 11.17 Simple electrical fire alarm system components

Some premises (e.g. a community centre) may only require a few interconnected manual call points and sounders. Other premises (e.g. a multilevel pub, or factory) may require a more complex warning system involving staff action. If a building has areas where a fire could develop undetected or where people work alone and might not see a fire (e.g. storerooms or warehouses), then it may be necessary to upgrade the fire warning system to incorporate some form of automatic fire detection (e.g. in some circumstances, this could be a series of interconnected smoke alarms, if shown to be acceptable in the risk assessment).

If the system fails, people in the premises still need to be warned and escape safely. A temporary arrangement, such as gongs, whistles or air horns, combined with suitable training, may be acceptable for a short period pending system repairs. All systems should be properly installed to recognised standards with an auxiliary power source, if temporary manual alarms would be unacceptable. Cables need protecting from fire by using mineral-insulated, copper-sheathed, silicone-rubber insulated or other approved fire-resistant cabling.

The fire warning sound levels should be loud enough to alert everyone, taking into account background noise, any other sound systems should be muted (automatically or manually) when the fire alarm sounds.

In areas with high background noise, or where people may be wearing hearing protectors, the audible warning should be supplemented, e.g. with visual or even vibratory alarms.

People with hearing difficulties

Where people have hearing difficulties, particularly those who are profoundly deaf, then simply hearing the fire warning is likely to be the major difficulty. If these persons are never alone while on the premises then this may not be a serious problem, as it would be reasonable for other occupants to let them know that the building should be evacuated. If a person with hearing difficulties is likely to be alone, then consider other means of raising the alarm. Among the most popular are visual beacons and vibrating devices or pagers that are linked to the existing fire alarm.

Voice alarms

Research has shown that some people and, in particular, members of the public, do not always react quickly to a conventional fire alarm. Voice alarms are therefore becoming increasingly popular and can also incorporate a public address facility. The message or messages sent must be carefully considered. It is therefore essential to ensure that voice-alarm systems are designed and installed by a person with specialist knowledge of these systems.

Schematic plan

In order to quickly determine where a fire has been detected, you should consider displaying a schematic plan showing fire alarm zones in a multi-zoned system adjacent to the control panel.

Manual call points

Manual call points, often known as 'break-glass' call points, enable a person who discovers a fire to immediately raise the alarm and warn other people in the premises of the danger.

People leaving a building because of a fire will normally leave by the way they entered. Consequently, manual call points are normally positioned at exits and storey exits that people may reasonably be expected to use in case of fire, not just those designated as fire exits. However, it is not necessary in every case to provide call points at every exit.

Manual call points should normally be positioned so that, after all fixtures and fittings, machinery and stock are in place, no one should have to travel more than 45 m to the nearest alarm point. This distance may be less if the premises cater for people of limited mobility or there are particularly hazardous areas. They should be conspicuous (red), fitted at a height of about 1.4 m (or less for premises with a significant number of wheelchair users), and not in an area likely to be obstructed.

Automatic fire detection

It is important to consider how long a fire is likely to burn before it is discovered. Fires are likely to be discovered quickly if they occur in places that are frequently visited by employees, or in occupied areas of a building. For example, employees are likely to smell/see burning or smoke. If it is thought that there might be some delay in fire being detected, automatic fire detection should be considered, linked into an electrical fire alarm system.

Automatic fire detection may be needed for a number of reasons. These can include:

▶ if there are areas where people are isolated or remote and could become trapped by a fire because they are unaware of its development, such as lone workers;
▶ sleeping accommodation is provided;
▶ if there are areas where a fire can develop unobserved (e.g. storerooms or warehouse);
▶ as a compensating feature, e.g. for inadequate structural fire protection, in dead-ends or where there are extended travel distances; and
▶ where smoke control and ventilation systems are controlled by the automatic fire detection system.

An automatic fire detection system should:

▶ be designed to accommodate the emergency evacuation procedure;
▶ give an automatic indication of the fire warning and its location. If the indicator panel is located in a part of the premises other than the control point (for example, an office) there should ideally be a repeater panel sited in the control point;
▶ be maintained and tested by a competent person; and
▶ communicate with a central control room (if there is one).

New automatic fire detection systems should be designed and installed by a competent person. Further guidance is given in BS 5839-116 or a more recent standard where applicable.

Where the public address system is part of the fire warning system it should be connected to an auxiliary power source to ensure the continued use of the system in the event of fire or other emergency.

Whichever warning or detection systems are in place, the fire and rescue service should always be called immediately if a fire occurs. When properly installed and maintained, these systems can be a significant factor in reducing the risk to life and limiting damage to your property in the event of a fire. These systems can help emergency services to respond very quickly to genuine emergencies.

Unfortunately, these systems can also produce 'unwanted' fire alarm signals, for example when the signals are triggered by mistake and not because of a fire or a test. The subsequent evacuation can cause unnecessary business disruption and they also mean that emergency services send fire appliances unnecessarily.

'Unwanted' fire alarm signals are a widespread problem and all fire and rescue services are trying to reduce the number of false alarms caused by automatic fire detection systems.

11.3.2 Extinguishing media

There are four main methods of extinguishing fires, which are explained as follows:

▶ **Cooling** – reducing the ignition temperature by taking the heat out of the fire – using water to limit or reduce the temperature.
▶ **Smothering** – limiting the oxygen available by smothering and preventing the mixture of oxygen and flammable vapour – by the use of foam or a fire blanket.
▶ **Starving** – limiting the fuel supply – by removing the source of fuel by switching off electrical power, isolating the flow of flammable liquids or removing wood and textiles, etc.
▶ **Chemical reaction** – by interrupting the chain of combustion and combining the hydrogen atoms with chlorine atoms in the hydrocarbon chain, for example with Halon extinguishers. (Halons have generally been withdrawn because of their detrimental effect on the environment, as ozone-depleting agents.)

Advantages and limitations of the main extinguishing media

Fire extinguishers are all red with 5% of the cylindrical area taken up with the colour code. The colour code (band) denotes on which class of fire the extinguisher can be used.

Water extinguishers (red band)

This type of extinguisher can only be used on Class A fires. They allow the user to direct water onto a fire from a considerable distance.

A 9-litre water extinguisher can be quite heavy and some water extinguishers with additives can achieve the same rating, although they are smaller and therefore considerably lighter. This type of extinguisher is not suitable for use on live electrical equipment, liquid or metal fires.

Water extinguishers with additives (red band)

This type of extinguisher is suitable for Class A fires. They can also be suitable for use on Class B fires and, where appropriate, this will be indicated on the extinguisher. They are generally more efficient than conventional water extinguishers.

Figure 11.18 Typical fire point in offices with extinguishers, fire notices and alarm break-glass call point

(a)

(b)

(c)

Figure 11.19 Various firefighting equipment: (a) fire blanket; (b) hose reel; (c) powder fire extinguisher

Foam extinguishers (cream band)

This type of extinguisher can be used on Class A or B fires and is particularly suited to extinguishing liquid fires such as petrol and diesel. They should not be used on free-flowing liquid fires unless the operator has been specially trained, as these have the potential to rapidly spread the fire to adjacent material. This type of extinguisher is not suitable for deep-fat fryers or chip pans. They should not be used on electrical or metal fires.

Powder extinguishers (blue band)

This type of extinguisher can be used on most classes of fire and achieve a good 'knock down' of the fire. They can be used on fires involving electrical equipment but will almost certainly render that equipment useless.

Because they do not cool the fire appreciably, it can re-ignite. Powder extinguishers can create a loss of visibility and may affect people who have breathing problems and are not generally suitable for confined spaces. They should not be used on metal fires.

Carbon dioxide extinguishers (black band)

This type of extinguisher is particularly suitable for fires involving electrical equipment as they will extinguish a fire without causing any further damage (except in the case of some electronic equipment, for example computers). As with all fires involving electrical equipment, the power should be disconnected if possible. These extinguishers should not be used on metal fires.

Wet chemical – class 'F' extinguishers

This type of extinguisher is particularly suitable for commercial catering establishments with deep-fat fryers. The intense heat in the fluid generated by fat fires means that when standard foam or carbon dioxide extinguishers stop discharging, re-ignition tends to occur.

Wet chemical extinguishers starve the fire of oxygen by sealing the burning fluid, which prevents flammable vapour reaching the atmosphere.

11.3.3 Portable firefighting equipment

Since 1 April 2006, employers or those who have control of non-domestic premises have had a statutory duty under fire safety and health and safety legislation to ensure that there are appropriate means of fighting fires. The employer or controller of non-domestic premises is known as the 'responsible person'. The responsible person also has to provide suitably trained people to operate non-automatic firefighting equipment.

Figure 11.20 Types of fire extinguishers and labels (Note: main colour of all extinguishers is red with 5% for label)

The responsible person must identify the firefighting requirement for their premises. If fire breaks out in the workplace and trained staff can safely extinguish it using suitable firefighting equipment, the risk to others will be removed. Therefore, all workplaces where people are at risk from fire should be provided with suitable firefighting equipment.

The most useful form of firefighting equipment for general fire risks is the water-type portable extinguisher or suitable alternative. One such extinguisher should be provided for around each 200 m² of floor space with a minimum of one per floor. If each floor has a hose reel, which is known to be in working order and of sufficient length for the floor it serves, there may be no need for water-type extinguishers to be provided.

Areas of special risks involving the use of oil, fats or electrical equipment may need carbon dioxide, dry powder or other types of extinguisher.

Fire extinguishers should be sited on exit routes, preferably near to exit doors or where they are provided for specific risks, near to the hazards they protect. Notices indicating the location of firefighting equipment should be displayed where the location of the equipment is not obvious or in areas of high fire risk where the notice will assist in reducing the risk to people in the workplace.

All Halon fire extinguishers should have been decommissioned as from December 2003 and disposed of safely.

Those carrying out hot work should have appropriate fire extinguishers with them and know how to use them.

The primary purpose of fire extinguishers is to tackle fires at a very early stage to enable people to make their escape. Putting out larger fires is the role of the fire and rescue services.

Extinguishers should conform to a recognised standard BS EN 3:7 Portable Fire Extinguishers. Characteristics, performance requirements and test methods.

Case study

It is very important to understand the limitations of portable firefighting equipment and ensure the safety of people as a first priority. At one cotton mill in the 1980s the author was involved with the investigation of a fire where they used **28 water extinguishers** fetched from all over the mill on various floors **before** sounding the alarm, evacuating the building and calling the fire and rescue services. Fortunately no one was seriously injured but the outcome could so easily have been very tragic indeed. Such equipment should only be used, if safe to do so, by trained people while the building is evacuated and the professionals are called.

11.3.4 Fixed firefighting equipment – sprinkler installations

Sprinklers should be considered as merely one component part of a total fire safety strategy, which is tailored to the existing and projected needs of a building. They have significant benefits to offer in suppressing fires until those best trained to deal with major incidents are on the scene to extinguish them.

The London Fire Brigade called for the following at the opening of the Grenfell Tower fire enquiry in 2017:

► The retrofitting of sprinklers in all residential high-rise tower blocks, as part of an appropriate package of fire safety measures.
► Sprinklers to be installed in all school new builds and major refurbishments.

As part of a £9.2m Grenfell Tower refit, sprinklers would have added about £200,000 to the cost and would in all probability have saved all the lives lost and the building.

Sprinklers, however, are an emotive topic. In some buildings, they have been used for a long time as the most significant element of a fire safety system. This situation is probably most prolific in warehouses and retail premises. They have unfortunately been resisted in most other buildings because of the initial capital cost and the perceived inherent risk of accidental water discharge.

It is anticipated that calls to fit sprinklers will continue and may be successful.

Figure 11.21 Various sprinkler heads designed to fit into a high-level water pipe system and spray water at different angles onto a fire below

Sprinkler systems can be very effective in controlling fires. They can be designed to protect life and/or property and may be regarded as a cost-effective solution for reducing the risks created by fire. If a building has a sprinkler installation, it may have been installed as a result of a business decision, for example for the protection of business assets, or it may have been installed as a requirement, for example imposed under a local Act, or an integral part of the building design.

Sprinkler systems should normally extend to the entire building. In a well-designed system, only those heads in the immediate vicinity of the fire will actually operate. Sprinkler installations typically comprise a water supply (preferably a stored water supply incorporating tanks), pumps, pipework and sprinkler heads. There are different types of sprinkler design; sprinklers can be operated to discharge water at roof or ceiling level or within storage racks. Other design types such as ESFR (early suppression fast response) and dry pipe may also be appropriate. In all cases, a competent person/contractor should be used to provide guidance.

The installation should be designed for the fire hazard, taking into account the building occupancy, the fire load and its burning characteristics and the sprinkler control characteristics. For each hazard the sprinkler installation design should take account of specific matters such as storage height, storage layout, ceiling clearance and sprinkler type (e.g. sprinkler orifice, sprinkler sensitivity).

There are some hazards where sprinklers should not be fitted, such as over salt baths and metal melt pans, because water will possibly cause an explosive reaction.

If any significant changes are being made to the premises, for example changing storage arrangements or material stored, the sprinkler installation should be checked to see that it is still appropriate and expert advice sought as necessary.

Sprinkler protection could give additional benefits, such as a reduction in the amount of portable firefighting equipment necessary, and the relaxation of restrictions in the design of buildings.

Guidance on the design and installation of new sprinkler systems and the maintenance of all systems is given in the Loss Prevention Council (LPC) Rules, BS EN 12845 or BS 5306 and should be carried out only by a competent person.

Routine maintenance by on-site personnel may include checking of pressure gauges, alarm systems, water supplies, any anti-freezing devices and automatic booster pump(s). For example diesel fire pumps should be given a test run for 30 minutes each week.

11.3.5 Inspection, maintenance and testing of fire equipment

It is important that equipment is fit for its purpose and is properly maintained and tested. One way in

which this can be achieved is through companies that specialise in the test and maintenance of firefighting equipment.

All equipment provided to assist escape from the premises, such as fire detection and warning systems and emergency lighting, and all equipment provided to assist with fighting fire, should be regularly checked and maintained by a suitably competent person in accordance with the manufacturer's recommendations. The alarm system should be tested every week, while the premises are in normal use. The test should be carried out by activating a different call point each week, at a fixed time. Table 11.1 gives guidance on the frequency of test and maintenance and provides a simple guide to good practice.

Table 11.1 Maintenance and testing of fire equipment

Equipment	Period	Action
Fire detection and fire warning systems including self-contained smoke alarms and manually operated devices	Weekly	• Check all systems for state of repair and operation • Repair or replace defective units • Test operation of systems, self-contained alarms and manually operated devices
	Annually	• Full check and test of system by competent service engineer • Clean self-contained smoke alarms and change batteries
Emergency lighting including self-contained units and torches	Weekly	• Operate torches and replace batteries as required • Repair or replace any defective unit
	Monthly	• Check all systems, units and torches for state of repair and apparent function
	Annually	• Full check and test of systems and units by competent service engineer • Replace batteries in torches
Firefighting equipment installation including hose reels	Weekly	• Check all extinguishers including hose reels for correct apparent function
	Annually	• Full check and test by competent service engineer

11.4 Fire evacuation

11.4.1 Means of escape in case of fire

(a) General

It is essential to ensure that people can escape quickly from a workplace if there is a fire. Normally the entrances and exits to the workplace will provide escape routes, particularly if staff have been trained in what to do in case of fire and if it is certain that an early warning will be given.

In modern buildings which have had Building Regulation approval, and where there have not been significant changes to the building or where the workplace has recently been inspected by the fire authorities and found to be satisfactory, it is likely that the means of escape will be adequate. It may occasionally be necessary to improve the fire protection on existing escape routes, or to provide additional exits. In making a decision about the adequacy of means of escape, the following points should be considered:

▶ People need to be able to turn away from a fire as they escape or be able to pass a fire when it is very small.

▶ If a single-direction escape route is in a corridor, the corridor may need to be protected from fire by fire-resisting partitions and self-closing fire doors.

▶ Stair openings can act as natural chimneys in fires. This makes escape from the upper parts of some workplaces difficult. Most stairways, therefore, need to be separated from the workplace by fire-resisting partitions and self-closing fire doors. Where stairways serve no more than two open areas, in shops for example, which people may need to use as escape routes, there may be no need to use this type of protection.

Figure 11.22 External fire escape from a multi-storey building. Extreme caution needed in winter and enclosure of staircase would be preferred for the UK climate

(b) Doors

Some doors may need to open in the direction of travel, such as:

▶ doors from a high-risk area, such as a paint spraying room or large kitchen;

▶ doors that may be used by more than 50 persons;

▶ doors at the foot of stairways where there may be a danger of people being crushed.

Some sliding doors may be suitable for escape purposes provided that they do not put people using them at additional risk, slide easily and are marked with the direction of opening.

Doors which only revolve and do not have hinged segments are not suitable as escape doors.

(c) Escape routes and travel distances

Escape routes should meet the following criteria:

▶ Where two or more escape routes are needed they should lead in different directions to places of safety.

▶ Escape routes need to be short and to lead people directly to a place of safety, such as the open air or an area of the workplace where there is no immediate danger.

▶ It should be possible for people to reach the open air without returning to the area of the fire. They should then be able to move well away from the building.

▶ Escape routes should be wide enough for the volume of people using them. A 750 mm door will allow up to 40 people to escape in 1 minute, so most doors and corridors will be wide enough. If the routes are likely to be used by people in wheelchairs, the minimum width will need to be 800 mm.

▶ While the workplace is in use, it must be possible to open all doors easily and immediately from the inside, without using a key or similar device. Doors must be readily opened in the direction of escape. Fire doors should be self-closing (fire doors to cupboards or lockers can be simply latched or locked).

▶ Make sure that there are no obstructions on escape routes, especially on corridors and stairways where people who are escaping could dislodge stored items or be caused to trip. Any fire hazards must be removed from exit routes as a fire on an exit route could have very serious consequences.

▶ Escape routes need regular checks to make sure that they are not obstructed and that exit doors are not locked. Self-closing fire-resisting doors should be checked to ensure doors close fully, including those fitted with automatic release mechanisms.

Figure 11.23 Fire escape route outside building

The maximum advisable travel distances from any area in a workplace to a fire exit door leading out to a relative place of safety should be in accordance with Table 11.2.

Table 11.2 Maximum travel distances

Fire hazard			
	Lower	**Normal**	**Higher**
Enclosed structures:			
Alternative	60 m	45 m	25 m
Dead-end	18 m	18 m	12 m
Semi-open structures:			
Alternative	200 m	100 m	60 m
Dead-end	25 m	18 m	12 m

Source: HSE.

(d) Lighting

Escape routes must be well lit. If the route has only artificial lighting or if it is used during the hours of darkness, alternative sources of lighting should be considered in case the power fails during a fire. Check the routes when it is dark as, for example, there may be street lighting outside that provides sufficient illumination. In small workplaces it may be enough to provide the staff with torches that they can use if the power fails. However, it may be necessary to provide battery-operated emergency lights so that if the mains lighting fails the lights will operate automatically. Candles, matches and cigarette lighters are not adequate forms of emergency lighting.

(e) Signs

Figure 11.24 International Fire Escape pictorial

Exit signs on doors or indicating exit routes should be provided where they will help people to find a safe escape route. Signs on exit routes should have directional arrows, 'up' for straight on and 'left', 'right' or 'down' according to the route to be taken. Advice on the use of all signs including exit signs can be found in Chapters 4 and 13.

(f) Escape times

Everyone in the building should be able to get to the nearest place of safety in between 2 and 3 minutes. This means that escape routes should be kept short. Where there is only one means of escape, or where the risk of fire is high, people should be able to reach a place of safety, or a place where there is more than one route available, in 1 minute.

The way to check this is to pace out the routes, walking slowly and noting the time. Start from where people work and walk to the nearest place of safety. Remember that the more people there are using the route, the longer they will take. People take longer to negotiate stairs and they are also likely to take longer if they have a disability.

Where fire drills are held, check how long it takes to evacuate each floor in the workplace. This can be used as a basis for assessment. If escape times are too long, it may be worth rearranging the workplace so that people are closer to the nearest place of safety, rather than undertake expensive alterations to provide additional escape routes.

Reaction time needs to be considered. This is the amount of time people will need for preparation before they escape. It may involve, for example, closing down machinery, issues of security or helping visitors or members of the public out of the premises. Reaction time needs to be as short as possible to reduce risk to staff. Assessment of escape routes should include this.

If reaction times are too long, additional routes may need to be provided. It is important that people know what to do in case of fire as this can lessen the time needed to evacuate the premises.

11.4.2 Evacuation procedures and emergency plans

(a) Introduction

Each workplace should have an emergency plan. The plan should include the action to be taken by staff in the event of fire, the evacuation procedure and the arrangements for calling the Fire and Rescue Authority.

For small workplaces, this could take the form of a simple fire action notice posted in positions where staff can read it and become familiar with it.

High fire-risk or larger workplaces will need more detailed plans, which take account of the findings of the risk assessment, for example the staff significantly at risk and their location. For large workplaces, notices giving clear and concise instructions of the routine to be followed in case of fire should be prominently displayed. The notice should include the method of raising an alarm in the case of fire and the location of an assembly point to which staff escaping from the workplace should report.

(b) Fire routines

Site managers must make sure that all employees are familiar with the means of escape in case of fire and their use, and with the routine to be followed in the event of fire.

To achieve this, routine procedures must be set up and made known to all employees, generally outlining the action to be taken in case of fire and specifically laying down the duties of certain nominated persons. Notices should be posted throughout the premises.

While the need in individual premises may vary, there are a number of basic components which should be considered when designing any fire routine procedures:

- the action to be taken on discovering a fire;
- the method of operating the fire alarm;
- the arrangements for calling the fire and rescue services;
- the stopping of machinery and plant;
- first-stage firefighting by employees;
- evacuation of the premises;
- assembly of staff, customers and visitors, and carrying out a roll call to account for everyone on the premises.

The procedures must take account of those people who may have difficulty in escaping quickly from a building because of their location or a disability. Insurance companies and other responsible people may need to be consulted where special procedures are necessary to protect buildings and plant during

or after people have been evacuated. For example there may be some special procedures necessary to ensure that a sprinkler system is operating in the event of fire.

(c) Supervisory duties/fire marshals

A member of the staff should be nominated to supervise all fire and emergency arrangements. This person should be in a senior position or at least have direct access to a senior manager. Senior members of the staff should be appointed as departmental fire marshals, with deputies for every occasion of absence, however brief. In the event of fire or other emergency, their duties would be, while it remains safe to do so, to ensure that:

- the alarm has been raised;
- the whole department, including toilets and small rooms, has been evacuated;
- the fire and rescue service has been called;
- fire doors are closed to prevent fire spread to adjoining compartments and to protect escape routes;
- plant and machinery are shut down wherever possible and any other actions required to safeguard the premises are taken where they do not expose people to undue risks;
- a roll call is carried out at the assembly point and the result reported to whoever is in control of the evacuation.

Under normal conditions, fire marshals should check that good standards of housekeeping and preventative maintenance exist in their department, that exits and escape routes are kept free from obstruction, that all firefighting appliances are available for use and fire points are not obstructed, that smoking is rigidly controlled, and that all members of staff under their control are familiar with the emergency procedure and know how to use the fire alarm and firefighting equipment.

(d) Assembly and roll call

Assembly points should be established for use in the event of evacuation. They should be in positions, preferably under cover, which are unlikely to be affected at the time of fire. In some cases, it may be necessary to make mutual arrangements with the occupiers of nearby premises.

In the case of small premises, a complete list of the names of all staff should be maintained so that a roll call can be made if evacuation becomes necessary.

In those premises where the number of staff would make a single roll call difficult, each departmental fire warden should maintain a list of the names of staff in their area. Roll call lists must be updated regularly.

(e) Fire notices

Printed instructions for the action to be taken in the event of fire should be displayed throughout the premises. The information contained in the instructions should be stated briefly and clearly. The staff and their deputies to whom specific duties are allocated should be identified.

Instruction for the immediate calling of the fire brigade in case of fire should be displayed at telephone switchboards, exchange telephone instruments and security lodges.

A typical fire notice is given in Appendix 11.2.

(f) Fire drills

Once a fire routine has been established, it must be tested at regular intervals in order to ensure that all staff are familiar with the action to be taken in an emergency.

The most effective way of achieving this is by carrying out fire drills at prescribed intervals. Drills should be held at least twice a year other than in areas dealing with hazardous processes, where they should be more

frequent. A programme of fire drills should be planned to ensure that all employees, including shift workers and part-time employees, are covered.

11.4.3 People with special needs

(a) General

Of all the people who may be especially at risk, employers will need to pay particular attention to people who have special needs, including those with a disability.

The Equality and Human Rights Commission estimates that 11 million people in this country have some form of disability, which may mean that they find it more difficult to leave a building if there is a fire. Under the Equalities Act 2010 if people with a disability could realistically expect to use premises, then employers or those in charge of premises must anticipate any reasonable adjustments that would make it easier for that right to be exercised.

The Equalities Act includes the concept of 'reasonable adjustments' and this can be carried over into fire

Figure 11.25 Fire evacuation diagram

safety law. It can mean different things in different circumstances. For a small business, it may be considered reasonable to provide contrasting colours on a handrail to help people with vision impairment to follow an escape route more easily. However, it might be unreasonable to expect that same business to install an expensive voice alarm system. Appropriate 'reasonable adjustments' for a large business or organisation may be much more significant.

Where people with special needs use or work on the premises, their needs should, so far as is practicable, be discussed with them. These will often be modest and may require only changes or modifications to existing procedures. You may need to develop individual 'personal emergency evacuation plans' (PEEPs) for disabled persons who frequently use a building. They will need to be confident of any plan/PEEP that is put in place after consultation with them. As part of your consultation exercise you will need to consider the matter of personal dignity.

If members of the public use the building, then those in control may need to develop a range of standard PEEPs which can be provided on request to a disabled person or others with special needs.

Guidance on removing barriers to the everyday needs of disabled people is in BS 8300. Much of this advice will also help disabled people during an evacuation.

Advice on the needs of people with a disability, including sensory impairment, is available from the organisations which represent various groups. (Names and addresses can be found on the internet.)

Many businesses have discovered the problem of reconciling their duties under the Equalities Act and RRFSO.

Employers need to consider the needs of all their staff and users of their building when fire evacuation strategies are being considered. This can include people with a surprisingly diverse range of access needs, not just those with perhaps more obvious disabilities such as wheelchair users. But access and egress for each of these groups can sometimes seem hard to reconcile with tightly controlled fire safety. For example Building Regulations require internal doors held open by electromagnetic devices to self-close when activated by smoke detectors, but a disabled person may not be able to open a heavy fire door once it has closed.

So what is more important, fulfilling fire safety requirements or fulfilling the requirements of the Equalities Act?

Essentially, fire safety concerns life-threatening incidents, the Equalities Act is about dignity and equal treatment. Basically, health and safety (including fire) overrides Equalities Act if there is a conflict. However, there really should not be a problem achieving both. In respect of fire, it is never acceptable to refuse someone

entry to an upper level on the basis of there not being an evacuation lift available. However, it is acceptable to pre-plan and decide to hold public meetings on the ground floor for this reason.

In other words, it should be possible to comply with fire safety legislation and comply with the Equalities Act, as long as there are management procedures in place to make sure both are adhered to successfully. In many cases, the answer lies in planning ahead, and implementing procedures which may involve other staff in personal evacuation plans, for example.

(b) Special needs, fire emergencies and precautions

If people with a disability are going to be in the premises, then employers must also provide a safe means for them to leave if there is a fire. Staff should be aware that people with a disability may not react, or can react differently, to a fire warning or a fire. You should give similar consideration to others with special needs such as parents with young children or the elderly.

In premises with a simple layout, a common-sense approach, such as offering to help lead a blind person or helping an elderly person down steps may be enough. In more complex premises, more elaborate plans and procedures will be needed, with trained staff assigned to specified duties.

Consider the needs of people with mental disabilities or spatial recognition problems. The range of disabilities encountered can be considerable, extending from mild epilepsy to complete disorientation in an emergency situation. Many of these can be addressed by properly trained staff, discreet and empathetic use of the 'buddy system' or by careful planning of colour and texture to identify escape routes.

People with special needs (including members of the public) need special consideration when planning for emergencies. But the problems this raises are seldom great. Employers should:

▶ identify everyone who may need special help to get out;
▶ allocate responsibility to specific staff to help people with a disability in emergency situations;
▶ consider possible escape routes;
▶ enable the safe use of lifts;
▶ enable people with a disability to summon help in emergencies;
▶ train staff to be able to help their colleagues;
▶ consider safe havens.

People with impaired vision must be encouraged to familiarise themselves with escape routes, particularly those not in regular use. A 'buddy' system would be helpful. But, to take account of absences, more than one employee working near anyone with impaired vision should be taught how to help them.

‘exitmaster’ stairway evacuation chairs for persons with disabilities or who cannot manage steps / stairs are located as follows:

Figure 11.26 Special addition to the fire notice for people with a disability

Figure 11.27 Stair evacuation chair for people with a disability

Where people have hearing difficulties, particularly those who are profoundly deaf, then simply hearing the fire warning is likely to be the major difficulty. If these persons are never alone while on the premises then this may not be a serious problem, as it would be reasonable for other occupants to let them know that the building should be evacuated. If a person with hearing difficulties is likely to be alone, then consider other means of raising the alarm. Among the most popular are visual beacons and vibrating devices or pagers that are linked to the existing fire alarm.

People with impaired hearing may not hear alarms in the same way as those with normal hearing but may still be able to recognise the sound. This may be tested during the weekly alarm audibility test. There are alternative means of signalling, such as lights or other visual signs, vibrating devices or specially selected sound signals.

Action on Hearing Loss, trading name of the Royal National Institute for Deaf People, can advise (see: www.actiononhearingloss.org.uk). Ask the fire brigade before installing alternative signals.

Wheelchair users or others with impaired mobility may need help to negotiate stairs, etc. Anyone selected to provide this help should be trained in the correct methods. Advice on the lifting and carrying of people can be obtained from the Fire and Rescue Service, Ambulance Service, British Red Cross Society, St John Ambulance Brigade or certain disability organisations.

Lifts should not be used as a means of escape in the event of a fire. If the power fails, the lift could stop between floors, trapping occupants in what may become a chimney of fire and smoke. But BS 5588: Part 8 provides advice on specially designed lifts for use by people with a disability in the event of a fire.

Employees with learning difficulties may also require special provision. Management should ensure that the colleagues of any employee with a learning difficulty know how to reassure them and lead them to safety.

11.4.4 Building plans and specifications

Plans and specifications can be used to assist understanding of a fire risk assessment or emergency plan. Even where not needed for this purpose they can help you and your staff keep your fire risk assessment and emergency plan under review and help the fire and rescue service in the event of fire. Any symbols used should be shown on a key. Plans and specifications could include the following:

▶ essential structural features such as the layout of function rooms, escape doors, wall partitions, corridors, stairways, etc. (including any fire-resisting structure and self-closing fire doors provided to protect the escape routes);
▶ location of refuges and lifts that have been designated suitable for use by people with a disability and others who may need assistance to escape in case of a fire;
▶ methods for fighting fire (details of the number, type and location of the firefighting equipment);
▶ location of manually operated fire alarm call points and control equipment for fire alarms;
▶ location of any control rooms and fire staff posts;
▶ location of any emergency lighting equipment and the exit route signs;
▶ location of any high-risk areas, equipment or process that must be immediately shut down by staff on hearing the fire alarm;

▶ location of any automatic firefighting systems, risers and sprinkler control valves;

▶ location of the main electrical supply switch, the main water shut-off valve and, where appropriate, the main gas or oil shut-off valves; and

▶ plans and specifications relating to all recent constructions.

This information should be passed on to any later users or owners of the premises.

11.5 Fire legislation in the UK – not part of the NEBOSH NGC syllabus

11.5.1 Introduction

Since the introduction of the Fire Services Act 1947, the fire authorities have had the responsibility for fighting fires in all types of premises. In 1971, the Fire Precautions Act gave the fire authorities control over certain fire procedures, means of escape and basic fire protection equipment through the drawing up and issuing of Fire Certificates in certain categories of building. The Fire Certification was mainly introduced to combat a number of serious industrial fires that had occurred, with a needless loss of life, where simple well-planned protection would have allowed people to escape unhurt.

Following a Government review in the 1990s, the Fire Precautions (Workplace) Regulations 1999 came into force in December of that year. These Regulations were made under the Health and Safety at Work (HSW) Act. They were amended in 1999 so as to apply to a wider range of premises including those already subject to the Fire Precautions Act 1971. In many ways these Regulations established the principles of fire risk assessment which would underpin a reformed legislative framework for fire safety. There remained a difference of opinion within Government, as to the right home for fire legislation. The Home Office believed that process fire issues should remain with the HSW Act, while general fire safety should have a different legislative vehicle.

In 2000 the Fire Safety Advisory Board was established to reform the fire legislation to simplify, rationalise and consolidate existing legislation. It would provide for a risk-based approach to fire safety allowing more efficient, effective enforcement by the fire and rescue service and other enforcing authorities.

The Regulatory Reform (Fire Safety) Order 2005 (RRFSO) SI No. 2005 1541 came into force on 1 October 2006. A summary of the Order is included in Chapter 13. The Fire Precautions Act 1971 was repealed and the Fire Precautions (Workplace) Regulations 1997 were revoked by the Order. When the CDM 2007 Regulations were introduced the RRFSO was amended, as the fire sections within the Construction (Health, Safety and Welfare) Regulations 1997 (now revoked) were transferred to the CDM 2007, and then retained in CDM 2015 which came into force in April 2015.

The RRFSO reforms the law relating to fire safety in non-domestic premises. The main emphasis of the changes is to move towards fire prevention. Fire Certificates under the Fire Precautions Act 1971 were abolished by the Order and cease to have legal status. The RRFSO replaces fire certification with a general duty to ensure, so far as is reasonably practicable, the safety of employees.

There is a duty to carry out fire risk assessments that should consider the safety of both employees and non-employees.

The Order imposes a number of specific duties in relation to fire precautions to be taken. The Order provides for the enforcement of the Order, appeals, offences and connected matters. The Order also gives effect in England and Wales to parts of a number of EU Directives including the Framework, Workplace, Chemical Agents and Explosive Atmospheres Directives.

11.5.2 The Regulatory Reform (Fire Safety) Order (RRFSO) – requirements

Outline of RRFSO

Part 1 General – The RRFSO applies to all non-domestic premises other than those listed in Article 6. The main duty-holder is the 'responsible person' in relation to the premises, defined in Article 3. The duties on the responsible person are extended to any person who has, to any extent, control of the premises to the extent of their control (Article 5).

Part 2 imposes duties on the responsible person in relation to fire safety in premises.

Part 3 provides for enforcement.

Part 4 provides for offences and appeals.

Part 5 provides for miscellaneous matters including firefighters' switches for luminous tube signs, maintenance measures provided to ensure the safety of firefighters, civil liability for breach of statutory duty by an employer, special requirements for licensed premises and consultation by other authorities.

Schedule 1 sets out the matters to be taken into account in carrying out a risk assessment (Parts 1 and 2), the general principles to be applied in implementing fire safety measures (Part 3) and the special measures to be taken in relation to dangerous substances (Part 4).

The Order is mainly enforced by the fire and rescue authorities.

PART 1 General

Meanings

The Order defines a responsible person as the person who is in control of the premises – this may be the owner or somebody else.

The meaning of general fire precautions is set out in the Order, which covers:

- reduction of fire risks and fire spread;
- means of escape;
- keeping means of escape available for use;
- firefighting;
- fire detection and fire warning;
- action to be taken in the event of fire;
- instruction and training of employees.

But they do not cover **process-related fire precautions.** These include the use of plant or machinery or the use or storage of any dangerous substances. Process-related fire precautions still come under either the Health and Safety Executive (HSE) or local authority, and are covered by the general duties imposed by the HSW Act 1974 or other specific Regulations like the Dangerous Substances and Explosive Atmosphere Regulations (DSEAR).

Duties

Duties are placed on a 'responsible person' who is:

- the employer in a workplace, to the extent they have control;
- any other person who has control of the premises; or
- the owner of the premises.

The obligations of a particular responsible person relate to matters within their control. It is therefore advisable that arrangements between responsible persons, where there may be more than one (such as contracts or tenancy agreements) should clarify the division of responsibilities.

Premises covered

The RRFSO does not cover domestic premises, offshore installations, a ship (normal shipboard activities under a master), remote fields, woods or other land forming part of an agricultural or forestry operation (it does cover the buildings), means of transport, a mine (it does cover the buildings at the surface) and a borehole site.

Other alternative provisions cover premises such as sports grounds.

PART 2 Fire Safety Duties

The responsible person must take appropriate general fire precautions to protect both employees and persons who are not employees to ensure that the premises are safe. Details are given in Chapter 13.

Risk assessment and arrangements

The responsible person must:

- make a 'suitable and sufficient' risk assessment to identify the general fire precautions required (usually known as a fire risk assessment) if a dangerous substance is or is liable to be present,

this assessment must include the special provisions of Part 1 of Schedule 1 to the Order;
- review the risk assessment regularly or when there have been significant changes; with regard to young persons, take into account Part 2 of Schedule 1;
- record the significant findings where five or more people are employed or where there is a licence or Alterations Notice in place for the premises;
- apply the principles of prevention Part 3 of Schedule 1;
- set out appropriate fire safety arrangements for planning, organisation, control, monitoring and review;
- eliminate or reduce risks from dangerous substances in accordance with Part 4 of Schedule 1.

Firefighting and fire detection

The responsible person must ensure that the premises are provided with appropriate:

- firefighting equipment (FFE);
- fire detectors and alarms;
- measures for firefighting which are adapted to the size and type of undertaking;
- trained and equipped competent persons to implement firefighting measures;
- contacts with external emergency services, particularly as regards firefighting, rescue work, first-aid and emergency medical care.

Emergency routes, exits and emergency procedures

The responsible person must ensure that routes to emergency exits and the exits themselves are kept clear and ready for use. See Chapter 13 for specific requirements for means of escape.

The responsible person must establish suitable and appropriate emergency procedures and appoint a sufficient number of competent persons to implement the procedures. This includes arrangements for undertaking fire drills and the provision of information and action regarding exposure to serious, imminent and unavoidable dangers. Where dangerous substances are used and/or stored, additional emergency measures covering the hazards must be set up. There should be appropriate visual or audible warnings and other communications systems to effect a prompt and safe exit from the endangered area.

Where necessary to protect persons' safety, all premises and facilities must be properly maintained and subject to a suitable system of maintenance.

Safety assistance

The responsible person must appoint (except in the case of the self-employed or partnerships where a person has the competence themselves) one or more people to assist in undertaking the preventative and protective measures. Competent persons must be given the time and means to fulfil their responsibilities. Competent persons must be kept informed of anything relevant to their role and have access to information on any dangerous substances present on the premises.

Provision of information

The responsible person must provide their own employees; and the employer of any employees of an outside undertaking; with comprehensible and relevant information on the risks, precautions taken, persons appointed for firefighting and fire drills, and appointment of competent persons. Additional information is also required for dangerous substances used and/or stored on the premises.

Before a child (16–18 years old) is employed, information on special risks to children must be given to the child's parent or guardian.

Capabilities, training and cooperation

The responsible person must ensure that adequate training is provided when people are first employed or exposed to new or increased risks. This may occur with new equipment, changes of responsibilities, new technologies, new systems of work and new substances used. The training has to be done in working hours and repeated periodically as appropriate.

Where two or more responsible persons share duties or premises, they are required to coordinate their activity and cooperate with each other, including keeping each other informed of risks.

Duties of employees

These are covered in Chapter 13 and are similar to requirements under the HSW Act. Employees must take care of themselves and other relevant persons. They must cooperate with the employer and inform them of any situation which they would reasonably consider to present a serious and immediate danger, or a shortcoming in the protection arrangements.

PART 3 Enforcement

Enforcing authorities

The fire and rescue authority in the area local to the premises is normally the enforcing authority. However, in stand-alone construction sites, premises which require a Nuclear license, and where a ship is under repair, the HSE is the enforcing authority, except for the general means of escape and fire emergency procedures under CDM 2015.

Fire and Rescue Authorities officers appointed under the RRFSO have similar powers to those under the HSW Act. These include the power to enter premises, make enquiries, require information, require facilities and assistance, take samples, require an article or substance to be dismantled or subjected to a process or test and the power to issue Enforcement Notices.

In addition to fire safety legislation, health and safety at work legislation also covers the elimination or minimisation of fire risks. As well as the particular and main general duties under the HSW Act, fire risks are covered by specific rules, such as for dangerous substances and explosive atmospheres, work equipment, electricity and other hazards. Thus, environmental health officers or HSE inspectors may enforce health and safety standards for the assessment and removal or control of process-related fire risks, where it is necessary, for the protection of workers and others.

Alterations and Enforcement Notices

Enforcing authorities can issue (see Chapter 13 for details):

Alterations Notices – where the premises constitute a serious risk to people either due to the features of the premises, or hazards present on the premises. The Notices require the responsible person to notify changes and may require them to record the significant findings of the risk assessment and safety arrangements and to supply a copy of the risk assessment before making changes.

Enforcement Notices – where there is a failure to comply with the requirements of the RRFSO. The Notice must specify the provisions concerned and MAY include directions on remedial action.

Prohibition Notices – where the risks are so serious that the use of the premises should be prohibited or restricted. The Notice MAY include directions on remedial action.

PART 4 Offences

Cases can be tried on summary conviction in a Magistrate's Court or on indictment in the Crown Court. The responsible person can be liable:

▶ on summary conviction to a fine not exceeding the statutory maximum;

▶ on indictment to an unlimited fine and/or imprisonment for up to 2 years for failure to comply with fire safety duties where there is a risk of death or serious injury and for failure to comply with an Alterations, Enforcement or Prohibition Notice.

Any person who fails to comply with their duties under the Order as regards fire risks can be prosecuted alongside or instead of a responsible person. Fines are limited to the statutory maximum (or levels 3 or 5 on the standard scale) on summary conviction but on indictment, the fine is unlimited.

Employees' Capabilities

The Fire Safety (Employees' Capabilities) (England) Regulations 2010 were introduced on 6 April 2010 to ensure that employers must take into account employees' capabilities as regards fire safety in entrusting tasks to them.

The making of these Regulations was a technical exercise to ensure that the EU legislation was properly transposed into English fire safety law. Not to do so may have resulted in EU infraction proceedings. Employers already had to comply with the regulation under wider health and safety legislation and the requirements of the 2010 Regulations were already implicit in the RRFSO.

11.5.3 Construction Design and Management Regulations 2015

Details of CDM 2015 are given in Chapter 13. Regulation 36 makes the Fire and Rescue authority responsible for the non-process fire-related sections of CDM 2015 on construction sites where other people are working within or as part of the site – for example, the work is being done in an occupied factory or office building.

The Regulations concerned are as follows.

Regulation 30: Emergency procedures

In so far as they concern fire emergency procedures and involve suitable procedures for evacuation, familiarisation of workers and testing the arrangements.

Regulation 31: Emergency routes and exits

In so far as they concern fire emergency routes and exits and involve the following:

▶ a sufficient number of emergency routes and exits;
▶ must lead directly as possible to an identified safe place;
▶ must be kept clear and free of obstruction;
▶ provided with emergency lights as necessary;
▶ suitably signed.

Regulation 32: Fire detection and firefighting

This involves the following:

▶ suitably located firefighting equipment (FFE);
▶ suitably located fire detection and alarm systems;
▶ proper maintenance, examination and testing;
▶ ease of access unless automatically activated;
▶ training for every person on the construction site in the operation of fire equipment that they are likely to use;
▶ where there is a particular risk of fire, people being instructed before starting the work;
▶ equipment suitably signed.

See Table 11.3 for the various enforcement authorities and their responsibilities on construction sites.

11.6 Fire risk assessment – not part of the NEBOSH NGC syllabus

11.6.1 General

Fire risk assessments come under the Regulatory Reform (Fire Safety) Order (RRFSO) and not under the Management of Health and Safety at Work Regulations 1999. See Section 11.5 for a summary of the RRFSO.

A fire risk assessment will indicate what fire precautions are needed. There are numerous ways of carrying out a fire risk assessment. The one described below is based on the method contained within Fire Safety Guides published by the Department of Communities and Local Government (see Appendix 11.1). A systematic approach, considered in five simple stages, is generally the best practical method. The RRFSO guides are downloadable on https://www.firesafe.org.uk/ministry-of-housing-communities-and-local-government-guides/

Table 11.3 Enforcement in respect of fire on construction sites

Type of premises fire issue		Fire and Rescue Authorities under RRFSO and CDM 2015	HSE under HSW Act, CDM 2015 and RRFSO	Local Authorities, Inspectors under HSW Act, and RRFSO	Defence fire service	A fire inspector or person appointed by the Secretary of State
Fire emergency procedures; escape routes and exits; fire detection and firefighting; and training issues on:	Stand-alone construction site		✓			
	Construction site where there are other activities (e.g. an occupied factory or office)	✓				
Process fire and explosion risks, where there are, e.g., tar boilers/ dangerous substances on:	Stand-alone sites		✓	✓		
	Sites where there are other activities (e.g. an occupied factory or office)					
Licensed nuclear sites and ships under construction or repair		✓				
Defence and Crown Armed Forces premises other than a ship					✓	✓
Sports grounds and sports stands			✓			
Crown premises except nuclear; UK Atomic Energy premises except nuclear						✓

397

11.6.2 **Stage 1 – identify fire hazards**

There are five main hazards produced by fire that should be considered when assessing the level of risk:

▶ oxygen depletion;
▶ flames and heat;
▶ smoke;
▶ gaseous combustion products;
▶ structural failure of buildings.

Of these, smoke and other gaseous combustion products are the most common cause of death in fires.

For a fire to occur, it needs sources of heat and fuel. If these hazards can be kept apart, removed or reduced, then the risks to people and businesses are minimised. **Identifying fire hazards** in the workplace is the first stage as follows:

(a) **Identify any combustibles** – Most workplaces contain combustible materials. Usually, the presence of normal stock in trade should not cause concern, provided the materials are used safely and stored away from sources of ignition. Good standards of housekeeping are essential to minimise the risk of a fire starting or spreading quickly.

The amount of combustible material in a workplace should be kept as low as is reasonably practicable. Materials should not be stored in gangways, corridors or stairways or where they may obstruct exit doors and routes. Fires often start and are assisted to spread by combustible waste in the workplace. Such waste should be collected frequently and removed from the workplace, particularly where processes create large quantities of it.

Some combustible materials, such as flammable liquids, gases or plastic foams, ignite more readily than others and quickly produce large quantities of heat and/or dense toxic smoke. Ideally, such materials should be stored away from the workplace or in fire-resisting stores. The quantity of these materials kept or used in the workplace should be as small as possible, normally no more than half a day's supply.

(b) **Identify any sources of heat** – All workplaces will contain heat/ignition sources; some will be obvious such as cooking sources, heaters, boilers, engines, smoking materials or heat from processes, whether in normal use or through carelessness or accidental failure. Others may be less obvious such as heat from chemical processes or electrical circuits and equipment.

Where possible, sources of ignition should be removed from the workplace or replaced with safer forms. Where this cannot be done, the ignition source should be kept well away from combustible materials or made the subject of management controls.

Particular care should be taken in areas where portable heaters are used or where smoking is permitted (now banned inside premises). Where heat is used as part of a process, it should be used carefully to reduce the chance of a fire as much as possible. Good security both inside and outside the workplace will help to combat the risk of arson.

Under smoke-free legislation, smoking is not permitted in enclosed or significantly enclosed areas. Outside designated safe areas should be provided for those who still wish to smoke. The smoking rules should be rigorously enforced (Figure 11.28).

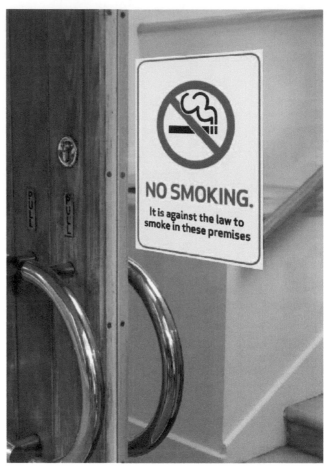

Figure 11.28 Smoke-free sign

Demolition work can involve a high risk of fire and explosion. In particular:

▶ Dismantling tank structures can cause the ignition of flammable residues. This is especially dangerous if hot methods are used to dismantle tanks before residues are thoroughly cleaned out. The work should only be done by specialists.
▶ Disruption and ignition of buried gas and electrical services is a common problem. It should always be assumed that buried services are present unless it is positively confirmed that the area is clear. A survey using service detection equipment must be carried out by a competent person to identify any services. The services should then be marked, competently purged or made dead, before any further work is done. A permit to excavate or dig is the normal formal procedure to cover buried services.

(c) **Identify any unsafe acts** – Persons undertaking unsafe acts such as smoking next to combustible materials, etc.

(d) **Identify any unsafe conditions** – These are hazards that may assist a fire to spread in the workplace, for example if there are large areas of hardboard or polystyrene tiles etc., or open stairs that can enable a fire to spread quickly, trapping people and engulfing the whole building.

An ideal method of identifying and recording these hazards is by means of a simple single-line plan, an example of which is illustrated in Figure 11.29. Checklists may also be used (see Appendix 11.1).

Figure 11.29 (a) Before fire risk assessment; (b) after fire risk assessment

399

11.6.3 Stage 2 – identify persons who are at significant risk

Consider the risk to any people who may be present. In many instances and particularly for most small workplaces the risk(s) identified will not be significant, and specific measures for persons in this category will not be required. There will, however, be some occasions when certain people may be especially at risk from the fire, because of their specific role, disability, sleeping, location or the workplace activity (see Section 11.4.3 for more information). Special consideration is needed if:

▶ sleeping accommodation is provided;
▶ persons are physically, visually or mentally challenged;
▶ people are unable to react quickly;
▶ persons are isolated.

People, such as visitors, the public or other workers, may come into the workplace from outside. The assessor must decide whether the current arrangements are satisfactory or if changes are needed.

Because fire is a dynamic event, which, if unchecked, will spread throughout the workplace, all people present will eventually be at risk if fire occurs. Where people are at risk, adequate means of escape from fire should be provided together with arrangements for detecting and giving warning of fire. Firefighting equipment suitable for the hazards in the workplace should be provided.

Some people may be at significant risk because they work in areas where fire is more likely or where rapid fire growth can be anticipated. Where possible, the hazards creating the high level of risk should be reduced. Specific steps should be taken to ensure that people affected are made aware of the danger and the action they should take to ensure their safety and the safety of others.

Young persons may not be employed unless risks to young persons have been addressed in the risk assessment. The issues to be covered are contained in Part 2 of Schedule 1 of the RRFSO 2005 (see Chapter 13).

11.6.4 Stage 3 – evaluate and reduce the risks

If the building has been built and maintained in accordance with Building Regulations and is being put to its designed use, it is likely that the means of escape provisions will either be adequate, or it will be easy to decide what is required in relation to the risk. Having identified the hazards and the persons at risk, the next stage is to reduce the chance of a fire occurring and spreading, thereby minimising the chance of harm to persons in the workplace. The principles of prevention laid down in the RRFSO should be followed at this stage. These are based on

EU Directive requirements and are therefore the same as those used in Regulation 4 of the Management Regulations (see Chapter 13).

Evaluate the risks

Attempt to classify each area as 'high', 'normal' or 'low risk'. If 'high risk', it may be necessary to reconsider the principles of prevention, otherwise additional compensatory measures will be required.

Low risk – Areas where there is minimal risk to persons' lives; where the risk of fire occurring is low; or the potential for fire, heat and smoke spreading is negligible and people would have plenty of time to react to an alert of fire.

Normal risk – Such areas will account for nearly all parts of most workplaces; where an outbreak of fire is likely to remain confined or spread slowly, with an effective fire warning allowing persons to escape to a place of safety.

High risk – Areas where the available time needed to evacuate the area is reduced by the speed of development of a fire, for example highly flammable or explosive materials stored or used (other than small quantities under controlled conditions); also where the reaction time to the fire alarm is slower because of the type of person present or the activity in the workplace, for example the infirm and elderly or persons sleeping on the premises.

Determine if the existing arrangements are adequate, or need improvement.

Matters that will have to be considered are:

▶ Means for detecting and giving warning in case of fire – can it be heard by all occupants? (see Section 11.3).
▶ Means of escape – are they adequate in size, number, location, well lit, unobstructed, safe to use, etc.? (see Section 11.4).
▶ Signs – for exits, fire routines, etc. (see Section 11.4).
▶ Firefighting arrangements – wall-mounted or in a cradle on fire exit routes, suitable types for hazards present and sufficient in number? (see Section 11.3).

11.6.5 Stage 4 – the findings (always recommended, see Stage 5 – review

The findings of the assessment and the actions (including maintenance) arising from it should be recorded. If five or more people are employed, or an Alterations Notice is required, a formal record of the significant findings and any measures proposed to deal with them must be recorded. (See summary of the Order in Section 11.5 and Chapter 13.) The record should indicate:

▶ the date the assessment was made;
▶ the hazards identified;

- any staff and other people especially at risk;
- what action needs to be taken, and by when (**action plan**);
- the conclusions arising.

The above guidelines are to be used with caution. Each part of the workplace must be looked at and a decision made on how quickly persons would react to an alert of fire in each area. Adequate safety measures will be required if persons are identified as being at risk. Where maximum travel distances (see Section 11.4 and Table 11.2) cannot be achieved, extra fire safety precautions will be needed.

Where persons are at risk or an unacceptable hazard still exists, additional fire safety precautions will be required to compensate for this, or alternatively repeat previous stages to manage risk to an acceptable level.

11.6.6 Stage 5 – monitor and review on a regular basis

The fire risk assessment is not a one-off procedure. It should be continually monitored to ensure that the existing fire safety arrangements and fire risk assessment remain realistic. The assessment should be reviewed if there is a significant change in the occupancy, work activity, the materials used or stored when building works are proposed, or when it is no longer thought to be valid.

11.6.7 Structural features

The workplace may contain features that could promote the rapid spread of fire, heat or smoke and affect escape routes. These features may include ducts or flues, openings in floors or walls, or combustible wall or ceiling linings. Where people are put at risk from these features, appropriate steps should be taken to reduce the potential for rapid fire spread by, for example, non-combustible automatic dampers fitted in ducts or to provide an early warning of fire so that people can leave the workplace before their escape routes become unusable.

Combustible wall or ceiling linings should not be used on escape routes and large areas should be removed wherever they are found. Other holes in fire-resisting floors, walls or ceilings should be filled in with fire-resisting material to prevent the passage of smoke, heat and flames.

11.6.8 Temporary workplaces, maintenance and refurbishment

Temporary workplaces such as construction sites, temporary buildings, festivals and fêtes all have requirements for fire precautions and means of escape in case of fire. The scale of the temporary workplace will dictate the requirements which will depend on:

- the number of persons working or visiting the site at any one time;
- the nature of the materials being used to construct the workplace or being used in the workplace. Are they flammable or highly flammable?
- height above or below the ground floor and how far it is to a place of safety;
- the location, whether in a remote area or close to water supplies and/or fire and rescue services;
- the size of the premises and whether audible warnings can be heard.

Risk assessments will be needed to determine the level of precautions necessary. Guidance has been given in Fire Safety in Construction 2nd edition HSG168 HSE Books downloadable at https://www.hse.gov.uk/pubns/books/hsg168.htm

Sources of heat or combustible materials may be introduced into the workplace during periods of maintenance or refurbishment. Where the work involves the introduction of heat, such as welding, this should be carefully controlled by a safe system of work, for example Hot Work Permit (see Chapter 4 for details). All materials brought into the workplace in connection with the work being carried out should be stored away from sources of heat and not obstruct exit routes.

11.6.9 Fire plans

Fire plans should be produced and attached to the fire risk assessment. A copy should be posted in the workplace. A single-line plan of the area or floor should be produced or an existing plan should be used which needs to show:

- escape routes, numbers of exits, number of stairs, fire-resisting doors, fire-resisting walls and partitions, places of safety, and the like;
- fire safety signs and notices including pictorial fire exit signs and fire action notices;
- the location of fire warning call points and sounders or rotary gongs;
- the location of emergency lights;
- the location and type of firefighting equipment.

See Figure 11.25 for an evacuation plan.

11.7 Further information

The Management of Health and Safety at Work Regulations 1999.

The Health and Safety (Safety Signs and Signals) Regulations 1996.

Regulatory Reform (Fire Safety) Order 2005.

Fire (Scotland) Act 2005.

Fire safety (Scotland) Regulations 2006.

Controlling fire and explosion risks in the workplace INDG370(rev1) 2013 ISBN 97807176 64856 www.hse.gov.uk/pubns/indg370.htm

The Dangerous Substances and Explosive Atmospheres Regulations 2002 SI 2002 No. 2776 ISBN 9780 11 042957 5.

Dangerous substances and explosive atmospheres, L138 second edition 2013. Dangerous Substances and Explosive Atmospheres Regulations 2002. Approved Code of Practice, 2003, HSE Books ISBN 978 0 7176 66164 www.hse.gov.uk/pubns/books/l138.htm

Fire Risk Assessment – Offices and Shops DCLG ISBN 978 1 85112 815 0.

Fire Risk Assessment – Factories and Warehouses DCLG ISBN 978 1 85112 816 7.

Fire Risk assessment – Sleeping accommodation DCLG ISBN 978 1 85112 817 4

Fire Risk assessment – Residential care premises DCLG ISBN 978 1 85112 818 1

Fire Risk assessment – Educational premises DCLG ISBN 978 1 85112 819 8

Fire Risk assessment – Small and medium places of assembly DCLG ISBN 978 1 85112 820

Fire Risk assessment – Large places of assembly DCLG ISBN 978 1 85112 821 1

Fire Risk assessment – Theatres, cinemas and similar places DCGL ISBN 978 1 85112 822

Fire Risk assessment – Open air events and venues DCGL ISBN 978 1 85112 823 5

Fire Risk assessment – Health care premises DCGL ISBN 978 1 85112 824 2

Fire Risk assessment – Transport premises and facilities DCGL ISBN 978 1 85112 825 9 www.communities.gov.uk/fire/firesafety/firesafetylaw

https://www.gov.uk/government/publications/fire-safety-approved-document-b

11.8 Practice questions

1. (a) **Outline** the differences between the three types of enforcement notice which may be issued under the Regulatory Reform (Fire Safety) Order.
   (b) **Identify** the principal fire safety duties of the 'responsible person' as defined by the Regulatory Reform (Fire Safety) Order.

2. (a) **Describe** the three sides of a 'fire triangle'.
   (b) **Outline THREE** methods of extinguishing fires.
   (c) **Identify FOUR** different types of ignition source that may cause a fire to occur, and give a typical workplace example of **EACH** type.
   (d) **Identify FIVE** classes of fire, the associated fuel sources and the appropriate fire extinguisher to use for each class.

3. **Outline THREE** methods of heat transfer (other than direct burning) and **explain** how each can cause the spread of fire in buildings.

4. (a) **Identify SIX** causes of fire in a workplace.
   (b) **Identify** the possible consequences on people of fires in a workplace.
   (c) **Outline** the main control measures to minimise the risk of fire in a workplace.

5. (a) **Identify** the five stages of a fire risk assessment in a workplace.
   (b) **Outline** the particular issues that should be considered in the fire risk assessment of temporary workplaces.
   (c) **Identify** additional factors to be taken into account during building refurbishment work.

6. **Outline** the contents of a typical fire plan that should be attached to a fire risk assessment.

7. (a) **Outline** effective control measures for the safe storage and handling of small containers containing flammable solvents.
   (b) **Identify EIGHT** types of unsatisfactory working practices using flammable gas cylinders that could increase the risk of a fire or explosion.

8. Fire protection is an important feature in the design of a new building.
   (a) **Identify SIX** structural measures that can help to prevent the spread of fire and smoke.
   (b) **Outline** the issues involved in the fire compartmentation of a building.

9. The failure of electrical equipment is one of the main causes of workplace fires.
   (a) **Outline** how fires could be caused by the failure of electrical cables and equipment.
   (b) **Identify TWO** types of extinguisher that can be used safely on fires involving electrical equipment and **identify** those extinguishers that should not be used in such fires.
   (c) **Outline** suitable control measures that should be taken to minimise the risk of fire from electrical equipment when used in a flammable atmosphere.

10. (a) **Identify TWO** types of emergency warning systems that can be installed in the building to ensure that all occupants can be made aware of the need to evacuate the building.

(b) **Outline** the main factors to be considered in the siting of fire extinguishers.

(c) **Outline** suitable arrangements for the inspection and maintenance of fire extinguishers in the workplace.

11. (a) **Outline ONE** advantage and **ONE** limitation of each of the main types of fire extinguishers.

(b) **Outline TWO** advantages and **TWO** disadvantages of using hose reels as a means of extinguishing fires.

(c) **Identify** the issues associated with the access for fire and rescue services to a workplace.

12. (a) **Outline** the principal requirements for a safe means of escape from a building in the event of a fire.

(b) **Outline** the role of fire marshals in the event of a fire emergency.

(c) **Outline** reasons that may delay the safe evacuation of occupants from a workplace during a fire.

13. (a) **Identify** the issues to consider for the location of an assembly point for use in a workplace evacuation.

(b) **Outline** the reasons for undertaking regular fire drills in the workplace.

(c) **Outline** the contents of a training programme for employees on the emergency action to take in the event of fire.

11

APPENDIX 11.1 Fire risk assessment checklist (as recommended in Fire Safety Guides published by the Department for Communities and Local Government in 2006)

Checklist

Follow the five key steps – Fill in the checklist – Assess your fire risk and plan fire safety.

1. Fire safety risk assessment

Fire starts when heat (source of ignition) comes into contact with fuel (anything that burns) and oxygen (air).

You need to keep sources of ignition and fuel apart.

How could a fire start?

Think about heaters, lighting, naked flames, electrical equipment, hot processes such as welding or grinding, cigarettes, matches and anything else that gets very hot or causes sparks.

What could burn?

Packaging, rubbish and furniture could all burn just like the more obvious fuels such as petrol, paint, varnish and white spirit. Also think about wood, paper, plastic, rubber and foam. Do the walls or ceilings have hardboard, chipboard or polystyrene? Check outside, too.

Check

▶ Have you found anything that could start a fire?

Make a note of it.

▶ Have you found anything that could burn?

Make a note of it.

2. People at risk

People at risk

Everyone is at risk if there is a fire. Think whether the risk is greater for some because of when or where they work, such as night staff, or because they're not familiar with the premises, such as visitors or customers.

Children, the elderly or disabled people are especially vulnerable.

Check

Have you identified:

▶ who could be at risk?
▶ who could be especially at risk?

Make a note of what you have found.

3. Evaluate and act

Evaluate

First, think about what you have found in Steps 1 and 2: what are the risks of a fire starting, and what are the risks to people in the building and nearby?

Remove and reduce risk

How can you avoid accidental fires? Could a source of heat or sparks fall, be knocked or pushed into something that would burn? Could that happen the other way round?

Protect

Take action to protect your premises and people from fire.

Check

▶ Have you assessed the risks of fire in your workplace?
▶ Have you assessed the risk to staff and visitors?
▶ Have you kept any source of fuel and heat/sparks apart?
▶ If someone wanted to start a fire deliberately, is there anything around they could use?
▶ Have you removed or secured any fuel an arsonist could use?
▶ Have you protected your premises from accidental fire or arson?

How can you make sure everyone is safe in case of fire?

▶ Will you know there is a fire?
▶ Do you have a plan to warn others?
▶ Who will make sure everyone gets out
▶ Who will call the fire service?
▶ Could you put out a small fire quickly and stop it spreading?

How will everyone escape?

▶ Have you planned escape routes?
▶ Have you made sure people will be able to safely find their way out, even at night if necessary?
▶ Does all your safety equipment work?
▶ Will people know what to do and how to use equipment?

Make a note of what you have found.

4. Record, plan and train

Record

Keep a record of any fire hazards and what you have done to reduce or remove them. If your premises are small, a record is a good idea. If you have five or more staff or have a licence, then you must keep a record of what you have found and what you have done.

Plan

You must have a clear plan of how to prevent fire and how you will keep people safe in case of fire. If you share a building with others, you need to coordinate your plan with them.

Train

You need to make sure your staff know what to do in case of fire, and if necessary, are trained for their roles.

Check

▷ Have you made a record of what you have found, and action you have taken?

▷ Have you planned what everyone will do if there is a fire?

▷ Have you discussed the plan with all staff?

Have you:

▷ informed and trained people (practised a fire drill and recorded how it went)?

▷ nominated staff to put in place your fire prevention measures, and trained them?

▷ made sure everyone can fulfil their role?

▷ informed temporary staff?

▷ consulted others who share a building with you, and included them in your plan?

5. Review

Keep your risk assessment under regular review. Over time, the risks may change.

If you identify significant changes in risk or make any significant changes to your plan, you must tell others who share the premises and, where appropriate, retrain staff.

Have you:

▷ made any changes to the building inside or out?

▷ had a fire or near miss?

▷ changed work practices?

▷ begun to store chemicals or dangerous substances?

▷ significantly changed your stock or stock levels?

▷ planned your next fire drill?

The checklist above can help you with the fire risk assessment but you may need additional information especially if you have large or complex premises.

Source: Department for Community and Local Government.

APPENDIX 11.2 Typical fire notice

Fire action
If you discover a fire

Operate nearest fire alarm point.

Call the Fire Brigade by telephoning ▪ 999

Leave the building by the nearest exit.

Report to your assembly point at ▪▬▬▬▬

Do not stop to collect personal belongings

11

CHAPTER 12

Electricity

> **Chapter summary**
>
> **This chapter covers the following topics in Element 12 of the NEBOSH NGC syllabus:**
> Hazards and risks
> Control measures
>
> **Learning outcomes**
>
> **These are the learning outcomes covered by this chapter:**
> **The learner will be able to:**
> 1. Outline the hazards, risks and dangers associated with the use of electricity in the workplace
> 2. Outline the control measures that should be taken when working with electrical systems or using electrical equipment in all workplace conditions

12.1 Principles, hazards and risks associated with the use of electricity at work

12.1.1 Introduction

Electricity is a widely used, efficient and convenient, but potentially hazardous method of transmitting and using energy. It is in use in every factory, workshop, laboratory and office in the country. Any use of electricity has the potential to be very hazardous with possibly fatal results. Legislation has been in place for many years to control and regulate the use of electrical energy and the activities associated with its use. Such legislation provides a framework for the standards required in the design, installation, maintenance and use of electrical equipment and systems and the supervision of these activities to minimise the risk of injury. Electrical work from the largest to the smallest installation must be carried out by people known to be competent to undertake such work. New installations always require expert advice at all appropriate levels to cover design aspects of the system and those of its associated equipment. Electrical systems and equipment must be properly selected, installed, used and maintained.

In the UK approximately 8% of all fatalities at work are caused by electric shock. Over the last few years, there have been 1,000 electrical accidents each year and about 30 people die of their injuries annually. The majority of the fatalities occur in the agriculture, extractive and utility supply and service industries, whereas the majority of the major accidents happen in the manufacturing, construction and service industries.

Only voltages up to and including **mains voltage** (220/240V) and the three principal electrical hazards – electric shock, electric burns and electrical fires and explosions – are considered in detail in this chapter.

Basic principles of electricity

In simple terms, electricity is the flow or movement of electrons through a substance which allows the transfer of electrical energy from one position to another. The substance through which the electricity flows is called a **conductor**. This flow or movement of electrons is known as the **electric current**. There are two forms of electric current – direct and alternating. **Direct current (DC)** involves the flow of electrons along a conductor from one end to the other. This type of current is mainly restricted to batteries, dynamos and similar devices. **Alternating current (AC)** is produced by a rotating magnetic alternator that produces a change in polarity every half turn causing an oscillation of the electrons rather than a flow of electrons so that energy is passed from one electron to the adjacent one and so on through the length of the conductor.

It is sometimes easier to understand the basic principles of electricity by comparing its movement with that of water in a pipe flowing downhill. The flow of water through the pipe (measured in litres per second) is similar to the current flowing through the conductor, which is measured in amperes, normally abbreviated to **amps (A)** Sometimes very small currents are used and these are measured in milliamps (mA).

Figure 12.1 Beware of electricity – typical sign

The higher the pressure drop is along the pipeline, the greater will be the flow rate of water and, in a similar way, the higher the electrical 'pressure difference' along the conductor, the higher the current will be. This electrical 'pressure difference' or potential difference is measured in **volts (V)**.

The flow rate through the pipe will also vary for a fixed pressure drop as the roughness on the inside surface of the pipe varies – the rougher the surface, the slower the flow and the higher the resistance to flow becomes. Similarly, for electricity, the poorer the conductor, the higher the resistance is to electrical current and the lower the current becomes. Electrical resistance is measured in **ohms**.

The voltage (V), the current (I) and the resistance (R) are related by the following formula, known as Ohm's law:

$$V = I \times R \text{ (volts)}$$

and, electrical power (P) is given by:

$$P = V \times I \text{ (watts)}$$

These basic formulae enable simple calculations to be made so that, for example, the correct size of fuse may be ascertained for a particular piece of electrical equipment.

Conductors and insulators

Conductors are nearly always metals, copper being a particularly good conductor, and are usually in wire form but they can be gases or liquids, water being a particularly good conductor of electricity. Superconductor is a term given to certain metals which have a very low resistance to electricity at low temperatures.

Very poor conductors are known as **insulators** and include materials such as rubber, timber and plastics. Insulating material is used to protect people from some of the hazards associated with electricity.

Short circuit

Electrical equipment components and an electrical power supply (normally the mains or a battery) are joined together by a conductor to form a **circuit**. If the circuit is broken in some way producing a fault so that the current flows directly to earth rather than to a piece of equipment, a **short circuit** is made. As the resistance is greatly reduced but the voltage remains the same, a rapid increase in current occurs which could cause significant problems if suitable protection were not available. A short circuit is thus a live to earth contact without resistance.

Earthing

The principle of **earthing** is that the Earth is basically a huge conductor which accepts any electrical current that is conducted to it. Any metal in an electrical device with which a person might come into contact (for example the metal casing of a food mixer) should be connected to earth so that if, due to a fault, the casing becomes live; the electricity will flow away to earth – an easier path than through the body. In the USA this process is called grounding.

The electricity supply company has one of its conductors solidly connected to earth and every circuit supplied by the company must have one of its conductors connected to earth. This means that if there is a fault, such as a break in the circuit, the current, known as the fault current, will return directly to earth, which forms the circuit of least resistance, thus maintaining the supply circuit. Other devices, such as fuses and residual current devices (RCDs), which will be described later, will also be needed within the circuit to interrupt the current flow to earth so as to protect people from electric shock and equipment from overheating. Good and effective earthing is absolutely essential and must be connected and checked by a competent person. Where a direct contact with earth is not possible, for example in a motor car, a common voltage reference point is used, such as the vehicle chassis.

Where other potential metallic conductors exist near to electrical conductors in a building, they must be connected to the main earth terminal to ensure **equipotential bonding** of all conductors to earth. This applies to gas, water and central heating pipes and other devices such as lightning protection systems. **Supplementary bonding** is required in bathrooms and kitchens where, for example, metal sinks and other metallic equipment surfaces are present. This involves the connection of a conductor from the sink to a water supply pipe which has been earthed by equipotential bonding. There have been several fatalities due to electric shocks from 'live' service pipes or kitchen sinks.

Some definitions

Certain terms are frequently used with reference to electricity and the more common ones are defined here.

Low voltage – This is a voltage normally not exceeding 1000V DC or 600V AC between conductors and earth or 1000V AC between phases. Mains voltage falls into this category.

High voltage – This is defined in national and international standards as a voltage exceeding 1000V DC or 600V AC between conductors and earth or 1000V AC between phases.

Mains voltage – The common voltage available in domestic premises and many workplaces and normally taken from three-pin socket points. In the UK, it is distributed by the national grid and is usually supplied between 220 and 240V AC and at 50 cycles/s.

Maintenance – A combination of any actions carried out to retain an item of electrical equipment in, or restore it to, an acceptable and safe condition.

Testing – A measurement carried out to monitor the conditions of an item of electrical equipment without physically altering the construction of the item or the electrical system to which it is connected.

12

Inspection – A maintenance action involving the careful scrutiny of an item of electrical equipment, using, if necessary, all the senses to detect any failure to meet an acceptable and safe condition. An inspection does not include any dismantling of the item of equipment.

Examination – An inspection together with the possible partial dismantling of an item of electrical equipment, including measurement and non-destructive testing as required, in order to arrive at a reliable conclusion as to its condition and safety.

Isolation – Involves cutting off the electrical supply from all or a discrete section of the installation by separating the installation or section from every source of electrical energy. This is the normal practice so as to ensure the safety of persons working on or in the vicinity of electrical components which are normally live and where there is a risk of direct contact with live electricity.

Competent electrical person – A person possessing sufficient electrical knowledge and experience to avoid the risks to health and safety associated with electrical equipment and electricity in general.

12.1.2 Hazards and risks

Electricity is a safe, clean and quiet method of transmitting energy. However, this apparently benign source of energy when accidentally brought into contact with conducting material, such as people, animals or metals, permits dangerous releases of energy which may result in serious damage or loss of life. Constant awareness is necessary to avoid and prevent danger from accidental releases of electrical energy.

The principal hazards associated with electricity are:

▶ electric shock
▶ electric burns
▶ electrical fires and explosions
▶ arcing
▶ secondary hazards.

The use of portable electrical equipment can lead to a higher likelihood of these hazards occurring.

Electric shock and burns

There are 1,000 workplace accidents and 30 fatalities involving electric shock and burns reported to the HSE each year.

Electric shock is the convulsive reaction by the human body to the flow of electric current through it. This sense of shock is accompanied by pain and, in more severe cases, by burning. The shock can be produced by low voltages, high voltages or lightning. Most incidents of electric shock occur when the person becomes the route to earth for a live conductor. The effect of electric shock and the resultant severity of injury depend upon the size of the electric current passing through the body which, in turn, depends on the voltage and the electrical resistance of the skin and body. If a person

comes into contact with a voltage above about 50 volts, they can receive a range of injuries, including those directly resulting from electrical shock (such as problems with breathing and heart function); and indirect effects resulting from loss of control (such as falling from height or coming into contact with moving machinery). The chance of being injured by an electric shock increases where it is damp or where there is a lot of metalwork.

If the skin is wet, a shock from mains voltage (220/240V) could well be fatal. The effect of shock is very dependent on conditions at the time but it is always dangerous and must be avoided. Electric burns are usually more severe than those caused by heat, since they can penetrate deep into the tissues of the body.

The effect of electric current on the human body depends on its pathway through the body (e.g. hand to hand or hand to foot), the frequency of the current, the length of time of the shock and the size of the current. Current size is dependent on the duration of contact and the electrical resistance of body tissue. The electrical resistance of the body is greatest in the skin and is approximately 100,000 ohm; however, this may be reduced by a factor of 100 when the skin is wet. The body beneath the skin offers very little resistance to electricity due to its very high water content and, while the overall body resistance varies considerably between people and during the lifetime of each person, it averages 1,000 ohm. Skin that is wounded, bruised or damaged will considerably reduce human electrical resistance and work should not be undertaken on electrical equipment if damaged skin is unprotected.

An electric current of 1mA is detectable by touch and one of 10mA will cause muscle contraction which may prevent the person from being able to release the conductor, and if the chest is in the current path, respiratory movement may be prevented, causing asphyxia. Current passing through the chest may also cause fibrillation of the heart (vibration of the heart muscle) and disrupt the normal rhythm of the heart, though this is likely only within a particular range of currents. The shock can also cause the heart to stop completely (cardiac arrest) and this will lead to the cessation of breathing. Current passing through the respiratory centre of the brain may cause respiratory arrest that does not quickly respond to the breaking of the electrical contact. These effects on the heart and respiratory system can be caused by currents as low as 25mA. It is not possible to be precise on the threshold current because it is dependent on the environmental conditions at the time, as well as the age, sex, body weight and health of the person.

Alternating current (AC) and direct current (DC) have slightly different effects on the human body, but both are dangerous above a certain voltage. The risk of injury changes according to the frequency of the AC, and it is common for DC to have an AC component (called ripple). The effect on a particular person is very difficult to predict as it depends upon a large number of factors, as mentioned above.

Burns of the skin occur at the point of electrical contact due to the high resistance of skin. These burns may be deep, slow to heal and often leave permanent scars. Burns may also occur inside the body along the path of the electric current, causing damage to muscle tissue and blood cells. These burns result from the direct contact with an electrical source. Burns can also be caused by indirect contact with an electrical source when such a source results in a fire. Successive annual UK fire statistics show that faulty appliances and leads continue to be the single most common cause of accidental fires in 'other buildings' (non-dwellings).

Burns associated with radiation and microwaves are dealt with in Chapter 6.

Treatment of electric shock and burns

There are many excellent posters available which illustrate a first-aid procedure for treating electric shock and such posters should be positioned close to electrical junction boxes or isolation switches (Figure 12.2). The

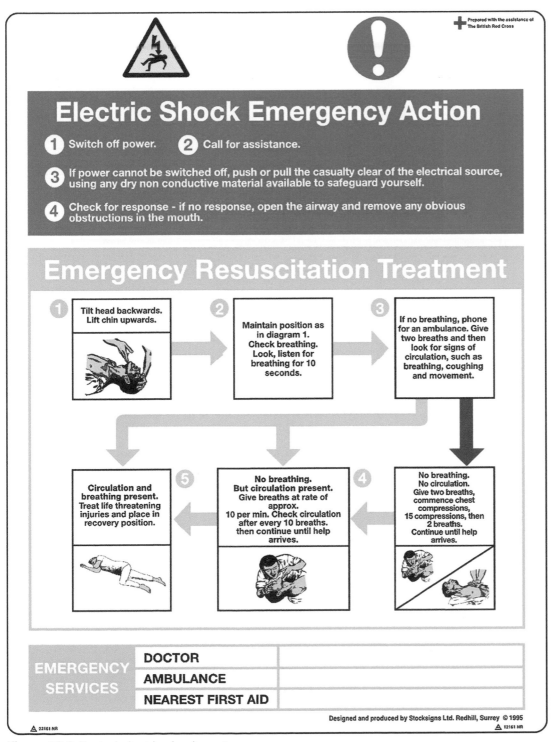

Figure 12.2 Typical electric shock poster

recommended procedure for treating an unconscious person who has received a **low-voltage** electric shock is as follows:

1. On finding a person suffering from electric shock, raise the alarm by calling for help from colleagues (including a trained first-aider).
2. Switch off the power if it is possible and/or the position of the emergency isolation switch is known.
3. Call for an ambulance.
4. If it is not possible to switch off the power, then push or pull the person away from the conductor using an object made from a good insulator, such as a wooden chair or broom. Remember to stand on dry insulating material, for example, a wooden pallet, rubber mat or wooden box. If these precautions are not taken, then the rescuer will also be electrocuted.
5. If the person is breathing, place him/her in the recovery position so that an open airway is maintained and the mouth can drain if necessary.
6. If the person is not breathing, apply mouth-to-mouth resuscitation and, in the absence of a pulse, chest compressions. When the person is breathing normally place them in the recovery position.
7. Treat any burns by placing a sterile dressing over the burn and secure with a bandage. Any loose skin or blisters should not be touched nor any lotions or ointments applied to the burn wound.
8. If the person regains consciousness, treat for normal shock.
9. Remain with the person until they are taken to a hospital or local surgery.

It is important to note that electrocution by high-voltage electricity is normally instantly fatal. On discovering a person who has been electrocuted by high-voltage electricity, the police and electricity supply company should be informed. If the person remains in contact with or within 18 m of the supply, then he/she should not be approached to within 18 m by others until the supply has been switched off and clearance has been given by the emergency services. High-voltage electricity can 'arc' over distances less than 18 m, thus electrocuting the would-be rescuer (Figure 12.3).

Electrical fires and explosions

Over 25% of all fires have a cause linked to a malfunction of either a piece of electrical equipment or wiring or both. Electrical fires are often caused by a lack of reasonable care in the maintenance and use of electrical installations and equipment. The electricity that provides heat and light and drives electric motors is capable of igniting insulating or other combustible material if the equipment is

18 m (20 yd)

Figure 12.3 Keep 18 m clear of high-voltage lines

misused, is not adequate to carry the electrical load or is not properly installed and maintained. The most common causes of fire in electrical installations are short circuits, overheating of cables and equipment, the ignition of flammable gases and vapours and the ignition of combustible substances by static electrical discharges.

Short circuits happen, as mentioned earlier, if insulation becomes faulty, and an unintended flow of current between two conductors or between one conductor and earth occurs. The amount of the current depends, among other things, upon the voltage, the condition of the insulating material and the distance between the conductors. At first the current flow will be low, but as the fault develops the current will increase and the area surrounding the fault will heat up. In time, if the fault persists, a total breakdown of insulation will result and excessive current will flow through the fault. If the fuse fails to operate or is in excess of the recommended fuse rating, overheating will occur and a fire will result. A fire can also be caused if combustible material is in close proximity to the heated wire or hot sparks are ejected. Short circuits are most likely to occur where electrical equipment or cables are susceptible to damage by water leaks or mechanical damage. Twisted or bent cables can also cause breakdowns in insulation materials.

Figure 12.4 Over 25% of fires are caused by electrical malfunction

Inspection covers and cable boxes are particular problem areas. Effective steps should be taken to prevent the entry of moisture as this will reduce or eliminate the risk. Covers can themselves be a problem especially in dusty areas where the dust can accumulate on flat insulating surfaces resulting in tracking between conductors at different voltages and a subsequent insulation failure. The interior of inspection panels should be kept clean and dust-free by using a suitable vacuum cleaner.

Overheating of cables and equipment will occur if they become overloaded. Electrical equipment and circuits are normally rated to carry a given safe current which will keep the temperature rise of the conductors in the circuit or appliance within permissible limits and avoid the possibility of fire. These safe currents define the maximum size of the fuse (the fuse rating) required for the appliance. A common cause of circuit overloading is the use of equipment and cables which are too small for the imposed electrical load. This is often caused by the addition of more and more equipment to the circuit, thus taking it beyond its original design specification. In offices, the overuse of multi-socket incorrectly fused outlet adaptors can create overload problems (sometimes known as the Christmas tree effect). The more modern multi-plugs are much safer as they lead to one fused plug and cannot be easily overloaded (see Figure 12.5). Another cause of overloading is

Figure 12.5 Ten-way surge-protected extension with telecoms

mechanical breakdown or wear of an electric motor and the driven machinery. Motors must be maintained in good condition with particular attention paid to bearing surfaces. Fuses do not always provide total protection against the overloading of motors and, in some cases, severe heating may occur without the fuses being activated.

Loose cable connections are one of the most common causes of overheating and may be readily detected (as well as overloaded cables) by a thermal imaging survey (a technique which indicates the presence of hot spots). The bunching of cables together can also cause excessive heat to be developed within the inner cable leading to a fire risk. This can happen with cable extension reels, which have only been partially unwound, used for high-energy appliances like an electric heater.

Ventilation is necessary to maintain safe temperatures in most electrical equipment and overheating is liable to occur if ventilation is in any way obstructed or reduced. All electrical equipment must be kept free of any obstructions that restrict the free supply of air to the equipment and, in particular, to the ventilation apertures.

Most electrical equipment either sparks in normal operation or is liable to spark under fault conditions. Some electrical appliances, such as electric heaters, are specifically designed to produce high temperatures. These circumstances create fire and explosion hazards, which demand very careful assessment in locations where processes capable of producing flammable concentrations of gas or vapour are used, or where flammable liquids are stored.

It is likely that many fires are caused by static electrical discharges. Static electricity can, in general, be eliminated by the careful design and selection of materials used in equipment and plant, and the materials used in products being manufactured. When it is impractical to avoid the generation of static electricity, a means of control must be devised. Where flammable materials are present, especially if they are gases or dusts, then there is a great danger of fire and explosion, even if there is only a small discharge of static electricity. The control and prevention of static electricity is considered in more detail later.

The use of electrical equipment in potentially flammable atmospheres should be avoided as far as possible. However, there will be many cases where electrical equipment must be used and, in these cases, the standards for the construction of the equipment should comply with the Equipment and Protective Systems Intended for Use in Potentially Explosive Atmospheres Regulations, known as ATEX. Details on the classification or zoning of areas are contained in the Dangerous Substances and Explosive Atmospheres Regulations and ACoPs.

12

Before electrical equipment is installed in any location where flammable vapours or gases may be present, the area must be zoned in accordance with the Dangerous Substances and Explosive Atmosphere Regulations, and records of the zoned areas must be marked on building drawings and revised when any zoned area is changed. The installation and maintenance of electrical equipment in potentially flammable atmospheres is a specialised task. It must only be undertaken by electricians or instrument mechanics who are trained to ATEX standards.

In the case of a fire involving electrical equipment, the first action must be the isolation of the power supply so that the circuit is no longer live. This is achieved by switching off the power supply at the mains isolation switch or at another appropriate point in the system. Where it is not possible to switch off the current, the fire must be attacked in a way which will not cause additional danger. The use of a non-conducting extinguishing medium, such as carbon dioxide or powder, is necessary. After extinguishing such a fire, careful watch should be kept for renewed outbreaks until the fault has been rectified. Re-ignition is a particular problem when carbon dioxide extinguishers are used, although less equipment may be damaged than is the case when powder is used.

Finally, the chances of electrical fires occurring are considerably reduced if the original installation was undertaken by competent electricians working to recognised standards, such as the Institution of Electrical Engineers' Code of Practice. It is also important to have a system of regular testing and inspection in place so that any remedial maintenance can take place.

Electric arcing

A person who is standing on earth too close to a high-voltage conductor may suffer flash burns as a result of arc formation. Such burns may be extensive and lower the resistance of the skin – so the electric shock may add to the ill-effects. Electric arc faults can cause temporary blindness by burning the retina of the eye and this may lead to additional secondary hazards. The quantity of electrical energy is as important as the size of the voltage since the voltage will determine the distance over which the arc will travel. The risk of arcing can be reduced by the insulation of live conductors.

Strong electromagnetic fields induce surface charges on people. If these charges accumulate, skin sensation is affected and spark discharges to earth may cause localised pain or bruising. Whether prolonged exposure to strong fields has any other significant effects on health has not been proved. However, the action of an implanted cardiac pacemaker may be disturbed by the close proximity of its wearer to a powerful electromagnetic field. The health effects of arcing and other non-ionising radiation are covered in Chapter 6.

Static electricity

Static electricity is produced by the build-up of electrons on weak electrical conductors or insulating materials. These materials may be gaseous, liquid or solid and may include flammable liquids, powders, plastic films and granules. Plastics have a high resistance that enables them to retain static charges for long periods of time. The generation of static may be caused by the rapid separation of highly insulated materials by friction or by transfer from one highly charged material to another in an electric field by induction (see Figure 12.6).

Dispensing Flammable Liquids

Figure 12.6 Prevention of static discharge; container connected to earthed drum

A static electric shock, perhaps caused by closing a door with a metallic handle, can produce a voltage greater than 10,000V. Since the current flows for a very short period of time, there is seldom any serious harm to an individual. However, discharges of static electricity may be sufficient to cause serious electric shock and are always a potential source of ignition when flammable liquid, dusts or powders are present. This is a particular problem in the parts of the printing industry where solvent-based inks are used on high-speed web presses. Flour dust in a mill has also been ignited by static electricity.

Static electricity may build up on both materials and people. When a charged person approaches flammable gases or vapours and a spark ignites the substance, the resulting explosion or fire often causes serious injury. In these situations, effective static control systems must be used.

Lightning strikes are a natural form of static electricity and result in large amounts of electrical energy being dissipated in a short time in a limited space with a varying degree of damage. The current produced in the vast majority of strikes exceeds 3,000 amps over a short period of time. Before a strike, the electrical potential between the cloud and earth might be about 100 million volts and the energy released at its peak might be about 100 million watts per metre of strike.

The need to provide lightning protection depends on a number of factors, which include:

► the risk of a strike occurring;
► the number of people likely to be affected;
► the location of the structure and the nearness of other tall structures in the vicinity;
► the type of construction, including the materials used;
► the contents of the structure or building (including any flammable substances);
► the value of the building and its contents.

Expert advice will be required from a specialist company in lightning protection, especially when flammable substances are involved. Lightning strikes can also cause complete destruction and/or significant disruption of electronic equipment.

Portable electrical equipment

Portable and transportable electrical equipment is defined by the Health and Safety Executive as *'not part of a fixed installation but may be connected to a fixed installation by means of a flexible cable and either a socket and plug or a spur box or similar means'*. It may be hand-held or hand-operated while connected to the supply, or is intended or likely to be moved while connected to the supply. The auxiliary equipment, such as extension leads, plugs and sockets, used with portable tools, is also classified as portable equipment. The term 'portable' means both portable and transportable (Figure 12.7).

Figure 12.7 Portable hand-held electric power tools

Almost 25% of all reportable electrical accidents involve portable electrical equipment (known as portable appliances). While most of these accidents were caused by electric shock, over 2,000 fires each year are started by faulty cables used by portable appliances, caused by a lack of effective maintenance. Portable electrical tools often present a high risk of injury, which is frequently caused by the conditions under which they are used. These conditions include the use of defective or unsuitable equipment and, indeed, the misuse of equipment. There must be a system to record the inspection, maintenance and repair of these tools.

Where plugs and sockets are used for portable tools, sufficient sockets must be provided for all the equipment, and adaptors should not be used. Many accidents are caused by faulty flexible cables, extension leads, plugs and sockets, particularly when these items become damp or worn. Accidents often occur when contact is made with some part of the tool which has become live (probably at mains voltage), while the user is standing on, or in contact with, an earthed conducting surface. If the electricity supply is at more than 50V AC, then the electric shock that a person may receive from such defective equipment is potentially lethal. In adverse environmental conditions, such as humid or damp atmospheres, even lower voltages can be dangerous. Portable electrical equipment should not be used in flammable atmospheres if it can be avoided and it must also comply with any standard relevant to the particular environment. Air-operated equipment should also be used as an alternative whenever it is practicable.

Some portable equipment requires substantial power to operate and may require voltages higher than those usually used for portable tools, so that the current is kept down to reasonable levels. In these cases, power leads with a separate earth conductor and earth screen must be used. Earth leakage relays and earth monitoring equipment must also be used, together with substantial plugs and sockets designed for this type of system.

Electrical equipment is safe when properly selected, used and maintained. It is important, however, that the environmental conditions are always carefully considered. The hazards associated with portable appliances increase with the frequency of use and the harshness of the environment (construction sites are often particularly hazardous in this respect). These factors must be considered when inspection, testing and maintenance procedures are being developed.

Secondary hazards

It is important to note that there are other hazards associated with portable electrical appliances, such as abrasion and impact, noise and vibration. Trailing leads used for portable equipment and raised socket points offer serious trip hazards and both should be used with great care near pedestrian walkways. Power drives from electric motors should always be guarded against entanglement hazards.

Secondary hazards are those additional hazards which present themselves as a result of an electrical hazard. It is very important that these hazards are considered during a risk assessment. An electric shock could lead to a fall from height if the shock occurred on a scaffold or it could lead to a collision with a vehicle if the victim collapsed on to a roadway.

Similarly, an electrical fire could lead to all the associated fire hazards outlined in Chapter 11 (e.g. suffocation, burns and structural collapse) and electrical burns can easily lead to infections.

High risks associated with the use of electricity

There are several sources of high risks associated with the use of electricity. These include:

▶ working with poorly maintained electrical equipment;
▶ using electrical equipment in adverse or hazardous environments such as wet, flammable or explosive atmospheres;
▶ working on mains electricity supplies;
▶ contact with underground cables during excavation work; and
▶ contact with live overhead power lines.

12.2 Control measures when working with electrical systems or using electrical equipment in all workplace conditions

12.2.1 Safe systems of work and competence

The principal control measures for electrical hazards are contained in the statutory precautionary requirements covered by the Electricity at Work Regulations, the main provisions of which are outlined in Chapter 13. They are applicable to all electrical equipment and systems found at the workplace, and impose duties on employers, employees and the self-employed.

The Regulations cover the following topics:

▶ the design, construction and maintenance of electrical systems, work activities and protective equipment;
▶ the strength and capability of electrical equipment;
▶ the protection of equipment against adverse and hazardous environments;
▶ the insulation, protection and placing of electrical conductors;
▶ the earthing of conductors and other suitable precautions;
▶ the integrity of referenced conductors;
▶ the suitability of joints and connections used in electrical systems;
▶ means for protection from excess current;
▶ means for cutting off the supply and for isolation;

▶ the precautions to be taken for work on equipment made dead;
▶ working on or near live conductors;
▶ adequate working space, access and lighting;
▶ the competence requirements for persons working on electrical equipment to prevent danger and injury.

Detailed safety standards for designers and installers of electrical systems and equipment are given a code of practice, published by the former Institution of Electrical Engineers. While these Regulations are not legally binding, they are recognised as a code of good practice and widely used as an industry standard.

BS 7671: Requirements for Electrical Installation – more commonly known as the IEE Wiring Regulations – is the national standard to which all domestic and industrial wiring has to conform. The recent edition includes substantial changes to harmonise with EU requirements. Among the changes to these Regulations is the inclusion of four regulations for the protection of people and livestock against voltage disturbances and electromagnetic influences. There is also a specific requirement for appropriate documentation for all installations. Seven special locations are covered to address the risk associated with certain environments or facilities, including exhibition areas, mobile units and temporary installations.

The standard wiring colours in the UK, shown in Table 12.1, are the same as elsewhere in Europe, Australia and New Zealand and follow the international standard IEC 60445 which defines basic safety principles for identifying electrical conductors by colours in electricity distribution wiring.

Table 12.1 Standard wiring colours

Wire	Colour
Protective earth (PE)	Green-and-yellow
Neutral (N)	Blue
Single phase: Line (L) Three phase: L1	Brown
Three phase: L2	Black
Three phase: L3	Grey

This edition of the Regulations requires that inspection and testing must be carried out by a 'competent person' to check that the electrical work meets required standards. It defines a competent person as someone *'who possesses sufficient technical knowledge and experience for the nature of the electrical work undertaken and is able at all times to prevent danger, and where appropriate, injury to themselves and others'*. After the initial testing, the Regulations recommend that every electrical installation is subject to periodic inspection and testing by a **competent person**.

Regulation 16 of the Electricity at Work Regulations states that

'No person shall be engaged in any work activity where technical knowledge or experience is necessary to prevent danger or, where appropriate, injury, unless he possesses such knowledge or experience, or is under such degree of supervision as may be appropriate having regard to the nature of the work'.

The Memorandum of Guidance on the Electricity at Work Regulations 1989 revised by the HSE in 2007 says that the scope of 'technical knowledge or experience' may include:

▶ adequate knowledge of electricity;
▶ adequate experience of electrical work;
▶ adequate understanding of the system to be worked on and practical experience of that class of system;
▶ understanding of the hazards which may arise during the work and the precautions which need to be taken;
▶ ability to recognise at all times whether it is safe for work to continue.

The risk of injury and damage inherent in the use of electricity can only be controlled effectively by the introduction of employee training, safe operating procedures (safe systems of work) and guidance to cover specific tasks.

Training is required at all levels of the organisation ranging from simple on-the-job instruction to apprenticeship for electrical technicians and supervisory courses for experienced electrical engineers. First-aid training related to the need for cardiovascular resuscitation and treatment of electric burns should be available to all people working on electrical equipment and their supervisors.

A **management system** should be in place to ensure that the electrical systems are installed, operated and maintained in a safe manner. All managers should be responsible for the provision of adequate resources of people, material and advice to ensure that the safety of electrical systems under their control is satisfactory and that **safe systems of work** are in place for all electrical equipment. For certain types of electrical work, such as working on high-voltage equipment or other work on electrical equipment which may give rise to danger, a permit to work will be required. The following items should be included in electrical permits to work:

▶ the permit issue number;
▶ the details of the work and its location;
▶ the significant hazards and risks involved in the work and the precautions to be taken including any personal protective equipment to be worn;
▶ details of the electrical items to be isolated and the points at which such isolations are to be made;
▶ the test procedures to be followed to confirm that circuits are dead;
▶ details of any special work tools required;

▶ details of the safety warning signs required;
▶ the emergency procedures required;
▶ the date and time of issue of the permit and the length of its duration and the cross-referencing with other permits issued;
▶ acceptance of the permit by the person carrying out the work; and
▶ on completion of the work, a signature of the person who authorised the work to ensure cancellation of the permit.

A permit should not be issued on equipment that is live. It is never absolutely safe to work on live electrical equipment. There are, however, a few circumstances where it is necessary to work live, but this must only be done after it has been determined that it is unreasonable for the work to be done dead. Even if working live can be justified, many precautions are needed to make sure that the risk is reduced 'so far as is reasonably practicable'.

For small factories and office or shop premises where the system voltages are normally at mains voltage, it may be necessary for an external competent person to be available to offer the necessary advice. Managers must set up a high-voltage permit-to-work system for all work at and above 600V. The system should be appropriate to the extent of the electrical system involved. Consideration should also be given to the introduction of a permit system for voltages under 600V. Chapter 4 gives more information on both safe systems of work and permits to work.

The additional control measures that should be taken when working with electricity or using electrical equipment are summarised by the following topics:

▶ the selection of suitable equipment;
▶ the use of protective systems;
▶ inspection and maintenance strategies.

These three groups of measures will be discussed in detail.

12.2.2 The selection and suitability of equipment

Many factors which affect the selection of suitable electrical equipment, such as flammable, explosive and damp atmospheres and adverse weather conditions, have already been considered. All conductors in a system that might be dangerous must be covered with an insulating material. Other issues include high or low temperatures, dirty or corrosive processes or problems associated with vegetation or animals (e.g. tree roots touching and displacing underground power cables, farm animals urinating near power supply lines and rats gnawing through cables). Temperature extremes will affect, for example, the lubrication of motor bearings and corrosive atmospheres can lead to the breakdown of insulating materials. The equipment selected must be suitable for the task demanded or either it will become overloaded or running costs will be too high.

The equipment should be installed to a recognised standard and capable of being isolated in the event of an emergency. It is also important that the equipment is effectively and safely earthed. Electricity supply failures may affect process plant and equipment. These are certain to happen at some time and the design of the installation should be such that a safe shutdown can be achieved in the event of a total mains failure. This may require the use of a battery-backed shutdown system or emergency standby electricity generators (assuming that this is cost-effective).

Finally, it is important to stress that electrical equipment must only be used within the rating performance given by the manufacturer and any accompanying instructions from the manufacturer or supplier must be carefully followed.

12.2.3 The advantages and limitations of protective systems

There are several different types of protective system and technique that may be used to protect people, plant and premises from electrical hazards, some of which, for example earthing, have already been considered earlier in this chapter. However, only the more common types of protection will be considered here.

Fuse

A fuse will provide protection against faults and continuous marginal current overloads. It is basically a thin strip of conducting wire which will melt when an excess of the rated current passes through it, thus breaking the circuit.

A fuse rated at 13A will melt when a current in excess of 13A passes through the fuse thus stopping the flow of current. A **circuit breaker** throws a switch off

when excess current passes and is similar in action to a fuse. Protection against overload is provided by fuses which detect a continuous marginal excess flow of current and energy. This overcurrent protection is arranged to operate before damage occurs, either to the supply system or to the load, which may be a motor or heater. When providing protection against overload, consideration needs to be made as to whether tripping the circuit could give rise to an even more dangerous situation, such as with firefighting equipment.

The prime objective of a fuse is to protect equipment or an installation from overheating and becoming a fire hazard. It is not an effective protection against electric shock due to the time that it takes to cut the current flow.

The examination of fuses is a vital part of an inspection programme to ensure that an item with the correct size or rating is fitted at all times.

Insulation

Insulation is used to protect people from electric shock, the short circuiting of live conductors and the dangers associated with fire and explosions. Insulation is achieved by covering the conductor with an insulating material. Insulation is often accompanied by the enclosure of the live conductors so that they are out of reach to people. A breakdown in insulation can cause electric shock, fire, explosion or instrument damage.

Isolation

The isolation of an electrical circuit involves more than 'switching off' the current in that the circuit is made dead and cannot be accidentally re-energised. It,

(a)

(b)

Figure 12.8 (a) Typical 240 volt fuses; (b) typical 240 volt mini circuit breaker therefore, creates a barrier between the equipment and the electrical supply which only an authorised person should be able to remove. When it is intended to carry out work, such as mechanical maintenance or a cleaning operation on plant or machinery, isolation of electrical equipment will be required to ensure safety during the work process. Isolators should always be locked off when work is to be done on electrical equipment.

Before working on an isolated circuit, checks must be made to ensure that the circuit is dead and that the isolation switch is 'locked off' and clearly labelled (Figure 12.9(a)).

(c)

Figure 12.9 A variety of electrical equipment: (a) electrical cabinet with multi-hasp and lock-off padlocks on the isolator; (b) 240 volt socket with RCD devices built in; (c) a lockable electrical control cabinet

(a)

(b)

Reduced low-voltage systems

When the working conditions are relatively severe, either due to wet conditions or heavy and frequent usage of equipment, reduced voltage systems should be used.

All portable tools used on construction sites, in vehicle washing stations or near swimming pools should operate on 120V or less, preferably with a centre-tapped to earth transformer at 55V. This means that while the full 120V are available to power the tool, only 55V are available should the worker suffer an electric shock. At this level of voltage, the effect of any electric shock should not be severe.

Safety Extra Low Voltage (SELV) – a voltage less than 50 volts – is used in low power tools, hand lights or soldering irons. Another way to utilise low-voltage equipment is to use **battery (cordless) operated hand tools**.

Residual current devices

If electrical equipment must operate at mains voltage, the best form of protection against electric shock is the residual current device (RCD). RCDs monitor and compare the current flowing in the live and neutral conductors supplying the protected equipment. Such

419

devices are very sensitive to differences of current between the live and neutral power lines and will cut the supply to the equipment in a very short period of time when a difference of only a few milliamperes occurs. It is the speed of the reaction which offers the protection against electric shock. For protection against electric shock, the RCD must have a rated residual current of 30mA or less and an operating time of 40 milliseconds or less at a residual current of 250mA. RCDs rated above 30mA provide very limited protection against harm from an electric shock.

The protected equipment must be properly protected by insulation and enclosure in addition to the RCD. The RCD will not prevent shock or limit the current resulting from an accidental contact with live electricity, but it will ensure that the duration of the shock is limited to the time taken for the RCD to operate. Electrical circuits must be securely isolated before any work is done on them. The RCD has a test button which should be tested frequently to ensure that it is working properly (Figure 12.9(b)). RCDs will also protect installations against fire, as they will interrupt the electrical supply before sufficient energy to start a fire has accumulated. However, this is not their prime purpose. Therefore the advantages of a RCD over a fuse are:

▶ a fuse is slower to act than a RCD and protects equipment rather than people;
▶ a blown fuse may be replaced by a fuse having an incorrect value: and
▶ a RCD, with its built-in test button, is easier to test and reset.

It is advisable to use a residual current device (RCD) whenever possible but particularly in wet or damp locations, such as outdoors. It is best practice to use an RCD that is incorporated into the switchboard of the electrical installation so that all circuits fed from that RCD are protected by it. A RCD that is incorporated into an ordinary mains socket, or plugged into it, will protect anything attached to that socket, but it is possible that equipment may have been accidentally plugged into another unprotected socket.

Double insulation

To remove the need for earthing on some portable power tools, double insulation is used. Double insulation employs two independent layers of insulation over the live conductors, each layer alone being adequate to insulate the electrical equipment safely. This prevents the exposure of live parts to exposed conductive parts such as the outer metal casing of the equipment. Consequently, an internal fault condition cannot make any part of the casing live.

Figure 12.10 shows the symbol which is marked on double-insulated portable power tools.

If the appliance has an insulating (such as plastic) cover or case and no exposed metal parts, it cannot

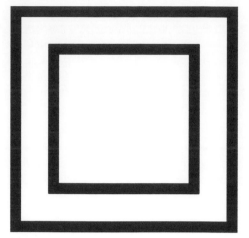

Figure 12.10 Double insulation sign

become live so cannot cause an electric shock. Thus the appliance is insulated in the normal way to avoid electric shock, then a second independent layer (such as the insulating plastic housing on a hand drill) is added. If part of the insulation system fails, the user is protected by the second insulation layer.

Double insulation is used on Class II appliances such as hand-held portable appliances and non-hand-held portable appliances such as desk fans and desk lamps. Many heating appliance – fan heaters, hair dryers or kettle – are double insulated. These devices are regularly handled and the cases must never become live. For a hair dryer this is especially important since these are sometimes used in a bathroom or with wet hands.

Double insulated tools and equipment must be inspected and maintained regularly and discarded if damaged.

12.2.4 Protection against contact with live overhead or buried power lines

Where possible all work likely to lead to contact with overhead power lines should be done in an area well clear of the line itself. It may be possible to alter the work and eliminate or reduce the risk. As a general rule, vehicles, plant or equipment should be brought no closer than:

▶ 15 m of overhead lines suspended from steel towers;
▶ 9 m of overhead lines supported on wooden poles.

Where a closer approach is likely either the lines should be made dead or barriers erected to prevent an approach. Permits to work are likely to be required if work close to the lines is necessary.

Where work is necessary directly beneath the lines, or blasting or other unusual activity has to be done adjacent to them, they may need to be made dead and a permit-to-work system operated. In certain situations induced AC voltages and even arcing can be created in fences and pipelines which run parallel to overhead lines. When there are concerns about nearby structures the advice of the safety adviser and/or the electrical

Figure 12.11 Multi-plug extension lead and special plugs and sockets: (a) reduced voltage normally 110 volts; (b) 240 volt lead with cut-out

supply company must be sought. Risk assessments should also be made and suitably recorded, the information being made available to all workers who are likely to be working near the power lines.

On many construction sites, vehicles will need to move beneath power lines. In these cases, the roadway should be covered by goalposts covered with warning tape. Bunting should be suspended, level with the top of the goalposts and just above ground level (often using empty oil drums), between poles across the site along the length of the power line. This will ensure that vehicles can only pass under the power lines by passing through the goalposts. Suitable warning signs should be placed on either side of the roadway on each side of the power line to warn of the overhead power line.

The presence of buried power lines is a common problem. Often official knowledge of the exact location of such services is less than accurate, particularly if the power lines have been in place for many years.

When underground cables are damaged, people can be injured or killed by electric shock or electrical arcs. Electric arcs can cause explosions and flames resulting in severe burns to hands, face and body, even if protective clothing is being worn. Such damage can be caused when a cable is:

▶ cut through by a sharp object such as the point of a tool; or
▶ crushed by a heavy object or powerful machine.

Cables that have been previously damaged but left unreported and unrepaired can also cause serious incidents.

The following precautions are suggested when there is uncertainty about the location of underground services in an area to be excavated:

▶ Check for any obvious signs of underground services, for example valve covers or patching of the road surface.

DANGER! OVERHEAD ELECTRICAL WIRING

(a)

Figure 12.12 Precautions for overhead lines: (a) 'goalpost' crossing points beneath lines to avoid contact by plant; (b) diagram showing normal dimensions for 'goalpost' crossing points and barriers

▶ Ensure that the excavation supervisor has the necessary service plans and is competent to use them to locate underground services.

▶ Ensure that all excavation workers are trained in safe digging practices and emergency procedures.

▶ Use locators to trace any services and mark the ground accordingly. A series of trial holes should be dug by hand to confirm the position of the pipes or cables. This is particularly important in the case of plastic pipes which cannot be detected by normal locating equipment.

▶ In areas where underground services may be present, only hand digging should be used with insulated tools. Spades and shovels should be used rather than picks and forks, which are more likely to pierce cables.

▶ Assume that all cables are 'live' unless it is known otherwise.

▶ Hand-held power tools should not be used within 0.5 m of the marked position of an electricity cable. Collars should be fitted to the tools so that initial penetration of the surface is restricted.

▶ Any suspected damage to cables must be reported to the service providers and the health and safety enforcement authority.

▶ All exposed cables should be backfilled with fine material such as dry sand or small gravel.

▶ The service plans must be updated when the new cables have been laid.

Excellent guidance is available in the HSE publication *Avoiding Danger from Underground Services* HSG47.

12.2.5 Emergency procedures following an electrical incident

General emergency procedures were covered in Chapter 4. However, even with the best of safety procedures in place, electrical incidents will occur so that specific and effective emergency procedures are essential. If someone is found unconscious as a result of electric shock or suffering from an electric burn, then the treatment procedure listed earlier in this chapter should be followed.

The emergency procedures should ensure that, for a serious electrical accident, the emergency services and key personnel in the organisation are notified and the emergency plan is activated. Such a plan should include the following points:

▶ The isolation of the electrical device that has caused the emergency by switching off the supply.
▶ If electric shock is the main emergency, the victim must not be touched until there is no longer a possibility of contact with the electrical current.
▶ If there is a fire as a result of electrical equipment malfunction, the fire procedures should also be activated.
▶ If the emergency could affect hazardous equipment or processes, essential actions, such as emergency plant shutdown, isolation or making processes safe, should be put into action. It is important that important items such as shut-off valves and electrical isolators are easily identifiable.
▶ Work must not resume after an emergency until a competent person has checked and declared that all electrical equipment affected by the emergency is safe and the electrical supply has been switched back on. If there are any doubts, then advice should be sought from the emergency services.
▶ Following the incident, an investigation should take place, a report produced and, if appropriate, a RIDDOR report submitted to the HSE.

12.2.6 Inspection and maintenance strategies

Inspection strategies

Regular inspection of electrical equipment is an essential component of any preventative maintenance programme and, therefore, regular inspection is required under Regulation 4(2) of the Electricity at Work Regulations, which has been quoted previously. Any strategy for the inspection of electrical equipment, particularly portable appliances, should involve the following considerations:

▶ a means of identifying the equipment to be tested;
▶ the number and type of appliances to be tested;
▶ the competence of those who will undertake the testing (whether in-house or brought in);
▶ the legal requirements for portable appliance testing (PAT) and other electrical equipment testing and the guidance available;

▶ organisational duties of those with responsibilities for PAT and other electrical equipment testing;
▶ test equipment selection and re-calibration;
▶ the development of a recording, monitoring and review system; and
▶ the development of any training requirements resulting from the test programme.

Maintenance strategies

Regulation 4(2) of the Electricity at Work Regulations requires that 'as may be necessary to prevent danger, all systems shall be maintained so as to prevent so far as is reasonably practicable, such danger'. Regular maintenance is, therefore, required to ensure that a serious risk of injury or fire does not result from installed electrical equipment. Maintenance standards should be set as high as possible so that a more reliable and safe electrical system will result. Inspection and maintenance periods should be determined by reference to the recommendations of the manufacturer, consideration of the operating conditions and the environment in which equipment is located. The importance of equipment within the plant, from the plant safety and operational viewpoint, will also have a bearing on inspection and maintenance periods. The mechanical safety of driven machinery is vital and the electrical maintenance and isolation of the electrically powered drives is an essential part of that safety.

The particular areas of interest for inspection and maintenance are:

▶ the cleanliness of insulator and conductor surfaces;
▶ the mechanical and electrical integrity of all joints and connections;
▶ the integrity of mechanical mechanisms, such as switches and relays;
▶ the calibration, condition and operation of all protection equipment, such as circuit breakers, RCDs and switches.

Safe operating procedures for the isolation of plant and machinery during both electrical and mechanical maintenance must be prepared and followed. All electrical isolators must, wherever possible, be fitted with mechanisms which can be locked in the 'open/off' position and there must be a procedure to allow fuse withdrawal wherever isolators are not fitted.

Working on live equipment with voltages in excess of 120V must not be permitted except where fault-finding or testing measurements cannot be done in any other way. Reasons such as the inconvenience of halting production are not acceptable.

Part of the maintenance process should include an appropriate system of visual inspection. By concentrating on a simple, inexpensive system of looking for visible signs of damage or faults, many of the electrical risks can be controlled, although more systematic testing may be necessary at a later stage.

12

Figure 12.13 Using a cable detector (CAT)

All fixed electrical installations should be inspected and tested periodically by a competent person, such as a member of the National Inspection Council for Electrical Installation Contracting (NICEIC).

Portable electrical appliances testing

Portable appliances should be subject to three levels of inspection – a user check, a formal visual inspection and a combined inspection and test.

User checks

When any portable electrical hand tool, appliance, extension lead or similar item of equipment is taken into use, at least once each week or, in the case of heavy work, before each shift, the following visual check and associated questions should be asked:

▶ Is there a recent PAT label attached to the equipment?
▶ Are any bare wires visible?
▶ Is the cable covering undamaged and free from cuts and abrasions (apart from light scuffing)?
▶ Is the cable too long or too short? (Does it present a trip hazard?)
▶ Is the plug in good condition (for example the casing is not cracked and the pins are not bent) (see Figure 12.14)?
▶ Are there no taped or other non-standard joints in the cable?
▶ Is the outer covering (sheath) of the cable gripped where it enters the plug or the equipment? (The coloured insulation of the internal wires should not be visible.)

▶ Is the outer case of the equipment undamaged or loose and are all screws in place?
▶ Are there any overheating or burn marks on the plug, cable, sockets or the equipment?
▶ Are the trip devices (RCDs) working effectively (by pressing the 'test' button)?

Figure 12.14 UK standard 3-pin plug wiring

Formal visual inspections and tests

There should be a **formal visual inspection** routinely carried out on all portable electrical appliances. Faulty equipment should be taken out of service as soon as the damage is noticed. At this inspection the plug cover (if not moulded) should be removed to check that the correct fuse is included, but the equipment itself should not be taken apart. This work can normally be carried out by a trained person who has sufficient information and knowledge.

Some faults, such as the loss of earth continuity due to wires breaking or loosening within the equipment, the breakdown of insulation and internal contamination (e.g. dust containing metal particles may cause short circuiting if it gets inside the tool), will not be spotted by visual inspections. To identify these problems, a programme of testing and inspection will be necessary.

This formal **combined testing and inspection** should be carried out by a competent person when there is reason to suspect the equipment may be faulty, damaged or contaminated, but this cannot be confirmed by visual inspection or after any repair, modification or similar work to the equipment, which could have affected its electrical safety. The competent person could be a person who has been specifically trained to carry out the testing of portable appliances using a simple 'pass/fail' type of tester. When more sophisticated tests are required, a competent person with the necessary technical electrical knowledge and experience would be needed. The inspection and testing should normally include the following checks:

▶ that the polarity is correct;
▶ that the correct fuses are being used;
▶ that all cables and cores are effectively terminated;
▶ that the equipment is suitable for its environment.

Testing need not be expensive in many low-risk premises like shops and offices, if an employee is trained to perform the tests and appropriate equipment is purchased.

Frequency of inspection and testing

The law requires an employer to ensure that their electrical equipment is maintained in order to prevent danger. It does not indicate how or how often this should be done.

Electrical equipment should be visually checked to spot early signs of damage or deterioration. The frequency of inspection and testing should be based on a risk assessment which is related to the usage, type and operational environment of the equipment. The harsher the working environment is, the more frequent the period of inspection. Thus tools used on a construction site should be tested much more frequently than a visual display unit which is never moved from a desk. Manufacturers or suppliers may recommend a suitable testing period. Table 12.2 lists the suggested intervals for inspection and testing derived from HSE publications *Maintaining Portable and Transportable Electrical Equipment* (HSG107 and INDG236 and 237).

It is very important to stress that there is no 'correct' interval for testing – it depends on the frequency of usage, type of equipment, and how and where it is used. A few years ago, a young trainee was badly scalded by a boiling kettle of water which exploded while in use. On investigation, an inspection report indicated that the kettle had been checked by a competent person and passed just a few weeks before the accident. Further investigation showed that this kettle was the only method of boiling water on the premises and was in use continuously for 24 hours each day. It was therefore unsuitable for the purpose and a plumbed-in continuous-use hot water heater would have been far more suitable.

HSE recommend that suppliers, who loan equipment, should formally inspect and test the equipment before each hire to ensure that it is safe to use. The person hiring the equipment should also take appropriate steps to ensure it remains safe to use throughout the hire period.

Records of inspection and testing

Schedules which give details of the inspection and maintenance periods and the respective programmes must be kept together with records of the inspection

Table 12.2 Suggested intervals for portable appliance inspection and testing

Type of business/ equipment	User checks	Formal visual inspection	Combined inspection and electrical tests
Equipment hire	Yes	Before issue and after return	Before issue
Construction	Yes	Before initial use and then every month	3 months
Industrial	Yes	Before initial use and then every 3 months	6–12 months
Hotels and offices, low-risk environments			
Battery operated (less than 20V)	No	No	No
Extra low voltage (less than 50V AC), e.g. telephone equipment, low-voltage desk lights	No	No	No
Desk-top computers, VDU screens	No	Yes 2–4 years	No if double insulated, otherwise up to 5 years
Photocopiers/fax machines/printers. Not hand-held; rarely moved	No	Yes 2–4 years	No if double insulated, otherwise up to 5 years
Double-insulated equipment: not hand-held. Moved occasionally, e.g. fans, table lamps, slide projectors	No	Yes 2–4 years	No
Double-insulated equipment: hand-held, e.g. some floor cleaners	Yes	Yes 6 months–1 year	No
Earthed equipment (class 1): e.g. electric kettles, some floor cleaners, portable electric heaters, some kitchen equipment and irons	Yes	Yes 6 months–1 year	Yes; 1–2 years
Cables (leads) and plug connected to the above. Extension leads (mains voltage)	Yes	Yes 6 months–4 years depending on the type of equipment it is connected to	Yes 1–5 years depending on the type of equipment it is connected to

Source: Derived from *HSE.*

Note

Operational experience may demonstrate that the above intervals can be reviewed.

findings and the work done during maintenance. Records must include both individual items of equipment and a description of the complete system or section of the system. They should always be kept up-to-date and with an audit procedure in place to monitor the records and any required actions. The records do not have to be paper based but could be stored electronically on a computer. It is good practice to label the piece of equipment with the date of the last combined test and inspection.

The effectiveness of the equipment maintenance programme may be monitored and reviewed if a record of tests is kept. It can also be used as an inventory of portable appliances and help to regulate the use of unauthorised appliances. The record will enable any adverse trends to be monitored and to check that suitable equipment has been selected. It may also give an indication as to whether the equipment is being used correctly.

Code of practice for portable appliance testing

The Institution of Engineering and Technology (IET) has published a revised edition of its 'Code of Practice for In-service Inspection and Testing of Electrical Equipment'. This guidance on portable appliance testing (PAT) suggests that UK companies are wasting more than £30 million annually by over-maintaining electrical appliances. It emphasises that it is not a legal requirement for companies to carry out PAT on every electrical appliance annually. It advises businesses to take a proportionate, risk-based approach to testing, which should consider the usage, type and environment of the equipment under question.

Advantages and limitations of PAT

The advantages of PAT include:

▶ an earlier recognition of potentially serious equipment faults, such as poor earthing, frayed and damaged cables and cracked plugs;
▶ discovery of incorrect or inappropriate electrical supply and/or equipment;
▶ discovery of incorrect fuses being used;
▶ a reduction in the number of electrical accidents;
▶ monitoring the misuse of portable appliances;
▶ equipment selection procedures checkable;
▶ an increased awareness of the hazards associated with electricity;
▶ a more regular maintenance regime should result.

The limitations of PAT include:

▶ some fixed equipment is tested too often leading to excessive costs;
▶ some unauthorised portable equipment, such as personal kettles, are never tested as there is no record of them;
▶ equipment may be misused or overused between tests due to a lack of understanding of the meaning of the test results;
▶ all faults, including trivial ones, are included on the action list, so the list becomes very long and the more significant faults are forgotten or overlooked;
▶ the level of competence of the tester can be too low;
▶ the testing equipment has not been properly calibrated and/or checked before testing takes place.

Most of the limitations may be addressed and the reduction in electrical accidents and injuries enables the advantages of PAT to greatly outweigh the limitations.

12.3 Further information

Statutory provisions

The Electricity at Work Regulations 1989

Other references

IET Wiring Regulations, 17th edition (first amendment), BS7671: 2008+A1:2012, ISBN: 978-1-84919-269-9

Electricity at Work – Safe Working Practices, HSG85, second edition 2007, HSE Books, ISBN: 978-0-7176-2164-4 www.hse.gov.uk/pubns/priced/hsg85.pdf

Health and Safety Toolbox, online resource, HSE, www.hse.gov.uk/toolbox/index.htm

Guidance on Safe Isolation procedures: www.select.org.uk/downloads/publications/Select%20-%20Safe%20Isolation%20Procedures.pdf

Memorandum of guidance on the Electricity at Work Regulation 1989, Guidance on Regulations, HSR25, second edition 2007, HSE Books, ISBN: 978-0-7176-6228-9 www.hse.gov.uk/pubns/priced/hsr25.pdf

Maintaining Portable and Transportable Electrical Equipment, HSG107, second edition 2004, HSE Books, ISBN: 978-0-7176-2805-6 www.hse.gov.uk/pubns/priced/hsg107.pdf

12.4 Practice questions

1. (a) **Explain** the relationship between voltage, current and resistance referring to a basic electric circuit.
   (b) **Outline** the main hazards associated with the use of electricity.
   (c) **Identify** the control measures that will reduce the risk of fire from electrical equipment.

2. (a) **Outline** the effects on the human body from a severe electric shock.
   (b) **Outline** the emergency action to take if a person suffers a severe electric shock.

3. **Explain** the following terms used in electrical work:
   (a) 'isolation'
   (b) 'earthing'

(c) 'conductors and insulators'

(d) 'overcurrent protection'

(e) 'electric arcing'

(f) 'static electricity'.

4. **Describe** how the following protective measures reduce the risk of electric shock and, in **EACH** case, **give** an example of their application.

(a) fuse

(b) reduced low voltage

(c) residual current devices

(d) double insulation.

5. Electrical plugs and cables can cause accidents in the workplace.

(a) **Identify FOUR** examples of faults and bad practices that could lead to such accidents.

(b) **Outline** the corresponding control measures that should be taken for **EACH** of the examples identified in (a) above.

6. A construction worker is working on a scaffold on the outside of a building. While using a portable 240V electrical drill to drill into the wall of the building, he makes direct contact with the electrical supply cable installed in the wall.

(a) **Identify FOUR** possible outcomes arising from this incident.

(b) **Identify FOUR** protective devices and procedures that could have reduced the risk of injury to this worker.

(c) **Describe** the types of inspection and/or test that should have been made on the drill.

7. (a) **Outline** the **THREE** levels of inspection that should be included in a maintenance and inspection strategy for portable electrical appliances.

(b) **Identify** the reasons for keeping records of the results of portable appliance testing within an organisation.

(c) **Outline** the issues to be considered when determining the frequency for the inspection and testing of a portable electrical tool.

8. (a) **Identify** the items that should be included on a checklist for the routine visual inspection of portable electrical appliances.

(b) **Identify EIGHT** examples of faults and bad practices that could contribute to accidents following the use of portable electrical appliances.

(c) **Identify** the advantages and limitations of portable appliance testing (PAT).

9. **Outline** the precautions to protect against contact with live electrical conductors when:

(a) excavating near underground cables

(b) working in the vicinity of overhead power lines.

12

CHAPTER 13

Summary of the main legal requirements

Chapter summary

1. An outline of the Health and Safety at Work (HSW) Act
2. Summaries of over 30 sets of regulations mainly made under the HSW Act which are covered by the NEBOSH (National Examination Board in Occupational Safety and Health) General Certificate syllabus
3. A summary of the Regulatory Reform (Fire Safety) Order
4. A very brief summary of over 17 other sets of regulations which are of interest to managers and students alike
5. Further information after each piece of legislation

13.1 Introduction

The achievement of an understanding of the basic legal requirements can be a daunting task. However, the health and safety student should not despair, because at NEBOSH National Certificate level they do not need to know the full history of UK health and safety legislation, the complete range of regulations made under the HSW Act 1974 or the details of the appropriate European Directives. These summaries, which are correct up to March 2020, cover those acts and regulations which are required for the Certificate student and will provide the essential foundation of knowledge. Students should check the latest edition of the NEBOSH Certificate guide to ensure that the regulations summarised here are the latest for the course being undertaken.

Many managers and other students will find these summaries a useful quick reference to health and safety legal requirements. However, anyone involved in achieving compliance with legal requirements should ensure they read the regulations themselves, any approved codes of practice and Health and Safety Executive (HSE) guidance.

13.2 The legal framework

13.2.1 General

The HSW Act is the foundation of British health and safety law. It describes the general duties that employers have towards their employees and to members of the public, and also the duties that employees have to themselves and to each other.

The term 'so far as is reasonably practicable' (sfairp) qualifies some of the duties in some sections, notably 2, 3, 4 and 6 of the HSW Act. In other words, the degree of risk in a particular job or workplace needs to be balanced against the time, trouble, cost and physical difficulty of taking measures to avoid or reduce the risk.

The law simply expects employers to behave in a way that demonstrates good management and common sense. They are required to look at what the hazards are and take sensible measures to control/reduce the risks from them.

The Management of Health and Safety at Work Regulations 1999 (MHSWR) (the Management Regulations) give employers specific requirements to manage health and safety under the HSW Act. Like the Act, they apply to every work activity.

The MHSW Regulations require every employer to carry out risk assessments. If there are five or more employees in the workplace, the significant findings of the risk assessments need to be recorded.

In a place like an office, risk assessment should be straightforward; but where there are serious hazards, such as those in a chemical plant, laboratory or an oil rig, it is likely to be more complicated.

The UK has now left the EU. The transition period is in place while the EU and UK negotiate new arrangements for a trade deal. It will end on 31 December 2020.

Your responsibility to protect the health and safety of people affected by your work activities remain the same during the transition period.

> 'The UK is committed to ensuring that only safe and compliant work equipment and machinery is placed on the market.
>
> Following the UK's exit from the EU HSE's priority is to maintain an effective regulatory system, which ensures that work equipment and machinery is safe and compliant.
>
> We will also respond to emerging risks and allow trade with the EU and the rest of the world that is as frictionless as possible.
>
> The UK is not a Member State of the EU. However, until the end of the transition period market access, and therefore UK–EU trade, should continue on the same terms as before.'

The HSE will publish the latest news on work equipment and machinery regulations on their website (https://www.hse.gov.uk/brexit/regulating-equipment.htm) or join their email newsletter.

13.2.2 The relationship between the regulator and industry

In April 2008 the HSC and the HSE merged into a single unified body called the Health and Safety Executive (HSE). The HSE consults widely with those affected by its proposals.

The HSE has many contact points with industry and work through:

► the Industry and Subject Advisory Committees, which have members drawn from the areas of work they cover, and focus on health and safety issues in particular industries (such as the textile industry, construction and education or areas such as toxic substances and genetic modification);

► intermediaries, such as small firms' organisations providing information and advice to employers and others with responsibilities under the HSW Act;

► guidance to enforcers, both the HSE inspectors and those of local authorities (LAs);

► the day-to-day contact which inspectors have with people at work.

The HSE consults with small firms through Small Firms Forums. It also seeks views in detail from representatives of small firms about the impact on them of proposed legislation.

13.2.3 Health and safety in Scotland

Although occupational health and safety is a reserved issue (i.e. not devolved to the Scottish Executive (SE)), there are some significant differences with some issues. The following extracts from the concordat between the HSE and the SE outline some of these issues.

13

Extracts from The Concordat between the Health and Safety Executive and the Scottish Executive

Introduction

1. This concordat is an agreement between the Scottish Executive and the Health and Safety Executive. It is intended to provide the framework to guide the future working relationship between the Health and Safety Executive (HSE) and the Scottish Executive (SE). The objective of the concordat is to ensure that the roles and responsibilities of HSE and the SE in the new constitutional structure are effectively translated into practical working

arrangements between the two organisations. The aim is to promote the establishment of close and harmonious working relationships and good communications at all levels between HSE and the SE, and in particular to foster constructive cooperation. Most day to day liaison between the SE and HSE will, in the first instance be handled by HSE's Director for Scotland, based in Edinburgh. [*Note*: where the concordat referred to HSC/E this has been changed to HSE meaning the new combined Authority.]

14 Nature of concordat

3. This concordat is a voluntary arrangement between HSE and the SE. It is not a binding agreement or contract and so does not create any legally enforceable rights, obligations, or restrictions. It is intended to be binding in honour only. The HSW Act provides for statutory consultation in connection with proposals for regulatory change; this concordat does not create any other right to be consulted or prevent consultation beyond that required by statute. Any failure to follow the terms of the concordat is not to be taken as invalidating decisions taken by HSE or the Scottish Executive.

15 Consultation

11. There are a significant number of areas where HSE and Scottish Ministers share or have closely related interests, as set out in the Annex. HSE and the SE recognise that the extent of these areas makes good communications essential in order to assist the process of policy formation and decision-making in each administration and to meet any consultation or other requirements connected with the exercise of a function. They also recognise that there will be mutual benefit from the exchange of information on scientific, technical and, where appropriate, policy matters.

12. Information obtained by use of HSE's statutory powers cannot be made available to SE where the release of that information is not for the purposes of HSE's functions. Subject to that, the HSE and the SE will:

▶ share relevant information, analysis and research;

▶ inform each other of any relevant information which comes to their attention which may require action or have resource consequences for the other party;

▶ seek to involve each other, as and when appropriate, in policy development on all topics where there is a reasonable

expectation that a policy initiative might affect the other's responsibilities, or be used or adapted by the other;

▶ inform each other at the earliest practicable stage of any emerging proposal to change primary or secondary legislation which might have an impact upon the other's responsibilities;

▶ inform each other at the earliest practicable stage of substantive new policy announcements which may be relevant to the other's responsibilities; and

▶ coordinate activities where appropriate.

16 Local authorities

16. The constitution of Scottish local authorities, which includes their establishment, dissolution, assets, liabilities, funding and receipts, is a devolved matter in which the Scottish Ministers have a policy interest as are most functions conferred upon such authorities. The HSE has an interest in the functions of local authorities under health and safety legislation, and under Section 18(4) of HSWA, the Commission may issue guidance to local authorities, as enforcing authorities, which they must follow. The HSE will consult the SE in advance where proposed changes in health and safety legislation, or directions under Section 18(4) might result in new burdens being imposed upon Scottish local authorities.

17 ANNEX – areas of common and closely related interest – extracts only

This annex lists subjects in which the Scottish Ministers and the HSE share an interest and other areas where there may be a particular need for consultation between HSE and the Scottish Executive.

18 Building control

This is a devolved matter, but the HSE has an interest in the related matter of construction safety.

19 Dangerous substances

The HSE have an interest under the HSW Act 1974 and the European Communities Act 1972 in the control of the keeping, notification, supply and use of explosive or highly flammable or otherwise dangerous substances which have the potential to create a major accident and in the prevention of the unlawful acquisition, possession and use of such substances. HSE also has an interest in the carriage by road and rail of dangerous and environmentally hazardous goods. The Scottish Ministers, however, have an interest in such matters for the purposes of protection of the environment and the planning system (by virtue of the Planning (Hazardous Substances) (Scotland) Act 1997). HSE will continue to advise what hazardous substances and in what quantities have significant risk off site. HSE involvement in individual hazardous substances consent/planning applications which are before Scottish Ministers for a decision will be on the basis of established planning and related public inquiry procedures.

20 Educational facilities; adventure activity centres

The safety of these premises is generally a devolved matter. However, the HSE has interests in the safety of adventure activity centres because of its responsibilities by virtue of the Activity Centres (Young Persons Safety) Act 1995 and subordinate legislation made under that Act. It also has an interest in the health and safety of workers and those affected by the activities of workers in these premises.

21 Fire safety

General fire safety is a devolved matter and the policy responsibility of the Scottish Ministers. However, HSE has an interest in the reserved matters of process fire precautions, fire precautions in relation to petroleum and petroleum spirit, fire safety on construction sites, ships under construction or repair by persons other than the master or crew, in mines, on offshore installations, and on any other premises which on 1 July 1999 are of a description specified in Part I of Schedule 1 to the Fire Certificates (Special Premises) Regulations 1976.

22 Food safety

This is generally a devolved matter and the policy responsibility of the Scottish Ministers. However, the HSE has an interest in the safety of workers in the food processing, manufacturing, cooking and food distribution industries and in related risks to the public.

23 Local authorities

The Scottish Ministers have an interest in the devolved matter of the constitution of Scottish local authorities, which includes their establishment, dissolution, assets, liabilities, funding and receipts. The HSE has an interest in the functions and performance of local authorities in relation to the enforcement of health and safety legislation.

24 Pesticides (including biocides and plant protection products)

The Scottish Ministers' interest is in the protection of the environment, public health, animal and plant health and food safety. HSE's interest is in the protection of workers and those affected by the activities of workers, protection of the environment, and product approvals.

25 Places of entertainment (cinemas, theatres, casinos, dance-halls, etc.), sports facilities, sports events and zoos)

The Scottish Ministers' interest is in public safety, fire safety, structural building standards, sanitation, food standards, pest control, security, etc. The HSE has an interest in the safety of workers and the safety of members of the public and others affected by work activity.

26 Ports, harbours and inland waterways

Scottish Ministers' interest is in ports policy and communications via the various ferry links between mainland Scotland and the islands. Responsibility for safety enforcement for the ships and their crews rests with the Maritime and Coastguard Agency. The HSE has an interest in the safety of workers and the safety of members of the public and others affected by work activity at inland waterways, the safety aspects of new harbour byelaws and the health and safety of shore side workers loading or unloading from berthed ships or the dock side.

27 Protection of the environment

This is a devolved matter and the policy responsibility of the Scottish Ministers. Responsibility for enforcement of much environmental legislation in Scotland rests with the Scottish Environment Protection Agency. However, the HSE has enforcement responsibilities for certain environmental legislation under agency arrangements under Section 13 of HSWA.

28 Public health, occupational health and health promotion

Public health is generally a devolved matter and the policy responsibility of the Scottish Ministers. Occupational Health is a reserved matter for HSE. General health promotion in Scotland is a devolved matter and the policy responsibility of the Scottish Ministers, but HSE has an interest in 'lifestyle' issues such

13

433

as misuse of drugs, alcohol and smoking (see also below), when these impinge on the workplace.

29 Rail safety

This is a reserved matter. However, the Scottish Ministers have an interest because of their executively devolved functions in relation to a number of aspects of the regulation of the rail industry in Scotland. HSE has an interest in any matter which could have an impact on its responsibility for the approval of new and altered works on railways, light rail and tramways.

30 Smoking

The regulation of smoking and passive smoking in workplaces is a reserved matter. However, the Scottish Ministers have an interest. Other aspects of smoking such as the regulation of tobacco advertising are generally devolved and are the policy responsibility of the Scottish Ministers.

31 Water and sewerage

This is generally a devolved matter and the policy responsibility of the Scottish Ministers. Water and sewerage services in Scotland are provided by three publicly owned water authorities. The HSE's interest is in matters relating to the safety of workers and those affected by the activities of workers.

13.2.4 Health and safety in Northern Ireland

The Health and Safety at Work (Northern Ireland) Order 1978 SI 1978/1039 (NI 9) www.legislation.gov.uk/nisi/1978/1039/contents effectively apply the Great Britain, Health and Safety at Work Act 1974 to Northern Ireland. The Health and Safety at Work (Amendment) (Northern Ireland) Order 1998 SI 1998/2795 (NI 18) established the Health and Safety Executive for Northern Ireland.

Much of the subordinate legislation made under the GB HSW Act has been transcribed into NI regulations made under the 1978 Order. For example, the Management of Health and Safety at Work Regulations (Northern Ireland) 2000. For a full list see www.hseni.gov.uk/resources/legislation.htm

This also applies to many ACoPs made under the GB HSW Act, which have been approved for use in Northern Ireland. A list is available at www.hseni.gov.uk/resources/legislation.htm. In addition, there are a number of NI ACoPS which have been adopted

separately from GB but these also follow closely GB legislation.

13.3 List of Acts, orders and regulations summarised

The following Acts and Regulations are covered in this summary:

- ▶ The HSW Act 1974
- ▶ Enterprise and Regulatory Reform Act 2013 (Strict liability issues only)
- ▶ The Health and Safety at Work etc. Act 1974 (Civil Liability) (Exceptions) Regulations 2013
- ▶ The Health and Safety (Fees) Regulations 2012

Most of the relevant regulations covering health and safety at work have been made under the HSW Act since 1974. Those relevant to the Certificate student are listed here and summarised in this chapter. The first list is alphabetical and some titles have been modified to allow an easier search. The second list is chronological with Statutory Instrument (SI) numbers.

The Classification, Labelling and Packaging of Substances and Mixtures Regulations are a EU Regulation which did not need member state regulations to be made.

The UK is not now a member state of the EU. However, until the end of the transition period, market access, and therefore UK–EU trade, should continue on the same terms as before. During the transition period:

- ▶ classification, labelling and packaging (CLP) requirements in place before the UK left the EU continue to be valid;
- ▶ the UK will recognise all new and revised harmonised classification and labelling adopted through EU arrangements which will have legal effect in the UK;
- ▶ HSE will not act as an EU CLP competent authority and will not participate in Risk Assessment Committee (RAC), CARACAL, CLP working group or REACH Committee meetings managed by the European Commission or an EU Agency. However, HSE may participate in EU/ECHA public consultations, commenting on new and revised harmonised classification and labelling proposals, and will offer, wherever possible, comments on any other proposed changes to CLP;
- ▶ UK-based suppliers have the same rights as EU-based suppliers to have their proposals for harmonised classification and labelling accepted and processed by a competent authority based in another EU member state provided the affected substance is placed on the market there;
- ▶ HSE's e-Bulletins and the CLP Helpdesk (send your enquiry to ukreachca@hse.gov.uk) will keep business up to date with developments;
- ▶ the UK's (HSE's) involvement with the UN GHS work continues.

The Regulatory Reform (Fire Safety) Order was made under the Regulatory Reform Act 2001.

13.3.1 Alphabetical list of regulations/orders summarised

Control of Artificial Optical Radiation at Work Regulations 2010;

Classification, Labelling and Packaging of Substances and Mixtures Regulations (European) adopting into EU UN Globally Harmonised System of Classification and Labelling of Chemicals(GHS);

Confined Spaces Regulations 1997;

Construction (Design and Management) Regulations 2015;

Health and Safety (Consultation with Employees) Regulations 1996;

Control of Substances Hazardous to Health Regulations 2002 and 2004 Amendment;

Dangerous Substances and Explosive Atmospheres Regulations 2002;

Health and Safety (Display Screen Equipment) Regulations 1992;

Electricity at Work Regulations 1989;

Employers Liability (Compulsory Insurance) Act 1969 and Regulations 1998;

The Regulatory Reform (Fire Safety) Order 2005 including:

Fire Safety (Employees' Capabilities) (England) Regulations 2010;

Fire Safety (Employees' Capabilities) (Wales) Regulations 2012

Fire (Scotland Act) 2005;

Police and Fire Reform (Scotland) Act 2012;

The Building Safety Bill;

Health and Safety (First Aid) Regulations 1981as amended 2013;

Hazardous Waste (England and Wales) Regulations 2005;

Health and Safety (Information for Employees) Regulations 1989;

Health and Safety (Amendment) (EU Exit) Regulations 2018

Ionising Radiations Regulations 2017;

Lifting Operations and Lifting Equipment Regulations 1998;

Management of Health and Safety at Work Regulations 1999 as amended;

Manual Handling Operations Regulations 1992 as amended.

Control of Noise at Work Regulations 2005;

Personal Protective Equipment at Work Regulations 1992 as amended;

Provision and Use of Work Equipment Regulations 1998 (except Part IV – Power Presses) as amended;

Reporting of Injuries, Diseases and Dangerous Occurrences Regulations 2013;

Safety Representatives and Safety Committees Regulations 1977;

Health and Safety (Safety Signs and Signals) Regulations 1996;

Supply of Machinery (Safety) Regulations 2008 and amendments;

Vibration at Work Regulations 2005;

Waste (England and Wales) Regulations 2011 as amended in 2012;

Workplace (Health, Safety and Welfare) Regulations 1992 as amended;

Work at Height Regulations 2005.

The following acts and regulations have also been included. With a few exceptions like the Road Traffic Acts, they are generally not in the NEBOSH General Certificate syllabus. Brief summaries only are given.

Building Regulations 2010 Approved Documents B & M Compensation Act 2006;

Control of Asbestos Regulations 2012;

Corporate Manslaughter and Corporate Homicide Act 2007;

Electrical Equipment (Safety) Regulations 1994;

Equality Act 2010 Disability Regulations 2010;

Gas Appliances (Safety) Regulations 1992;

Gas Safety (Installation and Use) Regulations 1998;

Health and Safety (Offences) Act 2008;

Lead at Work Regulations 2002;

Occupiers Liability Acts 1957 and 1984;

Personal Protective Equipment Regulations 2002;

Pesticides Regulations 1986;

Pressure Systems Safety Regulations 2000;

Road Traffic Acts 1988 and 1991;

Sharp Instruments in Healthcare Regulations 2013;

Smoke-free Regulations 2006;

Working Time Regulations 1998, 2001, 2003 and 2007.

13.3.2 Chronological list of regulations/orders summarised

The list below gives the correct titles, year produced and SI number.

Year	SI number	Title
1977	0500	Safety Representatives and Safety Committees Regulations
1981	0917	Health and Safety (First Aid) Regulations
1989	0635	Electricity at Work Regulations
1989	0682	Health and Safety (Information for Employees) Regulations
1992	2792	Health and Safety (Display Screen Equipment) Regulations
1992	2793	Manual Handling Operations Regulations
1992	2966	Personal Protective Equipment at Work Regulations
1992	3004	Workplace (Health, Safety and Welfare) Regulations
1996	0341	Health and Safety (Safety Signs and Signals) Regulations
1996	1513	Health and Safety (Consultation with Employees) Regulations
1997	1713	Confined Spaces Regulations

13

1998	2306	Provision and Use of Work Equipment Regulations (except Part IV – Power Presses)
1998	2307	Lifting Operations and Lifting Equipment Regulations
1998	2573	Employers Liability (Compulsory Insurance) Regulations
1999	437	Control of Substances Hazardous to Health Regulations
1999	3242	Management of Health and Safety at Work Regulations
2002	2677	Control of Substances Hazardous to Health Regulations
2002	2776	Dangerous Substances and Explosive Atmospheres Regulations
2004	3386	Control of Substances Hazardous to Health (Amendment) Regulations
2005	asp 5	The Fire Scotland Act
2005	735	Work at Height Regulations
2005	1541	The Regulatory Reform (Fire Safety) Order
2005	1093	Control of Vibration at Work Regulations
2008	1597	The Supply of Machinery (Safety) Regulations
2008	EC No. 1272	European Regulation on Classification, Labelling and Packaging of Substances and Mixtures (CLP Regulation)
2010	1140	The Control of Artificial Optical Radiation at Work Regulations
2013	1471	The Reporting of Injuries, Diseases and Dangerous Occurrences Regulations
2013	1667	The Health and Safety at Work etc. Act 1974 (Civil Liability) (Exceptions) Regulations
2015	51	Construction (Design and Management) Regulations
2015	363	The Health and Safety and Nuclear (Fees) Regulations
2017	1075	The Ionising Radiation Regulations
2018	1370	The Health and Safety (Amendment) (EU Exit) Regulations

2006	3368	Smoke-free (Premises and Enforcement) Regulations
2006	090	Scottish SI – The Prohibition of Smoking in Certain Premises (Scotland) Regulations
2007	W.68	The Smoke-free Premises etc. (Wales) Regulations
2007	Ch 19	Corporate Manslaughter and Corporate Homicide Act
2008	Ch 20	Health and Safety (Offences) Act
2010	471	The Fire Safety (Employees' Capabilities) (England) Regulations
2010	2214	The Building Regulations Approved Documents B and M
2010	Ch 15	Equality Act
2010	2128	The Equality Act 2010(Disability) Regulations
2012	Asp8	Police and Fire Reform (Scotland) Act
2012	632	The Control of Asbestos Regulations
2012	1085 (W133)	The Fire Safety (Employees' Capabilities) (Wales) Regulations
2013	645	The Health and Safety (Sharp Instruments in Healthcare) Regulations
2013	C24	Enterprise and Regulatory Reform Act (Strict Liability Changes only)

13.4 HSW Act 1974 as amended in 2015

The HSW Act was introduced to provide a comprehensive and integrated piece of legislation dealing with the health and safety of people at work and the protection of the public from work activities.

The Act imposes a duty of care on everyone at work related to their roles. This includes employers, employees, owners, occupiers, designers, suppliers and manufacturers of articles and substances for use at work. It also includes self employed people. The detailed requirements are spelt out in Regulations.

The Act basically consists of four parts.

Part 1 covers:

▶ the health and safety of people at work;
▶ protection of other people affected by work activities;
▶ the control of risks to health and safety from articles and substances used at work.

Part 2 sets up the Employment Medical Advisory Service.
Part 3 makes amendments to the safety aspects of building regulations.
Part 4 consists of general and miscellaneous provisions.

13.4.1 Duties of employers – Section 2

The employers' main general duties are to ensure, sfairp, the health, safety and welfare at work of all their employees, in particular:

▶ the provision of safe plant and systems of work;
▶ the safe use, handling, storage and transport of articles and substances;

Very brief summaries only for:

Year	SI number	Title
1957	Ch 31	Occupiers Liability Act
1984	Ch 3	Occupiers Liability Act
1986	1510	Control of Pesticides Regulations
1988	Ch52	Road Traffic Act
1991	Ch40	Road Traffic Act
1995	1629	Gas Appliances (Safety) Regulations
1998	2451	Gas Safety (Installation and Use) Regulations
1994	3260	Electrical Equipment (Safety) Regulations
1998	1833	Working Time Regulations
1999	743	The Control of Major Hazards Regulations and amendments
2000	128	Pressure Systems Safety Regulations
2002	1144	Personal Protective Equipment Regulations
2002	2676	The Control of Lead at Work Regulations
2003	1684	Working Time (Amendment) Regulations
2006	Ch 29	Compensation Act

- the provision of any required information, instruction, training and supervision;
- a safe place of work including safe access and egress; and
- a safe working environment with adequate welfare facilities.

When five or more people (The Employer's Health and Safety Policy Statements (Exception) Regulations 1975 (SI No. 1584) exempted an employer who employs less than 5 people) are employed the employer must:

- prepare a written general health and safety policy;
- set down the organisation and arrangements for putting that policy into effect;
- revise and update the policy as necessary;
- bring the policy and arrangements to the notice of all employees.

Employers must also:

- consult safety representatives appointed by recognised trade unions;
- consult safety representatives elected by employees;
- establish a safety committee if requested to do so by recognised safety representatives.

13.4.2 Duties of owners/occupiers – Sections 3 and 4

Every employer and self-employed person (self-employed person must be one who conducts an undertaking of a prescribed description, can be type of activity carried out or any other feature or whether person's health and safety may be affected by the undertaking) is under a duty to conduct their undertaking in such a way as to ensure, sfairp, that persons not in their employment (and themselves for self-employed), who may be affected, are not exposed to risks to their health and safety.

Those in control of non-domestic premises have a duty to ensure, sfairp, that the premises, the means of access and exit, and any plant or substances are safe and without risks to health. The common parts of residential premises are non-domestic.

Note: Section 5 was repealed by the Environmental Protection Act 1990.

13.4.3 Duties of manufacturers/ suppliers – Section 6

Persons who design, manufacture, import or supply any article or substance for use at work must ensure, sfairp, that:

- it is safe and without risks to health when properly used (i.e. according to manufacturers' instructions);
- they carry out such tests or examinations as are necessary for the performance of their duties;
- they provide adequate information (including revisions) to perform their duties;

- they carry out any necessary research to discover, eliminate or minimise any risks to health or safety;
- the installer or erector has done nothing regarding the way in which the article has been installed or erected to make it unsafe or a risk to health.

13.4.4 Duties of employees – Section 7

Two main duties are placed on employees:

- to take reasonable care for the health and safety of themselves and others who may be affected by their acts or omissions at work;
- to cooperate with their employer and others to enable them to fulfil their legal obligations.

13.4.5 Other duties – Sections 8 and 9

No person may misuse or interfere with anything provided in the interests of health, safety or welfare in pursuance of any of the relevant statutory provisions.

Employees cannot be charged for anything done, or provided, to comply with the relevant statutory provisions. For example, employees cannot be charged for personal protective equipment (PPE) required by health and safety legislation.

13.4.6 Powers of inspectors – Sections 20–25

Inspectors appointed under this Act have the authorisation to enter premises at any reasonable time (or any time in a dangerous situation), and to:

- take a constable with them if necessary;
- take with them another authorised person and necessary equipment;
- examine and investigate;
- require premises or anything in them to remain undisturbed for purposes of examination or investigation;
- take measurements, photographs and recordings;
- cause an article or substance to be dismantled or subjected to any test;
- take possession of or retain anything for examination or legal proceedings;
- if practical to do so take samples as long as a comparable sample is left behind;
- require any person who can give information to answer questions and sign a statement. Evidence given under this Act cannot be used against that person or their spouse;
- require information, facilities, records or assistance;
- do anything else necessary to enable them to carry out their duties;
- issue an **Improvement Notice(s)**, which is a notice identifying a contravention of the law and specifying a date by which the situation is remedied. If an appeal is to be made, the appeal procedure must be triggered within 21 days. The notice is suspended pending the outcome of the appeal;

13

437

▶ issue a **Prohibition Notice(s)**, which is a notice identifying and halting a situation which involves or will involve a risk of serious personal injury to which the relevant statutory provisions apply. A contravention need not have been committed. The notice can have immediate effect or be deferred, for example to allow a process to be shut down safely. Again there is provision to appeal against the notice but the order still stands until altered or rescinded by an employment tribunal;

▶ initiate prosecutions;

▶ seize, destroy or render harmless any article or substance which is a source of imminent danger to health and safety.

13.4.7 Offences – Section 33

The Health and Safety Offences Act 2008 increases penalties and provides courts with greater sentencing powers for those who flout health and safety legislation and the range of offences for which an individual can be imprisoned has also been broadened.

Further, under The Legal Aid, Sentencing and Punishment of Offenders Act 2012 Section 85(1) from 12 March 2012, Magistrates' Courts can issue unlimited fines.

There are strict guidelines which are observed by the regulators in their approach to the prosecution of health and safety offences. The HSE Enforcement Policy Statement makes it clear that prosecutions should be in the public interest and where one or more of a list of circumstances apply. These include where:

▶ death was a result of a breach of the legislation;

▶ there has been reckless disregard of health and safety requirements;

▶ there have been repeated breaches which give rise to significant risk, or persistent and significant poor compliance;

▶ false information has been supplied wilfully, or there has been intent to deceive in relation to a matter which gives rise to significant risk.

Summary of penalties under the HSW Act

Maxima (from 12 March 2015):

Unlimited fines in Magistrates courts for all summary offences, unlimited fines in higher courts

Imprisonment for nearly all offences – up to 6 months in Magistrates' Courts (until S154 of CJ Act 2003 is enacted, then 12 months) and two years in the Crown Court.

Table 13.1 Summary of maximum penalties in respect of offences committed after 12 March 2015

Section and contravention	Magistrates' Court	Crown Court
S.33(1)(a) – S2 – S6	Unlimited fine and/or 6 months imprisonment*	Unlimited fine and/or 2 years imprisonment
S.33(1)(a) – S7	Unlimited fine and/or 6 months imprisonment*	Unlimited fine and/or 2 years imprisonment
(b) – S8	Unlimited fine and/or 6 months imprisonment*	Unlimited fine and/or 2 years imprisonment
(b) – S9	Unlimited fine	Unlimited fine
(c) – any H & S regulations	Unlimited fine and/or 6 months imprisonment*	Unlimited fine and/or 2 years imprisonment
(d) – relating to S14 enquiries	Unlimited fine (Level 5)	Summarily only
(e) – any requirement imposed by Inspector under S20 & S25	Unlimited fine and/or 6 months imprisonment*	Unlimited fine and/or 2 years imprisonment
(f) – attempting to / preventing a person speaking to Inspector	Unlimited fine and/or 6 months imprisonment*	Unlimited fine and/or 2 years imprisonment
(g) – contravention of a PN or IN	Unlimited fine and/or 6 months imprisonment*	Unlimited fine and/or 2 years imprisonment
(h) – obstruction	Unlimited fine (Level 5) and/or 6 months imprisonment*	Summarily only
(i) – relating to S27	Unlimited fine	Unlimited fine
(j) – disclosure in contravention to S28 and S27(4)	Unlimited fine and/or 6 months imprisonment*	Unlimited fine and/or 2 years imprisonment
(k) – false statement	Unlimited fine and/or 6 months imprisonment*	Unlimited fine and/or 2 years imprisonment
(l) – false entry in register	Unlimited fine and/or 6 months imprisonment*	Unlimited fine and/or 2 years imprisonment
(m) – use a document with intent to deceive	Unlimited fine and/or 6 months imprisonment*	Unlimited fine and/or 2 years imprisonment
(n) – impersonating an inspector	Unlimited fine (Level 5)	Summarily only
(o) – relating to S42 orders	Unlimited fine and/or 6 months imprisonment*	Unlimited fine and/or 2 years imprisonment
Any other offence under existing statutory provisions	Unlimited fine and/or 6 months imprisonment*	Unlimited fine and/or 2 years imprisonment

Note: * The maximum term of imprisonment that may be imposed by a Magistrates' Court is currently 6 months. When s154(1) of the Criminal Justice Act 2003 is brought into force the maximum term will be increased to 51 weeks in England and Wales or 12 months in Scotland.

Prosecutions of individuals by health and safety regulators are not undertaken lightly. Any prosecutions of individuals are subject to the same strict considerations set out above and are only taken if warranted, and not in lieu of a case against their employer.

If a regulation has been contravened, failure to comply with an Approved Code of Practice is admissible in evidence as failure to comply. Where an offence is committed by a corporate body with the knowledge, connivance or neglect of a responsible person, both that person and the body are guilty of the offence. In proceedings the onus of proving the limits of what is reasonably practicable rests with the accused.

On conviction of directors for indictable offences in connection with the management of a company, the courts may also make a disqualification order as follows.

Lower court maximum	5 years disqualification
Higher court maximum	15 years disqualification

13.4.8 The merged HSE – Section 10

The majority of the proposals set out in the original Lord Robens' report were adopted in full and formed the basis of the HSW Act. However, contrary to Robens' recommendation, the HSW Act did not provide for a single authority, but two separate Crown Non-Department Public Bodies (NDPBs): the HSC and the HSE.

This recommendation was finally implemented and the HSC and the HSE merged on 1 April 2008 to form a single national regulatory body responsible for promoting the cause of better health and safety at work. The merged body is called the HSE and has provided greater clarity and transparency whilst maintaining its public accountability. Section 10 of the Act has been revised.

13.4.9 Civil Liability – Section 47

The Enterprise and Regulatory Reform Act 2013 removes the right of civil action against employers for breach of statutory duty in relation to health and safety at work regulations. This will address the potential unfairness that arises where an employer can be found liable to pay compensation to an employee despite having taken reasonable steps to protect them.

This unfairness was identified by Professor Löfstedt in his independent review of health and safety, 'Reclaiming Health and Safety for All' (November 2011) and arises where health and safety at work regulations impose a strict duty on employers, giving them no opportunity to defend themselves on the basis of having done all that was reasonable to protect their employees. The inability of employers to defend themselves in these cases helps fuel the perception of a compensation culture and the fear of being sued is driving businesses to over-comply with regulations, resulting in additional unnecessary costs.

Previously, compensation claims for workplace injuries or illnesses could be brought by two routes: breach of statutory duty in which failure to meet a particular standard in law has to be proved, or breach of common law duty of care in which negligence has to be proved.

The Act amends the law so that compensation claims can only be made where negligence or fault on the part of the employer can be proved.

This change will help redress the balance of the civil litigation system in respect of health and safety at work legislation. It will help employers' confidence, allowing them to focus on a sensible and practical approach to health and safety and keep costs down by avoiding over-compliance.

Employees will continue to have the same level of protection, as the standards set out in criminal law will not change and they will still be able to claim compensation where an employer has been negligent.

The Health and Safety at Work etc. Act 1974 (Civil Liability) (Exceptions) Regulations 2013. These Regulations create exceptions from the exclusion of civil liability for breach of statutory duty for pregnant workers and workers who have recently given birth or are breastfeeding ('new and expectant mothers').

13.4.10 Fees for Intervention

Fee for Intervention (FFI) is HSE's cost recovery regime implemented from 1 October 2012, under Regulations 23 to 25 of the Health and Safety (Fees) Regulations 2012.

These Regulations put a duty on HSE to recover its costs for carrying out its regulatory functions from those found to be in material breach of health and safety law.

Duty-holders who are compliant with the law, or where a breach is not material, will not be charged FFI for any work that HSE does with them.

A material breach is when, in the opinion of the HSE inspector, there is or has been a contravention of health and safety law that requires them to issue notice in writing of that opinion to the duty-holder.

Written notification from an HSE inspector may be by a notification of contravention, an improvement or prohibition notice, or a prosecution and must include the following information:

► the law that the inspector's opinion relates to;
► the reasons for their opinion; and
► notification that a fee is payable to HSE.

FFI applies to duty-holders where HSE is the enforcing authority. This includes employers, self-employed people who put others (including their employees or members of the public) at risk, and some individuals acting in a capacity other than as an employee, e.g. partners. It includes:

► public and limited companies;
► general, limited and limited liability partnerships; and
► Crown and public bodies.

The fee payable by duty-holders found to be in material breach of the law is £124 per hour (April 2015). The total amount to be recovered will be based on the amount of time it takes HSE to identify and conclude its regulatory action, in relation to the material breach (including associated office work), multiplied by the relevant hourly rate. This will include part hours.

13.4.11 Further information

An Introduction to Health and Safety. Guidance site on Health and Safety for Small Firms https://www.hse.gov.uk/toolbox/introduction.htm

Health and Safety made simple: The basics for your business INDG449(rev 1), HSE, 2014, www.hse.gov.uk/pubns/indg449.pdf

Health and Safety at Work etc. Act 1974, chapter 37, London, The Stationery Office www.legislation.gov.uk/ukpga/1974/37/contents

13.5 Control of Artificial Optical Radiation at Work Regulations 2010

13.5.1 Introduction

The Regulations impose duties on employers to protect both employees who may be exposed to risk from exposure to artificial optical radiation at work and other persons at work who might be affected by that work. AOR includes light emitted from all artificial sources in all its forms such as ultraviolet, infrared and laser beams, but excludes sunlight.

Regulation 1(2) defines exposure limit values as being those set out in Annexes I and II to the Directive, as amended from time to time, and these Annexes provide for exposure limit values for non-coherent radiation and laser radiation respectively.

13.5.2 Duties

The Regulations impose a duty to carry out a specific form of risk assessment where an employer carries out work which could expose its employees to levels of artificial optical radiation (i.e. artificial light) that could create a reasonably foreseeable risk of adverse health effects to the eyes or skin and where those risks have not already been eliminated or controlled (Regulation 3).

Where a risk assessment is necessary the Regulations also impose duties to:

(a) eliminate, or where this is not reasonably practicable, to reduce to as low a level as is reasonably practicable the risk of adverse health effects to the eyes or skin of the employee as a result of exposure to artificial optical radiation where this risk has been identified in the risk assessment (Regulation 4(1));

(b) devise an action plan comprising technical and organisational measures to prevent exposure to artificial optical radiation exceeding the exposure limit values where the risk assessment indicates that employees

are exposed to levels of artificial optical radiation that exceed the exposure limit values (Regulation 4(3));

(c) take action in the event that the exposure limit values are exceeded despite the implementation of the action plan and measures to eliminate or reduce so far as is reasonably practicable the risk of exposure (Regulation 4(5));

(d) demarcate, limit access to, and provide for appropriate signs in those areas where levels of artificial optical radiation are indicated in the risk assessment as exceeding the exposure limit values (Regulation 4(6) and (7));

(e) provide information and training if the risk assessment indicates that employees could be exposed to artificial optical radiation which could cause adverse health effects to the eyes or skin of the employee (Regulation 5); and

(f) provide health surveillance and medical examinations in certain cases (Regulation 6).

13.5.3 Safe light sources

The following information is taken from the HSE Guidance for Employers available at: www.hse.gov.uk/radiation/nonionising/employers-aor.pdf

The majority of light sources are safe, such as those described in List 1 below. If these sources, or similar, are being used workers are not at risk and there is no need to do anything further.

However, when making this decision, employers should consider the following points to ensure that all workers are protected:

▶ If there are any workers whose health is at particular risk (e.g. those with pre-existing medical conditions made worse by light).

▶ If workers use any chemicals (e.g. skin creams) which could react with light to make any health effects worse.

▶ If any workers are exposed to multiple sources of light at the same time.

▶ If exposure to bright light could present unrelated risks (e.g. temporary blindness could lead to mistakes being made in hazardous tasks).

List 1 Safe light sources

▶ All forms of ceiling-mounted lighting used in offices, etc. that have diffusers over bulbs or lamps.

▶ All forms of task lighting including desk lamps and tungsten-halogen lamps fitted with appropriate glass filters to remove unwanted ultraviolet light.

▶ Photocopiers.

▶ Computer or similar display equipment, including personal digital assistants (PDAs).

▶ Light emitting diode (LED) remote control devices.

▶ Photographic flashlamps – when used singly.

▶ Gas-fired overhead heaters.

▶ Vehicle indicator, brake, reversing and fog lamps.

▶ Any exempt or Risk Group 1 lamp or lamp system (including LEDs), as defined in British Standard BS EN 62471: 2008.

▶ Any Class 1 laser light product, as defined in British Standard BS EN 60825-1: 2007, for example laser printers and bar code scanners.

There are also some sources of light that, if used inappropriately, e.g. placed extremely close to the eyes or skin, have the potential to cause harm but which are perfectly safe under normal conditions of use. Examples include:

▶ Ceiling-mounted fluorescent lighting without diffusers over bulbs or lamps.

▶ High-pressure mercury floodlighting.

▶ Desktop projectors.

▶ Vehicle headlights.

▶ Non-laser medical applications such as:
  ▷ operating theatre and task lighting;
  ▷ diagnostic lighting such as foetal/neonatal transilluminators and X-ray light/viewing boxes.

▶ UV insect traps.

▶ Art and entertainment applications such as illumination by spotlights, effect lights and flashlamps (provided that any ultraviolet emissions have been filtered out).

▶ Multiple photographic flashlamps, for example in a studio.

▶ Any Risk Group 2 lamp or lamp system (including LEDs), as defined in British Standard BS EN 62471: 2008.

▶ Class 1M, 2 or 2M lasers, as defined in British Standard BS EN 60825-1: 2007, for example low-power laser pointers.

The above list is not exhaustive

13.5.4 Hazardous light sources

Some sources of light can cause a risk of ill health, such as: burns or reddening (erythema) of the skin or surface of the eye (photokeratitis); burns to the retina of the eye; so-called blue-light damage to the eye (photoretinitis) and damage to the lens of the eye that may bring about the early onset of cataract.

List 2 Hazardous light sources

Examples of hazardous sources of light that present a 'reasonably foreseeable' risk of harming the eyes and skin of workers and where control measures are needed include:

▶ Metal working – welding (both arc and oxy-fuel) and plasma cutting.

▶ Pharmaceutical and research – UV fluorescence and sterilisation systems.

▶ Hot industries – furnaces.

▶ Printing – UV curing of inks.

▶ Motor vehicle repairs – UV curing of paints and welding.

▶ Medical and cosmetic treatments – laser surgery, blue light and UV therapies, Intense Pulsed Light sources (IPLs).

▶ Industry, research and education, for example all use of Class 3B and Class 4 lasers, as defined in British Standard BS EN 60825-1: 2007.

▶ Any Risk Group 3 lamp or lamp system (including LEDs), as defined in British Standard BS EN 62471: 2008, for example search lights, professional projections systems.

Less common hazardous sources are associated with specialist activities – for example lasers exposed during the manufacture or repair of equipment, which would otherwise not be accessible.

13.5.5 Further information

Guidance for Employers on the Control of Artifical Optical Radiation at Work Regulations (AOR) 2010 www.hse.gov.uk/radiation/nonionising/employers-aor.pdf

The full text of the Regulations is available from the UK Government Information website at www.legislation.gov.uk/uksi/2010/1140/contents/made

13.6 Classification, Labelling and Packaging of Substances and Mixtures Regulation (European) adopting into EU UN Globally Harmonised System of Classification and Labelling of Chemicals (GHS) as amended from End of Implementation date of 31/12/2020 or later if changed

13.6.1 Introduction

Worldwide, there are many different laws on how to identify hazardous chemicals (classification) and how this information should be communicated to users (via labels and safety data sheets (SDSs)).

This is often confusing as the same chemical can have different hazard descriptions in different countries. The UN brought together experts from different countries to create the GHS.

The aim of GHS is to have, throughout the world, the same:

▶ criteria for classifying chemicals according to their health, environmental and physical hazards;

▶ hazard communication requirements for labelling and SDSs.

The GHS is not a law, it is an international agreement. To make the GHS legally apply, each country or bloc of countries must adopt the GHS through legislation. European Union (EU) member states agreed to adopt the GHS across the EU through a direct-acting Regulation, the European Regulation (EC) No. 1272/2008 on **C**lassification, **L**abelling and **P**ackaging of substances and mixtures. This is also known as the 'CLP Regulation' or just 'CLP'.

The UK has left the EU but remains strongly committed to the effective and safe management of chemicals.

13

Read the latest information on regulating chemicals during the transition period and beyond:

► Authorisation of biocidal substances and products (Biocides)

► Classification, Labelling and Packaging of substances and chemicals (CLP)

► Export and import of hazardous chemicals from and into Europe (PIC)

► Pesticides or Plant Protection Products (PPP)

► Registration, Evaluation, Authorisation and restriction of Chemicals (REACH)

The Health and Safety Executive (HSE) plays a key role in the UK's chemicals regulatory process working closely with the Department for Environment, Food and Rural Affairs (Defra) and the Environment Agency (EA).

As a responsible regulator their priorities are to:

► ensure the effective and safe management of chemicals to safeguard human health and the environment;

► respond to emerging risks;

► allow trade with the EU and the rest of the world that is as frictionless as possible.

13.6.2 The Classification, Labelling and Packaging (CLP) of substances and mixtures regulation

The CLP Regulation:

► adopted the majority of the GHS 'hazard classes' (nature of the hazard) and 'hazard categories' (severity of the hazard);

► kept a few aspects of the existing EU system, where these do not contradict the GHS, to maintain existing EU standards (e.g. harmful to the ozone layer);

► introduces newly designed hazard symbols, known as pictograms (see Figure 13.1);

► introduced labelling phrases, known as 'hazard statements' and 'precautionary statements';

	Example of hazard statement	Example of precautionary statement
	Heating may cause an explosion	Keep away from heat/sparks/open flames/hot surfaces – no smoking
	Heating may cause a fire	Keep only in original container
	May intensify fire; oxidizer	Take any precaution to avoid mixing with combustibles
	Causes serious eye damage	Wear eye protection
	Toxic if swallowed	Do not eat, drink or smoke when using this product
	Toxic to the aquatic life, with long lasting effects	Avoid release to the environment
	New pictogram, reflects serious longer term health hazards such as carcinogenicity and respiratory sensitization, e.g. may cause allergy or asthma symptoms or breathing difficulties if inhaled	In case of inadequate ventilation, wear respiratory protection
	New pictogram, refers to less serious health hazards such as skin irritancy/sensitization and replaces the CHIP [x] symbol, e.g. may cause an allergic skin reaction	Contaminated work clothing should not be allowed out of the workplace
	New pictogram, used when the containers hold gas under pressure, e.g. may explode when heated	None

Figure 13.1 GHS pictograms with examples of Hazard statements and Precautionary statements

- introduced provisions for a classification and labelling inventory;
- maintained the list of harmonised classifications.

13.6.3 Requirements of the CLP Regulation

The CLP Regulation requires:

- substances to be classified, labelled and packaged according to the CLP Regulation from 1 December 2010 onwards;
- mixtures to be classified, labelled and packaged according to the CLP Regulation from 1 June 2015 onwards.

The former system required that suppliers must use certain symbols and warning phrases on the label to inform users about the hazards of chemicals.

The CLP Regulation introduces new hazard pictograms (symbols), and new 'signal' words such as 'warning' and 'danger', together with new phrases for the different hazards.

Although many of the CLP pictograms are similar to the existing EU system, they have been redesigned. The pictograms are diamond shaped, white and black with a red border.

13.6.4 Chemical manufacture and storage – the European Commission's chemical strategy (REACH)

What is REACH?

REACH (Registration, Evaluation, Authorisation and restriction of Chemicals) is the system for controlling chemicals in the EU. It became law in the UK on 1 June 2007. In October 2006, Defra nominated HSE to be the UK Competent Authority for REACH, working closely with the Environment Agency and other partners to manage certain key aspects of the REACH system in the UK.

The UK is not now a member state of the EU. However, until the end of the transition period market access, and therefore UK–EU trade, should continue on the same terms as before.

During the transition period:

- EU REACH will continue to apply in the UK and businesses do not need to take any action to comply with UK REACH;
- there is no requirement to transfer UK-held registrations to an EU 27 legal entity in order to retain EU market access;
- registrations, authorisations and restrictions in place before the UK left the EU continue to be valid;
- the process for registering new chemicals under REACH remains unchanged, UK companies are still required to register with the European Chemicals Agency (ECHA);

- the UK will recognise all new decisions relating to REACH made by the EU.

The extent of the UK's ongoing participation in EU REACH following the transition period will be determined by the outcome of the UK's negotiations on its future relationship with the EU.

See www.gov.uk/guidance/how-to-comply-with-reach-chemical-regulations.

The UK REACH Helpdesk can be contacted at: ukreachca@hse.gov.uk.

Aims

REACH has several aims:

- to provide a high level of protection of human health and the environment from the use of chemicals;
- to allow the free movement of substances on the EU market;
- to make the people who place chemicals on the market (manufacturers and importers) responsible for understanding and managing the risks associated with their use;
- to promote the use of alternative methods for the assessment of the hazardous properties of substances;
- to enhance innovation in and the competitiveness of the EU chemicals industry.

No data, no market

A major part of REACH is the requirement for manufacturers or importers of substances to register them with a central European Chemicals Agency (ECHA). A registration package will be supported by a standard set of data on that substance. The amount of data required is proportionate to the amount of substance manufactured or supplied.

If substances are not registered, then the data on them will not be available and as a result, the company will no longer be able to manufacture or supply them legally, that is no data, no market.

Scope and exemptions

REACH applies to substances manufactured or imported into the EU in quantities of 1 tonne per year or more. Generally, it applies to all individual chemical substances on their own, in preparations or in articles (if the substance is intended to be released during normal and reasonably foreseeable conditions of use from an article). Some substances, e.g. radioactive substances, are specifically excluded. See REACH for details.

Registration

Registration is a requirement on industry (manufacturers/suppliers/importers) to collect and collate specified sets of information on the properties

13

443

of those substances they manufacture or supply. This information is used to perform an assessment of the hazards and risks that substance may pose and how those risks can be controlled. This information and its assessment are submitted to the ECHA in Helsinki.

Chemicals already existing (those on EINECS (European Inventory of Existing Commercial Chemical Substances) or manufactured in the EU prior to entry into force of REACH) are known as 'phase-in' substances under REACH. These will be registered in three phases according to their tonnage and/or hazardous properties.

Non-phase-in substances (i.e. those not on EINECS, or those which have never been manufactured previously or not pre-registered) were subject to registration by 1 June 2008.

Evaluation

Registration packages (dossiers) submitted under REACH can be evaluated for:

- ▶ **Compliance check**: A quality/accuracy check of the information submitted by industry.
- ▶ **Dossier evaluation:** A check that an appropriate testing plan has been proposed for substances registered at the higher tonnage levels ($100 tonnes/ annum).
- ▶ **Substance evaluation**: An evaluation of all the available data on a substance, from all registration dossiers. This is done by national Competent Authorities on substances that have been prioritised for potential regulatory action because of concerns about their properties or uses.

Substances of very high concern

Some substances have hazards that have serious consequences; for example they cause cancer, or they have other harmful properties and remain in the environment for a long time and gradually build up in animals. These are 'substances of high concern'. One of the aims of REACH is to control the use of such substances.

Authorisation

Authorisation is a feature of REACH that is new to the area of general chemicals management. As REACH progresses, a list of 'substances of very high concern' will be created. Substances on this list cannot be supplied or used unless an authorisation has been granted. A company wishing to market or use such a substance must apply to the ECHA in Helsinki for an authorisation, which may be granted or refused.

Restrictions

Any substance that poses a particular threat can be restricted. Restrictions take many forms, for example, from a total ban to not being allowed to supply it to the general public. Restrictions can be applied to any substance, including those that do not require registration. This part of REACH takes over the provisions of the Marketing & Use Directive.

Classification and labelling

An important part of chemical safety is clear information about any hazardous properties a chemical has. The classification of different chemicals according to their characteristics (e.g. those that are corrosive, or toxic to fish) follows an established system, which is reflected in REACH. REACH has been written with GHS in mind (see Section 13.6.1 for more information on GHS).

Information in the supply chain

The passage of information up and down the supply chain is a key feature of REACH. Users should be able to understand what manufacturers and importers know about the dangers involved in using chemicals and how to control risks. However, in order for suppliers to be able to assess these risks they need information from the downstream users about how the chemicals are used. REACH provides a framework in which information can be passed both up and down supply chains.

More detailed information about REACH can be found at the HSE REACH website.

Enforcement

Enforcing this very wide-ranging system presents new challenges to regulators across Europe. REACH places new duties on a range of different businesses. Mostly, the new duties will be on manufacturers and importers of chemicals, but there are also requirements for downstream users of chemicals to share information with their suppliers. HSE play a key role in enforcing REACH, both as the UK Competent Authority and more generally as the UK occupational health and safety regulator.

13.6.5 Further information

European Chemicals Agency information website for general and specific guidance: http://echa.europa.eu/web/guest/support/ guidance-on-reach-and-clp-implementation/ guidance-in-a-nutshell

Guidance in a nutshell – Compilation of safety data sheets, European Chemicals Agency 2013: http://echa.europa.eu/documents/10162/13643/ sds_nutshell_guidance_en.pdf

HSE Chemical Classification website: www.hse.gov.uk/ chemical-classification/index.htm

HSE REACH website: www.hse.gov.uk/reach/about.htm

HSE GHS website: www.hse.gov.uk/chemical-classification/index.htm

13.7 Confined Spaces Regulations 1997

13.7.1 Introduction

These Regulations concern any work that is carried on in a place which is substantially (but not always entirely) enclosed, where there is a reasonably foreseeable risk of serious injury from conditions and/or hazardous substances in the space or nearby. Every year about 15 people are killed and a number seriously injured across a wide range of industries (e.g. from simple open top pits to complex chemical plants). Rescuers without proper training and equipment, often become the victims.

13.7.2 Definitions

Confined space – means any place, including any chamber, tank, vat, silo, pit, trench, pipe, sewer, flue, well or similar space in which, by virtue of its enclosed nature, there arises a reasonably foreseeable specified risk.

Specified risk – means a risk to any person at work of:

▶ serious injury arising from a fire or explosion;
▶ loss of consciousness arising from an increase in body temperature;
▶ loss of consciousness or asphyxiation arising from gas, fume, vapour or the lack of oxygen;
▶ drowning arising from the an increase in the level of liquid;
▶ asphyxiation arising from a free-flowing solid or because of entrapment by it.

13.7.3 Employers' duties – Regulation 3

Duties are placed on employers to:

▶ comply regarding any work carried out by employees;
▶ ensure, sfairp, that other persons (e.g. use competent contractors) comply regarding work in the employer's control.

The self-employed also have duties to comply.

13.7.4 Work in confined space – Regulation 4

1. No person at work shall enter a confined space for any purpose unless it is not reasonably practicable to achieve that purpose without such entry.
2. Other than in an emergency, no person shall enter, carry out work or leave a confined space otherwise than in accordance with a safe system of work, relevant to the specified risks.

13.7.5 Risk assessment

Risk assessment is an essential part of complying with these Regulations and must be done (under the MHSWR 1999) to determine a safe system of work. The risk assessment needs to follow a hierarchy of controls for compliance. This should start with the measures – both in design and procedures – that can be adopted to enable any work to be carried out outside the confined space.

The assessment must be done by a competent person and will form the basis of a safe system of work. This will normally be formalised into a specific permit to work, applicable to a particular task. The assessment will involve the following.

(a) The general conditions to assess what may or may not be present. Consider:

▶ what have been the previous contents of the space;
▶ residues that have been left in the space, for example, sludge, rust, scale and what may be given off if they are disturbed;
▶ contamination which may arise from adjacent plant, processes, services, pipes or surrounding land, soil or strata;
▶ oxygen deficiency and enrichment. There are very high risks if the oxygen content differs significantly from the normal level of 20.8%; if it is above this level increased flammability exists; if it is below then impaired mental ability occurs, with loss of consciousness under 16%;
▶ that physical dimensions and layout of the space can affect air quality.

(b) Hazards arising directly from the work being undertaken. Consider:

▶ the use of cleaning chemicals and their direct effect or interaction with other substances;
▶ sources of ignition for flammable dusts, gases, vapours, plastics and the like.

(c) The need to isolate the confined space from outside services or from substances held inside, such as liquids, gases, steam, water, fire extinguishing media, exhaust gases, raw materials and energy sources.

(d) The requirement for emergency rescue arrangements including trained people and equipment.

13.7.6 Safe system of work

The detailed precautions required will depend on the nature of the confined space and the actual work being carried out. The main elements of a safe system of work which may form the basis of a 'permit-to-work' are:

▶ the type and extent of supervision;
▶ competence and training of people doing the work;
▶ communications between people inside, and from inside to outside and to summon help;

13

- testing and monitoring the atmosphere for hazardous gas, fume, vapour, dust, etc. and for concentration of oxygen;
- gas purging of toxic or flammable substances with air or inert gas such as nitrogen;
- good ventilation, sometimes by mechanical means;
- careful removal of residues using equipment which does not cause additional hazards;
- effective isolation from gases, liquids and other flowing materials by removal of pieces of pipe, blanked off pipes, locked off valves;
- effective isolation from electrical and mechanical equipment to ensure complete isolation with lock off and a tag system with key security. Need to secure against stored energy or gravity fall of heavy presses etc.;
- if it is not possible to make the confined space safe, the provision of personal and respiratory protective equipment;
- supply of gas via pipes and hoses carefully controlled;
- access and egress to give quick unobstructed and ready access and escape;
- fire prevention;
- lighting, including emergency lighting;
- prohibition of smoking;
- emergencies and rescue;
- limiting of working periods and the suitability of individuals.

13.7.7 Emergency arrangements – Regulation 5

Before people enter a confined space, suitable and sufficient rescue arrangements must be set up. These must:

- reduce the risks to rescuers sfairp;
- include the provision and maintenance of suitable resuscitation equipment designed to meet the specified risks.

To be suitable and sufficient, arrangements will need to cover:

- rescue and resuscitation equipment;
- raising the alarm, alerting rescue and watch keeping;
- safeguarding the rescuers;
- fire safety precautions and procedures;
- control of adjacent plant;
- first-aid arrangements;
- notification and consultation with emergency services;
- training of rescuers and simulations of emergencies;
- size of access openings to permit rescue with full breathing apparatus, harnesses, fall arrest gear and lifelines, which is the normal suitable respiratory protection and rescue equipment for confined spaces.

13.7.8 Training

Specific, detailed and frequent training is necessary for all people concerned with confined spaces, whether they are acting as rescuers, watchers or those carrying out the actual work inside the confined space. The training will need to cover all procedures and the use of equipment under realistic simulated conditions.

13.7.9 Further information

Guidance on Permit-to-Work Systems: A Guide for the Petroleum, Chemical and Allied Industries. HSG250, HSE, 2005, ISBN 9780717629435. www.hse.gov.uk/pubns/books/hsg250.htm

Respiratory Protective Equipment at Work: A Practical Guide HSG53 (fourth edition), 2013, HSE, ISBN 9780717664542. www.hse.gov.uk/pubns/books/hsg53.htm

Safe Work in Confined Spaces. Confined Spaces Regulations 1997. Approved Code of Practice, Regulations and Guidance L101 (third edition), 2014, HSE, ISBN 9780717666225. www.hse.gov.uk/pubns/books/l101.htm

Safe work in confined spaces INDG258(rev 1), 2013, HSE, www.hse.gov.uk/pubns/indg258.pdf

13.8 Construction (Design and Management) Regulations 2015 (CDM 2015)

13.8.1 Introduction and background

The CDM 2015 Regulations were introduced in April 2015.

The policy objectives behind the CDM 2015 are to:

- maintain or improve worker protection;
- simplify the regulatory package;
- improve health and safety standards on small construction sites;
- implement the Temporary or Mobile Construction Sites Directive (TMCSD) in a proportionate way;
- discourage bureaucracy; and
- meet better regulation principles.

The HSE commissioned extensive external research which was published in April 2012 and along with the other elements of the evaluation the broad conclusions were that:

- CDM 2007 was viewed more positively by duty-holders than the 1994 version;
- its broad structure was fit for purpose;
- problems generally arose through mis- and over-interpretation of the Regulations;
- significant concerns remained, however, in several areas:
  - ▷ the Regulations had not borne down on bureaucracy as hoped;

▷ the Regulations had led to an industry approach to competence which was heavy-handed and in many cases burdensome, particularly on SMEs;

▷ the coordination function in the pre-construction phase was not in many cases well embedded.

The balance of where serious and fatal injuries occur has shifted dramatically in the past 10–15 years. Two-thirds or more of fatalities now occur on small sites – sites where fewer than 15 people work – which is the reverse of the historical picture. The larger, more structured part of the industry has made significant progress in improving its management of health and safety risks over this time frame. Its motivation for achieving higher standards is often one of continuous improvement and innovation leading to best practice, rather than just meeting regulatory requirements.

13.8.2 Outline of regulations

The Regulations are divided into five parts. Part 1 deals with matters of interpretation and application. The Regulations apply to all construction work in Great Britain and its territorial waters, and apply to both employers and self-employed without distinction.

Part 2 covers general client duties which apply to all construction projects, including the requirements for notification.

Part 3 sets out health and safety duties and roles which apply to all construction projects.

Part 4 sets out the general requirements for all construction sites, and covers physical safeguards which need to be provided to prevent danger. Duties to achieve these standards are held by contractors who actually carry out the work, irrespective of whether they are employers or are self-employed. Duties are also held by those who do not do construction work themselves, but control the way in which the work is done. In each case, the extent of the duty is in proportion to the degree of control which the individual or organisation has over the work in question.

This does not mean that everyone involved with design, planning or management of the project legally must ensure that all the specific requirements in Part 4 are complied with. They only have duties if, in practice, they exercise sufficient control over the actual working methods, safeguards and site conditions. If, for example, a client specifies that a particular job is done in a particular way then the client will have a duty to make sure that their instructions comply with the requirements.

Contractors must not allow work to start or continue unless the necessary safeguards are in place,

whether they are providing the safeguards, or using something, for example a scaffold supplied by someone else.

Part 5 of the Regulations covers issues of enforcement in respect of fire; transitional provisions which will apply during the period when the regulations come into force; review; amendments and revocations of other legislation.

13.8.3 Definition of construction and client – Regulation 2

Construction work

'Construction work' means the carrying out of any building, civil engineering or engineering construction work and includes:

(a) the construction, alteration, conversion, fitting-out, commissioning, renovation, repair, upkeep, redecoration or other maintenance (including cleaning which involves the use of water or an abrasive at high pressure or the use of corrosive or toxic substances), decommissioning, demolition or dismantling of a structure;

(b) the preparation for an intended structure, including site clearance, exploration, investigation (but not site survey) and excavation (but not preconstruction archaeological investigations), and the clearance or preparation of the site or structure for use or occupation at its conclusion;

(c) the assembly on site of prefabricated elements to form a structure or the disassembly on site of prefabricated elements which, immediately before such disassembly, formed a structure;

(d) the removal of a structure or of any product or waste resulting from demolition or dismantling of a structure or from disassembly of prefabricated elements which immediately before such disassembly formed such a structure;

(e) the installation, commissioning, maintenance, repair or removal of mechanical, electrical, gas, compressed air, hydraulic, telecommunications, computer or similar services which are normally fixed within or to a structure.

This means that there are some things which may take place on a construction site, like tree planting and horticulture work, putting up marquees, positioning lightweight panels in an open plan office, archeological exploration and surveying, which are not construction work.

Client

'Client' means any person for whom a project is carried out; 'domestic client' means a client for whom a project is being carried out which is not in the course or furtherance of a business of that client.

13

> **Note:**
>
> Typical operating and owning costs of a building are in the ratio:
>
> ▶ 1 for construction costs
> ▶ 5 for maintenance and building operating costs
> ▶ 200 for business operating costs
>
> *Source:* Royal Academy of Engineering report on 'Long Term costs of Owning and Using Buildings'.

Part 2 – Client duties

13.8.4 Client duties for managing projects – Regulation 4

A client must:

▶ make arrangements for managing a project to ensure that construction work is carried out safely sfairp;
▶ the welfare requirements in Schedule 2 are complied with;
▶ take reasonable steps to ensure that the arrangements are maintained and reviewed throughout the project;
▶ provide pre-construction information to every designer and contractor appointed or being considered for appointment;
▶ ensure before the construction phase begins that the principal contractor (or contractor if only one) draws up a construction phase plan;
▶ ensure that the principal designer prepares an appropriate health and safety file for the project which:
  ▷ includes information which may be needed for subsequent projects;
  ▷ is revised from time to time;
  ▷ is kept available for inspection;
  ▷ is passed on if interest in the project passes to another client. If a client disposes of the client's interest in the structure, the client complies with the duty in paragraph 5(b)(iii) by providing the health and safety file to the person who acquires the client's interest in the structure and ensuring that that person is aware of the nature and purpose of the file;
▶ ensure that the principal designer complies with their duties under Regulations 11 and 12;
▶ ensure that the principal contractor complies with Regulations 12 to 14.

13.8.5 Appointment of principal designer and contractor – Regulation 5

Where it is foreseeable that more than one contractor will be working on the project at any time the client must appoint in writing:

▶ A principal designer for the responsibilities in Regulations 11 and 12.

▶ A principal contractor for the responsibilities in Regulations 12 to 14.

If an appointment is not made the client must fulfil either/or both of these roles. This part of Regulation 5(3) and (4) does not apply to domestic clients.

13.8.6 Notification – Regulation 6

The client must notify the HSE using the particulars in Schedule 1 (Figure 13.2) if the construction work on a site is scheduled to:

▶ last longer than 30 working days and have more than 20 workers working simultaneously at any point in the project; or
▶ exceed 500 person days.

The notice must be clearly displayed on the site in a form that can be read by any worker engaged in the construction work. Must be updated if necessary.

> **SCHEDULE 1**
> **PARTICULARS TO BE NOTIFIED TO THE EXECUTIVE**
>
> 1. Date of forwarding.
> 2. Exact address of the construction site.
> 3. The name of the local authority where the site is located.
> 4. A brief description of the project and the construction work which it includes.
> 5. Contact details of the client (name, address, telephone number and any e-mail address).
> 6. Contact details of the CDM co-ordinator (name, address, telephone number and any e-mail address).
> 7. Contact details of the principal contractor (name, address, telephone number and any e-mail address).
> 8. Date planned for the start of the construction phase.
> 9. The time allowed by the client to the principal contractor referred to in Regulation 15(b) for planning and preparation for construction work.
> 10. Planned duration of the construction phase.
> 11. Estimated maximum number of people at work on the construction site.
> 12. Planned number of contractors on the construction site.
> 13. Name and address of any contractor already appointed.
> 14. Name and address of any designer already engaged.
> 15. A declaration signed by or on behalf of the client that he is aware of his duties under these Regulations.

Figure 13.2 (a) CDM 2015 – Schedule 1

13.8.7 Application to domestic clients – Regulation 7

Unless there is a written agreement between the client and the principal designer, the duties in Regulations 4(1) to (7) and Regulation 6, must be carried out by:

▶ the contractor or principal contractor where there is more than one.
▶ the principal designer where there is a written agreement that they will fulfil those duties.

Where no appointment is made:

▶ the designer in control of the pre-construction phase is the principal designer; and

▶ the contractor in control of the construction phase is the principal contractor.

Regulation 5(3) and (4) does not apply (see 13.8.5).

Part 3 – Health and safety duties and roles

13.8.8 General duties – Regulation 8

Designers and contractors (including principal) appointed to work on a project must have: skills, knowledge and experience and if an organisation, the organisational capability, to fulfil the role they are given. They must not accept the appointment unless they fulfil these conditions.

Any person responsible for appointing a designer or contractor must take reasonable steps to satisfy themselves that the designer or contractor fulfils the conditions of the paragraph above.

Any duty-holders must cooperate with others on the same or adjoining construction sites.

A person involved in a project has a duty to report anything they are aware of which could endanger them or others.

Information must be comprehensible and provided as soon as is practicable.

13.8.9 Duties of designers – Regulations 9 and 10

A designer must:

▶ before starting work be satisfied that the client is aware of their duties under CDM 2015;
▶ take into account the general principles of prevention to eliminate risks for anyone:
  ▷ working or liable to be affected by construction work;
  ▷ maintaining or cleaning a structure; or
  ▷ using a structure designed as a workplace;
▶ if elimination is not possible the designer must sfairp:
  ▷ take steps to reduce and control the risks through the design process;
  ▷ provide information about the risks to the principal designer;
  ▷ ensure appropriate information is contained in the health and safety file;
▶ provide with the design sufficient information to help clients, other designer and contractors to fulfil their duties under CDM 2015.

Regulation 10 requires that if a design comes from outside Great Britain the person who commissions it or the client must ensure that Regulation 9 is complied with.

13.8.10 Duties of principal designer in relation to health and safety at the pre-construction phase – Regulation 11

A principal designer must plan, manage, monitor and coordinate the pre-construction phase of a project, taking into account the general principles of prevention

and the content of the construction phase plan and health and safety file to ensure:

▶ sfairp project is carried out without risks to health and safety with regard to design, technical, organisational and time required aspects;
▶ the foreseeable risks to people:
  ▷ doing or affected by construction work,
  ▷ maintaining or cleaning a structure; or
  ▷ using a structure designed as a workplace;
▶ all designers comply with their duties in Regulation 9;
▶ all people concerned with the pre-construction phase cooperate with the client, the principal designer and each other;
▶ that they assist the client to provide the pre-construction information and provide pre-construction information to every designer and contractor appointed or being considered, in a convenient form;
▶ liaison with the principal contractor for the duration of the project particularly over the preparation of the construction phase plan.

13.8.11 Construction phase plan and health and safety file

The principal contractor must draw up (or make arrangements for) a construction phase plan in the pre-construction phase period. This plan must set out:

▶ health and safety arrangements;
▶ site rules taking account of any industrial activity on the site;
▶ must include specific measures for work within Schedule 3 to the regulations.

And the:

▶ Principal designer must assist principal contractor to prepare the construction-phase plan by providing all relevant information from the client and designers;
▶ Principal contractor must review update and revise as necessary the construction-phase plan;
▶ Principal designer must prepare a health and safety file during the pre-construction phase;
▶ Principal designer must review, update and revise the health and safety file as necessary;
▶ Principal contractor must provide relevant information to the principal designer for the health and safety file.

And:

▶ If the appointment of the principal designer terminates before the end of the project the health and safety file must be passed on to the principal contractor who must then keep it reviewed, updated and revised as needed;
▶ At the end of the project the health and safety file must be passed to the client.

13

SCHEDULE 3
WORK INVOLVING PARTICULAR RISKS

1. Work which puts workers at risk of burial under earthfalls, engulfment in swampland or falling from a height, where the risk is particularly aggravated by the nature of the work or processes used or by the environment at the place of work or site;
2. Work which puts workers at risk from chemical or biological substances constituting a particular danger to the safety or health of workers or involving a legal requirement for health monitoring;
3. Work with ionizing radiation requiring the designation of controlled or supervised areas under regulation 16 of the Ionising Radiations Regulations 1999;
4. Work near high voltage power lines;
5. Work exposing workers to the risk of drowning;
6. Work on wells, underground earthworks and tunnels;
7. Work carried out by divers having a system of air supply;
8. Work carried out by workers in caissons with a compressed air atmosphere;
9. Work involving the use of explosives;
10. Work involving the assembly or dismantling of heavy prefabricated components.

Figure 13.2 (b) CDM 2015 – Schedule 3

13.8.12 Duties of the principal contractor in relation to health and safety at the construction phase – Regulation 13

The principal contractor must plan, manage, monitor and coordinate the construction phase taking into account the general principles of prevention to ensure that sfairp:

▶ project is carried out without risks to health and safety with regard to design, technical, organisational and time required aspects;
▶ cooperation between contractors is achieved;
▶ the implementation of legal requirement by contractors is coordinated;
▶ employers and self-employed apply general principles of prevention consistently and follow the construction phase plan;
▶ a suitable site induction is provided;
▶ access to unauthorised persons is controlled;
▶ Schedule 2 welfare facilities are provided throughout the construction phase;
▶ there is liaison with the principal designer throughout the project particularly regarding information for the health and safety file and management of the construction phased plan.

13.8.13 Principal contractor's duties to consult and engage with workers – Regulation 14

The principal contractor must:

▶ make and maintain arrangements to enable people on site to cooperate effectively to develop safety measures and check their effectiveness;

▶ consult workers or their representatives in good time on health and safety issues unless already consulted by their own employer;
▶ ensure that workers or their representatives can inspect and take copies of health and safety-related information for the project except:
  ▷ if disclosure would be against the interests of national security;
  ▷ if disclosure would contravene a legal prohibition;
  ▷ where the information relates specifically to an individual without their consent;
  ▷ where disclosure would cause substantial damage to the principal contractor's undertaking or other undertaking if they had supplied the information;
  ▷ where the information was obtained by the contractor in connection with legal proceedings.

13.8.14 Duties of contractors – Regulation 15

Contractors must:

▶ not carry out construction work on a project unless satisfied that the client is aware of their duties under CDM 2015;
▶ plan, manage and monitor the way construction work is done to ensure it is safe sfairp taking into account design, technical, organisational and time required aspects;
▶ where there is only one contractor working on the project, ensure that a construction phase plan is drawn up prior to setting up the site which must fulfil requirements of Regulation 12(2) which includes specific measures of Schedule 3 (see Figure 13.2(b));
▶ not employ or appoint a person unless they have or are obtaining the necessary skills, knowledge, training and experience to carry out the allocated tasks safely;
▶ must provide any employees or persons in their control, with appropriate supervision, information and instruction to ensure safety of the project including:
  ▷ suitable site induction;
  ▷ emergency procedures;
  ▷ information on risks to their health and safety either identified by their risk assessments or from another contractor's operations;
  ▷ any other information necessary;
▶ not start work until access to unauthorised persons is controlled;
▶ ensure that Schedule 2, welfare facilities are complied with for employees and others under their control.

13.8.15 Part 4 General requirements for all construction sites

The CHSW Regulations 1996 were revoked by CDM 2007 which has been retained by CDM 2015. Without the work at height provisions (Regulations 6, 7 and 8), their requirements form the basis of Part 4 and Schedule 2 (Welfare facilities) of the CDM 2015 Regulations.

Some issues already covered by the Workplace Regulations are particularly important in construction. These include traffic management and lighting. Because these are so important, CDM 2015 duplicates aspects of the Workplace Regulations requirements, although that was not legally necessary.

The issues covered by Part 4 apply to all construction projects and are as follows.

General application of regulations 17–35 – Regulation 16

Duties to comply with this part are placed on contractor or other person who controls the way in which construction work is carried on.

Duty placed on workers to report any defect they are aware of which may endanger the health and safety of themselves or another person.

Safe places of construction work – Regulation 17

This regulation requires sfairp:

- safe access and egress to places of work;
- maintenance of access and egress;
- that people are prevented from gaining access to unsafe access or workplaces;
- provision of safe places of work with adequate space and suitable for workers.

Good order and site security – Regulation 18

Requires that all parts of a construction be kept, sfairp, in good order and in a reasonable state of cleanliness.

Where necessary in the interests of health and safety the site's perimeter should be identified, signed and fenced off.

No timber or other materials with projecting nails or similar sharp objects shall be used or allowed to remain if they could be a source of danger.

Stability of structures – Regulation 19

All practical steps must be taken to ensure that:

- any new or existing structure which may be, or become, weak does not collapse accidentally;
- temporary supporting structures are designed, installed and maintained so as to withstand all foreseeable loads;
- a structure is not overloaded so as to be a source of danger.

Demolition or dismantling – Regulation 20

Demolition or dismantling must be planned and carried out so as to prevent danger or reduce it to as low a level as is reasonably possible.

Arrangements must be recorded in writing before the work takes place.

Explosives – Regulation 21

Explosives, sfairp, must be stored, transported and used safely and securely.

Explosives may only be fired if suitable and sufficient steps have been taken to ensure no one is exposed to risk of injury.

Excavations – Regulation 22

All practicable steps shall be taken to prevent danger to people including, where necessary, supports or battering to ensure that:

- excavations do not collapse accidentally;
- no material from side, roof or adjacent is dislodged or falls;
- a person is prevented from being buried or trapped by a fall of material;
- persons, work equipment or any accumulation of material are prevented from falling into the excavation;
- any part of an excavation or ground adjacent is not overloaded by work equipment or material.

No construction work may be carried out in an excavation where any supports or battering has been provided unless it has been inspected by a competent person:

- at the start of the shift (only one report required every 7 days);
- after any event likely to affect the strength or stability of the excavation;
- after any material accidentally falls or is dislodged;
- the person inspecting is satisfied that it is safe.

Where the person who carries out the inspection has informed the person on whose behalf the inspection was carried out, work must cease until the relevant matters have been corrected.

Cofferdams and caissons – Regulation 23

Every cofferdam or caisson must be:

- suitably designed and constructed;
- appropriately equipped so that people can gain shelter or escape if water or material enters;
- properly maintained.

13

A cofferdam or caisson may be used to carry out construction only if:

▶ it and any work equipment and materials which affect its safety have been inspected by a competent person:
  ▷ at the start of the shift (only one report required every 7 days);
  ▷ after any event likely to affect its strength or stability;
  ▷ the person inspecting is satisfied that it is safe.

Where the person who carries out the inspection has informed the person on whose behalf the inspection was carried out, work must cease until the relevant matters have been corrected.

Reports of inspection – Regulation 24

Before the end of the shift in which the report was completed, the person who carries out the inspection under Regulations 22 or 23 must:

▶ inform the person for whom they carried out the inspection if they are not satisfied that construction work can be carried out safely;

▶ prepare a report with the particulars shown in Figure 13.3.

A copy of the report must be provided within 24 hours of the relevant inspection. If the inspector is an employee or works under someone else's control, that person must ensure that these duties are performed.

Records must be kept at the site where the inspection was carried out until the work is complete and then for three months. Extracts must be provided for an inspector as and when they require. Only one report is required if more than one report is produced under each shift requirement of Regulation 22(4)(a)(i) or 23(2)(a)(i) within 7 days.

Inspection reports which must include—

(i) the name and address of the person on whose behalf the inspection was carried out;
(ii) the location of the place of construction work inspected;
(iii) a description of the place of construction work or part of that place inspected (including any work equipment and materials);
(iv) the date and time of the inspection;
(v) the details of any matter identified that could give rise to a risk to the safety of any person;
(vi) details of any action taken as a result of any matter identified in paragraph (v);
(vii) the details of any further action considered necessary; and
(viii) the name and position of the person making the report.

Figure 13.3 Content of inspection reports

Energy distribution installations – Regulation 25

Energy distribution installations must be suitably located, checked and clearly indicated.

Where there is a risk from electric power cables:

▶ they should be routed away from the risk areas; or
▶ the power should be cut off; or
▶ if the above safety measures are not reasonably practicable suitable warning notices, with barriers to exclude work equipment and suspended protections where vehicles have to pass underneath, should be provided; or
▶ something equally as safe must be done.

No construction work which is liable to create a risk to health or safety from an underground service, or from damage to or disturbance of it, shall be carried out unless suitable and sufficient steps have been taken to prevent such risk, sfairp.

Prevention of drowning – Regulation 26

Where persons could fall into water or other liquid, steps must be taken to:

▶ prevent, sfairp, people from falling;
▶ minimise the risk of drowning if people fall;
▶ provide, use and maintain suitable rescue equipment to ensure prompt rescue.

Transport to work by water must be suitable and sufficient and any vessel must not be overcrowded or overloaded.

Traffic routes – Regulation 27

Every construction site shall be organised in such a way that, sfairp, pedestrians and vehicles can move safely.

Traffic routes shall be suitable for the persons or vehicles using them, sufficient in number, in suitable positions and of sufficient size.

Steps must be taken to ensure that:

▶ pedestrians or vehicles may use them without causing danger to the health or safety of persons near them;
▶ any door or gate for pedestrians which leads onto a traffic route is sufficiently separated from it to enable them from a place of safety to see any approaching vehicle;
▶ there is sufficient separation between vehicles and pedestrians to ensure safety or, where this is not reasonably practicable:
  (i) there are provided other means for the protection of pedestrians;
  (ii) there are effective arrangements for warning any person liable to be crushed or trapped by any vehicle of its approach;
▶ any loading bay has at least one exit point for the exclusive use of pedestrians;
▶ where it is unsafe for pedestrians to use a gate intended primarily for vehicles, one or more doors for pedestrians provided in the immediate vicinity of the gate, is clearly marked and is kept free from obstruction.

Every traffic route must be:

► indicated by suitable signs;
► regularly checked;
► properly maintained.

No vehicle shall be driven on a traffic route unless, sfairp, that traffic route is free from obstruction and permits sufficient clearance.

Vehicles – Regulation 28

The unintended movement of any vehicle must be prevented or controlled.

Suitable and sufficient steps shall be taken to ensure that, where a person may be endangered by the movement of any vehicle, the person having effective control of the vehicle shall give suitable and sufficient warning.

Any vehicle being used for the purposes of construction work, shall, when being driven, operated or towed:

► be driven, operated or towed in a safe manner;
► be loaded so that it can be driven, operated or towed safely.

No person shall ride on any vehicle being used for the purposes of construction work except in a safe place provided.

No person shall remain on any vehicle during the loading or unloading of any loose material unless a safe place of work is provided and maintained.

Suitable measures must be taken to prevent any vehicle from falling into any excavation or pit, or into water, or overrunning the edge of any embankment or earthwork.

Prevention of risk from fire, flooding and asphyxiation – Regulation 29

Steps must be taken to prevent injury arising from:

► fires or explosions;
► flooding; or
► substances liable to cause asphyxiation.

Emergency procedures – Regulation 30

Where necessary, emergency procedures must be prepared and, where necessary, implemented to deal with any foreseeable emergency. They must include procedures for any necessary evacuation of the site or any part thereof.

The arrangements must take account of:

► the type of work for which the construction site is being used;
► the characteristics and size of the construction site and the number and location of places of work on that site;

► the work equipment being used;
► the number of persons likely to be present on the site at any one time;
► the physical and chemical properties of any substances or materials on or likely to be on the site.

Steps must be taken to make people on site familiar with the procedures and test the procedures at suitable intervals.

Emergency routes and exits – Regulation 31

Where necessary a sufficient number of suitable emergency routes and exits shall be provided to enable any person to reach a place of safety quickly in the event of danger.

An emergency route or exit (or, as appropriate, traffic route) provided must:

► lead as directly as possible to an identified safe area;
► be kept clear and free from obstruction and, where necessary, provided with emergency lighting so that such emergency route or exit may be used at any time;
► be indicated by suitable signs.

Fire detection and firefighting – Regulation 32

Where necessary, duty-holders must provide suitable and sufficient:

► firefighting equipment; and
► fire detection and alarm systems which shall be suitably located.

The equipment must be examined and tested at suitable intervals, properly maintained and suitably indicated by signs. If non-automatic, it must be easily accessible.

Every person at work on a construction site shall, sfairp, be instructed in the correct use of any firefighting equipment which it may be necessary for them to use. Where a work activity may give rise to a particular risk of fire, a person shall not carry out such work unless suitably instructed.

Firefighting equipment must be indicated by suitable signs.

Fresh air – Regulation 33

Steps must be taken to ensure, sfairp, that every place of work or approach has sufficient fresh or purified air to ensure that the place or approach is safe and without risks to health. Any plant used for the provision of fresh air must include, where necessary, an effective device to give visible or audible warning of any failure of the plant.

13

Temperature and weather protection – Regulation 34

Steps must be taken to ensure, that sfairp, during working hours the temperature at any place of work indoors is reasonable having regard to its purpose.

Every place of work outdoors must, where necessary, sfairp, and having regard to the purpose for which that place is used and any protective clothing or work equipment provided, be so arranged that it provides protection from adverse weather.

Lighting – Regulation 35

Every place of work and approach thereto and every traffic route shall be provided with suitable and sufficient lighting, which shall be, sfairp, by natural light.

The colour of any artificial lighting provided shall not adversely affect or change the perception of any sign or signal provided for the purposes of health and safety.

Suitable and sufficient secondary lighting shall be provided in any place where there would be a risk to the health or safety of any person in the event of failure of primary artificial lighting.

13.8.16 Welfare facilities – Schedule 2

Sanitary conveniences

1. (1) Suitable and sufficient sanitary conveniences must be provided or made available at readily accessible places.

(2) So far as is reasonably practicable, rooms containing sanitary conveniences must be adequately ventilated and lit.

(3) So far as is reasonably practicable, sanitary conveniences and the rooms containing them must be kept in a clean and orderly condition.

(4) Separate rooms containing sanitary conveniences must be provided for men and women, except where and so far as each convenience is in a separate room, the door of which is capable of being secured from the inside.

Washing facilities

2. (1) Suitable and sufficient washing facilities, including showers if required by the nature of the work or for health reasons, must, sfairp, be provided or made available at readily accessible places.

(2) Washing facilities must be provided—

(a) in the immediate vicinity of every sanitary convenience, whether or not also provided elsewhere; and

(b) in the vicinity of any changing rooms required by paragraph 4, whether or not provided elsewhere.

(3) Washing facilities must include—

(a) a supply of clean hot and cold, or warm, water (which must be running water sfairp);

(b) soap or other suitable means of cleaning; and

(c) towels or other suitable means of drying.

(4) Rooms containing washing facilities must be sufficiently ventilated and lit.

(5) Washing facilities and the rooms containing them must be kept in a clean and orderly condition.

(6) Subject to paragraph (7), separate washing facilities must be provided for men and women, except where they are provided in a room the door of which is capable of being secured from inside and the facilities in each room are intended to be used by only one person at a time.

(7) Sub-paragraph (6) does not apply to facilities which are provided for washing hands, forearms and the face only.

Drinking water

3. (1) An adequate supply of wholesome drinking water must be provided or made available at readily accessible and suitable places.

(2) Where necessary for reasons of health or safety, every supply of drinking water must be conspicuously marked by an appropriate sign.

(3) Where a supply of drinking water is provided, a sufficient number of suitable cups or other drinking vessels must also be provided, unless the supply of drinking water is in a jet from which persons can drink easily

Changing rooms and lockers

4. (1) Suitable and sufficient changing rooms must be provided or made available at readily accessible places if a worker—

(a) has to wear special clothing for the purposes of construction work; and

(b) cannot, for reasons of health or propriety, be expected to change elsewhere.

(2) Where necessary for reasons of propriety, there must be separate changing rooms for, or separate use of rooms by, men and women.

(3) Changing rooms must—

(a) be provided with seating; and

(b) include, where necessary, facilities to enable a person to dry any special clothing and any personal clothing or effects.

(4) Suitable and sufficient facilities must, where necessary, be provided or made available at readily accessible places to enable persons to lock away—

(a) any special clothing which is not taken home;

(b) their own clothing which is not worn during working hours; and

(c) their personal effects.

Facilities for rest

5. (1) Suitable and sufficient rest rooms or rest areas must be provided or made available at readily accessible places.

(2) Rest rooms and rest areas must—

(a) be equipped with an adequate number of tables and adequate seating with backs for the number of persons at work likely to use them at any one time;

(b) where necessary, include suitable facilities for any woman at work who is pregnant or who is a nursing mother to rest lying down;

(c) include suitable arrangements to ensure that meals can be prepared and eaten;

(d) include the means for boiling water; and

(e) be maintained at an appropriate temperature.

Part 5 General

13.8.17 Enforcement in respect of fire – Regulation 36

This Regulation is given in full.

(1) As regards Regulations 30 and 31 the enforcing authority in respect of a construction site which is contained within, or forms part of, premises which are occupied by persons other than those carrying out construction work or any activity related to this work, in so far as those regulations relate to fire are—

(a) in England and Wales the enforcing authority within the meaning of article 25 of the Regulatory Reform (Fire Safety) Order 2005(a) in respect of premises to which that Order applies; or

(b) in Scotland the enforcing authority within the meaning of Section 61 of the Fire (Scotland) Act 2005(b) in respect of premises to which Part 3 of that Act applies.

13.8.18 Further information

Asbestos Essentials HSG210 (fourth edition), 2018, ISBN 9780717666652 www.hse.gov.uk/pubns/books/hsg210.htm

Asbestos: The licensed contractors' guide HSG247, 2006, ISBN 9780717628742 www.hse.gov.uk/pubns/books/hsg247.htm

Electricity in Construction HSE website www.hse.gov.uk/electricity/information/construction.htm

Fire Safety in Construction Work HSG168rev, 2010, HSE, ISBN 9780717663453 www.hse.gov.uk/pubns/books/hsg168.htm.

Health and Safety in Construction HSG150 (rev 3), 2006, HSE, ISBN 9780717661824 (due for revision). www.hse.gov.uk/pubns/priced/hsg150.pdf

Health and Safety in Roof Work (fourth edition) HSG33, 2012, HSE, ISBN 9780717665273. www.hse.gov.uk/pubns/books/hsg33.htm

Safe Use of Vehicles on Construction Sites HSG144, 2009, HSE, ISBN 970717662913. www.hse.gov.uk/pubns/books/hsg144.htm

Protecting the public: Your next move HSG151, 2009, ISBN 9780717662944 www.hse.gov.uk/pubns/books/hsg151.htm

Managing health and safety in construction (Design and Management) Regulations 2015 L153 www.hse.gov.uk/pubns/books/l153.htm

Industry Guidance for clients, contractors, designers, principal designers, principal contractors and workers see www.citb.co.uk/health-safety-and-other-topics/health-safety/construction-design-and-management-regulations/

13.9 Health and Safety (Consultation with Employees) Regulations 1996

13.9.1 Application

These Regulations apply to all employers and employees in Great Britain except:

▶ where employees are covered by safety representatives appointed by recognised trade unions under the Safety Representatives and Safety Committees Regulations 1977 (note in workplaces where there is trade union recognition but either the trade union has not appointed a safety representative, or the union safety representative does not cover the whole workforce, the Health and Safety (Consultation with Employees) Regulations will apply);

▶ domestic staff employed in private households;

▶ crew of a ship under the direction of the master.

13.9.2 Employers' duty to consult – Regulation 3

The employer must consult relevant employees in good time with regard to:

▶ the introduction of any measure which may substantially affect their health and safety;

▶ the employer's arrangements for appointing or nominating competent persons under the Management of Health and Safety at Work Regulations 1999;

▶ any information required to be provided by legislation;

▶ the planning and organisation of any health and safety training required by legislation;

455

▶ the health and safety consequences to employees of the introduction of new technologies into the workplace.

The guidance emphasises the difference between informing and consulting. Consultation involves listening to employees' views and taking account of what they say before any decision is taken.

13.9.3 Persons to be consulted – Regulation 4

Employers must consult with either:

▶ the employees directly; or
▶ one or more persons from a group of employees, who were elected by employees in that group to represent them under these regulations. They are known as 'Representative(s) of Employee Safety' (ROES).

Where ROES are consulted, all employees represented must be informed of:

▶ the names of ROES;
▶ the group of employees represented.

Employers shall not consult a ROES if:

▶ the ROES does not wish to represent the group;
▶ the ROES has ceased to be employed in that group;
▶ the election period has expired;
▶ the ROES has become incapacitated from carrying out the necessary functions.

If an employer decides to consult directly with employees, they must inform them and ROES of that fact. Where no ROES is elected, employers will have to consult directly.

13.9.4 Duty to provide information – Regulation 5

Employers must provide enough information to enable ROES to participate fully and carry out their functions. This will include:

▶ what the likely risks and hazards arising from their work may be;
▶ reported accidents and diseases etc. under the Reporting of Injuries, Diseases and Dangerous Occurrences Regulations (RIDDOR) 2013;
▶ the measures in place, or which will be introduced, to eliminate or reduce the risks;
▶ what employees ought to do when encountering risks and hazards.

An employer need not disclose information which:

▶ could endanger national security;
▶ violates a legal prohibition;
▶ relates specifically to an individual without their consent;
▶ could substantially hurt the employer's undertaking or infringe commercial security;
▶ was obtained in connection with legal proceedings.

13.9.5 Functions of ROES – Regulation 6

ROES have the following functions (but no legal duties):

▶ to make representations to the employer on potential hazards and dangerous occurrences related to the group of employees represented;
▶ to make representations to the employer on matters affecting the general health and safety of relevant employees;
▶ to represent the group of employees for which they are the ROES in consultation with inspectors.

13.9.6 Training, time off and facilities – Regulation 7

Where an employer consults ROES, they shall:

▶ ensure that each ROES receives reasonable training at the employer's expense;
▶ allow time off with pay during ROES working hours to perform the duties of a ROES and while the ROES is a candidate for election;
▶ provide such other facilities and assistance that ROES may reasonably require.

13.9.7 Civil liability and complaints – Regulation 9 and Schedule 2

Breach of these Regulations does not confer any right of action in civil proceedings subject to Regulation 7(3) and Schedule 2 relating to complaints to industrial tribunals.

A ROES can complain to an industrial tribunal that:

▶ their employer has failed to permit time off for training or to be a candidate for election;
▶ their employer has failed to pay them as set out in Schedule 1 to the Regulations.

13.9.8 Elections

The guidance lays down some ideas for the elections, although there are no strict rules.

ROES do not need to be confined to consultation related to these Regulations. Some employers have ROES sitting on safety committees and taking part in accident investigation similarly to union Safety Representatives.

13.9.9 Further information

Consulting Workers on Health and Safety. Safety Representatives and Safety Committees Regulations 1977 (as amended) and Health and Safety (Consultation with Employees) Regulations 1996 (As amended). Approved Code of Practice and Guidance L146, HSE, 2014, ISBN 9780717664610. www.hse.gov.uk/pubns/books/l146.htm

Consulting Employees on Health and Safety. A Brief Guide to the Law INDG 232 (rev 2), HSE, 2014 ISBN 9780717663125. www.hse.gov.uk/pubns/indg232.pdf

Involving Your Workforce in Health and Safety: Good Practice for All Workplaces HSG263, HSE, 2015, ISBN 9780717662272. www.hse.gov.uk/pubns/books/hsg263.htm

13.10 Control of Substances Hazardous to Health Regulations (COSHH) 2002 and 2004 Amendment

13.10.1 Introduction

The 2002 COSHH Regulations updated the 1999 Regulations with a few changes which included: additional definitions like 'inhalable dust' and 'health surveillance'; clarify and extend the steps required under risk assessment; introduce a duty to deal with accidents and emergencies.

In 2004, the Amendment Regulations were passed, but they were not enforced until 2005. The 2004 Amendment Regulations replace Regulation 7(7) and (8) by substituting new requirements to observe principles of good practice for the control of exposure to substances hazardous to health introduced by Schedule 2A, to ensure that workplace exposure limits are not exceeded, and to ensure in respect of carcinogens and asthmagens that exposure is reduced to as low a level as is reasonably practicable. They also introduce a single new workplace exposure limit for substances hazardous to health which replaces occupational exposure standards and maximum exposure limits. The amendment regulations introduce a duty to review control measures other than the provision of plant and equipment, including systems of work and supervision, at suitable intervals.

The 6th edition of the ACoP was issued in 2013. The key changes were:

- takes account of regulatory changes following the introduction of the EU Regulations for REACH (Registration, Evaluation, Authorisation and restriction of Chemicals.) and CLP (European regulation (EC) No. 1272/2008 on classification, labelling and packaging of substances and mixtures);
- clarification of the requirements of Regulation 9, relating to thorough examination and test of control measures.

COSHH covers most substances hazardous to health found in workplaces of all types. The substances covered by COSHH include:

- substances used directly in work activities (e.g. solvents, paints, adhesives and cleaners);
- substances generated during processes or work activities (e.g. dust from sanding and fumes from welding);
- naturally occurring substances (e.g. grain dust).

But COSHH does **not** include:

- asbestos and lead, which have specific regulations;
- substances which are hazardous only because they:
  - are radioactive;
  - are simple asphyxiants;
  - are at high pressure;
  - are at extreme temperatures;
  - have explosive or flammable properties (separate Dangerous Substances and Explosive Atmospheres Regulations, DSEAR, cover these);
  - are biological agents if they are not directly connected with work and are not in the employer's control, such as catching flu from a workmate.

13.10.2 Definition of substance hazardous to health – Regulation 2

The range of substances regarded as hazardous under COSHH are:

- substances including preparations (mixtures of two or more substances) which are listed in Table 3.2 of part 3 of Annex VI of the CLP Regulation and for which an indication of danger specified for the substance is very toxic, toxic, harmful, corrosive; or
- substances with a workplace exposure limit as listed in EH40 published by the HSE;
- biological agents (bacteria and other micro-organisms) if they are directly connected with the work;
- any kind of dust in a concentration specified in the regulations, that is
  - (i) 10 mg/m^3, as a time-weighted average over an 8-hour period, of inhalable dust, or
  - (ii) 4 mg/m^3, as a time-weighted average over an 8-hour period, of respirable dust;
- any other substance which has comparable hazards to people's health but which, for technical reasons, is not covered by CLP.

Workplace exposure limit for a substance hazardous to health means the exposure limit approved by the HSC for that substance in relation to the specified reference period when calculated by a method approved by the HSC, as contained in HSE publication 'EH/40 Workplace Exposure Limits 2005' as updated from time to time.

13.10.3 Duties under COSHH – Regulation 3

The duties placed on employers under these Regulations are extended to (except for health surveillance, monitoring and information and training) persons who may be on the premises but not employed whether they are at work or not, for example visitors and contractors (see Table 13.2.)

457

Table 13.2 The employers' duties

Duty of employer relating to:	Duty for the protection of:		
	Employees	Other people on the premises	Other people likely to be affected by the work
Assessment (Regulation 6)	Yes	sfairp	sfairp
Prevention/control of exposure (Regulation 7)	Yes	sfairp	sfairp
Use of control measures and maintenance, examination and testing of control measures (Regulations 8 and 9	Yes	sfairp	sfairp
Monitoring exposure (Regulation 10)	Yes, where required	sfairp	No
Health surveillance (Regulation 11)	Yes, where appropriate	No	No
Information, training, etc. (Regulation 12)	Yes	sfairp	No
Arrangements to deal with accidents and emergencies (Regulation 13)	Yes	sfairp	No

Source: HSE ACoP 6th Edition.

13.10.4 General requirements

There are eight basic steps to comply with COSHH. They are:

1. Assess the risks to health;
2. Decide what precautions are needed;
3. Prevent or adequately control exposure;
4. Ensure that control measures are used and maintained;
5. Monitor the exposure of employees to hazardous substances;
6. Carry out appropriate health surveillance where necessary;
7. Prepare plans and procedures to deal with accidents, incidents and emergencies;
8. Ensure employees are properly informed, trained and supervised.

13.10.5 Steps 1 and 2: assessment of health risk – Regulation 6

No work may be carried where employees are liable to be exposed to substances hazardous to health unless:

▶ a suitable and sufficient risk assessment, including the steps need to meet COSHH, has been made.

The assessment must be reviewed and changes made regularly and immediately if:

▶ it is suspected that it is no longer valid;
▶ there has been a significant change in the work to which it relates;

▶ the results of any monitoring carried out in accordance with Regulation 10 shows it to be necessary.

The assessment includes:

▶ the hazardous properties of the substance;
▶ information on health effects provided by the supplier, including information contained in any relevant safety data sheet;
▶ the level, type and duration of exposure;
▶ the circumstances of the work, including the amount of the substance involved;
▶ activities, such as maintenance, where there is the potential for a high level of exposure;
▶ any relevant [workplace exposure limit] or similar occupational exposure limit;
▶ the effect of preventative and control measures which have been or will be taken in accordance with Regulation 7;
▶ the results of relevant health surveillance;
▶ the results of monitoring of exposure in accordance with Regulation 10;
▶ in circumstances where the work will involve exposure to more than one substance hazardous to health, the risk presented by exposure to such substances in combination;
▶ the approved classification of any biological agent; and
▶ such additional information as the employer may need in order to complete the risk assessment.

Where five or more employees are employed, the significant findings must be recorded and the steps taken to comply with Regulation 7 (this includes any revisions).

13.10.6 Step 3: prevention or control of exposure – Regulation 7

Every employer must ensure that exposure to substances hazardous to health is either:

▶ prevented or, where this is not reasonably practicable;
▶ adequately controlled.

Preference must be given to substituting with a safer substance. Where it is not reasonably practicable to prevent exposure, protection measures must be adopted in the following order of priority:

▶ the design and use of appropriate work processes, systems and engineering controls and the provision and use of suitable work equipment;
▶ the control of exposure at source, including adequate ventilation systems and appropriate organisational measures;
▶ where adequate control of exposure cannot be achieved by other means, the additional provision of suitable PPE.

The measures must include:

▶ safe handling, storage and transport (plus waste);
▶ reducing to a minimum required for the work the number of employees exposed, the level and duration of exposure, and the quantity of substance present at the workplace;
▶ the control of the working environment including general ventilation;
▶ appropriate hygiene measures.

Principles of good practice for the control of exposure to substances hazardous to health (Schedule 2A)

▶ Design and operate processes and activities to minimize emission, release and spread of substances hazardous to health.
▶ Take into account all relevant routes of exposure – inhalation, skin absorption and ingestion – when developing control measures.
▶ Control exposure by measures that are proportionate to the health risk.
▶ Choose the most effective and reliable control options which minimize the escape and spread of substances hazardous to health.
▶ Where adequate control of exposure cannot be achieved by other means, provide, in combination with other control measures, suitable PPE.
▶ Check and review regularly all elements of control measures for their continuing effectiveness.
▶ Inform and train all employees on the hazards and risks from the substances with which they work and the use of control measures developed to minimize the risks.
▶ Ensure that the introduction of control measures does not increase the overall risk to health and safety.

Figure 13.4 Principles of good practice

Control of exposure shall only be treated as adequate if:

▶ the principles of good practice as set out in Schedule 2A (see Figure 13.4) are applied;
▶ any workplace exposure limit approved for that substance is not exceeded;
▶ for a carcinogen (in Schedule 1 or with risk phrase R45, R46 or R49, likely to be Hazard Statements 340, 350, 351, 360, 361 under CLP Regulation) or sensitiser (Risk phrases R42 or R42/43, likely to be Hazard Statements 317, 334 under CLP Regulation) or asthmagen (Section C of HSE publication on Asthmagens ISBN 0 7176 1465 4) exposure is reduced to as low a level as is reasonably practicable.

PPE must conform to the Personal Protective Equipment (EC Directive) Regulations 2002.

Additionally, respiratory protective equipment must:

▶ be suitable for the purpose;
▶ comply with the EC Directive Regulations; or
▶ if there is no requirement imposed by these regulations, of a type that shall conform to a standard, approved by the HSE.

Where exposure to a carcinogen or biological agent is involved, additional precautions are laid down in Regulation 7.

13.10.7 Step 4: use, maintenance, examination and test of control measures – Regulations 8 and 9

Every employer must take all reasonable steps to:

▶ ensure that control measures, PPE or anything else provided under COSHH, are properly used or applied;
▶ carry out visual checks and observations at appropriate intervals;
▶ see that where more than one item of PPE is being worn, the different items are compatible with each other;
▶ supervise employees to ensure that the defined methods of work are being followed;
▶ monitor systems for the effectiveness of controls and prompt remedial action where necessary.

Employees must make full and proper use of control measures, PPE or anything else provided. Employees must also, as far as is reasonable:

▶ follow the defined methods of work;
▶ return items to accommodation and report defects immediately;
▶ remove any PPE, which could cause contamination, before eating, drinking or smoking;
▶ maintain a high standard of personal hygiene, and make proper use of facilities provided for washing, showering or bathing and for eating and drinking.

Employers shall also:

▶ properly maintain and keep clean plant and equipment, other engineering controls and PPE;
▶ in the case of the provision of systems of work and supervision or similar, review at suitable intervals and revise as necessary;
▶ carry out thorough examination and tests on engineering controls:
  ▷ in the case of local exhaust ventilation (LEV) at least once every 14 months (except those in Schedule 4 which covers blasting of castings – monthly; dry grinding, polishing or abrading of metal for more than 12 hours per week – 6 months; and jute manufacture – 6 months);
  ▷ in any other case, at suitable intervals;

13

▶ carry out thorough examination and tests where appropriate (except on disposable items) on respiratory protective equipment at suitable intervals;

▶ keep a record of examination and tests for at least 5 years;

▶ check, properly store, repair/replace and safely dispose of contaminated PPE as necessary.

13.10.8 Step 5: monitoring exposure – Regulation 10

Employers must monitor exposure:

▶ where this is necessary to ensure maintenance of control measures or the protection of health;

▶ specifically for vinyl chloride monomer or chromium plating as required by Schedule 5; and

▶ keep a record of identifiable personal exposures for 40 years and any other exposures for 5 years.

13.10.9 Step 6: health surveillance – Regulation 11

Employers must ensure employees, where appropriate, who are exposed or liable to be exposed, are under health surveillance. It is considered appropriate when:

▶ an employee is exposed (if significant) to substances or processes in Schedule 6;

▶ an identifiable disease or adverse health effect may be related to the exposure, and there is a reasonable likelihood that disease may occur and there are valid disease indication or effect detection methods.

Records of health surveillance containing approved particulars must be kept for 40 years or offered to HSE if trading ceases. If a person is exposed to a substance and/or process in Schedule 6, the health surveillance shall include medical surveillance by an employment medical adviser or appointed doctor at intervals not exceeding 12 months.

If a medical adviser certifies that a person should not be engaged in particular work they must not be permitted to carry out that work except under specified conditions. Health records must be available for the individual employee to see after reasonable notice.

Employees must present themselves, during working hours at the employer's expense, for appropriate health surveillance.

An employment medical adviser or appointed doctor has the power to inspect the workplace or look at records for the purpose of carrying out functions under COSHH.

13.10.10 Step 7: accidents and emergencies – Regulation 13

Prepare plans and procedures to deal with accidents, incidents and emergencies involving hazardous

substances beyond the risks of day-to-day use. This involves:

▶ first-aid provision and safety drills;

▶ information on emergency arrangements and specific hazards;

▶ suitable warning and other communications to enable appropriate response;

▶ ensuring suitable PPE and other equipment is available;

▶ making sure that the procedure is put into practice in an incident.

13.10.11 Step 8: information, instruction and training – Regulation 12

Where employees are likely to be exposed to substances hazardous to health, employers must provide:

▶ information, instruction and training on the risks to health and the precautions which should be taken (this duty is extended to anyone who may be affected);

▶ information on any monitoring of exposure (particularly if there is a workplace exposure limit where the employee or their representative must be informed immediately);

▶ information on collective results of health surveillance (designed so that individuals cannot be identified).

13.10.12 Defence – Regulation 21

It is a defence under these regulations for a person to show that they have taken all reasonable precautions and exercised all due diligence to avoid the commission of an offence.

13.10.13 Further information

A Step by Step Guide to COSHH Assessment HSG97, HSE , 2004, ISBN 9780717627851 www.hse.gov.uk/pubns/books/hsg97.htm

Control of Substances Hazardous to Health Approved Code of Practice and Guidance L5 (sixth edition),2013, ISBN 9780717665822 www.hse.gov.uk/pubns/books/l5.htm

Working with substances hazardous to health INDG136(rev5), 2012, HSE ISBN 9780717664863 www.hse.gov.uk/pubns/indg136.pdf.

COSHH Essentials guidance publications: www.hse.gov.uk/pubns/guidance/index.htm

COSHH HSE Micro website: www.hse.gov.uk/coshh/index.htm

EH40/2005 Workplace Exposure Limits, HSE Books, 2011, ISBN 978 0 7176 6446 7 www.hse.gov.uk/pubns/books/eh40.htm

Respiratory Protective Equipment at Work HSG53 (fourth edition), HSE, 2013, ISBN 9780717664542 www.hse.gov.uk/pubns/books/hsg53.htm

The Control of Substances Hazardous to Health Regulations 2002, SI 2002, 2677, 2004 Amendment www.legislation.gov.uk/uksi/2002/2677/contents/made

13.11 Dangerous Substances and Explosive Atmospheres Regulations (DSEAR) 2002

13.11.1 Introduction

DSEAR deals with the prevention of fires, explosions and similar energy-releasing events arising from dangerous substances.

In summary, the DSEAR Regulations require employers and the self-employed to:

▶ carry out an assessment of the fire and explosion risks of any work activities involving dangerous substances;

▶ provide measures to eliminate, or reduce sfairp, the identified fire and explosion risks;

▶ apply measures, sfairp, to control risks and to mitigate the detrimental effects of a fire or explosion;

▶ provide equipment and procedures to deal with accidents and emergencies;

▶ provide employees with information and precautionary training.

Where explosive atmospheres may occur:

▶ the workplaces should be classified into hazardous and non-hazardous places; and any hazardous places classified into zones on the basis of the frequency and duration of an explosive atmosphere, and where necessary marked with a sign;

▶ the equipment in classified zones should be safe so as to satisfy the requirements of the Equipment and Protective Systems Intended for Use in Potentially Explosive Atmospheres Regulations 1996;

▶ the workplaces should be verified by a competent person, as meeting the requirements of DSEAR.

The revised ACoP (August 2013) consolidates the contents of five existing ACoPs into a single new ACoP as proposed by the first consultation. As a consequence, the following four ACoPs have been withdrawn, L134 (Design), L135 (Storage), L136 (Control) and L137 (Maintenance).

Their contents have been consolidated into a revised and updated version of ACoP L138 – Dangerous Substances and Explosive Atmospheres Regulations 2002.

Changes in this edition include:

▶ Some guidance has been assigned ACoP status and vice versa. For example, previously no ACoP material

existed for Regulations 7 and 10. However these changes were introduced to provide clarification on how to comply and do not introduce any new requirements.

▶ Adjustments in light of European and other legislation introduced, amended or revoked after DSEAR came into force such as:
  ▷ General fire safety legislation
  ▷ Classification for labelling and packaging.

▶ Clarification on the scope of the regulations in relation to incompatible substances being outside the scope of DSEAR but within the scope of the Health and Safety at Work Act (HSWA).

13.11.2 Scope of regulations

Dangerous substance

The Regulations give a detailed interpretation of dangerous substance, which should be consulted. In summary it means:

▶ a substance or preparation that because of its chemical and sometimes physical properties and the way it is present and/or used at work, creates a fire or explosion risk to people; for example, substances like petrol, LPG, paints, cleaners, solvents and flammable gases;

▶ any dusts which could form an explosive mixture in air (not included in a substance or preparation); for example, many dusts from grinding, milling or sanding.

Explosive atmosphere

The Regulations give the definition as:

A mixture under atmospheric conditions, of air and one or more dangerous substances in the form of gases, vapours, mists or dusts, in which, after ignition has occurred, combustion spreads to the entire unburned mixture.

13.11.3 Application

The Regulations do not in general apply to:

▶ ships under the control of a master (DSEAR does apply when 'shoreside' workers and the ship's crew work together in a British port);

▶ areas used for medical treatment of patients;

▶ many gas appliances used for cooking, heating, hot water, refrigeration, etc. (except an appliance specifically designed for an industrial process), gas fittings;

▶ manufacture, use, transport of explosives or chemically unstable substances;

▶ mine, quarry or borehole activities;

▶ activity at an offshore installation;

▶ the use of means of transport (but the Regulations do cover means of transport intended for use in a potentially explosive atmosphere).

13

13.11.4 Risk assessment – Regulation 5

The risk assessment required by Regulation 5 must include:

- the hazardous properties of the dangerous substance;
- supplier information and Safety Data Sheet;
- the circumstances of the work including:
  - ▷ work processes and substances used and their possible interactions;
  - ▷ the amount of substance involved;
  - ▷ risks of substances in combination;
  - ▷ arrangements for safe handling, storage and transport and any waste which might contain dangerous substances;
- high-risk maintenance activities;
- likelihood of an explosive atmosphere;
- likelihood of ignition sources, including electrostatic discharges being present;
- scale of any possible fire or explosion;
- any places connected by openings to areas where there could be an explosive atmosphere;
- any additional information which may be needed.

The risk assessment should be reviewed regularly.

13.11.5 Elimination or reduction of risks – Regulation 6

Regulation 6 concerns the reduction of risks and tracks the normal hierarchy as follows:

- Substitute a dangerous substance by a substance or process which eliminates or reduces the risk, for example the use of water-based paints, or using a totally enclosed continuous process.
- Reduce the quantity of dangerous substance to a minimum; for example, only a half day supply in the workroom.
- Avoid releasing of a dangerous substance or minimise releases, for example keeping them in special closed containers.
- Control releases at source.
- Prevent the formation of explosive atmospheres, including the provision of sufficient ventilation.
- Ensure that any releases are suitably collected, contained and removed; suitable LEV in a paint spray booth is an example.
- Avoid ignition sources and adverse conditions, for example keep electrical equipment outside the area.
- Segregate incompatible dangerous substances, for example oxidising substances and other flammable substances.

Steps must also be taken to mitigate the detrimental effects of a fire or explosion by:

- keeping the number of people exposed to a minimum;
- avoidance of fire and explosion propagation;
- provision of explosion relief systems;

- provision of explosion suppression equipment;
- provision of very strong plant which can withstand an explosion;
- the provision of suitable PPE.

13.11.6 Classification of workplaces – Regulation 7 and Schedule 2

Where an explosive atmosphere may occur, workplaces must be classified into hazardous and non-hazardous places. Table 13.3 shows the zones specified by Schedule 2 of the Regulations.

Table 13.3 Classification zones

Classification Zones	
Zone 0	A place in which an explosive atmosphere consisting of a mixture with air of dangerous substances in the form of gas, vapour or mist is present continuously or for long periods.
Zone 1	A place in which an explosive atmosphere consisting of a mixture with air of dangerous substances in the form of gas, vapour or mist is likely to occur in normal operations occasionally.
Zone 2	A place in which an explosive atmosphere consisting of a mixture with air of dangerous substances in the form of gas, vapour or mist is not likely to occur in normal operations but, if it does, will persist for a short period only.
Zone 20	A place in which an explosive atmosphere in the form of a cloud of combustible dust in air is present continuously, or for long periods.
Zone 21	A place in which an explosive atmosphere in the form of a cloud of combustible dust in air is likely to occur in normal operations occasionally.
Zone 22	A place in which an explosive atmosphere in the form of a cloud of combustible dust in air is not likely to occur in normal operations but, if it does occur, will persist for a short period only.

Figure 13.5 Warning sign for places where explosive atmospheres may occur

13.11.7 Accidents, incidents and emergencies – Regulation 8

In addition to any normal fire prevention requirements employers must under these Regulations:

▶ ensure that procedures and first aid are in place with tested relevant safety drills;

▶ provide information on emergency arrangements including work hazards and those that are likely to arise at the time of an accident;

▶ provide suitable warning and other communications systems to enable an appropriate response, remedial actions and rescue operations to be made;

▶ where necessary, before any explosion condition is reached, provide visual or audible warnings and withdraw employees;

▶ provide escape facilities where the risk assessment indicates it is necessary.

In the event of an accident, immediate steps must be taken to:

▶ mitigate the effects of the event;

▶ restore the situation to normal;

▶ inform employees who may be affected.

Only essential persons may be permitted in the affected area. They must be provided with PPE, protective clothing and any necessary specialised safety equipment and plant.

13.11.8 Information instruction and training – Regulation 9

Under Regulation 9 where a dangerous substance is present, an employer must provide:

▶ suitable and sufficient information, instruction and training on the appropriate precautions and actions;

▶ details of the substances, any relevant data sheets and legal provisions;

▶ the significant findings of the risk assessment.

13.11.9 Contents of containers and pipes – Regulation 10

Regulation 10 requires that for containers and pipes (except where they are marked under legislation contained in Schedule 5 to the Regulations) the content and the nature of those contents and any associated hazards be clearly identified.

13.11.10 Further information

The Dangerous Substances and Explosive Atmosphere Regulations 2002, SI 2002, No. 2776, ISBN 9780110429575.

Dangerous Substances and Explosive Atmospheres, L138 (second edition). Dangerous Substances and Explosive Atmospheres Regulations 2002. Approved Code of Practice, 2013, HSE, ISBN 9780717666164 www.hse.gov.uk/pubns/priced/l138.pdf

Chemical warehousing: The storage of packaged dangerous substances, HSG71 (fourth edition), 2009, ISBN 9780717662371 https://www.hse.gov.uk/pubns/books/hsg71.htm

The storage of flammable liquids in containers. HSG51 (third edition), 2015, ISBN 9780717666089 https://www.hse.gov.uk/pubns/books/hsg51.htm

Fire and Explosions – How Safe is Your Workshop. A Short Guide to DSEAR, HSE INDG370(rev1), 2013, HSE, www.hse.gov.uk/pubns/indg370.htm

Safe Handling of Combustible Dusts; Precautions Against Explosion HSG103 (second edition), HSE Books, 2003, ISBN 9780717627264 www.hse.gov.uk/pubns/books/hsg103.htm

Safe Use and Handling of Flammable Liquids HSG140 (second edition), HSE, 2015, ISBN 9780717666096 www.hse.gov.uk/pubns/books/hsg140.htm

Safe Working with Flammable Substances INDG227 HSE, 2011, ISBN 978 0 7176 1154 6 www.hse.gov.uk/pubns/indg227.pdf

Safety in gas welding, cutting and similar processes, INDG297(rev 1), 2012, HSE, https://www.hse.gov.uk/pubns/indg297.pdf

13.12 Health and Safety (Display Screen Equipment) Regulations 1992 as amended in 2002

13.12.1 General

These Regulations cover the minimum health and safety requirements for the use of display screen equipment (DSE) and are accompanied by a guidance note. They typically apply to computer equipment with either a cathode ray tube or liquid crystal monitors. But any type of display is covered with some exceptions, for example, on board a means of transport, or where the main purpose is for screening a film or for a television. Multimedia equipment would generally be covered.

Equipment not specifically covered by these Regulations or where it is not being used by a defined 'user' is, nevertheless, covered by other requirements under the MHSWR and the Provision and Use of Work Equipment Regulations (PUWER).

13.12.2 Definitions – Regulation 1

(a) Display screen equipment (DSE) refers to any alphanumeric or graphic display screen, regardless of the display process involved.

(b) A user is an employee and an operator is a self-employed person, both of whom habitually use DSE as a significant part of their normal work. Both would be people to whom most or all of the following apply. A person:

  ▷ who depends on the DSE to do their job;

  ▷ who has no discretion as to use or non-use;

13

▷ who needs particular training and/or skills in the use of DSE to do their job;

▷ who uses DSE for continuous spell of an hour or more at a time;

▷ who does so on a more or less daily basis;

▷ for whom fast transfer of data is important for the job;

▷ of whom a high level of attention and concentration is required, in particular to prevent critical errors.

(c) A workstation is an assembly comprising:

▷ DSE with or without a keyboard, software or input device;

▷ optional accessories;

▷ disk drive, telephone, modem, printer, document holder, chair, desk, work surface, etc.;

▷ the immediate working environment.

13.12.3 Exemptions – Regulation 1(4)

Exemptions include DSE used in connection with:

▶ drivers' cabs or control cabs for vehicles or machinery;

▶ on board a means of transport;

▶ mainly intended for public operation;

▶ portable systems not in prolonged use;

▶ calculators, cash registers and small displays related to the direct use of this type of equipment;

▶ window typewriters.

13.12.4 Assessment of risk – Regulation 2

Possible hazards associated with DSE are physical (musculoskeletal) problems, visual fatigue and mental stress. They are not unique to display screen work, nor an inevitable consequence of it, and indeed research shows that the risk to the individual from typical display screen work is low. However, as in other types of work, ill health can result from poor work organisation, working environment, job design and posture, and from inappropriate working methods.

Employers must carry out a suitable and sufficient analysis of users' (regardless of who has provided them) and operators' (provided by the employer) workstations in order to assess the risks to health and safety. The guidance gives detailed information on workstation minimum standards and possible effects on health.

The assessment should be reviewed when major changes are made to software, hardware, furniture, environment or work requirements.

13.12.5 Workstations – Regulation 3

All workstations must meet the requirements laid down in the schedule to the Regulations. This schedule lays down minimum requirements for display screen

workstations, covering the equipment, the working environment and the interface between computer and user/operator.

13.12.6 Daily work routine of users – Regulation 4

The activities of users should be organised so that their daily work on DSE is periodically interrupted by breaks or changes of activity that reduce their workload at the equipment.

In most tasks, natural breaks or pauses occur from time to time during the day. If such breaks do not occur, deliberate breaks or pauses must be introduced. The guidance requires that breaks should be taken before the onset of fatigue and must be included in the working time. Short breaks are better than occasional long ones, for example, a 5–10-minute break after 50 to 60 minutes continuous screen and/or keyboard work is likely to be better than a 15-minute break every 2 hours. If possible, breaks should be taken away from the screen. Informal breaks, with time on other tasks, appear to be more effective in relieving visual fatigue than formal rest breaks.

13.12.7 Eyes and eyesight – Regulation 5

Initially on request, employees have the right to a free eye and eyesight test conducted by a competent person where they are:

▶ already users (as soon as practicable after the request);

▶ to become a user (before they become a user).

The employer must provide a further eye and eyesight test at regular intervals thereafter or when a user is experiencing visual difficulties which could be caused by working with DSE.

There is no reliable evidence that work with DSE causes any permanent damage to eyes or eyesight, but it may make users with pre-existing vision defects more aware of them.

An eye and eyesight test means a sight test as defined in the Opticians Act 1989. This should be carried out by a registered ophthalmic optician or medical practitioner (normally only those with an ophthalmic qualification do so).

Employers shall provide special corrective appliances to users where:

▶ normal corrective appliances cannot be used;

▶ the result of the eye and eyesight test shows that such provision is necessary.

The guidance indicates that the liability of the employer extends only to the provision of corrective appliances which are of a style and quality adequate for their function. If an employee chooses a more expensive design or multi-function correction appliances, the employer need only pay a proportion of the cost.

Employers are free to specify that users' eye and eyesight tests and correction appliances are provided by a nominated company or optician.

The confidential clinical information from the tests can only be supplied to the employer with the employee's consent.

Vision screen tests can be used to identify people with defects but they are not a substitute for the full eyesight test, and employees have the right to opt for the full test from the outset.

13.12.8 Training – Regulation 6

Employers shall ensure that adequate health and safety training is provided to users and potential users in the use of any workstation and refresh the training following any reorganisation. The guidance suggests a range of topics to be covered in the training. In summary these involve:

▶ the recognition of hazards and risks, including the absence of desirable features and the presence of undesirable ones;
▶ causes of risk and how harm may occur;
▶ what the user can do to correct them;
▶ how problems can be communicated to management;
▶ information on the regulations;
▶ the user's contribution to assessments.

13.12.9 Information – Regulation 7

Operators and users shall be provided with adequate information on all aspects of health and safety relating to their workstation and what steps the employer has taken to comply with the Regulations (insofar as the action taken relates to that operator or user and their work). Under Regulation 7, specific information should be provided as outlined in Regulation 3.

13.12.10 Further information

Display screen equipment Workstation Checklist, 2013, HSE, ISBN 9780717665211 www.hse.gov.uk/pubns/ck1.htm

Work with Display Screen Equipment, Health and Safety (Display Screen Equipment) Regulations 1992 as amended by the Health and Safety (Miscellaneous Amendments) Regulations 2002 – Guidance on Regulations L26 (second edition), 2003, HSE, ISBN 9780717625826 www.hse.gov.uk/pubns/books/l26.htm

Working with Display screen equipment (DSE) A brief guide, INDG36 (rev 4), 2013, HSE, ISBN 978071766227 www.hse.gov.uk/pubns/indg36.pdf

13.13 Electricity at Work Regulations 1989

The purpose of these Regulations is to require precautions to be taken against the risk of death or personal injury from electricity in work activities. The Regulations impose duties on persons ('duty-holders') in respect of systems, electrical equipment and conductors and in respect of work activities on or near electrical equipment. They apply to almost all places of work and electrical systems at all voltages.

Guidance on the Regulations is contained in the Memorandum of Guidance on the Electricity at Work Regulations 1989, published by the HSE.

13.13.1 Definitions

(a) **Electrical equipment** – includes anything used to generate, provide, transmit, rectify, convert, conduct, distribute, control, store, measure or use electrical energy.

(b) **Conductor** – means a conductor of electrical energy. It means any material (solid, liquid or gas), capable of conducting electricity.

Table 13.4 Provision of information under DSE Regulation 7

Information on							
		Risks from DSE workstation	Risk assessment and measures to reduce the risks Regs 2 and 3	Breaks and activity changes Reg. 4	Eye and eyesight texts Reg. 5	Initial training Reg. 6(1)	Training when workstation modified Reg. 6(2)
Does employer have to provide information to display screen workers who are:	Users employed by the undertaking	Yes	Yes	Yes	Yes	Yes	Yes
	Users employed by other employer	Yes	Yes	Yes	No	No	Yes
	Operators in the undertaking	Yes	Yes	No	No	No	No

13

(c) **System** – means an electrical system in which all the electrical equipment is, or may be, electrically connected to a common source of electrical energy, and includes the source and equipment. It includes portable generators and systems on vehicles.

(d) **Circuit conductor** – term used in Regulations 8 and 9 only, means a conductor in a system which is intended to carry electric current in normal conditions. It would include a combined neutral and earth conductor, but does not include a conductor provided solely to perform a protective connection to earth or other reference point and energised only during abnormal conditions.

(e) **Danger** – in the context of these Regulations means a risk of injury from any electrical hazard.

Every year about 30 people die from electric shock or electric burns at work. Each year several hundred serious burns are caused by arcing where the heat generated can be very intense. In addition, intense ultraviolet radiation from an electric arc can cause damage to the eyes – known as arc-eye. These hazards are all included in the definition of **Danger**.

13.13.2 Duties – Regulation 3

Duties are imposed on employers, self-employed and employees. The particular duties on employees are intended to emphasise the level of responsibility which many employees in the electrical trades and professions are expected to take on as part of their job. Employees are:

▶ to cooperate with their employer so far as is necessary to enable any duty placed on the employer to be complied with (this reiterates Section 7(b) of the HSW Act);

▶ to comply with the provisions of those Regulations insofar as they relate to matters which are within their control. (This is equivalent to duties placed on employers and self-employed where these matters are within their control.)

13.13.3 Systems, work activities and protective equipment – Regulation 4

Systems must, at all times, be of such construction as to prevent danger. Construction covers the physical condition, arrangement of components and design of the system and equipment.

All systems must be maintained so as to prevent danger.

Every work activity, including operation, use and maintenance or work near a system, shall be carried out in a way, which prevents danger.

Protective equipment shall be suitable, suitably maintained and used properly.

13.13.4 Strength and capability of equipment – Regulation 5

No electrical equipment may be put into use where its strength and capability may be exceeded in such a way as may give rise to danger, in normal transient or fault conditions.

13.13.5 Adverse or hazardous environments – Regulation 6

Electrical equipment which may be exposed to:

▶ mechanical damage;

▶ the effects of weather, natural hazards, temperature or pressure;

▶ the effects of wet, dirty, dusty or corrosive conditions;

▶ any flammable or explosive substances including dusts, vapours or gases;

shall be so constructed and protected that it prevents danger.

13.13.6 Insulation, protection and placing of conductors – Regulation 7

All conductors in a system which may give rise to danger shall either be suitably covered with insulating material and further protected as necessary, for example, against mechanical damage, using trunking or sheathing; or have precautions taken that will prevent danger, for example, being suitably placed like overhead electric power cables, or by having strictly controlled working practices.

13.13.7 Earthing, integrity and other suitable precautions – Regulations 8 and 9

Precautions shall be taken, either by earthing or by other suitable means, for example, double insulation, use of safe voltages and earth-free non-conducting environments, where a conductor, other than a circuit conductor, could become charged as a result of either the use of, or a fault in, a system.

An earth conductor shall be of sufficient strength and capability to discharge electrical energy to earth. The conductive part of equipment, which is not normally live but energised in a fault condition, could be a conductor.

If a circuit conductor is connected to earth or to any other reference point, nothing which could break electrical continuity or introduce high impedance, for example, fuse, thyristor or transistor, is allowed in the conductor unless suitable precautions are taken. Permitted devices would include a joint or bolted link, but not a removable link or manually operated knife switch without bonding of all exposed metal work and multiple earthing.

13.13.8 Connections – Regulation 10

Every joint and connection in a system shall be mechanically and electrically suitable for its use. This includes terminals, plugs and sockets.

13.13.9 Excess current protection – Regulation 11

Every part of a system shall be protected from excess current, for example, short circuit or overload, by a suitably located efficient means such as a fuse or circuit breaker.

13.13.10 Cutting off supply and isolation – Regulation 12

There should be suitably located and identified means of cutting off (switch) the supply of electricity to any electrical equipment and also isolating any electrical equipment. Although these are separate requirements, they could be affected by a single means. The isolator should be capable of being locked off to allow maintenance to be done safely.

Sources of electrical energy (accumulators, capacitors and generators) are exempt from this requirement, but precautions must be taken to prevent danger.

13.13.11 Work on equipment made dead – Regulation 13

Adequate precautions shall be taken to prevent electrical equipment that has been made dead from becoming live while work is carried out on or near the equipment. This will include means of locking off isolators, tagging equipment, permits to work and removing fuses.

13.13.12 Work on or near live conductors – Regulation 14

No persons shall work near a live conductor, except if it is insulated, unless:

▶ it is unreasonable in all the circumstances for it to be dead;

▶ it is reasonable in **all** the circumstances for them to be at work on or near it while it is live;

▶ suitable precautions (including where necessary the provision of suitable protective equipment) are taken to prevent injury.

13.13.13 Working space access and lighting – Regulation 15

Adequate working space, means of access and lighting shall be provided for all electrical equipment at which or near which work is being done in circumstances which may give rise to danger. This covers work of any kind. However, when the work is on live conductors, the access space must be sufficient for a person to fall back out of danger and if needed for persons to pass one another with ease and without hazard.

13.13.14 Competence – Regulation 16

Where technical knowledge or experience is necessary to prevent danger, all persons must possess such knowledge or experience or be under appropriate supervision.

13.13.15 Further information

Electrical Safety and You INDG231(rev1), 2012, HSE, www.hse.gov.uk/pubns/indg231.pdf

Electrical Switch Gear and Safety. A Concise Guide for Users INDG372(rev1), 2013, HSE, www.hse.gov.uk/pubns/indg372.pdf

Electricity at Work: Safe Working Practices HSG85, 2013 (third edition), HSE, ISBN 9780717665815 www.hse.gov.uk/pubns/books/hsg85.htm.

Guidance on Safe Isolation Procedures, SELECT, www.select.org.uk ISBN 9780717612724.

Maintaining Portable and Transportable Electrical Equipment HSG107 (third edition), 2013, HSE, ISBN 9780717666065 www.hse.gov.uk/pubns/books/hsg107.htm

Maintaining Portable Electrical Equipment in Low-risk Environments INDG236(rev 3), 2013, HSE, www.hse.gov.uk/pubns/indg236.pdf

Memorandum of Guidance on the Electricity at Work Regulations, Guidance on Regulations HSR25 (third edition), 2015, HSE, ISBN 9780717666362 www.hse.gov.uk/pubns/books/hsr25.htm

Safety in Electrical Testing at Work, General Guidance INDG354(rev1), 2013, HSE, www.hse.gov.uk/pubns/indg354.pdf

Avoidance of danger from overhead power-lines GS6 (fourth edition), 2013 www.hse.gov.uk/pubns/gs6.htm

13.14 Employers' Liability (Compulsory Insurance) Act 1969 and Regulations 1998 amended in 2002, 2004, 2008 and 2011

13.14.1 Introduction

Employers are responsible for the health and safety of employees while they are at work. Employees may be injured at work, or they or former employees may become ill as a result of their work while employed. They may try to claim compensation from the employer if they believe them to be responsible. The Employers' Liability Compulsory Insurance Act 1969 ensures that an employer has at least a minimum level of insurance cover against any such claims.

13

Employers' liability insurance will enable employers to meet the cost of compensation for employees' injuries or illnesses whether they are caused on or off site. However, any injuries or illnesses relating to motor accidents that occur while employees are working for them may be covered separately by motor insurance.

Public liability insurance is different. It covers for claims made against a person/company by members of the public or other businesses, but not for claims made by employees. While public liability insurance is generally voluntary, employers' liability insurance is compulsory. Employers can be fined if they do not hold a current employers' liability insurance policy which complies with the law.

Employers must use an authorised insurer working under the terms of the Financial Services and Markets Act 2000. The Financial Conduct Authority (FCA; formerly Financial Services Authority (FSA)) keeps a list of authorised insurance companies in the Financial Services Register, see www.fca.org.uk

13.14.2 Application

An employer needs employers' liability insurance unless they are exempt from the Employers' Liability Compulsory Insurance Act. The following employers are exempt:

▶ most public organisations including government departments and agencies, LAs, police authorities and nationalised industries;
▶ health service bodies, including National Health Service trusts, health authorities, Family Health Services Authorities and Scottish Health Boards and State Hospital Management Committees;
▶ some other organisations which are financed through public funds, such as passenger transport executives and magistrates' courts committees;
▶ family businesses, i.e. if employees are closely related to the employer (as husband, wife, father, mother, grandfather, grandmother, stepfather, stepmother, son, daughter, grandson, granddaughter, stepson, stepdaughter, brother, sister, half-brother or half-sister). However, this exemption does not apply to family businesses that are incorporated as limited companies except any employer which is a company that has only one employee who owns 50% or more of the share capital.

A full list of employers who are exempt from the need to have employers' liability insurance is shown at Schedule 2 of the Employers' Liability (Compulsory Insurance) Regulations 1998.

13.14.3 Coverage

Employers are only required by law to have employers' liability insurance for people whom they employ under a contract of service or apprenticeship. However, people who are normally thought of as self-employed may be considered to be employees for the purposes of employers' liability insurance.

Whether or not an employer needs employers' liability insurance for someone who works for them depends on the terms of the contract. This contract can be spoken, written or implied. It does not matter whether someone is usually called an employee or self-employed or what their tax status is. Whether the contract is called a contract of employment or a contract for services is largely irrelevant. What matters is the real nature of the employer/employee relationship and the degree of control the employer has over the work employees do.

There are no hard and fast rules about who counts as employee for the purposes of employers' liability insurance. The following paragraphs may help to give some indication.

In general, employers' liability insurance may be needed for a worker if:

▶ national insurance and income tax is deducted from the money paid to them;
▶ the employer has the right to control where and when they work and how they do it;
▶ most materials and equipment are supplied by the employer;
▶ the employer has a right to any profit workers make even though the employer may choose to share this with them through commission, performance pay or shares in the company; similarly, the employer will be responsible for any losses;
▶ that person is required to deliver the service personally and they cannot employ a substitute if they are unable to do the work;
▶ they are treated in the same way as other employees, for example if they do the same work under the same conditions as some other employee.

In some cases you will not need additional employers' liability insurance for volunteers or for:

▶ students who work unpaid;
▶ people who are not employed but are taking part in youth or adult training programmes;
▶ school children on work experience programmes.

In practice, insurance companies usually provide cover for people in these situations.

One difficult area is domestic help. In general, an employer will probably not need employers' liability insurance for people such as cleaners or gardeners if they work for more than one person. However, if they only work for one employer, that employer may be required to take out insurance to protect them.

13.14.4 Display of certificate

Under the Regulations, employers must display, in a suitable convenient location, a current copy of the certificate of insurance at each place of business where they employ relevant people. This can be done electronically as long as employees know how to access the information and have access to computers.

13.14.5 Retention of certificates

Since 1 October 2008 there has been no legal requirement for employers to keep copies of out-of-date certificates.

However, employers are strongly advised to keep, as far as is possible, a complete record of their employers' liability insurance. This is because some diseases can appear decades after exposure to their cause and former or current employees may decide to make a claim against their employer for the period they were exposed to the cause of their illness.

Employers that fail to hold the necessary insurance details risk having to meet the costs of such claims themselves.

13.14.6 Penalties

The HSE enforces the law on employers' liability insurance and HSE inspectors will check that employers have employers' liability insurance with an approved insurer for at least £5m. They will ask to see the certificate of insurance and other insurance details.

Employers can be fined up to £2,500 for any day they are without suitable insurance. If they do not display the certificate of insurance or refuse to make it available to HSE inspectors when they ask, employers can be fined up to £1,000.

13.14.7 Further information

Employers' Liability (Compulsory Insurance) Act 1969: A Guide for Employers HSE40 (rev 4), 2012, HSE, ISBN 9780717664801 www.hse.gov.uk/pubns/hse40.htm

13.15 Regulatory Reform (Fire Safety) Order 2005

13.15.1 Introduction

This Order, made under the Regulatory Reform Act 2001, reforms the law relating to fire safety in non-domestic premises. It replaces fire certification under the Fire Precautions Act 1971 (which it repeals) with a general duty to ensure, sfairp, the safety of employees, a general duty, in relation to non-employees to take such fire precautions as may reasonably be required in the circumstances to ensure that premises are safe and a duty to carry out a risk assessment.

The Fire Safety Order is the responsibility of the Department for Housing, Communities and Local Government and is enforced by the local fire and rescue authorities, with some exceptions which include:

▶ the HSE for:
  (i) nuclear installations;
  (ii) ships in construction or repair;
  (iii) construction sites other than a construction site which is contained within, or forms part of, premises which are occupied by persons other

than those carrying out the construction work or any activity arising from such work.
▶ In Crown-occupied and Crown-owned buildings, enforcement is carried out by the Fire Services Inspectorates appointed under the Fire Services Act 1947.
▶ LAs for Certificated Sports Grounds.

The order came into force on 1 October 2006 with a number of guidance documents.

In Scotland the fire safety legislation is enacted through the Fire (Scotland) Act 2005 which received Royal Assent on 1 April 2005 as amended by the Police and Fire Reform (Scotland) Act 2012.

Parts 1, 2, 4 and 5 of the Act commenced in August 2005. Part 3 introduced a new fire safety regime for non-domestic premises and came into force on 1 October 2006 and replaced the Fire Precautions Act 1971 and the Fire Precautions (Workplace) Regulations 1997, as amended. Fire certificates are no longer required after 1 October 2006 and the fire safety regime is based on the principle of risk assessment.

Ten guidance documents to complement the legislation and help those with fire safety responsibilities to understand their duties are available.

Part 1 General

13.15.2 Interpretation – Articles 1–7

Here are a few of the more important definitions from the articles. For a full list consult the Order directly.

(a) 'Premises' includes any place and, in particular, includes:
  ▶ any workplace;
  ▶ any vehicle, vessel, aircraft or hovercraft;
  ▶ any installation on land (including the foreshore and other land intermittently covered by water), and any other installation (whether floating, or resting on the seabed or the subsoil thereof, or resting on other land covered with water or the subsoil thereof);
  ▶ any tent or movable structure.

(b) 'Risk' means the risks to the safety of persons from fire. 'Safety' means the safety of persons in the event of fire. 'Workplace' means any premises or parts of premises, not being domestic premises, used for the purposes of an employer's undertaking and which are made available to an employee of the employer as a place of work and includes:
  ▶ any place within the premises to which such employee has access while at work;
  ▶ any room, lobby, corridor, staircase, road or other place;
  ▶ used as a means of access to or egress from that place of work; or
  ▶ where facilities are provided for use in connection with that place of work, other than a public road.

469

(c) 'Responsible person' means:

- ▶ in relation to a workplace, the employer, if the workplace is to any extent under their control;
- ▶ in relation to any premises not falling within paragraph (a):
  - ▶ the person who has control of the premises (as occupier or otherwise) in connection with the carrying on by them of a trade, business or other undertaking (for profit or not); or
  - ▶ the owner, where the person in control of the premises does not have control in connection with the carrying on by that person of a trade, business or other undertaking.

(d) 'General fire precautions' in relation to premises means:

- ▶ measures to reduce the risk of fire and the risk of the spread of fire;
- ▶ the means of escape from the premises;
- ▶ measures for securing that, at all material times, the means of escape can be safely and effectively used;
- ▶ measures in relation to fighting fires;
- ▶ the means for detecting fires and giving warning in case of fire;
- ▶ action to be taken in the event of fire, including:
- ▶ instruction and training of employees;
- ▶ measures to mitigate the effects of the fire.

These issues do not cover process-related fire risks including:

- ▶ the use of plant or machinery; or
- ▶ the use or storage of any dangerous substance.

Duties are placed on responsible persons in a workplace and in premises which are not workplaces to the extent that they have control over the premises. The order does not apply (Article 8) to:

- ▶ domestic premises;
- ▶ an offshore installation;
- ▶ a ship, in respect of the normal ship-board activities of a ship's crew which are carried out solely by the crew under the direction of the master;
- ▶ fields, woods or other land forming part of an agricultural or forestry undertaking but which is not inside a building and is situated away from the undertaking's main buildings;
- ▶ an aircraft, locomotive or rolling stock, trailer or semi-trailer used as a means of transport or a vehicle;
- ▶ a mine other than any building on the surface at a mine;
- ▶ a borehole site.

In addition certain provisions of the Order do not apply to groups of workers such as:

- ▶ occasional work which is not harmful to young people in a family undertaking;
- ▶ armed forces;
- ▶ members of police forces;
- ▶ emergency services.

Part 2 Fire Safety Duties

13.15.3 Duty to take general fire precautions – Article 8

The responsible person must:

(a) take such general fire precautions as will ensure, sfairp, the safety of any of his employees; and

(b) in relation to relevant persons who are not their employees, take such general fire precautions as may reasonably be required in the circumstances of the case to ensure that the premises are safe.

13.15.4 Risk assessment and fire safety arrangements – Articles 9 and 11

The responsible person must make a suitable and sufficient assessment of the risks to identify the general fire precautions that need to be taken.

Where a dangerous substance is or is liable to be present in or on the premises, the risk assessment must include consideration of the matters set out in Part 1 of Schedule 1, reproduced in Figure 13.6(a).

Risk assessments must be reviewed by the responsible person regularly and if it is no longer valid or there has been significant changes. The responsible person must not employ a young person unless risks to young persons have been considered in an assessment covering the following, which is Part 2 of Schedule 1, reproduced in Figure 13.6(b).

The responsible person must make and record (as per risk assessments) arrangements as are appropriate, for the effective planning, organisation, control, monitoring and review of the preventative and protective measures.

13.15.5 Principles of prevention to be applied and fire safety arrangements – Articles 10 and 11

Preventative and protective measures must be implemented on the basis of the principles specified in Part 3 of Schedule 1 as follows, which are broadly the same as those in the Management Regulations.

Principles of prevention

The principles are:

(a) avoiding risks;

(b) evaluating the risks which cannot be avoided;

(c) combating the risks at source;

(d) adapting to technical progress;

(e) replacing the dangerous substances by the non-dangerous or the less-dangerous substances;

(f) developing a coherent overall prevention policy which covers technology, organisation of work and the influence of factors relating to the working environment;

(g) giving collective protective measures priority over individual protective measures;

(h) giving appropriate instructions to employees.

MATTERS TO BE CONSIDERED IN RISK ASSESSMENT
IN RESPECT OF DANGEROUS SUBSTANCES

PART 1 OF SCHEDULE 1

The matters are;

(a) the hazardous properties of the substance

(b) information on safety provided by the supplier, including information contained in any relevant safety data sheet

(c) the circumstances of the work including:
   (i) the special, technical and organizational measures and the substances used and their possible interactions
   (ii) the amount of the substance involved
   (iii) where the work will involve more than one dangerous substance, the risk presented by such substances in combination
   (iv) the arrangements for the safe handling, storage and transport of dangerous substances and of waste containing dangerous substances.

(d) activities, such as maintenance, where there is the potential for a high level of risk

(e) the effect of measures which have been or will be taken pursuant to this Order

(f) the likelihood that an explosive atmosphere will occur and its persistance

(g) the likelihood that ignition sources, including electrostatic discharges, will be present and become active and effective

(h) the scale of the anticipated effects

(i) any places which are, or can be connected via openings to, places in which explosive atmospheres may occur

(j) such additional safety information as the responsible person may need in order to complete the assessment.

(a)

Matters to be taken into particular account in risk assessment in respect of young persons

The matters are:

(a) The inexperience, lack of awareness of risks and immaturity of young persons;

(b) The fitting-out and layout of the premises;

(c) The nature; degree and duration of exposure to physical and chemical agents;

(d) The form, range and use of work equipment and the way in which it is handled;

(e) The organization of processes and activities;

(f) The extent of the safety training provided or to be provided to young persons;

(g) Risks from agents, processes and work listed in the Annex to Council Directive 94/33/EC(a) on the protection of young people at work.

(b)

Figure 13.6 Fire safety order: (a) matters to be considered in dangerous substances; (b) in respect of young persons

Fire safety arrangements must be made and put into effect by the responsible person where five or more people are employed; there is a licence for the premises; or an Alterations Notice requires them to be recorded. The arrangements must cover effective planning, organisation, control, monitoring and review of the preventative and protective measures.

13.15.6 Elimination or reduction of risks from dangerous substances – Article 12

Where a dangerous substance is present in or on the premises, the responsible person must ensure that risk of

the substance is either eliminated or reduced, sfairp, by first replacing the dangerous substance with a safer alternative.

Where it is not reasonably practicable to eliminate risk the responsible person must, sfairp, apply measures including the measures specified in Part 4 of Schedule 1 to the Order (Figure 13.7).

Measures to be taken in respect of dangerous substances

1. In applying measures to control risks the responsible person must, in order of priority:
   (a) reduce the quantity of dangerous substances to a minimum
   (b) avoid or minimize the release of a dangerous substance
   (c) control the release of a dangerous substance at source
   (d) prevent the formation of an explosive atmosphere, including the application of appropriate ventilation
   (e) ensure that any release of a dangerous substance which may give rise to risk is suitably collected, safely contained, removed to a safe place, or otherwise rendered safe, as appropriate
   (f) avoid:
      (i) ignition sources including electrostatic discharges
      (ii) such other adverse conditions as could result in harmful physical effects from a dangerous substance
   (g) segregate incompatible dangerous substances.

2. The responsible person must ensure that mitigation measures applied in accordance with Article 12(3)(b) include:
   (a) reducing to a minimum the number of persons exposed
   (b) measures to avoid the propagation of fires or explosions
   (c) providing explosion pressure relief arrangements
   (d) providing explosion suppression equipment
   (e) providing plant which is constructed so as to withstand the pressure likely to be produced by an explosion
   (f) providing suitable personal protective equipment.

3. The responsible person must:
   (a) ensure that the premises are designed, constructed and maintained so as to reduce risk
   (b) ensure that suitable special, technical and organizational measures are designed, constructed, assembled, installed, provided and used so as to reduce risk
   (c) ensure that special, technical and organisational measures are maintained in an efficient state, in efficient working order and in good repair
   (d) ensure that equipment and protective systems meet the following requirements:
      (i) where power failure can give rise to the spread of additional risk, equipment and protective systems must be able to be maintained in a safe state of operation independently of the rest of the plant in the event of power failure
      (ii) means for manual override must be possible, operated by employees competent to do so, for shutting down equipment and protective systems incorporated within automatic processes which deviate from the intended operating conditions, provided that the provision or use of such means does not compromise safety
      (iii) on operation of emergency shutdown, accumulated energy must be dissipated as quickly and as safely as possible or isolated so that it no longer constitutes a hazard
      (iv) necessary measures must be taken to prevent confusion between connecting devices.
   (e) where the work is carried out in hazardous places or involves hazardous activities, ensure that appropriate systems of work are applied including:
      (i) the issuing of written instructions for the carrying out of work
      (ii) system of permits to work, with such permits being issued by a person with responsibility for this function prior to the commencement of the work concerned.

Figure 13.7 Measures to be taken in respect of dangerous substances

The responsible person must:

(a) arrange for the safe handling, storage and transport of dangerous substances and waste containing dangerous substances;

(b) ensure that any conditions necessary for ensuring the elimination or reduction of risk are maintained.

13.15.7 Firefighting and fire detection – Article 13

The premises must be equipped with appropriate firefighting equipment and with fire detectors and alarms; any non-automatic firefighting equipment provided must be easily accessible, simple to use and indicated by signs.

Appropriate measures must be taken for firefighting in the premises: to nominate and train competent persons to implement the measures; and arrange any necessary contacts with external emergency services.

A person is to be regarded as competent where they have sufficient training and experience or knowledge and other qualities to enable them to properly implement the measures referred to.

13.15.8 Emergency routes and exits – Article 14

Where necessary, routes to emergency exits from premises and the exits themselves must be kept clear at all times.

The following requirements must be complied with:

(a) emergency routes and exits must lead as directly as possible to a place of safety;

(b) in the event of danger, it must be possible for persons to evacuate the premises as quickly and as safely as possible;

(c) the number, distribution and dimensions of emergency routes and exits must be adequate having regard to the use, equipment and dimensions of the premises and the maximum number of persons who may be present there at any one time;

(d) emergency doors must open in the direction of escape;

(e) sliding or revolving doors must not be used for exits specifically intended as emergency exits;

(f) emergency doors must not be so locked or fastened that they cannot be easily and immediately opened by any person who may require to use them in an emergency;

(g) emergency routes and exits must be indicated by signs;

(h) emergency routes and exits requiring illumination must be provided with emergency lighting of adequate intensity in the case of failure of their normal lighting.

13.15.9 Procedures for serious and imminent danger and for danger areas – Article 15

The responsible person must:

(a) establish appropriate procedures, including safety drills;

(b) nominate a sufficient number of competent persons to implement evacuation procedures;

(c) provide adequate safety instruction for restricted areas.

Persons who are exposed to serious and imminent danger must be informed of the nature of the hazard and of the steps taken or to be taken to protect them from it. They must be able to stop work and immediately proceed to a place of safety in the event of their being exposed to serious, imminent and unavoidable danger; and procedures must require the persons concerned to be prevented from resuming work in any situation where there is still a serious and imminent danger.

13.15.10 Additional emergency measures in respect of dangerous substances – Article 16

In order to safeguard persons from an accident, incident or emergency related to the presence of a dangerous substance, the responsible person must (unless the risk assessment shows it is unnecessary) ensure that:

(a) information on emergency arrangements is available, including:
 (i) details of relevant work hazards and hazard identification arrangements;
 (ii) specific hazards likely to arise at the time of an accident, incident or emergency;

(b) suitable warning and other communication systems are established to enable an appropriate response, including remedial actions and rescue operations, to be made immediately when such an event occurs;

(c) where necessary, before any explosion conditions are reached, visual or audible warnings are given and relevant persons withdrawn;

(d) where the risk assessment indicates it is necessary, escape facilities are provided and maintained to ensure that, in the event of danger, persons can leave endangered places promptly and safely.

The information required must be:

(a) made available to accident and emergency services;

(b) displayed at the premises, unless the results of the risk assessment make this unnecessary.

In the event of a fire arising from an accident, incident or emergency related to the presence of a dangerous substance in or on the premises, the responsible person must ensure that:

(a) immediate steps are taken to:
(i) mitigate the effects of the fire;
(ii) restore the situation to normal;
(iii) inform those persons who may be affected.
(b) only those persons who are essential for the carrying out of repairs and other necessary work are permitted in the affected area and they are provided with:
(i) appropriate PPE and protective clothing; and
(ii) any necessary specialised safety equipment and plant which must be used until the situation is restored to normal.

13.15.11 Maintenance – Article 17

Any facilities, equipment and devices provided must be subject to a suitable system of maintenance and maintained in an efficient state, in efficient working order and in good repair.

13.15.12 Safety assistance – Article 18

The responsible person must (except a competent self-employed person) appoint one or more competent persons to assist them in undertaking the preventative and protective measures. If more than one person is appointed, they must make arrangements for ensuring adequate cooperation between them.

The number of persons appointed, the time available for them to fulfil their functions and the means at their disposal must be adequate having regard to the size of the premises, the risks to which relevant persons are exposed and the distribution of those risks throughout the premises.

13.15.13 Provision of information to employees and others – Articles 19 and 20

The responsible person must provide their employees with comprehensible and relevant information on:

▶ the risks to them identified by the risk assessment;
▶ the preventative and protective measures;
▶ the procedures for fire drills;
▶ the identities of persons nominated for firefighting or fire drills;
▶ the risks notified to them regarding shared premises.

Before employing a child, a parent (or guardian) of the child must be provided with comprehensible and relevant information on the risks to that child identified by the risk assessment; the preventative and protective measures; and the risks notified to them regarding shared premises.

Where a dangerous substance is present in or on the premises, additional information must be provided for employees as follows:

(a) the details of any such substance including:
(i) the name of the substance and the risk which it presents;
(ii) access to any relevant safety data sheets;
(iii) legislative provisions which apply to the substance;
(b) the significant findings of the risk assessment.

The responsible person must ensure that the employer of any employees from an outside undertaking who are working in or on the premises is provided with comprehensible and relevant information on:

(a) the risks to those employees;
(b) the preventative and protective measures taken by the responsible person.

The responsible person must ensure that any person working in their undertaking who is not their employee is provided with appropriate instructions and comprehensible and relevant information regarding any risks to that person.

13.15.14 Capabilities and training – Article 21

The responsible person must ensure that all employees are provided with adequate safety training:

(a) at the time when they are first employed;
(b) on their being exposed to new or increased risks because of:
(i) their being transferred or given a change of responsibilities;
(ii) the introduction of new work equipment or a change respecting work equipment already in use;
(iii) the introduction of new technology; or
(iv) the introduction of a new system of work, or a change respecting a system of work already in use.

The training must:

(a) include suitable and sufficient instruction and training on the appropriate precautions and actions to be taken by the employee;
(b) be repeated periodically where appropriate;
(c) be adapted to take account of any new or changed risks to the safety of the employees concerned;
(d) be provided in a manner appropriate to the risk identified by the risk assessment;
(e) take place during working hours.

Under the Fire Safety (Employee's Capabilities) Regulations for both England and Wales, Every employer must, in entrusting tasks to employees, take into account their capabilities as regards health and safety, so far as those capabilities relate to fire.

13

13.15.15 Cooperation and coordination – Article 22

Where two or more responsible persons share, or have duties in respect of, premises (whether on a temporary or a permanent basis) each such person must cooperate with, coordinate safety measures and inform the other responsible person concerned so far as is necessary to enable them to comply with the requirements and prohibitions imposed on them.

Where two or more responsible persons share premises (whether on a temporary or a permanent basis) where an explosive atmosphere may occur, the responsible person who has overall responsibility for the premises must coordinate the implementation of all the measures required.

13.15.16 General duties of employees at work – Article 23

Every employee must, while at work:

(a) take reasonable care for the safety of himself or herself and of other relevant persons who may be affected by acts or omissions at work;

(b) as regards any duty or requirement imposed on an employer by or under any provision of this Order, the employee must cooperate so far as is necessary to enable that duty or requirement to be performed or complied with;

(c) inform their employer or any other employee with specific responsibility for the safety of fellow employees:

(i) of any work situation which a person with the first-mentioned employee's training and instruction would reasonably consider represented a serious and immediate danger to safety;

(ii) of any matter which a person with the first-mentioned employee's training and instruction would reasonably consider represented a shortcoming in the employer's protection arrangements for safety.

Part 3

13.15.17 Enforcement – Articles 25–31

Part 3 of the Order is about enforcement and penalties. An enforcing authority may be the fire and rescue authority of the area (most cases), the HSE (nuclear sites, ships and construction sites without other operations) and fire Inspectors maintained by the Secretary of State (armed forces, UK Atomic Energy Authority and Crown premises).

Fire Inspectors and an officer of the fire brigade maintained by the fire and rescue authority, have similar powers (Articles 27 and 28) to inspectors under the HSW Act.

There are differences between the HSW Act and the Fire order in the notices which can be issued. Appeals can be made against these notices within 21 days and the notices are suspended until the appeal is heard. However, a Prohibition notice stands until confirmed or altered by the court. Fire Inspectors can issue the following notices.

Alterations Notices – Article 29

These may be issued where the premises concerned constitute a serious risk to relevant people or may constitute a serious risk if any change is made to the premises or its use.

Where an Alterations Notice has been served the responsible person must notify the enforcing authority before making any specified changes, which are as follows:

- a change to the premises;
- a change to the services, fittings or equipment in or on the premises;
- an increase in the quantities of dangerous substances which are present in or on the premises;
- a change in the use of the premises.

In addition, the Alterations Notice may also include the requirement for the responsible person to:

- record the significant findings of the risk assessment as per Article 9(7) and 9(6);
- record the fire safety arrangements as per Article 11(1) and 11(2);
- send a copy of the risk assessment to the enforcing authority, before making the above changes, and a summary of the proposed changes to the general fire precautions.

Enforcement Notices – Article 30

Where the enforcing authority is of the opinion that the responsible person has failed to comply with the requirements of this Order or any regulations made under it, they can issue an Enforcement Notice which must:

- state that the enforcing authority is of this opinion;
- specify the provisions which have not been complied with; and require that person to take steps to remedy the failure within a period (not less than 28 days) from the date of service of the notice.

An Enforcement Notice may include directions on the measures needed to remedy the failures. Choices of remedial action must be left open.

Before issuing an Enforcement Notice the enforcing authority must consult the relevant regulatory enforcing authorities including those under HSW Act and the Building Regulations.

Prohibition Notices – Article 31

If the enforcing authority is of the opinion that the risks, relating to escape from the premises, are so serious that the use of the premises ought to be prohibited or restricted they may issue a Prohibition Notice. The notice must:

▶ state that the enforcing authority is of this opinion;
▶ specify the provisions which give or may give rise to that risk;
▶ direct that the use to which the notice relates is prohibited or restricted as may be specified until the specified matters have been remedied.

A Prohibition Notice may include directions on the measures needed to remedy the failures. Choices of remedial action must be left open.

A Prohibition Notice takes effect immediately it is served.

Part 4

13.15.18 Offences and appeals – Articles 32–36

Cases can be tried in a Magistrates' Court or on indictment in the Crown Court.

The responsible person can be liable on conviction on indictment to a fine (not limited), or to imprisonment for a term not exceeding 2 years or both.

Any person can be liable to:

▶ on conviction on indictment to a fine (not limited) for an offence where that failure places one or more relevant people at risk of death or serious injury in the case of fire;
▶ on summary conviction to a fine at standard levels 3 or 5 depending on the particular offence.

In general, where an offence committed by a body corporate is proved to have been committed with the consent or connivance of any director, manager, secretary or other similar officer of the body corporate they, as well as the body corporate, are guilty of that offence – Article 32(9).

13.15.19 The Fire (Scotland) Act 2005

Fire is a devolved subject. Previous fire safety legislation in Scotland has been replaced by the Fire (Scotland) Act 2005, as amended, and by regulations made under that Act. Domestic premises do not generally fall within the scope of this Act, but there are exceptions which include those requiring a licence under the Houses in Multiple Occupation mandatory licensing scheme, and care home premises. However, the legislation does require any equipment or facilities provided for the protection or use of firefighters and located in common areas of private dwellings to be maintained.

The Police and Fire Reform (Scotland) Act 2012 part 2 – Fire Reform sets up the Scottish Fire and Rescue Service (SFRS). The 2005 Act is amended so that references to 'a relevant authority' is replaced by 'SFRS'. The SFRS also has the duty to secure 'best value' when carrying out its functions (see the SFRS website: www.firescotland.gov.uk).

Duties imposed by the legislation fall into seven general categories and are similar to those in England and Wales:

1. Carrying out a **fire safety risk assessment** of the premises;
2. Identifying the **fire safety measures** necessary as a result of the fire safety risk assessment outcome;
3. Implementing these fire safety measures using **risk reduction principles**;
4. Putting in place fire safety arrangements for the ongoing **control and review of the fire safety measures**;
5. Complying additionally with the specific requirements of the **fire safety regulations**;
6. Keeping the fire safety risk assessment and outcome under review; and
7. Record keeping.

The following is a summary of the main rights and responsibilities of those covered by the legislation. For a comprehensive list of rights and responsibilities, please refer to the legislation.

(a) Employers:
▶ are **entitled** to the cooperation of their employees in respect of fire safety responsibilities and for employees to take reasonable care for their own safety and others whom their actions could affect.
▶ **must** carry out (and review regularly) a fire safety risk assessment to identify the potential for fire to occur in the workplace and cause harm to employees and other persons in, or in the immediate vicinity of, the premises.
▶ **must** ensure, so far as is reasonably practicable, the safety of their employees from harm caused by fire in the workplace and take reasonable measures to ensure their safety and that of others there, or in the immediate vicinity. These measures should be aimed at avoiding or reducing any risks identified.
▶ **must** provide clear, appropriate information and instruction (and training where necessary) to employees and anyone else working on their premises, e.g. contractors and their employer(s), in relation to any risks identified and fire safety measures provided.
▶ **must** ensure that a record of a fire safety risk assessment is kept (electronically or paper-based) if they employ five or more employees (whether they are based in the premises or not), have a licence (such as a liquor licence) for the premises, are subject to registration (such as care home registration), or an alterations notice requiring this has been served in relation to the premises by the enforcing authority.

13

▶ **must** carry out (or review) a fire safety risk assessment before employing anyone under the age of 18, taking into account their youth and inexperience, and the particular risks they may be exposed to in the workplace. If a child is of school age, their parent must be informed of the risks and the measures taken to avoid or reduce them.

▶ **must** ensure that the premises are equipped to an appropriate level, with means of escape (ensuring these can be safely used), means of detecting fire and giving warning in the event of fire, means of fighting fires, and arrangements for action to be taken in the event of fire.

▶ **must** ensure that the premises, and any fire safety facilities, equipment and devices are maintained in good order.

▶ **must** cooperate with anyone else who has fire safety responsibilities for the same premises (including sharing information about the activities/measures undertaken, where appropriate) and take all reasonable steps to coordinate your fire safety measures regarding the premises.

(b) Employees:

▶ are **entitled** to the provision of adequate fire safety measures on the premises to ensure, so far as is reasonably practicable, their safety from harm caused by fire.

▶ are **entitled** to appropriate instruction (and training where necessary) about any risks identified on the premises, fire safety measures provided and what to do in the event of a fire.

▶ **must** take reasonable care to ensure the workplace is safe from harm caused by fire and do nothing that will place themselves or others at risk.

▶ **must** inform their employer (or a fellow employee with specific fire safety responsibilities), of anything relating to the premises which could represent a serious and immediate fire safety danger; of anything which they reasonably consider represents a shortcoming in the employer's fire safety protection arrangements; or in the event of fire.

▶ **must** cooperate with their employer, so far as is necessary, to allow them to comply with their fire safety responsibilities.

(c) Someone who is not an employer but has control of the premises or safety obligations to some extent, such as an owner, tenant, contractor, or occasional user (including voluntary groups):

▶ are **entitled** to cooperation from others with fire safety responsibilities for the premises.

▶ **must** carry out (and review regularly as necessary) a fire safety risk assessment of the premises, to the extent of their control, or obligations towards them, identifying any risks

to persons in, or in the immediate vicinity of, the premises in respect of harm caused by fire.

▶ **must** take reasonable fire safety measures on the basis of the results of their fire safety risk assessment to ensure, to the extent of the control or obligations they have, the safety of persons from harm caused by fire in the premises.

▶ **must** ensure that the premises are equipped to an appropriate level, relative to the extent of their control or obligations, with means of escape (ensuring they can be safely used), means of detecting fire and giving warning in the event of fire, means of fighting fires, and arrangements for action to be taken in the event of fire.

▶ **must** cooperate (including sharing information where appropriate) with others who have fire safety responsibilities for the premises, take all reasonable steps to coordinate fire safety measures with them, taking into account the extent of their control or obligation and others in respect of the premises.

▶ **must** ensure that the premises and any fire safety facilities, equipment and devices are maintained in good order, to the extent of their control or obligation towards the premises.

13.15.20 Fire and Rescue Services (Northern Ireland)

The law applies to almost all premises which are not private dwellings. Examples include:

▶ Offices and shops;
▶ Premises that provide care, including care homes and hospitals;
▶ Pubs, clubs and restaurants;
▶ Places of worship;
▶ Educational establishments including schools;
▶ Theatres and cinemas;
▶ Sports centres and other community premises;
▶ Hotels and hostels;
▶ Guest houses and B&B accommodation;
▶ Shared areas of properties common to several households;
▶ Houses in multiple occupation;
▶ Factories and warehouses;
▶ Tents and marquees;
▶ Transport premises and facilities;
▶ Animal premises and stables;
▶ Open air events and venues.

It does not apply to people's private homes, including individual flats in a block or house.

The main requirements under the legislation are very similar to other areas of the UK as follows:

▶ Carry out a fire risk assessment for the premises over which you have control;

► Identify the fire hazards and risks associated with the premises, materials/substances, activities, etc.;

► Identify the people, or groups of people at risk and anyone who may be especially at risk;

► Remove and reduce the risks as far as reasonably possible;

► Put in place general fire precautions to deal with any remaining risks;

► Implement additional preventative and protective measures if flammable or explosive substances are used or stored on the premises;

► Develop and implement appropriate emergency procedures in the event of fire;

► If you have five or more employees, or require a licence or registration, you must record the significant findings of the risk assessment and any actions you have taken to remove/reduce the risk;

► Review the risk assessment periodically or after significant changes in the workplace.

13.15.21 Fire Safety Bill

The following was included in the Queen's Speech 2019 and published by the Prime Minister's Office on 19 December 2019. In July 2020 this Bill is awaiting the Report Stage in the House of Commons:

'New measures will be brought forward ... to improve building safety.'

The purpose of the Bill is to:

► Deliver meaningful change to ensure an appalling tragedy like Grenfell can never happen again.

► Implement the relevant legislative recommendations of the Grenfell Tower Public Inquiry Phase 1 Report.

► Put beyond doubt that the Fire Safety Order will require building owners and managers of multi-occupied residential premises of any height to fully consider and mitigate the risks of any external wall systems and fire doors.

The main benefits of the Bill would be:

► Providing residents in these properties with reassurance that the Government is learning lessons from the Grenfell tragedy and taking steps to improve their safety.

► Ensuring that building owners and managers are clear that they are responsible for assessing the risks of external walls and fire doors in buildings of any height.

► Making clear Fire and Rescue Services' role to undertake enforcement against dangerous cladding and fire doors in residential buildings.

The main elements of the Bill are:

► Clarifying that the scope of the Fire Safety Order includes the external walls of the building, including cladding, and fire doors for domestic premises of multiple occupancy.

► Strengthening the relevant enforcement powers to hold building owners and managers to account.

► Providing a transitional period for building owners and managers (the 'responsible person') and Fire and Rescue Services to put in place the infrastructure for these changes.

Territorial extent and application

The Bill's provisions would extend and apply to England only. Fire safety policy and fire and rescue services are devolved to Scotland, Wales and Northern Ireland.

Key facts

► Dame Judith Hackitt's review found that the current regulatory system is not fit for purpose in relation to high-rise and complex buildings. The Government has agreed to take forward all 53 of Dame Judith's recommendations and in some areas is going further.

► Since the fire at Grenfell, the Government claim they have taken forward a wide range of improvements that have not required legislation.

13.15.22 Further information

Regulatory Reform, England and Wales. The Regulatory Reform (Fire Safety) Order 2005, SI 2005 No. 1541.

There are a number of guides produced by the Department of Communities and Local Government including the following subjects:

Regulatory Reform (Fire Safety) Order 2005 – A short guide to making your premises safe from fire. Product code: 05 FRSD 03546

Fire Risk Assessment – Offices and Shops ISBN 978185112815 0.

Fire Risk Assessment – Sleeping Accommodation ISBN 9781851128174.

Fire Risk Assessment – Residential Care Premises ISBN 9781851128181.

Fire Risk Assessment – Small and Medium Places of Assembly ISBN 978185112824.

Fire Risk Assessment – Large Places of Assembly ISBN 9781851128211.

Fire Risk Assessment – Factories and Warehouses ISBN 9781851128167.

Fire Risk Assessment – Theatres, Cinemas and Similar Premises ISBN 9781851128228.

Fire Risk Assessment – Educational Premises ISBN 9781851128198.

Fire Risk Assessment – Healthcare Premises ISBN 9781851128242.

Fire Risk Assessment – Transport Premises and Facilities ISBN 9781851128259.

Fire Risk Assessment – Open Air Events and Venues ISBN 9781851128235.

Fire Risk Assessment – Animal Premises and Stables ISBN 9781851128846.

13

For free downloads see website: https://www.gov.uk/workplace-fire-safety-your-responsibilities/fire-safety-advice-documents

Fire Safety Scotland see: www.firelawscotland.org/v2bbaa.html?pContentID=237

Northern Ireland Fire and Rescue Service's see: www.nifrs.org/firesafe/

Fire safety in construction, 2010, HSE, HSG168, ISBN 9780717663453 https://www.hse.gov.uk/pubns/books/hsg168.htm

Advice for Building Owners of Multi-storey, Multi-occupied Residential Buildings, January 2020, Ministry of Housing, Communities & Local Government

Fire Safety Bill 2019–2020 https://researchbriefings.parliament.uk/ResearchBriefing/Summary/CBP-8782

13.16 The Health and Safety (Amendment) (EU Exit) Regulations 2018 SI 2018 No. 1370

1. These Regulations ensure that European Union ('EU')-derived health and safety protections will continue to be available in domestic law after the United Kingdom ('UK') has left the EU. The regulations amended are detailed in Section 6 and 7 of this explanatory memorandum. These Regulations do not make any policy changes beyond the intent of ensuring continued operability of the relevant legislation.

2. The European Union Withdrawal Act ('EUWA') allows EU-derived legislation to be fixed to ensure it operates properly and effectively once the UK has left the EU. These amendments address deficiencies in health and safety legislation arising from the exit of the UK from the EU. This instrument amends provisions which will for example, become inappropriate or redundant.

3. This instrument amends the relevant legislation to ensure that existing protections and regulatory frameworks are maintained and continue to work in the same way once the UK has left the EU.

13.17 Health and Safety (First Aid) Regulations 1981 as amended

13.17.1 Introduction

These Regulations set out employers' duties to provide adequate first-aid facilities. They define first aid as:

▶ In cases where a person will need help from a medical practitioner or nurse, treatment for the purposes of preserving life and minimising the consequences of injury and illness until such help is obtained;

▶ Treatment of minor injuries which would otherwise receive no treatment or which do not need treatment by a medical practitioner or nurse.

From October 2013 the HSE have reissued L74 as Guidance only, taking into account the amendment of Regulation 3(2) to remove the requirement for the HSE to approve the training and qualifications of appointed first-aid personnel and other earlier amendments.

It was reissued with minor amendments in 2018. These were:

▶ update guidance on the use of automated external defibrillators and blended learning in first-aid training;

▶ incorporate some additional amendments to take account of other previous legislative changes.

13.17.2 Duty of the employer – Regulation 3

An employer shall provide, or ensure that there are provided:

▶ adequate and appropriate facilities and equipment;

▶ such number of suitable persons as is adequate and appropriate for rendering first aid; they must undergo such training and have such qualifications as may be appropriate in the circumstances;

▶ an appointed person, being someone to take charge of situations as well as first-aid equipment and facilities, where medical aid needs to be summoned. An appointed person will suffice where:

▷ the nature of the work is such that there are no specific serious hazards (offices, libraries, etc.); the workforce is small; the location makes further provision unnecessary;

▷ there is temporary (not planned holidays) or exceptional absence of the first-aider.

Employers must make an assessment of the first-aid requirements that are appropriate for each workplace. Where an employer provides first-aiders in the workplace, they should ensure they have undertaken suitable training, have an appropriate first-aid qualification and remain competent to perform their role. Typically, first-aiders will hold a valid certificate of competence in either First Aid at Work (FAW) or emergency First Aid at Work (EFAW). EFAW training enables a first-aider to give emergency first aid to someone who is injured or becomes ill while at work. FAW training includes EFAW and also equips the first-aider to apply first aid to a range of specific injuries and illness.

If an employer, after carrying out a needs assessment, decides a first-aider is not required in the workplace, a person should be appointed to take charge of the first-aid arrangements. The role of this appointed person includes looking after the first-aid equipment and facilities and calling the emergency services when required.

There must always be at least an appointed person in every workplace during working hours.

Any first-aid room provided under this Regulation must be easily accessible to stretchers and to any other equipment needed to convey patients to and from the room. It must be signposted according to the safety signs and signals regulations.

13.17.3 Employees information – Regulation 4

Employees must be informed of the arrangements for first aid, including the location of facilities, equipment and people.

13.17.4 Self-employed – Regulation 5

The self-employed shall provide such first-aid equipment as is appropriate to render first aid to themselves.

13.17.5 Further Information

Basic Advice on First Aid at Work INDG347, 2017, ISBN 9780717666683 www.hse.gov.uk/pubns/indg347.htm

First Aid at Work INDG214(rev2), HSE Books, 2014, ISBN 9780717664696 www.hse.gov.uk/pubns/indg214.htm

First Aid at Work. Health and Safety (First Aid) Regulations 1981. Guidance L74 (As amended 2018), HSE, 2013, ISBN 9780717665600 www.hse.gov.uk/pubns/books/l74.htm

13.18 Health and Safety (Information for Employees) Regulations 1989

13.18.1 General requirements

These Regulations require that the Approved Poster entitled, 'Health and safety – what you should know', is displayed or the Approved Leaflet is distributed.

This information tells employees in general terms about the requirements of health and safety law.

A modern and easy-to-read version of the poster and pocket cards were introduced on 6 April 2009.

The poster has three sections for completion by the employer. These are 1. Employee representative; 2. Management representative (e.g. appointed competent person to assist manager's health and safety appointee); 3. Details of how the Enforcing authority and the Employment Medical Advisory Service (EMAS) can be contacted.

13.18.2 Further information

Health and Safety (Information for Employees) Regulations 1989 (SI No. 682).

Health and Safety Information for Employees (Amendment) Regulations 2009, SI 2009/606.

Health and Safety Law poster A3 size, 2009, ISBN 9780717663699 www.hse.gov.uk/pubns/books/lawposter-a3.htm

Health and Safety Law What you need to know, 2009, HSE Books ISBN 9780717665013 www.hse.gov.uk/pubns/lawleaflet.pdf

13.19 Ionising Radiation Regulations 2017

13.19.1 Introduction

The main legal requirements enforced by HSE are the the Ionising Radiations Regulations 2017 (IRR17). They replace The Ionising Radiations Regulations 1999 (IRR99).

IRR17 applies to a large range of workplaces where radioactive substances and electrical equipment emitting ionising radiation are used. They also apply to work with natural radiation, including work in which people are exposed to naturally occurring radon gas and its decay products. Any employer who undertakes work with ionising radiation must comply with IRR17.

IRR17 requires employers to keep exposure to ionising radiations as low as reasonably practicable. Exposures must not exceed specified dose limits. Restriction of exposure should be achieved first by means of engineering control and design features. Where this is not reasonably practicable employers should introduce safe systems of work and only rely on the provision of personal protective equipment as a last resort.

The guidance which accompanies the Regulations and ACoP gives detailed advice about the scope and duties of the requirements imposed by IRR17. It is aimed at employers with duties under the Regulations but should also be useful to others such as radiation protection advisers, health and safety officers, radiation protection supervisors and safety representatives.

13.19.2 Changes from IRR99 to IRR17

They include:

▶ lowering the dose limit to the lens of the eye;
▶ flexibility for five-year averaging for dose limit to lens of the eye, subject to conditions specified by HSE;
▶ change to the radon reference level. The IRR99 radon reference level was over a 24-hour period, while EU Basic Safety Standards Directive (BSSD) expresses the reference level on an annual basis. Calculations show that the IRR99 reference level is broadly equivalent to the annual average reference level in IRR17;
▶ introducing a three-tier system of notification, registration and consent that replaces the IRR99 requirement for notification and prior authorisation;
▶ change to the requirement for notification, which for some radionuclides is at a lower threshold than in IRR99;
▶ broadening the scope of the definition of an outside worker so that it includes both classified and non-classified workers;
▶ change to the dose record retention period from 50 years to not less than 30 years after the last day of work;
▶ a requirement to put procedures in place to estimate doses to members of the public;

► a change to remove the requirement for a registered medical practitioner to be appointed 'in writing' for the purposes of these Regulations;

► introducing a requirement for authorisation of the annual whole-body dose limit in special cases – HSE or Office for Nuclear Regulation (ONR) may authorise the application of an effective dose limit of 100 mSv over five years (with no more than 50 mSv in a single year) rather than duty-holders only giving prior notification;

► recording and analysis of significant events, i.e. radiation accidents;

► removing the subsidiary dose limit for the abdomen of a woman of reproductive capacity;

► removing references to 'radiation employers', a term that has previously caused confusion, and replacing it where appropriate with reference to an employer who works with ionising radiation;

► those required to accommodate other legislative, standard or guidance changes.

13.19.3 Working with ionising radiation

Essentially, work with ionising radiation means:

(a) a practice which involves the production, processing, handling, use, holding, storage, transport or disposal of artificial radioactive substances and some naturally occurring sources, or the use of electrical equipment emitting ionising radiation at more than 5 kV (see definition of practice in Regulation 2(1));

(b) work in places where the radon gas concentration exceeds the values in Regulation 3(1)(b); or

(c) work with radioactive substances containing naturally occurring radionuclides not covered by the definition of a practice.

13.19.4 Duties of self-employed people

A self-employed person who works with ionising radiation will simultaneously have certain duties under these Regulations, both as an employer and as an employee.

For example, self-employed persons may need to take such steps as:

► carrying out assessments under Regulation 8;

► providing control measures under Regulation 9 to restrict exposure;

► designating themselves as classified persons under Regulation 21;

► making suitable arrangements under Regulation 22 with one or more approved dosimetry services (ADS) for assessment and recording of doses they receive;

► obtaining a radiation passbook and keeping it up to date;

► if they carry out services as an outside worker, making arrangements for their own training as required by Regulation 15;

► ensuring they use properly any dose meters provided by an ADS as required by Regulation 35.

13.19.5 General requirements

Some of the major considerations are:

► notification to HSE applies to any work with ionising radiations that does not require registration (Regulation 5) or consent (Regulation 7);

► carrying out a suitable and sufficient risk assessment before commencing a new activity (Regulation 8);

► restriction of exposure (Regulation 9);

► personal protective equipment (Regulation 10);

► maintenance and examination of engineering controls and PPE (Regulation 11);

► dose limitations (Regulation 12);

► contingency plans for emergencies (Regulation 13);

► radiation protection adviser appointment (Regulation 14);

► information instruction and training (Regulation 15).

13.19.6 Risk assessment

Where a radiation employer is required to undertake a risk assessment, the following matters need to be considered, where they are relevant:

► the nature of the sources of ionising radiation to be used, or likely to be present, including accumulation of radon in the working environment;

► estimated radiation dose rates to which anyone can be exposed;

► the likelihood of contamination arising and being spread;

► the results of any previous personal dosimetry or area monitoring relevant to the proposed work;

► advice from the manufacturer or supplier of equipment about its safe use and maintenance;

► engineering control measures and design features already in place or planned;

► any planned systems of work; estimated levels of airborne and surface contamination likely to be encountered;

► the effectiveness and the suitability of PPE to be provided;

► the extent of unrestricted access to working areas where dose rates or contamination levels are likely to be significant; possible accident situations, their likelihood and potential severity;

► the consequences of possible failures of control measures – such as electrical interlocks, ventilation systems and warning devices – or systems of work; steps to prevent identified accident situations, or limit their consequences.

This risk assessment should enable the employer to determine:

▶ what action is needed to ensure that the radiation exposure of all persons is kept as low as reasonably practicable (Regulation 9(1));

▶ what steps are necessary to achieve this control of exposure by the use of engineering controls, design features, safety devices and warning devices (Regulation 9(2)(a)) and, in addition, by the development of systems of work (Regulation 9(2)(b));

▶ whether it is appropriate to provide PPE and if so what type would be adequate and suitable (Regulation 9(2)(c));

▶ whether it is appropriate to establish any dose constraints for planning or design purposes, and if so what values should be used (Regulation 9(4));

▶ the need to alter the working conditions of any female employee who declares she is pregnant or is breastfeeding (Regulation 9(6));

▶ an appropriate investigation level to check that exposures are being restricted as far as are reasonably practicable (Regulation 9(8));

▶ what maintenance and testing schedules are required for the control measures selected (Regulation 11);

▶ what contingency plans are necessary to address reasonably foreseeable accidents (Regulation 13);

▶ the training needs of classified and non-classified employees (Regulation 15);

▶ the need to designate specific areas as controlled or supervised areas and to specify local rules (Regulations 17 and 18);

▶ the actions needed to ensure restriction of access and other specific measures in controlled or supervised areas (Regulation 19);

▶ the need to designate certain employees as classified persons (Regulation 21);

▶ the content of a suitable programme of dose assessment for employees designated as classified persons and for others who enter controlled areas (Regulations 19 and 22);

▶ the requirements for the leak testing of radioactive sources (Regulation 28);

▶ the responsibilities of managers for ensuring compliance with these Regulations;

▶ an appropriate programme of monitoring or auditing of arrangements to check that the requirements of these Regulations are being met.

13.19.7 Further information

The HSE have also produced a number of information sheets, which are available free at www.hse.gov.uk/radiation/ionising/publications.htm

Working with ionising radiation, L121 (second edition), 2018, HSE, ISBN 978071766621. This is a large document and need only be studied by those with a specific need to control IR. www.hse.gov.uk/pubns/books/l121.htm

Working safely with ionising radiation: Guidelines for expectant or breastfeeding mothers, INDG334(rev 1), 2015, www.hse.gov.uk/pubns/indg334.pdf

13.20 Lifting Operations and Lifting Equipment Regulations (LOLER) 1998 as amended in 2002

13.20.1 Introduction

This summary gives information about the LOLER 1998 which came into force on 5 December 1998.

The Regulations aim to reduce risks to people's health and safety from lifting equipment provided for use at work. In addition to the requirements of LOLER, lifting equipment is also subject to the requirements of the PUWER 1998.

Generally, the Regulations require that lifting equipment provided for use at work is:

▶ strong and stable enough for the particular use and marked to indicate SWL;

▶ positioned and installed to minimise any risks;

▶ used safely, i.e. the work is planned, organised and performed by competent people;

▶ subject to ongoing thorough examination and, where appropriate, inspection by competent people.

13.20.2 Definition

Lifting equipment includes **any equipment used at work for lifting or lowering loads**, including attachments used for anchoring, fixing or supporting it. The Regulations cover a wide range of equipment including cranes, fork-lift trucks, lifts, hoists, mobile elevating work platforms and vehicle inspection platform hoists. The definition also includes lifting accessories such as chains, slings, eyebolts, etc. LOLER **does not** apply to escalators; these are covered by more specific legislation, i.e. the Workplace (Health, Safety and Welfare) Regulations 1992.

If employees are allowed to provide their own lifting equipment, then this too is covered by the Regulations.

13.20.3 Application

The Regulations apply to an employer or self-employed person providing lifting equipment for use at work, or who has control of the use of lifting equipment. They do not apply to equipment to be used primarily by members of the public, for example, lifts in a shopping centre. However, such circumstances are covered by the HSW Act 1974.

13

LOLER applies to the way lifting equipment is used in industry and commerce. LOLER applies only to work activities, for example:

- a crane on hire to a construction site;
- a contract lift;
- a passenger lift provided for use of workers in an office block;
- refuse collecting vehicles lifting on a public road;
- patient hoist;
- fork-lift truck.

These Regulations add to the requirements of PUWER 98 and should be interpreted with them. For example, when selecting lifting equipment, PUWER Regulation 4, regarding suitability, should be considered in connection with:

- ergonomics;
- the conditions in which the equipment is to be used;
- safe access and egress;
- preventing slips, trips and falls;
- protecting the operator.

While employees do not have duties under LOLER, they do have general duties under the HSW Act and the MHSWR 1992, for example to take reasonable care of themselves and others who may be affected by their actions and to cooperate with others.

The Regulations cover places where the HSW Act applies – these include factories, offshore installations, agricultural premises, offices, shops, hospitals, hotels, places of entertainment, etc.

13.20.4 Strength and stability – Regulation 4

Lifting equipment shall be of adequate strength and stability for each load, having regard in particular to the stress induced at its mounting or fixing point.

Every part of a load and anything attached to it and used in lifting it shall be of adequate strength.

Account must be taken of the combination of forces to which the lifting equipment will be subjected, as well as the weight of any lifting accessories. The equipment should include an appropriate factor of safety against failure.

Stability needs to take into account the nature, load-bearing strength, stability, adjacent excavations and slope of the surface. For mobile equipment, keeping rails free of obstruction and tyres correctly inflated must be considered.

13.20.5 Lifting equipment for lifting persons – Regulation 5

To ensure safety of people being lifted, there are additional requirements for such equipment. The use of equipment not specifically designed for raising and lowering people should only be used in exceptional circumstances.

The Regulation applies to all lifting equipment used for raising and lowering people and requires that lifting equipment for lifting persons shall:

- prevent a person using it from being crushed, trapped or struck, or falling from the carrier;
- prevent, sfairp, persons using it while carrying out work from the carrier being crushed, trapped or struck or falling from the carrier;
- have suitable devices to prevent the risk of the carrier falling. If a device cannot be fitted, the carrier must have:
  - ▷ an enhanced safety coefficient suspension rope or chain;
  - ▷ the rope or chain inspected every working day by a competent person;
- be such that a person trapped in any carrier is not thereby exposed to any danger and can be freed.

13.20.6 Positioning and installation – Regulation 6

Lifting equipment must be positioned and installed so as to reduce the risks, sfairp, from:

- equipment or a load striking another person;
- a load drifting, falling freely or being released unintentionally;
- and it is otherwise safe.

Lifting equipment should be positioned and installed to minimise the need to lift loads over people and to prevent crushing in extreme positions. It should be designed to stop safely in the event of a power failure and not release its load. Lifting equipment, which follows a fixed path, should be enclosed with suitable and substantial interlocked gates and any necessary protection in the event of power failure.

13.20.7 Marking of lifting equipment – Regulation 7

Machinery and accessories for lifting loads shall be clearly marked to indicate their SWL, and:

- where the SWL depends on the configuration of the lifting equipment:
  - ▷ the machinery should be clearly marked to indicate its SWL for each configuration;
  - ▷ information which clearly indicates its SWL for each configuration should be kept with the machinery;
- accessories for lifting (e.g. hooks, slings) are also marked in such a way that it is possible to identify the characteristics necessary for their safe use (e.g. if they are part of an assembly);
- lifting equipment which is designed for lifting people is appropriately and clearly marked;
- lifting equipment not designed for lifting people, but which might be used in error, should be clearly marked to show it is not for lifting people.

13.20.8 Organisation of lifting operations – Regulation 8

Every lifting operation, that is lifting or lowering of a load, shall be:

▶ properly planned by a competent person;
▶ appropriately supervised;
▶ carried out in a safe manner.

The person planning the operation should have adequate practical and theoretical knowledge and experience of planning lifting operations. The plan will need to address the risks identified by the risk assessment and identify the resources, the procedures and the responsibilities required so that any lifting operation is carried out safely. For routine simple lifts a plan will normally be left to the people using the lifting equipment. For complex lifting operations, for example where two cranes are used to lift one load, a written plan may need to be produced each time.

The planning should take account of avoiding suspending loads over occupied areas, visibility, attaching/detaching and securing loads, the environment, location, overturning, proximity to other objects, lifting of people and pre-use checks of the equipment.

13.20.9 Thorough examination and inspection – Regulations 9 and 11

Before using lifting equipment for the first time by an employer, it must be thoroughly examined for any defect unless:

▶ the lifting equipment has not been used before;
▶ an EC declaration of conformity (where one should have been drawn up) has been received or made not more than 12 months before the lifting equipment is put into service;
▶ if it is obtained from another undertaking, it is accompanied by physical evidence of an examination.

A copy of this thorough examination report shall be kept for as long as the lifting equipment is used (or, for a lifting accessory, 2 years after the report is made) (Regulation 11).

Where safety depends on the installation conditions, the equipment shall be thoroughly examined:

▶ after installation and before being put into service;
▶ after assembly and before being put into service at a new site or in new location to ensure that it has been installed correctly and is safe to operate.

A copy of the thorough examination report shall be kept for as long as the lifting equipment is used at the place it was installed or assembled (Regulation 11).

Lifting equipment which is exposed to conditions causing deterioration that may result in dangerous situations, shall be:

▶ thoroughly examined at least every 6 months (for lifting equipment for lifting persons, or a lifting accessory); at least every 12 months (for other lifting equipment); or in accordance with an examination scheme; and each time that exceptional circumstances, liable to jeopardise the safety of the lifting equipment, have occurred and a copy of the report kept until the next report is made, or for 2 years (whichever is longer);
▶ inspected, if appropriate, by a competent person at suitable intervals between 'thorough examinations' (and a copy of the record kept until the next record is made).

All lifting equipment shall be accompanied by physical evidence that the last 'thorough examination' has been carried out before it leaves an employer's undertaking (or before it is used after leaving another undertaking).

The user, owner, manufacturer or some other independent party may draw up examination schemes, provided they have the necessary competence. Schemes should specify the intervals at which lifting equipment should be thoroughly examined and, where appropriate, those parts that need to be tested. The scheme should take account, for example, of its condition, the environment in which it is used, the number of lifting operations and the loads lifted.

The 'competent person' carrying out a thorough examination should have appropriate practical and theoretical knowledge and experience of the lifting equipment to be examined to enable them to detect defects or weaknesses and to assess their importance in relation to the safety and continued use of the lifting equipment. They should also determine whether a test is necessary and the most appropriate method for carrying it out.

13.20.10 Reports and defects – Regulation 10

The person making a 'thorough examination' of lifting equipment shall:

▶ notify the employer forthwith of any defect which, in their opinion, is or could become, dangerous; as soon as is practicable (within 28 days) write an authenticated report to:
  ▷ the employer;
  ▷ any person who hired or leased the lifting equipment, containing the information specified in Schedule 1;
▶ send a copy (as soon as is practicable) to the relevant enforcing authority where there is, in their opinion, a defect with an existing or imminent risk of serious personal injury (this will always be HSE if the lifting equipment has been hired or leased).

13

483

Every employer notified of a defect following a 'thorough examination' of lifting equipment should ensure that it is not used:

▶ before the defect is rectified;
▶ after a time specified in the schedule accompanying the report.

The person making an 'inspection' shall also notify the employer when, in his or her opinion, a defect is, or could become, dangerous and, as soon as is practicable, make a record of the inspection in writing.

13.20.11 Reports – Schedule 1

Schedule 1 lists the information to be contained in a report of a thorough examination. For example, name and address; identity of equipment; date of last thorough examination; SWL; appropriate interval; any dangerous or potentially dangerous defects; repairs required; date of next examination and test; details of the competent person.

13.20.12 Further information

HSE LOLER website: www.hse.gov.uk/work-equipment-machinery/loler.htm

Lifting Equipment and Lifting Operations Regulations 1998, SI No. 2307 1998, OPSI website: www.legislation.gov.uk/uksi/1998/2307/contents/made

Lifting Equipment at work. A brief guide, INDG290 (rev1), 2013, HSE, ISBN 9780717664832 https://www.hse.gov.uk/pubns/indg290.pdf

LOLER 1998 Lifting Operations and Lifting Equipment Regulations 1998, Open Learning Guidance. 1999, HSE Books, ISBN 978 0 7176 2464 5 www.hse.gov.uk/pubns/books/loler.htm

Rider-Operated lift trucks, L117, 2013, HSE, ISBN 9780717664412 www.hse.gov.uk/pubns/books/l117.htm

Safe Use of Lifting Equipment, Lifting Operations and Lifting Equipment Regulations, 1998, Approved Code of Practice and Guidance, L113 (second edition), 2014, HSE, ISBN 9780717665860 www.hse.gov.uk/pubns/books/l113.htm

Thorough Examination and Testing of Lifts. Simple Guidance for Lift Owners INDG339 (rev1), HSE, 2008, ISBN 9780717662555 www.hse.gov.uk/pubns/indg339.htm

Thorough examination of lifting equipment, A simple guide for employers, INDG422, HSE, 2008, ISBN 9780717663057 www.hse.gov.uk/pubns/indg422.pdf

13.21 Management of Health and Safety at Work Regulations 1999 as amended in 2003 and 2006

13.21.1 General

These Regulations give effect to the European Framework Directive on health and safety. They supplement the requirements of the HSW Act 1974

and specify a range of management issues, most of which must be carried out in all workplaces. The aim is to map out the organisation of precautionary measures in a systematic way, and to make sure that all staff are familiar with the measures and their own responsibilities. They were amended slightly in 2003, particularly in relation to civil liability (Regulation 22).

13.21.2 Risk assessment – Regulation 3

Every employer is required to make a 'suitable and sufficient' assessment of risks to employees, and risks to other people who might be affected by the organisation, such as visiting contractors and members of the public.

A systematic investigation of risks involved in all areas and operations is required, together with identification of the persons affected, a description of the controls in place and any further action required to reduce risks.

The risk assessments must take into account risks to new and expectant mothers and young people.

Significant findings from the assessments must be written down (or recorded by other means, such as on a computer) when there are five or more employees. The assessments need to be reviewed regularly and if necessary, when there have been significant changes, they should be modified.

13.21.3 Principles of prevention – Regulation 4

The following principles must be adopted when implementing any preventative and protective measures:

▶ avoiding risks;
▶ evaluating the risks which cannot be avoided;
▶ combating the risks at source;
▶ adapting the work to the individual, especially as regards the design of workplaces, the choice of work equipment and the choice of working and production methods, with a view, in particular, to alleviating monotonous work and work at a predetermined work rate and to reducing their effect on health;
▶ adapting to technical progress;
▶ replacing the dangerous substances by non-dangerous or less dangerous substances;
▶ developing a coherent overall prevention policy which covers technology, organisation of work, working conditions, social relationships and the influence of factors relating to the working environment;
▶ giving collective protective measures priority over individual protective measures;
▶ giving appropriate instruction to employees.

13.21.4 Effective arrangements for health and safety – Regulation 5

Formal arrangements must be devised (and recorded) for effective planning, organisation, control, monitoring and review of safety measures. This will involve an effective health and safety management system to implement the policy. Where there are five or more employees the arrangements should be recorded.

Planning involves a systematic approach to risk assessment, the selection of appropriate risk controls and establishing priorities with performance standards.

Organisation involves consultation and communication with employees; employee involvement in risk assessment; the provision of information; and securing competence with suitable instruction and training. Control involves clarifying responsibilities and making sure people adequately fulfil their responsibilities. It involves adequate and appropriate supervision.

Monitoring should include the measurement of how well the policy is being implemented and whether hazards are being controlled properly. It covers inspections of the workplace and management systems; and the investigation of incidents and accidents to ascertain the underlying causes and effect a remedy.

Review is essential to look at the whole of the health and safety management system to ensure that it is effective and achieving the correct standard of risk control.

13.21.5 Health surveillance – Regulation 6

In appropriate circumstances health surveillance of staff may be required – the ACoP describes more fully when this duty will arise. Health surveillance is considered relevant when: there is an identifiable disease or poor health condition; there are techniques to detect the disease; there is a reasonable likelihood that the disease will occur; and surveillance is likely to enhance the protection of the workers concerned. A competent person, who will range from a manager, in some cases, to a fully qualified occupational medical practitioner in others, should assess the extent of the surveillance.

13.21.6 Competent assistance – Regulation 7

Every employer is obliged to appoint one or more 'competent person(s)' to advise and assist in undertaking the necessary measures to comply with the relevant statutory requirements. They may be employees or outside consultants. The purpose is to make sure that all employers have access to health and safety expertise. Preference should be given to an in-house appointee who may be backed up by external expertise.

The competence of the person(s) appointed is to be judged in terms of their training, knowledge and experience of the work involved; it is not necessarily dependent upon particular qualifications. In simple situations, it may involve knowledge of relevant best practice, knowing one's limitations and taking external advice when necessary. In more complex situations or risks, fully qualified and appropriately experienced practitioners will be required.

Appointed competent persons must be provided with adequate information, time and resources to do their job.

13.21.7 Procedures for serious and imminent danger and contact with external services – Regulations 8 and 9

Procedures must be established for dealing with serious and imminent dangers, including fire evacuation plans and arrangements for other emergencies. A sufficient number of competent persons must be appointed to evacuate the premises in the event of an emergency. The procedures should allow for persons at risk to be informed of the hazards and how and when to evacuate to avoid danger. In shared workplaces employers must cooperate. Access to dangerous areas should be restricted to authorised and properly trained staff. Any necessary contact arrangements with external services for first aid, emergency medical care and rescue work must be set up.

13.21.8 Information for employees – Regulation 10

Information must be provided to staff on the risk assessment, risk controls, emergency procedures, the identity of the people appointed to assist on health and safety matters and risks notified by others.

The information provided must take into account the level of training, knowledge and experience of the employees. It must take account of language difficulties and be provided in a form that can be understood by everyone. The use of translations, symbols and diagrams should be considered. Where children under school-leaving age are at work, information on the risk assessments and control measures must be provided to the child's parent or guardians of children at work before the child starts work. It can be provided verbally or directly to the parent, guardians or school.

13

13.21.9 Cooperation and coordination – Regulations 11, 12 and 15

Where two or more employers share a workplace, each must:

▶ cooperate with other employers in health and safety matters;

▶ take reasonable steps to coordinate their safety precautions;

▶ inform the other employers of the risks to their employees, i.e. risks to neighbours' employees.

Where people from outside organisations are present to do work, they – and their employers – have to be provided with appropriate information on risks, health and the necessary precautions to be taken.

Temporary staff and staff with fixed-term contracts as well as permanent employees must be supplied with health and safety information before starting work (Regulations 12 and 15).

Regulation 11 does not apply to multi-occupied premises or sites where each unit, under the control of an individual tenant employer or self-employed person, is regarded as a separate workplace. In other cases, common areas may be shared workplaces, such as a reception area or canteen or they may be under the control of a person to whom Section 4 of HSW Act applies. Suitable arrangements may need to be put in place for these areas.

13.21.10 Capabilities and training – Regulation 13

When giving tasks to employees, their capabilities with regard to health and safety must be taken into account.

Employees must be provided with adequate health and safety training:

▶ on recruitment;

▶ on being exposed to new or increased risks;

▶ on the introduction of new procedures, systems or technology. Training must be repeated periodically and take place in working hours (or while being paid).

13.21.11 Duties on employees – Regulation 14

Equipment and materials must be used properly in accordance with instructions and training. Obligations on employees are extended to include certain requirements to report serious and immediate dangers and any shortcomings in the employer's protection arrangements.

13.21.12 New or expectant mothers – Regulations 16–18

Where work is of a kind that could present a risk to new or expectant mothers working there or their babies,

the risk assessments must include an assessment of such risks. When the risks cannot be avoided the employer must alter a woman's working conditions or hours to avoid the risks; offer suitable alternative work; or suspend from work on full pay. The woman must notify the employer in writing of her pregnancy, that she has given birth within the last 6 months or she is breastfeeding.

13.21.13 Young persons – Regulation 19

Employers must protect young persons at work from risks to their health and safety which are the result of lack of experience, or absence of awareness of existing or potential risks or which arise because they have not yet fully matured. Young persons may not be employed in a variety of situations which pose a significant risk to their health and safety that are a consequence of the following factors:

▶ physical or psychological capacity;

▶ pace of work;

▶ temperature extremes, noise or vibration;

▶ radiation;

▶ compressed air and diving;

▶ hazardous substances;

▶ lack of training and experience.

The exception to this is young persons over school-leaving age:

▶ where the work is necessary for their training;

▶ where they will be supervised by a competent person;

▶ where the risk will be reduced to the lowest level that is reasonably practicable.

The risk assessment must take the following specific factors into account:

▶ the fitting-out and layout of the workplace and the particular site where they will work;

▶ the nature of any physical, biological and chemical agents they will be exposed to, for how long and to what extent;

▶ what types of work equipment will be used and how this will be handled;

▶ how the work and processes involved are organised;

▶ the need to assess and provide health and safety training;

▶ risks from the particular agents, processes and work.

13.21.14 Provisions as to liability – Regulation 21

This provision is to prevent a defence for an employer by reason of any act or default by an employee or a competent person appointed under Regulation 7. Employees' duties to take reasonable care of their own health and safety and that of others affected by their work activities are unaffected by this provision.

13.21.15 Restriction of civil liability for breach of statutory duty – Regulation 22

Further restrictions have been imposed by the Enterprise and Regulatory Reform Act 2013. It amends the Health and Safety at Work etc. Act 1974 Section 47, to remove the right of civil action against employers and other duty-holders for breach of statutory duty in relation to certain health and safety legislation, other than where such a right is specifically provided for, which has been done by The Health and safety at Work etc. Act 1974 (Civil Liability) (Exceptions) Regulations 2013 which changed Regulation 22 (see below). This will address the potential unfairness that arises where an employer can be found liable to pay compensation to an employee despite having taken reasonable steps to protect them.

Regulation 22 of the Management of Health and Safety at Work Regulations ('the 1999 Regulations') excluded third parties from bringing claims for breach of a duty contained in those Regulations against employers and employees. This exclusion is no longer required because Section 47 of the 1974 Act now automatically excludes liability and express provision is only required for exceptions.

Regulation 22 has been changed to the following:

> 22.—(1) Breach of a duty imposed by Regulation 16, 16A, 17 or 17A shall, so far as it causes damage, be actionable by the new or expectant mother.
>
> (2) Any term of an agreement which purports to exclude or restrict any liability for such a breach is void.

13.21.16 Further information

Health and Safety made simple: The basics for your business, INDG 449(rev1),2014, HSE, INDG449 www.hse.gov.uk/pubns/indg449.pdf

Risk Assessment HSE INDG163 (rev 4), 2014, HSE, ISBN 9780717664634 www.hse.gov.uk/pubns/indg163.pdf

Leading Health and Safety at Work. Action for Directors and Board Members, business owners and organisations of all sizes INDG417(rev1), HSE, 2013 https://www.hse.gov.uk/pubns/indg417.pdf

The Management of Health and Safety at Work Regulations 1999, ISBN 9780110250516 www.opsi.gov.uk/si/si1999/uksi_19993242_en.pdf

Managing for Health and Safety HSG65 (third edition), 2013, HSE, ISBN 9780717664566 see: www.hse.gov.uk/pubns/books/hsg65.htm

The health and safety toolbox: How to control risks at work HSG268 2014, ISBN 9780717665877 www.hse.gov.uk/pubns/books/hsg268.htmm

13.22 Manual Handling Operations Regulations (MHO) 1992 as amended in 2002

13.22.1 General

The Regulations apply to the manual handling (any transporting or supporting) of loads, that is by human effort, as opposed to mechanical handling by fork-lift truck, crane, etc. Manual handling includes lifting, putting down, pushing, pulling, carrying or moving. The human effort may be applied directly to the load, or indirectly by pulling on a rope, chain or lever. Introducing mechanical assistance, like a hoist or sack truck, may reduce but not eliminate manual handling, as human effort is still required to move, steady or position the load.

The application of human effort for purposes other than transporting or supporting a load, for example pulling on a rope to lash down a load or moving a machine control, is not a manual handling operation. A load is a discrete movable object, but it does not include an implement, tool or machine while being used.

Injury in the context of these Regulations means to any part of the body. It should take account of the physical features of the load which might affect grip or cause direct injury, for example, slipperiness, sharp edges and extremes of temperature. It does not include injury caused by any toxic or corrosive substance which has leaked from a load, is on its surface or is part of the load. See Figure 13.8, which shows a flow chart for the Regulations.

13.22.2 Duties of employers: avoidance of manual handling – Regulation 4(1)(a)

Employers should take steps to avoid the need for employees to carry out MHO which involves a risk of their being injured.

The guidance suggests that a preliminary assessment should be carried out when making a general risk assessment under the MHSWR 1999. Employers should consider whether the operation can be eliminated, automated or mechanised.

13.22.3 Duties of employers assessment of risk – Regulation 4(1)(b)(i)

Where it is not reasonably practicable to avoid MHO, employers must make a suitable and sufficient risk assessment of all such MHO in accordance with the requirements of Schedule 1 to the Regulations (shown

13

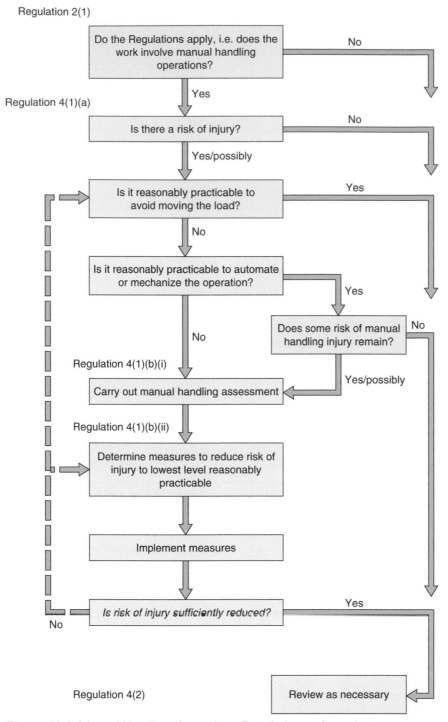

Figure 13.8 Manual Handling Operations Regulations – flow chart

in Table 13.5). This duty to assess the risk takes into account the task, the load, the working environment and individual capability.

13.22.4 Duties of employers: reducing the risk of injury – Regulation 4(1)(b)(ii)

Where it is not reasonably practicable to avoid the MHO with which there is a risk of injury, employers must take steps to reduce the risk of injury to the lowest level reasonably practicable.

The structured approach (considering the task, the load, the working environment and the individual capability) is

recommended in the guidance. The steps taken will involve ergonomics, changing the load, mechanical assistance, task layout, work routines, PPE, team working and training.

13.22.5 Duties of employers: additional information on the load – Regulation 4(1)(b)(iii)

Employers must take appropriate steps where manual handling cannot be avoided to provide general indications and, where practicable, precise information on:

- ▶ the weight of each load;
- ▶ the heaviest side of any load which does not have a central centre of gravity.

Table 13.5 Schedule 1 to the Manual Handling Operations Regulations

Factors	Questions
Factors to which the employer must have regard and questions they must consider when making an assessment of manual handling operations	
1. The tasks	*Do they involve:* Holding or manipulating loads at distance from trunk? Unsatisfactory bodily movement or posture, especially: Twisting the trunk? Stooping? Reaching upwards? Excessive movement of loads especially: Excessive lifting or lowering distances? Excessive carrying distances? Excessive pulling or pushing of loads? Risk of sudden movement of loads? Frequent or prolonged physical effort? Insufficient rest or recovery periods? A rate of work imposed by a process?
2. The loads	*Are they:* Heavy? Bulky or unwieldy? Difficult to grasp? Unstable, or with contents likely to shift? Sharp, hot or otherwise potentially damaging?
3. The working environment	*Are there:* Space constraints preventing good posture? Uneven, slippery or unstable floors? Variations in level of floors or work surfaces? Extremes of temperature or humidity? Conditions causing ventilation problems or gusts of wind? Poor lighting conditions?
4. Individual capability	*Does the job:* Require unusual strength, height, etc.? Create a hazard to those who might reasonably be considered to be pregnant or have a health problem? Require special information or training for its safe performance?
5. Other factors	Is movement or posture hindered by personal protective equipment or by clothing?

The information is probably best marked on the load. Sections 3 and 6 of the HSW Act may place duties on originators of loads, like manufacturers or packers.

13.22.6 Duties of employers – reviewing assessment – Regulation 4(2)

The assessment should be reviewed if there is reason to suspect that it is no longer valid or there have been significant changes in the particular MHO.

13.22.7 Individual capability – Regulation 4(3)(a)–(f)

A further requirement was added in 2002 by Amendment Regulations concerning the individual capabilities of people undertaking manual handling. Regard must be had in particular to:

(a) the physical suitability of the employee to carry out the operation;

(b) the clothing, footwear or other personal effects they are wearing;

(c) their knowledge and training;

(d) results of any relevant risk assessments carried out under Regulation 3 of the Management Regulations 1999;

(e) whether the employee is within a group of employees identified by that assessment as being especially at risk;

(f) the results of any health surveillance provided pursuant to Regulation 6 of the Management Regulations 1999.

13.22.8 Duty of employees – Regulation 5

Each employee, while at work, has to make proper use of any system of work provided for their use. This is in addition to other responsibilities under the HSW Act and the MHSWR.

The provisions do not include well-intentioned improvisation in an emergency, for example rescuing a casualty or fighting a fire.

13.22.9 Further information

Making the Best Use of Lifting and Handling Aids INDG398 (rev1), 2013, HSE, www.hse.gov.uk/pubns/indg398.pdf

Managing upper limb disorders in the workplace, INDG171 (rev3), 2020, www.hse.gov.uk/pubns/indg171.pdf

Manual Handling Assessment Chart INDG383 (rev3), 2019, HSE, www.hse.gov.uk/pubns/indg383.htm

Manual handling at work. A brief guide, INDG143 (rev4), 2020, HSE, www.hse.gov.uk/pubns/indg143.htm

Manual Handling, Manual Handling Operations Regulations 1992 (as amended), Guidance on Regulations Revised, L23 (fourth edition), 2016, HSE, ISBN 9780717666539 www.hse.gov.uk/pubns/books/l23.htm

Upper limb disorders in the workplace, HSG60, 2002, HSE, ISBN 9780717619788 www.hse.gov.uk/pubns/books/hsg60.htm

13.23 Control of Noise at Work Regulations 2005

13.23.1 Introduction

The Control of Noise at Work Regulations 2005 require employers to prevent or reduce risks to health and safety from exposure to noise at work. Employees also have duties under the Regulations.

The Regulations require employers to:

► assess the risks to their employees from noise at work;

► take action to reduce the noise exposure that produces those risks;

► provide their employees with hearing protection if they cannot reduce the noise exposure enough by using other methods;

► make sure the legal limits on noise exposure are not exceeded;

13

▶ provide their employees with information, instruction and training;

▶ carry out health surveillance where there is a risk to health.

The Regulations do not apply to:

▶ members of the public exposed to noise from their non-work activities, or making an informed choice to go to noisy places;

▶ low-level noise which is a nuisance but causes no risk of hearing damage.

The Control of Noise at Work Regulations came into force for the music and entertainment sectors on 6 April 2008 bringing them in line with all other sectors where the regulations have been in force since April

2006. The Control of Noise at Work Regulations now apply to pubs and clubs, amplified live music events, orchestras and other premises where live music or recorded music is played.

13.23.2 Exposure limit values and action levels – Regulation 4

The Control of Noise at Work Regulations require employers to take specific action at certain action values. These relate to:

▶ the levels of exposure to noise of their employees averaged over a working day or week;

▶ the maximum noise (peak sound pressure) to which employees are exposed in a working day.

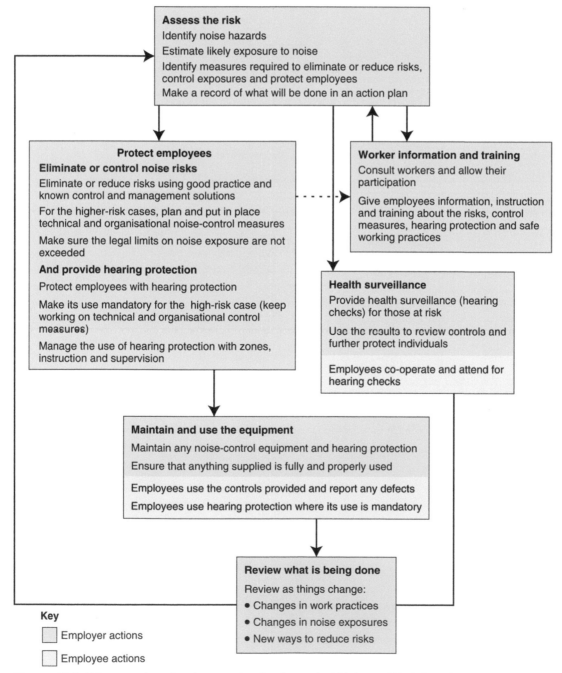

Figure 13.9 What needs to be done under the Control of Noise at Work Regulations 2005

The values are:

- ▶ lower exposure action values:
  - ▷ daily or weekly exposure of 80 dB (A-weighted);
  - ▷ peak sound pressure of 135 dB (C-weighted).
- ▶ upper exposure action values:
  - ▷ daily or weekly exposure of 85 dB (A-weighted);
  - ▷ peak sound pressure of 137 dB (C-weighted).

There are also levels of noise exposure which must not be exceeded:

- ▶ exposure limit values:
  - ▷ daily or weekly exposure of 87 dB (A-weighted);
  - ▷ peak sound pressure of 140 dB (C-weighted).

These exposure limit values take account of any reduction in exposure provided by hearing protection.

13.23.3 Risk assessment – Regulation 5

If employees are likely to be exposed to noise at or above the lower exposure value, a suitable and sufficient assessment of the risks must be made. Employers need to decide whether any further action is needed, and plan how to do it.

The risk assessment should:

- ▶ identify where there may be a risk from noise and who is likely to be affected;
- ▶ contain a reliable estimate of employees' exposures, and compare the exposure with the exposure action values and limit values;
- ▶ identify what needs to be done to comply with the law, e.g. whether noise-control measures or hearing protection are needed, and, if so, where and what type;
- ▶ identify any employees who need to be provided with health surveillance and whether any are at particular risk.

The risk assessment should include consideration of:

- ▶ the level, type and duration of exposure, including any exposure to peak sound pressure;
- ▶ the effects of exposure to noise on employees or groups of employees whose health is at particular risk;
- ▶ sfairp any effects on the health and safety of employees resulting from interaction, e.g. between noise and vibration;
- ▶ any indirect effects from the interaction between noise and audible warnings;
- ▶ manufacturers' information;
- ▶ availability of alternative equipment designed to reduce noise emissions;
- ▶ any extension to noise exposure due to extended hours or in supervised rest facilities;
- ▶ information following health surveillance;
- ▶ availability of personal hearing protectors with adequate attenuation characteristics.

It is essential that employers can show that their estimate of employees' exposure is representative of the work that they do. It needs to take account of:

- ▶ the work they do or are likely to do;
- ▶ the ways in which they do the work;
- ▶ how it might vary from one day to the next.

The estimate must be based on reliable information, for example measurements in their own workplace, information from other similar workplaces or data from suppliers of machinery.

Employers must record the significant findings of their risk assessment. They need to record in an action plan anything identified as being necessary to comply with the law, setting out what they have done and what they are going to do, with a timetable and saying who will be responsible for the work.

The risk assessment should be reviewed if circumstances in the workplace change and affect noise exposures. Also it should be reviewed regularly to make sure that the employer continues to do all that is reasonably practicable to control the noise risks. Even if it appears that nothing has changed, employers should not leave it for more than about 2 years without checking whether a review is needed.

13.23.4 Elimination or control of exposure – Regulation 6

Efforts to control noise risks and noise exposure should be concentrated on eliminating noise at source.

Wherever there is noise at work employers should be looking for alternative processes, equipment and/or working methods which would make the work quieter or mean people are exposed for shorter times. They should also be keeping up with what is good practice or the standard for noise control within their industry.

Where there are things that can be done to reduce risks from noise that are reasonably practicable, they should be done. However, where noise exposures are below the lower exposure action values, risks are low and so employers would only be expected to take actions which are relatively inexpensive and simple to carry out.

Where the assessment shows that employees are likely to be exposed at or above the upper exposure action values, employers must put in place a planned programme of noise control.

The risk assessment will have produced information on the risks and an action plan for controlling noise. Employers should use this information to:

- ▶ tackle the immediate risk, e.g. by providing hearing protection;
- ▶ identify what is possible to control noise, how much reduction could be achieved and what is reasonably practicable;

13

▶ establish priorities for action and a timetable (e.g. consider where there could be immediate benefits, what changes may need to be phased in over a longer period of time and the number of people exposed to the noise in each case);

▶ assign responsibilities to people to deliver the various parts of the plan;

▶ ensure the work on noise control is carried out;

▶ check that what has been done has worked.

Actions taken should be based on the general principles set out in the Management Regulations and should include consideration of:

▶ other working methods;

▶ choice of appropriate work equipment emitting the least possible noise;

▶ the design and layout of workplaces, workstations and rest facilities;

▶ suitable and sufficient information and training;

▶ reduction of noise by technical means;

▶ appropriate maintenance programmes;

▶ limitation of the duration and intensity of exposure;

▶ appropriate work schedules with adequate rest periods.

13.23.5 Hearing protection – Regulation 7

Hearing protection should be issued to employees:

▶ where extra protection is needed above what can been achieved using noise control;

▶ as a short-term measure while other methods of controlling noise are being developed.

Hearing protection should not be used as an alternative to controlling noise by technical and organisational means.

Employers should consult with their employees or their representatives on the type of hearing protection to be used.

Employers are required to:

▶ provide employees with hearing protectors if they ask for it and their noise exposure is between the lower and upper exposure action values;

▶ provide employees with hearing protectors and make sure they use them properly when their noise exposure exceeds the upper exposure action values;

▶ identify hearing protection zones, i.e. areas where the use of hearing protection is compulsory, and mark them with signs if possible. Restrict access to hearing protection zones where this is practicable and the noise exposure justifies it;

▶ provide employees with training and information on how to use and care for the hearing protectors;

▶ ensure that the hearing protectors are properly used and maintained.

13.23.6 Maintenance and use of equipment – Regulation 8

Employers need to make sure that hearing protection works effectively and to:

▶ check that all equipment provided in compliance with the regulations remains in good, clean condition and that there are no unofficial modifications;

▶ check that hearing protection is fully and properly used (except where it is provided for employees who are exposed at or above the lower exposure but below the upper exposure level). This is likely to mean that an employer needs to:

▷ put someone in authority in overall charge of issuing it and making sure replacements are readily available;

▷ carry out spot checks to see that the rules are being followed and that hearing protection is being used properly;

▷ ensure all managers and supervisors set a good example and wear hearing protection at all times when in hearing protection zones;

▷ ensure only people who need to be there enter hearing protection zones and do not stay longer than they need to.

13.23.7 Health surveillance – Regulation 9

Employers must provide health surveillance (hearing checks) for all their employees who are likely to be regularly exposed above the upper exposure action values, or are at risk for any reason – for example they already suffer from hearing loss or are particularly sensitive to damage.

The purpose of health surveillance is to:

▶ warn when employees might be suffering from early signs of hearing damage;

▶ give employers an opportunity to do something to prevent the damage getting worse;

▶ check that control measures are working.

The trade union safety representative, or employee representative and the employees concerned, should be consulted before introducing health surveillance. It is important that employees understand that the aim of health surveillance is to protect their hearing. Employers will need their understanding and cooperation if health surveillance is to be effective.

Health surveillance for hearing damage usually means:

▶ regular hearing checks in controlled conditions;

▶ telling employees about the results of their hearing checks (required after reasonable notice by employee);

▶ keeping health records;

▶ ensuring employees are examined by a doctor where hearing damage is identified.

If the doctor considers the hearing damage is likely to be the result of exposure, the employer must:

- ensure that the employee is informed by a suitably qualified person;
- review the risk assessment and control measures;
- consider assigning the employee to alternative work;
- ensure continued surveillance.

13.23.8 Information, instruction and training – Regulation 10

Where employees are exposed to noise which is likely to be at or above the lower exposure level, suitable and sufficient information, instruction and training must be provided and kept up to date. This includes the nature of the risks, compliance action taken, significant findings of the risk assessment, hearing protection, how to detect and report signs of hearing damage, entitlement to health surveillance, safe working practices and collective results of health surveillance.

13.23.9 Further information

Control of Noise at Work Regulations 2005 (SI No. 1643) www.legislation.gov.uk/uksi/2005/1643/contents/made

Controlling Noise at Work. The Control of Noise at Work Regulations 2005, L108 (second edition), 2005, HSE, ISBN 978017661640 www.hse.gov.uk/pubns/books/l108.htm

Noise – Don't lose your hearing INDG363 (rev 2), 2012, HSE, ISBN 9780717665105 www.hse.gov.uk/pubns/indg363.htm

Noise at Work: a brief guide to controlling the risks INDG362 (rev 2), 2012, HSE, ISBN 9780717664825 www.hse.gov.uk/pubns/indg362.htm

Sound Advice: Control of noise at work in music and entertainment, HSG260, 2008, HSE, ISBN 9780717663071 www.hse.gov.uk/pubns/books/hsg260.htm

13.24 Personal Protective Equipment at Work Regulations 1992 as amended in 2002 and 2013

13.24.1 Introduction

The effect of the PPE at Work Regulations is to ensure that certain basic duties governing the provision and use of PPE apply to all situations where PPE is required. The Regulations follow sound principles for the effective and economical use of PPE, which all employers should follow.

PPE, as defined, includes all equipment (including clothing affording protection against the weather) which is intended to be worn or held by a person at work and which protects them against one or more risks to their health and safety. Waterproof, weatherproof or insulated clothing is covered only if its use is necessary to protect against adverse climatic conditions.

Ordinary working clothes and uniforms, which do not specifically protect against risks to health and safety, and protective equipment worn in sports competitions, are not covered.

Where there is overlap in the duties in these Regulations and those covering lead, ionising radiations, asbestos, hazardous substances (COSHH) and noise, then the specific legislative requirements should prevail. For guidance on the use of PPE during the coronavirus pandemic, see https://www.hse.gov.uk/coronavirus/ppe-face-masks/index.htm.

13.24.2 Provision of PPE – Regulation 4

Every employer shall ensure that suitable PPE is provided to their employees who may be exposed to risks to their health and safety except where it has been adequately or more effectively controlled by other means. (Management Regulations require PPE to be the last choice in the principles of protection.)

PPE shall not be suitable unless:

- it is appropriate for the risks and the conditions of use including the period for which it is worn;
- it takes account of ergonomic requirements and the state of health of the wearer and the characteristics of the workstation of each person;
- it is capable of fitting the wearer correctly, by adjustments if necessary;
- it is, so far as is practicable, able to combat the risks without increasing overall risks;
- it complies with UK legislation on design or manufacture, i.e. it has a CE marking.

Where it is necessary that PPE is hygienic and otherwise free of risk to health, PPE must be provided to a person solely for their individual use.

13.24.3 Compatibility – Regulation 5

Where more than one health and safety risk necessitates the wearing of multiple PPE simultaneously then they shall be compatible and remain effective.

13.24.4 Assessment – Regulation 6

Before choosing any PPE, employers must ensure that an assessment is made to determine whether the PPE is suitable.

The assessment shall include:

- assessing risks which have not been avoided by other means;
- a definition of the characteristics that PPE must have to be effective, taking into account any risks created by the PPE itself;
- a comparison of available PPE with the required characteristics;
- information on whether the PPE is compatible with any other PPE which is in use, and which an employee would be required to wear simultaneously.

13

The assessment should be reviewed if it is no longer valid or there have been significant changes. In simple cases it will not be necessary to record the assessment but, in more complex cases, written records should be made and kept readily available for future reference.

13.24.5 Maintenance – Regulation 7

Every employer (and self-employed person) shall ensure that any PPE provided is maintained, including being replaced and cleaned, in an efficient state, in efficient working order and in good repair.

The guide emphasises the need to set up an effective system of maintenance for PPE. This should be proportionate to the risks and appropriate to the particular PPE. It could include, where appropriate, cleaning, disinfection, examination, replacement, repair and testing. For example, mechanical fall arrestor equipment or sub-aqua breathing apparatus will require planned preventative maintenance with examination, testing and overhaul. Records should be kept of the maintenance work. Gloves may only require periodic inspection by the user as necessary, depending on their use.

Spare parts must be compatible and be the proper part, suitably CE marked where applicable. Manufacturers' maintenance schedules and instructions should be followed unless alternative schemes are agreed with the manufacturer or agent.

In some cases these requirements can be fulfilled by using disposable PPE which can be discarded after use or when their life has expired. Users should know when to discard and replace disposable PPE.

13.24.6 Accommodation – Regulation 8

When an employer or self-employed person has to provide PPE they must ensure that appropriate accommodation is provided to store it when not in use.

The type of accommodation will vary and may just be suitable hooks for special clothing and small portable cases for goggles. It should be separate from normal outer clothing storage arrangements and protect the PPE from contamination or deterioration.

13.24.7 Information, instruction and training – Regulation 9

Employers shall provide employees with adequate and appropriate information, instruction and training on:

▶ the risks which the PPE will avoid or limit;
▶ the purpose for which and the manner in which PPE should be used;
▶ any action required of the employee to maintain the PPE.

Employers must ensure that the information is kept available. The employer shall where appropriate and at suitable intervals, organise demonstrations of the wearing of PPE. The guidance suggests the training should include:

▶ an explanation of the risks and why PPE is needed;
▶ the operation, performance and limitations of the equipment;
▶ instructions on the selection, use and storage of PPE;
▶ problems that can affect PPE relating to other equipment, working conditions, defective equipment, hygiene factors and poor fit;
▶ the recognition of defects and how to report problems with PPE;
▶ practice in putting on, wearing and removing PPE;
▶ practice in user cleaning and maintenance;
▶ how to store safely.

13.24.8 Use and reporting of defects – Regulations 10 and 11

Every employer shall take all reasonable steps to ensure that PPE is properly used.

Every employee shall:

▶ use PPE provided in accordance with training and instructions;
▶ return it to the accommodation provided after use;
▶ report any loss or obvious defect.

13.24.9 Further information

A Short Guide to the Personal Protective Equipment at Work Regulations 1992, INDG174 (rev2), HSE Books, 2013, ISBN 9780717661411 www.hse.gov.uk/pubns/indg174.pdf

Hard hats: What you need to know as a busy builder. Construction Information Sheet CIS70, HSE, 2013 https://www.hse.gov.uk/pubns/cis70.pdf

Personal Protective Equipment (EC Directive) Regulations 2002. SI 2002 No. 1144, www.legislation.gov.uk/uksi/2002/1144/contents/made

Personal Protective Equipment at Work. Personal Protective Equipment at Work Regulations 1992. Guidance on Regulations. L25 (second edition), HSE Books, 2005, ISBN 9780717665976 www.hse.gov.uk/pubns/books/l25.htm

Selecting Protective Gloves for Work with Chemicals. Guidance for Employers and Health and Safety Specialists INDG330, HSE, 2000. https://www.hse.gov.uk/pubns/indg330.pdf

Sun Protection Advice for Employers of Outdoor Workers INDG337, HSE, 2001, ISBN 9780717619825 www.hse.gov.uk/pubns/indg337.pdf

13.25 Provision and Use of Work Equipment Regulations 1998 (except Part IV) as amended in 2002 and 2013

13.25.1 Introduction

The PUWER Regulations 1998 are made under the HSW Act and their primary aim is to ensure that work equipment is used without risks to health and safety, regardless of its age, condition or origin. The requirements of PUWER that are relevant to woodworking machinery are set out in the Safe use of woodworking machinery Approved Code of Practice. PUWER has specific requirements for risk assessment which are covered under the Health and Safety Management Regulations 1999

Part IV of PUWER is concerned with power presses and is not part of the Certificate syllabus, and is therefore not covered in this summary.

13.25.2 Definitions

Work equipment means any machinery, appliance, apparatus, tool or installation for use at work.

Use in relation to work equipment means any activity involving work equipment and includes starting, stopping, programming, setting, transporting, repairing, modifying, maintaining, servicing and cleaning.

13.25.3 Duty-holders – Regulation 3

Under PUWER the following groups of people have duties placed on them:

▶ employers;
▶ the self-employed;
▶ people who have control of work equipment, for example plant hire companies.

In addition to all places of work, the Regulations apply to common parts of shared buildings, industrial estates and business parks; to temporary works sites including construction; to home working (but not to domestic work in a private household); to hotels, hostels and sheltered accommodation.

13.25.4 Suitability of work equipment – Regulation 4

Work equipment:

▶ has to be constructed or adapted so that it is suitable for its purpose;
▶ has to be selected with the conditions of use and the users' health and safety in mind;
▶ may only be used for operations for which, and under conditions for which, it is suitable.

This covers all types of use and conditions and must be considered for each particular use or condition. For example, scissors may be safer than knives with unprotected blades and should therefore be used for cutting operations where practicable; risks imposed by wet, hot or cold conditions must be considered.

13.25.5 Maintenance – Regulation 5

The Regulation sets out the general requirement to keep work equipment maintained in:

▶ an efficient state;
▶ efficient working order;
▶ good repair.

Compliance involves all three criteria. In addition, where there are maintenance logs for machinery, they must be kept up to date.

In many cases this will require routine and planned preventative maintenance of work equipment. When checks are made priority must be given to:

▶ safety;
▶ operating efficiency and performance;
▶ the equipment's general condition.

13.25.6 Inspection – Regulation 6

Where the safety of work equipment depends on the installation conditions, it must be inspected:

▶ after installation and before being put into service for the first time;
▶ after assembly at a new site or in a new location;
▶ to ensure that it has been installed correctly and is safe to operate.

Where work equipment is exposed to conditions causing deterioration which is liable to result in dangerous situations, it must be inspected:

▶ at suitable intervals;
▶ when exceptional circumstances occur.

Inspections must be determined and carried out by competent persons. An inspection will vary from a simple visual external inspection to a detailed comprehensive inspection which may include some dismantling and/or testing. However, the level of inspection would normally be less than that required for a thorough examination under, for example, LOLER, for certain items of lifting equipment.

Records of inspections must be kept with sufficient information to properly identify the equipment, its normal location, dates, faults found, action taken, to whom faults were reported, who carried out the inspection, when repairs were made, date of the next inspection.

When equipment leaves an employer's undertaking it must be accompanied by physical evidence that the last inspection has been carried out.

13.25.7 Specific risks – Regulation 7

Where the use of work equipment involves specific hazards, its use must be restricted to those persons given the specific task of using it, and repairs, etc. must be restricted to designated persons.

13

Designated persons must be properly trained to fulfil their designated task.

The ACoP requires that hazards must be controlled using a hierarchy of control measures, starting with elimination where this is possible, then considering hardware measures such as physical barriers and, lastly, software measures such as a safe system of work.

13.25.8 Information, instruction and training – Regulations 8 and 9

Persons who use work equipment must have adequate:

▶ health and safety information;
▶ where appropriate, written instructions about the use of the equipment;
▶ training for health and safety in methods which should be adopted when using the equipment, and for any hazards and precaution which should be taken to reduce risks.

Any persons who supervise the use of work equipment should also receive information, instruction and training. The training of young persons is especially important with the need for special risk assessments under the Management of Health and Safety at Work Regulations.

Health and safety training should take place within working hours.

13.25.9 Conformity with EU requirements – Regulation 10

The intention of this Regulation is to require that employers ensure that equipment, provided for use after 31 December 1992, conforms at all times with the relevant essential requirements in various European Directives made under Article 100A (now 95) of the Treaty of Rome. The requirements of PUWER 98 Regulations 11–19 and 22–29 only apply if the essential requirements do not apply to a particular piece of equipment.

However, PUWER Regulations 11–19 and 22–29 will apply if:

▶ they include requirements which were not included in the relevant product legislation;
▶ the relevant product legislation has not been complied with (e.g. the guards fitted on a machine when supplied were not adequate).

Employers using work equipment need to check that any new equipment has been made to the requirements of the relevant Directive, and has a CE marking, suitable instructions and a Certificate of Conformity.

The Machinery Directive was brought into UK law by the Supply of Machinery (Safety) Regulations 1992 as amended, which duplicate PUWER Regulations 11–19 and 22–29. On 29 December 2009 these regulations were replaced by the 2008 Regulations.

The employer still retains the duty to ensure that the equipment is safe to use.

In March 2020 the HSE made the following statement:

'The UK is committed to ensuring that only safe and compliant work equipment and machinery is placed on the market.

Following the UK's exit from the EU HSE's priority is to maintain an effective regulatory system, which ensures that work equipment and machinery is safe and compliant.

We will also respond to emerging risks and allow trade with the EU and the rest of the world that is as frictionless as possible.

The UK is not a Member State of the EU. However, until the end of the transition period market access, and therefore UK–EU trade, should continue on the same terms as before.

A transition period until 31 December 2020 is in place while the EU and UK negotiate new arrangements. The HSE will publish the latest news on work equipment and machinery regulations on their website.'

13.25.10 Dangerous parts of machinery – Regulation 11

Measures have to be taken which:

▶ prevent access to any dangerous part of machinery or to any rotating stock-bar;
▶ stop the movement of any dangerous part of machinery or rotating stock-bar before any person enters a danger zone.

The measures required follow the normal hierarchy and should be considered to the extent that they are practicable in each case. They consist of:

▶ the provision of fixed guards enclosing every dangerous part of machinery; then
▶ the provision of other guards or protection devices; then
▶ the provision of jigs, holders, push sticks or similar protection appliances used in conjunction with the machinery; then
▶ the provision of information, instruction, training and supervision as is necessary.

All guards and protection devices shall:

▶ be suitable for their purpose;
▶ be of good construction, sound material and adequate strength;
▶ be maintained in an efficient state, in efficient working order and in good repair;
▶ not give rise to increased risks to health and safety;
▶ not be easily bypassed or disabled;
▶ be situated at sufficient distance from the danger zone;

- not unduly restrict the view of the operating cycle of the machine where this is relevant;
- be so constructed or adapted that they allow operations necessary to fit or replace parts and for maintenance work, if possible without having to dismantle the guard or protection device.

13.25.11 Protection against specified hazards – Regulation 12

Exposure to health and safety risks from the following hazards must be prevented or adequately controlled:

- any article falling or being ejected from work equipment;
- rupture or disintegration of work equipment;
- work equipment catching fire or overheating;
- the unintended or premature discharge of any article, or of any gas, dust, liquid, vapour or other substance which is produced, used or stored in the work equipment;
- the unintended or premature explosion of the work equipment or any article or substance produced, used or stored in it.

13.25.12 High or very low temperature – Regulation 13

Work equipment and any article or substance produced, used or stored in work equipment which is at a high or very low temperature must have protection to prevent injury by burn, scald or sear.

This does not cover risks such as from radiant heat or glare.

Engineering methods of control such as insulation, doors, temperature control guards should be used where practicable but there are some cases, like cooker hot plates, where this is not possible.

13.25.13 Controls – Regulations 14–18

Where work equipment is provided with (Regulation 14):

- starting controls (including restarting after a stoppage); and
- controls which change speed, pressure or other operating condition which would affect health and safety;

it should not be possible to perform any operation except by a deliberate action on the control. This does not apply to the normal operating cycle of an automatic device.

Where appropriate, one or more readily accessible **stop controls** shall be provided to bring the work equipment to a safe condition in a safe manner (Regulation 15). They must:

- bring the work equipment to a complete stop where necessary;

- if necessary, switch off all sources of energy after stopping the functioning of the equipment;
- operate in priority to starting or operating controls.

Where appropriate, one or more readily accessible **emergency stop controls** (Regulation 16) must be provided unless it is not necessary:

- by the nature of the hazard;
- by the time taken for the stop controls to bring the equipment to a complete stop.

Emergency stop controls must have priority over stop controls. They should be provided where other safeguards are not adequate to prevent risk when something irregular happens. They should not be used as a substitute for safeguards or the normal method of stopping the equipment.

All controls for work equipment shall (Regulation 17):

- be clearly visible and identifiable including appropriate marking where necessary;
- not expose any person to a risk to their health and safety except where necessary.

Where appropriate, employers shall ensure that:

- controls are located in a safe place;
- systems of work are effective in preventing any person being in a danger zone when equipment is started;
- an audible, visible or other suitable warning is given whenever work equipment is about to start.

Persons in a danger zone as a result of starting or stopping equipment must have sufficient time and suitable means to avoid any risks.

Control systems (Regulation 18) must be safe and chosen so as to allow for failures, faults and constraints. They must:

- not create any increased risk to health and safety;
- not result in additional or increased risks when failure occurs;
- not impede the operation of any stop or emergency stop controls.

13.25.14 Isolation from sources of energy – Regulation 19

Work equipment must be provided with readily accessible and clearly identified means to isolate it from all sources of energy. Re-connection must not expose any person using the equipment to any risks.

The main purpose is to allow equipment to be made safe under particular circumstances, such as maintenance, when unsafe conditions occur, or when adverse conditions such as electrical equipment in a flammable atmosphere or wet conditions occur.

If isolation may cause a risk in itself, special precautions must be taken – for example a support for a hydraulic press tool which could fall under gravity if the system is isolated.

13

13.25.15 Stability – Regulation 20

Work equipment must be stabilised by clamping or otherwise as necessary to ensure health and safety.

Most machines used in a fixed position should be bolted or fastened so that they do not move or rock in use.

13.25.16 Lighting – Regulation 21

Suitable and sufficient lighting, taking account of the operations being carried out, must be provided where people use work equipment.

This will involve general lighting and in many cases local lighting, such as on a sewing machine. If access for maintenance is required regularly, permanent lighting should be considered.

13.25.17 Maintenance operations – Regulation 22

So far as is reasonably practicable, work equipment should be constructed or adapted to allow maintenance operations to be:

▶ conducted while it is shut down;
▶ undertaken without exposing people to risk;
▶ carried out after appropriate protection measures have been taken.

13.25.18 Markings and warnings – Regulations 23 and 24

Work equipment should have all appropriate **markings** for reasons of health and safety made in a clearly visible manner – for example, the maximum safe working load, stop and start controls, or the maximum rotation speed of an abrasive wheel.

Work equipment must incorporate **warnings** or **warning devices** as appropriate, which are unambiguous, easily perceived and easily understood.

They may be incorporated in systems of work, a notice, a flashing light or an audible warning. They are an active instruction or warning to take specific precautions or actions when a hazard exists.

13.25.19 Part III: mobile work equipment – Regulations 25–30

The main purpose of this section is to require additional precautions relating to work equipment while it is travelling from one location to another or where it does work while moving. If the equipment is designed primarily for travel on public roads, the Road Vehicles (Construction and Use) Regulations 1986 will normally be sufficient to comply with PUWER 98.

Mobile equipment would normally move on wheels, tracks, rollers, skids, etc. Mobile equipment may be self-propelled, towed or remote controlled and may incorporate attachments. Pedestrian-controlled work equipment such as lawn mowers is not covered by Part III.

Employees carried on mobile work equipment – Regulation 25

No employee may be carried on mobile work equipment unless:

▶ it is suitable for carrying persons;
▶ it incorporates features to reduce risks as low as is reasonably practicable, including risks from wheels and tracks.

Rolling over of mobile work equipment – Regulation 26

Where there is a risk of overturning it must be minimised by:

▶ stabilising the equipment;
▶ fitting a structure so that it only falls on its side (tip over protection structure such as the vertical mast of a fork-lift truck);
▶ fitting a structure which gives sufficient clearance for anyone being carried if it turns over further (rollover protective structures – ROPS);
▶ a device giving comparable protection;
▶ fitting a suitable restraining system for people if there is a risk of being crushed by rolling over.

This Regulation does not apply:

▶ to a fork-lift truck fitted with ROPS or vertical mast which prevents it turning on its side;
▶ where it would increase the overall risks;
▶ where it would not be reasonably practicable to operate equipment;
▶ to any equipment provided for use before 5 December 1998.

Overturning of fork-lift trucks – Regulation 27

Fork-lift trucks which carry an employee must be adapted or equipped to reduce the risk to safety from overturning to as low as is reasonably practicable.

Self-propelled work equipment – Regulation 28

Where self-propelled work equipment may involve risks while in motion, it shall have:

▶ facilities to prevent unauthorised starting;
▶ (with multiple rail-mounted equipment) facilities to minimise the consequences of collision;
▶ a device for braking and stopping;
▶ (where safety constraints so require) emergency facilities for braking and stopping, in the event of failure of the main facility, which have readily accessible or automatic controls;

▶ (where the driver's vision is inadequate) devices fitted to improve vision;

▶ (if used at night or in dark places) appropriate lighting fitted or otherwise it shall be made sufficiently safe for its use;

▶ if there is anything carried or towed that constitutes a fire hazard liable to endanger employees (particularly if escape is difficult such as from a tower crane), appropriate firefighting equipment carried, unless it is sufficiently close by.

Remote-controlled self-propelled work equipment – Regulation 29

Where remote-controlled self-propelled work equipment involves a risk while in motion it shall;

▶ stop automatically once it leaves its control range;

▶ have features or devices to guard against the risk of crushing or impact.

Drive shafts – Regulation 30

Where seizure of the drive shaft between mobile work equipment and its accessories or anything towed is likely to involve a risk to safety:

▶ the equipment must have means to prevent a seizure;

▶ where it cannot be avoided, every possible measure should be taken to avoid risks;

▶ the shaft should be safeguarded from contacting the ground and becoming soiled or damaged.

13.25.20 Part IV – power presses

Regulations 31–35 relate to power presses and are not included here as they are excluded from the NEBOSH Certificate syllabus. Details can be found in the Power Press ACoP.

13.25.21 Further information

Providing and using work equipment safely: A brief guide, INDG291 (rev1), 2013, HSE, ISBN 9780717624294 www.hse.gov.uk/pubns/indg291.pdf

Provision and Use of Work Equipment Regulations 1992, SI 1992 No. 2932. www.legislation.gov.uk/uksi/1998/2306/regulation/12/made

PUWER 1998, Provision and Use of Work Equipment Regulations 1998: Open Learning Guidance, HSE, 2008, ISBN 9780717662852 www.hse.gov.uk/pubns/books/puwer.htm

Safe use of woodworking machinery. Provision and Use of Work Equipment Regulations 1998 as applied to woodworking machinery. Approved Code of Practice and guidance L114 (second edition), 2014, HSE, ISBN 9780717666218 www.hse.gov.uk/pubns/books/l114.htm

Safe Use of Work Equipment, Provision and Use of Work Equipment Regulations 1998, Approved Code of Practice and Guidance, HSC L22 (fourth edition), 2014, HSE, ISBN 9780717666195 www.hse.gov.uk/pubns/books/l22.htm

The selection and management of mobile elevating work platforms, HSE information sheet, 2014, GEIS6. www.hse.gov.uk/pubns/geis6.htm

Using Work Equipment Safely INDG229(rev2), 2012, HSE, www.hse.gov.uk/pubns/indg229.pdf

13.26 The Reporting of Injuries, Diseases and Dangerous Occurrences Regulations 2013

13.26.1 Introduction

RIDDOR is the law that requires employers, and other people who are in control of work premises, to report and keep records of:

▶ work-related deaths

▶ certain serious injuries (reportable injuries);

▶ diagnosed cases of certain industrial diseases; and

▶ certain 'dangerous occurrences' (near-miss incidents).

In October 2013 the RIDDOR Regulations 2013 came into force.

The main changes simplify the reporting requirements in the following areas:

▶ The classification of 'major injuries' to workers is replaced with a shorter list of 'specified injuries'.

▶ The previous list of 47 types of industrial disease is replaced with eight categories of reportable work-related illness.

▶ Fewer types of 'dangerous occurrence' require reporting.

There are no significant changes to the reporting requirements for:

▶ Fatal accidents.

▶ Accidents to non-workers (members of the public).

▶ Accidents which result in the incapacitation of a worker for more than seven days.

Recording requirements remain broadly unchanged, including the requirement to record accidents resulting in the incapacitation of a worker for more than three days.

13.26.2 Definitions – Regulation 2

Accident includes an act of non-consensual physical violence done to a person at work.

Dangerous occurrence means an occurrence which arises out of or in connection with work and is of a class specified in Schedule 2 of the regulations.

13

Reportable incident means an incident giving rise to a notification or reporting requirement under these Regulations.

Disease includes a medical condition.

For the purposes of RIDDOR, an accident is a separate, identifiable, unintended incident that causes physical injury. This specifically includes acts of non-consensual violence to people at work.

Not all accidents need to be reported; a RIDDOR report is required only when:

▶ the accident is **work-related**; and
▶ it results in an injury of a type which is **reportable** (Regulations 4,5 and 6).

Examples of incidents that do and do not have to be reported are available at www.hse.gov.uk/riddor/do-i-need-to-report.htm

13.26.3 Responsible person

Reports under the RIDDOR Regulations should only be submitted by the **responsible person,** such as employers, the self-employed, and those in control of premises where incidents occur as follows:

Employers must report any work-related deaths, and certain work-related injuries, cases of disease, and near misses involving your employees wherever they are working.

Persons in control of premises, must report any work-related deaths, certain injuries to members of the public and self-employed people on their premises, and dangerous occurrences (some near-miss incidents) that occur on their premises.

Self-employed if working in someone else's work premises and suffer either a specified injury or an over-seven-day injury, then **the person in control of the premises** will be responsible for reporting. If a self-employed person has a reportable accident while working on their own premises or in domestic premises, or if a doctor tells them that they have a work-related disease or condition, then the **self-employed** person will need to report it.

Employment agencies are often the legal employer and are responsible for reporting like any other employer. In other cases such as self-employed people, the duty is on the host business. Agents should ensure that responsibilities under RIDDOR are clearly assigned.

There are special cases for mines, quarries, pipelines and wells (other than a water well).

13.26.4 Non-fatal injuries to workers – Regulation 4

Specified Injuries

Responsible persons must follow the reporting procedures where as a result of a work-related accident a person at work suffers a specified injury. The list

of 'specified injuries' in RIDDOR 2013 (Regulation 4) includes:

▶ a fracture, other than to fingers, thumbs and toes;
▶ amputation of an arm, hand, finger, thumb, leg, foot or toe;
▶ permanent loss of sight or reduction of sight;
▶ crush injuries leading to internal organ damage;
▶ any burns (covering more than 10% of the body, or damaging the eyes, respiratory system or other vital organs);
▶ scalpings (separation of skin from the head) which require hospital treatment;
▶ unconsciousness caused by head injury or asphyxia;
▶ any other injury arising from working in an enclosed space, which leads to hypothermia, heat-induced illness or requires resuscitation or admittance to hospital for more than 24 hours.

Over seven-day incapacitation of a worker

Accidents must be reported where they result in an employee or self-employed person being away from work, or unable to perform their normal work duties, for more than **seven** consecutive days as the result of their injury. This seven-day period does not include the day of the accident, but does include weekends and rest days. The responsible person must send a report in an approved manner, within 15 days of the accident.

13.26.5 Non-fatal injuries to non-workers – Regulation 5

Accidents to members of the public or others who are not at work must be reported if they result in an **injury** and the person is taken directly from the scene of the accident to hospital for treatment to that injury. Examinations and diagnostic tests do not constitute 'treatment' in such circumstances.

If the accident occurred at a hospital, the report only needs to be made if the injury is a 'specified injury'.

13.26.6 Work-related fatalities – Regulation 6

Where any person dies (up to a year after the date of the accident, regardless of whether it was reported at the time) as a result of a work-related accident or occupational exposure to a biological agent, the responsible person must report the death.

This does not apply to a self-employed person working on their own premises, although other reporting requirements to the police are likely to apply.

13.26.7 Dangerous occurrences

There are 27 categories of dangerous occurrences which must be reported by the responsible person following the procedure.

The list from **Schedule 2 Part 1 – General** is as follows:

Lifting machinery, etc.

1. The collapse, overturning, or the failure of any load-bearing part of lifts and lifting equipment.

Pressure systems

2. The failure of any closed vessel or of any associated pipework (other than a pipeline), forming part of a pressure system as defined by the Pressure Systems Safety Regulations 2000, where that failure could cause the death of any person.

Overhead electric lines

3. Any plant or equipment unintentionally coming into—
   (a) contact with an uninsulated overhead electric line in which the voltage exceeds 200 volts; or
   (b) close proximity with such an electric line, such that it causes an electrical discharge.

Electrical incidents causing explosion or fire

4. Any explosion or fire caused by an electrical short circuit or overload (including those resulting from accidental damage to the electrical plant) which either—
   (a) results in the stoppage of the plant involved for more than 24 hours; or
   (b) causes a significant risk of death.

Explosives

5. Any unintentional—
   (a) fire, explosion or ignition at a site where the manufacture or storage of explosives requires a licence or registration, as the case may be, under Regulation 9, 10 or 11 of the Manufacture and Storage of Explosives Regulations 2005; or
   (b) explosion or ignition of explosives (unless caused by the unintentional discharge of a weapon, where, apart from that unintentional discharge, the weapon and explosives functioned as they were designed to), except where a fail-safe device or safe system of work prevented any person being endangered as a result of the fire, explosion or ignition.

6. The misfire of explosives (other than at a mine or quarry, inside a well or involving a weapon) except where a fail-safe device or safe system of work prevented any person being endangered as a result of the misfire.

7. Any explosion, discharge or intentional fire or ignition which causes any injury to a person requiring first aid or medical treatment, other than at a mine or quarry.

8. (1) The projection of material beyond the boundary of the site on which the explosives are being used, or beyond the danger zone of the site, which caused or might have caused injury, except at a quarry.
   (2) In this paragraph, 'danger zone' means the area from which persons have been excluded or

forbidden to enter to avoid being endangered by any explosion or ignition of explosives.

9. The failure of shots to cause the intended extent of collapse or direction of fall of a structure in any demolition operation.

Biological agents

10. Any accident or incident which resulted, or could have resulted, in the release or escape of a biological agent likely to cause severe human infection or illness.

Malfunction of radiation generators and radiography

11. (1) The malfunction of—
    (a) a radiation generator or its ancillary equipment used in fixed or mobile industrial radiography, the irradiation of food or the processing of products by irradiation, which causes it to fail to de-energise at the end of the intended exposure period;
    (b) equipment used in fixed or mobile industrial radiography or gamma irradiation, which causes a radioactive source to fail to return to its safe position by the normal means at the end of the intended exposure period.
    (2) In this paragraph, 'radiation generator' means any electrical equipment emitting ionising radiation and containing components operating at a potential difference of more than 5kV.

Breathing apparatus

12. The malfunction of breathing apparatus—
    (a) where the malfunction causes a significant risk of personal injury to the user; or
    (b) during testing immediately prior to use, where the malfunction would have caused a significant risk to the health and safety of the user had it occurred during use, other than at a mine.

Diving operations

13. The failure, damaging or endangering of—
    (a) any life support equipment, including control panels, hoses and breathing apparatus; or
    (b) the dive platform, or any failure of the dive platform to remain on station, which causes a significant risk of personal injury to a diver.

14. The failure or endangering of any lifting equipment associated with a diving operation.

15. The trapping of a diver.

16. Any explosion in the vicinity of a diver.

17. Any uncontrolled ascent or any omitted decompression which causes a significant risk of personal injury to a diver.

Collapse of scaffolding

18. The complete or partial collapse (including falling, buckling or overturning) of—
    (a) a substantial part of any scaffold more than 5 metres in height;

501

(b) any supporting part of any slung or suspended scaffold which causes a working platform to fall (whether or not in use); or

(c) any part of any scaffold in circumstances such that there would be a significant risk of drowning to a person falling from the scaffold.

Train collisions

19. The collision of a train with any other train or vehicle, other than a collision reportable under Part 5 of this Schedule, which could have caused the death, or specified injury, of any person.

Wells

20. In relation to a well (other than a well sunk for the purpose of the abstraction of water)—
   (a) a blow-out (which includes any uncontrolled flow of well-fluids from a well);
   (b) the coming into operation of a blow-out prevention or diversion system to control flow of well-fluids where normal control procedures fail;
   (c) the detection of hydrogen sulphide at a well or in samples of well-fluids where the responsible person did not anticipate its presence in the reservoir drawn on by the well;
   (d) the taking of precautionary measures additional to any contained in the original drilling programme where a planned minimum separation distance between adjacent wells was not maintained; or
   (e) the mechanical failure of any part of a well whose purpose is to prevent or limit the effect of the unintentional release of fluids from a well or a reservoir being drawn on by a well, or whose failure would cause or contribute to such a release.

Pipelines or pipeline works

21. In relation to a pipeline or pipeline works—
   (a) any damage to, accidental or uncontrolled release from or inrush of anything into a pipeline;
   (b) the failure of any pipeline isolation device, associated equipment or system; or
   (c) the failure of equipment involved with pipeline works, which could cause personal injury to any person, or which results in the pipeline being shut down for more than 24 hours.

22. The unintentional change in position of a pipeline, or in the subsoil or seabed in the vicinity, which requires immediate attention to safeguard the pipeline's integrity or safety.

Part 2 Dangerous Occurences Reportable Except in Relation to an Offshore Workplace

Structural collapse

23. The unintentional collapse or partial collapse of—
   (a) any structure, which involves a fall of more than 5 tonnes of material; or

(b) any floor or wall of any place of work, arising from, or in connection with, ongoing construction work (including demolition, refurbishment and maintenance), whether above or below ground.

24. The unintentional collapse or partial collapse of any falsework.

Explosion or fire

25. Any unintentional explosion or fire in any plant or premises which results in the stoppage of that plant, or the suspension of normal work in those premises, for more than 24 hours.

Release of flammable liquids and gases

26. The sudden, unintentional and uncontrolled release—
   (a) inside a building—
      (i) of 100 kilograms or more of a flammable liquid;
      (ii) of 10 kilograms or more of a flammable liquid at a temperature above its normal boiling point;
      (iii) of 10 kilograms or more of a flammable gas; or
   (b) in the open air, of 500 kilograms or more of a flammable liquid or gas.

Hazardous escapes of substances

27. The unintentional release or escape of any substance which could cause personal injury to any person other than through the combustion of flammable liquids or gases.

Additional categories of dangerous occurrences apply to mines, quarries, relevant transport systems and offshore workplaces.

13.26.8 Reporting of cases of disease and occupational exposure – Regulations 8 and 9

Where a medical practitioner notifies the employer's responsible person that an employee suffers from a reportable work-related disease (i.e. receives a diagnosis), a completed disease report form (F2508A) should be sent to the enforcing authority. The full list is as follows:

(a) carpal tunnel syndrome, where the person's work involves regular use of percussive or vibrating tools;

(b) cramp in the hand or forearm, where the person's work involves prolonged periods of repetitive movement of the fingers, hand or arm;

(c) occupational dermatitis, where the person's work involves significant or regular exposure to a known skin sensitiser or irritant;

(d) hand-arm vibration syndrome, where the person's work involves regular use of percussive or vibrating tools, or the holding of materials which are subject to percussive processes, or processes causing vibration;

(e) occupational asthma, where the person's work involves significant or regular exposure to a known respiratory sensitiser; or

(f) tendonitis or tenosynovitis in the hand or forearm, where the person's work is physically demanding and involves frequent, repetitive movements.

For exposure to carcinogens, mutagens and biological agents:

(a) any cancer attributed to an occupational exposure to a known human carcinogen or mutagen (including ionising radiation); or

(b) any disease attributed to an occupational exposure to a biological agent.

Diseases offshore have separate criteria in Schedule 3 to the Regulations.

13.26.9 Gas incidents – Regulation 11

Distributors, fillers, importers and suppliers of flammable gas must report incidents where someone has died, lost consciousness, or been taken to hospital for treatment to an injury arising in connection with that gas. Such incidents should be reported using the online form.

Registered gas engineers (under the Gas Safe Register) must provide details of any gas appliances or fittings that they consider to be dangerous, to such an extent that people could die, lose consciousness or require hospital treatment. The danger could be due to the design, construction, installation, modification or servicing of that appliance or fitting, which could cause:

▶ an accidental leakage of gas;
▶ inadequate combustion of gas; or
▶ inadequate removal of products of the combustion of gas.

Unsafe gas appliances and fittings should be reported using the online form.

13.26.10 Recording and record-keeping – Regulation 12

A record of each incident reported (deaths, specified injuries, over 7-day incapacitations and dangerous occurrences) and each diagnosis received of a reportable disease or exposure must be kept at the place of business.

In addition records must be kept of **over 3-day** incapacitations.

Where an employer must keep an accident book under the Social Security (Claims and Payments) Regulations 1979, that record will be enough for the accident records. Records must be kept for at least 3 years.

13.26.11 Restrictions on reporting – Regulation 14

Reports are not required under RIDDOR in relation to the following:

▶ Accidents during medical or dental treatment, or during any examination carried out or supervised by a doctor or dentist.

▶ Accidents involving the movement of a vehicle on a public road (other than those associated with: loading or unloading operations; work alongside the road such as road maintenance; escapes of substances from the vehicle; and accidents involving trains).

▶ Accidents to members of the armed forces on duty.

▶ Reports are not required under RIDDOR where this would duplicate other similar reporting requirements as listed. These include reports required under:

▶ Nuclear Installations Act 1965
▶ Merchant Shipping Act 1988
▶ Ionising Radiations Regulations 1999
▶ Civil Aviation (Investigation of Military Air Accidents at Civil Aerodromes) Regulations 2005
▶ Civil Aviation (Investigation of Air Accidents and Incidents) Regulations 1996
▶ The Electricity Safety, Quality and Continuity Regulations 2002.

13.26.12 Reporting and recording procedures – Schedule 1

For most types of incident, including:

▶ accidents resulting in the **death** of any person;
▶ accidents resulting **in specified injuries** to workers;
▶ non-fatal accidents requiring hospital treatment to **non-workers**; and
▶ **dangerous occurrences**;

the responsible person must notify the enforcing authority without delay, in accordance with the reporting procedure (Schedule 1). This is most easily done by reporting online at: www.hse.gov.uk/riddor/report.htm#online.

Alternatively, **for fatal accidents and specified injuries only**, the responsible person can telephone the Incident Contact Centre on 0345 300 9923 (opening hours Monday to Friday 8.30 a.m. to 5 p.m.).

A report must be received within **10 days** of the incident.

For accidents resulting in the over seven-day incapacitation of a worker, the responsible person must notify the enforcing authority within **15 days** of the incident, using the appropriate online form F2508IE.

Cases of occupational disease, including those associated with exposure to carcinogens, mutagens or biological agents should be reported as soon as the responsible person receives a diagnosis, using the appropriate online form F2508AE at: https://extranet.hse.gov.uk/lfserver/external/F2508AE

There is no longer a paper form for RIDDOR reporting, since the online system is the preferred reporting mechanism. Should it be essential for the responsible person to submit a report by post, it should be sent to: RIDDOR Reports, Health and Safety Executive, Redgrave Court, Merton Road, Bootle, Merseyside, L20 7HS.

13

HSE has an out-of-hours duty officer. Circumstances where HSE may need to respond out of hours include:

▶ a work-related death or situation where there is a strong likelihood of death following an incident at, or connected with, work;

▶ a serious accident at a workplace so that HSE can gather details of physical evidence that would be lost with time; and

▶ following a major incident at a workplace where the severity of the incident, or the degree of public concern, requires an immediate public statement from either HSE or government ministers.

If the incident fits these descriptions ring the duty officer on 0151 922 9235. Further Information about contacting HSE out of hours can be found at www.hse.gov.uk/contact/outofhours.htm

13.26.13 Further information

Accident book BL510 (2018 edition) HSE 2018 ISBN 978 0 7176 6458 0 www.hse.gov.uk/pubns/books/accident-book.htm

HSE Website for RIDDOR www.hse.gov.uk/riddor/

Reporting accidents and incidents at work: A brief guide to RIDDOR INDG453(rev1), 2013, HSE www.hse.gov.uk/pubns/indg453-rev1.pdf

The Reporting of Injuries, Diseases and Dangerous Occurrences Regulations 2013, Si 2013, No. 1471, www.legislation.gov.uk/uksi/2013/1471/contents/made

13.27 Safety Representatives and Safety Committees Regulations 1977

These Regulations, made under the HSW ACT Section 2(4), prescribe the cases in which recognised trade unions may appoint safety representatives, specify the functions of such representatives and set out the obligations of employers towards them.

Employers' obligations to consult non-union employees are contained in the Health and Safety (Consultation with Employees) Regulations 1996.

13.27.1 Appointment – Regulation 3

A recognised trade union may appoint safety representatives from among employees in all cases where one or more employees are employed. When the employer is notified in writing the safety representatives have the functions set out in Regulation 4.

A person ceases to be a safety representative when:

▶ the appointment is terminated by the trade union;

▶ they resign;

▶ employment ceases.

A safety representative should have been with the employer for 2 years or have worked in similar employment for at least 2 years.

13.27.2 Functions – Regulation 4

These are functions and not duties. They include:

▶ representing employees in consultation with the employer;

▶ investigating potential hazards and dangerous occurrences;

▶ investigating the causes of accidents (*Note*: re accident book BI510 and Data Protection – if the injured person has ticked the box and signed the form the safety representative may see all details. If not the employer should conceal the injured person's identity and details when giving access to the safety representatives);

▶ investigating employee complaints relating to health, safety and welfare;

▶ making representations to the employer on health, safety and welfare matters;

▶ carrying out inspections;

▶ representing employees at the workplace in consultation with enforcing inspectors;

▶ receiving information;

▶ attending safety committee meetings.

Safety representatives must be afforded time off with pay to fulfil these functions and to undergo training.

13.27.3 Employers' duties – Regulation 4a

Every employer shall consult safety representatives in good time regarding:

▶ the introduction of any measure which may affect health and safety;

▶ the arrangements for appointing or nominating competent person(s) under the Management Regulations;

▶ any health and safety information required for employees;

▶ the planning and organising of any health and safety training for employees;

▶ the health and safety consequences of introducing new technology.

Employers must provide such facilities and assistance as safety representatives may reasonably require to carry out their functions.

13.27.4 Inspections – Regulations 5 and 6

Following reasonable notice in writing, safety representatives may inspect the workplace every quarter (or more frequently by agreement with the employer). They may inspect the workplace at any time, after consultation, when there have been substantial changes in the conditions of work or there is new information on workplace hazards published by the HSE.

Following an injury, disease or dangerous occurrence subject to Reporting of Injuries Diseases and Dangerous

Occurrences Regulations, and after notifying the employer, where it is reasonably practicable to do so, safety representatives may inspect the workplace if it is safe.

Employers must provide reasonable assistance and facilities, including provision for independent investigation and private discussion with employees. The employer may be present in the workplace during inspections.

13.27.5 Information – Regulation 7

Having given reasonable notice to the employer, safety representatives are entitled to inspect and take copies of any relevant statutory documents (except any health record of an identified person).

An exempt document is one:

▶ which could endanger national security;
▶ which could cause substantial commercial injury on the employer;
▶ that contravenes a prohibition;
▶ that relates to an individual without their consent;
▶ which has been obtained specifically for legal proceedings.

13.27.6 Safety committees – Regulation 9

When at least two safety representatives have requested in writing that a safety committee is set up, the employer has 3 months to comply. The employer must consult with the safety representatives and post a notice stating its composition and the workplaces covered by it, in a place where it can be easily read by employees. The guidance gives details on the composition and running of safety committees.

13.27.7 Complaints – Regulation 11

If the employer fails to permit safety representatives time off or fails to pay for time off, complaints can be made to an industrial tribunal within 3 months of the incident.

13.27.8 Further information

Consulting workers on health and safety: Safety representatives and safety Committees Regulations 1977 (as amended) and Health and Safety (Consultation with Employees) Regulations 1996 (as amended); Approved Codes of Practice and guidance. L146 (second edition), 2014, HSE, ISBN 9780717664610 www.hse.gov.uk/pubns/priced/l146.pdf

Consulting employees on health and safety. A brief guide to the law. INDG232(rev2), 2013, HSE, www.hse.gov.uk/pubns/indg232.pdf

Involving your workforce in health and safety: Good practice for all workplaces HSG263, HSE, 2008, ISBN 9780717662272 www.hse.gov.uk/pubns/books/hsg263.htm

13.28 Health and Safety (Safety Signs and Signals) Regulations 1996

13.28.1 Introduction

These Regulations came into force in April 1996. The results of the risk assessment made under the Management of Health and Safety at Work Regulations will have identified situations where there may be a residual risk when warnings or further information are necessary. If there is no significant risk there is no need to provide a sign.

13.28.2 Definitions – Regulation 2

'Safety sign' means a sign referring to a specific object, activity or situation and providing information or instruction about health and safety at work by means of a signboard, a safety colour, an illuminated sign, an acoustic sign, a verbal communication or a hand signal.

'Signboard' means a sign which provides information or instructions by a combination of geometric shape, colour and a symbol or pictogram and which is rendered visible by lighting of sufficient intensity.

'Hand signal' means a movement or position of the arms or hands or a combination, in coded form, for guiding persons who are carrying out manoeuvres which create a risk to the health and safety of people at work.

'Acoustic signal' means a coded sound signal which is released and transmitted by a device designed for that purpose, without the use of a human or artificial voice.

'Verbal communication' means a predetermined spoken message communicated by a human or artificial voice.

13.28.3 Provision and maintenance of safety signs – Regulation 4

The Regulations require employers to use and maintain a safety sign where there is a significant risk to health and safety that has not been avoided or controlled by other means, like engineering controls or safe systems of work, and where the use of a sign can help reduce the risk.

They apply to all workplaces and to all activities where people are employed, but exclude signs used in connection with transport or the supply and marketing of dangerous substances, products and equipment.

The Regulations require, where necessary, the use of road traffic signs in workplaces to regulate road traffic.

13.28.4 Information, instruction and training – Regulation 5

Every employer shall ensure that:

▶ comprehensible and relevant information on the measures to be taken in connection with safety signs is provided to each employee;
▶ each employee receives suitable and sufficient instruction and training in the meaning of safety signs.

13

13.28.5 Functions of colours, shapes and symbols in safety signs

Safety colours

Red

Red is a safety colour and must be used for any:
- ▶ prohibition sign concerning dangerous behaviour (e.g. the safety colour on a 'No-Smoking' sign). Prohibition signs must be round, with a black pictogram on a white background with red edging and a red diagonal line (top left, bottom right). The red part must take up at least 35% of the area of the sign;

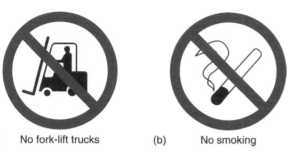

(a) No fork-lift trucks (b) No smoking

Figure 13.10 (a) No fork-lift trucks; (b) no smoking

- ▶ danger alarm concerning stop, shutdown, emergency cut-out devices, evacuate (e.g. the safety colour of an emergency stop button on equipment);
- ▶ firefighting equipment. Rectangular or square shape with white pictogram on a red background (red part must be at least 50% of the area of the sign).

(a) Hose Reel (b) Fire extinguisher

Figure 13.11 (a) Hose reel; (b) fire extinguisher

Red and white alternating stripes may be used for marking surface areas to show obstacles or dangerous locations.

Obstacles or dangerous locations

Yellow

Yellow (or amber) is a safety colour and must be used for any warning sign concerning the need to be careful, take precautions, examine or the like (e.g. the safety colour on hazard signs, such as for flammable material, electrical danger). Warning signs must be triangular, with a black pictogram on a yellow (or amber) background with black edging. The yellow (or amber) part must take up at least 50% of the area of the sign. Stripes should be of equal size and at 45 degrees.

(a) General danger (b) Explosive

Figure 13.12 (a) General danger; (b) explosive

Yellow and black alternating stripes may be used for marking surface areas to show obstacles or dangerous locations.

Obstacles or dangerous locations

Yellow may be used in continuous lines showing traffic routes.

Blue

Blue is a safety colour and must be used for any mandatory sign requiring specific behaviour or action (e.g. the safety colour on a 'Safety Helmet Must Be Worn' sign or a 'Pedestrians Must Use This Route' sign). Mandatory signs must be round, with a white pictogram on a blue background. The blue part must take up at least 50% of the area of the sign.

(a) Ear protection must be worn (b) Eye protection must be worn

Figure 13.13 (a) Ear protection must be worn; (b) eye protection must be worn

Green

Green is a safety colour and must be used for emergency escape signs (e.g. showing emergency doors, exits and routes) and first-aid signs (e.g. showing location of first-aid equipment and facilities). Escape and first-aid signs must be rectangular or square, with a white pictogram on a green background. The green part must take up at least 50% of the area of the sign. So long as the green takes up at least 50% of the area, it is sometimes permitted to use a green pictogram on a white background, for example where there is a green wall and the reversal provides a more effective sign than one with a green background and white border.

(a)   Means of Escape   (b)   First Aid

Figure 13.14 (a) Means of escape; (b) first aid

Other colours

White

White is not a safety colour but is used for: pictograms or other symbols on blue and green signs; in alternating red and white stripes to show obstacles or dangerous locations; in continuous lines showing traffic routes.

Black

Black is not a safety colour but is used for: pictograms or other symbols on yellow (or amber) signs and, except for fire signs, red signs; in alternating yellow and black stripes to show obstacles or dangerous locations.

Shapes

► Round signs must be used for: any prohibition (red) sign; mandatory (blue) sign.
► Triangular signs must be used for any warning (yellow or amber) sign.
► Square or rectangular signs must be used for any emergency escape sign and any first-aid sign.

13.28.6 Pictograms and other symbols

The meaning of a sign (other than verbal communication) must not rely on words. However, a sign may be supplemented with words to reinforce the message provided the words do not in fact distract from the message or create a danger.

A sign (other than verbal communication, acoustic signals or hand signals) should use a simple pictogram and/or other symbol (such as directional arrows, exclamation mark) to effectively communicate its message and so overcome language barriers.

Pictograms and symbols are included in the Regulations. Employee training is needed to understand the meaning of these as many are not inherently clear, some are meaningless to anyone who has not had their meaning explained and some can even be interpreted with their opposite meaning.

Pictograms and symbols included in the Regulations do not cover all situations for which graphic representation of a hazard or other detail may be needed. Any sign used for a situation not covered in the Regulations should include:

► the international symbol for general danger (exclamation mark!) if the sign is a warning sign and tests show that the sign is effective;
► in any other case a pictogram or symbol which has been tested and shown to be effective.

BS EN 7010:2012

These standards aim to bring about consistency in safety signage internationally. The date for adoption was January 2013, when it became a European Normative. Therefore, at that time, it replaced BS5499-5, the previous British Standard.

British Standards are not law. They are Codes of Practice, generally affecting only new products, as opposed to those previously produced. However, it is possible for standards to be given a type of legal status when they are referred to within legislation or government-issued guidance and Approved Codes of Practice.

The text of any words used to supplement a sign must convey the same meaning. For example, a round blue sign with a pictogram showing the white outline of a face with a solid white helmet on the head means 'Safety Helmet Must Be Worn' and so any text used must maintain the obligatory nature of the message.

13.28.7 Further information

The Health and Safety (Safety Signs and Signals) Regulations 1996, SI 1996/341. Stationery Office www.legislation.gov.uk/uksi/1996/341/contents/made

Safety Signs and Signals Guidance on the Regulations L64 (third edition), 2015, HSE, ISBN 9780717665983 www.hse.gov.uk/pubns/books/l64.htm

BS EN ISO 7010: 2012, Graphical symbols. Safety colours and safety signs. Registered safety signs British Standards Institute 2011 https://shop.bsigroup.com/SearchResults/?q=BS%20EN%20ISO%207010

13.29 The Supply of Machinery (Safety) Regulations 2008 as amended

13.29.1 Introduction

The Supply of Machinery (Safety) Regulations 2008 and the Supply of Machinery (Safety) (Amendment) Regulations 2011 implement Directives 2006/42/EC and 2009/127/EC on Machinery. Their scope extends to other products such as safety components, lifting

13

tackle and partly completed machinery, but excludes such items as domestic electrical machines and fairground equipment.

'The UK is committed to ensuring that only safe and compliant work equipment and machinery is placed on the market.

Following the UK's exit from the EU HSE's priority is to maintain an effective regulatory system, which ensures that work equipment and machinery is safe and compliant.

We will also respond to emerging risks and allow trade with the EU and the rest of the world that is as frictionless as possible.

The UK is not a Member State of the EU. However, until the end of the transition period market access, and therefore UK–EU trade, should continue on the same terms as before.

The latest on negotiations

A transition period until 31 December 2020 is in place while the EU and UK negotiate new arrangements. The HSE will publish the latest news on work equipment and machinery regulations [on our website] or join our email newsletter.'

The Regulations require that all machinery:

▶ is designed and constructed to be safe, meeting the essential health and safety requirements listed in the Regulations (these are supported by many harmonised standards)
▶ is CE marked
▶ is supplied with instructions in English
▶ has a Declaration of Conformity (or, in the case of partly completed machinery, a Declaration of Incorporation).

These Regulations apply to manufacturers or their authorised representatives.

In some cases, they apply to others – such as importers of non-CE-marked equipment from outside the EU, and those who design and construct machinery for their own use. However, the Regulations do not usually apply to intermediate suppliers of CE-marked machinery, which is covered by Section 6 of the HSW Act. (The non-application of the Regulations to intermediate suppliers may change in the next few years, when the Machinery Directive is brought into line with other EU legislation.) Both HSE and local trading standards enforce the provisions of these Regulations, depending on the field of use of the equipment (HSE leads where machinery is for use at work).

13.29.2 Main requirements for manufacturers

The Machinery Directives, and therefore the Supply of Machinery Regulations and amendments, require manufacturers of machinery (or their 'authorised representatives') to undertake the following activities

before they can place machinery on the European market or put it into service:

▶ carry out a detailed risk assessment by:
  ▷ identifying the health and safety hazards (trapping, crushing, electrical shock, dust or fumes, noise, vibration, etc.) that are likely to be present when the machinery is used. The essential health and safety requirements listed in the Regulations should be used by manufacturers as a guide to ensure all possible hazards have been considered, and that they have identified those relevant to the machine;
  ▷ assessing the likely level of risks involved;
▶ eliminate significant risks or, if that is not possible;
▶ provide safeguards (e.g. guarding dangerous parts of the machinery, providing noise enclosures, etc.) or, if that is not possible;
▶ provide information about any residual risks and place signs on the machinery to warn of risks that cannot be reduced in other ways (e.g. 'noisy machine' signs).

13.29.3 Suppliers requirements

If machinery is manufactured in the European Economic Area (EEA), suppliers should do all they can to ensure it is safe. If importing machinery from outside the EEA and it is CE marked with a Declaration of Conformity, suppliers will still need to follow the same checks as for machinery built inside the EEA. If it is not CE marked with a Declaration of Conformity, suppliers will need to make sure the machinery meets the requirements of the law in full as if they were the manufacturer. This means ensuring the machinery:

▶ meets the essential health and safety requirements;
▶ has (in certain cases) undergone conformity assessment by a notified body;
▶ has a technical file and Declaration of Conformity; and
▶ is CE marked.

Suppliers who are importing partly completed machinery should ensure it is supplied with instructions for its final incorporation with other machinery/equipment and a Declaration of Incorporation. Partly completed machinery should not be CE marked. The manufacturer of partly completed machinery should hold technical documentation in one of the EU languages, and must be able to produce it upon request to any relevant enforcing authority.

13.29.4 Further information

Buying New Machinery – A Short Guide to the Law INDG271(rev1), 2011, HSE, https://www.hse.gov.uk/pubns/indg271.pdf

Supplying New Machinery Advice to Suppliers INDG270, 2011, HSE, https://www.hse.gov.uk/pubns/indg270.pdf

The Supply of Machinery (Safety) Regulations 2008, OPSI: www.legislation.gov.uk/uksi/2008/1597/contents/made

13.30 Control of Vibration at Work Regulations 2005

13.30.1 Introduction

These Regulations came into force in July 2005 with some transitional arrangements for the exposure limits until 2010 (2014 for WBV the whole-body vibration exposure limit value for agriculture and forestry sectors).

The Regulations impose duties on employers to protect employees who may be exposed to risk from vibration at work, and other persons who may be affected by the work.

13.30.2 Interpretation – Regulation 2

'Daily exposure' means the quantity of mechanical vibration to which a worker is exposed during a working day, normalised to an 8-hour reference period, which takes account of the magnitude and duration of the vibration.

'Hand-arm vibration' (HAV) means mechanical vibration which is transmitted into the hands and arms during a work activity.

'Whole-body vibration' (WBV) means mechanical vibration which is transmitted into the body when seated or standing through the supporting surfaces, during a work activity or as described in Regulation 5(3)(f).

13.30.3 Application – Regulation 3

For work equipment first provided to employees for use prior to 6 July 2007 and where compliances with the exposure limit values are not possible, employers had until 2010 to comply and in the case of agriculture and forestry 2014 (for WBV).

However, action must be taken to use the latest technical advances and the organisational measures in accordance with Regulation 6(2).

13.30.4 Exposure limit values and action values – Regulation 4

HAV

(a) 8-hour daily exposure limit value is 5 m/s^2 A(8);
(b) 8-hour daily exposure action value is 2.5 m/s^2 A(8);
(c) daily exposure ascertained as set out in Schedule 2 Part I.

WBV

(a) 8-hour daily exposure limit value is 1.15 m/s^2 A(8);
(b) 8-hour daily exposure action value is 0.5 m/s^2 A(8);
(c) daily exposure ascertained as set out in Schedule 2 Part I.

13.30.5 Assessment of risk to health created by vibration at the workplace – Regulation 5

A suitable and sufficient risk assessment must be made where work liable to expose employees is carried on. The risk assessment must identify the measures which need to be taken to comply with these Regulations.

Assessment of daily exposure should be by means of:

▶ observation of specific working practices;
▶ reference to relevant work equipment vibration data;
▶ if necessary, measurement of the magnitude of vibration to which employees are exposed;
▶ likelihood of exposure at or above an exposure action value or above an exposure limit value.

The risk assessment shall include consideration of:

▶ the magnitude, type and duration of exposure, including any exposure to intermittent vibration or repeated shocks;
▶ the effects of exposure to vibration on employees whose health is at particular risk from such exposure;
▶ any effects of vibration on the workplace and work equipment, including the proper handling of controls, the reading of indicators, the stability of structures and the security of joints;
▶ any information provided by the manufacturers of work equipment;
▶ the availability of replacement equipment designed to reduce exposure to vibration;
▶ any extension of exposure at the workplace to WBV beyond normal working hours, including exposure in rest facilities supervised by the employer;
▶ specific working conditions such as low temperatures;
▶ appropriate information obtained from health surveillance including, where possible, published information.

The risk assessment shall be reviewed regularly and forthwith if:

▶ there is reason to suspect that the risk assessment is no longer valid; or
▶ there has been a significant change in the work to which the assessment relates;
▶ and where, as a result of the review, changes to the risk assessment are required, those changes shall be made.

The employer shall record:

▶ the significant findings of the risk assessment as soon as is practicable after the risk assessment is made or changed;
▶ the measures which they have taken and which they intend to take to meet the requirements of Regulations 6 and 8.

13

13.30.6 Elimination or control of exposure to vibration at the workplace – Regulation 6

Following the general principles of prevention in the Management Regulations, the employer shall ensure that risk from the exposure of his or her employees to vibration is either eliminated at source or reduced as low as reasonably practicable.

Where this is not reasonably practicable and the risk assessment indicates that an exposure action value is likely to be reached or exceeded, the employer shall reduce exposure to as low as reasonably practicable by establishing and implementing a programme of organisational and technical measures which is appropriate.

Consideration must be given to:

▶ other working methods which eliminate or reduce exposure to vibration;

▶ a choice of work equipment of appropriate ergonomic design which, taking account of the work to be done, produces the least possible vibration;

▶ the provision of auxiliary equipment which reduces the risk of injuries caused by vibration;

▶ appropriate maintenance programmes for work equipment, the workplace and workplace systems;

▶ the design and layout of workplaces, workstations and rest facilities;

▶ suitable and sufficient information and training for employees, such that work equipment may be used correctly and safely, in order to minimise their exposure to vibration;

▶ limitation of the duration and intensity of exposure to vibration;

▶ appropriate work schedules with adequate rest periods;

▶ the provision of clothing to protect employees from cold and damp.

Subject to implementation dates, the employer shall:

▶ ensure that no employees are exposed to vibration above an exposure limit value; or

▶ ensure if an exposure limit value is exceeded, the employer shall forthwith:

▷ take action to reduce exposure to vibration below the limit value;

▷ identify the reason for that limit being exceeded;

▷ modify the organisational and technical measures taken to prevent it being exceeded again.

This shall not apply where the exposure of an employee to vibration is usually below the exposure action value but varies markedly from time to time and may occasionally exceed the exposure limit value, provided that:

▶ any exposure to vibration averaged over 1 week is less than the exposure limit value;

▶ there is evidence to show that the risk from the actual pattern of exposure is less than the corresponding risk from constant exposure at the exposure limit value;

▶ risk is reduced to as low as reasonably practicable, taking into account the special circumstances;

▶ the employees concerned are subject to increased health surveillance, where such surveillance is appropriate.

Account must be taken of any employee whose health is likely to be particularly at risk from vibration.

13.30.7 Health surveillance – Regulation 7

If:

▶ the risk assessment indicates that there is a risk to the health of any employees who are, or are liable to be, exposed to vibration; or

▶ employees are exposed to vibration at or above an exposure action value;

then the employer shall ensure that such employees are under suitable health surveillance. Records must be kept, and employees given access to their own records and the enforcing authorities provided with copies, as may be required. A range of specified action is required if problems are found with the health surveillance results.

13.30.8 Information, instruction and training – Regulation 8

If:

▶ the risk assessment indicates that there is a risk to the health of employees who are, or who are liable to be, exposed to vibration; or

▶ employees are exposed to vibration at or above an exposure action value;

then the employer shall provide those employees and their representatives with suitable and sufficient information, instruction and training.

The information, instruction and training provided shall include:

▶ the organisational and technical measures taken in order to comply with the requirements of Regulation 6;

▶ the exposure limit values and action values;

▶ the significant findings of the risk assessment, including any measurements taken;

▶ the why and how to detect and report signs of injury;

▶ entitlement to appropriate health surveillance;

▶ safe working practices to minimise exposure to vibration;

▶ the collective results of any health surveillance undertaken in accordance with Regulation 7 in a form calculated to prevent those results from being identified as relating to a particular person.

The information, instruction and training required must be adapted to take account of significant changes in the type of work carried out or methods of work used by the employer, and cover all persons who carry out work in connection with the employer's duties under these Regulations.

13.30.9 Further information

Control Back-Pain Risks from Whole-Body Vibration Advice for Employers on the Control of Vibration at Work Regulations 2005 INDG242 (rev 1), HSE, 2005, ISBN 9780717661199 www.hse.gov.uk/pubns/indg242.pdf

Control of Vibration at Work Regulations 2005 SI No. 1093 www.legislation.gov.uk/uksi/2005/1093/contents/made

Drive Away Bad Backs Advice for Mobile Machine Operators and Drivers INDG404, HSE, 2005, ISBN 9780717661202 www.hse.gov.uk/pubns/indg404.pdf

Hand-Arm Vibration Advice for Employees Pocket Card INDG296 (rev 2), HSE, 2014, www.hse.gov.uk/pubns/indg296.pdf

Hand-arm vibration at work: A brief guide, INDG175(rev3), HSE, 2012, ISBN 9780717664887 www.hse.gov.uk/pubns/indg175.htm

Hand-Arm Vibration: Control of Vibration at Work Regulations L140, HSE, 2019, ISBN 9780717665655 www.hse.gov.uk/pubns/books/l140.htm

Vibration Solutions Practical Way to Reduce the Risk of Hand-Arm Vibration HSG170, HSE, 1997, ISBN 9780717609543 www.hse.gov.uk/pubns/books/hsg170.htm

Whole Body Vibration: The Control of Vibration at Work Regulations Guidance on Regulations L141, HSE, 2005, ISBN 9870717661268 www.hse.gov.uk/pubns/priced/l141.pdf

13.31 Workplace (Health, Safety and Welfare) Regulations 1992 as amended in 2002 and 2013

13.31.1 General

These Regulations were made to implement the European Directive on the minimum safety and health requirements for the workplace. A workplace for these purposes is defined very widely, to include any part of non-domestic premises to which people have access while at work, and any room, lobby, corridor, staircase or other means of access to or exit from them. The main exceptions to these rules are constructions sites, means of transport, mines and quarries and other mineral extraction sites. The stability and solidity of workplaces in a building were added in 2002. Ships in a British port not under the control of a master are now covered by the 2013 amendment to these Regulations.

Employers have a duty to ensure that workplaces under their control comply with the requirements of these Regulations. Any workplace and relevant equipment, devices and systems must be properly maintained. A maintenance system must be set up where appropriate, that is where a fault could result in a failure to comply and/or mechanical ventilation systems provided under Regulation 6.

The main requirements are summarised below. They are expressed in very general terms, and in each case it will be necessary to turn to the Approved Code of Practice associated with these Regulations for clarification of what is necessary to meet the objectives set.

13.31.2 Health: The Working Environment – Regulations 6–10

Ventilation

Ventilation must be effective in enclosed areas, and any plant used for this purpose must incorporate warning devices to signal breakdowns which might endanger health or safety.

A reasonable temperature

A reasonable temperature must be maintained during working hours and sufficient thermometers must be provided to enable people at work to determine the temperature in any workroom. The temperature should be comfortable without the need for special clothing. Special guidance is available for areas like food processing where it could be very hot or very cold.

Temperature should be at least 16°C, or where strenuous effort is involved, 13°C.

Lighting

Lighting must be suitable and sufficient. This should be natural lighting sfairp.

Emergency lighting

Emergency lighting shall be provided where persons are especially exposed to danger if artificial light fails. Lights should avoid glare and dazzle and should not themselves cause a hazard. They should not be obscured, and be properly maintained.

Workplaces, furniture and fittings

Workplaces, furniture and fittings should be kept sufficiently clean. Surfaces inside buildings shall be capable of being kept sufficiently clean.

Floors

Floors should not be slippery and wall surfaces should not increase fire risks.

Waste materials

Wastes should not be allowed to accumulate, except in suitable receptacles and should be kept free from offensive waste products and discharges.

511

Room dimensions

Room dimensions have to allow adequate unoccupied space to work in and to move freely; 11 m^3 minimum per person is required, excluding anything over 3 m high and furniture, etc.

Workstations

Workstations shall be suitable for any person in the workplace who is likely to work at that workstation and for any work that is likely to be done there.

Outside workstations

Outside workstations shall provide, sfairp, protection from adverse weather; provide adequate means of escape in emergencies; and ensure that no person is likely to slip or fall.

Seating

Seating shall be provided where work can or must be done sitting and shall be suitable for the person as well as the work. A footrest shall be provided where necessary.

13.31.3 Safety: Accident Prevention – Regulations 5, 12–19

Maintenance

The workplace and equipment, devices and systems shall be maintained (including cleaned as appropriate) in an efficient state, efficient working order and in good repair, and where appropriate subject to a system of maintenance. This generally means planned rather than breakdown maintenance. Systems include ventilation, emergency lighting, safety fences, window cleaning devices, and moving walkways.

Floors and traffic routes

Floors and the surface of traffic routes shall be suitably constructed for their intended purpose and free of slope and holes (unless fenced). They should not be uneven or slippery. The traffic routes should be of adequate width and height to allow people and vehicles to circulate safely with ease and they should be kept free of obstructions.

Additional precautions are necessary where pedestrians have to cross or share vehicle routes. Open sides of staircases should be fenced with an upper rail 900 mm or higher and a lower rail. Loading bays should have exits or refuges to avoid people getting crushed by vehicles.

Falls and falling objects

Now covered by Work at Height Regulations 2005 (see Section 13.32).

Tanks and pits

Where there is a risk of falling into a tank, pit or structure containing a dangerous substance that is likely to:

▶ scald or burn;
▶ be poisonous or corrosive;
▶ have an asphyxiating gas fume or vapour;
▶ have any granular or free-flowing substance likely to cause harm;

measures must be taken to securely fence or cover the tank, pit or structure.

Ladders and roofs

Now covered by Work at Height Regulations 2005 (see Section 13.32).

Glazing

Windows and transparent doors and partitions must be appropriately marked and protected against breakage.

Windows

Windows and skylights must open and close safely, and be arranged so that people may not fall out of them. They must be capable of being cleaned safely.

Traffic routes

Pedestrians and vehicles must be able to circulate safely. Separation should be provided between vehicle and people at doors, gateways and common routes. Workplaces should have protection from vehicles.

Doors and gates

Doors, gates and moving walkways have to be of sound construction and fitted with appropriate safety devices.

13.31.4 Welfare: provision of facilities – Regulations 20–25

Sanitary conveniences and washing facilities

Suitable and sufficient sanitary conveniences and washing facilities should be provided at readily accessible places. The facilities must be kept clean, adequately ventilated and lit. Washing facilities should have running hot and cold or warm water, soap and clean towels or other method of cleaning or drying. If necessary, showers should be provided. Men and women should have separate facilities unless each facility is in a separate room with a lockable door and is for use by only one person at a time.

Drinking water

An adequate supply of wholesome drinking water, with an upward drinking jet or suitable cups, should be provided. Water should only be supplied in refillable enclosed containers where it cannot be obtained directly from a mains supply.

Accommodation for clothing and facilities for changing

Adequate, suitable and secure space should be provided to store workers' own clothing and special clothing. The facilities should allow for drying clothing. Changing facilities should also be provided for workers who change into special work clothing.

Facilities to rest and to eat meals

Suitable and sufficient, readily accessible, rest facilities should be provided. Arrangements should include suitable facilities to eat meals; adequate seats with backrests and tables; and means of heating food (unless hot food is available nearby) and making hot drinks.

Canteens or restaurants

Canteens and restaurants may be used as rest facilities, provided there is no obligation to purchase food.

Suitable rest facilities

Suitable rest facilities should be provided for pregnant women and nursing mothers. They should be near sanitary facilities and, where necessary, include the facility to lie down.

Non-smokers

This is now covered by the smoke-free legislation which bans smoking in largely enclosed or enclosed work or public places. See Section 15.33.14.

13.31.5 Further information

How to Deal with Sick Building Syndrome: Guidance for Employers, Building Owners and Building Managers HSG132, HSE, 1995, ISBN 9780717608614 www.hse.gov.uk/pubns/books/hsg132.htm

Lighting at Work HSG38, HSE, 1997, ISBN 9780717612321 www.hse.gov.uk/pubns/books/hsg38.htm

Seating at Work HSG57, HSE, 1997, ISBN 9780717612314 www.hse.gov.uk/pubns/books/hsg57.htm

The Workplace (Health, Safety and Welfare) Regulations 1992, www.legislation.gov.uk/uksi/1992/3004/made/data.pdf

Workplace Health, Safety and Welfare. Workplace (Health, Safety and Welfare) Regulations 1992 Approved Code of Practice and Guidance L24 (second edition), 2013, HSE Books ISBN 9780717665839 www.hse.gov.uk/pubns/priced/124.pdf

Workplace Health, Safety and Welfare A Short Guide, INDG244 (rev 2), 2007, HSE Books, ISBN 9780717662777 www.hse.gov.uk/pubns/indg244.pdf

13.32 Work at Height Regulations 2005 as amended in 2007

13.32.1 Introduction

These Regulations bring together all current requirements on work at height into one goal-based set of Regulations. The Work at Height (Amendment) Regulations 2007, which came into force on 6 April 2007, apply to those who work at height providing instruction or leadership to one or more people engaged in caving or climbing by way of sport, recreation, team building or similar activities in Great Britain.

The Regulations require a risk assessment for all work conducted at height and arrangements to be put in place for:

▶ eliminating or minimising risks from working at height;
▶ safe systems of work for organising and performing work at height;
▶ safe systems for selecting suitable work equipment to perform work at height;
▶ safe systems for protecting people from the consequences of work at height.

13.32.2 Definitions – Regulation 2

'Work at height' means:

(a) work in any place, including a place at or below ground level;
(b) obtaining access to or egress from such place while at work except by a staircase in a permanent workplace where, if measures required by these Regulations were not taken, a person could fall a distance liable to cause personal injury.

'Working platform':

(a) means any platform used as a place of work or as a means of access to or egress from a place of work;
(b) includes any scaffold, suspended scaffold, cradle, mobile platform, trestle, gangway, gantry and stairway which is so used.

13.32.3 Organisation, planning and competence – Regulations 4 and 5

Work at height must be properly planned, appropriately supervised and carried out in a manner which is, sfairp, safe. The selection of appropriate work equipment is included in the planning. Work must not be carried out if the weather conditions would jeopardise safety or health (this does not apply where members of the police, fire, ambulance or other emergency services are acting in an emergency).

13

All people involved in work at height activity including planning, organising and supervising must be competent for such work, or if being trained, under competent supervision.

13.32.4 Avoidance of risk – Regulation 6

A risk assessment carried out under the Management Regulations must be taken into account when identifying the measures required by these Regulations. Work at height should be avoided if there are reasonably practicable alternatives.

Where work at height is carried out employers must take suitable and sufficient measures to prevent persons falling a distance liable to cause personal injury. The measures include:

(a) ensuring work is carried out:
  (i) from an existing workplace;
  (ii) using existing means of access and egress that comply with Schedule 1 of the Regulations (assuming it is safe and ergonomic to do so).

(b) where this is not reasonably practicable, providing work equipment (sfairp) for preventing a fall occurring.

Employers must take steps to minimise the distance and the consequences of a fall, if it is not prevented. Where the distance cannot be minimised (sfairp) the consequence of a fall must be minimised and additional training, instruction and other additional suitable and sufficient measures must be adopted to prevent a person falling a distance liable to cause personal injury.

13.32.5 General principles for selection of work equipment – Regulation 7

Collective protection measures must have priority over personal protection measures.

Employers must take account of:

▶ working conditions and the risks to the safety of persons at the place where the work equipment is to be used;
▶ the distance to be negotiated for access and egress;
▶ distance and consequences of a potential fall;

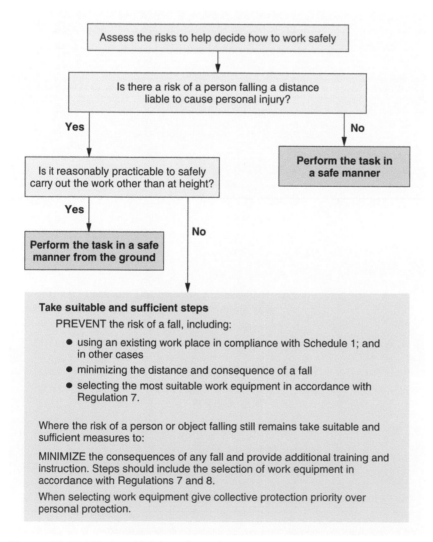

Figure 13.15 Work at Height – flow chart

- duration and frequency of use;
- need for evacuation and rescue in an emergency;
- additional risks of using, installing and removing the work equipment used or evacuation and rescue from it;
- other provisions of the Regulations.

Work equipment must:

- be appropriate for the nature of the work and the foreseeable loadings;
- allow passage without risk;
- be the most suitable work equipment having regard in particular to Regulation 6.

13.32.6 Requirements for particular work equipment – Regulation 8

Regulation 8 requires that various pieces of work equipment comply with the schedules to the Regulations as follows:

(a) A guard-rail, toe-board, barrier or similar means of protection comply with Schedule 2.

Schedule 2 in summary:

- they must be suitable, of sufficient dimensions, of sufficient strength and rigidity;
- be so placed, secured and used to prevent accidental displacement;
- prevent fall of persons and materials;
- supporting structure to be of sufficient strength and suitable for the purpose;
- no lateral opening save at point of access to a ladder or stairway where an opening is necessary and:
  - ▷ must be in place except for a time to gain access perform a particular task and then replaced;
  - ▷ compensatory safety measures must be provided if protection removed temporarily.

(b) A working platform, comply with Schedule 3 Part 1 and, in addition, where scaffolding is provided, Schedule 3 Part 2.

Schedule 3 Part 1 in summary:

Supporting structure must:

- be suitable, of sufficient strength and rigidity;
- if wheeled, be prevented, by appropriate device, from moving during work;
- not slip and have secure attachment;
- be stable while being erected, used, modified or dismantled.

Working platform to:

- be suitable and strong enough;
- have no accidental displacement of components;
- remain stable during dismantling;
- be dismantled so as to prevent accidental displacement;
- be of sufficient dimensions for safe passage and use;
- have a suitable surface and be constructed to prevent people falling through;

- have a suitable surface and be constructed to prevent material or objects falling through unless measures have been taken to protect other persons from falling objects;
- be erected, used and maintained so that risk of slipping and tripping is prevented and no person can be caught between working platform and adjacent structure;
- be not loaded so as to give risk of collapse or deformation.

Schedule 3 Part 2 additional for scaffolds:

- strength and stability calculations required unless they are available already or scaffold is assembled in conformity with a generally recognised standard configuration;
- an assembly, use and dismantling plan must be drawn up. Could be standard plan with supplements;
- copy of the plan and instructions to be available for persons doing the work;
- dimensions, form and layout to be suitable;
- when not available for use to be marked with warning signs and physical barrier preventing access;
- must be assembled, dismantled or significantly altered under supervision of competent person.

(c) a net, airbag or other collective safeguard for arresting falls which is not part of a personal fall protection system, comply with Schedule 4.

(d) any personal fall protection system to comply with Part 1 of Schedule 5.

- in the case of a work positioning system, comply with Part 2 of Schedule 5;
- in the case of rope access and positioning techniques, comply with Part 3 of Schedule 5;
- in the case of a fall arrest system, comply with Part 4 of Schedule 5;
- in the case of a work restraint system, comply with Part 5 of Schedule 5.

(e) a ladder to comply with Schedule 6.

Schedule 6 in summary:

- only to be used for work at height if the risk assessment demonstrates that the use of more suitable equipment is not justified because of the low risk and the short duration of use or existing features on site that cannot be altered;
- surface on which ladder rests must be stable, firm, of sufficient strength and suitable composition so its rungs remain horizontal;
- must be positioned to ensure stability;
- suspended ladder attached firmly without swing (except flexible ladder);
- portable ladder to be prevented from slipping by securing the stiles near their upper and lower ends and using an anti-slip device or by any other equivalent measures;
- when used for access must protrude above place of landing unless other firm handholds provided;

▶ interlocking or extension pieces must not move, relative to each other, during use;

▶ mobile ladder to be prevented from moving before being stepped on;

▶ where a ladder or run of ladders rises a vertical distance of 9 metres or more above its base, there shall, where reasonably practicable, be provided at suitable intervals sufficient safe landing areas or rest platforms;

▶ must be used so that a secure foothold and handhold are always available;

▶ must be used so that user can maintain a secure handhold while carrying a load (exceptions for using stepladders for low-risk, short-duration work).

13.32.7 Fragile surfaces – Regulation 9

Employers must take steps to prevent people falling through any fragile surface. Steps to be taken include:

▶ avoid, sfairp, passing across or near, or working on or near, a fragile surface;

▶ where this is not reasonably practicable:
  ▷ provide suitable and sufficient platforms, covering, guard-rails or other similar means of support or protection and use them;
  ▷ provide suitable and sufficient guard-rails to prevent persons falling through fragile surfaces;
  ▷ where a risk of falling remains, take suitable and sufficient measures to minimise the distances and consequences of a fall.

▶ provide prominent warning signs at approach to the fragile surface, or make people aware of the fragile surface if not reasonably practicable;

▶ where the risk of falling remains despite the precautions a suitable and sufficient, fall arrest system must be provided.

13.32.8 Falling objects and danger areas – Regulations 10 and 11

Suitable and sufficient steps must be taken:

▶ to prevent the fall of any material or object;

▶ if reasonably practicable, to prevent the fall of any material or object, prevent persons being struck by falling objects or material (if liable to cause personal injury) by for example providing covered walkways or fan scaffolds;

▶ to prevent material or objects being thrown or tipped from height where it is likely to cause injury;

▶ to store materials and objects in such a way as to prevent risk to any person arising from collapse, overturning or unintended movement;

▶ where danger of being struck exists, unauthorised persons must be kept out of the area by suitable devices and the area clearly indicated;

▶ to store materials and objects so as to prevent them collapsing, overturning or move in a way that could be a risk to people.

13.32.9 Inspection of work equipment – Regulation 12

This Regulation applies only to work equipment to which Regulation 8 and Schedules 2–6 apply. The requirements include the following:

▶ where work equipment used for work at height depends for safety on how it is installed or assembled, the employer must ensure that it is not used after installation or assembly in any position unless it has been inspected in that place;

▶ where work equipment is exposed to conditions causing deterioration which is liable to result in dangerous situations, it must also be inspected at suitable intervals and each time that exceptional circumstances which are liable to jeopardise the safety of the work equipment have occurred, for example a severe storm;

▶ in addition, a working platform used for construction and from which a person could fall more than 2 m has to be inspected in that position (mobile equipment on the site) before use and within the last 7 days. The particulars for this inspection are set out in Schedule 7;

▶ no work equipment (lifting equipment is covered under LOLER) may either leave the undertaking or be obtained from another undertaking without evidence of in-date inspection;

▶ results of inspections must be recorded and retained until the next due inspection is recorded. Reports must be provided within 24 hours and kept at the site until construction work is complete and thereafter at the office for 3 months.

13.32.10 Inspection of places of work at height – Regulation 13

The surface and every parapot, permanent rail or other such fall protection measure of every place of work at height must be checked on each occasion before the place is used.

13.32.11 Duties of persons at work – Regulation 12

Every person must, where working under the control of another person, report to that person any activity or defect relating to work at height which they know is likely to endanger themselves or others.

Work equipment must be used in accordance with training and instructions.

13.32.12 Further information

A Tool Box Talk on Leaning Ladders and Stepladders Safety INDG403, HSE, 2011, ISBN 978 0 7176 6106 0 www.hse.gov.uk/pubns/indg403.pdf

Health and Safety in Roof Work HSG33, 2012, HSE, ISBN 9780717665273 www.hse.gov.uk/pubns/priced/hsg33.pdf

Safe Use of Ladders and Stepladders: An Employers' Guide INDG455, 2014, HSE, www.hse.gov.uk/pubns/indg455.pdf

Safe use of ladders and step ladders – A brief guide, 2014 HSE, ISBN 9780717665181 www.hse.gov.uk/pubns/indg455.htm

The Work at Height (Amendment) Regulations 2007, SI 2007 No. 114, www.legislation.gov.uk/uksi/2007/114/contents/made

The Work at Height Regulations 2005, SI 2005/735, OPSI www.legislation.gov.uk/uksi/2005/735/contents/made

The Work at Height Regulations 2005 (as amended). A Brief Guide INDG401 (rev 2), 2014, HSE, www.hse.gov.uk/pubns/indg401.pdf

The Selection and Management of Mobile Elevating Work Platforms GEIS 6, HSE 2014, Web only. https://www.hse.gov.uk/pubns/geis6.pdf

Useful website – HSE's Work at Height: www.hse.gov.uk/work-at-height/index.htm

13.33 Other relevant legislation in brief

The following Regulations are included with brief summaries.

13.33.1 Building Regulations 2010 – Approved Documents B and M

These Regulations are not in the National General Certificate but are included in the National Fire Certificate and could be useful reference for Safety Practitioners.

The Building Regulations 2010 Fire Safety – Approved Document B Volumes 1 and 2

General introduction: Fire safety

Scope:

1. Approved Document B (Fire safety) has been published in two volumes. Volume 1 deals solely with dwelling houses (see Appendix E and Building Regulation 2(1)), while Volume 2 deals with all other types of building covered by the Building Regulations.

   Where very large (over 18 m in height) or unusual dwellinghouses are proposed some of the guidance in Volume 2 may be needed to supplement that given by Volume 1.

   Arrangement of Sections

2. The functional requirements B1 to B5 of Schedule 1 of the Building Regulations are dealt with separately in one or more Sections. The requirement is reproduced at the start of the relevant Sections, followed by an introduction to the subject.

3. The provisions set out in this document deal with different aspects of fire safety, with the following aims:

   B1: To ensure satisfactory provision of means of giving an alarm of fire and a satisfactory standard of means of escape for persons in the event of fire in a building.

   B2: To ensure fire spread over the internal linings of buildings is inhibited.

   B3: To ensure the stability of buildings in the event of fire; to ensure that there is a sufficient degree of fire separation within buildings and between adjoining buildings; to provide automatic fire suppression where necessary; and to inhibit the unseen spread of fire and smoke in concealed spaces in buildings.

   B4: To ensure external walls and roofs have adequate resistance to the spread of fire over the external envelope, and that spread of fire from one building to another is restricted.

   B5: To ensure satisfactory access for fire appliances to buildings and the provision of facilities in buildings to assist firefighters in the saving of life of people in and around buildings.

4. Whilst guidance appropriate to each of these aspects is set out separately in this document, many of the provisions are closely interlinked. For example, there is a close link between the provisions for means of escape (B1) and those for the control of fire growth (B2), fire containment and/or suppression (B3) and facilities for the fire and rescue service (B5). Similarly, there are links between B3 and the provisions for controlling external fire spread (B4), and between B3 and B5. Interaction between these different requirements should be recognised where variations in the standard of provision are being considered. A higher standard under one of the requirements may be of benefit in respect of one or more of the other requirements. The guidance in the document as a whole should be considered as a package aimed at achieving an acceptable standard of fire safety.

   Appendices: provisions common to more than one of Part B's requirements

5. Guidance on matters that refer to more than one of the Sections is in a series of Appendices, covering the following subjects:

   Appendix A – fire performance of materials, products and structures
   Appendix B – provisions regarding fire doors
   Appendix C – methods of measurement
   Appendix D – a classification of purpose groups
   Appendix E – definitions
   Appendix F – Standards and other publications referred to.

Fire performance of materials, products and structures

6. Much of the guidance throughout this document is given in terms of performance in relation to standard fire test methods. Details are drawn together in Appendix A to which reference is made where appropriate. In the case of fire protection systems, reference is made to standards for system design and installation. Standards referred to are listed in Appendix F.

Fire doors

7. Guidance in respect of fire doors is set out in Appendix 8.

Methods of measurement

8. Some form of measurement is an integral part of much of the guidance in this document and methods are set out in Appendix C.

Purpose groups

9. Much of the guidance in this document is related to the use of the building. The use classifications are termed purpose groups, and they are described in Appendix D. This document deals only with buildings in Purpose Groups 1b and 1c.

Definitions

10. The definitions are given in Appendix E.

The Building Regulations 2010 Access to and use of buildings – Approved Document M

The requirements of Part M will be met by making reasonable provision to ensure that buildings are accessible and usable. People, regardless of disability, age or gender, should be able to:

(a) gain access to buildings and to gain access within buildings and use their facilities, both as visitors and as people who live or work in them;

(b) use sanitary conveniences in the principal storey of a new dwelling.

The provisions are expected to enable occupants with disabilities to cope better with reducing mobility and to 'stay put' longer in their own homes. The provisions are not necessarily expected to facilitate fully independent living for all people with disabilities.

Where the requirements apply application of Part M

1. The requirements apply if:
   (a) a non-domestic building or a dwelling is newly erected;
   (b) an existing non-domestic building is extended, or undergoes a material alteration; or
   (c) an existing building or part of an existing building undergoes a material change of use to a hotel or boarding house, institution, public building or shop.

The terms 'institution', 'public building' and 'shop' are explained in Regulation 2.

It should be noted that, regardless of compliance with Building Regulations, there will be obligations under the Equality Act 2010 for service providers and employers to consider barriers created by physical features in buildings.

Extensions and material alterations: dwellings

2. Under Regulation 4(3), where any building is extended, or undergoes a material alteration, the building work must be carried out so that after it has been completed the building complies with the applicable requirements of Schedule 1, or, where it did not fully comply with any applicable requirement, it is no more unsatisfactory than before.

3. This rule applies to domestic as well as to non-domestic buildings. Under the Limits on Application in Part M, Part M does not apply to an extension of, or a material alteration of, a dwelling. However, an extension of a dwelling, or a material alteration of a dwelling, must not make the building less satisfactory in relation to Part M than it was before.

4. Under Regulation 3, the expression 'material alteration' is defined by reference to a list of 'relevant requirements' in Schedule 1. That list includes Part M. This means that an alteration of a dwelling is a material alteration if the work would result in the dwelling not complying with Part M where previously it did. Alternatively, if the dwelling did not previously comply with Part M, the dwelling should not be more unsatisfactory in relation to Part M after the material alteration.

It is irrelevant whether or not the dwelling was subject to Part M at the time of its construction. Under the general Limits on Application of Part M, the requirements of that Part do not apply to an extension of or a material alteration of a dwelling. This means that the extension or alteration work itself need not comply with Part M. However, a planned alteration to a dwelling that has the potential to reduce the compliance of the dwelling as a whole with Part M must be carried out in such a way that there is no reduction in the extent of Part M compliance. Similarly, an extension of a dwelling need not itself comply with Part M, but it must not result in the dwelling being less compliant with Part M.

Further information

Access to and use of Buildings Approved Document www.planningportal.gov.uk/buildingregulations/approveddocuments/partm/approved

Building regulations 2010 SI 2214 www.legislation.gov.uk/uksi/2010/2214/pdfs/uksi_20102214_en.pdf

Fire Safety Approved Document B Volume 1 – Dwelling Houses 2006 version with 2010 and 2013 amendments https://www.gov.uk/planning-permission-england-wales

13.33.2 Compensation Act 2006

The Compensation Act 2006 (c.29), was introduced in response to concerns about a growing compensation culture but conversely to ensure that the public received dependable service from claims management companies.

The Compensation Act contains provisions relating to the law of negligence and breach of statutory duty; damages for mesothelioma; and the regulation of claims management services.

Section 1 of the Act makes statutory provision that, in determining whether the omission of certain steps amounts to a breach of duty, the court MAY consider whether such steps, had they been performed, would prevent some desirable activity. For example, the court must consider whether precautionary and defensive measures might prevent something socially useful. Though this principle had often been observed by the common law (see *Miller v. Jackson* 1977, Denning's judgement), the Act places it on a statutory footing.

Section 2 stipulates that, in the event of an accident, an apology or offer or redress, such as paying for medical treatment, is not, of itself an admission of liability.

Sections 1 and 2 came into force on royal assent on 26 July 2006.

Section 3 reverses the common law on allocation of damages in mesothelioma claims arising from unlawful exposure to asbestos. In 2002, the House of Lords had controversially ruled that, where several parties had unlawfully exposed the claimant to asbestos and risk of mesothelioma, all were liable for his injury, even though the claimant could not prove which individual party had provided the critical single fibre believed to cause the disease. However, in *Barker v. Corus UK Ltd* the House of Lords held that the parties who contributed to the risk were severally but not jointly liable. This meant that a single defendant could only be held liable for a fraction of any damages proportional to the exposure for which they were responsible and that a claimant could only recover all their damages if they succeeded in actions against all such contributors. Section 3 reverses the House of Lords decision by making all such parties jointly and severally liable for the damage so that a claimant could recover the totality of their damages, even from a subgroup of potential defendants.

Section 3 is 'treated as having always had effect' (s.16(3)) (but does not apply to claims settled before 3 May 2006) and also extends to Scotland and Northern Ireland (s.17).

Part 2 of the Act seeks to regulate the provision of claims management services. As of 23 April 2007, an individual or a corporation may not provide claims management services by way of business unless authorised, exempt or otherwise in receipt of a waiver (s.4(1)). The Act creates a Claims Management Services Regulator to authorise and regulate claims management companies and to (s.5):

▶ Set and monitor standards of competence and professional conduct;
▶ Promote good practice, in particular as to the provision of information about charges and other matters to users;
▶ Promote practices likely to facilitate competition;
▶ Ensure that arrangements are made for the protection of users, including complaints handling.

It is a crime for an unauthorised person to provide or offer claims management services, or to pretend to be authorised. Offenders are punishable, on summary conviction, by a fine of up to level 5 on the standard scale or 51 weeks' imprisonment. If convicted on indictment in the Crown Court, offenders can be sentenced to an unlimited fine or two years' imprisonment (ss.7 and 11). Where a corporate crime is committed, the offender can only be fined and not imprisoned. The Regulator may investigate unauthorised trading and seek an injunction to prevent it or bring a criminal prosecution (s.8). It is a crime to obstruct the Regulator, punishable on summary conviction by a fine of up to level 5 on the standard scale (s.10).

Section 12 creates a Claims Management Services Tribunal to which a person may appeal a decision of the Regulator about authorisation (s.13(1)). There is a further route of appeal to the Court of Appeal (s.13(4)).

Part 2 came progressively into force from 1 December 2006.

13.33.3 Control of Asbestos Regulations 2012

The Control of Asbestos Regulations 2012 came into force on 6 April 2012 (Asbestos Regulations – SI 2012/632).

In practice the changes are fairly limited. They mean that some types of non-licensed work with asbestos now have additional requirements, i.e. notification of work, medical surveillance and record keeping. **All other requirements remain unchanged**.

Most work with asbestos needs to be carried out by a **licensed contractor**. This includes work on asbestos coating and asbestos lagging; and work on asbestos insulation or asbestos insulating board (AIB) where the risk assessment demonstrates that the fibre release will be high, e.g. the material is badly damaged, or the work is not short duration work.

'Short duration' means any one person doing this type of work for less than one hour, or more people doing the work for a total of less than two hours, in any seven consecutive days. The total time spent by all workers must not exceed two hours. This includes time spent setting up, cleaning and clearing up. Non-licensed work includes work on asbestos-containing textured coatings and asbestos cement.

13

From 6 April 2012, some non-licensed work needs to be **notified** to the relevant enforcing authority. Brief written records should be kept of non-licensed work, which has to be notified, e.g. copy of the notification with a list of workers on the job, plus the level of likely exposure of those workers to asbestos. This does not require air monitoring on every job, if an estimate of degree of exposure can be made based on experience of similar past tasks or published guidance.

By April 2015, all workers/self-employed doing **notifiable** non-licensed work with asbestos must be under health surveillance by a doctor. Workers who are already under health surveillance for licensed work need not have another medical examination for non-licensed work. BUT medicals for notifiable non-licensed work are not acceptable for those doing licensed work.

Some modernisation of language and changes to reflect other legislation, e.g. the prohibition section has been removed, as the prohibition of supply and use of asbestos is now covered by REACH.

The Asbestos Regulations also include the 'duty to manage asbestos' in non-domestic premises.

The duty to manage asbestos is contained in Regulation 4. It requires the person who has the duty (i.e. the 'duty-holder') to:

▶ take reasonable steps to find out if there are materials containing asbestos in non-domestic premises, and if so, its amount, where it is and what condition it is in;

▶ presume materials contain asbestos unless there is strong evidence that they do not;

▶ make, and keep up to date, a record of the location and condition of the asbestos-containing materials – or materials which are presumed to contain asbestos;

▶ assess the risk of anyone being exposed to fibres from the materials identified;

▶ prepare a plan that sets out in detail how the risks from these materials will be managed;

▶ take the necessary steps to put the plan into action;

▶ periodically review and monitor the plan and the arrangements to act on it so that the plan remains relevant and up to date; and

▶ provide information on the location and condition of the materials to anyone who is liable to work on or disturb them.

The Regulations require mandatory training for anyone liable to be exposed to asbestos fibres at work (see Regulation 10). This includes maintenance workers and others who may come into contact with or who may disturb asbestos (e.g. cable installers) as well as those involved in asbestos removal work.

Training must include detailed information on:

▶ recognising asbestos;

▶ how asbestos can affect your health;

▶ the added dangers of smoking;

▶ the uses and likely locations for asbestos in buildings;

▶ what work you are allowed to do by law;

▶ what the law requires you to do;

▶ procedures to protect yourself;

▶ what methods to use;

▶ what equipment you need to do the job properly;

▶ how to choose, use and look after personal protective equipment;

▶ recognising and dealing with other dangers, such as work at height;

▶ decontamination of yourself and work areas;

▶ emergency procedures; and

▶ waste disposal.

Refresher training is needed every year, or more often if:

▶ work methods change;

▶ the type of equipment used changes; or

▶ the type of work changes a lot.

Worker exposure must be below the airborne exposure limit (control limit). The Asbestos Regulations have a single control limit for all types of asbestos of 0.1 fibres per cm^3. A control limit is a maximum concentration of asbestos fibres in the air (averaged over any continuous 4-hour period) that must not be exceeded.

In addition, short-term exposures must be strictly controlled and worker exposure should not exceed 0.6 fibres per cm^3 of air averaged over any continuous 10-minute period using respiratory protective equipment if exposure cannot be reduced sufficiently using other means.

Respiratory protective equipment is an important part of the control regime but it must not be the sole measure used to reduce exposure and should only be used to supplement other measures. Work methods that control the release of fibres such as those detailed in the Asbestos Essentials Task Sheet (www.hse.gov.uk/asbestos/essentials/index.htm) for non-licenced work should be used. Respiratory protective equipment must be suitable, must fit properly and must ensure that worker exposure is reduced as low as is reasonably practicable.

Further information

A Comprehensive Guide to Managing Asbestos in Premises HSG227, HSE, 2002, ISBN: 9780717623815 www.hse.gov.uk/pubns/books/hsg227.htm

Asbestos Essentials HSG210 (fourth edition), HSE, 2018, ISBN: 9780717666652 www.hse.gov.uk/pubns/books/hsg210.htm

Asbestos Essentials task sheets are available on the Asbestos Essentials area of the HSE website www.hse.gov.uk/asbestos/essentials/index.htm.

Asbestos: The Analysts' Guide for Sampling, Analysis and Clearance Procedures HSG248, 2005, HSE, ISBN:

9780717628759, HSE Books, 2005 www.hse.gov.uk/pubns/books/hsg248.htm.

Asbestos: The Licensed Contractors Guide HSG247, HSE Books, 2006, ISBN: 9780 717628742 http://books.hse.gov.uk/hse/public/saleproduct.jsf?catalogueCode=9780717628742

Asbestos: The Survey Guide HSG264 (second edition), 2012, HSE, ISBN: 9780717665020 www.hse.gov.uk/pubns/books/hsg264.htm

Managing and working with asbestos. Control of Asbestos Regulations 2012. Approved Code of Practice and guidance. L143 (second edition) 2013, HSE, ISBN: 9780717666188 https://www.hse.gov.uk/pubns/books/l143.htm

Managing asbestos in buildings A brief guide INDG 223 (rev5) 2012, HSE Books, ISBN: 9780717664870 www.hse.gov.uk/pubns/indg223.htm

13.33.4 Corporate Manslaughter and Corporate Homicide Act 2007

The Act creates a new statutory offence of corporate manslaughter which will replace the common law offence of manslaughter by gross negligence where corporations and similar entities are concerned. In Scotland the new offence will be called 'corporate homicide'. An organisation will have committed the new offence if it:

▶ owes a duty of care to another person in defined circumstances;
▶ there is a management failure by its senior managers;
▶ it amounts to a gross breach of that duty resulting in a person's death.

On conviction the offence will be punishable by an unlimited fine and the courts will be able to make remedial orders requiring organisations to take steps to remedy the management failure concerned. It is important to note that the Act does not create a new individual liability. Individuals may still be charged with the existing offence of manslaughter by gross negligence. Crown immunity will not apply to the offence, although a number of public bodies and functions will be exempt from it (in defined circumstances). The Act came into force on 6 April 2008. Available at: www.legislation.gov.uk/ukpga/2007/19/contents

13.33.5 Electrical Equipment (Safety) Regulations 1994

These Regulations came into force in January 1995 and relate to the supply of electrical equipment with a working voltage between 50 and 1000 volts and are made under the Consumer Protection Act 1987. They apply to suppliers, which include both landlords and letting agents.

The Regulations apply to all mains voltage household electrical goods and require them to be safe so that there is no risk of injury or death to humans or pets, or risk of damage to property. They do not apply to fixed electrical wiring and built-in appliances like central heating systems. The Regulations also require that instructions be provided where safety depends on the user being aware of certain issues and equipment should be labelled with the CE marking.

There are other electrical consumer Regulations, like The Plugs and Sockets etc. (Safety) Regulations 1994 and the Low Voltage Electrical Equipment (Safety) Regulations 1989.

Available at: www.legislation.gov.uk/uksi/1994/3260/made

13.33.6 The Equality Act 2010 and (Disability) Regulations 2010

The Equality Act 2010 defines a person as disabled if they have a physical or mental impairment that has a substantial and long-term adverse effect on a person's ability to carry out normal day-to-day activities; such as reading, writing, using the telephone, having a conversation and travelling by public transport.

The key thing is not the impairment but its effect. Some people don't realise that impairments such as migraines, dyslexia, asthma and back pain can count as a disability if the adverse effect on the individual is substantial and long-term. Some conditions automatically count as disabilities for the purposes of the Equality Act 2010, from the point of first diagnosis – these are cancer, HIV and multiple sclerosis (MS).

It is discrimination to treat a disabled person unfavourably because of something connected with their disability (e.g. a tendency to make spelling mistakes arising from dyslexia). This type of discrimination is unlawful where the employer or other person acting for the employer knows, or could reasonably be expected to know, that the person has a disability.

Public authorities and public, private or voluntary organisations carrying out public functions have an Equality Duty. In summary, those subject to the duty must have due regard to the need to:

▶ Eliminate unlawful discrimination, harassment and victimisation and other conduct prohibited by the Act.
▶ Advance equality of opportunity and foster good relations between people who share a protected characteristic and those who do not.

Employers are required to make reasonable adjustments to jobs and workplaces for disabled workers. These may include:

▶ adjustments to the workplace to improve access or layout;

13

▶ giving some of the disabled person's duties to another person,

▶ transferring the disabled person to fill a vacancy;

▶ changing the working hours, e.g. flexi-time, job-share, starting later or finishing earlier;

▶ time off, e.g. for treatment, assessment, rehabilitation;

▶ training for disabled workers and their colleagues;

▶ getting new or adapting existing equipment;

▶ modifying instructions or procedures;

▶ improving communication, e.g. providing a reader or interpreter;

▶ providing alternative work (a last resort).

If a disabled person expects an employer to make a reasonable adjustment, they will need to provide the employer with enough information to carry out that adjustment. An employer must not disclose confidential details without explicit consent.

The Equality Act 2010 (Disability) Regulations 2010 came into force in August 2010.

These Regulations re-enact with amendments provisions which were previously made under the Disability Discrimination Act 1995 (c.50) and which are revoked by Regulation 15 of these Regulations.

Part 2 contains provisions which supplement those in the Act about when a person is disabled for the purposes of that Act. Regulation 3 excludes from the scope of the definition of disability addictions (other than those medically caused). Regulation 4 excludes certain conditions from being impairments for the purposes of the Act. Regulation 5 provides that severe disfigurements consisting of tattoos and certain body piercings are not to be treated as having a substantial adverse effect on a person's ability to carry out normal day-to-day activities. Regulation 6 contains provision for assessing the ability of a child under 6 years of age to carry out normal day-to-day activities. Regulation 7 deems a person who is certified by a consultant ophthalmologist as blind, severely sight-impaired, sight-impaired or partially sighted to be a disabled person. The various terms used in this provision reflect the fact that the terminology used in the certificates by consultant ophthalmologists has changed over time.

In Part 3, Regulation 8 sets out things which are to be treated as auxiliary aids or services for the purposes of paragraphs 2 to 4 of Schedule 4 to the Act.

Part 4 contains provisions about reasonable adjustments to physical features of premises. Regulation 9 provides that it is not reasonable for a provider of services, a public authority carrying out its functions or an association to have to remove or alter a physical feature which was provided in or in connection with a building to assist with access to the building or to use facilities and satisfies the design standard. The Schedule to these Regulations provides details of how to determine whether the design standard is satisfied.

Regulation 10 sets out the circumstances in which a relevant landlord is taken to have withheld consent for the purposes of Schedule 21 to the Act. Regulations 11 and 12 set out circumstances in which a relevant landlord is taken to have withheld consent for the purposes of Schedule 21 to the Act unreasonably and reasonably respectively.

Regulation 13 sets out circumstances in which a condition imposed by a landlord to consent to an alteration is reasonable for the purposes of Schedule 21 to the Act.

Regulation 14 provides modifications to Schedule 21 to the Act where the occupier occupies premises under a sub-tenancy.

Act available at: www.legislation.gov.uk/ukpga/2010/15/contents

Regulations available at: www.legislation.gov.uk/uksi/2010/2128/pdfs/uksi_20102128_en.pdf

13.33.7 Gas Appliances (Safety) Regulations 1995

The 1992 Regulations required the gas appliances and fittings to which they applied to conform with the essential requirements detailed in Schedule 3 and to be safe when normally used. Series-manufactured products were also required to undergo EU type-examination together with one of a number of prescribed options for production control, at the choice of the manufacturer, conducted by third-party notified bodies (Regulations 8, 10–18, 20–23). Single appliances or small numbers may be subject to EU unit verification (Regulations 8 and 19). The 1992 Regulations are revoked.

These Regulations re-enact those requirements (with certain minor amendments) (Regulations 4(1)(c) and 4(5), 8 and 23).

Second-hand appliances are not covered by these Regulations but they retain in force, for second-hand gas appliances only, the Gas Cooking Appliances (Safety) Regulations 1989 and the Heating Appliances (Fireguard) (Safety) Regulations 1991 Regulation 1(3).

The Gas Appliance Regulations cover the safety standards on new gas appliances which have to:

▶ satisfy safety and efficiency standards;

▶ undergo type-examination and supervision during manufacture;

▶ carry the CE mark and specified information;

▶ be accompanied by instructions and warnings in the language of destination.

Available at: www.legislation.gov.uk/uksi/1995/1629/contents/made

13.33.8 Gas Safety (Installation and Use) Regulations 1998

The Installation and Use Regulations place duties on gas consumers, installers, suppliers and landlords to ensure that:

▶ only competent people work on gas installations;

▶ no one is permitted to use suspect gas appliances;

▶ landlords are responsible, in certain cases, to make sure that fittings and flues are maintained;

▶ with the exception of room-sealed appliances there are restrictions on gas appliances in sleeping accommodation;

▶ instantaneous gas water heaters must be room-sealed or fitted with appropriate safety devices

Available at: www.legislation.gov.uk/uksi/1998/2451/contents/made

From April 2009 a new gas installer registration scheme replaced the former CORGI gas scheme. The 'Gas Safe Register' is run by the Capita Group Plc on behalf of the HSE. A Gas Safety engineer must be used for any type of gas work, including installation, maintenance and servicing. See: www.gassaferegister.co.uk/default.aspx

See: Safety in the installation and use of gas systems and appliances: Gas Safety (Installation and use) Regulations 1998 (GSIUR) as amended. Approved Code of Practice and guidance. L56 (fifth edition), 2018, HSE. www.hse.gov.uk/pubns/priced/l56.pdf

13.33.9 Control of Lead at Work Regulations 2002

These Regulations came into force in November 2002 and impose requirements for the protection of employees who might be exposed to lead at work and others who might be affected by the work. The Regulations:

▶ require occupational exposure levels for lead and lead alkyls;

▶ require blood-lead action and suspension levels for women of reproductive capacity and others;

▶ re-impose a prohibition for women of reproductive capacity and young persons in specified activities;

▶ require an employer to carry out a risk assessment;

▶ require employers to restrict areas where exposures are likely to be significant if there is a failure of control measures;

▶ impose requirements for the examination and testing of engineering controls and RPE and the keeping of PPE;

▶ impose new sampling procedures for air monitoring;

▶ impose requirements in relation to medical surveillance;

▶ require information to be given to employees;

▶ require the keeping of records and identification of containers and pipes.

Information available at: www.legislation.gov.uk/uksi/2002/2676/contents/made

and www.hse.gov.uk/lead/

13.33.10 Occupiers Liability Acts 1957 and 1984 – Civil Law

The 1957 Act concerns the duty that the occupier of premises has towards visitors in relation to the condition of the premises and to things which have or have not been done to them. The Act imposes the following:

▶ there is a duty to take reasonable care to see that a visitor is reasonably safe in using the premises for the purpose for which they were invited or permitted by the occupier to be there;

▶ the common duty of care will differ depending on the visitor, so a greater duty is owed to children;

▶ an occupier can expect that a person in the exercise of their trade or profession will appreciate and guard against normal risks, for example, a window cleaner;

▶ no duty of care is owed to someone exercising a public right of way.

The 1984 Act extends the duty of care to persons other than visitors, i.e. trespassers. The occupier has to take reasonable care in all the circumstances to see that non-visitors do not get hurt on the premises because of its condition or the things done or not done to it. The occupier must cover perceived dangers and must have reasonable grounds to know that the trespassers may be in the vicinity.

Available at: www.legislation.gov.uk/ukpga/Eliz2/5-6/31/contents

www.legislation.gov.uk/ukpga/1984/3/contents https://www.hse.gov.uk/voluntary/when-it-applies.htm

13.33.11 Personal Protective Equipment Regulations 2002

These Regulations relate to approximation of the laws of EU member states. They place duties on persons who place PPE on the market to comply with certain standards. These requirements are that PPE must satisfy the basic health and safety requirements which are applicable to that class or type of PPE, the appropriate conformity assessment procedures must be carried out, CE marking must be correctly affixed and the PPE must not compromise the safety of individuals, domestic animals or property when properly maintained and used.

Available at: www.legislation.gov.uk/uksi/2002/1144/contents/made

13.33.12 Control of Pesticides Regulations 1986 as amended 1997 (COPR)

These Regulations made under the Food and Environment Protection Act 1985 all came into force by 1 January 1988. The Regulations apply to any pesticide or any substance used generally for plant control and protection against pests of all types, including anti-fouling paint used on boats. They do not apply to substances covered by other Acts like the Medicines Act 1968, The Food Safety Act 1990, those used in laboratories and a number of other specific applications.

13

No person may advertise, sell, supply, store, or use a pesticide unless it has received ministerial approval and the conditions of the approval have been complied with. The approval may be experimental, provisional or full, and the Minister has powers to impose conditions.

The Regulations also cover the need for users to be competent and have received adequate information and training. Certificates of competence (or working under the supervision of a person with a certificate) are required where pesticides approved for agricultural use are used commercially.

Defra issued a consultation document in August 2007 which covers the update of these Regulations as follows.

At present there are four Regulations (plus their amendments) to control and monitor marketing and use of pesticides in England and Wales.

▶ The Control of Pesticides Regulations 1986 and its amendment in 1997;
▶ The Plant Protection Products Regulations 2005 and its amendment in 2006;
▶ The Plant Protection Products (Basic Conditions) Regulations 1997;
▶ The Plant Protection Products (Fees) Regulations 2007.

Defra's proposal is to consolidate these four Regulations (plus amendments) into one Regulation.

Available at: www.legislation.gov.uk/uksi/1986/1510/contents/made

13.33.13 Pressure Systems Safety Regulations 2000 (PSSR)

These Regulations came into effect in February 2000 and replace the Pressure Systems and Transportable Gas Containers Regulations 1989. Transportable gas containers are covered by the Use of Transportable Pressure Receptacles Regulations 1996 (SI 1996, No. 2092).

The aim of PSSR is to prevent serious injury from the hazard associated with stored energy as a result of a pressure system or one of its parts failing. The Regulations cover:

▶ steam at any pressure;
▶ gases which exert a pressure in excess of 0.5 bar above atmospheric pressure;
▶ fluids which may be mixtures of liquids, gases and vapours where the gas or vapour phase may exert a pressure in excess of 0.5 bar above atmospheric pressure.

With the exception of scalding from steam, the Regulations do not consider the effects of the hazardous contents being released following failure. The stored contents are of concern where they can accelerate wear and cause more rapid deterioration and an increased risk of failure.

Available at: www.legislation.gov.uk/uksi/2000/128/contents/made

www.hse.gov.uk/pressure-systems/about.htm

13.33.14 Road Traffic Acts 1988 and 1991

Road rules from Road Traffic Acts 1988 and 1991 are contained in the Highway Code.

Many references to legislation can be found abbreviated throughout the Code. The Code contains a table of legislation under the 'Road User and the law' section. It is not intended to be a comprehensive guide, but a guide to some of the important points of law. For the precise wording of the law, please refer to the various Acts and Regulations (as amended) indicated in the Code.

Most of the provisions apply on all roads throughout Great Britain, although there are some exceptions. The definition of a road in England and Wales is 'any highway and any other road to which the public has access and includes bridges over which a road passes' (Road Traffic Act (RTA) 1988 sect 192(1)). In Scotland, there is a similar definition which is extended to include any way over which the public have a right of passage (Road (Scotland) Act (R(S)A) 1984 sect151(1)).

It is important to note that references to 'road' therefore generally include footpaths, bridleways and cycle tracks, and many roadways and driveways on private land (including many car parks). In most cases, the law will apply to them and there may be additional rules for particular paths or ways. Some serious driving offences, including drink-driving offences, also apply to all public places, for example public car parks.

This Highway Code applies to England, Scotland and Wales. The Highway Code is essential reading for everyone who uses the public highways, which is virtually the whole population.

The most vulnerable road users are pedestrians, particularly children, older or disabled people, cyclists, motorcyclists and horse riders. It is important that all road users are aware of the Code and are considerate towards each other. This applies to pedestrians as much as to drivers and riders. Many of the rules in the Code are legal requirements, and if you disobey these rules you are committing a criminal offence. You may be fined, given penalty points on your licence or be disqualified from driving. In the most serious cases you may be sent to prison. Such rules are identified by the use of the words 'MUST'. In addition, the rule includes an abbreviated reference to the legislation which creates the offence.

Although failure to comply with the other rules of the Code will not, in itself, cause a person to be prosecuted, The Highway Code may be used in evidence in any court proceedings under the Traffic Acts to establish liability. This includes rules which use advisory wording such as 'should/should not' or 'do/do not'. Knowing and applying the rules contained in The Highway Code could significantly reduce road casualties. Cutting the number of deaths and injuries that occur on our roads every day is a responsibility we all share. The Highway Code can help us discharge that responsibility.

The Highway Code covers;

- ▶ Rules for pedestrians;
- ▶ Rules for animals;
- ▶ Rules for cyclists;
- ▶ Rules for motorcyclists;
- ▶ Rules for drivers and motorcyclists;
- ▶ General rules, techniques and advice for all drivers and riders;
- ▶ Using the road;
- ▶ Road users requiring extra care;
- ▶ Driving in adverse weather conditions;
- ▶ Waiting and parking;
- ▶ Motorways;
- ▶ Breakdowns and accidents;
- ▶ Road works;
- ▶ Railway level crossings;
- ▶ Tramways;
- ▶ Light signals controlling traffic;
- ▶ Signals to other road users;
- ▶ Signals by authorised persons;
- ▶ Traffic signs;
- ▶ Road markings;
- ▶ Vehicle markings;
- ▶ Annexes
    1. You and your bicycle
    2. Motorcycle licence requirements
    3. Motor vehicle documentation and learner driver requirements
    4. The road user and the law
    5. Penalties
    6. Vehicle maintenance, safety and security
    7. First aid on the road
    8. Safety code for new drivers;
- ▶ Supplementary notes.

Here we have picked out the legal requirements from the Code applying to drivers of motor vehicles and motorcyclists as an example of the Code. All those who drive for work should be familiar with the latest copy of the Code, which is available from all good book stores and at https://www.gov.uk/browse/driving/highway-code

1. **Vehicle condition**. You must ensure your vehicle and trailer comply with the full requirements of the Road Vehicles (Construction and Use) (CUR) Regulations and Road Vehicles Lighting Regulations.
2. **Fitness to drive.** Make sure that you are fit to drive. You must report to the Driver and Vehicle Licensing

Agency (DVLA) any health condition likely to affect your driving. Law RTA 1988 sect 94.

3. **Vision**. You must be able to read a vehicle number plate, in good daylight, from a distance of 20 metres (or 20.5 metres where the old-style number plate is used). If you need to wear glasses (or contact lenses) to do this, you must wear them at all times while driving. The police have the power to require a driver to undertake an eyesight test. Laws RTA 1988 sect 96 & MV(DL)R reg 40 & sch 8.
4. You must not drive with a breath alcohol level higher than 35 microgrammes/100 millilitres of breath or a blood alcohol level of more than 80 milligrammes/100 millilitres of blood. Law RTA 1988 sects 4, 5 & 11(2).
5. You must not drive under the influence of drugs or medicine. Law RTA 1988 sect 4.
6. **Vehicle towing and loading**. As a driver:
    (a) you must not tow more than your licence permits. If you passed a car test after 1 Jan 1997 you are restricted on the weight of trailer you can tow:
    (b) you must not overload your vehicle or trailer;
    (c) You must secure your load and it must not stick out dangerously;
    Laws CUR reg 100 & MV(DL)R reg 43.
7. **Seat belts and child restraints.** You must wear a seat belt in cars, vans and other goods vehicles if one is fitted. Adults, and children aged 14 years and over, must use a seat belt or child restraint, where fitted, when seated in minibuses, buses and coaches. Laws RTA 1988 sects 14 & 15, Motor Vehicle (Wearing of Seat Belts)[MV(WSB)] Regulations, MV(WSB by Children in Front Seats) Regulations & MV(WSB)(Amendment)R.
8. The driver must ensure that all children under 14 years of age in cars, vans and other goods vehicles wear seat belts or sit in an approved child restraint where required. If a child is under 1.35 metres (approx 4 feet 5 inches) tall, a baby seat, child seat, booster seat or booster cushion must be used suitable for the child's weight and fitted to the manufacturer's instructions.
9. A rear-facing baby seat must not be fitted into a seat protected by an active frontal airbag, as in a crash it can cause serious injury or death to the child. Laws RTA 1988 sects 14 & 15, MV(WSB)R, MV(WSBCFS) R & MV(WSB)(A)R.

Electric bikes

You can ride an electric bike in England, Scotland and Wales if you are 14 or over, as long as it meets certain requirements.

These electric bikes are known as 'electrically assisted pedal cycles' (EAPCs). You do not need a licence to ride one and it does not need to be registered, taxed or insured.

13

What counts as an EAPC

An EAPC must have pedals that can be used to propel it.

It must show either:

- ▶ the power output
- ▶ the manufacturer of the motor.

It must also show either:

- ▶ the battery's voltage
- ▶ the maximum speed of the bike.

Its electric motor:

- ▶ must have a maximum power output of 250 watts
- ▶ should not be able to propel the bike when it is travelling more than 15.5 mph.

An EAPC can have more than 2 wheels (for example, a tricycle).

Where you can ride

If a bike meets the EAPC requirements it is classed as a normal pedal bike. This means you can ride it on cycle paths and anywhere else pedal bikes are allowed.

Other kinds of electric bike

Any electric bike that does not meet the EAPC rules is classed as a motorcycle or moped and needs to be registered and taxed. You will need a driving licence to ride one and you must wear a crash helmet.

The bike must also be type approved if either:

- ▶ it does not meet the EAPC rules
- ▶ it can be propelled without pedalling (a 'twist and go' EAPC).

This should have been done by the manufacturer or importer before you bought it. If it has been type approved, it will have a plate showing its type approval number.

15.35.15 The Health and Safety (Sharp Instruments in Healthcare) Regulations 2013

The Health and Safety (Sharp Instruments in Healthcare) Regulations came into force on 11 May 2013. Northern Ireland introduced equivalent regulations on the same date.

The Regulations were introduced to implement a European Directive. They supplement the existing health and safety legislation that already requires employers across all sectors to take effective action to control the risk from sharps injuries.

Employers and contractors working in the healthcare sector are required to:

- ▶ Have effective arrangements for the safe use and disposal (including using 'safer sharps' where reasonably practicable, restricting the practice of recapping of needles and placing sharps bins close to the point of use).

- ▶ Provide the necessary information and training to workers.
- ▶ Investigate and take action in response to work-related sharps injuries.

Guidance for healthcare employers and employees, HSIS7, HSE, 2013, www.hse.gov.uk/pubns/hsis7.htm

13.33.16 Smoke-free legislation

Five sets of smoke-free Regulations set out the detail of the smoke-free legislation. Within the UK, smoking law is a devolved issue and therefore similar legislation has been enacted in all parts of the UK.

The smoke-free law was introduced to protect employees and the public from the harmful effects of second-hand smoke.

Key points are:

- ▶ It is against the law to smoke in virtually all 'enclosed' and 'substantially enclosed' public places and workplaces.
- ▶ Public transport and work vehicles used by more than one person must be smoke-free at all times.
- ▶ No-smoking signs must be displayed in all smoke-free premises and vehicles.
- ▶ Staff smoking rooms and indoor smoking areas are no longer allowed, so anyone who wants to smoke has to go outside.
- ▶ Managers of smoke-free premises and vehicles have legal responsibilities to prevent people from smoking.
- ▶ If anyone is uncertain where they can or cannot smoke, they need to look for the No-smoking signs or ask someone in charge.

Available at www.smokefreeengland.co.uk/what-do-i-do/quick-guide/ www.legislation.gov.uk/all?title=Smoke%20free%20

13.33.17 Working Time Regulations 1998 as amended

These Regulations came into force in October 1998 and, for specified workers, restrict the working week to 48 hours per 7-day period. Individuals can voluntarily agree to disapply the weekly working hours limit. Employers must keep a copy of all such individual agreements. Workers whose working time is not measured or predetermined, or who can themselves determine the duration of their working day, are excepted from weekly working time, night work, rest periods and breaks.

The Regulations were amended, with effect from 1 August 2003, to extend working time measures in full to all non-mobile workers in road, sea, inland waterways and lake transport, to all workers in the railway and offshore sectors, and to all workers in aviation who are not covered by the sectoral Aviation Directive. The Regulations applied to junior doctors from 1 August 2004.

Mobile workers in road transport have more limited protections. Those subject to European Drivers' hours rules 3820/85 are entitled to 4 weeks' paid annual leave and health assessments if night workers, with effect from 1 August 2003. Mobile workers not covered by European drivers' hours rules will be entitled to an average 48 hours per week, 4 weeks' paid holiday, health assessments if night workers and adequate rest.

The Regulations were previously amended, with effect from 6 April 2003, to provide enhanced rights for adolescent workers.

The basic rights and protections that the Regulations provide are a:

▶ limit of an average of 48 hours a week which a worker can be required to work (though workers can choose to work more if they want to);

▶ limit of an average of 8 hours' work in 24 hours which night workers can be required to work;

▶ right for night workers to receive free health assessments;

▶ right to 11 hours' rest a day;

▶ right to a day off each week;

▶ right to an in-work rest break if the working day is longer than 6 hours;

▶ right to 28 days' paid leave per year from April 2009.

Further amendment in 2009 applied to some Doctors in training.

1998 Regulations available at: www.legislation.gov.uk/uksi/1998/1833/contents/made

2003 amendment available at: www.legislation.gov.uk/uksi/2003/1684/contents/made

And for HSE guidance see: www.hse.gov.uk/contact/faqs/workingtimedirective.htm

Climate change, environmental management, international and other aspects of health and safety

14.1 Introduction

This chapter covers some topics that are no longer in the current syllabus for the NEBOSH National General Certificate but may well be useful to many health and safety managers. These include climate change, other environmental management responsibilities in brief and international issues.

Additional sections on health and safety in the home and safe cycling have also been included.

In October 2018, the Intergovernmental Panel on Climate Change (IPCC) issued its Special Report on the impacts of global warming of 1.5°C above pre-industrial levels and related global greenhouse gas emission pathways, a follow-up on their report of 1992. Since then many governments and dozens of local authorities have declared a climate emergency. The world has slowly begun to wake up to the greatest environmental challenge it has ever faced.

Since March 2020, Covid-19 has become a major topic of concern for many governments, some of which have declared an emergency. Italy effectively shut down most of its socialising, education, sports and travel, advising people to stay at home. Milan, a city of 3.1 million inhabitants was described in *The Guardian*, one day after the shutdown, as a city with no one at work, with restaurants, schools, university, cinemas, places of entertainment, pubs, many shops all closed. Supermarket shelves empty and children out playing on the empty streets free of traffic, as they are not at school. Other countries in Europe, including the UK, introduced lockdowns in an attempt to limit the effects of Covid-19. Countries are really acting as if it was an emergency with the appropriate advice and tough restrictions. Leaders are saying their actions are led by the experts. Italy and no doubt many other countries appear to be on a war footing. The economic effects are going to be very significant, but politicians are

What the UN are saying:

https://www.ipcc.ch/2018/10/08/summary-for-policymakers-of-ipcc-special-report-on-global-warming-of-1-5c-approved-by-governments/

➤ The IPCC report finds - limiting global warming to 1.5°C would require "rapid and far-reaching" transitions in land, energy, industry, buildings, transport and cities.

➤ Global net human-caused emissions of CO₂ need to fall by ~45% (from 2010 levels) by 2030

➤ Net zero needs to be reached around 2050.

➤ Remaining emissions need to be balanced by removing CO₂ from the air.

➤ "Unprecedented change needed to limit warming to 1.5°C within the laws of chemistry and physics" **Jim Skea**, Co-Chair IPCC Working Group III

IPCC, policymakers summary, 2018

Figure 14.1 The IPCC report 2018 summary

saying they are doing what is necessary to protect their citizens and manage their way through this crisis.

The effects of Covid-19 are potentially serious. It is estimated that 80% of the population may get the virus and between 1% and 3% could die. In the UK this means about 500,000–1,500,000 may die from Covid-19, which is extremely serious and rightly has the urgent attention of the government. One owner of a care home believed if Covid-19 takes hold they will lose 90% of their residents. Relatives and other activity providers are all being denied access.

The climate emergency is of a completely different order of magnitude. It is reliably estimated that the next year, 2020/21, is a fast-closing window in which humanity needs to take action to limit global warming and keep the world within livable temperatures. In 2020 we are already seeing more and more extremes in our weather patterns, resulting in droughts, floods, melting ice, sea level rises, fires, high temperatures, very strong winds and much more. This was predicted by the experts and there is much more to come. Serious results are likely to take place within the next decade. The results will be catastrophic for humans and all living things on the Earth.

There have been five mass extinctions in the Earth's history, an event which means that 75% of species die out. Extinction events can be triggered by volcanoes, asteroids and/or cosmic events. These triggers lead to spikes in atmospheric carbon dioxide (CO_2) and tens of millennia of climate change. The five mass extinctions saw temperature rises of 4–7°C over 20,000 to 200,000-year periods. In the 700-million-year history of complex life on Earth, the arrival of the 'human super predator' is unprecedented in terms of the effects on climate and biodiversity and it represents:

▶ the fastest ever sustained rate of extinction (200 species a day);
▶ the fastest ever sustained rise in atmospheric CO_2 levels;
▶ possibly the fastest ever rise in temperatures;
▶ the fastest ever rate of ocean acidification.

At current rates the Holocene Extinction (the sixth mass extinction) will have occurred by the end of the century (75% by 2100). Between 1850 and 2150 humans could raise the Earth's temperature by 5–10°C, resulting in civilisation collapse or human extinction.

The Holocene Extinction is the first time that a species has caused this themselves in the full knowledge, since at least 1992, of what we are doing. Our trajectory at the moment predicts that we may have over 2 degrees of warming above pre-industrial levels already locked in and this level is continuing to rise. Our carbon footprint has to be drastically reduced by 60% by 2030 if we are to stand any chance of reaching our net zero carbon levels by 2050 which is required by UK law. Given global warming of 3.5 degrees or so, by the end of the century or early 22nd century, the world is predicted to only be

able to support about 4bn humans at the outside. Given that world population is predicted to increase this century from 7.5bn to 11bn there is an alarming and catastrophic gap between what the world can support and the predicted population. The effects on plant life, insects and the animal world has already been very severe with 25% of the world's species at threat level of extinction.

The scale of these challenges is unprecedented in human history and yet the actions we are currently taking are more concerned with economic growth, a major cause of our predicament, than emergency action to meet the challenge. The UN said in 2018 that we need to scale up our efforts five-fold to meet the challenge and we must go on a war footing like the Second World War. This level of action has only been seen to meet Covid-19, which is not an existential threat to life on Earth. Clearly there is much, much more that has to be done to meet the climate emergency.

14.2 Climate change

14.2.1 Introduction – position as seen in March 2020

As reported in *The Guardian* on 10 March 2020:

The UN Secretary General António Guterres has delivered the UN's assessment of the global climate in 2019. He said, 'The world is "way off track" in dealing with the climate emergency and time is fast running out.' Scientists said the threat was greater than that from the coronavirus, and world leaders must not be diverted away from climate action.

The climate assessment is led by the UN's World Meteorological Organization (WMO), with input from the UN's agencies for environment, food, health, disasters, migration and refugees, as well as scientific centres.

In 2019 the oceans were at the hottest on record, with at least 84% of the seas experiencing one or more marine heatwaves. Surface air temperatures around the world were the hottest ever recorded, after a natural El Niño event boosted figures in 2016.

The report says results from the World Glacier Monitoring Service indicate 2018-19 was the 32nd year in a row in which more ice was lost than gained. The melting of land ice combined with thermal expansion of water pushed sea levels up to the highest mark since records began.

The long-term decline of Arctic sea ice also continued in 2019, with the September average extent – usually the lowest of the year – the third worst on record.

"Climate change is the defining challenge of our time. We are currently way off track to meeting either the 1.5C or 2C targets that the Paris agreement calls for," said Guterres. 2019 ended with a global average temperature of 1.1C above pre-industrial levels.

14

"Time is fast running out for us to avert the worst impacts of climate disruption and protect our societies."

He added: "We need more ambition on [emission cuts], adaptation and finance in time for the climate conference, Cop26, in Glasgow, UK, in November. That is the only way to ensure a safer, more prosperous and sustainable future for all people on a healthy planet."

[...] A heatwave in Europe was made five times more likely by global heating, and the scorching summer led to 20,000 emergency hospital admissions and 1,462 premature deaths in France alone. India and Japan also sweltered and Australia started and ended the year with severe heat and had its driest year on record. Australia had "an exceptionally prolonged and severe fire season", the WMO noted.

Floods and storms contributed most to displacing people from their homes, particularly Cyclone Idai in Mozambique and its neighbours, Cyclone Fani in south Asia, Hurricane Dorian in the Caribbean, and flooding in Iran, the Philippines and Ethiopia. The number of internal displacements from such disasters is estimated to have been close to 22 million people in 2019, up from 17 million in 2018.

The US saw heavy rains, with the total from July 2018 to June 2019 being the highest on record. Total economic losses in the US for the year were estimated at $20bn, the WMO said.

Unpredictable climate and extreme weather was a factor in 26 of the 33 nations that were hit by food crises in 2019, and was the main driver in 12 of the countries. "After a decade of steady decline, hunger is on the rise again – over 820 million suffered from hunger in 2018, the latest global data available," the report says.

The WMO said unusually heavy precipitation in late 2019 was also a factor in the severe desert locust outbreak in the Horn of Africa, which is the worst for decades and expected to spread further by June 2020 in a severe threat to food security.

Yet despite these dire warnings, the world is very slow to get the necessary plans and actions set up to mitigate the worst effects of the climate changes. We are all slow to shed our reliance on consumerism and governments are scared to measure development in anything but GDP. But there is hope in the UK as the central government plus the two regional governments of Scotland and Wales, along with many councils, at both town and county levels, have declared a climate emergency. Governments and counties have yet to follow this up with adequate plans to meet the declaration. However at least two town/city councils have produced really ground-breaking action plans which have been approved. These are:

1. Helston Climate action plan, September 2019 (https://transitionhelston.org.uk/2019/10/05/climate-action-plan-published). A town of 11,700 inhabitants which is aiming to be carbon neutral by 2030; and

2. Bristol net zero by 2030 (https://www.cse.org.uk/projects/view/1363) an evidence-based report produced for the Bristol One City Climate Strategy

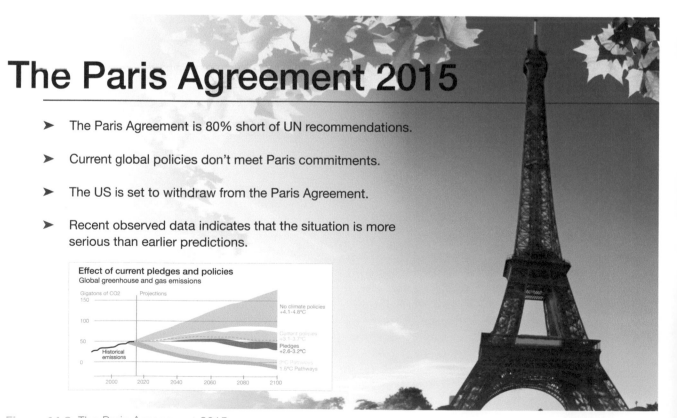

The Paris Agreement 2015

➤ The Paris Agreement is 80% short of UN recommendations.

➤ Current global policies don't meet Paris commitments.

➤ The US is set to withdraw from the Paris Agreement.

➤ Recent observed data indicates that the situation is more serious than earlier predictions.

Effect of current pledges and policies
Global greenhouse and gas emissions

Figure 14.2 The Paris Agreement 2015

published in February 2020 (https://www.cse.org.uk/news/view/2436). A city of 470,000 people.

A number of other county councils are making a significant effort to decarbonise but their plans may only involve direct emissions by the council and not the total emissions of the county.

14.2.2 Some carbon maths

A big part of the challenge is that as a society we have not committed sufficiently to reduce emissions. One conclusion we've reached is that we all need to learn – and get real – about 'carbon maths'. This is the basic mathematical concepts that are important to understanding how the carbon issue applies to each of us, whether as individuals, families, businesses, or other organisations.

One aspect of this is relatively simple but quite important. Scientists account for carbon emissions by classifying them into three categories, or 'scopes':

▶ Scope 1 emissions are the direct emissions that your activities create – like the exhaust from the car you drive, or for a business, the trucks it drives to transport its products from one place to another or the generators it might run.

▶ Scope 2 emissions are indirect emissions that come from the production of the electricity or heat you use, like the traditional energy sources that light up your home or power the buildings owned by a business.

▶ Scope 3 emissions are the indirect emissions that come from all the other activities in which you're engaged, including the emissions associated with producing the food you eat, or manufacturing the products that you buy. For a business, these emission sources can be extensive, and must be accounted for across its entire supply chain, the materials in its buildings, the business travel of its employees, and the full life cycle of its products, including the electricity customers may consume when using the product. Given this broad range, a company's scope 3 emissions are often far larger than its scope 1 and 2 emissions put together.

There is another aspect of carbon maths that is also essential. This is the difference between being 'carbon neutral' and being 'net zero'. While they sound similar, in fact they are different.

▶ Given common usage, companies have typically said they are 'carbon neutral' if they offset their emissions with payments *either* to avoid a reduction in emissions *or* remove carbon from the atmosphere. But these are two very different things. For example, one way to avoid a reduction in emissions is to pay someone not to cut down the trees on the land they own. This is a good thing, but in effect it pays someone *not* to do something that would have a negative impact. It doesn't lead to planting more trees that would have a positive impact by removing carbon.

▶ In contrast, 'net zero' means that a company actually removes as much carbon as it emits. The reason the phrase is 'net zero' and not just 'zero' is because there are still carbon emissions, but these are equal to carbon removal. And 'carbon negative' means that a company is removing more carbon than it emits each year.

14.2.3 Climate safety responsibility

Organisations have a duty of care for their employees, including in the area of occupational health and safety. Increasingly, this is understood as extending beyond the workplace to a range of locations the employee works from or passes through to attend work. Recently, climate change is being regarded as an issue of employee health and safety both within the workplace and more broadly.

Workers in many occupations already experience climate and weather-related hazards in their jobs. The increasing frequency, disruption and severity of extreme events and the increasingly hostile normal working conditions due to climate change will make these hazards more extreme and affect more and more workers. As economies have to adjust to higher temperatures, extremes of precipitation and rising sea levels this will impact on working hours, locations and conditions. This will affect the injury risk from, for example, firefighting, clearing debris from hurricanes, heat-related illnesses in agricultural, fishery, construction, landscape and other outdoor workers. It will also make the community challenges much more severe for workers, their families and those they seek to help or who live nearby.

Increased air pollution, biological hazards and pesticides are all on the increase. Increased ambient temperature due to climate change may also increase occupational exposure to hazardous chemicals and their adverse effects on workers. Psychological stresses and productivity losses are also on the increase, and these can cause conflict between employees and employers.

Multiple strategies will be required to reduce the risks of work-related illnesses and injuries from climate change. Employer and worker preparedness includes recognising, anticipating and controlling potential occupational safety and health hazards resulting from both climate change and from implementation of mitigation and adaptation measures to address climate change. For example, policies to mitigate and adapt to climate change may lead to changes in infrastructure and energy use that may positively or negatively impact workers' health and safety. Occupational health surveillance will need to be broadened to include the health and safety impacts of climate change on workers.

Action from employers will need to include:

1. Management policies and planning to enhance risk assessment and eliminate or control unacceptable risks;

2. Emergency response with suitable equipment and trained and protected workforce;
3. Long-term adaptation to a zero-carbon world;
4. Sustainable transport;
5. Sustainable energy sources and conservation;
6. Passive solutions like building design and insulation;
7. Occupational health to ensure workers are best protected from extremes and identify and support vulnerable people;
8. Sympathetic regeneration and support needs of the workforce facing a climate emergency at work while their families are trying to protect their home and children (a first responder firefighter battling to save a house while their own is burning to the ground and making their family homeless);
9. Engineering solutions for carbon capture and manufacture in a zero-carbon world.

14.2.4 Summary of potential zero-carbon action to meet the climate emergency

According to Zero Carbon Britain: Rising to the Climate Emergency report published in 2019 by the Centre for Alternative Technology (https://cat.org.uk) we can make major changes to decarbonise our lives.

Energy

We could cut energy demands by 50% if we reduce the amount we travel, change our modes of travel, retrofit all existing buildings and improve internal temperature control. Greater use of public transport, walking, cycling, switching to efficient electric vehicles and two-thirds less flying would reduce energy demand for transport by 75%.

Heat loss from buildings can be reduced by:

▶ sealing and reducing draughts;
▶ improving insulation in floors, roofs, doors, windows, etc.;
▶ recovering heat leaving the building through ventilation.

Some new buildings designed to Passivhaus standards only demand about 10% of the energy of most existing buildings. Much improved heat controls could reduce demand by only heating to the temperature required and when the rooms are in use.

Solar can be used for hot water heating. Room heating can use geothermal or heat pumps taking heat from the air or the ground. More efficient and smarter appliances can minimise heat used. LED lighting is now readily available and dramatically reduces the energy used. The use of batteries or large hot water tanks can store energy produced when the sun shines, for example, and deliver it when needed.

It is possible to supply 100% of the UK's energy needs using sources such as solar, geothermal, hydro, onshore and offshore wind, hydrogen and tidal.

Transport

About 41% of UK energy demand and 37% of UK greenhouse gas emissions are from transport. Surface passenger transport is about 50% of demand with aviation and freight about a quarter each. Radical changes are required to reduce the amount we travel and the way we do it to reduce energy demands. We also need to make urban environments healthy and clean places to live and work.

We need to:

▶ improve the use of communication technology like Zoom or Skype;
▶ reduce commutes to work and recreation;
▶ improve our infrastructure for cycling and walking;
▶ improve the use of public transport through bus, coach and rail;
▶ increase the number of occupants in cars through car sharing.

We will need to consider alternative fuels as petrol and diesel-fuelled vehicles are phased out by 2035 or perhaps 2032 in the UK. Electric vehicles are much more efficient and, as charging points and battery technology improve, will cease to have a range challenge in the future. The concept of using hydrogen to power a fuel cell which then drives an electric propulsion system is perhaps the most promising technology in 2020. It increases the vehicle range to 400–500 miles and reduces the refuelling time to 10 minutes. It also massively reduces the use of precious metals for battery manufacture making hydrogen fuel-cell vehicles a real possibility.

A remarkable, unintended consequence of the Covid-19 pandemic, is the massive reduction in air travel that occurred in just over two weeks. Italy was down by 90%, Virgin 85%, British Airways 80%. The airports were groaning under the financial shocks and some may close. As at early March 2020, no one fully knows what the outcome will be but it is highly likely that the CO_2 emissions in 2020 will be significantly reduced. In the medium term, major modifications to this industry are needed to change aircraft propulsion methods and reduce the amount of flying. The effects of emissions high up in the atmosphere means that flying is particularly detrimental.

Moving goods around requires some fundamental rethink in order to:

▶ reduce the amount of goods we move;
▶ reduce the distance travelled by goods by using local sources;
▶ improve the efficiency of goods vehicles;
▶ limit the use of air freight to all but essential items;
▶ improve the efficiency of ships; and
▶ overall reduce the amount of goods that we use.

Land use

Land use accounts for about 10% of the UK emissions. Land management includes restoring essential biodiversity, providing a good mix of food crops, reducing unnecessary food imports through a change in diets and growing local with less land used for grazing livestock. Farmers would have space for a range of other uses including plant-based sources of protein like beans, nuts, cereals and vegetables. It will involve doubling the forest area to 24% of the land used in the UK. Capturing large tonnage of CO_2 and providing more room for wild habitat and conserved areas.

The UK would become more self-reliant in food, reducing imports from 42% to about 17% and improve our health with a better and more balanced diet.

14.2.5 Environmental management issues in brief

The current NEBOSH National General Certificate no longer includes general environmental considerations. However, many health and safety managers also have an 'environmental' responsibility. This section attempts to give some background information on environmental issues. NEBOSH also offers a Certificate in Environmental Management and a Diploma in Environmental Management that covers these issues in much more detail.

Organisations must also be concerned with aspects of the environment. There will be an interaction between the health and safety policy and the environmental policy which many organisations are now developing. Many of these interactions will be concerned with good practice, the reputation of the organisation within the wider community and the establishment of a good health and safety culture. The health and safety data sheet, used for a COSHH assessment, also contains information of an environmental nature covering ecological information and disposal considerations. An important component in the management of environmental considerations is the Environmental Management System (EMS).

The Environmental Management System should cover all aspects of an organisation's activities, products and services that could interact with the environment. Such aspects or elements include;

▶ the raw materials used and their source;
▶ by-products from processes;
▶ waste material produced;
▶ the energy and water used;
▶ emissions to air and water;
▶ lifespan of products and end-of-life disposal;
▶ packaging and transport issues; and
▶ environmental hazards associated with any materials used.

Environmental assessments need to be reviewed if there are significant changes to some of the elements, such as changes in raw material source or change of product. However, changes in plant or location, the availability of new processes and technologies or recycling opportunities could also be reasons. In addition, new legislation or market and public pressures, evidence of environmental damage or accidents could also prompt a review. Reviews of assessments may also arise as a result of audits or from annual reviews as part of an EMS or striving for continuous improvement.

There are three environmental issues which place statutory duties on employers and are directly related to the health and safety function. These are:

▶ air pollution;
▶ water pollution;
▶ waste disposal.

The statutory duties are contained in the Environmental Protection Act (EPA) 1990 and several of its subsequent Regulations. The Act is enforced by various state agencies (the Environment Agency, local authorities) and these agencies have very similar powers to the HSE (e.g. enforcement, prohibition notices and prosecution). In addition to serving notices, they can enter premises at any reasonable time or at any time if they have reason to believe that there is immediate risk of serious pollution to the environment. They may be accompanied by other authorised persons or a police constable and take with them any equipment or materials required to exercise their powers.

Once there they can make any examinations or inspections necessary and take samples, photographs, measurements and copies of documents. They may cause premises or parts of premises to be left undisturbed and ask for articles to be dismantled, tested or rendered harmless. Articles and substances may be taken and examined or ensured that they are secure from tampering and available as evidence. Any persons involved are required to offer assistance or facilities as necessary and may be required to answer questions and sign a declaration that their statements are true. The end result may be the initiation of legal proceedings.

Environmental managers should have a clear understanding of the harm that can be done to a business when regulatory controls are breached. The potential liabilities arising from non-compliance with environmental regulatory controls can be significant leading to criminal prosecutions by the enforcing authorities or through private prosecutions. The results of a successful prosecution could result in:

▶ the significant fines that can be given for serious breaches of the law;
▶ the potential for prison sentences and for personal prosecution for directors and managers;
▶ administrative sanctions including serving notices such as revocation notices, suspension notices and enforcement notices;
▶ orders requiring works to be undertaken to prevent pollution;
▶ award of clean-up costs for rectifying environmental damage;
▶ loss of value of property, land or a business; civil liability, resulting in claims for damages or injunctions;
▶ adverse publicity and its effect on relations with key stakeholders, including shareholders, customers, neighbours, insurers and banks; and

14

▶ insurance issues due to the fact that most standard business policies exclude liability for environmental damage unless sudden and unforeseen.

The Act is divided into nine parts but this chapter will only be concerned with Part 1 (the control of pollution into the air, water or land) and Part 2 (waste disposal).

Pollution is a term that covers more than the effect on the environment of atmospheric emissions, effluent discharges and solid waste disposal from industrial processes. It also includes the effect of noise, vibration, heat and light on the environment. This was recognised in the 1996 EU Directive on Integrated Pollution Prevention and Control, and the Pollution Prevention and Control Act 1999 replaced the Integrated Pollution Control Regulations made under Part 1 of the EPA by extending those powers to cover waste minimisation, energy efficiency, noise and site restoration. Thus the Pollution Prevention and Control Act has now replaced Part 1 of the EPA.

Figure 14.3 Example of industrial air pollution

The Environmental Damage (Prevention and Remediation) Regulations came into force in 2009 (in England only) to implement the EU Environmental Liability Directive. They cover environmental damage caused by economic activity of any public or private organisation. Under the Regulations such organisations are obliged to report to the Environment Agency any activity that has or could cause environmental damage. Such damage could be to protected species, natural habitats, surface or ground water, sites of special

scientific interest or land in general. If the Environment Agency considers that such damage has occurred, it can serve a remediation notice on the organisation outlining any remedial action required. If the organisation fails to remedy the problem, the Environment Agency is empowered to undertake the remediation itself and order any associated costs from the organisation.

The Regulatory Enforcement and Sanctions Act has given the Environmental Agency powers to enforce environmental law using civil sanctions rather than criminal prosecutions. The powers include the imposition of fixed and variable fines and prohibition notices to prevent the continuation of an activity until compliance has been assured. Other powers include the use of compliance notices that require remedial action to be taken; restoration notices that require the restoration of particular situation and enforcement undertakings. These allow corrective actions to be undertaken before any enforcement is taken. Appeals against any of these measures are possible to an independent tribunal.

14.3 International issues

The authors of this book have published 'International Health and Safety at Work' which covers the syllabus for the NEBOSH International General Certificate and gives a detailed account of the operation of occupational health and safety internationally. A basic introduction is given here to three aspects of the international scene – the International Labour Organization (ILO), the differences between fault and no-fault compensation schemes that exist in various parts of the world and working abroad. Many countries have either fault or no-fault compensation schemes for workers involved in accidents. Knowledge of these schemes is important for those who work in several different countries The hazards of working in unfamiliar countries and/or climates are important health and safety issues and include snake bites, diseases (such as malaria and yellow fever) and sunstroke.

The International Labour Organization (ILO), World Health Organization (WHO) and United Nations (UN) have estimated that there are 270 million occupational accidents and 160 million occupational diseases every year throughout the world – and these are recognised as relatively conservative estimates due to probable underreporting. The ILO estimates that 2 million women and men die each year as a result of occupational accidents and work-related diseases. Table 14.1 shows the global numbers in more detail.

In the USA in 2002, approximately 2 million workers were victims of workplace violence. In the UK, 1.7% of working adults (357,000 workers) were the victims of one or more incidents of workplace violence.

Some 10% of all skin cancers are estimated to be attributable to workplace exposure to hazardous substances. In Latin America, 37% of miners have

Table 14.1 Numbers of global work-related adverse events

Event	Average (daily)	Annually
Work-related death	5,000	2,000,000
Work-related deaths to children	60	22,000
Work-related accidents	740,000	270,000,000
Work-related disease	438,000	160,000,000
Hazardous substance deaths	1,205	440,000
Asbestos-related deaths	274	100,000

silicosis, rising to 50% among miners over 50. In India, 54.6% of slate pencil workers and 36.2% of stone cutters have silicosis.

In the course of the 20th century, industrialised countries saw a clear decrease in serious injuries, not least because of real advances in making the workplace healthier and safer. The challenge is to extend the benefits of this experience to the whole working world. However, 1984 saw the worst chemical disaster ever when 2,500 people were killed and over 200,000 injured in the space of a few hours at Bhopal. Not only were the workers affected but their families, their neighbours and whole communities. Decades later, many people are still affected by the disaster and are dying as a result. The rusting remains of a once magnificent plant remain as a reminder of the disaster.

According to the ILO, deaths due to work-related accidents and illnesses represent 3.9% of all deaths, and 15% of the world's population suffers a minor or major occupational accident or work-related disease in any one year. A large number of the unemployed – up to 30% – report that they suffer from an injury or disease dating from the time at which they were employed. The number of fatal occupational accidents, especially in Asia and Latin America, is increasing.

The main preventable factors for accidents are:

▶ lack of a preventative safety and health culture;
▶ poor management systems;
▶ poor supervision and enforcement by the government.

Experience has shown that a strong safety culture is beneficial for workers, employers and governments alike. Various prevention techniques have proved themselves effective, both in avoiding workplace accidents and illnesses and improving business performance. Today's high standards in some countries are a direct result of long-term policies encouraging tripartite social dialogue, collective bargaining between trade unions and employers, and effective health and safety legislation backed by potent labour inspection. The ILO believes that safety management systems like ILO-OSH 2001 provide a powerful tool for developing a sustainable safety and health culture at the enterprise level and mechanisms for the continual improvement of the working environment.

(a)

(b)

Figure 14.4 (a) Dangerous means of access with platform tied to bucket of excavator. A local expediency accepted as normal, Morocco; (b) good quality scaffold with protection near water, Belgium

14.3.1 The role and function of the ILO

The ILO is a specialised agency of the United Nations that seeks to promote social justice through establishing and safeguarding internationally recognised human and labour rights. It was founded in 1919 by the Treaty of Versailles at the end of the First World War.

The motivation behind the creation of such an organisation was primarily humanitarian. Working conditions at the time were becoming unacceptable to a civilised society. Long hours, unsafe, unhygienic and dangerous conditions were common in low-paid manufacturing careers. Indeed, in the wake of the Russian Revolution, there was concern that such working conditions could lead to social unrest and even other revolutions. The ILO was created as a tripartite organisation with governments, employers and workers represented on its governing body.

The ILO formulates international labour standards and attempts to establish minimum rights including freedom of association, the right to organise, collective bargaining, abolition of forced labour, equality of opportunity and treatment and other standards that regulate conditions across all work-related activities.

Representatives of all ILO member states meet annually in Geneva for the International Labour Conference, which acts as a forum where social and labour questions of importance to the entire world are discussed. At this conference, labour standards are adopted and decisions made on policy and future programmes of work.

The ILO has 178 member states but if a country is not a member, the ILO still has influence as a source of guidance when social problems occur.

The main principles on which the ILO is based are:

1. labour is not a commodity;
2. freedom of expression and of association are essential to sustained progress;
3. poverty anywhere constitutes a danger to prosperity everywhere;
4. the 'war against want' requires to be carried on with unrelenting vigour within each nation, and by continuous and concerted international effort in which the representatives of workers and employers, enjoying equal status with those of governments, join with them in free discussion and democratic decision with a view to the promotion of the common welfare.

A campaign launched by the ILO has been to seek to eliminate child labour throughout the world. In particular, the ILO is concerned about children who work in hazardous working conditions, bonded child labourers and extremely young working children. It is trying to create a worldwide movement to combat the problem by:

▶ implementing measures which will prevent child labour;

▶ withdrawing children from dangerous working conditions;
▶ providing alternatives;
▶ improving working conditions as a transitional measure towards the elimination of child labour.

ILO Conventions and Recommendations

The international labour standards were developed for four reasons. The main motivation was to improve working conditions with respect to health and safety and career advancement. The second motivation was to reduce the potential for social unrest as industrialisation progressed. Thirdly, the member states want common standards so that no single country has a competitive advantage over another due to poor working conditions. Finally, the union of these countries creates the possibility of a lasting peace based on social justice.

International labour standards are adopted by the International Labour Conference. They take the form of Conventions and Recommendations. At the present time, there are 187 Conventions and 198 Recommendations, some of which date back to 1919.

International labour standards contain flexible measures to take into account the different conditions and levels of development among member states. However, a government that ratifies a Convention must comply with all its articles. Standards reflect the different cultural and historical backgrounds of the member states as well as their diverse legal systems and levels of economic development.

ILO occupational safety and health standards can be divided into four groups, and an example is given in each case:

1. Guiding policies for action

   The Occupational Safety and Health Convention, 1985 (No. 155) and its accompanying Recommendation (No. 164) emphasise the need for preventative measures and a coherent national policy on occupational safety and health. They also stress employers' responsibilities and the rights and duties of workers.

2. Protection in given branches of economic activity

   The Safety and Health in Construction Convention, 1988 (No. 167) and its accompanying Recommendation (No. 175) stipulate the basic principles and measures to promote safety and health of workers in construction.

3. Protection against specific risks

   The Asbestos Convention, 1986 (No. 162) and its accompanying Recommendation (No. 172) gives managerial, technical and medical measures to protect workers against asbestos dust.

4. Measures of protection

   Migrant Workers (Supplementary Provisions) Convention, 1975 (No. 143) aims to protect the safety and health of migrant workers.

ILO Conventions are international treaties signed by ILO member states and each country has an obligation to comply with the standards that the Convention establishes.

In contrast, **ILO Recommendations** are non-binding instruments that often deal with the same topics as Conventions. Recommendations are adopted when the subject, or an aspect of it, is not considered suitable or appropriate at that time for a Convention. Recommendations guide the national policy of member states so that a common international practice may develop and be followed by the adoption of a Convention.

ILO standards are the same for every member state and the ILO has consistently opposed the concept of different standards for different regions of the world or groups of countries.

The standards are modified and modernised as needed. The Governing Body of the ILO periodically reviews individual standards to ensure their continuing relevance.

Supervision of international labour standards is conducted by requiring the countries that have ratified Conventions to periodically present a report with details of the measures that they have taken, in law and practice, to apply each ratified Convention. In parallel, employers' and workers' organisations can initiate contentious proceedings against a member state for its alleged non-compliance with a convention it has ratified. In addition, any member country can lodge a complaint against another member state which, in its opinion, has not ensured in a satisfactory manner the implementation of a Convention which both of them have ratified. Moreover, a special procedure exists in the field of freedom of association to deal with complaints submitted by governments or by employers' or workers' organisations against a member state whether or not the country concerned has ratified the relevant Conventions. Finally, the ILO has systems in place to examine the enforcement of international labour standards in specific situations.

Figure 14.5 World Cup stadium, under construction, Cape Town, 2009

The ILO also publishes Codes of Practice, guidance and manuals on health and safety matters. These are often used as reference material by either those responsible for drafting detailed Regulations or those who have responsibility for health and safety within an organisation. They are more detailed than either Conventions or Recommendations and suggest practical solutions for the application of ILO standards. Codes of Practice indicate 'what should be done'. They are developed by tripartite meetings of experts and the final publication is approved by the ILO Governing Body.

For example, the construction industry has a Safety and Health in Construction Convention, 1988 (No. 167) that obliges signatory ILO member states to comply with the construction standards laid out in the Convention – the Convention is a relatively brief statement of those standards. The accompanying Recommendation (No. 175) gives additional information on the Convention statements. The Code of Practice gives more detailed information than the Recommendation. This can best be illustrated by contrasting the coverage of scaffolds and ladders by the three documents shown in Appendix 14.1.

Important ILO Conventions (C) and Recommendations (R) in the field of occupational safety and health include:

▶ C 115 Radiation Protection and (R 114), 1960;
▶ C 120 Hygiene (Commerce and Offices) and (R 120), 1964;
▶ C 139 Occupational Cancer and (R 147), 1974;
▶ C 148 Working Environment (Air, Pollution, Noise and Vibration) and (R 156), 1977;
▶ C 155 Occupational Safety and Health and (R 164), 1981;
▶ C 161 Occupational Health Services and (R 171), 1985;
▶ C 162 Asbestos and (R 172), 1986;
▶ C 167 Safety and Health in Construction and (R 175), 1988;
▶ C 170 Chemicals and (R 177), 1990;
▶ C 174 Prevention of Major Industrial Accidents and (R 181), 1993;
▶ C 176 Safety and Health in Mines and (R 176), 1995;
▶ C 184 Safety and Health in Agriculture and (R 192), 2001;
▶ C 187 Promotional Framework for Occupational Safety and Health and (R 197), 2006;
▶ R 97 Protection of Workers' Health Recommendation, 1953;
▶ R 102 Welfare Facilities Recommendation, 1956;
▶ R 31 List of Occupational Diseases Recommendation, 2002.

Copies of Conventions, Recommendations and Codes of Practice can be obtained from the ILO website: www.ilo.org/safework

14

Figure 14.6 Good quality access on a MEWP for pointing brickwork in Belgium

14.3.2 Compensation and insurance issues

Accidents/incidents arising out of the organisation's activities resulting in injuries to people and incidents resulting in damage to property can lead to compensation claims. There are two facets to this – employers' liability insurance, and fault and no-fault compensation.

Employers' liability insurance

In many countries, employers are required to take out employers' liability insurance to cover their liability in the event of accidents and work-related ill-health to employees and others who may be affected by their operations. This ensures that any employee, who successfully sues his/her employer following an accident, is assured of receiving compensation irrespective of the financial position of the employer.

Fault and no-fault injury compensation

In the UK, compensation for an injury following an accident is achieved by means of a successful legal action in a civil court (as discussed in Chapter 1). In such cases, injured employees sue their employer for negligence and the employer is found liable or at fault. This approach to compensation is adversarial, costly and can deter injured individuals of limited means from pursuing their claim. In a recent medical negligence claim in Ireland, costs were awarded against a couple who were acting on behalf of their disabled son, and they were faced with a bill for £3m.

The spiralling cost of insurance premiums to cover the increasing level and number of compensation awards, despite the Woolf reforms (see Chapter 5), has led to another debate on the introduction of a no-fault compensation system. It has been estimated that in medical negligence cases, it takes on average six years to settle a claim and only 10% of claimants ever see any compensation.

No-fault compensation systems are available in many parts of the world, in particular New Zealand and several states in the USA. In these systems, amounts of compensation are agreed centrally at a national or state level according to the type and severity of the injury. The compensation is often in the form of a structured continuous award rather than a lump sum and may be awarded in the form of a service, such as nursing care, rather than cash.

The no-fault concept was first examined in the 1930s in the USA to achieve the award of compensation quickly in motor car accident claims without the need for litigation. It was introduced first in the State of Massachusetts in 1971 and is now mandatory in nine other states although several other states have either repealed or modified no-fault schemes.

In 1978, the Pearson Commission in the UK rejected a no-fault system for dealing with clinical negligence even though it acknowledged that the existing tort system was costly, cumbersome and prone to delay. Its principal reasons for rejection were given as the difficulties in reviewing the existing tort liability system and in determining the causes of injuries.

In New Zealand, there was a general dissatisfaction with the workers' compensation scheme, which was similar to the adversarial fault-based system used in Australia and the UK. In 1974, a no-fault accident compensation system was introduced and administered by the Accident Compensation Corporation (ACC). This followed the publication of the Woodhouse Report in 1966 which advocated 24-hour accident cover for everybody in New Zealand.

The Woodhouse Report suggested the following five principles for any national compensation system:

- ▶ community responsibility;
- ▶ comprehensive entitlement irrespective of income or job status;
- ▶ complete rehabilitation for the injured party;
- ▶ real compensation for the injured party;
- ▶ administrative efficiency of the compensation scheme.

The proposed scheme was to be financed by channelling all accident insurance premiums to one national organisation (the ACC).

The advantages of a no-fault compensation scheme include:

1. Accident claims are settled much quicker than in fault schemes.
2. Accident reporting rates will improve.
3. Accidents become much easier to investigate because blame is no longer an issue.
4. Normal disciplinary procedures within an organisation or through a professional body are unaffected and can be used if the accident resulted from negligence on the part of an individual.
5. More funds are available from insurance premiums for the injured party and less used in the judicial and administrative process.

The possible disadvantages of a no-fault compensation system are:

1. There is often an increase in the number of claims, some of which may not be justified.
2. There is a lack of direct accountability of managers and employers for accidents.
3. Mental injury and trauma are often excluded from no-fault schemes because of the difficulty in measuring these conditions.
4. There is more difficulty in defining the causes of many injuries and industrial diseases than in a fault scheme.
5. The monetary value of compensation awards tend to be considerably lower than those in fault schemes (although this can be seen as an advantage).

No-fault compensation schemes exist in many countries including Canada and the Scandinavian countries. However, a recent attempt to introduce such a scheme in New South Wales, Australia, was defeated in the legislative assembly.

14.3.3 Working abroad

Introduction

In March 2020 we are living in unprecedented times when most travel has been suspended and hotels shut down. It is very difficult to know how this will end and what resulting business world will emerge. The advice here is given for what we have come to believe was normal with extensive business and social travel throughout the world. With Covid-19 and climate change this may never return to the position up to the end of 2019. Staff may travel and work abroad in the course of the organisation's business. Most of this work is no more risky than the equivalent activity in their home country and takes place in countries that are considered safe for travellers. Increasingly though, work takes place in countries or parts of countries where the relevant government office advises against travel. With Covid-19 most governments are advising against travel and many countries have shut their borders.

This section outlines the steps needed for safe management of these activities. It applies to work carried out by staff (working alone or with colleagues).

General safety advice

Safety begins when you pack. To help avoid becoming a target, do not dress so as to mark yourself as an affluent tourist. Expensive-looking jewellery, for instance, can draw the wrong attention.

▶ Always try to travel light. You can move more quickly and will be more likely to have a free hand. You will also be less tired and less likely to set your luggage down, leaving it unattended.
▶ Carry the minimum number of valuables, and plan places to conceal them. Your passport, cash and credit cards are most secure when locked in a hotel safe. When you have to carry them on your person, you may wish to put them in various places rather than all in one wallet or pouch. Avoid handbags, fanny packs and outside pockets that are easy targets for thieves. Inside pockets and a sturdy shoulder bag with the strap worn across your chest are somewhat safer. One of the safest places to carry valuables is in a pouch or money belt worn under your clothing.
▶ If you wear glasses, pack an extra pair. Bring them and any medicines you need in your carry-on luggage.

To avoid problems when passing through customs, keep medicines in their original, labelled containers. Bring copies of your prescriptions and the generic names for the drugs. If a medication is unusual or contains narcotics, carry a letter from your doctor attesting to your need to take the drug. If you have any doubt about the legality of carrying a certain drug into a country, consult the embassy or consulate of that country before you travel.

▶ Bring traveller's cheques and one or two major credit cards instead of cash.
▶ Pack an extra set of passport photos along with a photocopy of your passport's information page to make replacement of your passport easier in the event it is lost or stolen.
▶ Put your name, and organisation's address and telephone numbers inside and outside of each piece of luggage. Use covered luggage tags to avoid casual observation of your identity or nationality. If possible, lock your luggage.

Risks involved in overseas work

Risks to health and safety that arise from overseas work can be:

1. work-related (e.g. physical risks from the fieldwork itself);

541

2. health-related (e.g. exposure to tropical diseases);
3. related to personal security (e.g. associated with local criminal or political activity, or civil unrest).

The following aspects concentrate on the third point above as the health and safety aspects of the work itself are covered throughout this book.

Risk assessment

Risk assessments must be made for all work abroad. This need not necessarily be onerous – the nature and complexity of the assessment should reflect the risks involved in the work. In many cases the work itself is not hazardous but it takes place in hazardous surroundings. In these cases the risk assessment should concentrate on the travel-associated risks. For instance, no written risk assessment would be needed for low-risk activities such as attending conferences or visiting colleagues in Australia or the European Union, but similar activities in more dangerous locations would require one.

Travel advice

Many national foreign office departments or consulates will have information about safe travel. The UK's Foreign and Commonwealth Office (FCO) has a website (www. fco.gov.uk/travel) which carries up-to-date travel advice for over 200 countries. It could be used as the basis for informing all those travelling on business or leisure of the risks that they might face. Although the advice pertains to the safety of British travellers, it is relevant to other nationalities as British insurance companies use it to determine the extent or validity of insurance cover.

Figure 14.7 Driving outside the UK

Health matters

Staff travelling on business should take advice on the potential health risk associated with travel. This should be sought in good time so that any recommended immunisations can be given and/or prophylactic drugs prescribed.

Appendix 14.2 has some useful safe travel recommendations.

14.3.4 **ILO recommendations for road transport drivers**

The ILO covers some aspects of driver health and safety in the Hours of Work and Rest Periods (Road Transport) Convention 1979 (No. 153) and its accompanying Hours of Work and Rest Periods (Road Transport) Recommendation 1979 (No. 153).

The Convention states that no driver shall be allowed to drive continuously for more than four hours without a break, although this may be exceeded by no more than one hour under certain circumstances. The maximum total driving time, including overtime, must not exceed nine hours per day or 48 hours per week. The length of the break and, as appropriate, the way in which the break may be split shall be determined by the competent authority or body in each country.

The Convention also states that the daily rest of drivers shall be at least ten consecutive hours during any 24-hour period starting from the beginning of the working day. The daily rest may be calculated as an average over periods to be determined by the competent authority or body in each country. During the daily rest the driver shall not be required to remain in or near the vehicle if he has taken the necessary precautions to ensure the safety of the vehicle and its load.

The Recommendation adds the following detail:

▶ The normal hours of work should not exceed eight per day as an average.
▶ The length of the break after the four-hour driving period and, as appropriate, the way in which the break may be split should be determined by the competent authority or body in each country.
▶ The maximum total driving time, including overtime, should exceed neither nine hours per day nor 48 hours per week (averaged, if necessary over a maximum period of four weeks).
▶ The daily rest of drivers should be at least 11 consecutive hours during any 24-hour period starting from the beginning of the working day.
▶ The daily rest may be calculated as an average over periods to be determined by the competent authority or body in each country; provided that the daily rest should in no case be less than eight hours (exceptions to the recommended duration of the daily rest periods and the manner of taking such rest periods may be provided in the cases of vehicles having a crew of two drivers and of vehicles using a ferry-boat or a train).
▶ The minimum duration of the weekly rest should be 24 consecutive hours, preceded or followed by the daily rest.
▶ In long-distance transport, it should be possible to accumulate weekly rest over two consecutive weeks. In appropriate cases, the competent authority or body in each country may approve the accumulation of this rest over a longer time.

14.4 Health and safety in the home

Why does a book on health and safety at work include a section on home safety? Recent quotes from practitioners in emergency medicine, based on previously unpublished NHS research, show that more accidents occur in the home than anywhere else.

Dr Cliff Mann, president of the College of Emergency Medicine, is quoted in *The Guardian* (12 December 2014), saying that leisure pursuits such as trampolining, horse riding, mountain biking and DIY are more dangerous than previously thought. Because workplace and road safety have improved, home and leisure pursuits now cause more injury and death, averaging 1,000 cases a month. These accidents cost accident and emergency (A&E) units approximately £1bn a year.

'The reality is that you are more likely to die sorting out the Christmas lights by taking them out of the loft or trying to fix the faulty plug or flex, than an electrician is on a building site. You are safer being at work or driving to and from work than you are being at home or doing a leisure activity with your family and friends. People think the kitchen is the most dangerous room in the house but it's actually the living room. Children get scalded from their parents having a hot cup of tea, people get burned by the fire and elderly people can trip on rugs.

With DIY the use of drills and chainsaws can lead to very severe injuries. You can half-amputate a limb with a chainsaw. Falling off a ladder … some people die and some never work again. Falls from lofts, typically a drop of 10 ft, can result in broken limbs and bleeding in the brain, and such falls are more common than is generally realised.'

Figures from A&E at the John Radcliffe Hospital Oxford and the Horton General hospital in Banbury show that over an 18-month period 26,310 patients were treated for injuries: 41% of these were people hurt at home, 25% were hurt during leisure activities, 15% were road accidents, 10% were education related such as PE lessons and only 8% were at work.

Here are some facts and figures taken from RoSPA's research. Unfortunately there are no government statistics after 2002 because the former Department of Trade and Industry (DTI) ceased to collect injury data for its Home Accident Surveillance System (HASS) and Leisure Accident Surveillance System (LASS). (See www.rospa.com)

Home safety facts and figures

- More accidents happen at home than anywhere else.
- Every year there are approximately 4,000 deaths as the result of a home accident. Around 120 of these are individuals aged below 15 years and 1,300 are over 75.
- During 2002 there were 2.7 million home accidents requiring hospital treatment, of which 477,500 involved children under 5.
- Children under the age of 5 and people over 65 (particularly those over 75) are most likely to have an accident at home.
- Falls are the most common accidents, which can cause serious injury at any time of life. Some 55% of accidental injuries in the home involve falls.
- More women than men over the age of 65 die as the result of an accident in the home.
- Approximately 1,500 people aged over 75 die annually as the result of a fall.
- Every year around 120 children under 15 die as the result of an accident in the home.
- Around 25,000 under-fives attend A&E departments each year after being accidentally poisoned.
- Some 26,000 under-fives are burnt or scalded in the home every year. A hot drink can still scald a small child up to 15 minutes after it is made.
- More accidents happen in the lounge/living room than anywhere else in the home.
- Every year more than 4,200 children are involved in falls on the stairs and 4,000 children under the age of 15 are injured falling from windows.
- Boys have more accidents than girls.
- The cost to society of UK home accident injuries has been estimated at £45.63 billion (£45,630 million) annually.

Even if you buy a ticket every week in a year, you are 20 times more likely to suffer a fatal home accident than you are to win the National Lottery!

Figure 14.8 Gas explosion during the night but man and his daughter escaped unharmed – house destroyed, Southampton, 2014

Prevention – a few hints

Fire safety

▶ Keep fires and heaters well guarded. Nursery guards with side clips are essential if there are young children around.

▶ Portable heaters and candles should be positioned so that they cannot be knocked over and away from furniture and curtains.

▶ Clothes should not be dried and aired over or near a fire or cooker.

▶ Never smoke in bed.

▶ Never leave a deep fat fryer or pan of fat unattended on a lit stove.

▶ Keep matches and lighters out of the reach of children.

▶ Fit approved smoke detectors on each floor, either mains operated or with long-life batteries.

▶ Plan your escape route.

Electrical safety

▶ If you buy an older property, get the wiring checked straight away.

▶ Worn and damaged flexes should not be used and flexes should not be wired together.

▶ Portable mains operated devices should not be used in the bathroom.

▶ Have electric blankets serviced and checked regularly.

▶ If an appliance seems to be faulty stop using it.

▶ Residual current devices (RCDs) protect the whole house and make the use of power tools safer.

▶ Buy electrical equipment with a CE mark, BEAB mark, BS mark or British Standard number.

▶ Never overload an electrical socket.

Heating and cooking

▶ Never block or lean anything against air vents or air bricks as they help waste gases to escape. The same applies to outside grilles.

▶ Be aware that carbon monoxide poisoning causes drowsiness and flu-like symptoms.

▶ Chimneys in use should be swept annually, more often if burning wood.

▶ Check the pilot light on a water heater and gas cooker to make sure it is lit.

▶ Beware of secondhand appliances and cowboy installers when buying gas appliances. Look for the safety mark or seal of approval only use Gas Safe approved installers.

▶ Young children and elderly people are especially at risk from fires and heaters.

▶ If a gas leak is suspected turn off the supply, open the windows, call the gas supplier and avoid operating any switches as they can spark.

▶ Bioethanol and fuel gel burners should be used with great care as the flame from them is virtually invisible and the fuel is highly flammable.

▶ Kitchen planning is important – heavy items should not be kept on high shelves. Things in daily use should be easy to reach:
  ▷ Knives need to be kept sharp and out of the reach of children.
  ▷ Pan handles need to be turned inwards on the stove.
  ▷ Hot liquids need special care if there are children or vulnerable adults around.
  ▷ A cordless kettle is the safest option.

Safety glass

▶ Ordinary glass at low level is dangerous but less so if the area is well lit and people are made aware that it is there. Children should never be allowed to play near it.

▶ Safety glass is much better, either laminated or toughened. In new buildings it is a requirement in areas below 800 mm.

Medicines and cleaners

▶ Have a lockable medicine cabinet and keep medicines clearly labelled.

▶ Make sure that chemicals are not kept where children can see or reach them.

▶ Return old medicines to the pharmacy.

DIY and garden safety

▶ Gas and electrical work should always be done by a competent qualified person.

▶ Keep tools clean and in good repair.

▶ Plan ahead and allow plenty of time. Being unprepared and in a hurry is more likely to result in an accident.

▶ Tools, paint, chemicals, anything sharp, should be kept out of sight and reach of children.

▶ Be extra careful with adhesives, especially instant ones.

▶ Disconnect the electricity before working on electrical appliances and tools.

▶ Wear strong shoes and appropriate clothing when mowing, strimming, brushcutting. Keep hands and feet away from blades and twine. Children and pets are at risk here too.

▶ Keep products in original containers.

▶ Barbecues should be well away from trees, fences and buildings. Never pour petrol on a barbecue or bonfire that is alight.

▶ A residual current device (RCD) is essential for outdoor electrical equipment unless the property is already wired for this.

Accidents to older people

Higher living standards, better health care and awareness of the need to pay attention to diet and exercise have led to people living and working longer, so we need to think about health and safety in respect of older people. People over 65 years of age are most at risk of an accident in and around the home. Most of these accidents happen on the stairs or in the kitchen. Falls from stairs or steps are the most common, but people also fall from a bed or chair, or trip over a rug or mat. As people age they become less stable for a variety of reasons. Awareness of this fact means that areas of risk can be identified to make the older person's environment safer.

Main areas of risk that should be targeted are:

▶ trips and fall from poor eyesight, frailty and balance, negotiating stairs, no or insecure handrails, poor flooring and unsuitable shoes;

▶ fire-related accidents from cookers, flammable materials, candles, coal fires, heaters and electric blankets;

▶ poisoning accidents from medicines, carbon monoxide and piped gas;

▶ burns from radiators, electric fires and cookers; scalds from kettles, hot water bottles and uncontrolled hot water;

▶ hypothermia from poor heating, sedentary living, inadequate nutrition and clothes.

Finally a word from Prof. Keith Willett, NHS England's director for acute care:

'The number of people who come through Accident and Emergency with DIY, leisure and gardening injuries is striking. Working as a trauma surgeon for over 30 years, it has been frustrating to see how many people come to A&E with injuries that could have been avoided with a bit more care. However I would much rather people remained active and fit for their general mental and physical well-being than that they avoid sport.'

14.5 Safe cycling

Cycling has gained in popularity over recent years and various initiatives have encouraged people to cycle to work, school and college as well as taking up cycling as a hobby and to improve their fitness. So of course there are many more bikes out there on the road and although local and national authorities are improving cycling conditions by making new cycle routes, there are still a lot of accidents. In fact according to Sustrans (the sustainable transport charity and organisers of the National Cycle Network) in Britain around 19,000 cyclists are killed or injured every year in reported road accidents.

But there are many ways to cycle safely. Here are some guidelines:

Road cycling

▶ follow the Highway Code; cyclists are road users so red traffic lights apply to them and unless designated, pavements are not for cycling;

▶ adjust your speed – wet and icy surfaces alter the way that tyres and brakes behave;

▶ claim your space on the road and be decisive. Give parked cars a wide berth in case someone opens a door;

▶ wear a helmet;

▶ keep your bike in good condition.

Figure 14.9 Tour de France for professionals, French Alps, 2014

Be visible

▶ ride where you can be seen especially when passing trucks and buses;

▶ use lights in poor visibility and at night and wear bright and reflective clothing;

▶ make eye contact with other road users to be sure that they have seen you;

▶ signal clearly;

▶ use your bell;

▶ be aware that some new LED lights are so bright that they can dazzle oncoming cyclists. There is usually a lower intensity setting to use in town.

Awareness of other vehicles

Many collisions occur when a cyclist is on the inside of a vehicle that is going left. Avoid 'undertaking' large vehicles and never cycle along the inside of trucks and buses. When a large vehicle turns left, it often swings out to the right first to get round the corner and this creates a large empty space. Cyclists are tempted to ride into this space, but the gap will disappear as the vehicle swings round to the left. A lot of accidents have happened this way.

Cycle training

People who are new to cycling or have not cycled for some time will find that cycle training helps to build confidence and develop skills. The CTC (Cyclists' Touring Club) can give advice about courses and trainers (www.ctc.org.uk).

14

Figure 14.10 Tour de France for softies, France, 2014

Motorists, be aware!

▶ when turning left watch for cyclists coming up the nearside;

▶ give cyclists a wide berth when overtaking;

▶ dip your headlights at night when approaching cyclists;

▶ in wet or icy weather be extra vigilant when passing cyclists;

Cyclists and motorists have equal rights to the same road space. On shared pavements the same applies to cyclists and pedestrians. Respect for each other will benefit all of us.

14.6 Further information

The Health and Safety at Work etc. Act 1974

The Management of Health and Safety at Work Regulations 1999 (as amended)

The Environmental Protection Act 1990

The Hazardous Waste (England and Wales) Regulations 2005

ILOLEX (ILO database of International Law) www.ilo.org/ilolex/index.htm

Occupational Safety and Health Convention (C155), ILO www.ilo.org/ilolex/cgi-lex/convde.pl?C155

Occupational Safety and Health Recommendation (R164), ILO www.ilo.org/ilolex/cgi-lex/convde.pl?R164

Safety in the Global Village, IOSH Information Sheet, 1999 www.iosh.co.uk/index.cfm?go=technical.guidance

IPCC 2019 Special Report on Climate Change and Land https://www.scimex.org/newsfeed/ipcc-2019-special-report-on-climate-change-and-land

Zero Carbon Britain: Rising to the Climate Emergency, Centre for Alternative Technology, 2019 https://www.cat.org.uk/info-resources/zero-carbon-britain/

APPENDIX 14.1 Scaffolds and ladders

1.1 Convention (Safety and Health in Construction) (167)

Article 14

1. Where work cannot safely be done on or from the ground or from part of a building or other permanent structure, a safe and suitable scaffold shall be provided and maintained, or other equally safe and suitable provision shall be made.

2. In the absence of alternative safe means of access to elevated working places, suitable and sound ladders shall be provided. They shall be properly secured against inadvertent movement.

3. All scaffolds and ladders shall be constructed and used in accordance with national laws and Regulations.

4. Scaffolds shall be inspected by a competent person in such cases and at such times as shall be prescribed by national laws or Regulations.

1.2 Recommendation (Safety and Health in Construction) (175)

Scaffolds

16. Every scaffold and part thereof should be of suitable and sound material and of adequate size and strength for the purpose for which it is used and be maintained in a proper condition.

17. Every scaffold should be properly designed, erected and maintained so as to prevent collapse or accidental displacement when properly used.

18. The working platforms, gangways and stairways of scaffolds should be of such dimensions and so constructed and guarded as to protect persons against falling or being endangered by falling objects.

19. No scaffold should be overloaded or otherwise misused.

20. A scaffold should not be erected, substantially altered or dismantled except by or under the supervision of a competent person.

21. Scaffolds as prescribed by national laws or Regulations should be inspected, and the results recorded, by a competent person:

    (i)   before being taken into use

    (ii) at periodic intervals thereafter

    (iii) after any alteration, interruption in use, exposure to weather or seismic condition or any other occurrence likely to have affected their strength or stability.

1.3 Code of Practice – Safety and Health in Construction

The Code of Practice covers scaffolds and ladders under the following topics over five pages:

1. general provisions
2. materials
3. design and construction
4. inspection and maintenance
5. lifting appliances on scaffolds
6. prefabricated scaffolds
7. use of scaffolds.

APPENDIX 14.2 International travel tips

1. General international travel tips

▷ Leave copies of your passport, visa and itinerary with your office and home. If you should lose any documents you would have easy access to a copy.

▷ Ask the relevant authorities/travel experts for a country profile if you are travelling to a new destination. They can provide extensive data on weather, local customs, food, political climate and much more.

▷ Your luggage may be searched upon entering a country, so pack accordingly. Know what items are taboo in the country you are visiting.

▷ Be wary of anyone loitering around doors to your room, and notify the desk of your suspicions.

▷ Do not swap or sell any personal effects, such as jewellery or religious items, in countries where it is illegal to do so.

▷ Avoid photographing military installations, border protection points and sensitive government buildings.

▷ Avoid any area of the city which has been the target of a terrorist attack or is a place where they are known to congregate.

▷ Know the location of the embassy of your passport country, and have that phone number with you at all times.

▷ If you go out, let another member of your group know when you expect to return and the general location you are visiting. You may wish to provide a contact number as well.

▷ If you are in a situation where someone starts shooting a gun, drop to the floor or get down as low as possible and don't move until you are sure the danger has passed. If possible, shield yourself behind or under a solid object.

14

2. Hotel travel tips

▶ Reserve rooms above first floor.

▶ Keep doors and windows locked, and check them before retiring.

▶ Put valuables in the hotel room safe.

▶ Keep curtains/blinds closed after dark.

▶ Verify workmen with the front desk before admitting them into your room.

▶ Locate your nearest emergency exit and fire extinguisher.

▶ Be careful answering the door. Use the peephole, or leave the chain attached when opening the door to talk.

▶ Don't accept packages or items unless you know the source.

▶ Don't answer the phone with your name.

▶ Remember that the doorknob signs tell an outsider whether you are in the room (Do Not Disturb) or out of the room (Please Make Up This Room).

▶ Take your passport and other forms of identification with you when leaving the room.

▶ Do not indiscriminately give out names and room numbers of others in your group.

▶ Know your hotel phone number.

3. Taxi travel tips

▶ If possible, pick your own taxi.

▶ Have the address of your destination and hotel written in the local language.

▶ If you call a taxi, wait for it indoors.

▶ Compare the face of the taxi driver with the posted licence.

▶ When in a metered taxi, make sure the meter is activated.

▶ When boarding a bus or limo, do so promptly.

▶ Keep an eye on the loading and unloading of your baggage, especially if you are sharing a taxi or limo which makes an intermediate stop.

4. Lost luggage travel tips

▶ If your luggage is lost, find an airline representative and report the problem.

▶ You will be given a form to complete. DO THIS IMMEDIATELY, BEFORE YOU LEAVE THE AIRPORT. This form will describe the luggage and its contents.

▶ Obtain a copy of the form before giving it to an airline representative.

▶ If you are asked to surrender your claim checks, make sure you note this on the form.

▶ 98% of lost luggage is found within several hours, and will be put on the next flight to your destination.

▶ If you cannot wait, make sure the airline has your address to forward your luggage.

▶ If you need to buy or rent replacement items, most carriers will front you some cash.

▶ If luggage is never found, you will be asked to estimate the value (depreciated) of its contents. If you claim anything new or costly (i.e. camera), you may be asked to provide receipts and other documentation.

5. Walking travel tips from the travel experts at Safe Harbours

▶ Avoid walking alone or travelling to remote places after dark; avoid poorly lit streets and narrow alleys.

▶ Do not wear flashy jewellery.

▶ Avoid public demonstrations, accidents or civil disturbances.

▶ Do not engage in loud conversations or arguments.

▶ Avoid discussing personal matters with people you do not know.

▶ Keep a phrase book handy.

▶ Know how to use the pay phones in the country you are visiting, and carry exact change in local currency.

▶ Avoid bringing any unnecessary attention to yourself.

CHAPTER 15

Study skills

Learning outcomes

These are the learning objectives covered by this chapter:

▶ How to plan and organise self-study work

▶ How to manage study and revision time

▶ Understanding and organising revision

▶ Understanding the concept of memory techniques

▶ Understanding how to tackle examinations in the exam room and afterwards

▶ Understanding about examiners' reports and what some of the latest reports are saying

▶ Understanding the marks allocated to each NEBOSH question.

15.1 Introduction

This chapter speaks directly to you, the student. For the NEBOSH certificate you need study skills and exam techniques and this means careful planning from the beginning of the course.

Think in terms of:

▶ clear and realistic goals, both short and long term;

▶ techniques for studying and passing exams;

▶ a well-organised approach;

▶ strong motivation.

People say that genius is 99% application and only 1% inspiration, so good organisation and exam technique are what really matters when it comes to passing exams. Studying is an activity in its own right and there are ways in which you can make your study more effective and give yourself the greatest possible chance of success. The place where you study, the way you plan your study time, how you adapt your study to suit your lifestyle, the way that you take notes, memory and concentration skills, revision techniques, and being clear about the contents of the syllabus and what is expected by the examiner – attention to all these details will give you the best chance of passing the exam.

15.2 Find a place to study

The skill of studying usually has to be learnt, so it's worth thinking carefully about the basics. If you can find somewhere quiet and free from distractions, and keep that place just for your studies, it will be easier to get down to work. You need a place with good ventilation to help you to stay alert, and somewhere that has a comfortable temperature. Make sure there is enough light – a reading lamp helps to prevent eyestrain and tiredness.

An upright chair is better than an armchair. Make sure that the workspace is large enough. The size recommended by Warwick University, for example, is a minimum of 60 cm x 1.5 m. If you can, find a place where your study materials do not have to be cleared away – it's easy to put off starting work if they are not instantly accessible.

15.3 Make a study plan

Start by making a study timetable. First of all, put aside some time for study so that it does not get squeezed out of your working week. Your timetable will need to include time for carrying out assignments set by tutors, time for going over lecture notes and materials, any further reading rated as 'essential' and, if possible, reading rated as 'desirable'. You should allocate time for revision as the course progresses because the more regularly you revise, the more firmly the information will become fixed in your memory. This will make it much easier to remember when you take the exams.

After every hour of study, take a short break and, if possible, have a change of atmosphere. Physical exercise will help to increase your concentration. Even a short, brisk walk round the block will improve your attention span.

Vary your activities while you are studying. For example, you could spend some time gathering information from books, the internet and so on, then maybe work on a diagram or a graph, followed by writing, reading and so on.

15.4 Time management

Finding time to study and take exams when your life is already filled with work, family and social commitments is clearly not going to be easy. The Open University (OU) has been working with adult learners since 1969 and they provide a lot of useful, free information about this at www.open.edu/openlearn/free-courses. Here are some ideas from the section called 'manage your time effectively' to help you to plan.

The aim is to develop good work habits and time-management practices. These are often a matter of developing the right attitudes towards your work and towards your time.

Ten tips to help you manage your time:

1. establish goals and targets;
2. work smarter, not harder – value your time;
3. avoid attempting too much;
4. schedule activities, allow time for emergencies, minimise interruption, identify and use your peak energy times, put similar tasks together;
5. stay in control of paper work and electronic work;
6. minimise day dreaming;

7. use a log to track time wasting;
8. put a time limit on some tasks;
9. be decisive, finish things;
10. work at a steady pace and review your progress regularly.

Keep a log of your learning so that you can reflect on how you use your time – are you managing it effectively? Think about how you deal with the unexpected, with emergencies. How do you prioritise your tasks and your time? Can you set deadlines? Can you meet them? If not, how do you deal with the situation?

There is a section on time management in the Guide to the NEBOSH General Certificate. This subject is also mentioned in 15.13.1 Planning and revision.

15.5 Blocked thinking

Sometimes your thinking can become blocked. If this happens, try leaving the problem area of study alone for a few days and tackle a different part of the subject, since if you concentrate too hard on something that is too difficult you are likely to lose confidence. Usually, after a break, the difficulty vanishes and the problem clears. This could be because the solution has emerged from another perspective, or because you have learnt something new that has supplied the answer. Psychologists think it's possible that some problems can be solved during sleep when your mind has a chance to wander and think laterally.

15.6 Taking notes

The most efficient way to store notes is in a loose-leaf folder, because you can easily add extra information. Write only on one side of the sheet and use margined paper. The facing pages can be used to make summaries or extra points. If notes are clearly written and well spaced out, they are much more straightforward to work from and more attractive to return to later. Use colour, highlighting and underlining too; notes that look good are more appealing and easier to revise.

During revision it will be important to be able to identify subjects quickly, so use headings, numbering, lettering, bullet points, indentation and so on. It is worth spending a bit of extra time to make revision easier.

Key words and phrases are better than continuous prose when you are taking notes. While you are writing down information it is easy to miss essential points that are being made by the tutor. Try to read through your notes within 36 hours to make sure that they are completely clear, so that when you revise you will not be puzzling over what you have written. At this point you may be able to add more information, while it is fresh in your mind. Reading through notes in this way also helps to fix the information in your memory and the level of recall will be further improved by reading them again a week later. Although it seems time-consuming, this technique will save you a lot of time in the long run.

Many people find that they can understand and remember more effectively by making 'mind maps' (also known as 'pattern notes'). These will be discussed in more detail in the revision section.

15.7 Reading for study

Most people will be working as well as studying, so you will need techniques to help you make the best use of your time. For example, you don't need to read a whole book for study purposes, apart from books written specifically to a syllabus, such as this one. Use the index and be selective. In this book an example would be Chapter 13 on summaries of legislation.

Where a book is not specifically written to a syllabus, time-saving techniques such as skim reading can be used. First of all, flip through the book to see what it contains, looking at the contents, summaries, introduction, tables, diagrams and so on. It should be possible to identify areas that are necessary to the syllabus. These can be marked with removable adhesive strips or 'Post-it' notes.

Secondly, quickly read through the parts of the book that you have identified as useful. Two ways of doing this are:

(a) run a finger slowly down the centre of a page, watching it as it moves. You will pick up relevant words and phrases as your finger moves down the page;
(b) read in phrases, rather than word by word, which will increase your reading speed.

Both these methods need practice but they save a lot of time and effort. It is better to read through a piece two or three times, quickly, than to read it once slowly; but if there is a really difficult piece of information that is absolutely essential to understand, read it aloud, slowly.

Finally, at the end of each section or chapter, make a few brief notes giving the essential points.

It is worth learning to speed-read if a course involves a lot of reading. Tony Buzan's series The Mind Set explains the technique.

15.8 Free learning resources from the Open University

The OpenLearn section of the Open University is a very useful resource which is free and available worldwide. There is an extensive study skills section containing units which range from three to fifty hours of study.

Just create an account and sign in. A free statement of participation is available on completion.

Here are a few examples of what is available:

▶ Am I ready to study in English? (5 hours)
This unit provides a series of exercises to help you to reflect on the use of English at an academic (as opposed to everyday) level. It is designed for people who have been educated in a language other than English, or for those who have studied in a country where the conventions are different from those used

15

in the British educational system. It's a free course at Level 1 (introductory).

▶ Revision and examinations (6 hours)
This unit is for people who are unsure about exams. Perhaps you have not taken an exam for a long time, or have had a bad experience with exams in the past; or maybe you have never taken an exam? This unit aims to help you to develop techniques for revision and exam taking and to reassure you.

▶ Learning how to learn (6 hours)
This unit looks at learning as an active process and provides activities that will help you to learn more effectively.

You can access these free resources on www.open. edu/openlearn/free-courses

15.9 Organising for revision

As the exam approaches, you will need to organise all the information from the course in preparation for revision. Ideally, of course, you should have been doing this throughout, but it is not too late to start. It will just mean extra work. Information from lectures, reading, practical experience and assignments will have to be organised into a form that makes it easier for you to remember. As well as this, you will already possess some knowledge that is relevant to the syllabus. Don't underestimate the importance of this. It is worth spending some time thinking about how this store of information

relates to and can be used to expand and back up the new information that you have learnt on the course. Examiners' Reports have, in the past, referred specifically to the advantages to be gained from this. For example:

> 'Some candidates, perhaps from their work experience, showed a good knowledge of a permit-to-work system and produced reasonable answers.'

But be warned. The examiners also pointed out that:

> 'Answers based solely on "what we do in our organisation" … do not earn high marks.'

15.10 Organising information

You can organise the information in preparation for an exam in several ways:

▶ read back through the notes on a regular basis while the course is still in progress; this will help to fix the information in your memory;

▶ make revision cards: condense the information down so that it fits on a set of postcards. This makes it possible to carry a lot of information around in your pocket. The activity of condensing the information is revision in itself. This way you can keep your revision material available and spend a few spare moments from time to time, reading it through.

For example: 'Outline the factors that may increase risks to pregnant employees'. On a card, the information could be condensed as shown in Figure 15.1.

Figure 15.1 Revision notes

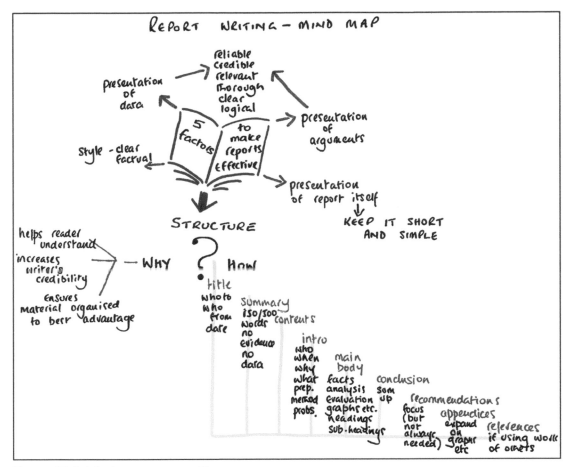

Figure 15.2 Mind map report writing

Make mind maps (also known as pattern notes): like revision cards, the act of making them is revision in itself. Because they are based on visual images some people find them easier to remember and they can be made as complex or as simple as you like. Use colour and imagery. It will be easier to follow and to recall the information (Buzan). Figure 15.2 shows a mind map based on the report writing section of this book.

Use key words as an aid to memory. For example, a list of key words for 'ways of reducing the risk of a fire starting in the workplace' could be reduced to an eight-letter nonsense word, as follows:

FLICSHEV:

Flammable liquids – provide proper storage facilities
Lubrication – regular lubrication of machinery
Incompatible chemicals – need segregation
Control of hot work
Smoking – of cigarettes and materials, to be controlled
Housekeeping (good) – prevent accumulation of waste
Electrical equipment – needs frequent inspection for damage
Ventilation – outlets should not be obstructed.

The effort involved in making the list and inventing the word helps to fix the information in your memory.

Learning by rote is a popular learning method in many countries and is very useful for specific learning tasks, such as tables and formulae. But it is not appropriate to this type of syllabus. Rote learning can prevent you from applying what you have learnt to a wide range of situations. If you limit yourself in this way you will be in danger of not having sufficient understanding of the depth of the subject, particularly if you learn set answers to exam questions. The NEBOSH examiners' reports specifically warn against rote learning of set answers. They point out that 'Candidates should prepare themselves for this vocational examination by ensuring their understanding.' Working through old exam papers is very helpful and you can learn a lot by doing revision in this way but don't be tempted to learn a 'perfect' version of an answer as they are all designed to make you think about specific problems that you might encounter in your working life.

It is useful to have a range of techniques for memorising, since what works for one person may not work for another (see 15.11). There are several other memory techniques available, but the ones described in this paragraph are the most commonly used.

15.11 Being aware of your learning style

All of us are different and we learn in different ways, so it makes sense to be able to identify your own preferred style of learning. The main learning styles are

15

visual, auditory and **kinaesthetic** and it helps if you can choose ways of learning and remembering that fit in with your own personal style.

For example, if you are a person who learns in a **visual** way, you may find it easier to understand and recall information when it is presented in the form of diagrams, the use of flip charts, handouts, demonstrations and videos. When you are taking notes and later in revision, you will perhaps find it more effective to:

▶ Use colour
▶ Organise in columns and categories
▶ Use mnemonics in a visual way (see the 'key word' tip in 15.10)
▶ Use mind mapping (see Figure 15.2)
▶ Use index cards
▶ Highlight, circle and underline
▶ Make posters and/or use Post-it notes.

If you are a person who leans towards learning in an **auditory** way, you will probably find lectures, discussions, dialogue and music are useful, so here are a few ideas to help you to learn more effectively:

▶ Discussion of new ideas and information
▶ Recording information as well as taking notes
▶ Using word association
▶ Repeating facts with your eyes closed
▶ Watching videos
▶ Using rhyme, rhythm and music to help you to retain information
▶ Using mnemonics aurally (see 15.10).

Maybe you are a person who works best with a **kinaesthetic** approach to learning, and find that touching, feeling and actively doing are the best ways for you to learn. Hands-on activity, role play, simulation and practising skills are likely to be the methods you are most comfortable with. Here are a few possibilities:

▶ Copying notes repeatedly to organise them in your mind
▶ Making on-site visits
▶ Using colour and highlighting
▶ Making posters and models to aid memory
▶ Skim reading, then reading in detail
▶ Taking study breaks
▶ Revise on an exercise bike.

(Arp, 2014)

This is not to say that people only conform to the visual, the auditory or the kinaesthetic model. Most people will be on a spectrum and use more than one type of learning, but you have probably already identified the one that is prevalent in your learning style and you can use that information to make both the learning process and revision a bit easier.

15.12 How does memory work?

Understanding how your memory works will help you to use it more effectively. In common with other scientific subjects, there is still a lot to be discovered about the workings of the mind, and a considerable amount of disagreement about what has been discovered so far; but psychologists generally seem to agree about the way that memory works.

The process of remembering is divided, roughly, into two sections – short-term and long-term memory. Items that go into short-term memory, if no further attention is paid to them, will fade away and be forgotten. If they are rehearsed they will stay in short-term memory for a while. Think, for example, of how you remember a phone number when you make a call. Storing something in long-term memory requires a greater 'depth of processing', that is to say, more mental activity is required so that:

(a) information stays in the memory (storage); and
(b) information can be found when it is needed (retrieval).

If you are studying on the NEBOSH course, it is likely that you are a mature student. This will give you some advantages when it comes to learning. Because your life and work experience will almost always be more comprehensive and of a higher complexity than that of younger students, you are likely to possess more 'schemas' (areas of knowledge) to which new information can be attached, and you will have more experiences already stored in your brain that can provide an explanation for the new pieces of knowledge that you are acquiring. Add to this, a very high level of motivation, a stronger level of incentive towards success and more determined application, and it becomes clear that you, as a mature student, have many advantages.

Research shows that, as learners, we take in 10% of what we read; 20% of what we hear; 30% of what we see; 50% of what we see and hear; 70% of what we ourselves say; and 90% of what we ourselves do (Northedge, 2012). You can see from this that people who are simultaneously working at and studying a subject have a marked advantage.

But, it is also true that mature students tend to be very nervous about exams and assignments and often need a good deal of reassurance and support to enable them to realise their full potential.

15.13 How to deal with exams

There are three stages to taking an exam:

▶ planning and revision;
▶ the exam room;
▶ after the exam.

For people who require special consideration, there are provisions within NEBOSH to allow extra time or the use of special equipment. If you think you may be eligible for reasonable adjustments/special consideration, you should apply to NEBOSH several weeks before the date of the exam. Have a look at the NEBOSH Guide if you think any of this may apply to you.

15.13.1 Planning and revision

1. It is absolutely essential that you know what you are going to be examined on and what form the exam will take. You should read through the syllabus and if you are concerned about any area of it, this should be raised with your course tutor well before the date of the exam.

2. Read the Guide to the NEBOSH General Certificate in Occupational Health and Safety and the examiners' reports. After the exams every year, the examiners highlight the most common mistakes made by students (see Section 15.14 on examiners' reports). They also provide useful information about, for example, pass rates, levels attained (distinction, credit, pass, fail), time management and other hints on exam technique.

3. You should work through some recent past papers, against the clock, to get used to the 'feel' of the exam. If possible, ask your tutor to set up mock exams and make sure that the papers are marked. Some of the shorter courses will not be able to provide this service, and if this is the case, do try to do at least one or two timed questions as part of your revision. You may find it difficult at first to write an answer as a first draft. Now that everyone is used to the facilities provided by word processing, you may need to practise writing an answer without being able to edit the text. Past papers, examiners' reports and syllabuses are available from NEBOSH. The website is at www.nebosh.org.uk.

4. You will need to know where the exam is to be held and the date and time. If possible, visit the building beforehand to help build confidence about the location, availability of parking and so on.

5. Make a chart of the time leading up to the exam. Include all activities, work, leisure, and social, as well as the time to be used for revision, so that the schedule is realistic.

6. Try to eat and sleep well and take some exercise.

7. Revision techniques were covered in the sections on revision and memory earlier in this chapter.

Set realistic targets, then achieve them. Make plans and stick to them.

15.13.2 In the examination room

- Read through the exam paper very carefully.
- Check the instructions – how many questions have to be answered? From which section(s)?
- Make a time plan.
- Underline **command words**, e.g. 'identify' (see Table 15.1). Using the words in the question when writing the answers will help to keep the answer on track.
- Stick to the instructions given in the question. A working knowledge of the command words is vital to your success. NEBOSH no longer uses the command word 'sketch', but candidates may use sketches to illustrate their answers where appropriate.
- Write clearly. Illegible answers don't get marked.
- Look at the number of marks allocated to a question to pick up clues as to how much time should be spent on it.
- Mark questions which look possible and identify any that look impossible.
- It is rarely necessary to answer exam questions in a particular order. Start with the question that you feel most comfortable with since it will help to boost your confidence. Make sure it is clearly identified by number for the examiner.
- Answer the question that is set, not the one you wish was on the paper.
- If ideas for other answers spring to mind while you are writing, jot down a reminder on a separate piece of paper. It is easy to forget that bit of information when you are concentrating on something else.
- Plan the use of time and plan the answers. Include some time to check over each answer.
- Stick to the time plan; stick to the point; make points quickly and clearly.

Table 15.1 Terminology used in NEBOSH exams

Command word	Definition
Identify	To give reference to an item, which could be its name or title. NB: normally a word or phrase is sufficient provided the reference is clear.
Give	To provide short, factual answers. NB: normally a single word, phrase or sentence will be sufficient.
Outline	To indicate the principle features or different parts of. NB: an exhaustive description is not required. What is sought is a brief summary of the major aspects of whatever is stated in the question.
Describe	To give a detailed written account of the distinctive features of a subject. The account should be factual, without any attempt to explain. When describing a subject (or object) a test of sufficient detail would be that another person would be able to visualise what you are describing.
Explain	To provide an understanding. To make an idea or relationship clear. NB: this command word is testing the candidate's ability to know or understand why or how something happens. Is often associated with the words 'how' or 'why'.

Early marks in an exam question are easier to pick up than the last one or two, so make sure that all the questions are attempted within the time plan. No marks are given for correct information that is not relevant to the question.

Don't be distracted by the behaviour of other students. Someone who is requesting more paper has not necessarily written a better answer; they may simply have larger handwriting. People who start to scribble madly as soon as they turn over the question sheet are not in possession of some extra ability – they simply haven't planned their exam paper properly.

Keep calm, plan carefully, don't panic.

15.13.3 After the exam

If there are several exams to be taken, you need to stay calm, relaxed and confident. It is not a good idea to get into discussion about other people's experiences of the exam. After one exam, focus on the next. If something went wrong during the exam (for example, illness or severe family problems), the tutor and the examining board should be alerted immediately.

15.13.4 Unit NG2 Risk Assessment

NEBOSH has produced a clear and straightforward booklet on how to complete your risk assessment. Before you start the assessment you must have completed your studies of the whole NG syllabus (elements 1 to 11). It is essential that you read and follow the procedure laid out in the National General Certificate in Occupational Health and Safety: Unit NG2: Risk Assessment: Part 1: Guidance and information for learners.

The practical assessment time is 3 hours

15.14 Open book examinations

NEBOSH is adapting to the challenges and opportunities presented by the Covid-19 pandemic. They have therefore decided to introduce an open book examination (OBE) that can be taken at home as a replacement for the invigilated paper-based examinations for Unit 1 of the National and International General Certificates.

The first open book examination took place on 6 August 2020 with learners registering via their Learning Partner. At first it is in the English language only with other languages being planned at a later date. Open book examinations are common practice and UK Regulators welcome this method of assessment.

NEBOSH have produced several guides:

1. **NEBOSH Open Book Examinations: Learner Guide**
   The document focuses on general guidance to prepare for the open book examination for NG1, IG1, NGC1 and IGC1 units.

2. **NEBOSH Open Book Examinations: Technical Learner Guide**
   This document focuses on the practicalities of accessing, completing and submitting open book examination using NEBOSH's online examination platform.

3. **NEBOSH Open Book Examinations: App Guide**
   This document focuses on the practicalities of accessing, and submitting an open book

examination (OBE) using the NEBOSH online examination platform's Moodle app.

These open book examinations for NG1 and IG1 will not affect the method of assessments for NG2 and IG2.

NEBOSH have given the following guidance to learners on their website: www.NEBOSH.org.uk/obe

Give yourself the greatest chance of success and make sure you are prepared before the day of your examination:

▶ *Study and revise - Make sure you thoroughly understand the topic so that you apply your knowledge to the unfamiliar scenario presented in the examination*

▶ *Prepare an environment that:*
   ▷ *is comfortable with a desk and chair*
   ▷ *has got room for your notes, books and other resources*
   ▷ *is free from distractions*
   ▷ *is well lit*
   ▷ *has a good internet connection available*
   ▷ *is not too hot or too cold*

▶ *Read through the resources provided on the website (1,2,3 above plus sample papers)*

▶ *Familiarise yourself with the online examination platform before the day of your examination*

15.15 The examiners' reports

15.15.1 A few points from the examiners' reports

The examiners' reports provide information to help candidates and tutors in future exams. They aim to be constructive and to help you towards a better understanding of the syllabus and how you are assessed.

The examiners note that many candidates are well prepared and that their answers are relevant and comprehensive and show that they not only have the necessary knowledge, but that they are able to apply it to the workplace situation.

On the other hand, there are some candidates who appear to be unprepared and the examiners emphasise that good preparation is essential. You will need to study the content of the syllabus, while at the same time understanding how the concepts contained in it apply in the workplace. The examiners also stress the importance of applying the information that you have learnt, to the question that is asked on the paper. It is especially important to understand that rote-learning will not lead to success in this type of exam.

Here is a quote from the latest examiners' reports:

'In order to meet the pass standard for this assessment, acquisition of knowledge and understanding across the syllabus are prerequisites. However candidates need to demonstrate their knowledge and understanding in answering the

questions set. Referral of candidates in this unit is invariably because they are unable to write a full, well-informed answer to one or more of the questions asked.

Some candidates find it difficult to relate their learning to the questions and as a result offer responses reliant on recalled knowledge and conjecture and fail to demonstrate a sufficient degree of knowledge and understanding. Candidates should prepare themselves for this vocational examination by ensuring their understanding, not rote-learning pre-prepared answers.'

The section on 'common pitfalls' warns about common mistakes that are made by people taking these exams. Here are seven things which often cause people to fail to reach their full potential:

▶ Failure to apply the basic principles of exam technique can make the difference between passing the exam and being referred.

▶ Some candidates do not attempt all the required questions or they fail to provide complete answers. Even if your mind goes blank, do make an attempt to answer a question that is compulsory. Applying basic health and safety management principles can generate points that will add to your mark.

▶ If you provide information that is relevant to the topic but not relevant to the question that is set, that information cannot gain any marks. Always answer the question that is set.

▶ Knowledge of the command words is essential (see Table 15.1). They are the instructions that guide you on the depth of answer that is required. For example, if a question asks you to 'describe' something, you will not get many marks if you 'outline' it.

▶ Where a question is divided into sub-sections, you need to indicate clearly which part of the question you are answering. Use the numbering from the question in your answer. If your answer is structured to address the different parts of the question, that will help to draw out the points you need to make in your answer.

▶ Time planning is absolutely vital. If you spend a lot of time on one answer and give a lot of unnecessary information, you may be left with insufficient time to address all the questions.

▶ If you have doubts about the legibility of your handwriting, practise writing at speed. The examiner has to be able to read it to mark it.

The examiners also point out that you do not need to start a new page for each section of a question.

15.15.2 Marks for practice revision questions

We have been asked about the allocation of marks to the practice revision questions shown at the end of each chapter. However, since the marks awarded to each part of a question can vary, only general guidance can be given.

All one- or two-part questions are given 8 marks with a minimum of 2 marks awarded to any part.

Questions with three or more parts are given 20 marks with a minimum of 4 marks for each part.

NEBOSH exam questions have a similar format and allocation of marks.

15.16 Conclusion

Passing health and safety exams and assessments has a lot in common with any other subject being examined. Read the questions carefully and plan your answers. The old carpenter's saying 'Measure twice, think twice, cut once' can be applied to exam technique:

'Read twice, think twice and write once.'

15.17 References

Arp, D. Standards Manager NEBOSH (2014) Presentation to course providers 'Preparing Students for Assessment'.

Buzan, T. A wide range of study skills books by this author can be found at www.thinkbuzan.com

Guide to the NEBOSH General Certificate.

Leicester, Coventry and Nottingham Universities Study Skills booklets.

NEBOSH Examiners' Reports.

Northedge, A. 2012 The Good Study Guide. The Open University. ISBN 978 0 7492 5974 7.

Free learning resources from the Open University www.open.edu/openlearn/free-courses

NEBOSH Guidance on command words used in learning outcomes and question papers Version 4 April 2017 online.

National General Certificate in Occupational Health and Safety: Unit NG2: Risk Assessment.

15

CHAPTER 16

Specimen answers to practice questions

16.1 Introduction

The NEBOSH National General Certificate is assessed by one written examination paper (NGC1 – The management of health and safety) only and NGC2 – a Risk Assessment in the workplace of the candidate. Both the examination paper and the risk assessment must be passed so that the NEBOSH National General Certificate may be awarded.

In this chapter, specimen answers are given to one long and two short questions for the written paper. It is important to stress that there are no unique answers to these questions, but the answers presented should provide a useful guide to the depth and breadth expected by the NEBOSH examiners. Candidates are strongly advised to read past examiners' reports which are published by NEBOSH. These are very useful documents because they not only give an indication of the expected answers to each question of a particular paper but they also indicate some common errors made by candidates.

NEBOSH is adapting to the challenges and opportunities presented by the Covid-19 pandemic. They have therefore decided to introduce an open book examination (OBE) that can be taken at home as a replacement for the invigilated paper-based examinations for Unit 1 of the National and International General Certificates.

The first open book examination took place on 6 August 2020 with learners registering via their Learning Partner. At first it is in the English language only with other languages being planned at a later date. Open book examinations are common practice and UK Regulators welcome this method of assessment.

See Section 15.14 for more information and www. nebosh.org.uk/obe

16.2 The written examination

The previous chapter on study skills gives some useful advice on tackling examinations and should be read in conjunction with this chapter. NEBOSH has a commendably thorough system for question paper preparation to ensure that no candidates are disadvantaged by question ambiguity. Candidates should pay particular attention to the meaning of action verbs, such as 'outline', used in the questions.

Candidates have always had most difficulty with the action verb '**outline**' and, for this reason, several of the specimen questions chosen use '**outline**' so that some guidance on the depth of answer expected by the examiners can be given.

NEBOSH no longer uses the command word 'sketch; however, candidates may use sketches to illustrate their answers, when appropriate.

It is difficult to give a definitive guide on the exact length of answer to the examination questions because some expected answers will be longer than others and candidates answer in different ways. As a general guide, for the long answer question on the examination paper, it should take about 25 minutes to write about one and a half pages (550–620 words). Each of the 10 short answer questions requires about half a page of writing (170–210 words).

Although candidates may not be guaranteed 100% by following the guidance in this chapter, they should obtain a comfortable pass.

NGC1 – The management of health and safety

Paper 1 – Question 1

(a) **Outline** the factors that should be considered when selecting individuals to assist in carrying out risk assessments in the workplace. **(5)**

(b) **Describe** the key stages of a general risk assessment. **(5)**

(c) **Outline** a hierarchy of measures for controlling exposures to hazardous substances. **(10)**

Answer:

(a) The most important factor is the competence and experience of the individuals in hazard identification and risk assessment. Some training in these areas should offer evidence of the required competence. They should be experienced in the process or activity under assessment and have technical knowledge of any plant or equipment used. They should have knowledge of any relevant standards, Health and Safety Executive (HSE) guidance and Regulations relating to the activity or process.

Figure 16.1 Select a competent and experienced person to carry out a risk assessment

(b) There are five key stages to a risk assessment suggested by the HSE as follows:

The first stage is hazard identification, which involves looking at significant hazards which could result in serious harm to people. Trivial hazards should be ignored. This will involve touring the workplace concerned looking for the hazards in consultation with workers themselves and also reviewing any accidents, ill health or incidents that have occurred.

Stage 2 is to identify the persons who could be harmed – this may be workers, visitors, contractors, neighbours or even the general public. Special groups at risk, like young persons, nursing or expectant mothers and people with a disability, should also be identified.

Stage 3 is the evaluation of the risks and deciding if existing precautions or control measures are adequate. The purpose is to reduce all residual risks after controls have been put in to as low as is reasonably practicable. It is usual to have a qualitative approach and rank risks as high, medium or low after looking at the severity of likely harm and the likelihood of it happening. A simple risk matrix can be used to get a level of risk.

The team should then consider whether the existing controls are adequate and meet any guidance or legal standards using the hierarchy of controls.

Stage 4 of the risk assessment is to record the significant findings which should be done. The findings should include any action that is necessary to reduce risks and improve existing controls, preferably set against a timescale. The information contained in the risk assessment must be disseminated to workers and discussed at the next health and safety committee meeting.

Stage 5 is a timescale set to review and possibly revise the assessment which must also be done if there are significant changes in the workplace or the equipment and materials being used.

(c) The various stages of the usual hierarchy of risk controls are in bold in this answer.

Elimination or **substitution** is the best and most effective way of avoiding a severe hazard and its associated risks. Elimination occurs when a process or activity is totally abandoned because the associated risk is too high. Substitution describes the use of a less hazardous form of the substance. There are many examples of substitution, such as the use of water-based rather than oil-based paints and the use of asbestos substitutes.

In some cases it is possible to **change the method of working** so that exposures are reduced, such as the use of rods to clear drains instead of strong chemicals. It may be possible to use the substance in a safer form; for example, in liquid or pellets to prevent dust from powders. Sometimes the pattern of work can be changed so that people can do things in a more natural way; for example, by encouraging people in offices to take breaks from computer screens by getting up to photocopy or fetch documents.

Reduced or limited time exposure involves reducing the time that the employee is exposed to the hazardous substance by giving the employee either other work or rest periods.

If the above measures cannot be applied, then the next stage in the hierarchy is the introduction of **engineering controls**, such as isolation (using an enclosure, a barrier or guard), insulation (used on any electrical or temperature hazard) or ventilation (exhausting any hazardous fumes or gases either naturally or by the use of extractor fans and hoods). If ventilation is to be used, it must reduce the exposure level for employees to below the workplace exposure limit (WEL).

Housekeeping is a very cheap and effective means of controlling risks. It involves keeping the workplace clean and tidy at all times and maintaining good storage systems for hazardous substances.

A **safe system of work** is a requirement of the Health and Safety at Work (HSW) Act and describes the safe method of performing the job.

Training and information are important but should not be used in isolation. Information includes such items as signs, posters, systems of work and general health and safety arrangements.

Personal protective equipment (PPE) should only be used as a last resort. There are many reasons for this. It relies on people wearing the equipment at all times and it must be used properly.

Where necessary **health surveillance** should be introduced to monitor the effects on people, and air quality may need to be monitored to check exposure.

Welfare facilities, which include general workplace ventilation, lighting and heating and the provision of drinking water, sanitation and washing facilities, are the next stage in the hierarchy.

All risk control measures, including **training and supervision** must be **monitored** by competent people to check on their continuing effectiveness. Periodically the risk control measures should be **reviewed**. Monitoring and other reports are crucial for the review to be useful. Reviews often take place at safety committee and/or at management meetings. A serious accident or incident should lead to an immediate review of the risk control measures in place.

Finally, special control requirements are needed for carcinogens.

Paper 1 – Question 2

Outline ways in which employers may motivate their employees to comply with health and safety procedures. **(8)**

Answer:

Motivation is the driving force behind the way a person acts or the way in which people are stimulated to act. The best way to motivate employees to comply with health and safety procedures is to improve their understanding of the consequences of not working safely, their knowledge of good safety practices and the promotion of their ownership of health and safety. This can be done by effective training (induction, refresher and continuous) and the provision of information showing the commitment of the organisation to safety and by the encouragement of a positive health and safety culture with good communications systems. Managers should set a good example by encouraging safe behaviour and obeying all the health and safety rules themselves, even when there is a difficult conflict between production schedules and health and safety standards. A good working environment and welfare facilities will also encourage motivation. Involvement

in the decision-making process in a meaningful way, such as regular team briefings, the development of risk assessments and safe systems of work, health and safety meetings and effective joint consultation arrangements, will also improve motivation as will the use of incentive schemes. However, there are other important influences on motivation such as recognition and promotion opportunities, job security and job satisfaction. Self-interest, in all its forms, is a significant motivator.

Although somewhat negative, it is necessary sometimes to resort to disciplinary procedures to get people to behave in a safe way. This is rather like speed cameras on roads with the potential for fines and points on your licence.

Paper 1 – Question 3

(a) **Explain** why young persons may be at a greater risk from accidents at work. **(4)**
(b) **Outline** the measures that could be taken to minimize the risks to young employees. **(4)**

Answer:

(a) Young workers have a lack of experience, knowledge and awareness of risks in the workplace. They tend to be subject to peer pressure and behave in a boisterous manner. They are often willing to work hard and want to please their supervisor and can become over-enthusiastic.

(b) The Management of Health and Safety at Work Regulations require that a special risk assessment must be made before a young person is employed. This should help to identify the measures which should be taken to minimize the risks to young people.

There should be clear lines of communication and regular appraisals. A health surveillance programme should also be in place.

16.3 NGC2: Risk Assessment

NEBOSH has published a detailed Guidance document entitled 'NGC2: Risk Assessment, Guidance and Information for Learners and Learning Partners'. It covers the four stages, the location and the expected time (3 hours) of the assessment. It is recommended that candidates and course providers read this document.

CHAPTER 17

Sources of information and guidance

This chapter covers:

1. Guidance on searching the internet
2. Websites for principal sources of information and guidance on health and safety issues

It will also give you access to:

3. A range of forms for use in the management of health, safety and fire issues which are illustrated at the end of the chapter

---

17.1 Introduction

The authors and publishers are not responsible for the contents or reliability of the linked websites and do not necessarily endorse the views expressed within them. Listing should not be taken as endorsement of any kind. We cannot guarantee that these links will work all the time and we have no control over the availability of the linked pages. For this chapter of the book, as in Chapter 15, a less formal style has been chosen, speaking directly to you, the student.

Today using the internet for access to legislation, standards and guidance is universal and of great benefit to the international OH&S practitioner or person seeking information. Clearly there are many organisations who give small amounts of information free in order to gain your membership or custom. As internet users know, documents, guidance, reports, etc. can remain sitting on websites for years. So it is very important to ensure that they are up to date and authoritative. Guidance from good reliable sources is usually kept up to date and often states the date on which the site was last updated.

Fortunately in the OH&S field there are many organisations that freely allow the download of information because they are either governments, government funded or have a strong wish to make good guidance widely available. These include IOSH, RoSPA and many similar OSH organisations around the world; government departments in the USA, UK, Europe and beyond; universities and local authorities. In the UK, the HSE now allows free download of most of their extensive, authoritative and well-produced guidance. This is a significant change of policy in the last few years and provides a wealth of information for the OH&S practitioner and others.

Finding these sources of information and getting the best results from your searches can be a daunting task. The following general guidance may help when searching for occupational, safety, health, fire, chemical and environmental information on the internet. The websites are mainly in English but some are also in several other languages or a local language and English.

There is also a very useful free, UK Open University, 9-hour introductory course, Finding information in information technology and computing, LIB 5 at: http://openlearn.open.ac.uk/course/view.php?id=2370

17.2 How to search the internet effectively

17.2.1 The key to successful searching

Remember, you are smarter than a computer. Use your intelligence. Search engines are fast, but dumb.

A search engine's ability to understand what you want is very limited. It will obediently look for occurrences of your keywords all over the Web, but it doesn't understand what your keywords mean or why they're important to you. To a search engine, a keyword is just a string of characters. It doesn't know the difference between cancer the crab and cancer the disease … and it doesn't care.

But *you* know what your query means (at least, we hope you do!). Therefore, you must supply the brains. The search engine will supply the raw computing power. While searching, bookmark any other sites of interest. Here are a few tips.

17.2.2 Use unique, specific terms

It is simply amazing how many Web pages are returned when performing a search. You might guess that the terms "hard hat" are relatively specialised. A Google search of those terms returned 65,700,000 results! To reduce the number of pages returned, use *unique* terms that are *specific* to the subject you are researching. "Safety hard hat" reduces this to 93,000. Enter a short question, such as "what time is it?"

17.2.3 Use the minus sign (–) to narrow the search

How many times have you searched for a term and had the search engine return something totally unexpected? Terms with multiple meanings can return a lot of unwanted results. The rarely used but powerful minus sign, equivalent to a Boolean NOT, can remove many unwanted results. For example, when searching for the insect caterpillar, references to the company Caterpillar, Inc. will also be returned. Use Caterpillar -Inc to exclude references to the company or Caterpillar -Inc -Cat to further refine the search.

17.2.4 Use of + sign

Place a **plus sign** (+) prior to each word to view each "word separately" within your search results, such as +writer+grammar+punctuation.

17.2.5 Use quotation marks for exact phrases

I often remember parts of phrases I have seen on a Web page or part of a quotation I want to track down. Using quotation marks around a phrase will return only those exact words in that order. It's one of the best ways to limit the pages returned. Example: "Safe as houses". Of course, you must have the phrase exactly right – and if your memory is as good as mine, that can be problematic.

17.2.6 Don't use common words and punctuation

Common terms like *a* and *the* are called stop words and are usually ignored. Punctuation is also typically ignored. But there are exceptions. Common words and punctuation marks *should* be used when searching for a specific phrase inside quotes. There are cases when common words like *the* are significant. For instance, Raven and The Raven return entirely different results.

17.2.7 Capitalisation

Most search engines do not distinguish between uppercase and lowercase, even within quotation marks. The following are all equivalent:

▶ technology
▶ Technology
▶ TECHNOLOGY
▶ "technology"
▶ "Technology"

17.2.8 Drop the suffixes

It's usually best to enter the base word so that you don't exclude relevant pages. For example, *bird* and not *birds*, *walk* and not *walked*. One exception is if you are looking for sites that focus on the act of walking, enter the whole term *walking*.

17.2.9 Maximise AutoComplete

Ordering search terms from general to specific in the search box will display helpful results in a drop-down list and is the most efficient way to use AutoComplete. Selecting the appropriate item as it appears will save time typing.

The standard Google start page will display a drop-down list of suggestions supplied by the Google search engine. This option can be a handy way to discover similar, related searches. For example, typing in *health and safety* will not only bring up the suggestion *health and safety at work* but also *health and safety legislation* and *health and safety jobs*.

17.2.10 Specific file type search

In Google you can search by specific type of file by writing **filetype:<name_of_type>**, for example:

"filetype:pdf" and write the subject that you are going to search then you will only get pdf files on that subject in the results.

17.2.11 Authors, institutions and other sources

Do you know the name of any author(s) working in this subject? By using the author's name, you may retrieve other references to similar work on the subject of your choice.

Is there an institution or competent authority(ies) known to have done some work in this area? Again try using the name; you may retrieve even more references.

Are any journals/indexing/abstracting service(s) specialising in the subject which are known to you? Again you can add these to your search.

Are there any information centre(s) specialising in the subject? This is similar to author searches because these information centres may well have produced a publication on the subject.

17.2.12 Don't forget the phone book

Telephone directories (white pages and yellow pages) are widely available now on the Web. This means that if you know someone's name and what town they live in, you can access their address, phone number, even their age. There are also databases ('reverse look-up') that allow you to type in a phone number and get the name and address of the person who owns the phone.

If you know the address of the person you are seeking, you can easily get a map of their town, street, and neighbourhood on the many web map sites. Some maps are precise enough to show the exact location of the person's home.

17.2.13 Privacy issues

The same things that you can find out about other people, other people can find out about you.

Here's a list of some of the databases someone might access when researching you:

▶ property sales and ownership;
▶ land registry;
▶ postal service change-of-address records;
▶ telephone books, past and present;
▶ automobile registration records;
▶ other vehicle registration, i.e. boats, private airplanes;
▶ subscription lists for magazines and periodicals;
▶ national marketing databases;
▶ national email directories;
▶ voter registration lists;
▶ bankruptcy filings;
▶ court proceedings/judgements;
▶ professional licences;
▶ tax records;

17

▶ credit bureau records;

▶ domain name registration records.

If you are concerned about your privacy, you can ask to have your personal information removed from web databases. It is difficult to remove all trace of yourself, though. Some events and transactions are legitimately matters of public record, and more of these public records are becoming available every year via the Web.

17.3 Some useful websites

1. **British Institute for Occupational Hygienists:** www.bohs.org/
   BIOH is the professional body for Occupational Hygienists in the UK and organises examinations and awards in Occupational Hygiene.

2. **CIS:** www.ilo.org/safework/cis/WCMS_CON_TXT_SAF_INF_STT_EN/lang--en/index.htm
   International Occupational Safety and Health Information Centres (CIS) and its National and Collaborating Centres which are located all over the world. The goal of this Network is to help its members and the rest of the world to find information from the participating countries on subjects related to occupational safety and health (OSH). The CIS Centres Information Network can be accessed in three languages (English, French and Spanish).

3. **Chartered Institution of Waste Management (CIWM):** www.ciwm.co.uk
   The CIWM is the professional body which represents over 4,000 waste management professionals – predominantly in the UK but also overseas. The CIWM sets the professional standards for individuals working in the waste management industry and has various grades of membership.

4. **Department for Communities and Local Government (DCLG):** https://www.gov.uk/government/collections/fire-safety-law-and-guidance-documents-for-business
   DCLG has responsibility for Fire and has published a series of guides on the Regulatory Reform Fire Safety Order 2005 which can be downloaded free from their website.

5. **European Agency for Safety and Health at Work:** http://osha.europa.eu/en
   This Agency was set up by the European Union to promote the exchange of technical, scientific and economic information between member states. Has links to the European Commission legislation and also to each of the member states' legislation. Also contains links to Australia, Canada, Iceland, Switzerland, Norway, USA legislation. Note that the legislation appears in the language of that country.

6. **European Commission EUR-LEX portal:** http://eur-lex.europa.eu/en/index.htm
   Details of European legislation, publications including the Official Journal of the European Union which is updated daily, links to other Commission sites.

7. **FIRA International Ltd:** www.fira.co.uk/
   FIRA has driven the need for higher standards through testing, research and innovation for the furniture and allied industries. A non-government funded organisation, FIRA is supported by all sections of the industry through the Furniture Industry Research Association, ensuring ongoing research programmes which bring benefits to all. With unrivalled support from across the whole industry, FIRA also has the influence and capability to help shape legislation and regulations.

8. **Fire Protection Association**: www.thefpa.co.uk/
   The Fire Protection Association is the UK's national fire safety organisation. The association works to identify and draw attention to the dangers of fire and the necessary fire prevention measures.

9. **Glossary:** www.ilo.org/public/english/protection/safework/cis/products/safetytm/glossary.htm
   Glossary of Terms on Chemical Safety from the International Programme on Chemical Safety (IPCS) which is a joint venture between the United Nations Environment Programme (UNEP), the International Labour Organization (ILO) and the World Health Organization (WHO). The language of chemical safety is drawn from many sources. These include medicine, toxicology, pharmacology, epidemiology, ecotoxicity and environmental pollution. Its terminology has developed in an unstructured manner with proliferation into multiple terms, some with overlapping, alternative, or even ambiguous meanings. To facilitate international communication and comprehension this Glossary of Terms on Chemical Safety has been produced. It does not claim to be an exhaustive compilation nor a definitive list of approved terms, but because language and terminology are not static the compilers welcome suggestions for additions and improvements.

10. **Hazards Forum:** http://hazardsforum.org.uk
    The Hazards Forum was established in 1989 by the four major engineering institutions, the Institutions of Civil, Electrical, Mechanical and Chemical Engineers. The Forum was set up because of concern about several major disasters, both technological and natural. The Forum believes that there is widespread public misunderstanding of the nature of hazards and the risks they pose. The public reaction to the risks from genetically modified foods and those from smoking are two examples of differing attitudes. These attitudes may result from media attention; suspicion of assurances of safety bodies with a vested interest; or just ignorance. Whatever the reason the Forum believes there is a need to fill an educated but unbiased role and this the Forum aspires to do.

11. **Health and Safety Executive (HSE)**: www.hse.gov.uk
    The Health and Safety Executive's main website containing general information on their areas of work and the latest news.

12. **Health Development Agency (HDA):**
www.hda-online.org.uk
The HDA is a special health authority that aims to improve the health of people in the UK – in particular to reduce inequalities in health between those who are well off and those on low incomes or reliant on state benefits.

13. **Institution of Occupational Safety and Health (IOSH):** www.iosh.org.uk
IOSH is the chartered organisation representing occupational safety and health practitioners predominately in the UK but elsewhere also. IOSH has a valuable website with technical services for members, training courses and many local branch meetings for practitioners.

14. **IOSH International Links**
IOSH has a comprehensive international site. International development is an integral part of IOSH's vision: "a world of work which is safe, healthy and sustainable". We need to improve safety and health at work not only within the UK, but also in developing and newly industrialised countries. The site gives access to linked sites like the ILO, WHO and local OSH organisations in many countries on all continents.

15. **International Labour Organization (ILO):**
www.ilo.org/global/lang–en/index.htm
The ILO is the global body responsible for drawing up and overseeing international labour standards. Working with its member states, the ILO seeks to ensure that labour standards are respected in practice as well as principle. The International Labour Organization's site contains the Conventions and Recommendations which are continually being developed. Contains the full text of these documents.
ILO Safework bookshelf: www.ilo.org/safework_bookshelf/english?d&nd=170000102&nh=0
The *SafeWork Bookshelf* is a collection of key OSH documents produced, in whole or in part, by the ILO. It was compiled by CIS, the information arm of the SafeWork Programme of the ILO.

16. **ISO International Organization for Standardization:** www.iso.org/iso/home.htm
ISO is the world's largest developer and publisher of International Standards.
ISO is a network of the national standards institutes of 162 countries, one member per country, with a Central Secretariat in Geneva, Switzerland, that coordinates the system.

17. **Legislation – UK:** www.legislation.gov.uk
This website is managed by the National Archives on behalf of HM Government. Publishing all UK legislation is a core part of the remit of Her Majesty's Stationery Office (HMSO), part of the National Archives, and the Office of the Queen's Printer for Scotland. These are freely available in full text. All Acts from 1987 and Statutory Instruments from 1988 are available on the website. There are separate pages for Scottish, Northern Ireland and Welsh legislation.

18. **Legislation – Other countries:**
www.gksoft.com/govt/en
Governments around the world on the Web. Lists government websites from over 220 countries. Classified under Worldwide Governments, European Governments, African Governments, American Governments, Asian Governments, Oceanian Governments. There are some multi-government institutions also listed.

19. **National Examination Board in Occupational Safety and Health (NEBOSH):** www.nebosh.org.uk
NEBOSH is the main accredited awarding body for Level 6 and Level 3 health and safety qualifications. Their awards are run by over 400 training organisations.

20. **The National Safety Council USA:**
www.nsc.org/Pages/Home.aspx
The National Safety Council saves lives by preventing injuries and deaths at work, in homes and communities, and on the roads, through leadership, research, education and advocacy.

21. **OSH World:** www.oshworld.com
List of organisations, governments, associations, trade and technical associations, trade unions listed under country and then alphabetically by name. Major organisations listed in each country including the ILO Health and Safety Information Centres.

22. **OSH Links** (formerly known as the Internet Directory): www.ccohs.ca/oshlinks/about.html
An extensive resource that can be used to find primarily Canadian websites that cover occupational health and safety subjects.

23. **Royal Society for the Prevention of Accidents (RoSPA):** www.rospa.com/
The Royal Society for the Prevention of Accidents is a registered charity established more than 100 years ago that aims to campaign for change, influence opinion, contribute to debate, educate and inform – for the good of all. By providing information, advice, resources and training, RoSPA is actively involved in the promotion of safety and the prevention of accidents in all areas of life – at work, in the home, on the roads, in schools, at leisure and on (or near) water. RoSPA's mission is to save lives and reduce injuries.

24. **Specifications and Standards**
American National Standards Institute:
www.ansi.org
British Standards Institution: www.bsi-global.com
European Committee for Standardisation:
www.cenorm.be
ILI – Informe London Information: www.ili.co.uk
International Standards Organisation: www.iso.ch

17

25. **Worksafe Western Australia:**
www.commerce.wa.gov.au/WorkSafe/
WorkSafe is a division of the Department of Commerce, the Western Australian State Government agency responsible for the administration of the Occupational Safety and Health Act 1984. The principal objective of the Act is to promote and secure the safety and health of people in the workplace.

26. **World Health Organization (WHO):**
www.who.int/about/en
WHO is the directing and coordinating authority for health within the United Nations system. It is responsible for providing leadership on global health matters, shaping the health research agenda, setting norms and standards, articulating evidence-based policy options, providing technical support to countries and monitoring and assessing health trends.

17.4.1 Management

M1 GENERAL HEALTH & SAFETY RISK ASSESSMENT EXAMPLE 1		No.	
Firm/Company		Department	
Contact name		Nature of Business	
Telephone no.			

Principal Hazards

1.
2.
3.
4.
5.

Persons at Risk

Main Legal Requirements

1.
2.
3.
4.
5.

Significant Risks	Consequences
1.	
2.	
3.	
4.	
5.	

Existing Control Measures

1.
2.
3.
4.
5.

Residual risk, i.e. after controls are in place

Severity		Likelihood		Residual Risk = SxL	

Information - relevant HSE and trade publications

Comments from Line Manager	Comments from Risk Assessor

Signed		Date		Signed		Date	
Review Date							

M1 – Risk assessment 1

17

M2 RISK ASSESSMENT REPORT FORM EXAMPLE 2

| Firm/Company | | Date of Assessment | |
| Assessor | | Date of review | |

Hazards	Persons Affected	Risks	Initial Risk Level	Existing Controls	Additional Controls	Action by when?	Action by whom?	Done

M3 CONTRACTORS' RISK ASSESSMENT EXAMPLE FOR CONFINED SPACES

INITIAL RISK ASSESSMENT DATE

	Significant Hazards	Persons at Risk	Low	Medium	High	Residual risk
1	Poisoning from toxic gases	Employees			√	Low
2	Asphyxiation – lack of oxygen	Employees			√	Low
3	Explosion	Employees			√	Low
4	Fire	Employees			√	Low
5	Excessive heat	Employees		√		Medium
6	Drowning	Employees			√	Low
7						
8						

ACTION ALREADY TAKEN TO REDUCE THE RISKS

Compliance with:

HSE Guidance L101 - Entry into Confined Spaces
Local Authority/client safety standards, e.g. on sewer entry.
Entry into Confined Spaces Regulations 1997
The Construction (Design and Management) Regulations 2015
Dangerous Substances and Explosive Atmospheres Regulations 2002

Planning:

Eliminate need for entry or use of hazardous materials by selection of alternative methods of work or materials.
Assessment of ventilation available and possible local exhaust ventilation requirements, potential presence of hazardous gases/atmosphere, process by-products, need for improved hygiene/welfare facility.
Prepare a method statement for the work.

Physical:

Documented entry system will apply, with a Permit to Work and method statement. Adequate ventilation will be present or arranged. Detection equipment will be present before entry to check on levels of oxygen and presence of toxic or explosive substances. The area will be tested before entry and continually during the presence of persons in the confined space. Breathing apparatus or airlines will be provided if local ventilation is not possible. Where no breathing apparatus is assessed as being required, emergency BA and rescue harnesses will be provided. Rescue equipment including lifting equipment, resuscitation facilities, safety lines and harnesses will be provided. A communication system with those in confined space will be established. Air will not be sweetened with pure oxygen. Precautions for safe use of any plant or heavier-than-air gases in the confined space must be established before entry. Necessary PPE and hygiene facilities will be provided for those entering sewers.

Managerial/Supervisory:

The management role is to decide on nature of the confined space and to put a safe system into operation, including checking the above and preparing a method statement. Flood potential and isolations must be checked.

Training:

Full training required for all entering and managing confined spaces. Rescue surface party to be trained, including first aid and operation of testing equipment. All operatives must be certified as trained and supervisory staff trained to the same standard. All persons entering a confined space should be physically fit.

Site Manager's Comments

RISK RE-ASSESSMENT DATE	

M3 – Risk assessment – example confined space

17

M4 CONTRACTORS' RISK ASSESSMENT EXAMPLE FOR WORK ON FRAGILE ROOFS

INITIAL RISK ASSESSMENT

	Significant Hazards	Persons at Risk	Low	Medium	High	Residual risk
1	Falls of persons through materials	Employees and Public below			√	Medium
2	Access across fragile material	Employees		√		Low
3						
4						
5						
6						

ACTION ALREADY TAKEN TO REDUCE THE RISKS

Compliance with:

Lifting Operations & Lifting Equipment Regulations 1998 (LOLER)
Provision and Use of Work Equipment Regulations 1998 (PUWER)
Work at Height Regulations 2005
HSE Guidance HSG 150 – Health and Safety in Construction
HSE Guidance HSG 33 – Safety in Roof work
Construction (Design and Management) Regulations 2015.

Planning:

Fragile materials will be identified before work begins. In each case, a risk assessment will be made to provide a safe system of work taking account of the work to be done, access/egress requirements and protection of the area beneath the work area. A method statement will be prepared.

Physical:

Suitable means of access will be provided, such as roof ladder, crawling boards, scaffolding or mobile elevating work platform. Handrails will be fitted. Where access is possible alongside fragile materials such as roof lights, covers will be provided or the fragile material will be fenced off. Soft landing airbags or catch nets will be provided as appropriate. Barriers and signs will be provided so as to isolate the area below fragile materials while work is in progress. No person is permitted to walk on suspected fragile materials for any purpose, including access and surveying.

Managerial/Supervisory:

The role of management is to prepare a method statement prior to commencement of the work, and to arrange for provision of suitable access equipment and trained personnel as required by the safe system devised. Managers must check risk assessments and method statements supplied by subcontractors and others, including the self-employed, to ensure that the proposed work method is safe.

Training:

All operatives must be given specific instructions on the system of work to be used in each case. Selection may be required of operatives who have experience of the work and are physically fit. MEWP operators must be fully trained with a suitable certificate of competence.

Site Manager's Comments

RISK RE-ASSESSMENT DATE	

M4 – Risk assessment – example fragile roof

M5 WORKPLACE INSPECTION REPORT FORM

Company/Organisation	
Work area covered by this inspection	
Activity carried out in work place	
Person carrying out inspection	Date of inspection

Observations	Priority/risk (H,M,L)	Actions to be taken (if any)	Time Scale
List hazards, unsafe practices and good practices		List all immediate and longer term actions required	Immediate, 1 week etc.

M5 – Workplace inspection

17

M6 WORKPLACE INSPECTION CHECKLIST

PREMISES

1	**Work at Height**	Ladders & Step ladders	Right equipment for the job? Level base? Correct angle? Secured at top and bottom? Equipment in good condition?
		Working platforms/ temporary scaffolds	Regularly inspected? Suitable for the task? Properly erected? Maintained and inspected?
		Use of mobile elevating work platforms (MEWPs)	Suitable for task? Operators properly trained? Properly maintained?
2	**Access**	Access ways	Adequate for people, machinery and work in progress? Unobstructed? Properly marked? Stairs in good condition? Handrails provided?
3	**Working Environment**	Housekeeping	Tidy, clean, well organized?
		Flooring	Even and in good condition? Non-slippery? Crowded?
		Comfort /health	Too hot/cold? Ventilation? Humidity? Dusty? Lighting? Slip risk controlled?
		Cleaning	Hygienic conditions? Normal conversation possible?
		Noise	Noise assessment needed/not needed? Noise areas designated? Tasks require uncomfortable postures or actions?
		Ergonomics	Frequent repetitive actions accompanying muscular strain? Workstation assessments needed/not needed?
		Visual display units	Chairs adjustable/comfortable/maintained properly? Cables properly controlled? Lighting OK? No glare? Washing and toilet facilities satisfactory?
4	**Welfare**	Toilets/Washing	Kept clean, with soap and towels/ Adequate changing facilities? Clean and adequate/Means of heating food?
		Eating facilities	For pregnant or nursing mothers?
		Rest room	Kept clean? Suitably placed and provisioned?
		First aid	Appointed person? Trained first aider? Correct signs and notices? Eye wash bottles as necessary? Portable equipment tested?
5	**Services**	Electrical equipment	Leads tidy not damaged? Fixed installation inspected? Equipment serviced annually?
		Gas	Hot and cold water provided?
		Water	Drinking water provided?
6	**Fire Precautions**	Fire extinguishers	In place? Full? Correct type? Maintenance contract?
		Fire instructions	Posted up? Not defaced or damaged?
		Fire alarms	Fitted and tested regularly?
		Means of escape/ Fire exits	Adequate for the numbers involved? Unobstructed? Easily Opened? Properly Signed?
7	**Work Equipment**	Lifting equipment	Thoroughly examined? Properly maintained? Slings etc properly maintained? Operators properly trained? Written schemes for inspection?
		Pressure systems	Safe working pressure marked? Properly maintained?

M6 – Workplace inspection checklist

M6 WORKPLACE INSPECTION CHECKLIST (Cont.)

PREMISES

7	Work Equipment (cont.)	Sharps	Safety knives used? Knives/needles/glass properly used/disposed of?
		Vibration	Any vibration problems with hand held machinery or with whole body from vehicle seats etc?
		Tools and equipment	Right tool for the job? In good condition?
8	Manual & Mechanical Handling	Manual handling	Moving excessive weight? Assessments carried out? Using correct technique? Could it be eliminated or reduced?
		Mechanical handling	Forklifts and other trucks properly maintained? Drivers authorized and properly trained? Passengers only where specifically intended with suitable seat?
9	Vehicles	On site	Speeding limits? Following correct route? Properly serviced? Drivers authorized
		Road risks	Suitable vehicles used? No use of mobile phones when driving? Properly serviced? Schedules managed properly?
10	Dangerous substances	Flammable liquids and gases	Stored properly? Used properly/minimum quantities in workplace? Sources of ignition? Correct signs used?
11	Hazardous substances	Chemicals	COSHH assessments OK? Exposures adequately controlled? Data sheet information available? Spillage procedure available? Properly stored and separated as necessary?
		Exhaust ventilation	Properly disposed of? Suitable and sufficient? Properly maintained? Inspected regularly?

PROCEDURES

12	Risk assessments		Carried out? General and fire? Suitable and sufficient?
13	Safe systems of Work		Provided as necessary? Kept up to date? Followed?
14	Permits to work		Used for high risk maintenance? Procedure OK? Properly followed?
15	Personal Protective Equipment		Correct type? Worn correctly? Good condition?
16	Contractors		Is their competence checked thoroughly? Are there control rules and procedures? Are they followed?
17	Notices, Signs and Posters	Employers' liability insurance	Notice displayed? In date?
		Health and Safety law poster	Displayed? Latest design?
		Safety Signs	Correct type of sign used? Signs in place and maintained?

PEOPLE

18	Health surveillance		Specific surveillance required by law? Stress or fatigue?
19	People's behaviour		Are behaviour audits carried out? Is behaviour considered in the safety programme?
20	Training and supervision		Suitable and sufficient? Induction training? Refresher training?
21	Appropriate authorised person		Is there a system for authorising people for certain special tasks like permits to work, dangerous machinery, entry into confined spaces?
22	Violence		Any violence likely in workplace? Is it controlled? Are there policies in place?
23	Especially at risk categories	Young persons	Employed? Special risk assessments?
		New or expectant mothers	Employed? Special risk assessments?

17

M7 JOB SAFETY ANALYSIS

Job		Date	
Department		Carried out by	

Description of job

Legal requirements and guidance

Task steps	Hazards	Likelihood	Severity	Risk L x S	Controls

Safe systems of work

Job Instruction

Training requirements

Review Date	

M7 – Job safety analysis

M8 ESSENTIAL ELEMENTS — PERMIT TO WORK

1. Permit Title		2. Permit No.	

3. Job Title

4. Plant Identification

5. Description of Work

6. Hazard Identification

7. Precautions Necessary	7. Signatures

8. Protective Equipment

9. Authorization	
10. Acceptance	
11. Extension/Shift Handover	
12. Hand back	
13. Cancellation	

NOTES REFER TO BOX NO ON FORM
2. Should have any reference to other relevant permits or isolation certificates.
5. Description of work to be done and its limitations.
6. Hazard identification including residual hazards and hazards introduced by the work.
7. Precautions necessary – person(s) who carry out precautions, e.g. isolations, should sign that precautions have been taken.
8. Protective equipment needed for the task.
9. Authorization signature confirming that isolations have been made and precautions taken, except where these can only be taken during the work. Date and time duration of permit.
10. Acceptance signature confirming understanding of work to be done, hazards involved and precautions required. Also confirming permit information has been explained to all workers involved.
11. Extension/Shift handover signature confirming checks have been made that plant remains safe to be worked on, and new acceptance/workers made fully aware of hazards/precautions. New time expiry given.
12. Hand back signed by acceptor certifying work completed. Signed by issuer certifying work completed and plant ready for testing and re-commissioning.
13. Cancellation certifying work tested and plant satisfactorily recommissioned

Source HSE

17

M9 WITNESS STATEMENT FORM		
Type of event		Date of occurrence	
Location of event			
Surname			
Forename(s)			
Address			
Telephone No.			
Employer			
Position (if applicable)			

Please provide an account of the event, as you saw or heard it arise, that may be of use in the incident/near miss investigation

Signed		Date	

M9 – Witness statement

M10 ACCIDENT/INCIDENT REPORT

Injured Person		Date		Time
Position		Place of Accident		
Department		Details of Injury		
Investigation carried out				
Position		Date	Estimated absence	

Brief details of Accident (Diagrams, photographs and any witness statements should be attached where necessary)

Immediate Causes	Underlying or Root causes

Conclusions (How can this kind of incident/accident be prevented?)

Action to be taken		Completion Date

Please ensure that an accident investigation and report is completed and forwarded to Personnel within 48 hours of the accident occurring. Some Incidents e.g. involving death or major injuries or dangerous occurrences may have to be notified to the Enforcing Authority. (Mandatory in the UK)

Signature of Manager making Report		Date	
Copies to go to	Personnel Manager	Health & Safety Manager	Payroll Controller

INJURED PERSON

Surname		Forenames	

Home address

Age		Male	Female	Occupation

By ticking this box I give consent to my employer and Best Practice Co(if not employer) to disclose my personal information and details of the incident which appear on this form to safety representatives and representatives of employee safety for them to carry out the health and safety functions given to them by law. (May not apply outside Europe)

Signature of injured person		Date	
Employee	Agency Temp	Contractor	Visitor	Youth Trainee

M10 – Accident report 1

17

KIND OF ACCIDENT Indicate what kind of accident led to injury or condition (tick one box)

Contact with moving machinery or material being machined	1	Injured whilst handling or lifting	5	Drowning or Asphyxiation	9	Contact with electricity or an electrical discharge	13
Struck by moving, including flying, or falling object	2	Slip, trip or fall on same level	6	Exposure to or contact with harmful substance	10	Injured by an animal	14
Struck by moving vehicle	3	Fall from height. Indicate approx. distance of fall……mtrs	7	Exposure to fire	11	Violence	15
Struck against something fixed or stationery	4	Trapped by something collapsing or overturning	8	Exposure to an explosion	12	Other kind of accident	16

Detail any machinery, chemicals, tools etc. involved

Accident first reported to:

Position & Dept.		Name	
First Aid/medical attention by	First Aider Name		Dept.
	Doctor Name	Medical Centre	Hospital

WITNESSES

Name	Position and Department	Statement obtained (attach all statements)	
		Yes	No
		Yes	No
		Yes	No
		Yes	No
		Yes	No

For Office Use Only (if relevant)

Date reported to Enforcing Authority	by telephone		On form F2508		by email	
Date reported to Company Insurers		Were the recommendations effective?				

If No say what further action should be taken:

First Aid Treatment and Accident Record

BEST PRACTICE COMPANY
logo goes here

Date: _____ **FIRST AIDER AND/OR INJURED PERSON** **No:** _____

Part A: FIRST AID TREATMENT/EXAMINATION RECORD

1. About the person who was seen/treated

Name ....   Address ....

....   Postcode ....

Occupation ....   Employer ....

2. About you, the person filling in this record

▼ If the incident **DID NOT** happen to you fill in your name, address and occupation

Name ....   Address ....

....   Postcode ....

Occupation ....

3. Treatment given (if any)

please provide details ....

....

....

....

Part B: INCIDENT AND INJURY DETAILS (if believed to be a work related incident complete the following)

4. Details of incident

▼ Say when it happened: Date _____ / _____ / _____   Time _____

▼ Say where it happened. State which room or area ....

....

▼ Say how the incident happened – give the cause if you can ....

....

▼ If the person who had the incident suffered an injury, say what it was ....

....

▼ Please sign the record and date it

Signature ....   Date _____ / _____ / _____

5. Further details of injury (if any)

☐ Amputation	☐ Burn - Heat	☐ Electric shock	☐ Puncture
☐ Break/Fracture	☐ Concussion/headache	☐ Foreign body	☐ Scald
☐ Bruises (contusions)	☐ Crush	☐ Graze	☐ Splinters and blisters
☐ Burn - Chemical	☐ Cuts (laceration)	☐ Multiple Injury	☐ Sprain
			☐ Strain

M11 – First aid treatment and accident record 2 revised Jan 2015

17

6. For the Person Involved in the Incident Only

By ticking this box I give consent to my employer and Best Practice (if not employer) to disclose my personal information and details of the incident which appear on this form to safety representatives and representatives of employee safety for them to carry out health and safety functions given to them by law. ☐

Signature .... Date .... / .... / ....

FIRST AIDER PLEASE _IMMEDIATELY_ INFORM RELEVANT TEAM LEADER/LINE MANAGER AFTER FIRST AID TREATMENT IS COMPLETED

name of person notified _____ date _____

7. BP Co Team Leader/Line manager Information - Please complete immediately and inform your line manager

Was a risk assessment carried out? YES/NO   If yes who did it? ....

If no, why not? ....

Was correct protective clothing worn?   YES/NO

Why did the incident occur? What was its cause? ....

What action has been taken to prevent a recurrence? ....

....

....

Type of incident: First Aid Injury ☐   *HSE Reportable - 7 days ☐   Other (give details) ☐
* must also notify Factory
Manager/Health & Safety   *Hospital Visit ☐   *HSE Specified ☐
Manager immediately   *Lost time injury ☐   Serious Injury

Level of Investigation required (tick box)   NONE [ ]   MINIMAL [ ]   LOW [ ]   MEDIUM [ ]   HIGH [ ]

Above must be completed within 24 hours of incident

8. BP Co Management Review - by Line Manager & Health and Safety Manager or Deputy

Comments ....

Name   **Signed**   **Date**

Type of Incident is correct ☐   **Level of Investigation is correct** ☐

Name   **Signed**   **Date**

9. H&S Management System Entry - ONLY FOLLOWING COMPLETION OF 1-8 ABOVE

Entry by ....   Number ....

Action Allocated to ....   Date .... / .... / ....

17.4.2 Safety

S1 MACHINERY RISK ASSESSMENT							
MACHINE NO	Model		Manufacturer			Risk Assessment No	
	Model no.		Other			Department ID	

HAZARDS	Hazard	Yes	No	Hazard		Yes	No
	Trapping/drawing in			Stored energy/vacuum/ kinetic			
	Impact/crushing/mobility			Electrical/radiation			
	Cutting/severing/sharp edges			Thermal/noise/vibration			
	Entanglement/ejection			Material/sunstance			
	Rough/slippery surface/falling			Environment/ergonomic			

WHO MIGHT BE HARMED?	Operatives		Cleaners		Maintenance		Visitors	Others

GUARDING	Fixed Guards		Interlocked Movable Guards		Adjustable Guards		Fixed Distance Guard	
	Adequate Enclosure of drives and motors		Fitted to machine		Fitted to machine		Fitted to machine	
			Closed to run		Readily Adjustable		Securely Fixed	
	Securely fixed in position		Design OK with positive switches and robust		Prevent Ejection		Prevents access to danger zone	
	Tools required to remove		Safe to open		Maintained OK		Tool to remove	
			Maintained OK				Maintained OK	

CONTROLS/ WARNINGS/ INSTRUCTIONS/ TRAINING	Controls				Instruction Training		Area	
	Clearly identified		Shrouded start		Warning signs		Lighting OK	
	Function Clear		Warning device		Signs clear		Stability OK	
	Easy to use		Isolator nearby		Safety sheet OK		Ventilation OK	
	Emergency stop		Isolator lockable		SOP available		LEV needed	
	Machine stops OK		Permit required		Training OK		Access operators	
	Safe at stop		Seating needed				Access maintenance	

ACTION REQUIRED

Severity		Frequency		Residual Risk (SxE)	

S1 – Risk assessment machinery revised Jan 2015

17

ASSOCIATED PERMITS (specify type, if none state NA)		ON THE JOB COPY	**BEST PRACTICE COMPANY** logo goes here	GENERAL WORK PERMIT	Number GWP 00000000
Type	Prefix & No.				
Type	Prefix & No.				
Type	Prefix & No.				

SECTION 1 ISSUE   PLEASE PRINT DETAILS

Permit Receiver/Competent Person in charge of work	Location of work/equipment to be worked on
Name of person detailed to carry out work	Details of work to be done

| Risk Assessment attached? Yes ☐ No ☐ Reference | Safety method Statement or Safe System of Work attached? | Yes ☐ | No ☐ | NA ☐ | Reference |

SECTION 2 ISOLATION of electrical or mechanical plant, liquid or gas pipeline or other energy source – give details

Item	Lock location & reference	Isolated by: Print Name

| Attached Isolation sheet? Yes ☐ No ☐ Reference | Location of Keys |

SECTION 3 PREPARATIONS/PRECAUTIONS tick box

	Yes	No	NA		Yes	No	NA
1. Has every source of energy been isolated?				5. Are vessels/pipes free of toxic/flammable gas			
2. Have all isolations been tagged?				6. Any asbestos containing materials present?			
3. Have all isolations been tested?				7. Are the risk assessment control measures in place?			
4a. Does the standard of pipeline isolation meet minimum Corporate standards?				8. Other precautions/control measures required (specify)			
4b. If not, specify additional precautions							

SECTION 4 PERSONAL PROTECTIVE EQUIPMENT tick box

Site Standard	Other specify
Safety Goggles	
Hearing Protection	
Respiratory Protection	Specify Type

SECTION 5 TIME LIMITS

From hrs On / /	To hrs On / /

Max 24 hours See overleaf for details of time extensions allowed

SECTION 6 AUTHORISATION Permit Issuer/Authorised Person	SECTION 7 RECEIPT Permit Receiver/Competent Person in charge
I certify that it is safe to work in the area/on the equipment detailed in Section 1 above and that all safety measures detailed in Section 2-4 have been carried out/complied with. ALL OTHER PARTS ARE DANGEROUS Print name Date Time	I have read, understood and accept the requirements of this permit. I will ensure that everyone working under my supervision will strictly follow the requirements of this permit. I have checked the isolations. Print Name Date Time

SECTION 8 SUSPENSION OF GENERAL WORK PERMIT this is an exception and must be signed on the front of 'on the job'

I certify that the task for which this Permit was issued has now been suspended. We have agreed and implemented a procedure which complies with the criteria noted in the checklist in Section 13 overleaf

Signed Permit receiver/Competent person in charge	Signed Authorised person
Date Time	Date Time

SECTION 9 CLEARANCE Permit Receiver/Competent person in charge	SECTION 10 CANCELLATION Permit Issuer/Authorised Person
I certify that the work for which the permit was issued is now COMPLETED and that all persons at risk have been WITHDRAWN and WARNED that it is NO LONGER SAFE to work on the plant specified on this permit and that GEAR TOOLS and EQUIPMENT are all CLEAR Print name Date Time	This permit to work is hereby CANCELLED, all plant is restored to safe operating conditions, including the replacement of guards. Print name Date Time
Signed	Signed

S2a – General work permit

PERMIT TIME EXTENSION/TRANSFER

Time Extensions

General Work Permits are normally only valid for a maximum of 24 hours.
When a permit is extended beyond 24 hours a Permit issuer/Authorised Person must check ALL the conditions in the Permit. When satisfied there have been no changes and it is safe to continue the work specified in Section 1 Time Extension/Transfer Section 11 must be completed. The permit may be continued beyond a shift and be transferred to another Permit Receiver/Competent Person, who must check that all the precautions are in place and that they understand all of the requirements of the permit. Section 11 must be signed by the new Permit Receiver.

General Note

Except for Section 8 Suspension, the front sheet must not be altered or changed in any way. If there is a need to change the permit conditions (except Section 8 overleaf) then a new permit must be used. Extension sheet Mxxx must be used when Section 11 below is full.

SECTION 11 TIME EXTENSION/TRANSFER OPERATIONAL PERIODS

Permit is extended from	Print Name	Date as FRONT of Permit
____ hrs to ____ hrs	Permit receiver	Signed
____ hrs to ____ hrs	Permit receiver	Signed
Permit is extended from		**Date**
____ hrs to ____ hrs	Authorised Person	Signed
____ hrs to ____ hrs	Permit receiver	Signed
____ hrs to ____ hrs	Permit receiver	Signed
____ hrs to ____ hrs	Permit receiver	Signed
Permit is extended from		**Date**
____ hrs to ____ hrs	Authorised Person	Signed
____ hrs to ____ hrs	Permit receiver	Signed
____ hrs to ____ hrs	Permit receiver	Signed
____ hrs to ____ hrs	Permit receiver	Signed
Permit is extended from		**Date**
____ hrs to ____ hrs	Authorised Person	Signed
____ hrs to ____ hrs	Permit receiver	Signed
____ hrs to ____ hrs	Permit receiver	Signed
____ hrs to ____ hrs	Permit receiver	Signed
Permit is extended from		**Date**
____ hrs to ____ hrs	Authorised Person	Signed
____ hrs to ____ hrs	Permit receiver	Signed
____ hrs to ____ hrs	Permit receiver	Signed
____ hrs to ____ hrs	Permit receiver	Signed
Permit is extended from		**Date**
____ hrs to ____ hrs	Authorised Person	Signed
____ hrs to ____ hrs	Permit receiver	Signed
____ hrs to ____ hrs	Permit receiver	Signed
____ hrs to ____ hrs	Permit receiver	Signed

SECTION 12 ADDITIONAL PEOPLE WORKING ON PLANT

SECTION 13 SUSPENSION OF GENERAL WORK PERMIT check list

Operation of machinery or plant with safeguards removed must be considered a measure of last resort and may only be carried out where it is not practicable to perform tasks, such as examinations or adjustments, in any other manner with the safeguards in place. Should it be necessary to re-energise part of the machinery to which this permit applies before the task is complete a Permit Issuer/Authorised Person must complete the following checklist & sign Section 8.

CHECK LIST	tick box	Yes	NA	CHECK LIST	tick box	Yes	NA
Is number of people involved as small as possible?				Is area free of obstructions & slipping hazards?			
Have precautions been taken to keep people away?				Are all other parts suitably protected?			
Has loose clothing been avoided?				Are suitable precautions against hazards maintained?			
Is safe access with foothold & handhold provided?				Are people competent, briefed and supervised as necessary?			
Is there a means of stopping equipment?							
Check list completed by: print name				Signed		Date	

S2b – General work permit back of form

17.4.3 Health

	Site	Best Site	Work Area	Laboratory	COSHH Assessment No.	001
BEST PRACTICE COMPANY						

Substance / Material: Task / Activity : Hazardous Content: Health Hazard Groups: A / B / C / D / E	FORMALDEHYDE Laboratory Work Formaldehyde* (40%), Methanol (10%). Group "C"	No of Personnel involved in Activity: Frequency of Operation: Duration of Exposure: Quantity / Amount Exposed To: Specific Work Area:	One Person Once per day Up to ½ Hour Daily Use of small quantities under laboratory conditions. Inside well ventilated Area

Procedures Reference: Corporate Minimum Standard, H&SMan-002 Section 2 H&SMan-02 Section 3	Routes of Entry to Body:	INHALATION	✓	SKIN	✓
		INGESTION	✓	EYES	✓

Hazard Classification:	Residual Risk: HIGH HAZARD LIQUID	(WEL) Workplace Exposure Limits: Formaldehyde *2.5mg/m3TWA, 2.5mg/m3STEL WEL [15mins]

Monitoring Arrangements

If using engineering controls, ensure maintenance (Regulation 9.)

RECORDS   MAINTENANCE

Health Surveillance Requirements

Material contains MEL and skin sensitising assigned substance, consider monitoring and skin checks (Regulation 10&11)

BLOOD SAMPLING   LUNG FUNCTION   SKIN CHECK

Hazards from Activity / Health Risks

AVOID CONTACT WITH SKIN & EYES
CAUSES BURNS
TOXIC IF SWALLOWED, BY INHALATION AND IN CONTACT WITH SKIN
LIMITED EVIDENCE OF A CARCINOGENIC EFFECT
MAY CAUSE SENSITIZATION BY SKIN CONTACT
RISK OF SERIOUS DAMAGE TO EYES
Do not breathe in vapour
When using do not eat, drink or smoke

Control Measures / Storage

Personnel must wear lab coat or protective overalls when working with substance
Chemical Resistant Gloves must be worn [Nitrile]
Eye Protection to BS EN 166 Grade 3
Wear rubber boots
General or local exhaust ventilation should be used to control airborne levels
Store away from oxidising agents, stored at a temperature sufficiently high to avoid the formation of polymers

PPE / Workplace Environment Control Measure Symbols

Chemical Overalls   Glasses   Chemical   Dust Mask

First Aid

INHALATION - REMOVE TO FRESH AIR AND REST
IN THE EVENT OF SIGNIFICANT EXPOSURE CALL FOR MEDICAL ASSISTANCE IMMEDIATELY
INGESTION - DO NOT INDUCE VOMITING
INGESTION - GIVE PLENTY OF WATER IN SIPS
INGESTION - GET IMMEDIATE MEDICAL ATTENTION
EYE - IRRIGATE WITH WATER FOR AT LEAST 15 MINUTES
EYE CONTACT - GET IMMEDIATE MEDICAL ATTENTION
SKIN - REMOVE CLOTHING & WASH CONTAMINATED AREA WITH PLENTY OF WATER
SKIN CONTACT - GET IMMEDIATE MEDICAL ATTENTION

Spillage and Disposal

VENTILATE AREA
DO NOT ALLOW SPILLAGE TO ENTER MAINS DRAIN/SEWERS/WATER COURSES
MARK THE AREA AND WARN ALL PERSONNEL
WEAR NITRILE GLOVES
WEAR EYE GOGGLES (GRADE 3)
WEAR RPE WITH ORGANIC FILTER (A)
WEAR PROTECTIVE OVERALLS & CHEMICAL/SAFETY FOOTWEAR
ABSORB IN SAND OR INERT ABSORBENT MATERIAL
COLLECT INTO A CONTAINER, CLOSE LID
DISPOSE OF FOLLOWING SUITABLE SITE PROCEDURE OR SEEK L.A. GUIDANCE

Fire and Emergency

ISOLATED SMALL SCALE FIRE:
DRY POWDER – WATER SPRAY OR MIST – FOAM – CARBON DIOXIDE (CO2)
LARGE FIRE: EVACUATE AREA, CALL FIRE BRIGADE OR FOLLOW SITE PROCEDURE
WEAR SELF-CONTAINED BREATHING APPARATUS AND PROTECTIVE CLOTHING
TOXIC FUMES ARE PRODUCED WHEN SUBSTANCE IS INVOLVED IN A FIRE
KEEP CONTAINERS COOL WITH WATER SPRAY

Supplier: **Chemical distribution Ltd**
Address: **Bridge Way, Anytown**

Post Code: **AN19 1SW**
Telephone: **+44 (0) 1206 543210**
Emergency Tel: **01234 567891**
MSDS Reference:
Link to Manufactures Safety Data Sheet

Assessor Name	A.Another
Date of Assessment	10.01.11
Approved By	A.Another
Date Re-Assessment Due	10.01.13
Assessment Printed Date	10.01.12
Assessment Revision No	01

H1 – COSHH assessment 1

BEST PRACTICE COMPANY logo goes here	**Site**		**Work Area**		**COSHH** Assessment No.	

Substance / Material:
Task / Activity :
Hazardous Content:

Health Hazard Groups:
A / B / C / D / E

No of Personnel involved in Activity:
Frequency of Operation:
Duration of Exposure:
Quantity / Amount Exposed To:
Specific Work Area:

Procedures Reference:	Routes of Entry to Body:	INHALATION		SKIN	
		INGESTION		EYES	

Hazard Classification:	Residual Risk:	(WEL) Workplace Exposure Limits:

Monitoring Arrangements:

Hazards from Activity / Health Risks

Control Measures / Storage

First Aid

PPE / Workplace Environment Control Measure Symbols

Spillage and Disposal

Supplier:
Address:

Post Code:
Telephone:
Emergency Tel:
MSDS Reference:
Link to Manufactures Safety Data Sheet

Fire and Emergency

Assessor Name	
Date of Assessment	
Approved By	
Date Re-Assessment Due	
Assessment Printed Date	
Assessment Revision No	

H1a – COSHH assessment example

17

COSHH 1 – DETAILS OF SUBSTANCES USED OR STORED

Name of Manager:

Name of Department Area:

SUBSTANCE DETAILS

1. Information from the label

Trade name:	Manufacturer's name:

Name of many chemical constituents listed:

Hazard marking...	corrosive		irritant		harmful		toxic		very toxic	

Risk phrases or Hazard Statements noted on label (eg Harmful in contact with skin)

Safety phrases or Precautionary Statements noted on label (eg Avoid contact with skin)

PRECAUTIONS noted on label (eg Use in well ventilated area)

2. Have you got Safety Data Sheet or MSDS for this product	YES		NO	

DETAILS OF USE

3. What is it used for?

4. By whom?

5. How often?

6. Where?

7. What CONTROL measures (precautions) are used?
  (e.g. local ventilation, goggles, respirator, protective gloves etc)

8. Is it ABSOLUTELY ESSENTIAL to keep/use this substance?	YES		NO	
9. Can it be DISPOSED OF NOW?	YES		NO	

H2 – COSHH assessment 2

588

COSHH 2 – ASSESSMENT OF A SUBSTANCE

1. Name of Substance:

2. The process or description of job where the substance is used:

3. Location of the process where the substance is used:

4. Health & Safety information on substance:

a) Hazards to health:

b) Precautions required:

5. Number of persons exposed:

6. Frequency and duration of exposure:

7. Control measures that are in use:

8. The assessment, an evaluation of the risks to health:

9. Details of steps to be taken to reduce the exposure:

10. Action to be taken (name):	**Date:**

11. Date of next assessment/review:

12. Name and position of person making this statement:

13. Date of assessment:

17

EXAMPLE OF A WORKSTATION SELF ASSESSMENT CHECKLIST

Name	Department	Date

The completion of this checklist will enable you to carry out a self-assessment of your own workstation. Your views are essential in order to enable us to achieve our objective of ensuring your comfort and safety at work. Please circle the answer that best describes your opinion, for each of the questions listed.
The form should be returned to……………………………………………………………as soon as it has been completed.

ENVIRONMENT

1. Lighting

Describe the lighting at your workstation.	About right		Too bright		Too dark	
Do you get distracting reflections on your screen?	Never		Sometimes		Constantly	
What control do you have over local lighting?	Full control		Some control		No control	

2. Temperature and Humidity

At your workstation, is it usually:	Comfortable		Too warm		Too cold	
Is the air around your workstation:	Comfortable		Too dry			

3. Noise

Are you distracted by noise from work equipment?	Never		Occasionally		Constantly	

4. Space

Describe the amount of space around your workstation.	Adequate		Inadequate			

FURNITURE

5. Chair

Can you adjust the height of the seat?	Yes		No	
Can you adjust the height and angle of the backrest? Is the chair stable?	Yes		No	
Is the chair stable?	Yes		No	
Does it allow movement?	Yes		No	
Is the chair in a good state of repair?	Yes		No	
If your chair has arms, do they get in the way?	Yes		No	

6. Desk

Is the desk surface large enough to allow you to place all your equipment where you want it?			Yes		No	
Is the height of the desk suitable?	Yes		Too high		Too low	
Does the desk have a matt surface (non-reflectant)?			Yes		No	

7. Footrest

If you cannot place your feet flat on the floor whilst keying, has a footrest been supplied?			Yes		No	

H3 – Workstation assessment

EXAMPLE OF A WORKSTATION SELF ASSESSMENT CHECKLIST (cont.)

8. Document Holder

If it would be of benefit to use a document holder, has one been supplied?	Yes		No	
If you have a document holder, is it adjustable to suit your needs?	Yes		No	

DISPLAY SCREEN EQUIPMENT

9. Display Screen

Can you easily adjust the brightness and the contrast between the characters on screen and the background	Yes		No	
Does the screen tilt and swivel freely?	Yes		No	
Is the screen image stable and free from flicker?	Yes		No	
Is the screen at a height, which is comfortable for you?	Yes		No	

10. Keyboard

Is the keyboard separate from the screen?	Yes		No	
Can you raise and lower the keyboard height?	Yes		No	
Can you easily see the symbols on the keys?	Yes		No	
Is there enough space to rest your hands in front of the keyboard?	Yes		No	

11. Software

Do you understand how to use the software?	Yes		No	

12. Training

Have you been trained in the use of your workstation?	Yes		No	
Have you been trained in the use of software?	Yes		No	
If you were to have a problem relating to display screen work, would you know the correct procedures to follow?	Yes		No	
Do you understand the arrangements for eye and eyesight tests?	Yes		No	

Any other comments

17

EXAMPLE OF A NOISE ASSESSMENT RECORD FORM

Name of Department			Date of Survey	

Lower Exposure Action Level: 80 dBA daily or weekly	Upper Action Level: 85dBA daily or weekly	Peak Pressure: 135 dB(C)/137dB(C)

Workplace	Number of persons exposed	Noise level (leg) dB (A)	Daily exposure period	$L_{EP,d}$ dB (A)	Peak pressure dB (C)	Comments

General Comments:

Instrument used		Date of last calibration		
Signature		Position		Date

H4 – Noise assessment

MANUAL HANDLING OF LOADS: ASSESSMENT CHECKLIST

Section A – Preliminary	Department:	Date:

TASK DESCRIPTION: Factors beyond the limits of the guidelines?	IS AN ASSESSMENT NEEDED? (ie is there a potential risk of injury, and are the factors beyond the limits of the guidelines?) YES/NO

If 'YES 'continue. If 'NO' the assessment need go no further.

Tasks covered by this assessment (detailed description): Locations: People involved: Date of assessment:	Diagrams and other information:

SECTION B – See separate sheet for detailed analysis

SECTION C – Overall assessment of the risk of injury	Low		Med		High	

SECTION D – Remedial action needed

Remedial steps that should be taken, in priority order:

a

b

c

d

e

f

g

h

Date by which action should be taken:

Date for reassessment:

Assessor's name:	Signature

H5 – Manual handling assessment 1

MANUAL HANDLING RISK ASSESSMENT: EMPLOYEE CHECKLIST

Task Description		Employee's ID no.			
RISK FACTORS					
A. Task Characteristics	Yes/No	**Risk Level**			**Current Controls**
1. Loads held away from trunk?		H	M	L	
2. Twisting?					
3. Stooping?					
4. Reaching upwards?					
5. Extensive vertical movements?					
6. Long carrying distances?					
7. Strenuous pushing or pulling?					
8. Unpredictable movements of loads?					
9. Repetitive handling operations?					
10. Insufficient periods of rest/recovery?					
11. High work rate imposed?					
B. Load characteristics					
1. Heavy?					
2. Bulky?					
3. Difficult to grasp?					
4. Unstable/unpredictable?					
5. Harmful (sharp/hot)?					
C. Work environment characteristics					
1. Postural constraints?					
2. Floor suitability?					
3. Even surface?					
4. Thermal/humidity suitability?					
5. Lighting suitability?					
D. Individual characteristics					
1. Unusual capability required?					
2. Hazard to those with health problems?					
3. Hazard to pregnant workers?					
4. Special information/training required?					
Any further action needed?					
Details:					

H6 – Manual handling assessment 2

17.4.4 Fire

F1 FIRE SAFETY MAINTENANCE CHECKLIST

A fire safety maintenance checklist can be used as a means of supporting your fire safety policy. This example list is not intended to be comprehensive and should not be used as a substitute for carrying out a fire risk assessment. You can modify the example where necessary, to fit your premises and may need to incorporate the recommendations of manufacturers and installers of the fire safety equipment/ systems that you may have installed in your premises. Any ticks in the "No" boxes should result in further investigation and appropriate action as necessary. In larger and more complex premises you may need to seek the assistance of a competent person to carry out some of the checks.

Source: Department for Communities and Local Government Fire Safety Guides

Fire Safety Maintenance Checklist	Yes	No	N/A	Comments
DAILY CHECKS (not normally recorded)				
Escape routes				
Can all fire exits be opened immediately and easily?				
Are fire doors clear of obstructions?				
Are escape routes clear?				
Fire warning systems				
Is the indicator panel showing 'normal'?				
Are whistles, gongs or air horns in place?				
Escape lighting				
Are luminaires and exit signs in good condition and undamaged?				
Is emergency lighting and sign lighting working correctly?				
Fire-fighting equipment				
Are all fire extinguishers in place?				
Are fire extinguishers clearly visible?				
Are vehicles blocking fire hydrants or access to them?				
WEEKLY CHECKS				
Escape routes				
Do all emergency fastening devices to fire exits (push bars and pads, etc.) work correctly?				
Are external routes clear and safe?				
Fire warning systems				
Does testing a manual call point send a signal to the indicator panel? (Disconnect the link to the receiving centre or tell them you are doing a test.)				
Did the alarm system work correctly when tested?				
Did staff and other people hear the fire alarm?				
Did any linked fire protection systems operate correctly? (e.g. magnetic door holder released, smoke curtains drop)				
Do all visual alarms and/or vibrating alarms and pagers (as applicable) work?				
Do voice alarm systems work correctly? Was the message understood?				
Escape lighting				
Are charging indicators (if fitted) visible?				
Fire-fighting equipment				
Is all equipment in good condition?				
Additional items from manufacturer's recommendations.				
MONTHLY CHECKS				
Escape routes				
Do all electronic release mechanisms on escape doors work correctly? Do they 'fail safe' in the open position?				

F1 – Fire safety maintenance checklist

17

Fire Safety Maintenance Checklist	Yes	No	N/A	Comments
Do all automatic opening doors on escape routes 'fail safe' in the open position?				
Are fire door seals and self-closing devices in good condition?				
Do all roller shutters provided for fire compartmentation work correctly?				
Are external escape stairs safe?				
Do all internal self-closing fire doors work correctly?				
Escape lighting				
Do all luminaires and exit signs function correctly when tested?				
Have all emergency generators been tested? (Normally run for one hour.)				
Fire-fighting equipment				
Is the pressure in 'stored pressure' fire extinguishers correct?				
Additional items from manufacturer's recommendations.				
THREE-MONTHLY CHECKS				
General				
Are any emergency water tanks/ponds at their normal capacity?				
Are vehicles blocking fire hydrants or access to them?				
Additional items from manufacturer's recommendations.				
SIX-MONTHLY CHECKS				
General				
Has any fire-fighting or emergency evacuation lift been tested by a competent person?				
Has any sprinkler system been tested by a competent person?				
Have the release and closing mechanisms of any fire-resisting compartment doors and shutters been tested by a competent person?				
Fire warning system				
Has the system been checked by a competent person?				
Escape lighting				
Do all luminaires operate on test for one third of their rated value?				
Additional items from manufacturer's recommendations.				
ANNUAL CHECKS				
Escape routes				
Do all self-closing fire doors fit correctly?				
Is escape route compartmentation in good repair?				
Escape lighting				
Do all luminaires operate on test for their full rated duration?				
Has the system been checked by a competent person?				
Fire-fighting equipment				
Has all fire-fighting equipment been checked by a competent person?				
MISCELLANEOUS				
Has any dry/wet rising fire main been tested by a competent person?				
Has the smoke and heat ventilation system been tested by a competent person?				
Has external access for the fire service been checked for ongoing availability?				
Have any firefighters' switches been tested?				
Has the fire hydrant bypass flow valve control been tested by a competent person?				
Are any necessary fire engine direction signs in place?				

F1a Fire safety maintenance annual checks

F2 FIRE RISK ASSESSMENT RECORD – SIGNIFICANT FINDINGS

Risk Assessment for		
Assessment undertaken by		
Company	Date	
Address	Completed by	
	Sign	
Sheet no.	Floor/Area	Use

Step 1 – Identify Fire Hazards

Sources of ignition	Sources of fuel	Sources of oxygen

Step 2 – People at risk

Step 3 – Evaluate, remove, reduce and protect from risk

Evaluate the risk of the fire occurring

Evaluate the risk to people from a fire starting in the premises

Remove and reduce the hazards that may cause a fire

Remove and reduce the risks to people from a fire

Assessment review

Assessment review date	Completed by	Sign

Review outcome (where substantial changes have occurred a new record sheet should be used)

F2 – Fire risk assessment record

17

17.4.5 Construction

C1 CONSTRUCTION INSPECTION REPORT

1. Name and address of person for whom inspection was carried out.

2. Site address.	3. Date and time of inspection.

4. Location and description of place of work or work equipment inspected.

5. Matters which give rise to any health and safety risks.

6. Can work be carried out safely?	Yes		No	
7. If not, name of person informed.				

8. Details of any other action taken as a result of matters identified in 5 above.

9. Details of any further action considered necessary.

10. Name and position of person making the report.	
11. Date and time report handed over.	
12. Name and position of person receiving report.	

C1 – Construction inspection report

C2 EXAMPLE RISK ASSESSMENT FOR CONTRACT BRICKLAYERS

IMPORTANT REMINDER

Company name:
TVW Contract Bricklayers

Date of risk assessment:
6/3/2006

This example risk assessment shows the kind of approach a small business might take. Use it as a guide to think through some of the hazards in your business and the steps you need to take to control the risks. Please note that it is not a generic risk assessment that you can just put your company name on and adopt wholesale without any thought. This would not satisfy the law – and would not be effective in protecting people.
Every business is different – you need to think through the hazards and controls required in your business for yourself. Source HSE

What are the hazards?	Who might be harmed and how?	What are you already doing?	What further action is necessary?	Action by who?	Action by when?	Done
Falling from height	Serious injury or even fatal injury could occur if a worker fall	■ **Agree** scaffolding requirements at contract stage, including appropriate load rating and provision of loading bays. ■ **Bricklayers'** supervisor to check with the site manager that the correct scaffold is provided and inspected. ■ **Workers** instructed not to interfere with or misuse scaffold – supervisor to keep an eye out for problems. ■ **Ladders** in good condition, adequately secured (lashed) and placed on firm surface. ■ **Band** stands with handrails to be used for work on internal walls. ■ **Workers** trained to put up bandstands	■ **Scaffold** requirements agreed, including loading bays and appropriate load rating. ■ **Supervisor** to speak regularly to site manager to arrange scaffold alterations and ensure that weekly inspections have been carried out.	TB LG	20/3/06 from 1/5/06	20/3/06
Collapse of scaffold	Operatives on scaffold may incur crush injuries, or worse, if the scaffold collapses on top of them.	■ **Agree** scaffolding requirements at contact stage, including appropriate loading bays. ■ **Bricklayers'** supervisor to check with the site manager that the correct scaffold is provided and inspected.	■ **Supervisor** to keep a check to make sure that scaffold is not overloaded with materials.	LG	from1/5/06	
Falling objects hitting head or body, including feet	Serious head and other injuries to workers, others on site and members of the public.	■ **Brick guards** kept in position on scaffold lifts. ■ **Waste materials** removed from scaffolding and placed in skip. ■ **Safety helmets** and protective footwear (with steel toecaps and mid-soles) supplied and worn at all times.	■ **Supervisor** to monitor use of safety hats and protective footwear.	LG	from1/5/06	
Manual handling	All workers could suffer from back injury and long term pain if regularly lifting/carrying heavy or awkward objects	■ **Bricks** mortar etc. to be transported and lifted to scaffolding using telehandler provided by principal contractor. ■ **Provision** of lifting bay agreed with principal contractor. ■ **Bricks/blocks** to be covered with tarpaulin when stored on site to prevent taking up water. ■ **Spot boards** to be raised with blocks to easy working height. ■ **Trolley** to be used for moving loads of bricks around the scaffold lift. ■ **Check** at tender stage for any blocks or lintels over 20kg and make arrangements.	■ Heaviest blocks are 15kg no special arrangements necessary. ■ Concrete lintels are well over 20kg, to be positioned using telehandler (all are accessible) ■ All workers to be instructed not to carry materials up by hand.	VP VP LG	20/3/06 from1/5/06 from1/5/06	

C2 – Risk assessment bricklayer example

17

What are the hazards?	Who might be harmed and how?	What are you already doing?	What further action is necessary?	Action by who?	Action by when?	Done
Workers stuck or crushed by moving vehicles on site	Workers could suffer serious or even fatal injuries from vehicles and machines on site – particularly when reversing	■ **Manager** to agree safe route to work area with principal contractor based upon the construction phase health and safety plan. ■ **Induction** to each site to be carried out for all workers on first day.	■ Safe route agreed with principal contractor based. ■ Supervisor to liaise with site ■ Manager to ensure safe route stays clear. ■ Instruct staff that they must never drive vehicles and plant on this site. ■ High-visibility vests to be provided. ■ Supervisor to check vests are worn on all sites where the principal contractor requires them.	TB LG LG LG LG LG	20/3/06 from1/5/06 from1/5/06 from1/5/06 from1/5/06 from1/5/06	20/3/06
Slips and trips	All workers may suffer sprains or fractures if they trip over waste including brick bands and pallet debris. Slips at height could result in a serious fall.	■ **Good** housekeeping maintained at all times. ■ **Waste** including brick bands and pallet debris disposed of in skip. ■ **Safety** footwear provided to all workers. ■ **Safe** route to workplace agreed with principal contractor based upon the construction phase health and safety plan.	■ Temporary storage locations to be agreed with site manager ■ Supervisor to ensure that workers wear safety footwear whenever on site.	TB LG	20/2/06 from1/5/06	20/3/06
Stepping on nails and sharp objects	All workers could suffer foot injuries.	■ **Safety boots** with steel toecaps and mid-soles provided to all workers. ■ **Waste** disposed of in skips.	■ Explain the need to wear safety boots and dispose of waste in skips – repeat annually.	LG	1/5/06	1/5/06
			■ Supervisor to check that safety boots are always worn and waste disposed of properly.	LG	from 1/5/06	
Hazard to eyes, cutting bricks	Bricklayers could suffer eye injury through flying brick fragments.	■ **Safety goggles** (EN 166 B standard) worn when breaking bricks.	■ Use of goggles to be monitored by supervisor.	LG	from 1/5/06	
Hazardous substances, mortar	Direct skin contact with the mortar could also cause bricklayer contact dermatitis and burns.	■ **Risk of dermatitis** or cement burns and precautions explained to all workers. ■ **Use cement** or cement containing products within the use-by date. ■ **Direct skin contact** to be avoided, CE marked PVC gloves used when handling mortar. ■ **Good washing** facilities on site, with hot and cold water, soap and basins large enough to wash forearms. ■ **Principal** contractor's first-aid includes emergency eyewash	■ Training on how to treat exposure to be given to all operatives. ■ Supervisor to be aware of anyone with early signs of dermatitis.	TB LG	17/4/06 from 1/5/06	26/4/06

What are the hazards?	Who might be harmed and how?	What are you already doing?	What further action is necessary?	Action by who?	Action by when?	Done
Dust from cutting bricks	Dust exposure could cause silicosis.	■ **Angle grinders** replaced with block splitter, removing the risk of significant dust exposure. ■ **The use of** a grinder for chasing etc. is not needed on this job.				
Operating cement mixer	Workers could be crushed or cut if the mixer topples or they get caught in moving parts. Damage to electrics could result in a shock.	■ **Cement mixer** located on firm, level ground. ■ **Mixer** is fully guarded and guards in place during operation. ■ **Mixer** is 110 volt and PAT tested every three months.	■ Supervisor to check mixer daily for obvious damage.	LG	from 1/5/06	
Noise from use of equipment, eg angle grinder	Workers using grinders or working near people who may suffer hearing loss.	■ **Angle grinders** replaced with block splitter, removing high noise levels from our work. ■ **Construction** phase plan show other trades using grinders etc. should not be working close enough to cause problems.	■ Supervisor to monitor and talk to site manager if noisy work does start close by.	LG	from 1/5/06	
Vibration from use of equipment such as angle grinder	Exposure to vibration can lead to the development of 'vibration white finger' (VWF).	■ **Angle grinders** replaced with block splitter. No significant vibration left.				
Fire/explosion	All operatives in the vicinity could suffer from smoke inhalation or burns.	■ **Suitable** fire extinguisher kept in site office and welfare block. ■ **Good housekeeping** monitored by supervisor.	■ Supervisor to brief all workers on first day on emergency arrangements agreed with principal contractor.	LG	1/5/06	1/5/06
Welfare/ first-aid	Good facilities help prevent dermatitis etc.	■ **Principal contractor** will have facilities on site by the time bricklaying starts, including: - ○ flushing toilet; ○ hot and cold running water, soap, towels and full-size washbasins; ○ heated canteen with kettle etc; ○ first-aid equipment; ○ principal contractor will arrange clearing and ensure the necessary electrical and heating safety checks are made; ○ and site agent is appointed person for first-aid.	■ Supervisor to brief workers on facilities and keeping them clean.	LG	1/5/06	1/5/06

17

C3 EXAMPLE RISK ASSESSMENT FOR WOODWORK

Company name:	IMPORTANT REMINDER
The Woodworking Company	This example risk assessment shows the kind of approach a small business might take. Use it as a guide to think through some of the hazards in your business and the steps you need to take to control the risks. Please note that it is not a generic risk assessment that you can just put your company name on and adopt wholesale without any thought. This would not satisfy the law – and would not be effective in protecting people. **Every business is different – you need to think through the hazards and controls required in your business for yourself. Source HSE**
Date of risk assessment: 28/09/07	

What are the hazards?	Who might be harmed and how?	What are you already doing?	What further action is necessary?	Action by who?	Action by when?	Done
Exposure to wood dust	Staff risk lung diseases, such as asthma, from inhaling wood dust. Hardwood dust can cause cancer, particularly of the nose	■ Local exhaust ventilation (LEV) provided at machines and staff are trained in using it properly. Bricklayers' supervisor to check with the site manager that the correct scaffold is provided and inspected. ■ LEV maintained to keep it in good condition and working effectively. LEV inspected every 14 months by a competent person. ■ Wood dust cleared up using a suitable vacuum cleaner, fitted with an appropriate filter. ■ Suitable respiratory protective equipment (RPE) as well as LEV for very dusty jobs, and staff trained in how to use it. ■ Staff do health surveillance questionnaire before starting, then annually. ■ Any affected staff referred to a medical professional.	■ Remind staff of the risks of wood dust, and why these controls are necessary.	Manager	7/10/07	1/10/07
			■ Remind staff never to dry sweep wood dust, which just spreads the dust around.	Manager	7/10/07	1/10/07
Machinery	Staff risk serious and possibly fatal cut injuries following contact with moving parts of machinery, particularly saw blades	■ All machines guarded according to manufacturers' instructions. ■ Guards inspected regularly and maintained as necessary to ensure their good condition. ■ Staff have sufficient space at machines to work safely. ■ Staff monitored by manager to ensure guards always used. ■ All staff trained in safe use of machines by a competent person. ■ All machines braked and fitted with necessary safety features, eg chip limited tooling etc.	■ Download information sheets on the safe use of the machines used in the workshop from HSE website and pin them up in mess room.	Manager	4/10/07	1/1/07
Manual Handling	Staff may suffer musculoskeletal disorders, such as back pain, from handling heavy/bulky objects, eg timber boards and machinery parts. Also risk cuts when handling tooling, or splinters when handling pallets.	■ Staff trained in manual handling. ■ Workbenches and machine tables set at a comfortable height. ■ Strong, thick gloves provided for handling tooling and pallets. ■ Panel trolley and lifting hooks available for moving boards. ■ Systems of work in place for the safe and careful handling of assembled furniture.	■ Where possible, store tooling next to the machine to reduce carrying distance. ■ Remind staff to ask for a new set of gloves when old ones show wear and tear, and not to try to lift objects that appear too heavy.	Manager Manager	30/9/07 4/4/07	29/9/07 4/4/07

C3 – Risk assessment woodwork example

What are the hazards?	Who might be harmed and how?	What are you already doing?	What further action is necessary?	Action by who?	Action by when?	Done
Noise	Staff and others may suffer temporary or permanent hearing damage from exposure to noise from woodworking machinery	■ Noise enclosures used where practicable, and maintained in good condition. ■ Low-noise tooling used where possible. ■ Planned maintenance programme for machinery and LEV systems. ■ Suitable hearing protectors provided for staff and staff trained how to use them. Check and maintain them according to advice given by supplier. ■ Staff trained in risks of noise exposure. ■ Staff trained in systems of work to reduce noise exposure (eg suitable feed rates for certain jobs, timber control etc).	■ Consider if certain machines could be safely mounted on anti-vibration mountings. ■ Include noise emission in specification for new vertical spindle moulder, to be purchased next year.	Manager Manager	30/9/07 30/9/07	29/9/07 29/9/07
Vehicles	Staff may suffer serious, possibly fatal, injuries if struck by a vehicle such as a lift truck or a delivery lorry.	■ Fork-lift truck maintained and inspected as per lease contract. ■ Lift truck operated only by staff who have been trained to use it. ■ Pedestrian walkways marked. ■ Only authorised people allowed in yard for deliveries/dispatch.	■ Ensure drivers get out of their vehicle and stand in a safe area while it is being loaded/unloaded	Manager and all other staff	31/9/07	15/9/07
Slips, trips and falls	Staff could suffer injuries such as bruising or fractures if they trip over objects, or slip, eg on spillages, and fall.	■ Generally good housekeeping – off-cuts cleared away promptly, dust cleared regularly etc. ■ Staff wear strong safety shoes that have a good grip. ■ Good lighting in all areas.	■ Remind staff to clear up spillages of wax or polish immediately, even very minor spillages.	Manager	31/9/07	15/9/07
Electrical	Staff could get electrical shocks or burns from using faulty electrical equipment, eg machinery, or a faulty installation. Electrical faults can also lead to fires	■ Residual current device (RCD) built into main switchboard. ■ Staff trained to spot and report any defective plugs, discoloured sockets or damaged cable/equipment to manager. ■ No personal electrical appliances, eg toasters or fans, allowed.	■ Ask landlord when the next safety check of the electrical installation will be done. ■ Confirm with landlord the system for making safe any damage to building installation electrics, eg broken light switches or sockets.	Company secretary Company secretary	31/9/07 31/9/07	15/9/07 15/9/07
Work at height	Falls from any height can cause bruising and fractures	■ Strong stepladder, in good condition, provided. ■ Only trained, authorised staff allowed to work at height.	■ Condition of stepladder to be checked periodically.	Manager	31/9/07	as required
Fire	If trapped, staff could suffer fatal injuries from smoke inhalation/burns	■ Fire risk assessment done, see www.communities.gov.uk/fire and necessary action taken.	■ Ensure the actions identified as necessary by the fire risk assessment are completed.	Manager	from now	

C4 Job safety analysis – construction

603

Index

Page numbers in *italics* denote an illustration, **bold** indicates a table or box